LAYERED INTRUSIONS

Series

Developments in Petrology

1. K.R. Mehnert
MIGMATITES AND THE ORIGIN OF GRANITIC ROCKS

2. V. Marmo
GRANITE PETROLOGY AND THE GRANITE PROBLEM

3. J. Didier
GRANITES AND THEIR ENCLAVES
The Bearing of Enclaves on the Origin of Granites

4. J.A. O'Keefe
TEKTITES AND THEIR ORIGIN

5. C.J. Allègre and S.R. Hart (Editors)
TRACE ELEMENTS IN IGNEOUS PETROLOGY

6. F. Barker (Editor)
TRONDHJEMITES, DACITES, AND RELATED ROCKS

7. Ch.J. Hughes
IGNEOUS PETROLOGY

8. R.W. Le Maitre
NUMERICAL PETROLOGY
Statistical Information of Geochemical Data

9. M. Suk
PETROLOGY OF METAMORPHIC ROCKS

10. C.E. Weaver and Associates
SHALE-SLATE METAMORPHISM IN SOUTHERN APPALACHIANS

11A. J. Kornprobst (Editor)
KIMBERLITES. I:KIMBERLITES AND RELATED ROCKS

11B. J. Kornprobst (Editor)
KIMBERLITES. II: THE MANTLE AND CRUST-MANTLE RELATIONSHIPS

12. D.C. Smith (Editor)
ECLOGITES AND ECLOGITE-FACIES ROCKS

13. J. Didier and B. Barbarin (Editors)
ENCLAVES AND GRANITE PETROLOGY

14. J.N. Boland and J.D. Fitzgerald (Editors)
DEFECTS AND PROCESSES IN THE SOLID STATE: GEOSCIENCE APPLICATIONS
THE McLAREN VOLUME

Developments in Petrology 15

LAYERED INTRUSIONS

Edited by:

RICHARD GRANT CAWTHORN

Department of Geology, University of the Witwatersrand, Johannesburg, P.O. WITS 2050, South Africa

1996

ELSEVIER

Amsterdam – Lausanne – New York – Oxford – Shannon – Tokyo

ELSEVIER SCIENCE B.V.
Sara Burgerhartstraat 25
P.O. Box 211, 1000 AE Amsterdam, The Netherlands

ISBN Hardbound 0 444 81768 9
ISBN Paperback 0 444 82518 5

This book is printed on acid-free paper.

Printed in The Netherlands

Contents

Preface

The book by Lawrence Wager and Malcolm Brown on *Layered Igneous Rocks* has become a milestone in the geological literature, and it was with some trepidation that I set out to try to update this classic treatise. As mentioned in the *Foreword* by Malcolm Brown, their intention was to stimulate interest in this field of petrology, and their success is reflected by the list of nearly 800 names in the Author Index at the back of this book, who have contributed to our understanding of layered intrusions.

It was a personal dilemma deciding which intrusions to include in this book, and which of the many other intrusions would have to be excluded because of space limitations. Some of these bodies have been comprehensively researched in the recent literature, whereas others, perhaps due to lack of exposure or geographical isolation still require more detailed study with modern techniques, and yet others display specific unique or intriguing features, but do not justify an entire chapter. Wager and Brown included observations on sills and other intrusions in which modal variation or fractionation was recorded, notably the Palisades Sill, but there are now so many examples that it is not possible to include those here.

Their book in 1968 marked a quantum leap from a more descriptive approach to an attempt to quantify the physical and chemical processes in magma chambers. While basic observation is still the cornerstone to any study, modern analytical techniques permit far more detailed evaluation of these processes.

By enlisting the support of 24 different authors in fourteen chapters, I hope that this book will present an overview of what we know about *Layered Intrusions*, their differences, as well as similarities. In each chapter I asked each author to present sufficient observation and information content to make this a useful reference volume, even if some of the current ideas become superseded. The first four chapters summarize our understanding of layering processes, the relevance of fluid dynamics, the textures observed in these slowly cooled rocks and some of their mineral deposits. The remaining ten chapters review the geology of some of the intrusions which have moulded our ideas about processes in magma chambers.

I am grateful to the many colleagues in South Africa with whom I have explored and experienced the Bushveld Complex, and the many associates elsewhere who have shown me other intrusions and shared other concepts, which provide the variety and challenges in interpreting these spectacular geological phenomena.

I should like to thank the following people:

LD Ashwal	LA Larsen	SA Morse	JH Bedard
CE Lesher	AR Philpotts	AE Boudreau	B Lipin
DL Reid	CH Donaldson	S Maaløe	D Shelley
RP Hall	AA Mitchell	RA Weibe	

who, together with several of the authors of other chapters in this book, reviewed the manuscripts.

I also thank Drs Berlinda Kerkhoff of Elsevier Science for guidance during the planning and preparation of this book, and for accepting delays in deadlines with such understanding. Dr Feodor Walraven undertook all the type-setting, and his care and willingness to make repeated corrections and changes are greatly appreciated.

Finally, but most importantly, very many thanks go to my wife, Pat, for editing many of these chapters, and for her continued support and patience while I was pre-occupied with the production of this book.

Grant Cawthorn
Johannesburg
1996

Cover Photograph

The photograph on the front cover shows one of the many enigmatic features of the interlayered anorthosite-chromitite sequence at Dwars River in the Upper Critical Zone of the Bushveld Complex. Chromitite layers in anorthosite frequently bifurcate, but preserve a constant thickness of chromitite in vertical sections. Aspects of these and other features are described on pages 7, 8 and 129.

Foreword

G.M. Brown

Oxford, U.K.

The book on *Layered Igneous Rocks*, which the late Lawrence R. Wager and I published in 1968, was intended not only to present available information and ideas, but also to stimulate widespread interest in the subject. The achievement of the latter ambition, as exemplified by this new book on *Layered Intrusions* and demonstrated by more than 25 years of preceding global researches of high calibre, is a most welcome outcome.

Nowadays it would be difficult for only two authors to write the sort of comprehensive research review that we attempted, because of the greatly expanded scale of current data and ideas. Grant Cawthorn has made the right decision by encouraging a wide range of experts to deal with critical topics and types of layered igneous intrusion, for which in 1993 he provided a stimulus through organizing the Johannesburg Symposium and Bushveld Field Excursion.

There have been many occasions when I have wondered whether the significance even of igneous layering would survive the sophisticated probings that have been applied to most of the Earth Sciences over the past two or three decades. But the subject remains alive and healthy, not least because many of the observations and hypotheses which we presented have since been questioned rigorously and, where found wanting, replaced by more defensible alternatives. That applies particularly to the processes responsible for rhythmic and cyclic layering.

The concept of crystal settling and sorting within magma bodies, given exceptional support through Wager and Deer's classic 1939 memoir on the Skaergaard Intrusion, was thereafter recognized as a major influence on the differentiation of basaltic magmas (as envisaged in principle by earlier workers such as Charles Darwin and N.L. Bowen). Subsequent work by H.H. Hess and E.D. Jackson (Stillwater) and B.V. Lombaard (Bushveld) revealed problems in the application of the Skaergaard model to very thick anorthositic, dunitic or pyroxenitic layers. Since then, thanks especially to the persistence of Alex McBirney in seeking solutions to many layering anomalies, and the application by several additional researchers of experimental, electron-probe, trace-element, stable-isotopic, textural, and fluid-dynamic studies, there are now attractive new hypotheses as well as confirmation of some aspects of the earlier ones. Much evidence of these developments is contained in this book, with a commendable emphasis on the need for further research and the recognition that detailed differences between layered igneous intrusions require as much attention as their shared properties.

In 1968 we tended to emphasize those properties which were striking in their similarities, such as certain layered structures and textures, mineral compositional trends, and chemical fractionation patterns. However, we had been alerted to significant differences from a study of the Rum layered ultrabasic intrusion in the 1950s. There, an "open system" was proposed (compared with a Skaergaard-type "closed system"), in which periodically the crustal magma chamber was partially drained, and replenished from its basalt source. Of attendant significance, but less emphasized in subsequent research, was the view that such layered "crystal subtraction reservoirs" could have fed central-type volcanoes and therefore played a key role in the conversion of primary to derivative magmas. Thus a complex series of "integration stages" coupled with "differentiation stages" was envisaged, first for the Rum intrusion and later thought to be applicable to the Bushveld intrusion. Now that electron-probe, ion-probe, and refined trace-element analyses are available, it is proving possible to distinguish some of those events in relation, for example, to reversals in mineral fractionation trends.

Major differences between types of layered igneous intrusion are also evident from sophisticated textural studies, where great advances have been made by R.H. Hunter and others. When, in 1960, we first developed an "igneous cumulate" terminology in collaboration with W.J. Wadsworth, the initial aim was to overcome the use of separate names for each contrasted, thin layer within a single rock specimen. Our analogy was with metamorphic petrography where the rock name (*e.g.* schist) was pre-fixed by the main mineral assemblage in order of relative abundance. That aim has proven too cumbersome to apply, which is not surprising when one attempts to substitute andesine-rich ferrodiorite by one such as andesine-ferrohortonolite-ferrohedenbergite cumulate! However, the general term "cumulate" has survived, although it is clearly causing problems in regard to an inferred process.

I feel that there is no pressing need to seek an alternative name if cumulates are allowed to embrace a wider variety of products, all from the *accumulation (concentration) of crystalline material.* Hence sunken cumulates would be only one variant, along with flotation, floor-growth, and other types of cumulate. So far as the role of interstitial liquid is concerned, it is nowadays clear that what we identified as orthocumulates and adcumulates depended on assumptions regarding sedimentation and solidification processes that are over-ruled by the likelihood of more complex processes operating at crystal boundaries and between interstitial and main-body liquids.

Wager and I believed that early rock nomenclature was a dull subject stemming from a profusion of place-oriented names, whereas process-oriented names were a more lively prospect. That has certainly proved true, but more as a “hornet's nest” than a solution! Nobody can doubt that dunite occurs at Mount Dun in New Zealand, but I suspect that should it be called, say, a floor-growth olivine cumulate it would become a very controversial subject for many years to come. That seems to be a reasonable microcosm of geology. Good field observations are hard to refute, whereas the fun begins with the interpretations.

Every researcher on Layered Intrusions will find one or more aspects with an enduring appeal. Additional to understanding the exposed intrusions themselves, there are the broader implications for volcanology, economic deposits, ocean-crust evolution and lunar crust-upper mantle evolution, as evidenced in numerous publications other than in the field covered by this book. To end on a personal note, I have been fortunate in at least two respects, separated by many years of addiction to the subject. First, to have been a student of Bill Wager, and now to be invited by Grant Cawthorn, once one of my students, to write this Foreword. In doing so, I welcome the company of so many fellow-addict contributors, who together guarantee the continued strength and vigour of our subject.

LAYERED INTRUSIONS
R.G. Cawthorn (editor)

Mechanisms of Formation of Igneous Layering

H.R. Naslund[a] and A.R. McBirney[b]

[a]Department of Geological Sciences, State University of New York, Binghamton, New York, 13902-6000, U.S.A.

[b]Department of Geology, University of Oregon, Eugene, Oregon, 97403, U.S.A.

Abstract

Layering is a common, almost ubiquitous, feature of gabbroic and syenitic intrusions. Individual layers, or layered sequences, however, vary greatly in such features as thickness and length, the nature of layer boundaries, internal vertical and lateral variations within layers, and the relationships to other nearby layers. Their modal proportions, grain-sizes, mineral compositions, whole-rock compositions, and textures present in layers and their surrounding host rock, are also quite varied. Given the wide range of these characteristics, it is unlikely that any single layer-forming mechanism can explain all or even most of the known occurrences of igneous layering.

A wide variety of layer-forming mechanisms has been proposed. Some operate during the initial filling of a magma chamber, as a result of the settling of crystals carried in suspension, flow segregation during magma transport, magma chamber recharge, or magma mixing. Other proposed mechanisms operate in response to continuous, intermittent, or double-diffusive convection. Layering may also form as the result of mechanical processes, such as gravity settling, crystal sorting by magma currents, magmatic deformation, compaction, seismic shocks, or tectonic deformation. Variations of intensive parameters and kinetic factors, such as fluctuations of rates of nucleation and growth of crystals, oxygen fugacity, pressure, and rates of separation of immiscible liquids, may also be responsible for certain types of layering. During late-stage crystallization and cooling, layering may form in response to porous flow of interstitial liquids, metasomatism, constitutional zone refining, solidification contraction, Ostwald ripening, or contact metamorphism.

The simple concept of a magma chamber undergoing differentiation as a result of early-formed crystals settling out of the magma and accumulating in layers on the floor of the chamber, has been discarded by most petrologists in favor of models involving *in situ* crystallization, in which magma chambers are thought to have the general form of a central mass of nearly crystal-free magma, that gradually loses heat and crystallizes inwards from its margins. The transition from crystal-free magma in the central part of the chamber to completely solidified rock in the outer parts is thought to occur through a marginal zone of crystal-liquid mush. As magmas crystallize and differentiate, components included in early-crystallizing minerals are depleted, while those excluded from these phases are enriched. It is unclear, however, how the latter are effectively transferred through the crystal mush zone, so that crystallization at margins results in differentiation of the body as a whole. It is also not clear what non-steady-state or non-equilibrium processes are responsible for the formation of layering during the crystallization process. Because these two problems are interrelated, an understanding of the formation of igneous layering should eventually lead to a better understanding of the processes responsible for igneous differentiation. The time scales and length scales involved in the formation of igneous layering preclude direct experimentation on silicate melts at magmatic temperatures, and as a result, the origin of these features must be largely deduced from field observations and theoretical considerations. The challenge for the igneous petrologist is to determine which features of igneous layering are diagnostic of a particular mechanism, which reflect subsequent compositional or textural modifications, and which can best discriminate between the plethora of possible mechanisms that have been proposed.

1. INTRODUCTION

Countless studies of layered intrusions have drawn heavily on evidence deduced from layering to interpret basic processes of crystallization and differentiation. The simple model of a magma chamber undergoing differentiation as a result of early-formed crystals settling out of the magma and accumulating in layers on the floor of the chamber, has been discarded by many petrologists in favour of *in situ* crystallization with or without contributions from crystal settling and/or current flow in the magma adjacent to the crystallization front. According to this view, magma chambers have the general form of a central mass of nearly crystal-free magma, either convecting or stagnant, which gradually loses heat and crystallizes inwards from its margins. The transition from crystal-free magma in the central part of the chamber to completely solidified rock in the outer parts is thought to occur through a marginal zone of crystal-liquid mush with the percentage of liquid decreasing systematically in the direction of falling temperature (Figure 1).

Two important problems that remain to be resolved are:

1) As magmas crystallize and differentiate, components included in early crystallizing minerals are depleted in the remaining magma, while those excluded from these phases are enriched. It is not clear, however, how the latter are effectively transferred through the crystal mush zone, so that crystallization at the margins results in differentiation of the body as a whole.
2) Layers are a very common feature in slowly cooled mafic intrusions. It is not known, however, what non-steady-state or non-equilibrium processes are responsible for the formation of these inhomogeneities. It is also not known when, during the transition from liquid magma to solid rock, layering develops.

Because these two problems are closely inter-related, an understanding of igneous layering should lead to a better understanding of the processes responsible for igneous differentiation, and *vice versa*.

A number of distinct phenomena are described as igneous layering. A layer can be defined as a sheet-like inhomogeneity resulting from variations in the composition, modal proportions, or textures of minerals. Individual layers differ greatly in thickness, lateral extent, boundary characteristics, internal structures, and the textural, grain-size, and/or modal variation between the layer and its host rock. Layers also differ in their relationships to other near-by layers. They may be isolated, intermittent, or cyclic. Some have regular, parallel spacing, while others are cross stratified.

A wide variety of layer-forming mechanisms has been proposed (Table 1), and although many are applicable to specific occurrences, no single process can explain all types of igneous layering. Some operate during the initial filling of a magma chamber, some during the initial stages of crystallization when the system is dominated by silicate liquid, others during later stages of crystallization in a crystal-liquid mush, and still others during sub-solidus cooling or reheating. Some mechanisms may operate at more than one stage of the solidification process. Many layers, perhaps most, appear to have formed by a combination of mechanisms.

The references given in this chapter are meant only to illustrate the particular layer-forming mechanisms under discussion; no attempt has been made to cite or evaluate every reference for a particular mechanism. This discussion will not consider the details of "cryptic layering", for to do so would lead us into the much broader realm of igneous differentiation.

2. MAGMA EMPLACEMENT

2.1. Crystals carried in suspension

Because many lavas are erupted as phenocryst-liquid mixtures, it is likely that many of the magmas filling intrusions are also emplaced as crystal-rich liquids. The distribution of phenocrysts in many thick sills and ponded lavas (*c.f.* the Shonkin Sag, Tasmanian Dolerite, and Makoapuhi Lava Lake) have broad S-shaped vertical profiles as a result of the settling of crystals carried in suspension at the time of magma emplacement (Marsh, 1989). Large, dense crystals in the upper parts of these bodies settle faster than the rate of advance of the upper

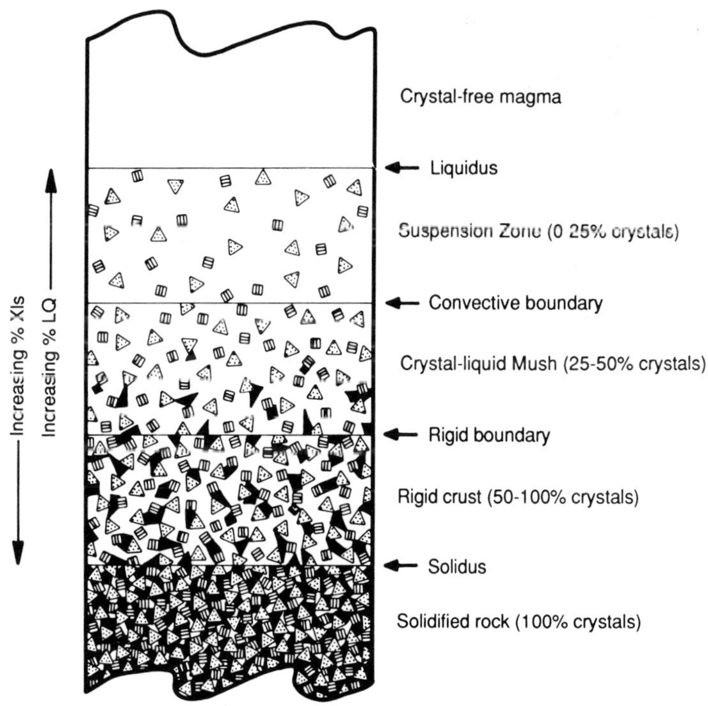

Figure 1. *Schematic profile of the interface between crystal-free magma in the centre of a magma chamber and solidified rock at the margin. Neither the absolute nor the relative dimensions of the zones are known. Modified from Marsh (1989).*

Table 1
Mechanisms for the formation of igneous layers

Mechanisms
Mechanisms that operate during magma emplacement.
Crystals carried in suspension
Flow segregation
Magma chamber recharge
Magma mixing
Mechanisms that operate in response to magma convection patterns.
Continuous convection
Intermittent convection
Double diffusive convection
Mechanisms that are the result of mechanical processes.
Gravity settling
Magma currents
Magmatic deformation
Compaction
Seismic shocks
Tectonic deformation
Mechanisms that result from variations in intensive parameters.
Nucleation rate fluctuations
Diffusion-controlled nucleation and growth
Crystal growth in thermal gradients
Oxygen fugacity fluctuations
Pressure fluctuations
Immiscibility
Mechanisms that occur during late-stage crystallization and cooling.
Interstitial crystal growth
Metasomatism
Constitutional zone refining
Solidification contraction
Ostwald ripening
Contact metamorphism

capture front and accumulate in the lower parts when they reach the upward-advancing accumulation front at the floor. This process results in broad phenocryst-poor zones or layers in the upper part and broad phenocryst-rich zones or layers in the lower part. Layers formed by this mechanism are generally thick units with gradational upper and lower boundaries, and may have bimodal grain-size distributions.

2.2. Flow segregation

The movement of phenocryst-rich magmas through conduits can result in flow segregation and concentration of crystals into specific parts of the flowing magma. This Bagnold effect causes suspended solids within a moving fluid to migrate towards regions with minimum shear stress. Large variations in phenocryst abundance, that have been attributed to flow segregation, are common in dykes and sills (*c.f.* Simkin, 1967; Gibb, 1968; Blake, 1968; Komar, 1972; Bebien and Gaghy, 1978; Ross, 1986), and in some cases these variations can be described as modal layering.

The well-known olivine horizon of the Palisades sill is a layer of olivine-rich dolerite ranging from 1 to 10 m in thickness. It is located 10 to 13 m above the basal contact of the sill and is traceable for over 40 km along strike (Walker, 1969). The origin of this unit has been debated for almost 100 years, and was cited by Bowen (1928, p.71) as a classic example of crystal settling. Recent interpretations, however, argue against gravity settling, suggesting instead that the olivine horizon is the result of either a separate pulse of olivine-rich magma (Husch, 1990) or an initially inhomogeneous magma (Gorring and Naslund, 1995). Both interpretations suggest that olivine was concentrated in the olivine-rich zone by flow segregation. Irregular cm- to m-scale layering within the olivine horizon appears to be the result of minor variations in the degree of flow segregation. Geochemical evidence from the lower part of the Palisades Sill indicates that, although plagioclase/augite and augite/orthopyroxene ratios are relatively constant, olivine/(plagioclase + pyroxene) in the olivine horizon is quite varied, suggesting that the olivine has been mechanically sorted (Gorring and Naslund, 1995).

An origin by flow segregation of a phenocryst-rich magma has also been proposed for a basal tongue of bronzite-rich dolerite that thins away from the inferred feeder system in the York Haven Diabase Sheet over a lateral distance of ~10 km (Mangan *et al.*, 1993). Discontinuous zones of weakly developed modal layering with cross-bedding in the bronzite-rich tongue may be the result of small differences in shear stress within the flowing magma.

2.3. Magma chamber recharge

Earlier suggestions that individual igneous layers were the result of separate injections of magma have been largely discounted, because the bulk compositions of many of the layers could not have been liquid at any reasonable igneous temperature. Formation of layers by separate magma injections may be a viable mechanism, however, for layers with bulk compositions comparable to those of lavas, or for layers that represent only limited differentiation of the injected magma followed by removal of the residual liquid. In either case, the injected liquid, or crystal-liquid mixture, should have a bulk composition, viscosity, density, and liquidus temperature appropriate for magmas at the depth of emplacement and in the tectonic setting in which the intrusion was formed.

In the Muskox intrusion, cyclic layered units have been attributed to repeated influxes of new magma into the chamber (Irvine and Smith, 1967). An ideal cycle has a basal dunite with 1 to 2% chromite, followed upward by a harzburgite with ~1% chromite, and an upper-most orthopyroxenite with only a trace of chromite. Within each cyclic unit whole-rock and mineral compositions typically become progressively more Fe-rich upward. Compatible trace elements, such as Ni in olivine, show a progressive decrease upwards as well. Chromitite layers are present within the dunite subunit in many of these cycles. Similar cyclic units are present in the Stillwater, Great Dyke, Bushveld, Rum, Jimberlana, and other intrusions (Jackson 1970; Campbell, 1977; Dunham and Wadsworth, 1978). The base of each cycle is thought by some to represent the influx of new primitive magma into the chamber, because it is marked by an abrupt shift to more primitive mineral and whole-rock compositions (Huppert and Sparks, 1980). This interpretation has been questioned, because it requires implausible regularity in the injection of precisely the required volumes and compositions of magma to produce the observed trends. Moreover, as Brandeis (1992) has shown, it is inconsistent with mass balance relations; the amounts of magma needed to satisfy the compositional and density requirements are far too large to be accommodated in the intrusion. Alternatively, the base of each cycle could represent a period of convective overturn in an otherwise relatively stagnant magma (Jackson, 1961).

Alternating peridotite and troctolite (allivalite) layers in the Rum intrusion have been attributed to repeated injections of picritic magma that ponded beneath cooler, lighter residual magma already in the chamber (Emeleus, 1987; Volker and Upton, 1990). Each pulse of picritic, partly-crystallized magma is thought to have formed a peridotite layer and then mixed with the resident magma in the chamber to form a troctolite layer. Alternatively, the peridotite layers may have formed from picritic magma injected as sills into a partly crystallized, layered troctolite (Bédard *et al.*, 1988). The Eastern Layered Series of the Rum intrusion has 16 such peridotite/troctolite units. Isotopic analyses confirm that the peridotites crystallized from a primitive magma and the troctolites from a more evolved, contaminated magma (Palacz and Tait, 1985). A similar model has been proposed for peridotite and troctolite layers in the Cuillin Igneous Complex, Skye (Claydon and Bell, 1992).

In the Kap Edvard Holm intrusion, layers of fine-grained, equigranular "gabbro" are thought to have formed by "intraplutonic quench" as hot fresh magma was injected into the chamber and chilled against the chamber floor (Tegner *et al.*, 1993). In the Klokken gabbro-syenite complex of Southern Greenland, alternating "granular" and "laminated" syenite layers have been attributed to lateral tongues of laminated syenite injected into pre-existing granular chilled roof rocks causing "layers" of granular textured rock to spall off and settle into the magma (Parsons, 1979). Although the granular sheets appear to have maintained coherency despite their extreme aspect ratios, some granular layers can be traced laterally into planes of autoliths. In the Isle au Haut Igneous Complex, Maine, a sequence of alternating gabbroic and dioritic layers appear to have formed from repeated injections of small batches of gabbroic magma into an evolving dioritic magma chamber (Chapman and Rhodes, 1992). Density relationships caused the gabbroic magma to be injected sill-like between the crystalline floor of the chamber and the overlying dioritic magma.

Multiple injection of magma into solidified or nearly solidified rocks has also been proposed to explain alternating layers of aplite and pegmatite centimetres to metres thick, with sharp intrusive contacts (Jahns and Tuttle, 1963). In some cases, aplite layers have injected pegmatite and in other cases pegmatite layers appear to have injected aplite. Alternatively, pegmatite-aplite layer pairs may form from injection of a homogeneous magma lens or sill that separates *in situ* into an upper pegmatitic layer and a lower aplitic layer, as has been suggested for the Calamity Peak intrusion, South Dakota (Duke *et al.*, 1988).

2.4. Magma mixing

A great deal of attention has been given to the origin of chromitite layers in layered intrusions. Since chromium is a trace element in magmas, the formation of a layer with >90% chromite must involve a column of magma hundreds of times thicker than the layer formed from it. The most common explanation for these chromitite layers is that a magma precipitating both olivine and chromite, ceased to crystallize olivine for a period of time, while chromite remained the only liquidus phase (Lipin, 1993). In the Bushveld intrusion, individual chromitite layers can be traced for hundreds of kilometres along strike with little change in thickness or stratigraphic position, suggesting that some chamber-wide process was responsible for layer formation. Irvine (1975) proposed that chromitite layers form as a result of contamination of a

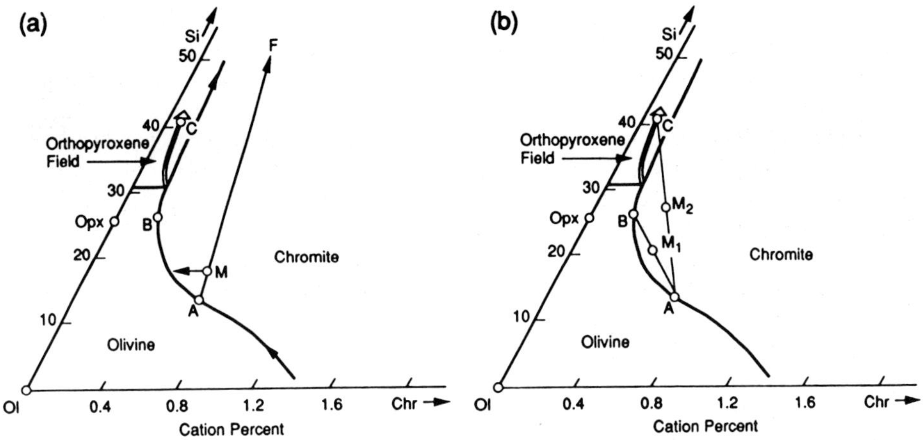

***Figure 2.** (a). Part of the system SiO_2–Mg_2SiO_4–Cr_2FeO_4 showing the fields of olivine, orthopyroxene, and chromite. Note the difference in scales between the Ol–Si and the Ol–Chr sides. Primitive magma of composition A differentiates along the curve A–B precipitating a dunite with 1.5 to 0.5% chromite. Continued differentiation from B to C moves the magma into the orthopyroxene field where first olivine and then chromite cease to crystallize. The magma path leaves the pyroxene–chromite peritectic curve and follows the heavy arrow in the orthopyroxene field, because Cr is an included element in orthopyroxene. Contamination of primitive magma at A with felsic crust (F) results in magma with composition M that will crystallize only chromite until it returns to the olivine-chromite cotectic. (b). Mixing differentiated magma at B or C with primitive magma at A results in hybrid magmas M1 or M2 that will crystallize only chromite until they return to the olivine–chromite cotectic. Figures modified from Irvine (1977).*

magma with felsic crustal rocks which forces the magma off the cotectic, and into the chromite stability field. Olivine will cease to crystallize, and the magma will precipitate only chromite until the composition of the magma returns to the cotectic (Figure 2a). It is difficult, however, to imagine how a viscous liquid of low density produced by melting of felsic crustal rocks could be efficiently mixed with a large body of underlying denser magma to produce uniform layers of chromite extending for tens or hundreds of kilometres.

Alternatively, a magma which has partly differentiated, could be forced into the chromite stability field if mixed with a more primitive magma during magma chamber recharge (Irvine, 1977) (Figure 2b). Owing to the relatively greater ease with which a basaltic magma will mix with a more primitive magma, and the evidence of magma chamber recharge associated with many chromitite layers, the second model is the more widely cited. Sequences containing numerous, sharply-bounded layers that alternate between >99% chromite and <1% chromite (Figure 3) would require numerous, abrupt, mixing episodes and the almost complete expulsion of interstitial melt, if attributed to a magma mixing mechanism. These characteristics, and the apparent replacement features associated with some chromitite layers (Figure 3), suggest that the chromite may have been redistributed and concentrated by late-stage

Figure 3. *Alternating chromitite and anorthosite layers, Dwars River, Bushveld Complex, South Africa. The extreme degree of fractionation between layers suggests that processes other than mechanical sorting must have been involved in the formation of these layers. Note the partly resorbed block of anorthosite in the upper chromitite layer.*

metasomatic processes, such as those proposed by Boudreau (1994). These processes are discussed in greater detail in a later section.

The origin of layers containing magmatic sulphides, such as the Merensky and J-M reefs in the Bushveld and Stillwater intrusions respectively, has also received much attention, mainly because of their economic importance. Like some of the chromitite layers just mentioned, these units can be traced over great distances along strike, suggesting that they are produced by events that affected the entire magma chamber. Mixing of primitive and evolved magmas along a curved phase boundary is one of the many mechanisms that have been proposed for the origin of these sulphide-rich layers (Naldrett *et al.*, 1987, 1990). Alternatively, mixing of anorthositic and mafic magmas, neither of which is saturated in sulphides, could result in a hybrid melt oversaturated in sulphides (Todd *et al.*, 1982; Irvine *et al.*, 1983).

Similar processes may be responsible for other examples of phase layering where there appears to be an abrupt change in the liquidus phase assemblage during crystallization. Any phase with a convex saturation surface, could potentially become oversaturated in a hybrid magma formed as a mixture of two liquids on or near that saturation surface, and either begin to precipitate or if already a liquidus phase greatly increase in modal proportion. Likewise, any phase with a concave saturation surface could potentially become undersaturated in a hybrid magma formed as a mixture of two magmas both of which were previously saturated in that phase. In a magma chamber with a stagnant boundary layer adjacent to the crystallization front, mixing of a differentiated boundary layer with less differentiated magma from the main reservoir could also result in a hybrid magma oversaturated in one or more phases. Mixing along a saturation surface that is convex with respect to temperature as well as composition, may cause supersaturation of one or more phases. Such a mechanism has been proposed for the formation of layers rich in skeletal magnetite, skeletal ilmenite, and hopper apatite crystals in the Skaergaard Upper Border Series (Naslund, 1984b; Keith and Naslund, 1987).

3. PATTERNS OF MAGMA CONVECTION

3.1. Continuous convection

Wager attributed the uniform "host" rock between graded rhythmic layers to deposition from "a gentle and fairly continuous convective circulation" (Wager, 1963; Wager and Brown, 1968), and discarded the earlier idea (Wager and Deer, 1939) that rhythmic layers were the result of convection currents, that "like the wind, would be variable in velocity". Although the importance of convection in magma chambers has been the subject of recent debate (Martin *et al.*, 1987; Marsh, 1989; 1991; Gibb and Henderson, 1992), in most shallow magma chambers heat is lost mainly through the roof, while most accumulation takes place at the floor. Both processes may create buoyancy fluxes that could promote "highly unsteady chaotic convection in the magma" (Martin *et al.*, 1987). As pointed out by Jackson (1961), crystallization is facilitated at the floor as a result of a 3° per km increase in average liquidus temperatures with increasing depth. Adiabatic decompression, however, lowers the temperature of a convecting liquid ~0.3° per km as it convects upwards resulting in an effective liquidus temperature differential of 2.7° per km. Convection in a well-mixed column of magma 4 km thick could result in magma at the roof being at or above its liquidus temperature, while magma at the floor is 10° below its liquidus temperature. In large magma chambers (in particular, those with a large vertical dimension, such as the Bushveld or the Muskox) heat transfer from the floor to the roof is thought to have resulted in crystallization at the floor and melting of country rock at the roof. In smaller bodies, such as the Skaergaard, Kiglaplait, and Palisades intrusions, thick floor sequences formed simultaneously with thin roof sequences attesting to the efficient transfer of heat from the base of the magma to its top. There is a general tendency for the proportion of rocks crystallized under the roof to be an inverse function of the total thickness of intrusions. An alternative explanation for thin roof sequences is that the crystal mush under the roof becomes unstable and sinks to the floor.

In dilute crystal suspensions, convection can keep particles in suspension as long as their settling velocities are small relative to turbulent fluid velocities (Marsh and Maxey, 1985; Marsh, 1988; Sparks *et al.*, 1993). Above a critical concentration, however, the particles will settle out in mass leaving behind a nearly crystal-free liquid. In convecting magmas, critical concentrations for silicate minerals are as low as 0.002 to 0.03%. In a slowly cooled, convecting magma, the concentration of crystals in suspension will increase at a steady rate, triggering discrete sedimentation events each time the concentration exceeds the critical value (Sparks *et al.*, 1993). In a multiply-saturated system containing minerals with different settling velocities, and different critical concentrations, complex sequences of layers can result from steady convection and steady cooling.

In the Khibina alkaline massif of the Kola Peninsula, massive feldspathic urtites have been attributed to crystal settling from a steadily convecting, multiply saturated eutectic magma, while layers of fine-grained ijolite, and monomineralic layers of fine-grained and coarse-grained apatite, and coarse-grained nepheline have been attributed to irregular convection with changing intensities of flow (Kogarko and Khapaev, 1987). In the Skaergaard intrusion, intervals of massive gabbro separating modally-graded rhythmic layers have been attributed to crystallization from steady convection currents (Wager and Brown, 1968; Irvine, 1987). In the Ploumanac'h subalkaline granite, Brittany, curved, truncated, and reversely-graded rhythmic layers have been attributed to Bagnold sorting in convection currents within an outer annular zone (Barrière, 1981).

3.2. Intermittent convection

A mechanism has been proposed for the formation of cyclic units in the Stillwater intrusion in which each cyclic unit begins with a brief episode of convective overturn followed by a long period of stagnation (Hess, 1960). Kakortokite layering in the Ilimaussaq intrusion has also been attributed to periodic episodes of convective overturn followed by periods of relative stagnation (Engell, 1973).

Layering in the Middle Zone of the Skaergaard intrusion is characterized by alternating plagioclase-rich and pyroxene-rich layers with sharp upper and lower boundaries, that range in thickness from 0.3 to 6 m and can be traced in outcrop for 2 km or more along strike with little obvious change in thickness (Naslund *et al.*, 1991). Individual layers appear to cross the entire intrusion. Pyroxene-rich layers have greater abundances of excluded trace elements and coarser grain sizes, than do plagioclase-rich layers. The layers are striking when viewed from a distance, because trace amounts of olivine in the pyroxene-rich layers give them a brown stain on weathered surfaces. Because the layering is difficult to follow when standing on the outcrop, one gains the erroneous impression that the layer boundaries are gradational (Irvine, 1987). It has been proposed that the plagioclase-rich layers form during periods of stagnation, and that pyroxene-rich layers form during periods of convection (Naslund *et al.*, 1991). The mineral proportions in the pyroxene-rich layers, however, more closely approximate cotectic abundances than do the proportions in the plagioclase-rich layers. Plagioclase in the plagioclase-rich layers has a low K_2O content similar to those of the Upper Border Series suggesting that they have accumulated excess plagioclase transported from the roof zone of the intrusion (Jang and Naslund, 1994). If such is the case, the plagioclase-rich layers may form during periods of convection, while the pyroxene-rich layers form during periods of stagnation. The relatively sharp layer boundaries, the variable layer thicknesses, and the intrusion-wide extent of individual layers, suggest a mechanism that is randomly and abruptly turned on and off. A model based on periodic convection and non-convection fits well with these observations.

3.3. Double diffusive convection

Double diffusive convection is a natural phenomenon that can be readily demonstrated in tank experiments and is commonly observed in oceans and saline lakes. The basic requisites for double diffusive stratification are, firstly, vertical differences in the concentration of two components having different diffusivities, and secondly, opposing effects of these components on the density distribution in the vertical dimension. The mechanism is best illustrated by the conditions at the surface of the sea, where temperature and the concentration of salt tend to be highest at the surface, and to decrease downward. Stratification is stabilized by the thermal profile, but destabilized by the concentration gradient of salt. Because heat diffuses much faster than salt, the diffusion of heat downwards tends to increase the density of the top layer causing it to descend to a depth where it finds its own density level. In the ocean, double diffusive convection effects are observed when heat and salinity are diffusing in the same direction because they have opposite effects on liquid density. In a basaltic magma, double diffusive convection effects would be expected when heat diffuses in the same direction as Fe, Mg, and/or Ca, or when heat diffuses in a direction opposite from that of Si, Na, and/or K.

Although the two components in a double diffusive system are normally thought of as temperature and composition, they could just as well be two chemical species, such as mafic and felsic components of a magma, so long as the one tends to increase density and the other to decrease it. In the case of a crystallizing magma, fractionation of the liquid adjacent to a front of crystallization at the floor would produce gradients of heat and chemical concentrations, some tending to stabilize the liquid and others causing it to overturn. Whether a liquid layer turns over or not, depends on the combined effects of these factors (McBirney, 1985).

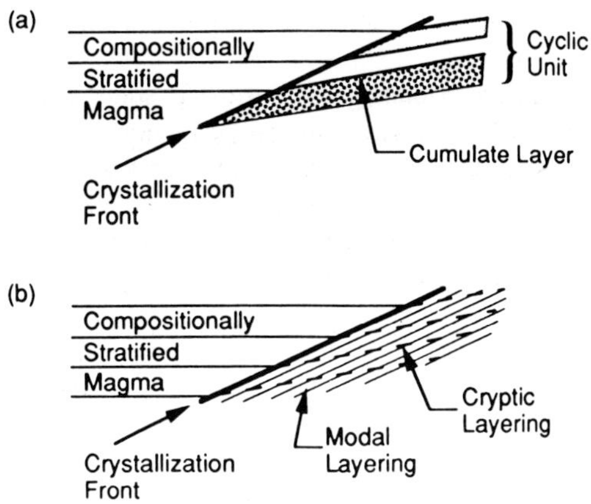

Figure 4. (a). Thick cumulate layers growing laterally into a gravity stratified magma at an angle to the crystallization front as proposed by Irvine et al. (1983). (b). Cumulate layers growing parallel to a sloping chamber floor into a gravity stratified magma such that individual layers crystallize from magmas that vary in composition along the strike of the layer as proposed by Wilson and Larsen (1985), Wilson et al. (1987), and Robins et al. (1987). Figures modified from Huppert et al. (1987).

Numerous workers have proposed double diffusive convection as a mechanism for the formation of igneous layering (McBirney and Noyes, 1979; Chen and Turner, 1980; Kerr and Turner, 1982; Irvine *et al.*,1983; Huppert and Sparks, 1984; Wilson and Larsen, 1985; Wilson *et al.*, 1987; Robins *et al.*, 1987). On closer examination, however, questions have been raised as to the effectiveness of the mechanism in magmas. It is not clear, for example, how a layered liquid can be transformed into a layered solid. As crystallization proceeds, convection cells associated with double diffusive convection are likely to migrate away from the crystallization front as fast as the front propagates upwards, and as a result, the layered liquid will remain ahead of the crystallizing solid. It has also been pointed out (McBirney, 1985) that horizontal double-diffusive convection cells will not develop in magmas near their liquidus, because temperature and composition are not independent variables (as they are in tank experiments), and the compositional effects on density are so great, that the thermal gradient required to exceed the critical Rayleigh number necessary for convection would require severe undercooling or superheating. Such wide excursions from liquidus conditions seem unlikely in slowly crystallizing magmas. As Sparks and Huppert (1987) have pointed out, however, convection theory is inadequate to predict when double diffusive convection will occur in low-temperature brines, so convection in high-temperature magmas may not conform to theoretical predictions.

In the Honningsvåg intrusive suite, mineral compositions and modal proportions vary systematically along the strike of individual layers or units. It has been suggested that these layers may have crystallized parallel to the sloping floor of a chamber containing a compositionally and thermally stratified magma (Robins *et al.*, 1987). The stratification is thought to have developed during magma recharge, when hot dense magma pooled beneath a cooler, more-differentiated, magma. Diffusion of heat and chemical components across the interface between these magmas is said to have resulted in a series of stacked, horizontal, double-diffusive layers. Crystallization along the sloping floor of the chamber is thought to have produced individual layers from a series of liquids that became progressively more Mg-rich downward in the stratified magma (Figure 4). A similar explanation has been proposed for the Fongen-Hyllingen complex (Wilson and Larsen, 1985; Wilson *et al.*, 1987) in which even more extreme compositional variations are observed along strike within individual layers. Olivine and plagioclase vary systematically from Fo_{75} and An_{63}, to Fo_{13} and An_{42}, over a strike distance of 7 km. The growth of such a layer by the proposed mechanism requires a magma chamber which contains a series of liquids covering an extreme range of igneous differentiation. In the Great Dyke, Zimbabwe, lateral variations in minor element content in pyroxene within the P1 pyroxenite have also been attributed to crystallization from different liquid layers within a diffusively stratified magma (Wilson, 1992).

In the Stillwater intrusion, it has been proposed that individual layers of different mineralogies crystallize simultaneously from separate magma layers by lateral accretion from the margins inward (Irvine *et al.*, 1983). Individual layers metres to tens of metres thick, are said to have formed at an angle to the inclined crystallization front which was propagating laterally into a double-diffusively stratified magma (Figure 4). The heat transfer mechanism by which such a large magma chamber could crystallize horizontal layers from the margins inward, rather than from the floor upwards, is not known.

4. PROCESSES OF MECHANICAL SORTING

4.1. Gravity Settling

Solid spherical particles in a liquid will settle under the influence of gravity according to the well-known Stokes' law equation:

$$V = 2r^2 g\left(\rho_1 - \rho_2\right) / 9\eta$$

where V is the velocity of a sphere of radius r and density ρ_1, settling through a Newtonian liquid of density ρ_2 and viscosity η, under the influence of gravity g. Polymerized liquids, such as magmas, are not Newtonian, however, so before settling can begin, the downward force on a crystal must overcome the yield strength of the magma (σ_y) such that:

$$\left[r g\left(\rho_1 - \rho_2\right) / 3\right] - \sigma_y > 0$$

(McBirney and Noyes, 1979). Experimental studies and field measurements on lava flows (Murase and McBirney, 1973; McBirney and Murase, 1984) indicate that yield strengths increase with increasing time and SiO_2 content, and decrease with increasing temperature and H_2O content. Measurements on crystallizing lava lakes indicate yield strengths of the order of

700 to 1200 dynes per cm^2 for a basalt at about 1130°C (Shaw *et al.*, 1968). Hulme (1974) estimated even greater values from the morphology of flow fronts of more silica-rich lava flows. In a stagnant magma with a yield strength between 500 and 1000 dynes per cm^2 (typical of a basic magma), olivine and pyroxene crystals would have to attain a size of 3 to 5 cm in

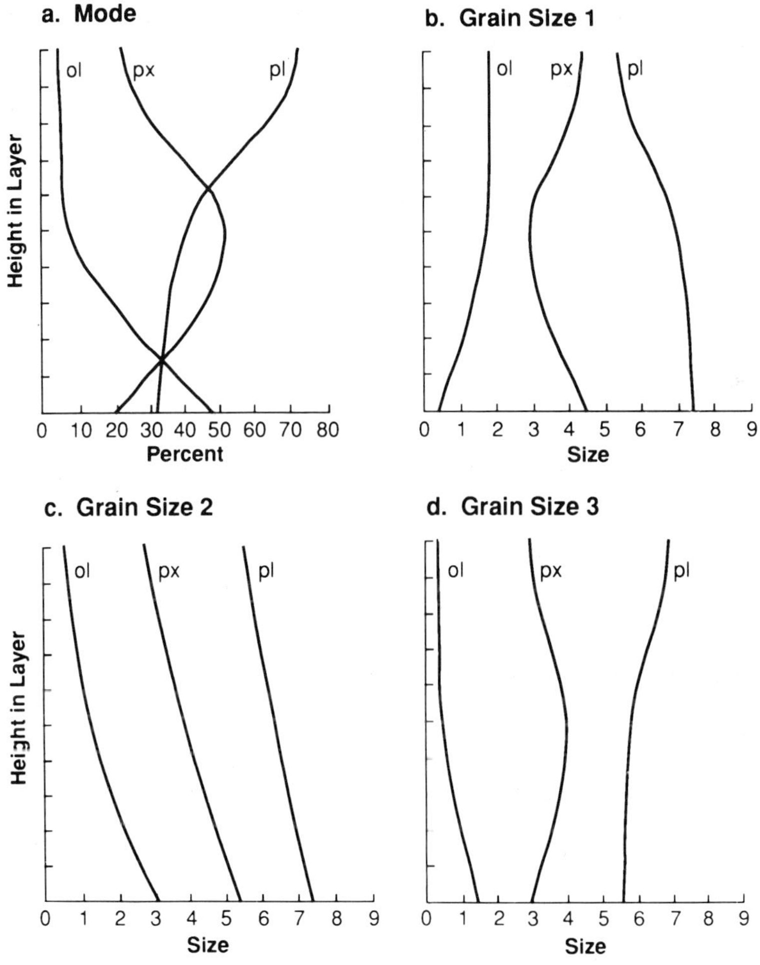

Figure 5. *(a). Theoretical profiles through a graded layer containing three minerals: olivine (ol); pyroxene (px); and plagioclase (pl), showing modal variation through the layer. (b). Theoretical grain-size distribution through the layer in (a) in which grain size is inversely correlated with mode as would be expected by a nucleation controlled process. (c) Theoretical distribution in which grain size for all phases increases downward as would be expected for a gravity-controlled process. (d). Theoretical distribution in which grain size is positively correlated with mode as would be expected for a flow segregation or an Ostwald ripening process.*

order to overcome the yield strength and initiate settling (McBirney and Noyes, 1979). The yield strength of a magma is greatly decreased during viscous flow, however, suggesting that crystal settling may be more effective in moving magmas.

If grain-size variations in layers produced by crystal settling follow Stokes' law, the coarsest grain sizes for each phase in a layer should be concentrated at the base and become finer upwards (Figure 5). Although examples of layers formed by crystal settling have been proposed in a wide variety of rock types, the best evidence for crystal settling comes from magmas that had very low viscosities and low yield strengths. Graded layers in the Imilik gabbro (Brown and Farmer, 1971), and in the Vesturhorn eucritic gabbro (Roobol, 1972) are graded in terms of mineral density and grain-size with the densest minerals and the largest grain sizes concentrated at the base of each layer. In the Duke Island peridotite (Irvine, 1974), grain-size sorting dominates the crystal distribution (Figure 6a), and density sorting is weakly developed, or in some layers even reversely graded. This pattern can be attributed to the fact that the pyroxene is, on average, coarser than the olivine, so that many layers have coarse pyroxene-rich bases and finer olivine-rich tops. In the Skaergaard intrusion graded layers are generally density-sorted, but show little or no size sorting (Figure 6b).

Goode (1976) has suggested that crystal settling in a system with continuous crystal nucleation and growth will result in massive unlayered rock. He proposed that density-sorted graded layers in the Kalka layered intrusion, Australia, resulted from "repeated bursts of

Figure 6. *(a). Size grading of olivine- and pyroxene-rich layers in the Duke Island ultramafic intrusion. (b). Density grading of pyroxene, plagioclase, and oxides in the Skaergaard intrusion.*

discontinuous nucleation, followed by differential gravity settling" (p. 379). Depending on the thickness of the nucleation zone, the height of the nucleation zone above the accumulation front, and the time interval between nucleation bursts, differential crystal settling might produce isomodal layers, graded layers, or reversely graded layers (see section 5.1). Owing to the complex thermal and rheological structure of crystallizing boundary layers, it is difficult to say whether these mechanisms could produce layering (Mangan and Marsh, 1992).

4.2. Magma currents

The apparent similarity between modally graded layers and certain types of sedimentary bedding has led many petrologists to ascribe both to deposition from turbidity currents. In the Duke Island ultramafic complex of Alaska graded layering is associated with scour-and-fill structures, slumping, angular unconformities, and layer truncations (Irvine, 1974). The "obvious similarity to graded bedding in clastic sediments leaves little doubt that" layering in the Duke Island Complex "is due to sedimentation from currents in a highly fluid medium" (Irvine, 1974, p.13). In the Fongen-Hyllingen complex, however, Thy (1983) argues against current formation of layers, even though scour-and-fill, and slump structures are common, because the plagioclase:pyroxene ratio is relatively constant within layers, and the rhythmic layering is discordant to cryptic layering. Field relations suggesting that currents have acted on partly consolidated layers do not necessarily imply that the layering was formed by gravity settling or current deposition.

Modally graded layers are a widespread, almost ubiquitous feature of the Layered Series of the Skaergaard intrusion from Lower Zone *a* through Upper Zone *a*, but are not well developed in the roof or wall sequences. In the latter, the layers tend to be more bimodal with mafic minerals more abundant in the outer part of the layers, and felsic minerals concentrated in the inner part. Layers are density-graded with olivine, ilmenite, and magnetite concentrated at the base, pyroxene in the middle, and plagioclase at the top. The lower contacts are sharp but mafic to felsic boundaries are generally gradational. They range in thickness from a few centimetres to tens of centimetres and in lateral extent from tens to hundreds of metres. They typically occur in irregular or random sequences in which individual graded layers are separated by unlayered gabbro. Some graded layers are associated with sedimentary-like features such as cross-bedding, scour-and-fill structures, and lateral grading.

These layers have been attributed to crystal-rich density currents that broke away from the walls of the intrusion and moved out across the floor leaving a density-sorted layer behind (Wager and Brown, 1968; Irvine, 1987; Conrad and Naslund, 1989). The material in the layer may have been derived primarily from the current (Wager and Brown, 1968; Irvine, 1987) or may have a substantial contribution from a stagnant zone of *in situ* crystallization on the floor that was stirred and sorted by the passing current (Conrad and Naslund, 1989). The absence of discontinuities in the wall sequence makes the second interpretation more likely. Density sorting in stagnant liquids or in laminar flow should result in grain-size sorting in which the largest grains of each mineral occur at the base of the layer, while grain-size sorting in a turbulent flow (elutriation) should result in the largest grains of each mineral occurring where that mineral is most abundant (Figure 5). Although the Skaergaard modally graded layers do not show obvious size sorting, detailed grain-size measurements on six modally graded layer sequences indicate a strong correlation between grain size and mineral mode (Conrad and Naslund, 1989). To date, however, the origin of modally graded layers has not been rigorously examined in terms of what is now known about deposition of mixed solids from suspensions. It is known from industrial experience, for example, that when two or more particle types of differing sizes and densities settle from slurries, they may be deposited in a variety of ways depending on their relative size distribution, shape distribution, and concentrations, and on the physical properties of the liquid.

Near the top of Upper Zone *a* in the Skaergaard, modally graded layering is replaced by remarkable trough structures composed of stacks of synformal layers 10 to 50 m wide, up to 100 m high, and 450 m or more in length. Some troughs form broad, shallow, linear depressions while others are distinctly U-shaped with steep sides dipping up to 80° toward the trough axis. Over 21 principal troughs and 23 subsidiary troughs have been mapped (Irvine, 1987). The troughs are subparallel and are spaced at approximately 30 to 50 m intervals, separated by ridges of more massive ferrogabbro. The trough structures have been attributed to intermittent density currents that became "canalized" during the later stages of crystallization of the intrusion (Wager and Brown, 1968). Their forms, however, appear to be depositional and not erosional. If they formed from density currents and were the sites of increased deposition, it is not clear why they did not fill in within a short vertical sequence. Irvine (1987) has proposed a complex model in which the trough form is maintained by elongate, subparallel roller convection cells, and layering within the troughs is deposited by density currents much like those proposed for the modally graded Skaergaard layering. The layers in many of the trough structures, however, are of more extreme composition than any other Skaergaard layers. Some are nearly pure anorthosites, while others consist almost entirely of olivine, pyroxene, and Fe-Ti oxides. In addition, most troughs are surrounded by halos of anorthositic gabbro. These features suggest that some process other than, or in addition to, magmatic sedimentation, must have been involved. Sonnenthal (1992) and McBirney and Nicolas (in review) have suggested that the structural and geochemical features of the trough structures may be best explained as a result of compaction.

4.3. Magmatic deformation

Layering can also be produced by various types of deformation, including viscous flow, slumping, and compaction. Deformation of crystallizing magmas differs from that of liquids in that the former are normally anisotropic. In this sense they have much in common with metamorphic rocks, but they differ from solids in that almost all of the strain is taken up by the liquid matrix, and individual crystals show much less evidence of mechanical deformation. The way a partly crystallized magma responds to stress is very sensitive to the proportions of solids and liquids. The deformation features produced in magmas with less than a critical melt fraction of 20 to 30% differ from those in which enough liquid is present to prevent extensive grain-to-grain contact of the suspended solids (Nicolas, 1992).

The most distinctive feature of layering produced by simple shear of crystallizing magma is a linear orientation of crystals within a plane of foliation defined by tabular crystals. Foliation alone is not necessarily the result of magmatic flow; it may have any one of a variety of origins. The strong foliation commonly referred to as “igneous lamination”, for example, has been attributed to compaction, but in the Skaergaard intrusion no relationship can be found between strongly laminated, plagioclase-rich rocks and deformation (McBirney and Hunter, 1995). Although the preferred orientation of platey plagioclase crystals may be very marked (Figure 7), the lamination in some units crosses lithologic boundaries and may vary by as much as 90° over distances of a few tens of centimetres. As Higgins (1991) has shown, mechanical rotation of grains alone is not sufficient to generate strong fabrics. Thus, this type of strong foliation

***Figure** 7. Layer-parallel igneous lamination in the Layered Series of the Skaergaard intrusion produces a planar schistosity in the gabbro.*

without lineation is unlikely to be primarily the product of deformation even though it may be associated with it.

A distinctive type of layering produced by deformation results from the segregation of liquids into zones of minimum stress to form lenses and schlieren. Layers of this kind are common in zones of disturbance, particularly near the margins of intrusions. They are characterized by sharply defined dark and light layers that in extreme cases may be nearly monomineralic (Figure 8). Some are more mafic than the host rock, others are more felsic, and some have both mafic and felsic rocks in close association.

Layered gabbros from the lower crustal section of the Oman ophiolite have strong magmatic foliations and lineations that are at an oblique angle to the compositional layering. These fabrics have been interpreted as due to imbrication and laminar flow within the ophiolite magma chamber (Benn and Allard, 1989). It has been suggested that the imbrication direction of these fabrics can be used to determine the shear sense during magmatic flow. An origin by shear during magmatic flow has also been proposed for layers of laminated anorthosite within a massive anorthosite host rock in the Sept Iles intrusion (Higgins, 1991).

4.4. Compaction

The processes by which crystals are consolidated into solid layered rocks are complex and poorly understood. Whether crystals settle out of suspension during viscous flow or from a dispersed state during slow growth in an advancing zone of crystallization, the ensuing

Figure 8. *Schlieren of mafic and felsic gabbro in Lower Zone of the Skaergaard intrusion developed as a result of segregation of liquids into zones of minimum stress during magmatic deformation. Note that the white anorthosite cuts across a mafic layer at an angle close to 45°. The latter is parallel to the planes of slumping near the steep margins of the Layered Series.*

compaction may develop some form of layering as a result of mechanical sorting, recrystallization, or some combination of the two.

Coats (1936) was probably the first to point out that crystals of differing sizes and densities tend to sort themselves in crude layers as they consolidate under the force of gravity. The forces responsible for this sorting are not well understood, but seem to be related to a self-organization of particles according to their drag coefficients in a viscous fluid. Layering that is thought to have been caused by an effect of this kind is found in coarse, pyroxene-rich zones of the York Haven Diabase and in some of the large sills of Antarctica (B.D. Marsh, pers. comm.). Until the process can be evaluated more quantitatively, however, it is difficult to judge its importance to igneous layering.

Even when crystals form a self-supporting framework they can continue to compact, reducing the pore space and driving out interstitial liquids. Textural evidence shows that crystals may be deformed during compaction (McBirney and Hunter, 1995), and that pressure solution at the contacts of grains may be at least equally important (Dick and Sinton, 1979). Because the surface energy of a crystal increases with stress, points where stress is concentrated tend to dissolve while those under less stress tend to grow (Fyfe, 1976). The presence of a liquid or fluid medium is essential for effective transfer of mass from one site to the other. Liquids expelled from deeper in the zone of crystallization would greatly facilitate this process. Because these liquids are not in equilibrium with the crystals at higher, hotter levels, they tend to dissolve the crystal matrix through which they are percolating, absorbing heat and moderating the chemical and thermal gradients (McBirney, 1987; 1995).

Pressure solution can produce layering in a rock undergoing simple shear (Dick and Sinton, 1979). The mechanism is based on the principle that, if two mineral species differ in their ability to deform under stress, the more readily deformed species will be preferentially concentrated in zones of greatest shear by selective dissolution and reprecipitation of the more rigid phase. This same mechanism can operate under pure shear associated with compaction. Any initial modal variations will result in the less deformable mineral being under greater stress in a layer where it is the subordinate phase than in one where it is more abundant. Once the relative size of grains is reduced by pressure solution, the chemical potential difference is further increased by the size-dependent difference of surface energy (see section 6.5). As a result, the mineral will preferentially reprecipitate in the layer where it is most abundant, and an initially weak inhomogeneity can develop into layers that are increasingly mono-mineralic. Liquid expelled by compaction and rising through the crystal mush helps surmount the limitations of diffusive transfer and increases the vertical dimensions of the layering.

Magma expelled during compaction may move through the crystal pile as waves or pulses (Richter and McKenzie, 1984) which may contribute to layering formed in the manner just described. Expelled liquid may also collect along shear planes forming layers with evolved compositions. Alternatively, liquids expelled by compaction may pond on the floor of the magma chamber and crystallize as “adcumulus layers” at the crystallization front (Kanaris-Sotiriou, 1974). Discontinuous pegmatitic layers of granophyre in massive anorthosites of the Sept Iles intrusion have been attributed to the expulsion of interstitial liquids during compaction (Higgins, 1991).

4.5. Seismic shocks

Experimental studies show that spontaneous nucleation and growth can be triggered in supersaturated liquids by agitation. Seismic shock waves may cause layering by intermittent agitation of a supersaturated magma resulting in changes in the rates of crystal nucleation, growth, or settling (Hoffer, 1965). Alternatively, seismic shock waves might result in disruption and crystal sorting within the suspension zone of an *in situ* crystallization front along the floor of a magma chamber. In the Klokken gabbro-syenite complex, Greenland, granular layers overlying some graded layers have been attributed to "the spalling off of a granular sheet from the roof initiated by minor earth movements" (Parsons, 1979, p. 691).

Aftershocks and local deformation occurred in Long Valley in apparent response to the 1992 Landers earthquake which had an epicentre 400 km away. These local events were attributed to the development of approximately 0.1 kbars of overpressure in the magma chamber beneath the Long Valley Caldera, as the result of the rise and expansion of gas bubbles dislodged by the distant Landers earthquake (Linde *et al.*, 1994). Pressure fluctuations of this magnitude could result in the formation of layering by either triggering a burst of nucleation, or shifting phase boundaries in a multiply-saturated system (see section 5.5). Alternatively, syn-magmatic deformation may result in fractures and sudden vapour loss (Lofgren and Donaldson, 1975) which can trigger layer formation. Seismically induced layers should be laterally continuous over the entire chamber, and because such events are short lived relative to the cooling times of intrusions, they might be characterized by thin abrupt layers in otherwise homogeneous rock. If such layers could be identified in an intrusion, they might provide a record of seismicity during solidification (Hoffer, 1965).

4.6. Tectonic deformation

Thayer (1963) suggested that "flow-layering" forms in alpine peridotite-gabbro complexes during emplacement as crystal-liquid mushes. Such flow layers may be monomineralic or polymineralic, *i.e.* dunitic, anorthositic, or gabbroic; their contacts can range from sharp to gradational; and they may have foliation, lineation, or both. Although some flow layers appear to be parallel and uniform over distances of tens of metres, careful examination usually reveals that they are lenticular and pinch out within a few metres. Boudinage and fold structures are common. Similar flow layering in the Gosse Pile intrusion of Australia has been attributed to sub-solidus, syn-tectonic annealing (Moore, 1973), while flow layering in the Josephine and Red Mountain peridotites has been attributed to metamorphic differentiation accompanying deformation, pressure solution, and anatexis under mantle conditions prior to emplacement in the crust (Dick and Sinton, 1979). Petrofabric studies of olivine in the dunites of Almklovdalen, Norway suggest that textural layering in these bodies formed at sub-solidus temperatures during deformation and recrystallization (Lappin, 1967).

5. VARIATIONS OF INTENSIVE PARAMETERS

5.1. Nucleation rate fluctuations

Magmas must be supersaturated in order to nucleate and grow crystals, because, by definition, the nucleation rate and the growth rate of any crystal at equilibrium is zero. Supersaturation of a crystal-liquid system can be obtained by cooling the system below the equilibrium temperature, by shifting the liquid away from the equilibrium composition, or by changing the intensive parameters (T, P, P_{H_2O}, f_{O_2}). As a result, all crystallization in intrusions occurs under supersaturated conditions.

The growth rate of a crystal in a melt is dependent primarily on the volume free energy change associated with transferring components from the melt to the crystal, while the nucleation rate of a phase is dependent upon both the volume free energy term and a surface free energy term:

$$G_n = (G_s \cdot S_n) + (G_v \cdot V_n)$$

where G_n is the free energy of a crystal nucleus, G_s is the surface free energy term, G_v is the volume free energy term, S_n is the surface area of the nucleus, and V_n is the volume of the nucleus. Owing to their small size, crystal nuclei have large surface areas relative to their volumes. Surface free energy terms are uniformly positive and increase as a function of the surface area as a crystal nucleus grows. In order for nucleation to occur, the volume free energy term, which increases as the volume of the nucleus grows, must be negative and must increase in magnitude at a faster rate than the surface free energy (Figure 9). The nucleus size at which this occurs is called the critical radius. Because the volume free energy term increases with supersaturation, whereas the surface free energy term remains relatively constant, the critical radius for a given phase decreases with increasing supersaturation. The likelihood that random collisions might create molecular clusters that exceed the critical radius, therefore, will increase with supersaturation. The increase in growth rate as a function of increasing supersaturation generally exceeds the increase in nucleation rate (Figure 10), because for

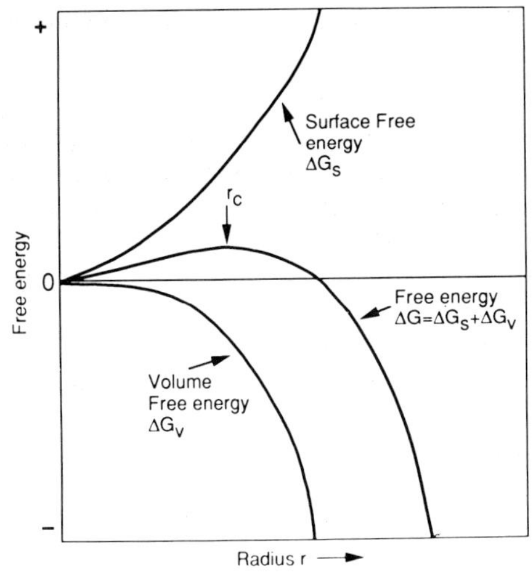

***Figure 9**. Plot of free energy versus radius for small nuclei. The surface free energy increases as a function of r^2 while the volume free energy increases as a function of r^3. The critical radius (r_c) marks the point where continued growth of the nuclei decreases the total free energy.*

crystals orders of magnitude larger than the critical radius, the surface area (and the surface free energy term) increases at a much slower rate than the volume (and the volume free energy term). Both the nucleation rate and the growth rate eventually fall off with increasing supersaturation of a melt, because at high degrees of supersaturation the melt undergoes transformation into a glass in which molecular motion is greatly retarded.

Numerous investigators have proposed layer-forming mechanisms based on the difference between increasing nucleation rates and increasing growth rates in supersaturated systems (*c.f.* Harker, 1909). Wager and Brown (1968) attributed the growth of crescumulate layers in the Marginal Border Series of the Skaergaard intrusion to delayed nucleation and rapid growth in stagnant magma before convection began. Wager (1959) suggested that cyclic layering in the Bushveld, characterized by graded units with basal chromitite, followed upwards by orthopyroxenites, and finally by plagioclase-rich rocks, is the result of the order in which the phases nucleated, which was controlled by the complexity of their crystal structures. Hawkes (1967) proposed a similar mechanism for rhythmic layers in the Freetown Complex, Sierra Leone, in which layers rich in olivine or pyroxene at their base and rich in plagioclase at their tops, form because olivine and pyroxene nucleate at lower degrees of undercooling than does plagioclase. Wager (1959) also suggested that within the Skaergaard intrusion, the large-scale, intrusion-wide layers, such as the "Triple Group", were difficult to explain "solely on a specific gravity and winnowing basis" (p. 79) and that variations in crystal nucleation rates probably played a role. Maaløe (1978) suggested that both macro-rhythmic layering and modally-graded rhythmic layering in the Skaergaard intrusion may be the result of an interplay between nucleation rates and growth rates within the Skaergaard magma chamber. In this model, supersaturation develops until one phase nucleates, after which growth of the nucleated phase decreases supersaturation and, hence, the nucleation rate of that phase, and increases supersaturation and, hence, the nucleation rate of the other phases. As a given phase nucleates and grows it causes a compositional shift in the magma under relatively isothermal conditions, that results in the nucleation and growth of additional phases. Hort *et al.* (1993) have examined this phenomenon and conclude that layering due to oscillatory nucleation can occur only in intrusions of more than a certain thickness and also depends on the viscosity of the magma and the growth rate of crystals.

Sørensen and Larsen (1987) proposed a model for the Ilimaussuaq intrusion in which increasing vapour pressure caused an increase in the nucleation rates of feldspar and nepheline relative to pyroxene, and hence produced normally graded layers, while decreasing vapour pressure caused a decrease in the nucleation rates of feldspar and nepheline, and hence produced inversely graded layers (see section 5.5). Parsons and Becker (1987) proposed a similar model for the Klokken intrusion. Goode (1976) proposed an explanation for density-graded layers in the Kalka intrusion, Australia, involving repeated bursts of simultaneous crystal nucleation followed by differential settling of pyroxene and olivine relative to plagioclase (see section 4.1). Lofgren and Donaldson (1975) suggested that alternating layers of crescumulate (comb-layered) plagioclase and pyroxene result from nucleation and growth in a supersaturated (compositionally supercooled) boundary layer. The nucleation of one phase results in rapid growth of a crescumulate layer outwards into the supersaturated melt, and results in the eventual buildup of rejected components to the point where a second phase nucleates and forms a second layer (Figure 11).

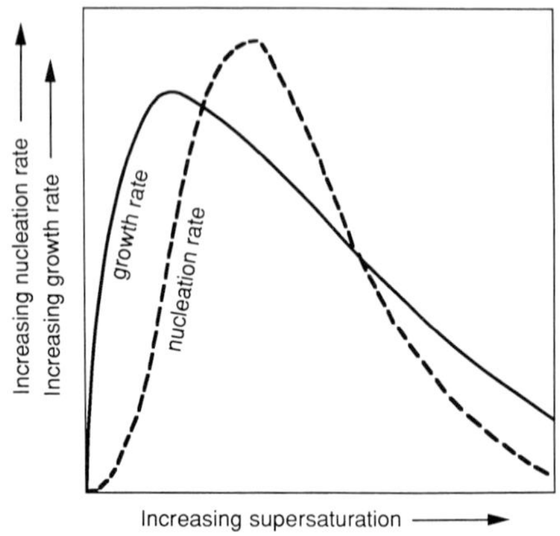

Figure 10. *Plot of nucleation rate and crystal growth rate vs. increasing supersaturation. The growth rate curve (solid line) peaks at lower degrees of supersaturation than does the nucleation curve (dashed curve). In a crystal-free system, supersaturation will increase until sufficient nucleation occurs to allow crystal growth to decrease the degree of supersaturation. Crystal growth will continue at low degrees of supersaturation inhibiting further nucleation. In a situation where a thermal gradient or a compositional gradient is migrating into a magma body, layering may form in response to cycles of increased supersaturation, nucleation, crystal growth, and reduced supersaturation.*

In any mechanism dependent upon differences in nucleation rates to cause differences in modal abundances, layers with greater abundances of a phase should have more nuclei and, hence, a smaller average grain size than layers with smaller abundances and fewer nuclei. Samples from layered sequences in which modal layering has been formed by differences in nucleation rates, should therefore, demonstrate a negative correlation between mode and average grain size for individual phases (see Figure 5). Maaløe (1978; 1987) suggested that there is a strong negative correlation between mode and grain size of individual minerals in Skaergaard rhythmic layering. He used "crystallinity" (C) and "crystal index" (n) to calculate an average grain volume and an average nucleation density from the number of crystals (N) in a given area of thin section (A) and the per cent mode of the mineral (M) as follows:

$C = (N / A)^{3/2}$ and $n = (N / A)^{3/2} / (0.01M)$

This procedure has been shown to be incorrect (Conrad and Naslund, 1989); it results in a negative correlation between mode and average grain size even in sequences in which the reverse is true. Direct measurements of average grain sizes in Skaergaard layered sequences suggest that within the intrusion-wide macro-rhythmic sequences, the pyroxene-rich layers have coarser pyroxene and plagioclase than do the more plagioclase-rich layers (Naslund *et al.*, 1991), and that within the more locally developed modally graded layers there is a positive correlation between mode and grain size (Conrad and Naslund, 1989). These results suggest that variations in nucleation rate did not play an important role in the formation of either of these types of layering.

5.2. Diffusion controlled nucleation and growth

The phenomenon of Liesegang banding (Liesegang, 1896) is a well-known process of oscillatory crystallization and rhythmic layering. The effect can be demonstrated at low temperatures by simple experiments (McBirney and Noyes, 1979). Liesegang banding in sedimentary rocks consists of fine-scale mineral layering formed during diagenesis, often at high angles to the original sedimentary layering. Knopf (1908) and Liesegang (1913) suggested that orbicular textures in granitic rocks formed as a result of the Liesegang phenomena operating in partly solidified magmas. Ray (1952), Leveson (1966), and McBirney *et al.* (1990) have described other examples of orbicular structures that may have formed in this way.

Taubeneck and Poldervaart (1960) and McBirney and Noyes (1979) proposed a mechanism involving diffusion of heat and chemical components in the boundary layer at the margin of a magma chamber to form rhythmic layering. In this model, if crystals of a mineral nucleate and begin to grow, the components that make up that mineral will diffuse towards the growing crystals forming a zone of depletion adjacent to the crystallization front that inhibits further nucleation. If nucleation requires a significant degree of supersaturation, initial crystal growth will be rapid and the depletion zone will rapidly advance towards the main magma reservoir.

As the system approaches equilibrium temperatures, the growth rate becomes slower, and the rate of advance of the depletion zone decreases. Because the rate of diffusion of heat remains relatively constant, the advancing cooling front eventually overtakes the edge of the depletion zone and initiates a new pulse of nucleation (Figure 12). Although the rate of heat diffusion is relatively constant, the rate of advance of the cooling front acts

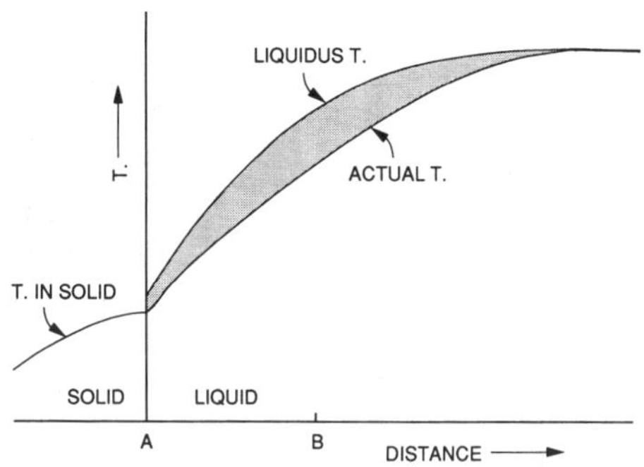

Figure 11. *A plot of temperature vs. distance for a profile through a crystallization front in which the liquidus temperature in the adjacent magma is depressed by the addition of rejected components at the crystal-liquid interface. The hachured area represents a zone of constitutional supercooling. Nucleation of a layer at A on the horizontal axis lowers the liquidus temperature curve and raises the actual temperature curve as a result of the release of rejected components and the heat of crystallization at the crystallization front. As the crystallization rate decreases, the liquidus temperature curve rises and the actual temperature curve falls, resulting in a sufficient degree of supersaturation at B to nucleate a new layer. Because the liquid is oversaturated beyond the interface, any crystal that happens to extend into that region will grow rapidly in that direction and produce long acicular crystals oriented normal to the front of crystallization.*

antithetically to the rate of chemical diffusion. After each new pulse of nucleation, crystal growth and the resultant diffusion of components towards the growing crystals accelerates. The sudden release of the latent heat of crystallization that accompanies accelerated crystal growth acts to slow or temporarily halt the advance of the cooling front. The same principles apply if the two diffusing components are chemical species of differing diffusivities.

In a multiply saturated system, the supersaturation of each phase is affected by the nucleation and growth of other phases, so that the formation of a layer rich in one mineral component may act to trigger formation of a following layer rich in another. Like layering formed by changes in nucleation rates, layering formed by diffusion-controlled nucleation

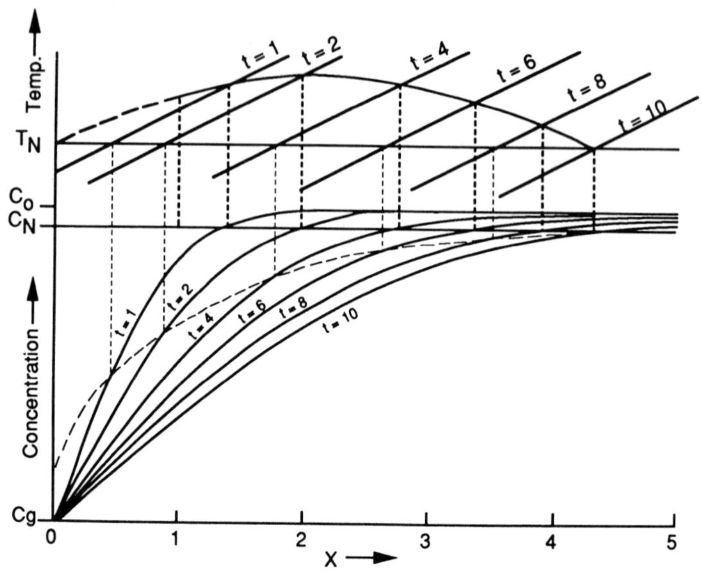

Figure 12. *Changes following nucleation and rapid crystal growth at position x=0 and time t=0. The lower part of the diagram shows concentration in the magma vs. distance profiles for time t=1, 2, 4, 6, 8, and 10. The upper part of the diagram shows temperature vs. distance profiles for t=1, 2, 4, 6, 8, and 10. For simplicity, temperature profiles are shown as straight lines, whereas in reality, they would be complex functions of heat loss to the walls, heat loss to convecting magma, and heat gain from crystallization. Time units and distance units are arbitrary.* C_o *denotes the initial concentration,* C_N *denotes the concentration necessary for nucleation at temperature* T_N*, and* C_g *denotes the concentration following rapid growth. The upper solid curve (constructed with heavy vertical dashed lines) indicates the temperature in the magma at a given position of x when the concentration profile falls below* C_N*. The lower dashed curve (constructed with the light vertical dashed lines) indicates the concentration in the magma at a given position of x when the temperature reaches TN. Following initial nucleation at position x=0 and time t=0, nucleation is inhibited until position x=4.3 and t=10 when the concentration is again above* C_N *and the temperature is below* T_N*. Figure modified from McBirney and Noyes (1979).*

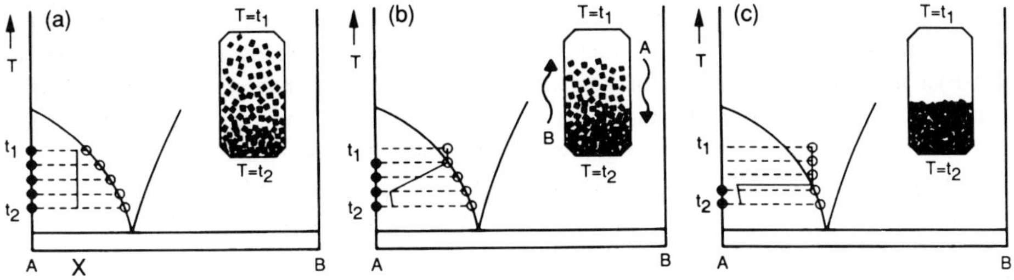

Figure 13. *(a). Initial crystallization across a zone with a temperature gradient results in 10% crystallization at the hot end and 50% crystallization at the cooler end for an initial uniform bulk composition at X denoted by the solid vertical line. The composition of the interstitial liquid (denoted by open circles) will follow the liquidus curve, and the composition of the solid (denoted by filled circles) will be pure A. (b). Migration of component A down its compositional gradient towards the cooler end and component B down its compositional gradient toward the hot end will promote increased crystallization at the cooler end and dissolution at the hot end. The bulk composition will shift towards component B at the hot end and towards A at the cooler end as shown by the solid line. (c). If allowed to go to completion, the end result will be solid A with a minimum of liquid at the cold end and all liquid at the hot end. The final bulk composition profile is shown by the solid line. Figures modified from Lesher and Walker (1988).*

should demonstrate a negative correlation between modal proportion and average grain size for individual phases.

5.3. Crystal growth in thermal gradients

Experimental studies (Lesher and Walker, 1988) have demonstrated that chemical migration in thermal gradients might act as a potential driving force for cumulate compaction and layer formation in slowly cooled plutonic bodies. In a multiply saturated melt, individual crystal solubilities change as a function of temperature, setting up gradients in interstitial melt composition wherever there is a gradient in temperature. Mass transport in response to this thermal and compositional gradient, referred to as thermal migration (Lesher and Walker, 1988), acts to promote additional growth in the cooler regions of a crystal mush and migration of interstitial melt towards the warmer regions. In Figure 13a, a thermal gradient applied across an originally homogeneous interval of melt results in a few crystals (~10%) in equilibrium with a melt enriched in component A at the high temperature end, and many more crystals (~50%) in equilibrium with a melt enriched in B at the cooler end. As long as the crystal - liquid mush remains permeable, component A will diffuse down its compositional gradient towards the cooler end, and component B will diffuse down its compositional gradient towards the hotter end (Figure 13b). If the process is allowed to go to completion, the final result will be a layer of coarse crystals with a minimum amount of trapped liquid at the cooler end, and a homogeneous expelled liquid at the hotter end (Figure 13c).

Heat loss to the country rock promotes the migration of interstitial liquids back into the main magma reservoir, while heat loss to convection within the chamber promotes trapping of interstitial liquids within the crystal mush. Because rates of thermal diffusion greatly exceed those of compositional diffusion (*i.e.* the Lewis number = thermal diffusion / chemical diffusion = ~10^4) chemical migration cannot keep pace with solidification in a steady-state system. Layering, by definition, however, is not a steady-state process, but rather one that requires some intermittent fluctuation of conditions. Thermal migration in a cumulate pile 10 m thick could cause mass reorganization on a scale of *ca.* 1 mm, while thermal migration in a crystallizing zone 1 km thick could result in mass reorganization on a scale of 10 cm to 1 m (Lesher and Walker, 1988). If a thickness on the scale of 10 to 100 m is assumed, interruption of the solidification process at the appropriate intervals could result in layers on the scale of mms to cms as a result of thermal migration.

5.4. Fluctuations of oxygen fugacity

The liquidus phases in equilibrium with a magma are controlled by composition, temperature, and oxygen fugacity. In systems that co-precipitate silicate and oxide minerals, oxygen fugacity can control the phases crystallized, the liquid differentiation path, and the compositions of the phases in equilibrium. In the system Mg_2SiO_4-FeO-Fe_2O_3-$CaAl_2Si_2O_8$-SiO_2 (Figure 14) a liquid in equilibrium with plagioclase, pyroxene, and olivine at low f_{O_2} (10^{-11}) will be in equilibrium with only pyroxene at higher f_{O_2} (10^{-9}). Experimental studies also indicate that pyroxenes and spinels precipitated at higher f_{O_2} are more Mg-rich than those precipitated from the same magma at lower f_{O_2}. Pulsating or fluctuating f_{O_2} in these systems could result in sequences of silicate-rich and oxide-rich layers with complex variations in mineral composition (Ulmer, 1969). Oxygen fugacity variations in a magma could be caused by assimilation of water-rich or CO_2-rich country rocks, gas release through vents to the surface, loss of gases by diffusion, temperature fluctuations, convection, or fractionation of oxide-rich phases. Layered sequences with alternating chromite-rich and silicate-rich layers (such as those in the Lower Zones of the Stillwater and the Bushveld), or with magnetite-rich and silicate-rich layers (such as those in the Bushveld and Skaergaard) may have formed as a result of variations in f_{O_2} within the crystallizing magma (*c.f.* Cameron, 1975; 1977). Reynolds (1985a) has suggested that extensive magnetite-rich layers in the upper zone of the Bushveld intrusion formed as a result of variations in f_{O_2}, T, f_{H_2O}/f_{H_2}, and Fe_2O_3/FeO in an iron-enriched liquid, formed by the local precipitation of plagioclase, that ponded on the floor of the intrusion. He attributed the conversion of an intial oxide-rich layer, into a nearly mono-mineralic layer, to subsolidus annealing and densification.

Oxygen fugacity fluctuations may also affect the relative stabilities of silicate phases and result in modal layering. In the Norite I subzone of the Stillwater intrusion, plagioclase in anorthosite has higher Fe and lower Eu contents than does plagioclase from norite, suggesting that anorthosite layers may have formed as a result of a reduction in pyroxene stability during intervals of increased oxygen fugacity (Ryder, 1984). Unclear in any of the oxygen fugacity driven models is how a change in f_{O_2} can be propagated over great distances through an intrusion to produce laterally-extensive layers.

5.5. Pressure fluctuations

Repeated variations of either total pressure or vapour pressure have been proposed to explain alternating layers of aegirine, arfvedsonite, and eudialyte in the Ilimaussaq intrusion, Greenland (Ussing, 1911; Ferguson and Pulvertaft, 1963). "Inversely" graded layers within the Ilimaussaq intrusion may have formed during periods of gradually increasing vapour pressure, while "normally" graded layers formed during periods of gradually decreasing vapour pressure

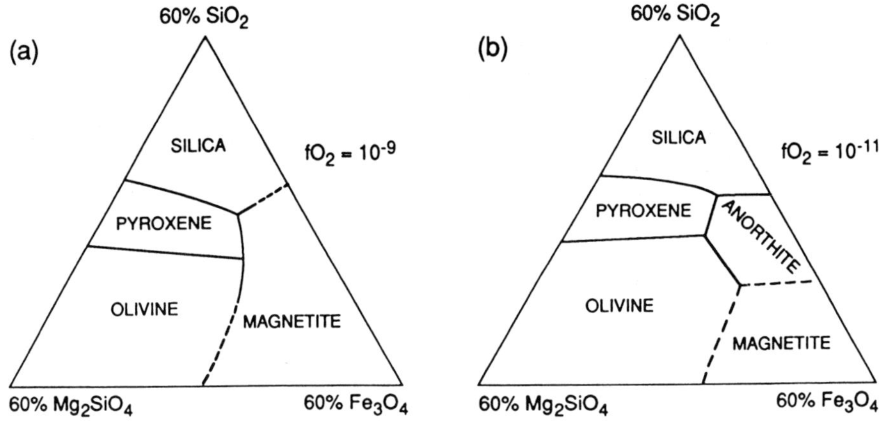

Figure 14. *(a). Phase relations on the 40% anorthite join in the system $MgSi_2O_4$-FeO-Fe_2O_3-$CaAl_2Si_2O_8$-SiO_2 at an oxygen fugacity of 10^{-9}. (b). The same join at an oxygen fugacity of 10^{-11}. Oxidation of a liquid in equilibrium with pyroxene, anorthite, and olivine at an oxygen fugacity of 10^{-11} will result in a liquid saturated only in pyroxene. Figures modified from Ulmer (1969).*

(Sørensen and Larsen, 1987). Inversely-graded layers in the Klokken gabbro-syenite complex, Greenland have been attributed to rhythmic pressure build-up followed by sudden release (Parsons, 1979). Rhythmic textural and modal layering in the Calamity Peak pluton, South Dakota, has been attributed to repetitive episodes of water vapour exsolution triggered by the precipitation of tourmaline (Rockhold *et al.*, 1987). The depletion of boron in the melt by the crystallization of tourmaline lowers the solubility of water, and results in the exsolution of a volatile phase. Partitioning of boron into the released vapour causes tourmaline crystallization to cease. Slight fluctuations in confining pressure on a magma saturated in volatiles has been proposed to explain mm- to cm-thick layers of garnet, tourmaline, and muscovite in some pegmatite-aplite associations (Jahns and Tuttle, 1963; Jahns, 1982). A sudden release of pressure has also been proposed as a mechanism for rapidly inducing the supersaturation conditions necessary for crescumulate layers in plutonic environments where rapid heat loss is unlikely (Lofgren and Donaldson, 1975).

Changes in total pressure within a crystallizing magma chamber could change the equilibrium liquidus assemblage and result in phase layering (Cameron, 1977; Lipin, 1993). In the systems Mg_2SiO_4-$CaAl_2Si_2O_8$-SiO_2 (Sen and Presnall, 1984) and Mg_2SiO_4-Fe_2O_3-$CaAl_2Si_2O_8$-SiO_2 (Osborn, 1978) the fields of spinel and orthopyroxene expand with increasing pressure, over the range of 1 bar to 10 kbars, at the expense of the olivine and plagioclase fields (Figure 15). Pressure increases within a magma chamber could result in chromite, magnetite, or orthopyroxene-rich layers, while pressure decreases could result in anorthositic or dunitic layers. Laterally continuous chromitite layers in the Stillwater Complex have also been attributed to such changes in pressure (Lipin, 1993).

The effects of a pressure change would be felt nearly simultaneously over the entire magma chamber, and as a result, a pressure-change mechanism for layer formation is particularly

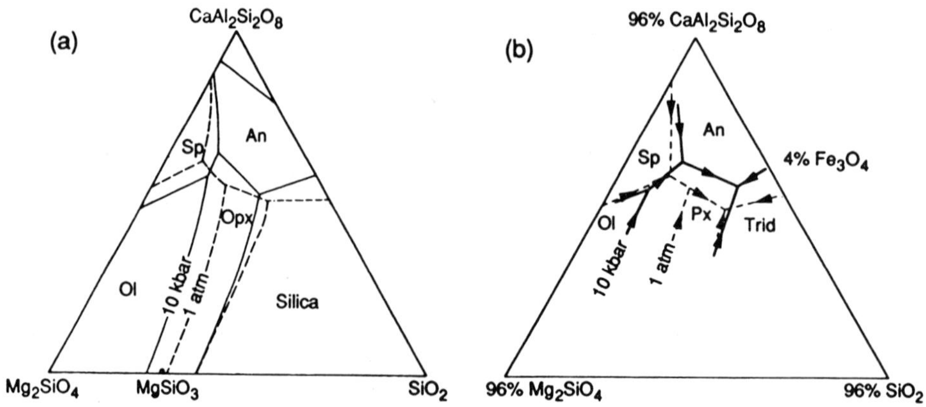

Figure 15. *(a). Phase relations in the system $CaAl_2Si_2O_8$-Mg_2SiO_4-SiO_2 at 1 atm. and 10 kbars. A liquid in equilibrium with olivine, spinel, and anorthite at high pressure will precipitate only olivine at lower pressure. Figure modified from Sen and Presnall (1984). (b). Phase relations on the 4% Fe_3O_4 join in the system $CaAl_2Si_2O_8$-Mg_2SiO_4-SiO_2-Fe_3O_4 at 1 atm. and 10 kbars. A liquid in equilibrium with spinel, anorthite, and orthopyroxene at high pressure will precipitate only anorthite at lower pressure. Figure modified from Osborn (1978). In both phase diagrams, a liquid in equilibrium with olivine, orthopyroxene, and plagioclase at low pressure will precipitate only orthopyroxene at higher pressure.*

attractive for explaining layers of great lateral extent (Cameron, 1977; Lipin, 1993). Possible mechanisms for pressure fluctuations within a magma chamber include exsolution and expansion of a vapour phase (Lipin, 1993), emplacement of a new magma into an existing chamber, convective overturn (Jackson, 1961), volcanic eruptions from the chamber (Sørensen and Larsen, 1987), tectonic stress (Cameron, 1977), and fracturing of the overlying crust. The country rocks enclosing a magma chamber will fracture or deform in response to large or long-term pressure changes within the magma. Small, temporary pressure changes are possible, however, as long as they do not exceed the tensile strength of the country rock. Calculated pressure fluctuations in the summit chambers of Kilauea and Krafla volcanoes reach a maximum of 0.2 to 0.25 kbars (Pollard *et al.*, 1983), and the rise and expansion of bubbles in the magma beneath Long Valley Caldera, may have produced temporary overpressures within the chamber on the order of 0.1 kbars (Linde *et al.*, 1994). Even minor shifts in phase equilibria can produce large variations in modal abundances if a large thickness of magma is shifting its bulk composition by precipitating a thin layer of crystals. Shifting the phase boundary in a 100 m thick column of magma 0.1% away from plagioclase could result in a 10 cm thick layer of anorthosite. Alternatively, in a well-mixed system, 10 cm of anorthosite distributed over a 50 cm interval would increase the apparent modal percentage of plagioclase by 20%.

5.6. Immiscibility

Mafic magmas that differentiate to extreme degrees of iron-enrichment may separate into two immiscible liquids, one rich in silica, alumina, and alkalies, and the other rich in iron and other mafic cations (McBirney, 1975; Philpotts, 1976; Roedder, 1978). Conditions that may promote immiscibility include high concentrations of Fe_2O_3, FeO, P_2O_5, and TiO_2; low concentrations of MgO, CaO, and Al_2O_3; and large ratios of Fe_2O_3/FeO, K_2O/Na_2O, and $(Na_2O + K_2O)/Al_2O_3$ (Naslund, 1983). Immiscible silicate liquid pairs should possess some or all of the following characteristics: identical liquidus mineral assemblages and temperatures; similar FeO/MgO and MnO/FeO ratios; larger Na_2O/K_2O and $Al_2O_3/(Na_2O + K_2O)$ ratios and greater P_2O_5, TiO_2, MgO, MnO, Zr, and REE contents in the more iron-rich liquid; and greater K_2O, Na_2O, Al_2O_3, and Rb contents in the more silica-rich liquid (Watson, 1976; Naslund, 1983). In layers formed from immiscible crystal-liquid mixtures, however, the proportions and compositions of the crystals in each liquid must be considered before the bulk compositions of layers can be compared to experimental immiscible liquids.

In Upper Zone *c* and Upper Border Series γ of the Skaergaard intrusion, pods, sills, and layers of melanogranophyre appear to have formed as a result of liquid-liquid separation during the final stages of crystallization of the intrusion (McBirney and Nakamura, 1974; McBirney, 1975; Naslund, 1984a). Dykes, sills, layers, and pods of Fe-Ti oxide- and apatite-rich rocks (nelsonites) associated with anorthosites and diorites in a variety of localities may also have formed as a result of liquid immiscibility (Philpotts, 1967; Kolker, 1982). Reynolds (1985b) has suggested that three zones of apatite- and oxide-rich rocks in the Bushveld Complex may have formed from immiscible liquids. One of the zones contains a 2 m thick layer of almost pure apatite, magnetite, and ilmenite with the proportions ~70% Fe-Ti-oxides and ~30% apatite, similar to the proportions reported from other nelsonites.

Immiscibility between sulphide and silicate liquids has been proposed as a mechanism for the formation of ore horizons or layers rich in Pt and Pd (Naldrett *et al.*, 1987; 1990). The exceptionally large values for the distribution coefficients D^{Pt} sul./sil. and D^{Pd} sul./sil. (where D^X sul./sil. = concentration of X in the sulphide liquid / concentration of X in the silicate liquid) may explain why these horizons have platinum group element contents several orders of magnitude greater than other parts of their host intrusions. The thin yet laterally extensive nature of these ore layers suggests that immiscibility was abruptly induced over wide areas of the crystallizing magma chamber.

6. LATE-STAGE PROCESSES

6.1. Interstitial crystal growth

The pore spaces between crystals formed during the initial phase of solidification are ultimately filled by overgrowths on the original crystals and by new, late-crystallizing minerals. The growth of crystals of nearly constant composition requires that components expelled from the growing crystals be removed from the crystallization site and that components included in the growing crystals be transported to the crystallization site. This may occur at the crystal-magma interface when the solidification rate is very slow, or within the crystal-liquid mush if convective transfer can effectively move components through the crystal pile (Sparks *et al.*, 1985). Thick monomineralic layers in some intrusions attest to the efficiency of the exchange process.

Morse (1979) suggested that anorthosite layers in the Kiglapait intrusion formed as a result of "adcumulus growth" on the floor of the magma chamber. Goode (1977) reported layers several metres thick in the Kalka intrusion, Australia, that form from alternating intergranular mineral assemblages, one pyroxene-rich and one plagioclase-rich, suggesting that layering formed during crystallization of the interstitial melt. In the Rum intrusion, granular-textured layers and laminae cut across the contacts between pyroxene-rich and pyroxene-poor units, suggesting that they formed during late-stage crystallization within the crystal liquid mush (Young and Donaldson, 1985).

6.2. Metasomatism

Irvine (1980) suggested that a process of infiltration metasomatism acts in layered intrusions to re-equilibrate cumulus minerals with intercumulus liquids migrating upwards as a result of compaction. The main effects of such a process are to displace upwards geochemical discontinuities associated with phase layering, and in some cases, to produce a vertical alignment of crystals (Irvine, 1980).

Boudreau (1982) suggested that olivine layers and the J-M Pt-Pd horizon in the Banded Zone of the Stillwater intrusion formed as a result of late-stage metasomatism. These olivine layers are characterized by coarse to pegmatoidal textures, and some contain unusual amounts

Figure 16. *Mafic pegmatite layers replacing the leucocratic parts of modally-graded rhythmic layers in Upper Zone* a *of the Skaergaard intrusion. Individual pegmatitic layers may follow the leucocratic part of one modally graded layer for some distance, and then cut at an angle across the stratigraphy, before following the leucocratic part of a parallel, but stratigraphicly higher, second modally graded layer.*

of biotite. Anorthosites with few if any mafic minerals form halos on both sides of the more olivine-rich layers, and the anorthosite layers thicken and thin along strike as the olivine layers thicken and thin sympathetically. The Pt-Pd sulphide mineralization of the J-M reef is most commonly found within these olivine-rich rocks or their associated anorthosites. Boudreau (1982) proposed a process of bimetasomatism in which materials are transported in two directions. Volatile components and SiO_2 diffuse outwards while mafic components diffuse inwards to form troctolitic and anorthositic layers from rock that was originally of gabbroic or noritic composition. The end result of such a process may be monomineralic layers with sharp contacts. Nicholson and Mathez (1991) proposed a similar process to explain features of the Merensky Reef of the Bushveld intrusion, but suggested that magmatic volatiles interacted with a zone of interstitial melt to produce the reef.

In the Duke Island complex (Irvine, 1987), dunite and pyroxenite have metasomatically replaced olivine clinopyroxenite through large volumes of rock, sometimes with no obvious channeling of the metasomatic fluids. There is little evidence, however, to indicate that metasomatism has produced layering. Metasomatism and recrystallization appear to have either modified or destroyed pre-existing layers. Similar features are common in ophiolites (Dick and Sinton, 1979).

In the Skaergaard intrusion, coarse-grained gabbroic pegmatite with abundant interstitial granophyre has replaced the leucocratic parts of some graded layers. Many of these pegmatitic zones follow one graded layer for some distance and then abruptly cut across the sequence to follow another layer. In other places, two or more pegmatitic zones join and continue as one unit (Figure 16). With the exception of excess quartz, K-feldspar, and apatite, the modal abundances in the pegmatitic replacements are similar to those found in the leucocratic parts of unaltered layers. Olivine in the pegmatite is more Fe-rich than that in the host rock, and the plagioclase is more anorthitic. Field relations suggest that these pegmatite "layers" are the result of recrystallization in response to fluid metasomatism. Alternatively, the mafic pegmatites may be the result of upward-migrating, water-rich, low-density, interstitial Skaergaard liquids in the final stages of crystallization (Sonnenthal, 1992; Larsen and Brooks, 1994).

In the Gars-bheinn ultramafic sill on the Isle of Skye, coarse-grained feldspathic layers have been attributed to metasomatism by silica-rich fluids (Beran and Hutchinson, 1984). The feldspathic layers become more abundant upward, and at the top of the section make up half of the rock. Although generally concordant, some coarse-grained veins are transgressive.

In Lower Zone *a* of the Skaergaard intrusion, discontinuous layers of anorthosite and iron-rich pyroxenites appear to have formed by metasomatic replacement of Lower Zone *a* gabbros. Some of these discontinuous layers may represent smeared out roof autoliths which were re-equilibrated and partially remobilized after settling to the floor of the magma chamber (Naslund, 1986), but others are clearly the result of volume-for-volume replacement (McBirney, 1995).

6.3. Constitutional zone refining

An additional mechanism of layer formation that could conceivably occur during melt migration through the cumulus pile is based on a process of constitutional zone refining (McBirney, 1987). Thermal zone refining is a well understood process in the field of metallurgy where it is used for the purification of metals. During thermal zone refining, a solid bar of metal is passed through a furnace so that only a small section of the bar will be partly molten at any given time. A zone of melt forms on the leading edge of the bar, and subsequently passes through the bar as it slowly moves through the furnace. As the zone of melt passes through the bar it is continuously melting at one boundary and crystallizing at the other. Impurities in the metal, for which the distribution coefficient (concentration in the solid / concentration in the liquid) is less than 1.0, will be preferentially retained in the melt, and after repeated passages, will be swept to the trailing end of the bar.

Figure 17. *Inch-scale layering in the Stillwater intrusion, Montana. The layers consist of doublets of pyroxene-rich rock in an anorthosite host. Note hammer for scale.*

Constitutional zone refining can occur under relatively isothermal conditions if a zone of flux migrates through a crystal-liquid mixture causing a depression of the melting temperature and, therefore, an increase in the proportion of partial melt. As the zone of flux melting migrates through the crystal-liquid pile, components with low-melting temperatures (*i.e.* components with solid / liquid distribution coefficients less than 1.0) will be concentrated in the melt. Water and alkalies are likely fluxing agents that are excluded during the crystallization of typical layered intrusions. Flux migration in a crystal-liquid pile is likely to be accelerated by diffusion of the fluxing agents down a geochemical potential gradient, by compaction of the crystals under their own weight, by the buoyancy effect of concentrating water and alkalies in the residual magma, and/or by separation of a vapour phase. Because the proportion of melt steadily increases as the zone migrates through the pile, it is not a steady state process, but rather one that passes through the crystal-liquid mush as a series of pulses or waves. The effects of water on the position of phase boundaries could shift cotectic proportions and lead to layers with significantly different modal proportions. Alternatively, the stopping and starting of the constitutional zone refining process could lead to interfaces where minerals are crystallized in the order of their ease of nucleation, and therefore, result in modally graded layers. In normal zone refining, the transfer of trace elements is strictly limited by the maximum concentration set for the liquid by the distribution coefficient; once the liquid is saturated, the moving zone can no longer extract more of an element as it advances through new rock. This is not true, however, if the excluded components have the effect of lowering melting temperatures and thereby increasing the proportion of liquid. Boudreau (1988) and Nicholson and Mathez (1991) have suggested that certain features of the Merensky reef of the Bushveld intrusion and Stillwater can be best explained by magmatic vapour migrating upwards through the cumulate pile, and causing an increase in the proportion of interstitial liquid at the level of the reef.

Figure 18. *Outcrop of finely banded orbicules in a rhyolite dyke near the eastern margin of the Skaergaard intrusion.*

6.4. Solidification contraction

Petersen (1987) has suggested that instead of being expelled by compaction, interstitial liquids will be drawn into partially solidified crystal-liquid mixtures in response to a volume contraction of 7 to 10% during solidification. During crystallization the rejected solute will continue to flow from the crystallization front deeper into the accumulating crystal pile leaving the main magma reservoir unfractionated. Layering may form in response to variations in percolation rates. High percolation rates encourage crystal growth by effectively removing rejected solute from the crystallization front, and may result in adcumulate layers that act to seal off underlying liquids. Low percolation rates result in uniform mesocumulates. The flow of interstitial liquids downward into the crystal pile in response to solidification contraction results in thick sequences in which there is little or no geochemical evidence of progressive fractionation, but which appear to have very large contents of trapped liquid. In general, intrusions with well-developed layering do not fit these criteria.

6.5. Ostwald ripening

An assemblage of crystals of mixed grain sizes is inherently unstable, in that larger grains can grow at the expense of smaller ones in order to minimize the total surface free energy of the system (Boudreau, 1987). Such a process of Ostwald grain ripening, can occur under isothermal and isochemical conditions in which the heat absorbed and components released as the smaller grains dissolve is exactly balanced by the release of heat and uptake of components

Figure 19. *Rheomorphic layering in the contact aureole of the Basistoppen Sill produced by contact metamorphism. Originally homogeneous Upper Zone* c *ferrodiorites of the Skaergaard intrusion, have been partially melted to produce dark Fe-rich ultramafic layers that represent the solidified partial melt, and light andesine-anorthosite layers that represent the residual crystals. Note tip of ice axe for scale.*

as the larger grains grow. The volumetric free energy terms for both small and large grains are negative, while their surface energy terms are positive. As a result, larger grains with small ratios of surface area to volume have less total free energy per mole than do smaller grains. The resulting chemical potential gradient aids in the transfer of components between grains, because the chemical potential at which a small grain dissolves exceeds that at which a large grain grows. A mathematical treatment of Ostwald ripening called "the competitive particle growth model" or "geochemical self-organization" has been proposed by P.J. Ortoleva and his co-workers (Feinn *et al.*, 1978; Lovett *et al.*,1978; Feeney *et al.*, 1983; Ortoleva *et al.*, 1987).

Inch-scale layering in the Stillwater intrusion consists of parallel, evenly-spaced, pyroxene-rich layers in a host of almost pure anorthosite. In some sequences the layers are evenly-spaced doublets (Figure 17). The pyroxene within the gabbroic anorthosite layers has a interstitial texture suggesting that the layers, which are defined by the presence or absence of pyroxene, must have formed by a late-stage process. There is a crude mosaic or honeycomb pattern to the distribution of pyroxene within the plane of the layering, similar to that observed in experimental gels produced by Ostwald ripening, and a positive correlation between pyroxene grain-size and layer spacing (Boudreau, 1987), suggesting that the layers formed in response to grain-size coarsening of pyroxene within an anorthositic crystal mush. Any zone or layer where grains are marginally larger than those in their surroundings, will be energetically favoured and will grow by diffusion of components from the surroundings where grains are dissolving (Boudreau, 1987). In slowly cooled intrusions, the process may continue to the extreme situation where growth of a coarse grained pyroxene-rich layer has depleted the surrounding rock of pyroxene creating an almost pyroxene-free anorthositic host rock. Dissolving crystals above a layer are also at a chemical potential disadvantage with respect to crystals at higher levels, and the latter may begin to grow and generate a new layer at some set distance from the first. In this way, a series of regularly spaced layers may be produced. The exact spacing of the layers would be controlled by the interplay between the growth rate and the diffusion rate. Layering formed by Ostwald ripening should show a positive correlation between mode and grain size (see Figure 5). Rocks that have undergone extensive Ostwald ripening should also have predictable grain-size distributions on a size *vs.* frequency plot (Chai, 1974; Baronnet, 1982).

A remarkable example of layer formation by Ostwald ripening has developed under sub-solidus conditions during devitrification of a siliceous dyke (McBirney *et al.*, 1990). Layers two to three millimetres thick consisting of quartz alternating with albite and K-feldspar, have formed spherical clots 25 to 30 cm in diameter within a metre-wide rhyolitic dyke near the eastern margin of the Skaergaard intrusion (Figure 18). Although neither the dyke nor the host rocks show conspicuous evidence of hydrothermal alteration, the formation of the layering may have been related to, or assisted by fluid flow along a small fault that cuts the dyke.

Figure 20. *Layering within the Mikis Fjord Macrodyke, East Greenland, produced as a result of contact metamorphism of a roof pendant of zeolite-rich, hydrothermally altered basalts.*

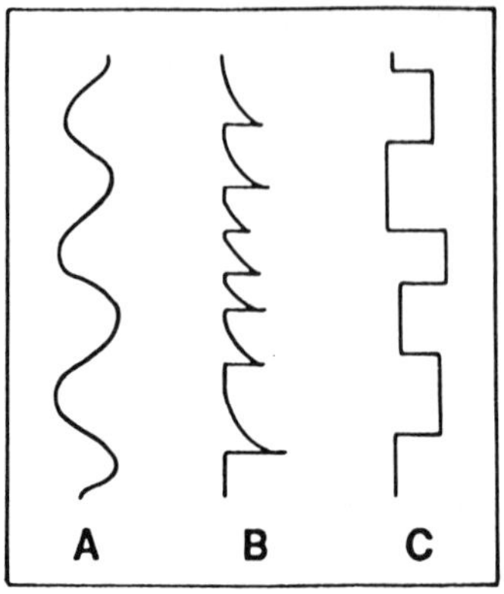

Figure 21. Three styles of rhythmic layering. In A the system varies gradually between two extreme sets of conditions. In B, the system is abruptly disturbed by a sudden change in conditions followed by a gradual return to the original conditions. In C, the system abruptly changes from one set of stable conditions to another set of stable conditions, then after a period of stability, the system abruptly reverses back to the original conditions. Figure modified from Naslund et al. (1991).

6.6. Contact metamorphism

The Basistoppen sill was intruded into the Skaergaard intrusion shortly after the latter solidified and before regional tilting (Wager and Brown, 1968). Where the sill cuts rocks of Upper Zone *c* and Upper Border Zone γ, the ferrodiorites of these zones have been partly remelted (Naslund, 1986). Owing to kinetic effects, the partial melting process has preferentially melted and remobilized the mafic components leaving a residue of plagioclase. As a result of contact metamorphism, partial melting, and rheomorphism, the original unlayered ferrodiorites adjacent to the contact of the Basistoppen sill have been transformed into alternating layers of andesine anorthosites and Fe-rich olivine pyroxenites (Figure 19).

In the Mikis Fjord Macrodyke, a distinctive layered division 100 to 200 m thick, composed of rocks ranging from metabasalt to medium-grained, olivine gabbro, formed adjacent to the roof. Well-developed layering in these rocks has been interpreted (Lesher *et al.*, 1992) to have formed by thermal metamorphism and partial melting of a large roof pendant of hydrothermally altered basalts (Figure 20). Although the layers have many features in common with layers in larger intrusions, the rocks are granular in texture, and individual layers can be traced along strike into metabasalts with amygdules filled with plagioclase and zeolites. Isotopic studies suggest that the layered rocks are not cogenetic with the underlying unlayered gabbros of the Macrodyke, but rather are isotopically similar to the surrounding host lavas of the Mikis Formation.

7. CONCLUSIONS

Owing to the wide variety of igneous layering that has been recognized, it is unlikely that any single layer-forming mechanism can explain all or even most of the known occurrences. Indeed, some types of layering may be the result of multiple mechanisms operating at different stages of crystallization. The different mechanisms that have been proposed should result in layered sequences with a variety of patterns (Figure 21). Important characteristics to consider are thickness and length, the nature of boundaries, any internal vertical or lateral variations, and the relationships to nearby layers. Modal proportions, grain-size, mineral composition, whole-rock composition, and textural patterns within layers are also likely to reflect the mechanism responsible for their formation. The challenge for the igneous petrologist is to determine which features are diagnostic of a particular mechanism, which reflect subsequent compositional or textural modifications, and which can best discriminate between the plethora of possible mechanisms.

8. ACKNOWLEDGEMENTS

The authors wish to thank Dr. A.E. Boudreau and Dr. C.I. Chalokwu for constructive comments on earlier drafts of this manuscript. Anne Hull prepared the illustrations and David Tuttle assisted with photography.

9. REFERENCES

Baronnet, A., 1982. Ostwald ripening in solution: the case of calcite and mica. *Estudios Geol.* **38**, 185-98.

Barriere, M., 1981. On curved laminae, graded layers, convection currents, and dynamic crystal sorting in the Ploumanac'h (Brittany) subalkaline granite. *Contr. Miner. Petrol.* **77**, 214-24.

Bebien, J., & Gaghy, C.L., 1978. Importance of flow differentiation in magmatic evolution: an example from an ophiolitic sheeted complex. *J. Geol.* **87**, 579-82.

Bédard, J.H., Sparks, R.S.J., Renner, R., Cheadle, M.J., & Hallworth, M.A., 1988. Peridotite sills and metasomatic gabbros in the Eastern Layered series of the Rhum complex. *J. Geol. Soc. London* **145**, 207-24.

Benn, K., & Allard, B., 1989. Preferred mineral orientations related to magmatic flow in ophiolitic layered gabbros. *J. Petrology* **30**, 925-46.

Beran, J.C., & Hutchinson, R., 1984. Layering in the Gars-bheinn ultrabasic sill, Isle of Skye: A new interpretation and its implications. *Scott. J. Geol.* **20**, 329-41.

Blake, D.H., 1968. Gravitational sorting of phenocrysts in some Icelandic intrusive sheets. *Geol. Mag.* **105**, 140-8.

Boudreau, A.E., 1982. The main platinum zone, Stillwater complex, MT - evidence for bimetasomatism and a secondary origin for olivine. In: Walker, D. & McCallum, I.S. (eds.) *Workshop on Magmatic Processes of Early Planetary Crusts: Magma Oceans and Stratiform Layered Intrusions LPI Tech. Rpt. 82-01.* Houston: Lunar and Planetary Institute, 59-61.

Boudreau, A.E., 1987. Pattern formation during crystallization and the formation of fine-scale layering. In. Parsons, I. (ed.) *Origins of Igneous Layering*. Dordrecht: Reidel, 453-71.

Boudreau, A.E., 1988. Investigations of the Stillwater Complex. IV. The role of volatiles in the petrogenesis of the J-M Reef, Mineapolis Adit section. *Can. Miner.* **26**, 193-208.

Boudreau, A.E., 1994. Mineral segregation during crystal aging in two-crustal, two-component systems. *S. Afr. J. Geol.* **97**, 473-85.

Bowen, N.L., 1928. *The Evolution of the Igneous Rocks*. Princeton, NJ: Princeton University Press, 332 pp.

Brandeis, G., 1992. Constraints on the formation of cyclic units in ultramafic zones of large basaltic chambers. *Contr. Miner. Petrol.* **112**, 312-28.

Brown, P.E., & Farmer, D.G., 1971. Size-graded layering in the Imilik gabbro, East Greenland. *Geol. Mag.* **108**, 465-76.

Cameron, E.N., 1975. Postcumulus and subsolidus equilibration of chromite and coexisting silicates in the Eastern Bushveld Complex. *Geochim. Cosmochim. Acta* **39**, 1021-33.

Cameron, E.N., 1977. Chromite in the central sector, eastern Bushveld Complex, South Africa, *Am. Miner.* **62**, 1082-96.

Campbell, I.H., 1977. A study of macro-rhythmic layering and cumulate processes in the Jimberlana intrusion, western Australia. Part 1: The Upper Layered Series. *J. Petrology* **18**, 185-215.

Chai, B.H.T., 1974. Mass transfer of calcite during hydrothermal recrystallization. In: Hofmann, A.W., Giletti, B.J., Yoder, H.S. Jr., & Yund, R.A. (eds.) *Geochemical Transport and Kinetics.* Washington, D.C.: Carnegie Institution of Washington, 205-18.

Chapman, M., & Rhodes, J.M., 1992. Composite layering in the Isle au Haut igneous complex, Maine: evidence for periodic invasion of a mafic magma into an evolving magma reservoir. *J. Volc. Geotherm. Res.* **51**, 41-60.

Chen, C.F., & Turner, J.S., 1980. Crystallization in a double diffusive system. *J. Geophys. Res.* **85**, 2573-93.

Claydon, R.V. & Bell, B.R., 1992. The structure and petrology of ultrabasic rocks in the southern part of the Cuillin Igneous Complex, Isle of Skye. *Trans. Roy. Soc. Edin.: Earth Sci.* **83**, 635-53.

Coats, R.R., 1936. Primary banding in basic plutonic rocks. *J. Geol.* **44**, 407-419.

Conrad, M.E., & Naslund, H.R., 1989. Modally-graded rhythmic layering in the Skaergaard intrusion. *J. Petrology* **30**, 251-69.

Dick, H.J.B., & Sinton, J.M., 1979. Compositional layering in alpine peridotites: evidence for pressure solution creep in the mantle. *J. Geol.* **87**, 403-16.

Duke, E.F., Redden, J.A., & Papike, J.S., 1988. Calamity Peak layered granite-pegmatite complex, Black Hills, South Dakota: structure and emplacement. *Geol. Soc. Am. Bull.* **100**, 825-40.

Dunham, A.C., & Wadsworth, W.J., 1978. Cryptic variation in the Rhum layered intrusion. *Miner. Mag.* **42**, 347-56.

Emeleus, C.H., 1987. The Rhum Layered complex, Inner Hebrides, Scotland. In: Parsons, I. (ed.) *Origins of Igneous Layering*. Dordrecht: Reidel, 263-86.

Engell, J., 1973. A closed system crystal-fractionation model for the agpaitic Ilimaussaq intrusion, South Greenland, with special reference to the lujavrites. *Bull. Geol. Soc. Denmark* **22**, 334-62.

Feeney, R., Schmidt, S.L., Stricholm, P., Chadam, J., & Ortoleva, P., 1983. Perioditic precipitation and coarsening waves: application of the competitive growth model. *J. Chem. Phys.* **78**, 1293-311.

Feinn, D., Ortoleva, P., Scalf, W., & Wolff, M., 1978. Spontaneous pattern formation in precipitating systems. *J. Chem. Phys.* **69**, 27-39.

Ferguson, J., & Pulvertaft, T.C.R., 1963. Contrasted styles of igneous layering in the Gardar Province of South Greenland. *Miner. Soc. Am. Spec. Pap.* **1**, 10-21.

Fyfe, W.S., 1976. Chemical aspects of rock deformation. *Roy. Soc. London Phil. Trans. Ser. A* **283**, 221-8.

Gibb, F.G.F., 1968. Flow differentiation in xenolithic ultrabasic dykes of the Cuillins and the Strathaird Peninsula, Isle of Skye, Scotland. *J. Petrology* **9**, 411-43.

Gibb, F.G.F., & Henderson, C.M.B., 1992. Convection and crystal settling in sills. *Contr. Miner. Petrol.* **109**, 538-45.

Goode, A.D.T., 1976. Small-scale primary cumulus igneous layering in the Kalka layered intrusion, Giles Complex, central Australia. *J.Petrology* **17**, 379-97.

Goode, A.D.T., 1977. Intercumulus igneous layering in the Kalka layered intrusion, central Australia. *Geol. Mag.* **114**, 215-8.

Gorring, M.L., & Naslund, H.R., 1995. Geochemical reversals within the lower 100 m of the Palisades sill, New Jersey. *Contr. Miner. Petrol.* **119**, 263-76.

Harker, A., 1909. *Natural History of Igneous Rocks.* New York: Macmillan Company, 384 pp.

Hawkes, D.D., 1967. Order of abundant crystal nucleation in a natural magma. *Geol. Mag.* **104**, 473-86.

Hess, H.H., 1960. Stillwater igneous complex. *Mem. Geol. Soc. Am.* **80**, 1-230.

Higgins, M.D., 1991. The origin of laminated and massive anorthosite, Sept Iles layered intrusion, Quebec, Canada. *Contr. Miner. Petrol.* **106**, 340-54.

Hoffer, A., 1965. Seismic control of layering in intrusions. *Am. Miner.* **50**, 1125-8.

Hort, M., Marsh, B.D., & Spohn, T., 1993. Igneous layering through oscillatory nucleation and crystal settling in well-mixed magmas. *Contr. Miner. Petrol.* **114**, 425-40.

Hulme, G., 1974. The interpretation of lava flow morphology. *Geophys. J. Roy. Astronom. Soc.* **39**, 361-83.

Huppert, H.E., & Sparks, R.S.J., 1980. The fluid dynamics of a basaltic magma chamber replenished by influx of hot, dense ultrabasic magma. *Contr. Miner. Petrol.* **75**, 279-89.

Huppert, H.E., & Sparks, R.S.J., 1984. Double diffusive convection due to crystallization in magmas. *Ann. Rev. Earth Planet. Sci.* **12**, 11-37.

Huppert, H.E., Sparks, R.S.J., Wilson, J.R., Hallworth, M.A., & Leitch, A.M., 1987. Laboratory experiments with aqueous solutions modeling magma chamber processes - II. Cooling and crystallization along inclined planes. In: Parsons, I. (ed.) *Origins of Igneous Layering.* Dordrecht: Reidel, 539-68.

Husch, J.M., 1990. Palisades sill: origin of the olivine zone by separate magmatic injection rather than gravity settling. *Geology* **18**, 699-702.

Irvine, T.N., 1974. Petrology of the Duke Island ultramafic complex southeastern Alaska, Boulder, Co: *Geol. Soc. Am. Mem.* **138**, 240 pp.

Irvine, T.N., 1975. Crystallization sequences in the Muskox intrusion and other layered intrusions: II Origin of chromitite layers and similar deposits of other magmatic ores. *Geochim. Cosmochim. Acta* **39**, 991-1020.

Irvine, T.N., 1977. Origin of chromite layers in the Muskox intrusion and other stratiform intrusions: a new interpretation. *Geology* **5**, 273-7.

Irvine, T.N., 1980. Magmatic infiltration metasomatism, double-diffusive fractional crystallization, and adcumulus growth in the Muskox intrusion and other layered intrusions. In: Hargraves, R.B. (ed.) *Physics of Magmatic Processes.* Princeton, NJ: Princeton University Press, 325-84.

Irvine, T.N., 1987. Layering and related structures in the Duke Island and Skaergaard intrusions: similarities, differences, and origins. In: Parsons, I. (ed.) *Origins of Igneous Layering.* Dordrecht: Reidel, 185-245.

Irvine, T.N., & Smith, C.H.,1967. The ultramafic rocks of the Muskox intrusion, Northwest Territories, Canada. In: Wyllie, P.J. (ed.) *Ultramafic and Related Rocks.* New York: John Wiley & Sons, Inc., 38-49.

Irvine, T.N., Keith, D.W., & Todd, S.G., 1983. The J-M platinum-palladium reef of the Stillwater complex, Montana: II Origin by double diffusive convective magma mixing and implications for the Bushveld complex. *Econ. Geol.* **78**, 1287-334.

Jackson, E.D., 1961. Primary textures and mineral associations in the Ultramafic zone of the Stillwater complex, Montana. *U.S. Geol. Sur. Prof. Paper* **358**, 1 106.

Jackson, E.D., 1970. The cyclic unit in layered intrusions - a comparison of the repetitive stratigraphy in the ultramafic parts of the Stillwater, Muskox, Great Dyke and Bushveld Complexes. *Spec. Publ. Geol. Soc. S. Afr.* **1**, 391-424.

Jahns, R.H., 1982. Internal evolution of pegmatite bodies. In: Cèrny, P. (ed.) *Miner. Assoc. Canada Short Course Handbook* **8**, 293-327.

Jahns, R.H., & Tuttle, O.F., 1963. Layered pegmatite-aplite intrusions. *Miner. Soc. Am. Spec. Paper* **1**, 78-92.

Jang, Y.D., & Naslund, H.R., 1994. Compositional variations within graded layers in the Skaergaard intrusion. *Geol. Soc. Am. Abst. with Prog.* **26**, no. 2, 25.

Kanaris-Sotiriou, R., 1974. Fine-scale layering in igneous intrusions: A possible mechanism for a non-depositional origin. *Geol. Mag.* **111**, 157-62.

Keith, D.W., & Naslund, H.R., 1987. Petrographic and chemical characteristics of a layered sequence in the Upper Border Zone of the Skaergaard intrusion, East Greenland. *Geol. Soc. Am. Abst. with Prog.* **19**, 723.

Knopf, A., 1908. Geology of the Seward Peninsula tin deposits. *Bull. U.S. Geol. Surv.* **358**, 1-71.

Kerr, R.C., & Turner, J.S., 1982. Layered convection and crystal layers in multicomponent systems. *Nature* **298**, 731-3.

Kogarko, L.N., & Khapaev, V.V., 1987. The modeling of formation of apatite deposits of the Khibina massif (Kola Peninsula). In: Parsons, I. (ed.) *Origins of Igneous Layering*. Dordrecht: Reidel, 589-611.

Kolker, A., 1982. Mineralogy and geochemistry of Fe-Ti oxide and apatite (Nelsonite) deposits and evaluation of the liquid immiscibility hypothesis. *Econ. Geol.* **77**, 1146-58.

Komar, P.D., 1972. Mechanical interactions of phenocrysts and flow differentiation in igneous dikes and sills. *Geol. Soc. Am. Bull.* **83**, 973-88.

Larsen, R.B., & Brooks, C.K., 1994. Origin and evolution of gabbroic pegmatites in the Skaergaard intrusion, East Greenland. *J. Petrology* **35**, 1651-80.

Lappin, M.A., 1967. Structural and petrofabric studies of the dunites of Almklovadalen, Nordfjord, Norway. In: Wyllie, P.J. (ed.) *Ultramafic and Related Rocks*. New York: John Wiley & Sons, Inc., 183-90.

Lesher, C.E., & Walker, D., 1988. Cumulate maturation and melt migration in a temperature gradient. *J. Geophys. Res.* **93**, 10295-311.

Lesher, C.E., Rosing, M.T., & Bird, D.K., 1992. Metasomatic transformation of host lavas of the Miki Fjord macrodyke, East Greenland. *EOS* **73**, n.44, 640.

Leveson, D.J., 1966. Orbicular rocks - A review. *Geol. Soc. Am. Bull.* **77**, 409-26.

Liesegang, R.E., 1896. Über einige Eigenschaften von Gallerten. *Naturw. Wochschr.* **11**, 353-62.

Liesegang, R.E., 1913. *Geologische Diffusionen.* Dresden: T. Steinkopff, 180 pp.

Linde, A.T., Sacks, I.S., Johnson, M.J.S., Hill, D.P., & Bilham, R.G., 1994. Increased pressure from rising bubbles as a mechanism from remotely triggered seismicity. *Nature* **371**, 408-10.

Lipin, B.R., 1993. Pressure increases, the formation of chromite seams, and the development of the ultramafic series in the Stillwater Complex, Montana. *J. Petrology* **34**, 955-76.

Lofgren, G.E., & Donaldson, C.H., 1975. Curved branching crystals and differentiation in comb-layered rocks. *Contr. Miner. Petrol.* **49**, 309-19.

Lovett, R., Ortoleva, P., & Ross, J., 1978. Kinetic instabilities in first order phase transitions. *J. Chem. Phys.* **69**, 947-55.

Maaløe, S., 1978. The origin of rhythmic layering. *Miner. Mag.* **42**, 337-45.

Maaløe, S., 1987. Rhythmic layering of the Skaergaard intrusion. In: Parsons, I. (ed.) *Origins of Igneous Layering.* Dordrecht: Reidel, 247-62.

Mangan, M.T., & Marsh, B.D., 1992. Solidification front fractionation in phenocryst-free sheet-like magma bodies. *J. Geol.* **100**, 605-20.

Mangan, M.T., Marsh, B.D., Froelich, A.J., & Gottfried, D., 1993. Emplacement and differentiation of the York Haven Diabase Sheet, Pennsylvania. *J. Petrology* **34**, 1271-302.

Marsh, B.D., 1988. Crystal capture, sorting, and retention in convecting magma. *Geol. Soc. Am. Bull.* **100**, 1720-37.

Marsh, B.D., 1989. On convective style and vigor in sheet-like magma chambers. *J. Petrology* **30**, 479-530.

Marsh, B.D., 1991. Reply to comments on "Convective styles and vigor in sheet-like magma chambers". *J. Petrology* **32**, 855-60.

Marsh, B.D., & Maxey, M.R., 1985. On the distribution and separation of crystals in convecting magma. *J. Volc. Geotherm. Res.* **24**, 95-150.

Martin, D., Griffiths, R.W., & Campbell, I.H., 1987. Compositional and thermal convection in magma chambers. *Contr. Miner. Petrol.* **96**, 465-75.

McBirney, A.R., 1975. Differentiation of the Skaergaard intrusion. *Nature* **253**, 691-4.

McBirney, A.R., 1985. Further considerations of double-diffusive stratification and layering in the Skaergaard intrusion. *J. Petrology* **26**, 993-1001.

McBirney, A.R., 1987. Constitutional zone refining of layered intrusions. In: Parsons, I. (ed.) *Origins of Igneous Layering*. Dordrecht: Reidel, 437-52.

McBirney, A.R., 1995. Mechanisms of differentiation in the Skaergaard intrusion. *J. Geol. Soc. London* **152**, 421-35.

McBirney, A.R., & Hunter, R.H., 1995. The cumulate paradigm reconsidered. *J. Geol.* **103**, 114-22.

McBirney, A.R., & Murase, T., 1984. Rheological properties of magmas. *Ann. Rev. Earth Planet. Sci.* **12**, 337-57.

McBirney, A.R., & Nakamura, Y., 1974. Immiscibility in late-stage magmas of the Skaergaard intrusion. *Yrbk. Carnegie Inst. Wash.* **73**, 348-52.

McBirney, A.R., & Nicolas, A., In Review. The Skaergaard Layered Series, Part II Magmatic flow and dynamic layering.

McBirney, A.R., & Noyes, R.M., 1979. Crystallization and layering of the Skaergaard intrusion. *J. Petrology* **20**, 487-554.

McBirney, A.R., White, C.M., & Boudreau, A.E., 1990. Spontaneous development of concentric layering in a solidified siliceous dike, East Greenland. *Earth-Sci. Rev.* **29**, 321-30.

Moore, A.C., 1973. Studies of igneous and tectonic textures and layering in the rocks of the Gosse Pile Intrusion, Central Australia. *J. Petrology* **14**, 49-80.

Morse, S.A., 1979. Kiglapait geochemistry - II. Petrography. *J. Petrology* **20**, 591-624.

Murase, T., & McBirney, A.R., 1973. Properties of some common igneous rocks and their melts at high temperatures. *Geol. Soc. Am. Bull.* **84**, 3563-92.

Naldrett, A.J., Cameron, G., von Gruenewaldt, G., & Sharpe, M.R., 1987. The formation of stratiform PGE deposits in layered intrusions. In: Parsons, I. (ed.) *Origins of Igneous Layering*. Dordrecht: Reidel, 313-97.

Naldrett, A.J., Brügmann, G.E., & Wilson, A.H., 1990. Models for the concentration of PGE in layered intrusions. *Can. Miner.* **28**, 389-408.

Naslund, H.R., 1983. The effect of oxygen fugacity on liquid immiscibility in iron-bearing silicate melts. *Am. J. Sci.* **283**, 1034-59.

Naslund, H.R., 1984a. The petrology of the Upper Border Series of the Skaergaard intrusion. *J. Petrology* **25**, 1-28.

Naslund, H.R., 1984b. Supersaturation and crystal growth in the roof-zone of the Skaergaard magma chamber. *Contr. Miner. Petrol.* **86**, 89-93.

Naslund, H.R., 1986. Disequilibrium partial melting and rheomorphic layer formation in the contact aureole of the Basistoppen sill. *Contr. Miner. Petrol.* **93**, 359-67.

Naslund, H.R., Turner, P.A., & Keith, D.W., 1991. Crystallization and Layer Formation in the Middle Zone of the Skaergaard intrusion. *Bull. Geol. Soc. Denmark* **38**, 165-71.

Nicolas, A., 1992. Kinematics in magmatic rocks, with special reference to gabbro. *J. Petrology* **33**, 891-915.

Nicholson, D.M., & Mathez, E.A., 1991. Petrogenesis of the Merensky Reef in the Rustenburg section of the Bushveld Complex. *Contr. Miner. Petrol.* **107**, 293-309.

Ortoleva, P., Merino, E., Moore, C., & Chadam, J., 1987. Geochemical self-organization, I. Reaction-transport feedbacks and modeling approach. *Am. J. Sci.* **287**, 979-1007.

Osborn, E.F., 1978. Changes in phase relations in response to change in pressure from 1 atm. to 10 kbar for the system Mg_2SiO_4-iron oxide-$CaAl_2Si_2O_8$-SiO_2. *Yrbk. Carnegie Inst. Wash.* **77**, 784-90.

Palacz, Z.A., & Tait, S.R., 1985. Isotopic and geochemical investigation of unit 10 from the Eastern Layered Series of the Rhum intrusion, Northwest Scotland. *Geol. Mag.* **122**, 485-90.

Parsons, I., 1979. The Klokken gabbro - syenite complex, South Greenland: Cryptic variation and origin of inversely-graded layering. *J. Petrology* **20**, 653-94.

Parsons, I., & Becker, S.M., 1987. Layering, compaction and post-magmatic processes in the Klokken intrusion. In: Parsons, I. (ed.) *Origins of Igneous Layering*. Dordrecht: Reidel, 29-92.

Petersen, J.S., 1987. Solidification contraction: another approach to cumulus processes and the origin of igneous layering. In: Parsons, I. (ed.) *Origins of Igneous Layering*. Dordrecht: Reidel, 505-26.

Philpotts, A.R., 1967. Origin of certain iron-titanium oxide and apatite rocks. *Econ. Geol.* **62**, 303-15.

Philpotts, A.R., 1976. Silicate liquid immiscibility: its probable extent and petrogenetic significance. *Am. J. Sci.* **276**, 1147-77.

Pollard D.D., Delany, P.T., Duffield, W.A., Endo, E.T., & Okamura, A.T., 1983. Surface deformation in volcanic rifts. *Tectonophysics* **94**, 541-84.

Ray, R.G., 1952. Orbicular diorite from southern Alaska. *Am. J. Sci.* **250**, 57-70.

Reynolds, I.M., 1985a. The nature and origin of titaniferous magnetite-rich layeres in the Upper Zone of the Bushveld Complex: a review and synthesis. *Econ. Geol.* **80**, 1089-108.

Reynolds, I.M., 1985b. Contrasting mineralogy and textural relationships in the uppermost titaniferous magnetite layers of the Bushveld Complex in the Bierkraal area north of Rustenburg. *Econ. Geol.* **80**, 1027-48.

Richter, R.M., & McKenzie, D.P., 1984. Dynamical models for melt segregation from a deformable matrix. *J. Geol.* **92**, 729-40.

Robins, B., Haukvik, L., & Jansen, S., 1987. The organization and internal structure of cyclic units in the Honningsvåg intrusive suite, North Norway: Implications for intrusive mechanisms, double-diffusive convection and pore-magma infiltration. In: Parsons, I. (ed.) *Origins of Igneous Layering*. Dordrecht: Reidel, 287-312.

Rockhold, J.R., Nabelek, P.I., & Glasscock, M.D., 1987. Origin of rhythmic layering in the Calamity peak satellite pluton of the Harney Peak Granite, South Dakota: the role of boron. *Geochim. Cosmochim. Acta* **51**, 487-96.

Roedder, E., 1978. Silicate liquid immiscibility in magmas and in the system K_2O-FeO-Al_2O_3-SiO_2: an example of serendipity. *Geochim. Cosmochim. Acta* **42**, 1597-617.

Roobol, M.J., 1972. Size-graded igneous layering in an Icelandic intrusion. *Geol. Mag.* **109**, 393-403.

Ross, M.E., 1986. Flow differentiation, phenocryst alignment, and compositional trends within a dolerite dike at Rockport, Massachusetts. *Geol. Soc. Am. Bull.* **97**, 232-40.

Ryder, G., 1984. Oxidation and layering in the Stillwater intrusion. *Geol. Soc. Am. Abstr. with Prog.* **16**, 642.

Sen, G., & Presnall, D.C., 1984. Liquidus phase relationships on the join anorthite-forsterite-quartz at 10 kbar with applications to basalt genesis. *Contr. Miner. Petrol.* **85**, 404-8.

Shaw, H.R., Peck, D.L., Wright, T.L., & Okamura, R., 1968. The viscosity of basaltic magma: an analysis of field measurements in Makaopuhi lava lake, Hawaii. *Am. J. Sci.* **266**, 225-64.

Simkin, T., 1967. Flow differentiation in the picritic sills of North Skye. In: Wyllie, P.J. (ed.) *Ultramafic and Related Rocks*. New York: John Wiley, 64-9.

Sonnenthal, E.L., 1992. Geochemistry of dendritic anorthosites and associated pegmatites in the Skaergaard intrusion, East Greenland: evidence for metasomatism by a chlorine-rich fluid. *J. Volc. Geotherm. Res.* **52**, 209-30.

Sørensen, H., & Larsen, L.M., 1987. Layering in the Ilimaussaq Alkaline intrusion. South Greenland. In: Parsons, I. (ed.) *Origins of Igneous Layering*. Dordrecht: Reidel, 1-28.

Sparks, R.S.J., & Huppert, H.E., 1987. Laboratory experiments with aqueous solutions modeling magma chamber processes. I. discussion of their validity and geologic application. In: Parsons, I. (ed.) *Origins of Igneous Layering*. Dordrecht: Reidel, 527-38.

Sparks, R.S.J., Huppert, H.E., Kerr, R.C., McKenzie, D.P., & Tait, S.R., 1985. Postcumulus processes in layered intrusions. *Geol. Mag.* **122**, 555-68.

Sparks, R.S.J., Huppert, H.E., Koyaguchi, T., & Hallworth, M.A., 1993. Origin of modal and rhythmic igneous layering by sedimentation in a convecting magma chamber. *Nature* **361**, 246-9.

Taubeneck, W.H., & Poldervaart, A., 1960. Geology of the Elkhorn mountains, northeastern Oregon: part II Willow Lake intrusion. *Geol. Soc. Am. Bull.* **71**, 1295-1322.

Tegner, C., Wilson, J.R., & Brooks, C.K., 1993. Intraplutonic quench zones in the Kap Edvard Holm layered gabbro complex, East Greenland. *J. Petrology* **34**, 681-710.

Thayer, T.P., 1963. Flow-layering in alpine peridotite-gabbro complexes. *Miner. Soc. Am. Spec. Paper* **1**, 55-61.

Thy, P., 1983. Cumulate chemistry and its bearing on the origin of layering: evidence from the Fongen-Hyllingen basic complex, Norway. *Tschermaks Min. Petr. Mitt.* **32**, 1-24.

Todd, S.G., Keith, D.W., LeRoy, L.W., Schissel, D.J., Mann, E.L., & Irvine, T.N., 1982. The J-M platinum-palladium reef of the Stillwater Complex, Montana. *Econ. Geol.* **77**, 1454-80.

Ulmer, G.C., 1969. Experimental investigations of chromite spinels. In: Wilson, H.D.B. (ed.) *Magmatic Ore Deposits. Econ. Geol. Monograph 4*, 114-31.

Ussing, N.V., 1911. Geology of the country around Julianehaab, Greenland. *Medd. Grønland* **169**, 1-60.

Volker, J.A., & Upton, B.G.J., 1990. The structure and petrogenesis of the Trallval and Ruinsival areas of the Rhum ultrabasic complex. *Trans. Roy. Soc. Edin.: Earth Sci.* **81**, 69-88.

Wager, L.R., 1959. Differing powers of crystal nucleation as a factor producing diversity in layered igneous intrusions. *Geol. Mag.* **96**, 75-80.

Wager, L.R., 1963. The mechanism of adcumulus growth in the layered series of the Skaergaard intrusion. *Miner. Soc. Am. Spec. Paper* **1**, 1-9.

Wager, L.R., & Brown, G.M., 1968. *Layered Igneous Rocks*. San Francisco, CA: W.H. Freeman & Co., 587 pp.

Wager, L.R., & Deer, W.A., 1939. Geologic investigations in East Greenland, Part III, The petrology of the Skaergaard intrusion, Kangerdlugssuaq, East Greenland. *Medd. Grønland* **105**, 1-352.

Walker, K.R., 1969. The Palisades sill, New Jersey: A reinvestigation. *Geol. Soc. Am. Spec. Paper* **111**, 1-178.

Watson, E.B., 1976. Two-liquid partition coefficients: experimental data and geochemical implications. *Contr. Miner. Petrol.* **56**, 119-34.

Wilson, A.H., 1992. The geology of the Great Dyke, Zimbabwe: Crystallization, layering, and cumulate formation in the P1 pyroxenite of Cyclic Unit 1 of the Darwendale subchamber. *J. Petrology* **33**, 611-63.

Wilson, J.R., & Larsen, S.B., 1985. Two dimensional study of a layered intrusion: the Hyllingen series, Norway. *Geol. Mag.* **122**, 97-124.

Wilson, J.R., Menuge, J.F., Pedersen, S., & Engell-Sørensen, O., 1987. The southern part of the Fongen-Hyllingen layered mafic complex, Norway: Emplacement and crystallization of compositionally stratified magma. In: Parsons, I. (ed.) *Origins of Igneous Layering*. Dordrecht: Reidel, 287-312.

Young, I.M., & Donaldson, C.H., 1985. Formation of granular-textured layers and laminae within the Rhum crystal pile. *Geol. Mag.* **122**, 519-28.

LAYERED INTRUSIONS
R.G. Cawthorn (editor)

Fluid Dynamic Processes in Basaltic Magma Chambers

I.H. Campbell

Research School of Earth Sciences, Australian National University, Canberra, A.C.T. 0200, Australia.

Abstract

Convection in magma chambers is driven by small density differences that originate at the margins of a magma chamber or when a new pulse of magma enters a chamber. Buoyancy anomalies at the margins of magma chambers result from cooling or crystallization at the floor, roof or walls of the intrusion. Cooling produces a thermal boundary layer which is typically between 10 cm and 1 m wide with the temperature drop across the layer between 0.05 and 1°C. Compositional boundary layers, produced by crystallization, are much thinner than thermal boundary layers and are no more than a few millimetres wide. The compositional step across them normally lies between 0.6 to 12 wt%. Calculated thermal and compositional flux Rayleigh numbers, assuming convection over the full depth range of the chamber, are typically greater than 10^{12} and 10^{19} respectively, well above the critical value of 10^6 that marks the transition from laminar to turbulent convection. Laminar or cellular convection is only possible in a convecting layer if its depth is less than 10 cm.

A new pulse of magma entering a chamber may have a density that is less than or greater than the fractionated magma in the chamber. If it is light it will rise to the top of the chamber as a plume. If it is dense it will form a fountain. In both cases the flow will be turbulent and the input magma will mix extensively with the fractionated magma in the chamber, leading to stratification. If the input magma is hotter than the fractionated magma, the stratified hybrid zone produced at the floor of the chamber by a fountain will consist of hot, compositionally dense magma overlain by cooler, compositionally lighter magma. Because the distribution of heat is unstable the hybrid layer will break up into double-diffusive convecting layers. A plume of hot, light magma will produce a hybrid zone at the top of the chamber that is stably stratified with respect to both temperature and composition. The magma will remain stably stratified until heat loss to the surroundings can overcome the stable density gradient and convection can recommence.

Crystallization, dissolving, or melting at the floor or roof of the chamber can also lead to stratification of the magma. A light melt released at the floor, by any of these processes, has an homogenizing influence on the overlying magma whereas light magma released at the roof stratifies the magma at the top of the chamber. The release of a dense magma has the reverse effect; it stratifies the magma if formed near the base of the chamber but has an homogenizing influence on magma if formed near the roof. Where melting or dissolving of the roof produces a light magma it will pond against the roof and the chamber will divide into two layers separated by a double-diffusive interface. Much of the heat required to melt the roof is provided by the latent heat released by crystallization at the floor and it is transmitted across the interface by diffusion. However, little mass is transferred across the interface. That is, assimilation-fractional crystallization is not an important process in basaltic magma chambers.

1. INTRODUCTION

It is now recognized that much of the diversity seen in layered mafic intrusions results from convective processes in the magma chamber. However, because we can only observe the crystallization products of magmas and not the magmas themselves, the form of convection in magma chambers must be determined by inference and not by direct observation. Petrologists working in the field or making measurements in the laboratory often appeal to fluid dynamical processes to explain their observations. The physics of convection is well understood. Less well-understood is how a magma chamber, which may be chemically zoned, crystallizes to produce layered rocks. Many of the convective processes that occur in magma chambers will not be recorded in the crystallized rock record. That is, working backwards from the solidification products of magma chambers to interpret convective processes that may have been operating in magma chambers is rarely straightforward.

In this paper, I review the convective processes that are likely to occur in basaltic magma chambers. I will, for the most part, avoid the more difficult step of relating the convective processes described to the observed features in layered igneous rocks although some generalizations will be drawn. Emphasis will be placed on describing the physical principles that underlie convective processes rather than providing mathematical descriptions in the form of equations.

2. FUNDAMENTALS OF CONVECTION

Fluid dynamicists make extensive use of non-dimensional ratios in which one force acting on a fluid is balanced against another. These ratios are used to quantify the flow characteristics of a dynamic fluid. The advantage of dimensionless numbers is that they are independent of scale or the type of fluid under consideration. For example, the Reynolds number can be used to predict the transition from laminar to turbulent flow in a plume whether the flow takes place in aqueous solutions in small laboratory experiments, in large oil fires that rise over 10 km into the atmosphere, or in the mantle.

There are two basic forms of convection: convection from an extended source and convection from a point or line source. Convection from an extended source occurs when a magma chamber is cooled from above (or the side) or heated from below. It is driven by small buoyancy differences that develop in narrow boundary layers at the margins of the intrusion. These buoyancy differences can be thermal or compositional. Thermal buoyancy is produced by cooling at the margins or by the release of latent heat at the crystal-liquid interface. Compositional buoyancy is produced by crystallization at the floor, walls, or roof of the chamber or by melting (or dissolving) of the roof.

Convection from point or line sources takes place when a new pulse of magma enters the chamber through a pipe (point source) or dyke (line source). Again the convection may be driven by thermal or compositional buoyancy: thermal when the new pulse has a different temperature from the magma in the chamber and compositional when its composition is different. Generally the new pulse will be both compositionally and thermally different from the magma in the chamber and these differences can produce buoyancy differences that are in the same or opposite sense.

2.1. Convection from small sources

Convection from point or line sources occurs when a new pulse of magma is injected into the chamber as a jet, plume or fountain. The term 'jet' is used to describe a forced flow of fluid of the same temperature and composition as the ambient fluid, emitted from an isolated source which can be either a small, nearly circular hole (point source) or a narrow slit (line source). The properties of the flow are determined by the momentum flux at the source or by the Reynolds number *Re* defined by:

$$\mathrm{Re} = \frac{wd}{\nu}, \tag{1}$$

the value of which determines whether or not the jet is turbulent (symbols are defined in Table 1). The Reynolds number is an example of a dimensionless number and, in this case, it expresses the balance between the inertial forces which drive the flow and tend to make it unsteady, and the viscous forces that retard and stabilize the flow. The flow is laminar when the Reynolds number is small, but as *Re* approaches 30, the flow starts to become unsteady and the fluid begins to entrain or mix with its surroundings. As *Re* rises above ~30 the flow becomes increasingly turbulent and mixing is progressively more efficient until, at approximately *Re* = ~400, the flow becomes fully turbulent and further increases in *Re* have little influence on the efficiency of mixing. The mixed fluid spreads out as a cone or wedge away from the source. The Reynolds number can also be used to characterize flow in a pipe but in that case the criterion for turbulence is $Re > 2000$. The higher value for pipes is due to the stabilizing influence of the pipe walls.

A 'plume' is the flow produced by an isolated source of buoyancy and here the buoyancy flux, which may be due to heat or compositional differences, is the fundamental parameter. Again plumes may be laminar or turbulent, depending on a similar Reynolds number criterion to that for a jet. The momentum flux of a plume increases with distance above the source through the action of the buoyancy force, and so does the Reynolds number; thus an originally laminar plume may become turbulent at greater heights. A jet of dense fluid projected upwards with excess momentum will eventually be brought to rest by negative buoyancy forces and turn back to form what we have called a 'fountain'; this, too, is turbulent at $Re > 100$ and mixes vigorously with its surroundings when the properties of the inflowing and ambient fluids are not very different.

2.2. Convection from extended sources

Convection in magma chambers is driven by small buoyancy differences in boundary layers at the margin of the chamber. There are two types: thermal boundary layers normally produced by diffusion of heat into the cool walls or roof, and compositional boundary layers produced by diffusion of mass at a boundary where crystallization or melting is occurring.

At a vertical boundary layer, convective motion starts as soon as a buoyancy anomaly develops. In the case of a cool thermal boundary layer in a basaltic magma chamber, heat diffuses relatively slowly into the wall but the buoyancy of the thin, cool boundary layer drags much more fluid into motion through the action of viscosity (*i.e.* the viscous boundary layer is much thicker than the thermal boundary layer). The Prandtl number *Pr*, defined by

$$\mathrm{Pr} = \frac{\nu}{\kappa_T}, \tag{2}$$

is a measure of the relative thicknesses of the viscous and thermal boundary layers during laminar flow; the ratio of the thickness of the layers being roughly proportional to Pr.

Another parameter arises during crystallization or melting, when molecular diffusion from a boundary is the process producing a compositionally buoyant boundary layer (instead of the diffusion of heat). This parameter is the Schmidt number where $Sc = \nu/\kappa_S$ (*i.e.* Sc is the equivalent of Pr, using κ_S, instead of κ_T). The ratio of the viscous to the compositional boundary layer scales is $Sc^{1/2}$; the latter, both in aqueous solutions and magmas, is even thinner than the thermal boundary layer. The relative widths of the compositional, thermal and momentum boundaries are proportional to $\kappa_S^{1/2}$, $\kappa_T^{1/2}$, and $\nu^{1/2}$, as shown in Figure 1.

When a buoyancy anomaly exists across a horizontal layer of fluid, it may remain static, with motions being opposed by viscosity and by the action of thermal or compositional diffusion, which smooths out buoyancy anomalies. Under these conditions buoyancy anomalies are dispersed by diffusion. Instability and convective motion set in only when the Rayleigh number, based on the thickness of the fluid layer, reaches a minimum value of order 10^3. For thermally produced buoyancy the Rayleigh number, Ra is:

$$Ra = \frac{g\alpha\,\Delta T h^3}{\nu\kappa_T}. \tag{3}$$

It expresses the balance between the driving buoyancy forces and the two diffusive processes, viscosity (ν) and the thermal diffusivity (κ_T) which retard the motion and tend to stabilize it. If the total depth h of the fluid is much larger than the boundary layer thickness and Ra is greater than the critical value, the boundary layer becomes unstable when Ra, based on its thickness ($\delta \ll h$), reaches 10^3, and breaks away to form a plume which feeds buoyancy into the overlying convecting magma. That is, plumes intermittently break away from horizontal boundary layers when they acquire enough local buoyancy to overcome the viscous forces that oppose their rise or fall.

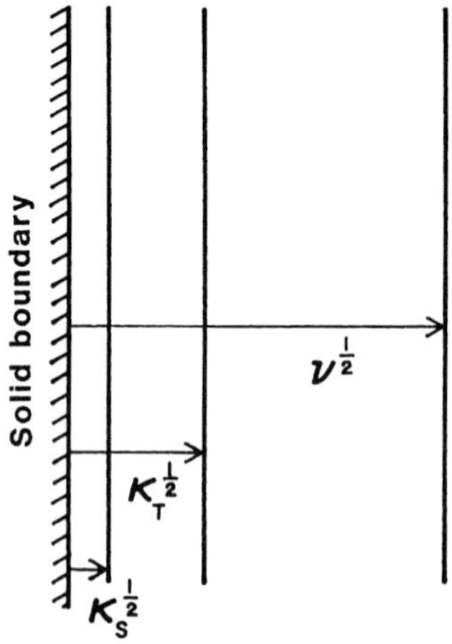

Figure 1. *Diagrammatic representation of the relative thicknesses of the boundary layers formed by molecular diffusion ($\kappa_S^{1/2}$), heat ($\kappa_T^{1/2}$) and momentum ($\nu^{1/2}$) away from a solid boundary (after Turner and Campbell, 1986).*

Boundary layer flow at an inclined floor or roof is obviously intermediate between the vertical and horizontal cases. The fluid flows laterally along the boundary layer until the local Ra exceeds $\sim 10^3$ when it breaks away to form a plume. If the angle of the boundary is shallow, lateral flow is unimportant but it becomes increasingly important as the angle steepens.

The same flow patterns can be expected in two convecting systems only if the systems have the same geometric form and the same values of both Ra and Pr. In this case h in (3) is the height of the chamber. For a given fluid (*i.e.* Pr) the value of Ra (or the Grashof number defined by $Gr = Ra/Pr$) indicates the type of flow to be expected, and determines whether it will be laminar or turbulent, and thus plays a similar role to the Reynolds number for a plume. When the Prandtl number is large, as is the case for a basaltic magma, the transition from laminar (cellular) to turbulent convection takes place at $Ra > 10^6$.

Table 1
List of symbols used in text and values used in calculations

Symbol	Units	Values	Description
w	$m\ s^{-1}$		Mean fluid velocity
d	m		Diameter of source or dyke width
ρ	$kg\ m^{-3}$		Magma density
η	$kg\ m^{-1}\ s^{-1}$		Viscosity of magma
ν	$m^2\ s^{-1}$		Kinematic viscosity of magma $\left(=\dfrac{\eta}{\rho}\right)$
C_p	$J\ kg^{-1}\ °C^{-1}$	1.1×10^3	Specific heat capacity
C	–	0.10	A constant
d_s	m		Width of compositional boundary layer
d_L	m		Width of latent heat boundary layer
d_T	m		Width of thermal boundary layer
f	$kg\ m^{-2}\ s^{-1}$		Mass flux out of magma due to crystallization
g	$m\ s^{-1}$		Acceleration due to gravity
h	m		Depth of magma
L	$J\ kg^{-1}$	8.4×10^5	Latent heat of crystallization
q	Wm^{-2}		Heat flux out of chamber
q_L	Wm^{-2}		Flux of latent heat released by crystallization
ΔS	(weight fraction)		Change in composition across chamber
T	°C		Temperature
ΔT	°C		Drop in temperature across chamber
α	$°C^{-1}$		Thermal expansion coefficient
β	$(\text{weight fraction})^{-1}$		Compositional "expansion" coefficient
κ_S	$m^2\ s^{1}$	10^{-11}	Diffusivity of mass in magma
κ_T	$m^2\ s^{-1}$	8×10^{-7}	Diffusivity of heat in magma
$\left.\dfrac{\partial T}{\partial S}\right\|_{liq}$	°C		Slope of liquidus in T-S space
Ra	–		Conventional thermal Rayleigh number
Ra_f	–		Flux-based thermal Rayleigh number
Rs	–		Conventional compositional Rayleigh number
Rs_f	–		Flux-based compositional Rayleigh number

3. QUANTIFYING CONVECTION IN BASALTIC MAGMA CHAMBERS

3.1. Thermal convection

Unfortunately (3) cannot be used directly to calculate the Rayleigh number for magma chambers because ΔT, the temperature step across the convecting magma in the chamber, is not known. However, this problem can be avoided if the flux Rayleigh number Ra_f is used instead of the thermal Rayleigh number (Martin *et al.*, 1987):

$$Ra_f = \frac{g\alpha q h^4}{\nu \kappa_T^2 \rho C_P}, \tag{4}$$

$$Ra = \left(\frac{Ra_f}{C}\right), \tag{5}$$

where C is a numerical constant with a value of approximately 0.1.

Magma chambers lose most of the heat through their roof (Irvine, 1970; Turner and Campbell, 1986). If heat loss is assumed to be entirely by conduction the heat flux can be calculated from the equations of Carslaw and Jaeger (1959). For a magma chamber which has already cooled for tens of thousands of years and is buried deep in the crust the minimum likely heat flux is 0.4 Wm^{-2}, whereas 4 Wm^{-2} is a more reasonable value for a shallow chamber. Higher heat fluxes are also likely during the early stages of cooling or where cooling is enhanced by hydrothermal circulation in the roof, which is likely to be an important factor in the cooling of most chambers (*e.g.* Skaergaard: Taylor and Forester, 1979).

Martin *et al.* (1987) have calculated thermal Rayleigh numbers for heat fluxes varying between 4×10^{-1} Wm^{-2} to 4×10^{3} Wm^{-2} assuming convection occurs over the full depth of the chamber (Figure 2). The minimum value obtained for the lowest plausible heat flux is 10^{12} for a magma chamber >1 km thick, well above the value of 10^{6} that marks the transition from laminar to turbulent convection. Convection must therefore be turbulent. Thermally driven laminar convection is only possible in basaltic chambers for convecting layers that are less than 10 m deep and, even then, only if the minimum plausible heat flux is assumed.

3.2. Compositional convection

Although most of the heat is lost through the roof of the chamber two factors make the floor the major site of crystallization during the early and middle stages of the evolution of a chamber. First, most basaltic magmas melt the roof of the chamber, creating a ponded layer of felsic melt with the liquidus temperature well below that of the remainder of the magma in the chamber (Campbell and Turner, 1987). The second factor is the well-known pressure effect on the liquidus temperature which increases the supersaturation in a homogeneous magma by about 1.2°C km^{-1} for olivine and 3.4°C km^{-1} for orthopyroxene. This property implies a large degree of supersaturation at the bottom of the chamber and consequently more rapid crystallization at the floor than the roof if, indeed, any crystallization is occurring at the roof.

The crystallization of dense minerals such as olivine and/or pyroxene at the floor of a basaltic magma chamber leaves the melt adjacent to the crystal-liquid interface depleted in dense components. The depleted fluid is less dense than the remainder of the magma and convects upwards away from the growing crystals. This type of convection is called compositional convection and can be described in terms of a compositional Rayleigh number which is analogous to its thermal equivalent. The compositional Rayleigh number Rs and flux Rayleigh number Rs_f are defined as:

$$Rs = \frac{g\beta\Delta S h^3}{\nu\kappa_S}, \tag{6}$$

$$Rs_f = \frac{g\beta\Delta f h^4}{\rho\nu\kappa_S^2}, \tag{7}$$

where β is a compositional "expansion" coefficient such that the liquid density ρ obeys $\rho = \rho_0 (1+\beta\Delta S)$ and ΔS is a concentration difference between the top and bottom boundaries analogous to ΔT in the thermal Rayleigh number. f is the mass flux out of the liquid due to crystallization of the heavier component – the "solute" S. The corresponding compositional version of equation (5) is

$$Rs = \left(\frac{Rs_f}{C}\right)^{3/4}, \tag{8}$$

where the constant C is again approximately equal to 0.1 and the ratio of the conventional Rayleigh numbers is given by

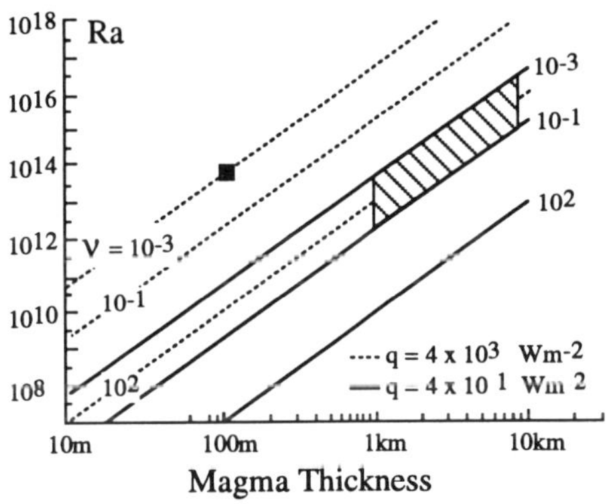

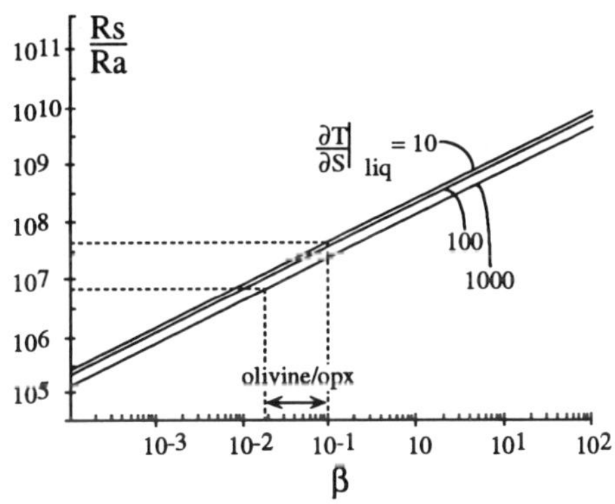

Figure 2. *(left) Ra plotted against h for rapidly cooled chambers (broken lines) and slowly cooled chambers (solid lines). Lines are plotted for different kinematic viscosities (in $m^2\ s^{-1}$) and are labelled accordingly. The field for mafic intrusions (e.g. Bushveld, Great Dyke, Stillwater, Jimberlana), assuming the magmas to be homogeneous over the depth h, is shown by the hatched box. The solid square shows the approximate position of a 100 m layer of picrite emplaced under a cooler, more fractionated magma (modified after Martin et al., 1987).*

Figure 3. *(right) The ratio of conventional Rayleigh numbers Rs/Ra against β (in 'per weight fraction') for various values of $\left.\frac{\partial T}{\partial S}\right|_{liq}$ (°C per weight fraction using $\rho = 2{,}500$ kg m^{-3} and $L = 8.4 \times 10^5$ J kg^{-1} (modified after Martin et al., 1987).*

$$\frac{Rs}{Ra} = \left(\frac{Rs_f}{Ra_f}\right)^{3/4}. \tag{9}$$

Martin *et al.* (1987) have calculated typical values for *Rs/Ra* while olivine or orthopyroxene are crystallizing from a melt. There is some uncertainty regarding the appropriate value of κ_S to use in equations 6 and 7. Martin *et al.* (1987) used a value of 10^{-10} m^2 s^{-1} but more recent work by Kress and Ghiorso (1995) and Zhang *et al.* (1989) has shown that κ_S for an andesite is $\sim 2\times10^{-12}$ m^2 s^{-1} at 1200°C and $\sim 6\times10^{-12}$ m^2 s^{-1} at 1300°C. The diffusion coefficient can be expected to be slightly higher in a low viscosity basaltic magma so a value of 1×10^{-11} m^2 s^{-1} is taken as a reasonable estimate for κ_S in a basaltic magma at 1250°C but the value could be as low as 1×10^{-12} m^2 s^{-1} in an evolved basalt at 1150°C or as high as 1×10^{-10} m^2 s^{-1} for a picritic magma at 1400°C.

Figure 3 shows how this ratio varies as a function of the density coefficient β for various values of the liquidus slope. Note that variations in the liquidus slope which is about 700°C per weight fraction for olivine and 70°C per weight fraction for orthopyroxene have only a small influence on *Rs/Ra*. It is apparent from Figure 3 that for olivine or orthopyroxene crystallization the compositional Rayleigh number is a factor of about 10^7 greater than the thermal Rayleigh number. Taking the value of *Ra* from Figure 2 and the ratio of *Rs/Ra* from Figure 3 we find that the compositional Rayleigh number for mafic magmas, crystallizing olivine or orthopyroxene lies between 10^{19} and 10^{23} for layer depths between 1 and 10 kms. Laminar or cellular convection is only possible if the heat flux is low and if the layer depth is less than 10 cm. It is therefore unlikely to occur in basaltic magma chambers.

4. BOUNDARY LAYERS

As already noted, convection in magma chambers is driven by buoyancy anomalies that build up in boundary layers at the roof, wall or floor of the intrusion. These boundary layers grow by diffusion until, at a critical time t^*, the local Rayleigh number, based on the boundary layer thickness δ, reaches a critical value of approximately 10^3. At this time the boundary layer becomes unstable and breaks away as a plume of buoyant material to join the convection in the interior of the magma chamber.

4.1. Upper thermal boundary layer without melting or crystallization

If it is assumed that no crystallization or melting occurs at the top of the chamber, the heat flux q through the roof must pass through the upper thermal boundary layer. We can approximate the intermittently unstable boundary layer by a layer of constant thickness across which heat is transported solely by steady-state conduction, so that

$$d_T = \kappa_T \rho C_P \frac{\Delta T}{q}. \tag{10}$$

Because convective stirring keeps the magma well mixed outside the boundary layers the temperature drop across the thermal boundary layer in (10) is equal to the overall temperature difference, ΔT, available to drive convection. As a consequence, it can be shown (Martin *et al.*, 1987) that the thickness of the boundary layer is given by

$$d_T = \left[\frac{\rho C_P \nu \kappa_T^2}{C^3 g \alpha q} \right]^{1/4}, \tag{11}$$

and that the temperature difference across the boundary layer is given by

$$\Delta T = (\nu / g\alpha\kappa_T^2)^{1/4} (q / \rho C_P C)^{3/4}. \tag{12}$$

Figure 4 shows the values for the thermal boundary layer thickness and temperature drop as a function of heat flux and magma viscosity. The upper thermal boundary layer thickness is generally between 10 cm and 1 m and the temperature drop lies between 0.05 and 1°C. This implies that temperature differences, within a convecting layer in a basaltic magma chamber, will typically be 0.1°C and always less than 1°C for conductive cooling.

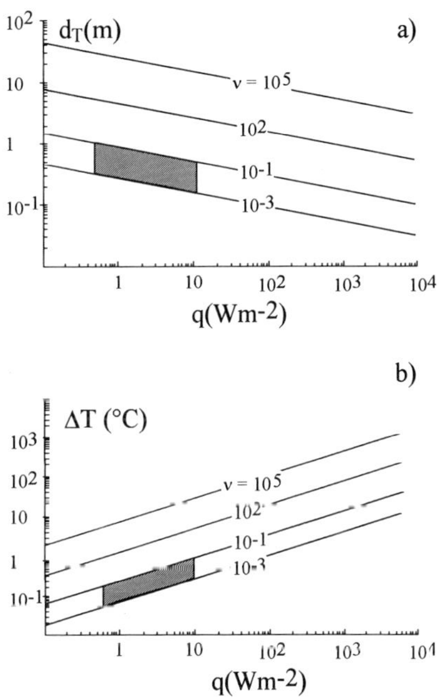

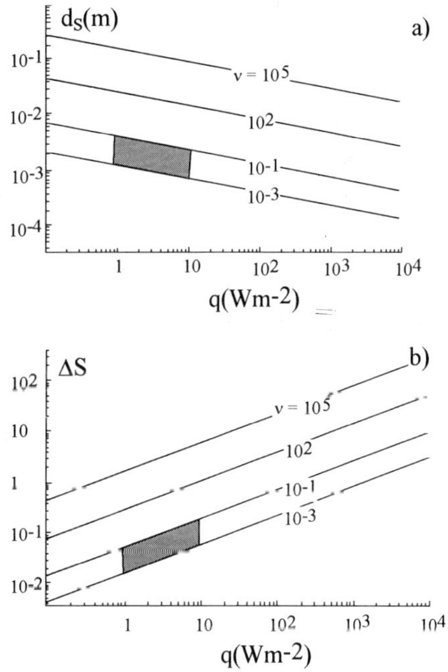

Figure 4. (left) (a) The thickness of the thermal boundary layer and (b) the temperature difference across the thermal boundary layer as functions of cooling rate for different viscosities. The field for normal basaltic magma chambers, cooled by conduction, is indicated by the hatched box (modified after Martin et al., 1987).

Figure 5. (right) (a) The thickness of the compositional boundary layer and (b) the compositional difference ΔS across the compositional boundary layer as a function of cooling rate for different viscosities of magmas crystallizing olivine. Note that ΔS is expressed as a weight fraction. Multiply by 100 to convert into weight percent. Again the field for normal basaltic chambers is indicated by the hatched box (after Martin et al., 1987).

When crystallization occurs at the roof, some of the conductive heat loss is consumed by latent heat of crystallization. The effective heat flux in equations 11 and 12 is therefore lowered, which results in an increase in d_T and a decrease in ΔT. The importance of crystallization at the roof has been quantified in a series of papers by Kerr *et al.* (1989, 1990a, b, c) and Worster *et al.* (1990). *Their results are consistent with the field observation that, in large magma chambers of the type considered in this volume, most of the crystallization occurs at the floor even though most of the heat is lost through the roof.* The effect of crystallization at the roof on d_T and ΔT can therefore be neglected except in thin sills and during the final stages of crystallization of large chambers.

Melting at the roof absorbs latent heat and because heat is transferred between the zone of melting and the underlying basaltic magma by convection, roof melting can lead to a dramatic increase in the effective heat flux. For example, calculations similar to those carried out by Huppert and Sparks (1988) show that, under ideal conditions, melting of the roof of a 5 km thick magma chamber in the hot, lower crust can increase the heat flux by up to 10^3. An increase in q leads to an increase in ΔT and a decrease in d_T that can be predicted from equations 11 and 12. For upper crustal magma chambers, like the Bushveld, Skaergaard and Stillwater Complex, emplaced into relatively cool crust, the effect will be far less important but could still be significant.

4.2. Combined compositional and latent heat boundary layers

If a dense phase such as olivine crystallizes at the floor of a magma chamber a light magma is released giving rise to a buoyant compositional boundary layer. In addition, the release of latent heat of crystallization produces a thermal boundary layer which is also buoyant. As a consequence two buoyant boundary layers develop at the floor of the chamber and, because heat diffuses faster than mass, the thermal boundary layer is wider than the compositional boundary layer. However, in spite of this difference, the critical time scale for the boundary layers to break away and become unstable is similar, which means that the two boundary layers can be treated independently (Martin *et al.*, 1987).

Following a procedure similar to that described in connection with equations 10 and 12 Martin *et al.* have shown that the width of the compositional boundary layer is given by:

$$d_S = d_T\left(\frac{Ra_f}{Rs_f}\right)^{1/4} = \left[\frac{\rho\nu\kappa_S^2}{C^3 g\beta q}\left(L + C_P \left.\frac{\partial T}{\partial S}\right|_{liq}\right)\right]^{1/4} \tag{13}$$

and the compositional change across the boundary layer by

$$\Delta S = \frac{f d_S}{\rho\kappa_S} = \left(\frac{\nu}{g\beta\kappa_S^2}\right)^{1/4}\left[\frac{q}{pC\left(L + C_P \left.\frac{\partial T}{\partial S}\right|_{liq}\right)}\right]^{3/4} \tag{14}$$

Figure 5 shows the range of values in d_s and ΔS for $\kappa_s = 10^{-11}$ m^2 s^{-1} as a function of magma viscosity and cooling rate when olivine is crystallizing from a mafic magma. The thickness of the compositional boundary layer is approximately 30 times smaller than the upper thermal boundary layer with values of d_s lying between ~0.6 mm and 6 mm. Corresponding values of ΔS are between 0.6 and 12 wt% for conductive cooling.

Martin *et al.* (1987) have illustrated the nature of the lower boundary layer by considering a specific example. The example they chose is a magma of Great Dyke composition crystallizing olivine. The results of their calculations are illustrated in Figure 6. It was assumed that the total

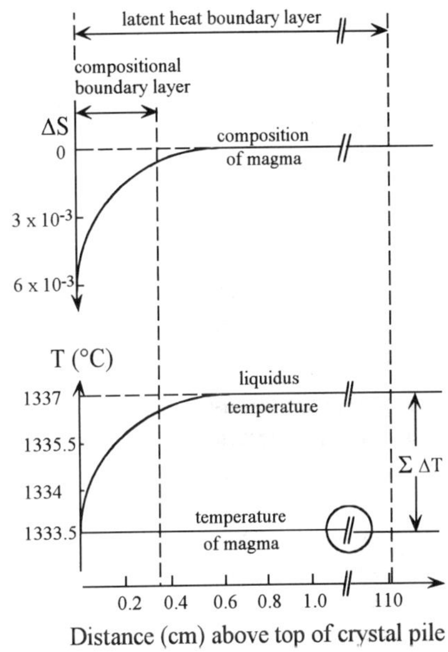

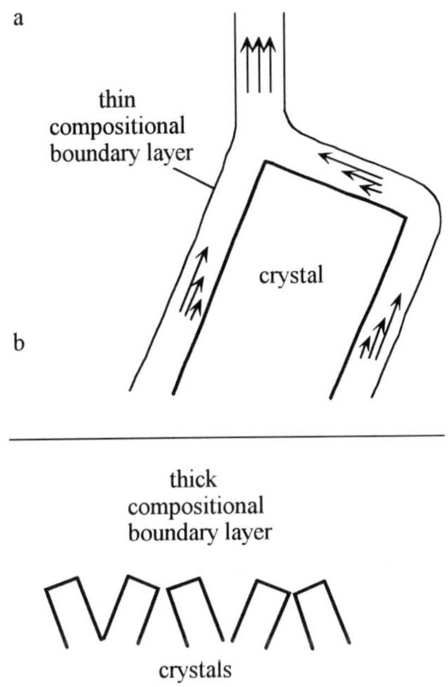

Figure 6. *(left) A sketch of chemical and thermal profiles in the compositional and latent heat boundary layers at the base of a magma chamber. The numbers shown are those calculated for the parental magma of the Great Dyke (Wilson, 1982) cooling at 4×10^{-1} Wm^{-2} in a 3 km deep chamber, but the qualitative form of the diagram is appropriate to all magmas which release a light fluid on crystallization. Note that the depletion of the magma in the crystallizing components across the compositional boundary layer leads to a sharp decrease in the liquidus temperature close to the growing crystals. Also note that the total amount of supercooling ($\Sigma\Delta T$) always increases away from the crystal/liquid interface, and that compositional effects contribute by far the largest portion of this supercooling. The slope of the line representing magma temperature is reflected in the small (~ 0.02°C) drop across the break between 3 cm and 110 cm (circled in the diagram) (modified after Martin et al., 1987).*

Figure 7. *(right) Sketches of the two possible configurations of compositional boundary layer and growing crystals. (a) At high cooling rates for the least viscous magmas the crystals are large compared with the thickness of the boundary layer. The size of the arrows is intended to indicate the speed of fluid moving in the boundary layer. (b) In most situations the crystals are small compared with the thickness of the boundary layer, which in this case grows by diffusion and breaks down periodically (modified after Martin et al., 1987).*

thickness of the magma was 3 km, that q = 0.4 Wm^{-2}, and that the temperature of crystallization was 1337°C. The following points are noteworthy.

1) The temperature increases slightly (by 0.02°C) from the interior of the magma chamber towards the top of the crystal pile (here considered to be a planar surface) because crystallization at the floor results in the release of a significant amount of latent heat.
2) The thickness of the latent heat boundary layer is 110 cm compared with 3.5 mm for the compositional boundary layer.
3) The calculated temperature drop across the latent heat boundary layer is 0.02°C, resulting in a *very* small thermal gradient.
4) The change in the olivine content of the melt across the compositional boundary layer is 0.6 wt%, resulting in a strong compositional gradient close to the growing crystals.
5) The depletion of the olivine component of the melt within the compositional boundary layer lowers the liquidus temperature of the magma adjacent to the growing crystals by 4.5°C compared with the liquidus temperature of the far field magma.
6) If the melt adjacent to the crystals at the top of the crystal pile is assumed to be in equilibrium with the crystals, the far field magma is supercooled by 4.5°C, almost all of which (4.48°C) is due to compositional depletion in the boundary layer and only 0.02°C is due to thermal differences. That is, the supercooling of the magma outside the boundary layers (*i.e.* the bulk of the magma) is dominated by compositional effects and is more appropriate described as supersaturation (or, equivalently, as constitutional supercooling).
7) Supersaturation (supercooling) increases with distance from the crystal-liquid interface.

4.3. Discussion

The calculated compositional boundary layer thickness of 0.6 mm to 6 mm for crystallization in a basaltic magma chamber is comparable with the crystal size, which is generally less than 3 mm. Similar calculations for the crystallization of salts from an aqueous solution show that, under these conditions, the compositional boundary layer is appreciably smaller than the size of the crystals. This difference between the thickness of the boundary layer and crystal size has important implications for nucleation and crystal growth.

The case of a thin boundary layer is illustrated in Figure 7a. Crystal growth leads to the formation of a thin boundary layer which clings to the growing crystal and is held in place by viscous stresses. If the fluid released is light it flows up the sides of the crystal to the highest corner where it breaks away as a plume. The system rapidly evolves to a steady state with compositional profiles within the depleted boundary layer controlled by a balance between diffusion and advection of the buoyant layer, with the composition of the fluid adjacent to the growing crystals being strongly depleted in those components that are required to form the solid. Heterogeneous nucleation of new crystals within the boundary layer requires the degree of supersaturation to exceed some critical level. Because the compositional profile in the boundary layer around the growing crystals evolves to a steady state this rarely happens and crystal growth occurs through the continual growth of existing crystals rather than by the nucleation of new crystals. As a consequence, crystals evolve to an elongate habit with crystals growing at right angles to the crystal liquid interface being favoured over those growing at an acute angle because the former are growing into free space whereas the growth of the latter is impeded by adjacent crystals. This is the type of crystal growth normally observed when a salt such as sodium carbonate crystallizes from an aqueous solution. Such textures are, however, rare in layered intrusions. An important exception is the harrisites of the Rum Intrusion (Wager and Brown, 1968). The narrow boundary layer required for the development of this texture is most likely to occur in melts of low viscosity and/or when the rate of heat loss is unusually high (see Figure 5). These conditions may be met when a new pulse of dense picritic magma is injected into a magma chamber containing cooler gabbroic melt. Under these circumstances the chamber can divide into two layers separated by a double-diffusive interface and this can lead to a rate of heat loss for the picritic layer which is orders of magnitude higher than normal (Huppert and Sparks, 1980; see below).

The case of the compositional boundary layer being thicker than the size of the growing crystals is illustrated in Figure 7b. The boundary layer is initially stagnant and there is no steady flow of fluid away from the crystals. Crystal growth depletes the boundary layer adjacent to the growing crystals which grows by molecular diffusion until it acquires enough buoyancy to overcome the viscosity of the overlying liquid, *i.e.* the boundary layer grows until the local Rayleigh number, based on its thickness, exceeds approximately 10^3. At this point it breaks away to form a plume. As a consequence, the boundary layer is continually being built up and swept away. The composition of the fluid adjacent to the growing crystals is continually changing in response to the irregular fluctuations in the thickness of the boundary layer. When the boundary layer is thin the composition of the melt adjacent to the growing crystals is similar to the far field magma outside the boundary layer. However, as the boundary layer thickens, the melt adjacent to the growing crystal becomes progressively depleted. These fluctuations allow the supersaturation at the crystal liquid interface to occasionally exceed that required for the heterogeneous nucleation of new crystals even though the average supercooling may be less than that in the narrow boundary layer case. The result is that new crystals can nucleate, within the boundary layer, against existing crystals leading to the more equant crystal habit characteristic of layered intrusions.

If boundary layers are continually building up, then being swept away by convection, this will result in oscillations in the composition of the melt at the crystal-liquid interface which should produce oscillatory zoning in the cumulate crystals. The process will be most effective when the boundary layer thickness is comparable to the crystal size as is normally the case for layered intrusions. Unfortunately, the diffusive time scale for minerals such as olivine and orthopyroxene is greater than the cooling time scale for large igneous bodies, and any zoning which forms during primary crystallization is annealed during the slow cooling of large intru-

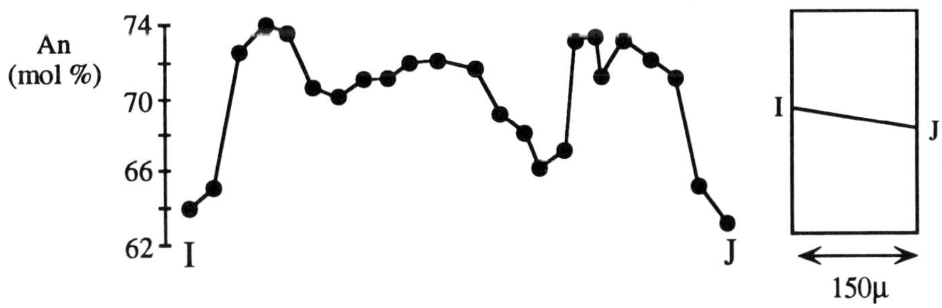

Figure 8. *Compositional variations determined by electron microprobe along a traverse IJ across a zoned plagioclase crystal from the Jimberlana intrusion (after Campbell, 1973).*

sions. Diffusion rates in plagioclase, however, are much slower and plagioclase crystals do preserve their primary zoning, even in layered intrusions. Although the sweeping cycles are unsteady and chaotic, the average time-scale for a cycle is 3 or 4 days, which compares with approximately 7 days for the time to grow a 2 mm crystal (Martin *et al.*, 1987). A typical plagioclase crystal in a layered intrusion should therefore contain 1 or 2 oscillatory zones. Campbell (1973) made a detailed study of zoning in cumulus plagioclase crystals of gabbros from the Jimberlana Intrusion. He found that they showed complex zoning (Figure 8) and that the magnitude of the oscillations was comparable to that predicted in Figure 5b (albeit for olivine).

Brandeis and Jaupart (1986) and Morse (1986), following Wager and Deer (1939), have developed models for crystallizing magmas in which crystals nucleate and grow in the upper thermal boundary layer and periodically cascade to the floor in a gravity current. This requires the temperature drop across the boundary layer to be large enough to enable homogeneous nucleation to occur there in preference to heterogeneous nucleation against the walls and floor of the chamber (Campbell, 1978). The temperature drop across the thermal boundary layer in mafic magma chambers is usually of the order 0.05° to 1°C (Figure 4b) except at extremely high cooling rates. The predicted temperature drop is too low to enable homogeneous nucleation to occur within the upper thermal boundary layer of basaltic magma chambers. Nucleation, if it occurs at all within the upper boundary layer, will be heterogeneous against pre-existing crystals at the roof of the chamber (see section 9).

5. DOUBLE-DIFFUSIVE CONVECTION

Convection driven by compositional or temperature differences has been considered briefly above. Now consider the combined effect of several properties, at least two of which influence the density of the fluid. Magmas are multicomponent systems, and the application of purely thermal or purely compositional convection theory to them ignores a variety of novel convective phenomena that have been described under the general headings of double-diffusive convection or (more recently) multicomponent convection (Turner, 1973, 1985; Turner and Gustafson, 1978; Huppert and Turner, 1981a; Sparks *et al.*, 1984). A characteristic feature of double-diffusive convection is that a flux of one property, imposed either from below or at the side on a gradient of another property with a different molecular diffusivity, tends to produce a series of convecting layers, rather than a single, large-scale overturning. For example, if a magma with a compositional density gradient is cooled from above it will break up into a set of horizontal, convecting layers, each of them well mixed in temperature and composition and separated by interfaces across which heat and composition are transported by molecular diffusion. Heat, which diffuses faster than mass, is transferred across these diffusive interfaces and causes instability and convection in the layers above, while the composition of the layer changes little and preserves the "stable" density steps between the layers.

A fluid which has opposing vertical gradients of two solutes or components with different diffusivities can also break up into double-diffusive convection layers. Consider again the case where the density contribution of the faster diffusing component is gravitationally unstable (Figure 9a) (which has been called the "diffusive" regime). It is convenient to describe the behaviour in terms of the heat-solute system, but it is important to realize that the same ideas apply to properties with much closer molecular diffusivities. Because the destabilizing property (say heat) diffuses faster than the stabilizing one (say salt), local non-uniformities of the vertical gradient are amplified, and the system begins to convect. As in the previous case of cooling

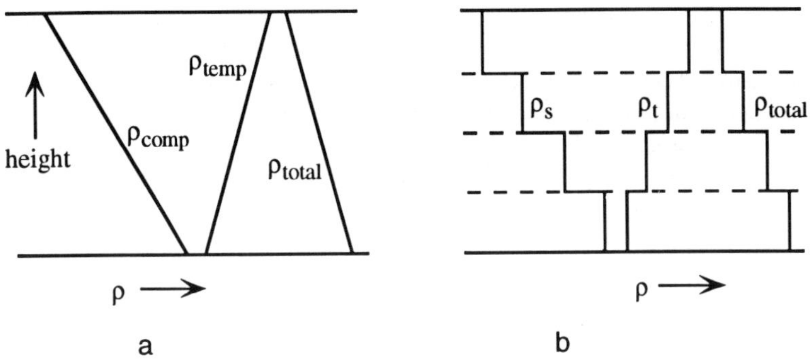

Figure 9. *(a) The distribution of density due to composition and temperature in a layer of fluid before it breaks up into double-diffusive convection cells. Note that the density contribution of the faster diffusing component (in this case heat) is destabilizing but that the overall density gradient is stable. (b) The system depicted in 9a will break up into a series of double-diffusive convecting layers, driven by the release of the potential energy in the unstable distribution of the faster diffusing property (heat). Individual convecting layers are well mixed and are separated by thin diffusive interfaces (indicated by dashed lines), across which heat and composition are transferred by molecular diffusion (heat more rapidly than composition) (after Turner and Campbell, 1986).*

from above, the fluid cannot convect as a single unit but breaks up into a series of well-mixed convecting layers, separated by double-diffusive interfaces (Figure 9b). The essential physical characteristic of a double-diffusive interface is that it is a stagnant boundary layer, with marked thermal and compositional gradients, separating two well-mixed convecting layers. Note that the density is always greater at the bottom. In fact the density difference between the top and bottom of a series of convecting layers will increase with time, so long as convection continues to be driven by double diffusion, since the energy for convection comes from a decrease in the gravitational potential energy of the system.

When the more slowly diffusing component is "destabilizing" (*e.g.* heavy at the top) another phenomenon is observed, either when hot salty solution lies above colder fresher water, or when there are two different chemical components. Consider, for example, the two-layer case where a light layer of sugar solution is placed above denser salt solution. Long, narrow, vertical convection cells, called "salt fingers", are formed rapidly and extend through the interface, which thickens with time.

A finger interface forms in a magma chamber when hot compositionally dense magma overlies cooler compositionally lighter magma. That is, the upper liquid has a lower density only because it is hotter than the underlying one. If both liquids were at the same temperature the density relations would be reversed. The rapid horizontal diffusion of heat between adjacent fingers makes it possible to release the potential energy stored in the unstable compositional density difference, and it is this release of potential energy that drives convection. The flow in adjacent fingers is in opposite directions. The downward moving fingers are losing heat and their density is increasing, whereas the reverse is true for the upward moving fingers (Figure 10). The fingers act as tubes transporting fluid from one convecting layer to the other. As fluid moves through the downward moving fingers it loses heat and its density increases so that by the time it reaches the bottom of the interface its density is greater than that of the underlying liquid. This results in the release of a flux of dense fluid at the top of the lower convecting layer which drives convection in that layer. Conversely, upward moving fingers release a flux of light fluid at the base of the upper layer. As a consequence, the density difference between the two convecting layers increases with time and, in this respect, finger interfaces are similar to diffusive interfaces. An important difference between finger and diffusive interfaces, however, is that finger interfaces are driven by the unstable distribution of compositions whereas diffusive interfaces are driven by the unstable distribution of heat. Finger interfaces result in extensive mixing between the fluids on opposite sides of the interface whereas, in the case of a diffusive interface, composition mixing is more limited.

In addition to the Rayleigh number and Prandtl number two further dimensionless numbers are required to characterize the motions in double-diffusive convection: the diffusivity ratio τ (or its inverse, the Lewis number, which is used in engineering literature) and the density ratio R_ρ. They are defined by

$$\tau = \kappa_T / \kappa_S, \tag{15}$$

$$R_\rho = \beta \Delta S / \alpha \Delta T \cdot \tag{16}$$

R_ρ is the ratio of the separate contributions of composition and heat to the density difference across an interface. It is an overall measure of the extent to which the system is double-diffusive; for a given value of $\beta \Delta S$, double-diffusive convection becomes most important as R_ρ

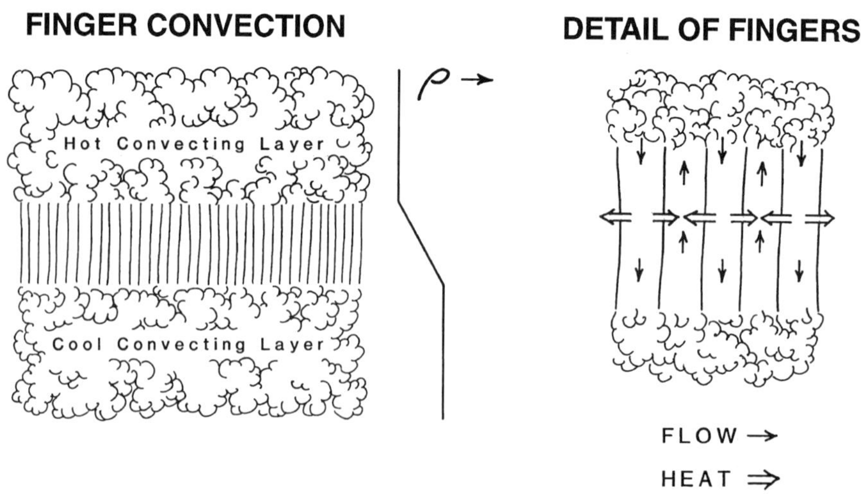

Figure 10. Diagrammatic cross section through a finger interface. The single arrows represent vertical fluid flow and the double arrows horizontal heat flow.

approaches unity (*i.e.* as the property gradients approach each other) for both the "diffusive" and finger configurations.

6. FILLING OF MAGMA CHAMBERS

Now consider the fluid dynamical processes associated with the injection of a new pulse of magma into a chamber, a problem first addressed by Sparks *et al.* (1980). Assume, as a starting condition for several of the processes discussed below, that the chamber is filled with a homogeneous magma, *i.e.* the magma has the same composition and temperature throughout the chamber. Such a chamber could remain well mixed during further smaller inflows only when the physical properties of the new magma (composition, temperature, density) are identical to those of the resident magma. A jet of fluid at a high Reynolds number will mix rapidly with surroundings having the same properties, and it will be immaterial whether the jet is directed upwards or at an angle to the vertical. At low *Re*, there will be less mixing with the surroundings, but again the new and old magmas will be indistinguishable. If any property differences exist, however, complete mixing is very unlikely to occur, and in the following sections the various ways in which compositional and thermal stratification can be produced as a result of the filling process itself are discussed.

6.1. Replenishment by a dense input without mixing

This case was considered by Huppert and Sparks (1980) and it was also the subject of a laboratory investigation by Huppert and Turner (1981). In the laboratory a denser layer of hot KNO_3 solution was injected below colder $NaNO_3$ solution, taking care to minimize the mixing. In the case of a magma chamber a similar end-result could be obtained if the flow enters the chamber with little momentum (*e.g.* if the ascending magma is driven upward by a very small density difference between the magma and the surrounding rocks). An important criterion, however, is that the filling rate should be fast enough for a substantial layer to be built up at the bottom, without mixing and before significant cooling or crystallization can occur.

The application is to a magma chamber replenished from below with hot, dense, more primitive magma. The striking feature is the establishment and maintenance of a sharp horizontal "double-diffusive" interface between the two layers, with hot picritic magma overlain by cooler fractionated magma. Because this interface is thin (of the order of only tens of centimetres thick), heat is transferred rapidly from the lower layer to the upper layer while compositional properties are transported very slowly. In fact calculations by Huppert and Sparks (1980) show that the rate of cooling of the lower layer by heat loss to the upper layer is orders of magnitude faster than the normal rate of cooling of a magma chamber by conduction through its walls. Rapid heat loss by the lower layer leads to rapid crystallization of olivine (if the new pulse of magma is picritic in composition) which lowers the density of the lower layer and this can lead to overturning and complete mixing between the layers.

6.2. Fountains

The case considered above assumes no mixing between the input magma and the fractionated magma in the chamber. If, however, the input is turbulent extensive mixing can occur and this case has been studied by Campbell and Turner (1989) in a series of experiments in which a salt solution was injected into fresh water. If a fluid or magma which is denser than the fluid in the chamber enters the chamber with upward momentum it will initially be carried upwards into the overlying magma, forming a fountain. The momentum within the ascending fluid decreases with increasing height due to the influence of buoyancy forces which act in the opposite direction to the motion. Eventually the negative buoyancy forces overcome the initial upward momentum of the fountain and the new pulse falls back to the floor of the chamber in an annular ring about the feeder. The motion within the fountain is highly turbulent, resulting in the host fluid being drawn into or entrained by the added fluid. The resulting mixture collects at the base of the fountain then spreads out across the bottom of the tank to form a hybrid layer (Figure 11a).

During the early stages of an experiment the fountain entrains mainly the host fluid, and entrainment, expressed as the percentage of host fluid entrained into the fountain, is highly

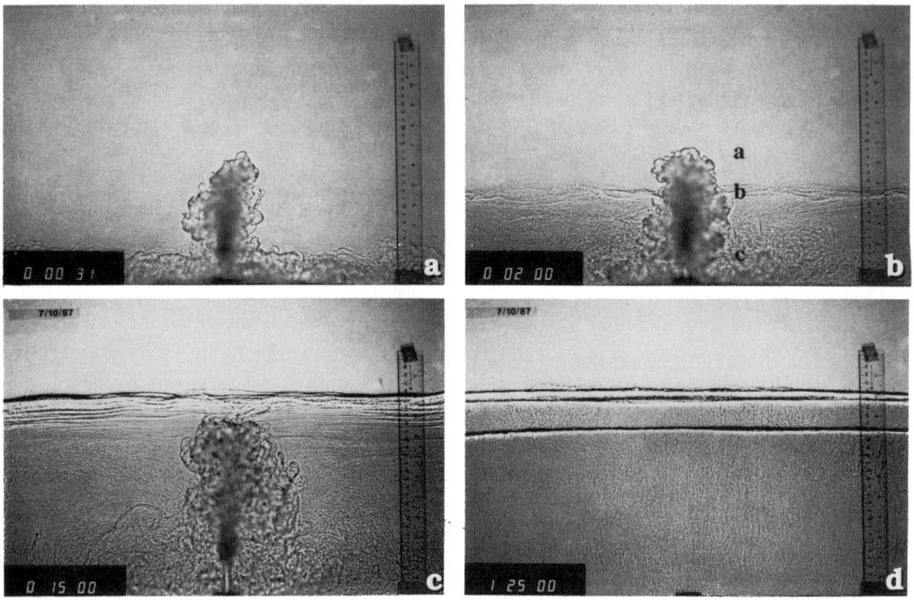

Figure 11. *The fountain produced when a dense sugar solution is injected into a tank containing lighter salt water. (a) The momentum of the dense fluid initially carries it upwards into the overlying fluid, forming a fountain. Eventually, when negative buoyancy forces overcome the initial upward momentum, the new pulse falls back to the floor and collects, forming a hybrid layer. (b) As additional fluid is added to the tank the hybrid layer advances up the side of the fountain. Entrainment of host fluid is restricted to that part of the fountain that rises above the top of the hybrid layer (between a and b). (c) The filling stage, when the top of the hybrid layer is above the level of the fountain so that no new host fluid can be entrained into the fountain. (d) The hybrid layer breaks up into three double-diffusive convecting layers. Convection is being driven by the unstable distribution of heat, the faster diffusing of the two components. Note that the thickness of the layers decreases towards the top of the hybrid zone where the density gradient is highest (after Campbell and Turner, 1989).*

efficient (>97%). As a consequence the thickness of the hybrid layer increases rapidly compared with the rate of filling of the tank. As filling continues and the hybrid layer grows (Figure 11b), the fountain begins to entrain both the host fluid (between a and b) and the hybrid layer (between b and c). This has two effects: (i) the rate of entrainment of host fluid into the fountain is continually decreasing, so that the rate of advance of the hybrid layer is also decreasing; and (ii) since this hybrid layer is denser than the host magma, the density of the fluid falling back down the sides of the fountain is continually increasing. This dense fluid, when it reaches the bottom of the tank, flows laterally away from the fountain in a turbulent flow, displacing lighter fluid in its path and resulting in the development of a stable density stratification in the hybrid layer (Figure 12a).

Eventually the hybrid layer rises above the top of the fountain and host fluid can no longer be entrained into the fountain. From this point, the rate of advance of the hybrid layer is controlled only by the rate at which fluid is added to the tank. Thus two distinct stages can be recognized during the development of the hybrid layer. (1) The "entrainment" stage during which the fountain penetrates the top of the hybrid layer, allowing entrainment of the host fluid (Figure 11b). (2) The 'filling' stage, when the top of the hybrid layer is above the level of the fountain so that no new host fluid can be entrained into the fountain (Figure 11c).

The factors controlling the average mixing ratio in the hybrid layer are the amount of magma added to the chamber, and the height of rise of the fountain, which controls the amount of entrainment and therefore the degree of dilution of the hybrid layer. The average steady height (h_1) reached by a line fountain, fed by a dyke of width d, is given by:

$$h_1 = 1.3\left(\frac{\Delta\rho_1}{\Delta\rho_0}\frac{1}{f}\right)^{-2/3} d \qquad (17)$$

where $\Delta\rho_0$ is the density difference between the input magma and the host magma in the chamber, $\Delta\rho_1$ is the mean density contrast between the magma and the wall rocks and f is a friction factor with a value of ~0.03.

Observations suggest that the feeder dykes to continental magma chambers are likely to lie within the range 10-100 m, although most probably towards the lower end of that range. Dykes narrower than 1 m would freeze while passing through cool continental crust (Bruce and Huppert, 1990a, b). Calculated fountain heights for a range of values of $\Delta\rho_0$ and dyke widths between 1 and 30 m, assuming that the average density difference between the ascending magma and wall rocks is 300 kg m^{-3}, are given in Table 2. Taking 3 m as the minimum likely width of the feeder dyke to a major layered intrusion and $\Delta\rho_0$ = 30 kg m^{-3} as the maximum likely density difference between the input magma and the fractionated host magma, the minimum height of rise of a fountain is 186 m. Furthermore, the Reynolds number is well in excess of the critical number for turbulent flow so that the input magma can be expected to mix extensively with the host magma to produce a zoned hybrid layer at the bottom of the chamber. The highest Mg/Fe ratio for olivines and pyroxenes from cyclic units from the ultramafic zones of layered intrusions is often well below the value expected for minerals crystallizing from a melt derived directly from the mantle, supporting the hypothesis that new pulses of magma can mix extensively with the fractionated magma in the chamber.

Table 2
The height of rise h_l (in metres) of line fountains calculated from equation (17) and the corresponding Reynolds number for a range of magma densities and viscosities and for different dyke widths.

d(m)	30	10	3	1
h_l (for $\Delta\rho_0 = 30$)	1860	620	186	62
h_l (for $\Delta\rho_0 = 10$)	3900	1300	390	130
Re ($\eta = 1$ Pa s)	2.7×10^6	5.3×10^5	8.7×10^4	1.7×10^4
Re ($\eta = 10$ Pa s)	2.7×10^5	5.3×10^4	8.7×10^3	1.7×10^3

6.3. Hot fountains

If the input magma is hotter than the fractionated magma in the chamber, as would normally be the case during filling of a magma chamber, the hybrid layer becomes thermally stratified with cooler, compositionally lighter magma overlying hotter, denser magma, *i.e.* the hybrid layer becomes stratified with opposing gradients one stabilizing (composition), the other destabilizing (heat). Because the destabilizing property, heat, diffuses faster than the stabilizing property, composition, the hybrid layer breaks up into a number of well-mixed double-diffusive convecting layers.

Campbell and Turner (1989) carried out a series of experiments in which hot salty water was injected into cooler fresh water. The temperature and density difference between the input fluid and the host fluid and the amount of fluid added were varied systematically. Temperature profiles through the hybrid layer, for a typical experiment, are illustrated in Figure 13. The first of the temperature profiles was taken 15 min. after the completion of filling, before the onset of double-diffusive convection. With the exception of a narrow zone of cool fluid at the bottom, produced by cooling against the floor of the tank, the form of the profile is similar to that produced in the constant temperature experiments. The fluid in this narrow zone was cooled from below and could not convect until the temperature of the Perspex floor of the tank rose above the temperature of the lower fluid layer. Similarly, the lowermost layer in a magma chamber can be expected to remain stagnant, following the injection of a new pulse of magma, until the temperature of the lower layer falls below the temperature of the temporary floor of the chamber. The second profile in Figure 13 was taken 2 hrs. after filling, by which time the hybrid layer had broken up into four double-diffusive convecting layers. It shows the well-mixed layers separated by double-diffusive interfaces. Note that the temperature difference between the lower layers is small but that there is a pronounced temperature step at the top of the hybrid layer.

The hot fountain experiments allowed Campbell and Turner to draw two qualitative generalizations regarding the influence of the various fountain parameters on the number of layers that form in the hybrid zone. First, raising the density of the input fluid at constant temperature increases the density gradient relative to the thermal gradient through the hybrid layer. This has a stabilizing influence on convection. It restricts the size of the layers that form and leads to the formation of a thick upper interface. Second, increasing the temperature of the input fluid without changing the density increases the temperature gradient relative to the density gradient. This has a destabilizing influence and leads to the formation of fewer, thicker

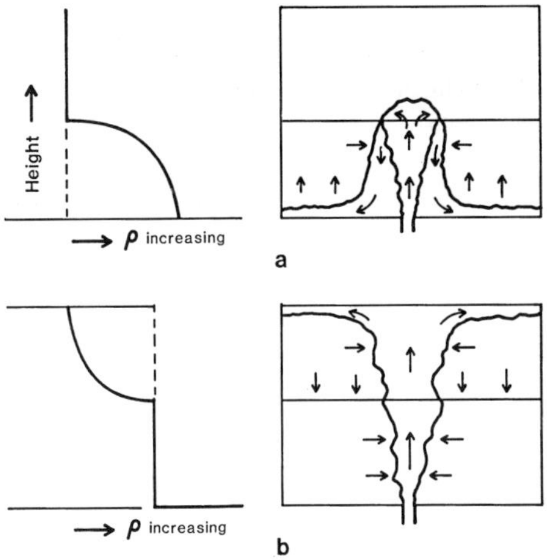

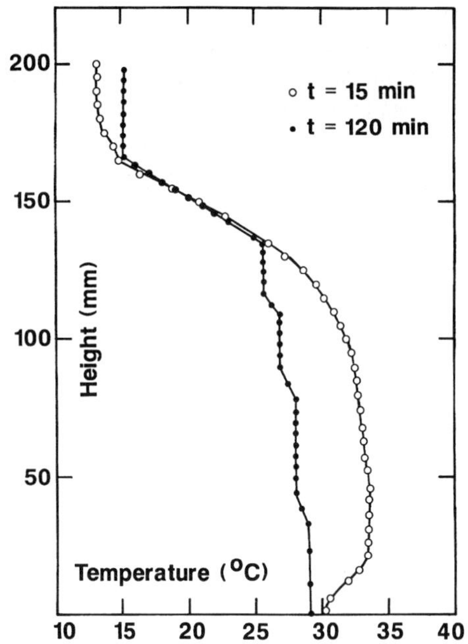

Figure 12. *(left) The flows and schematic density profiles produced in a rectangular box originally containing homogeneous fluid, when it is partly filled, (a) by a dense turbulent fountain and (b) by a light turbulent plume (after Campbell and Turner, 1986).*
Figure 13. *(right) Temperature profiles through the hybrid layer taken 15 min. and 120 min. after the start of an experiment in which hot salty water at 90°C was injected into a tank containing fresh water at 12°C. $\Delta\rho_0$ for the experiment was 41.5 kg m^{-3} (after Campbell and Turner, 1989).*

layers and to the formation of a thin upper interface. Although we are confident that these qualitative predictions also apply to magma chambers it is not possible, at present, to predict quantitatively the number or size of the layers even for simple laboratory experiments.

6.5. Finger fountains

The injection of U- and A-type magmas into large magma chambers, such as the Bushveld and Stillwater Complexes, gives rise to the possibility that a new pulse of magma that enters the chamber may be denser than the fractionated magma in the chamber because it is colder, although it is compositionally lighter. This situation is described by Campbell and Turner (1986b). The new pulse will form a fountain which will result in the development of a hybrid layer as described before but, in this case, hot, compositionally dense magma will overly cool compositionally light magma. The distribution of composition is unstable and it drives convection, leading to the formation of one or more finger interfaces within the hybrid layer, depending on the R_ρ of the system. If R_ρ is high, a single finger interface forms at the contact between the hybrid layer (which may extend to the floor of the chamber) and the fractionated magma above (the first front). However, if R_ρ is closer to 1.0 and if the compositional difference between the magmas is large, the hybrid zone may break up into a series of convective layers separated by finger interfaces. The important difference between finger and diffusive fountains is that the finger interface that forms at the first front in the former allows more rapid mixing between the hybrid layer and the fractionated magma above and this mixing can continue until the compositional density difference between the layers is negligible. Diffusive interfaces, on the other hand, inhibit mixing between adjacent layers, which will retain different compositions unless crystallization in the lower layer reduces its density so that the layers can overturn and mix as discussed above.

6.5. Fountains with fluid of different viscosity

Campbell and Turner (1985, 1986a) have reported a series of experiments designed to test the influence of viscosity on magma mixing when a new pulse of low viscosity magma is injected into a chamber containing a magma of higher viscosity. The reported experiments are for fountains but the results should be equally applicable to other kinds of turbulent flows such as plumes or jets. Campbell and Turner found that if the host fluid is much more viscous than the injected fluid, very little mixing occurs at moderate flow rates, even if flow within the fountain is fully turbulent. The amount of mixing is related to the ability of the incoming fluid to distort the outer surface of the fountain (the contact between the fluids). If the outer surface is smooth no mixing occurs, but if it is rough the fluids mix efficiently. Hence the input fluid must enter the chamber or tank with sufficient momentum to entrain the outer fluid. This happens when:

$$wd > k\nu_2 \qquad (18)$$

where ν_2 is the viscosity of the outer fluid and k is a constant. If $wd/\nu_2 > 70$, the inflowing fluid mixes with the host fluid as if there was no viscosity difference between them but, if $wd/\nu_2 < 7$, little or no mixing occurs even if motion within the fountain is fully turbulent. Alternatively, equation (18) can be expressed in terms of the Reynolds number of the inflow as:

$$Re_1 = \frac{wd}{\nu_1} > k\frac{\nu_2}{\nu_1} \qquad (19)$$

where ν_1 is the viscosity of the input fluid.

6.6. Stability of layering produced by fountains

In section 6.2 it was shown that, following the entry of a new pulse of hot dense magma into a chamber, a hybrid layer develops near the floor that breaks up into double-diffusive convecting layers. These layers, once formed, can persist for an extended period of time.

Consider the case of a magma chamber that is fed by a number of pulses of a picritic magma that crystallizes olivine followed by orthopyroxene. Crystallization of olivine and pyroxenes from a picritic magma lowers its density so that, in principle, the crystallization of these minerals from the lowermost layer could decrease its density until it becomes the same as the layer above, leading to overturning and mixing (Huppert and Sparks, 1980; Huppert and Turner, 1981; Huppert *et al.*, 1982). However, in a system of stacked double-diffusive convecting layers the temperature of the layers must decrease upwards and, if crystallization decreases the density of the fluid, the density of the cooler upper layers must always be less than that of the warmer, less fractionated, lower layers. Overturning therefore appears to be impossible in such a system.

The above discussion ignores two factors that assist overturning. First, if crystallization occurs *in situ* at the margins of the chamber, crystallization can occur at the floor and walls of the lower layer whereas it is confined to the walls of the upper layers. This can result in more extensive crystallization in the lower layer, causing it to become more evolved and thus more fractionated than the upper layers and this can lead to overturning (Campbell and Turner, 1989).

The second factor that affects overturning is the influence of pressure on the olivine and orthopyroxene liquidus which increases by about 1.2°C km^{-1} for olivine and 3.4°C km^{-1} for orthopyroxene. In a series of stacked convecting layers 2 km thick, the liquidus temperature for olivine at the bottom of the lowermost layer will be 2.4°C lower than it is at the top of the uppermost layer and 6.8°C lower for orthopyroxene. This can result in the lowermost layer being more fractionated and therefore lighter than the overlying layers. That is, olivine (or orthopyroxene) crystallization can produce overturning in series of stacked double-diffusive layers, produced from a single parent magma, but the time scale for overturning will probably be much greater than envisaged by Huppert and Sparks (1980) and Huppert and Turner (1981b).

6.7. Light inputs

When light fluid is injected at the bottom of a homogeneous layer of comparable viscosity, and the Reynolds number of the input is high, a turbulent plume will form (Sparks *et al.*, 1980). This fluid will vigorously entrain the host fluid, so that the mixture arriving at the top of the tank or chamber will contain a large proportion of the latter. In a deep, narrow tank where the plume becomes as wide as the tank, complete mixing results. A tank or chamber which is much wider than it is deep can, on the other hand, be treated by the "filling" box model of Baines and Turner (1969). They showed, theoretically and experimentally, how a continuing inflow of this kind will build up a stratified layer at the top, bounded below by a sharp front that moves downwards. The fluid that has already spread out along the boundary and become part of the environment will lower the density of the subsequent plume fluid mixing with it, so that the lightest fluid will always be deposited at the top, pushing the previously accumulated layers downwards. A typical density profile produced by a light turbulent input into a rectangular box is illustrated in Figure 12b.

If the input magma is also hotter than the host magma the chamber will become thermally stratified with hot magma overlying cooler magma. The system will become doubly stable, that is stably stratified with respect to both temperature and composition. The magma will remain stagnant until heat loss to the surroundings can overcome the stable density gradient and convection, driven by heat fluxes through the boundaries, can recommence. The chamber may then break up into a series of double-diffusive convecting layers. Note that there will be a hiatus in crystallization at the floor of the chamber while the magma is stagnant because, during this period, there is no heat transfer between the cooling roof and the zone of crystallization at the floor.

6.8. Plumes in a stratified environment

To this point it has been assumed that the new pulses enter a chamber which is homogeneous and well-mixed. There are however many processes that result in the development of a stable density stratification in the chamber. It has already been seen that the filling process itself can lead to stratification. Other processes which can stratify a chamber are the release of dense fluid by crystallization at the floor, or the release of light fluid at an inward sloping boundary. The latter effect can be due either to crystallization at the wall or roof of the chamber (Turner, 1980; McBirney, 1980) or to the melting of less dense wall rocks.

A turbulent plume entering a stratified chamber from below begins to entrain the host fluid as soon as it enters the chamber. Since the surrounding fluid is denser than the input fluid, entrainment increases the density of the fluid in the rising plume until at some height it becomes equal to that of the environment. At this level it spreads out laterally at a level that is dependent on the buoyancy flux at the source and the density gradient of the environment. If the input fluid is hotter than the host fluid, the intrusion will also be hotter than the fluid above or below the intrusion. A diffusive interface will form at the top of the intrusion and a finger interface at the bottom.

7. ZONED MAGMA CHAMBERS

Crystallization experiments have shown that the release of a flux of light fluid at a vertical boundary can lead to zoning of magma chambers (Turner, 1980; McBirney, 1980). Whether crystallization leads to zoning or homogenization of magma chambers depends on the shape of the chamber and on whether the fluid released is lighter or denser than the host magma (Sparks and Huppert, 1984). In practice the walls of magma chambers will rarely be vertical and it is more relevant to consider crystallization at a sloping boundary. Two cases need to be considered: crystallization leading to the release of a dense fluid, and crystallization leading to the release of a light fluid. Only convection at a sloping roof will be described because, from a fluid dynamic point of view, crystallization at the floor is the same problem inverted.

If crystallization at the roof releases a magma that is denser than the magma in the chamber, a boundary layer of dense magma will develop at the roof of the chamber. Magma within this boundary layer will immediately start to flow downwards but will cling to the roof, held in place by the viscous forces exerted by the underlying magma. Eventually, when the local Rayleigh number exceeds 10^3, it will acquire enough buoyancy to break away and form a plume that will sink into the magma below. The flux of dense melt sinking through the chamber will have an homogenizing influence on the underlying magma. Similarly, the release of light magma from the sloping floor of a chamber will have an homogenizing influence (Martin and Campbell, 1988).

If the magma released at the sloping roof is light it will continue to flow up the roof as a laminar boundary layer. This phenomena has been studied by Worster and Leitch (1985) and by Nilson *et al.* (1985). They have shown that there is a significant difference in the stratification produced by a laminar boundary layer, compared with that set up by a turbulent plume. In the turbulent filling box case already discussed, a sharp "first front" or density step is set up, moving in the opposite direction to the plume, and the largest density gradients are immediately behind this. With a laminar boundary layer, however, the magnitude of the density gradient produced in the interior fluid increases instead of decreasing in the direction of flow of the boundary layer, and it varies smoothly so that there is no density front. This is because the laminar boundary layer, which is lightest near the wall, can be carried around the corner at the top of the region it is stratifying without mixing with the adjacent fluid. As a consequence, the vertical density profile of the stratified region at the top of the chamber has the same qualitative features as a density profile through the boundary layer, *i.e.* a larger density gradient near the top boundary. As flow continues the inner, buoyant part of the boundary layer (which is much thinner than the whole, viscously-driven layer at high *Pr*) is carried up into the stratified region and is "detrained" there, each part at its own density level, while the outer viscous layer flows out into the environment below the stratified region. This leads eventually to stratification of the upper part of the chamber. Similarly, release of a dense fluid at the sloping floor of the chamber in the interior will lead to stratification of the lower part of the chamber.

7.1. Two chamber geometries compared

With these simple principles in mind two basic shapes need to be considered to understand the influence of chamber geometry on compositional convection; a funnel-shaped intrusion depicted in Figure 14a and an inverted funnel depicted in Figure 14b (Turner and Campbell, 1986). For simplicity it will be assumed that crystallization occurs simultaneously at all boundaries but that the crystallization rate increases with depth due to the pressure effect. The form of compositional convection for chambers with more complex geometries can be predicted from the principles that will now be illustrated using these two basic forms.

Consider first the case of a flux of light fluid generated by crystallization at the boundaries of a funnel-shaped intrusion (Figure 14a1). The light fluid released from the sloping floor will tend to move away from the boundary and mix convectively with the overlying magma. This will have an homogenizing influence on the magma in the chamber, which will be well mixed at all levels with the possible exception of a narrow zone at the roof (McBirney *et al.,* 1985). Here there will be two competing processes. First, local crystallization will tend to stratify the top of the chamber. Second, tending to destroy that stratification as it forms is a flux of buoyant fluid released by crystallization at the floor. Since most crystallization occurs at the bottom of basaltic chambers, it seems probable that the convection due to the light fluid released from below will dominate, and that stable stratification will not develop at the top of the chamber. However, during the final stages of crystallization, when the distance between the roof and floor is small and the pressure effect is less important, there may be sufficient crystallization at the roof to produce stable stratification at the top of the chamber.

With a flux of light fluid released in a reversed funnel (Figure 14b1), there are again two competing processes, but this time the stratifying fluid flowing up the sides of the intrusion is likely to dominate and produce stable stratification at the apex of the chamber. This is because the light fluid released by crystallization at the sloping roof will flow along the roof and collect at the apex of the chamber concentrating the light fluid into a small volume and helping to stabilize the developing stratification.

The principles for a dense flux are similar to those discussed in connection with a light flux. This time the dense fluid ponds at the bottom of the chamber (Figures 14a2 and 14b2). Stable stratification is likely to develop in both geometries because of the importance of bottom crystallization, but it is likely to be better developed in the case of the normal funnel (Figure 14a2) due to the channelling effect of the inward sloping walls. However, where heat loss through the floor is important, for example in thin sills and during the early stages of the crystallization of large magma chambers, the magma at the bottom of the chamber may be stably stratified with respect to both composition and temperature. Under these conditions it will remain stagnant and cool by conduction until heat loss to the surroundings can overcome the stable density gradient and convection can recommence.

It should be apparent from the above discussion that crystallization will often occur at more than one surface of a magma chamber simultaneously, producing fluxes of buoyant fluid which have opposing effects. Whether convection stratifies or homogenizes the chamber depends on which flux dominates. It has been argued, for example, that a light flux in a normal funnel will

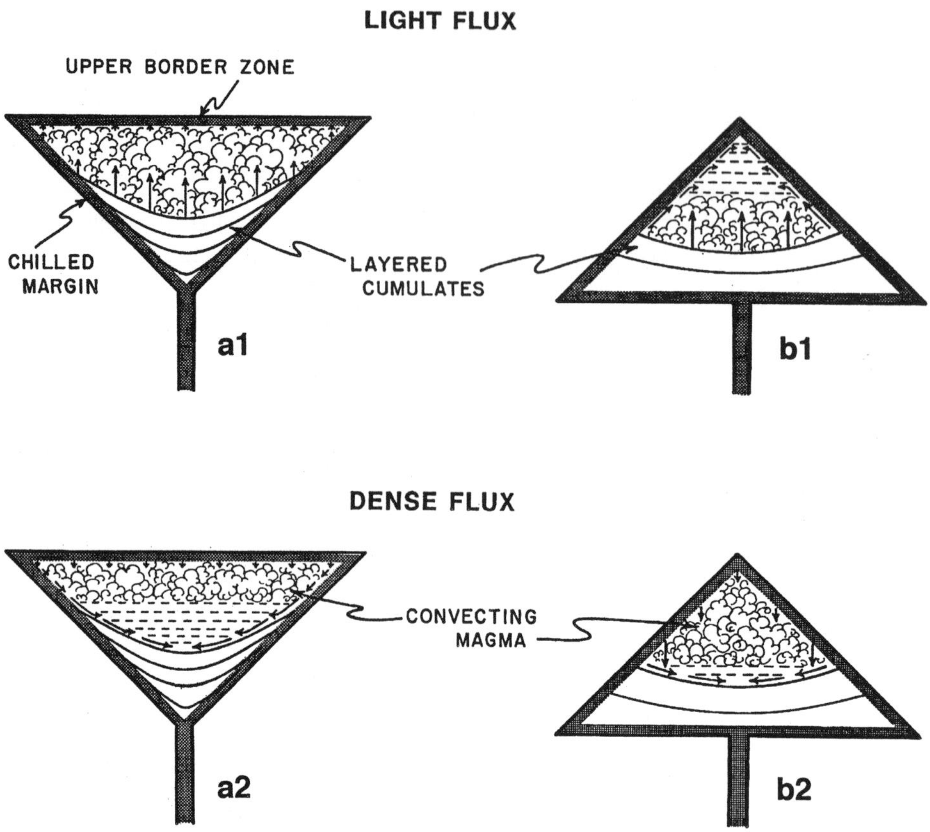

Figure 14. Diagrammatic representation of convection in a funnel-shaped (a1 and a2) and inverted funnel-shaped intrusion (b1 and b2). In a1 and b1 a light flux of magma is released by crystallization, in a2 and b2 a dense flux is released. Dashed lines represent zoned magma, and swirls convection. Vertical scale exaggerated. See text for further explanation (after Turner and Campbell, 1986).

generally not produce stratification at the top of the chamber. In a small chamber the pressure effect will be less important than it is in large chambers and, as a consequence, top crystallization will be more important. Stable stratification may therefore develop at the top of a small chamber whereas it may not in a larger chamber.

8. ASSIMILATION IN MAGMA CHAMBERS

At any contact where the melting point of the country rock is less than the temperature of the magma or if the country rocks can dissolve in the magma, the magma will begin to assimilate the walls of the chamber and, if the melt convects away from the boundary, this will continue until the onset of crystallization at that contact. Once crystallization begins the contact becomes protected by a layer of crystalline rock which must melt (or dissolve) before further assimilation can occur. This is only possible if the chamber receives a fresh input of magma.

The fate of magma generated by melting of the chamber walls depends on its density relative to the magma in the chamber. The melts produced may, of course, be lighter or denser than the magma in the chamber, but light melts will normally predominate because the average composition of the crust lies between andesite and granodiorite.

Light magmas, produced by melting of the floor of the chamber, will rise away from the contact and be assimilated into the overlying melt. Melting consumes latent heat which will lower the temperature of the basaltic melt and eventually lead to crystallization. Once crystallization commences at a contact, assimilation will normally stop although melting of low melting points rocks in the footwall may lead to some disruption of the contact. Melting at the floor is therefore not considered to be an important factor during the crystallization of most magma chambers (Campbell and Turner, 1987; Kerr, 1994).

Light melts generated at the roof will rise and collect in cupolas, high points in the roof structure (Figure 15). The upper sections of the roof will therefore be in contact with a low melting-point felsic magma and no crystallization will occur at this contact during the early stages of the evolution of the chamber. The chamber will stratify, with the bulk of the chamber being filled with basaltic magma, but the upper part will contain felsic magma. This upper layer may become compositionally zoned especially if the roof is heterogeneous and, if this is the case, it will remain stably stratified as long as a substantial flux of light magma is being released at the top of the chamber by melting of the roof. Later, when the rate of melting slows and the cooling through the roof becomes more important than the compositional flux, the upper layer may break up into double-diffusive convecting layers. A double-diffusive interface will form at the base of the upper zone, across which heat, but little mass, will be transferred. The heat acquired by the upper layer will maintain it in a superheated state and, in so doing, prevent crystallization occurring in that layer. Thus no chilled margin will form at the upper contact of the cupola, and as long as the melt remains superheated, assimilation will continue. At the same time crystallization will continue in the lower layer, with most of the latent heat released being transmitted to the upper

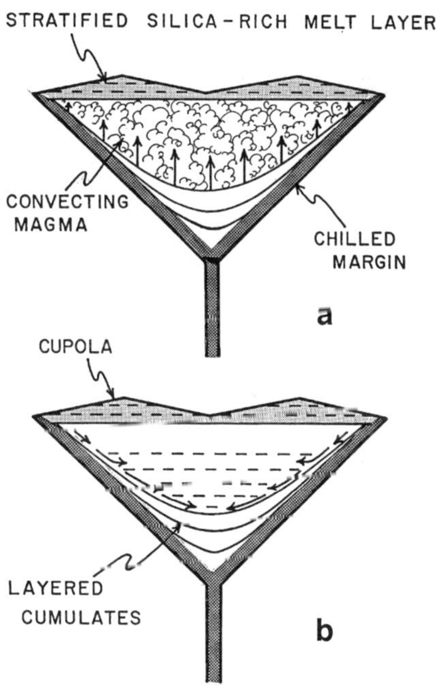

Figure 15. *Diagrammatic representation of convection due to bottom crystallization in a funnel-shaped intrusion which melts its roof. (a) light magma released; (b) dense magma released. Dashed lines represent zoned magma, swirls convection. Vertical scale exaggerated (after Turner and Campbell, 1986).*

layer through the double-diffusive interface. In this way the heat required for assimilation of the roof is acquired from crystallization in the lower layer and in this respect it is similar to previous assimilation models. The important difference between this model and previous suggestions is that assimilation and crystallization are required to occur simultaneously at different levels in the magma chamber (Campbell and Turner, 1987; Huppert and Sparks, 1988). Melting rates have been calculated by Huppert and Sparks (1988) and by Kerr (1994) who have obtained values of a few metres per year.

An important prediction of the roof melting hypothesis, that little mass is transferred across the interface between the felsic magma at the roof and the basaltic magma below (Campbell and Turner, 1987), has recently been confirmed by detailed isotopic studies of cumulates from the Muskox and Skaergaard intrusions by Stewart and DePaolo (1990, 1992, 1996).

If the roof of the chamber slopes, as will normally be the case, the light magma released by assimilation will flow along the roof and pond at the top of the chamber. This flux of light magma along the boundary is directly analogous to the release of light fluid by side wall crystallization and can lead to zoning of the chamber for exactly the same reasons.

8.1. Potholes

A special case of assimilation in magma chambers occurs following the injection of a new pulse of magma into a chamber which can result in erosion of the cumulate pile. If the products of erosion increase the density of the melt, the contaminated magma ponds at the base of the chamber, and erosion is rapidly arrested. If, however, the products are light, they are swept away by compositional convection and replaced by uncontaminated magma allowing erosion to continue.

An example of this type of assimilation in layered intrusions occurs at the levels of the Merensky Reef and UG-2 chromitite layer of the Bushveld Complex. At each of these stages in the evolution of the Bushveld new pulses of olivine or bronzite-saturated magma have entered the chamber and flowed out across the floor. This magma dissolved the underlying plagioclase cumulates. The principle is illustrated using the system diopside-anorthite (Figure 16). If a

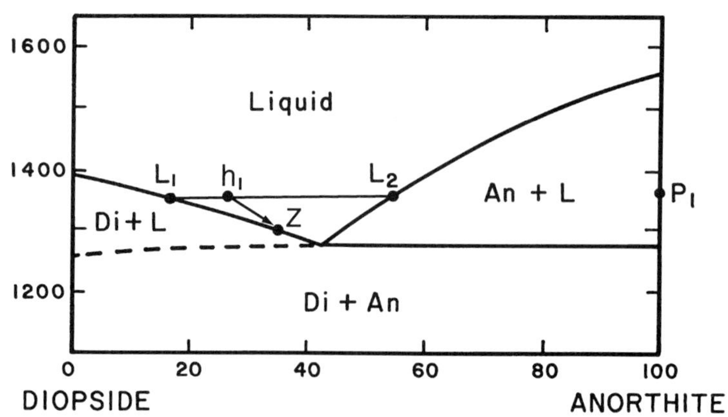

Figure 16. *The system diopside-anorthite. See text for further explanation (Campbell, 1986).*

pyroxene-saturated magma L_1 enters the chamber and mixes with a plagioclase-saturated magma L_2 to form a mixed magma h_1, the hybrid magma will be undersaturated with respect to both plagioclase and pyroxene. It will dissolve plagioclase-rich cumulates at the floor of the chamber to produce a contaminated magma that is light and is swept away by compositional convection, so preventing the build-up of plagioclase-saturated liquid (L_2) at the floor of the chamber. Plagioclase assimilation will drive the liquid in the direction of the arrow (taking both the specific and latent heats involved into account) and will continue until diopside starts to crystallize at Z (Campbell, 1986).

The dissolution of a plagioclase cumulate by an olivine or bronzite-saturated magma has been modelled by Campbell (1986) in a series of experiments in which ice, held at the bottom of a tank, was dissolved by an overlying salt solution. Square holes placed in the ice before the start of an experiment rapidly become rounded in both plan and section and assumed a shape similar to that of the smaller potholes of the Merensky Reef and UG-2. Furthermore, the surface of the ice developed a pitted texture, similar to the dimpled surface at the base of the Merensky Reef. The problem has been quantified by Kerr (1994), whose analysis predicts a dissolution rate of 25 cm y^{-1} for $\kappa_S = 10^{-11}$ m^2 s^{-1}.

9. CRYSTAL SETTLING

It has been assumed, in this review, that crystallization in magma chambers occurs *in situ* at the floor, walls and roof of the chamber. The field evidence, summarized by Campbell (1978) and McBirney and Noyes (1979), suggests that *in situ* crystallization is the dominant mechanism of crystallization in layered intrusions. However, crystal settling may be important under some circumstances.

Whether crystals form *in situ* at the floor of the chamber or settle through the magma depends on the mechanism of nucleation. Crystal settling requires the crystals to nucleate homogeneously within the chamber, whereas *in situ* crystallization implies heterogenous nucleation at the floor, walls and roof of the intrusion. Because the activation energy for homogeneous nucleation is much higher than for heterogenous nucleation, the amount of supercooling required for homogeneous nucleation is appreciably greater than for heterogeneous nucleation (Campbell, 1978). This principle has been illustrated experimentally by Martin (1990) who showed that an aqueous solution of potassium nitrate, cooled from above, crystallizes heterogeneously at the floor of the tank if the cooling rate is low (low supercooling) but by a mixture of heterogeneous and homogeneous nucleation at high cooling rates (higher supercooling). He also found that the likelihood of homogeneous nucleation is increased by raising the viscosity of the fluid.

The amount of supercooling, in a crystallizing magma chamber that loses heat by conduction through its wall rocks, will normally be between 1° and 20°C (Martin *et al.*, 1987; Martin, 1990). The supercooling required to produce homogeneous nucleation in a basaltic magma is not known but is unlikely to be less than for pure metals which vary between 77°C for mercury and 319°C for nickel (Campbell, 1978). Crystal settling is therefore unlikely to be an important factor in large magma chambers. It may, however, be important in thin sills and lava lakes where high rates of cooling may produce the level of supercooling required for homogeneous nucleation. Crystal settling may also occur if steeply dipping cumulates at the margin of a large intrusion become unstable and slump into the centre of the intrusion, forming a density current.

Crystal settling in the convective regime relevant to large magma chambers, that is when the Stoke's Law settling velocity (v_S) is less than the root mean square vertical component of the convective velocity at mid-depth (w), has been considered by Martin and Nokes (1989). If $w > v_S$ settling is still possible because convective velocities are height-dependent and must decrease to zero at the boundaries of the magma chamber. Vigorous convection within the main body of magma ensures that suspended crystals are evenly distributed within the chamber so that the convective process brings a continuous supply of crystals into the zone of reduced convective velocities at the bottom of the chamber. Here the crystals can settle out at a velocity that reaches the full Stokes velocity at the chamber floor. Martin and Nokes (1989) showed that the number of suspended crystals decay exponentially with time and that the decay constant is equal to v_s/h, where h is the depth of the fluid.

10. ACKNOWLEDGEMENTS

I wish to thank Stewart Turner and Ross Kerr for reviewing the manuscript and Jan Bitmead and Ross Wylde-Browne for helping with the diagrams.

11. REFERENCES

Baines, W.D., & Turner, J.S., 1969. Turbulent buoyant convection from a source in a confined region. *J. Fluid Mech.* **37,** 51-80.

Brandeis, G., & Jaupart, C., 1986. On the interaction between convection and crystallization in cooling magma chambers. *Earth Planet. Sci. Lett.* **77**, 345-61.

Bruce, P.M., & Huppert, H.E., 1990a. Thermal controls of basaltic fissure eruptions. *Nature* **342,** 665-7.

Bruce, P.M., & Huppert, H.E., 1990b. Solidification and melting along dykes by the laminar flow of basaltic magma. *Magma Transport and Storage.* New York: Wiley, 87-101.

Campbell, I.H., 1973. Aspects of the petrology of the Jimberlana Layered Intrusion of Western Australia. *PhD Thesis,* London University.

Campbell, I.H., 1978. Some problems with the cumulus theory. *Lithos* **11,** 311-23.

Campbell, I.H., 1986. A fluid dynamic model for the potholes of the Merensky Reef. *Econ. Geol.* **81,** 1118-25.

Campbell, I.H., & Turner, J.S., 1985. Turbulent mixing between fluids with different viscosities. *Nature* **313,** 39-42.

Campbell, I.H., & Turner, J.S., 1986a. The influence of viscosity on fountains in magma chambers. *J. Petrology* **27,** 1-30.

Campbell, I.H., & Turner, J.S., 1986b. The role of convection in the formation of platinum and chromitite deposits in layered intrusions. *Miner. Assoc. Can. Short Course in Silicate Melts*, 236-78.

Campbell, I.H., & Turner, J.S., 1987. A laboratory investigation of assimilation at the top of a basaltic magma chamber. *J. Geology* **95,** 155-72.

Campbell, I.H., & Turner, J.S., 1989. Fountains in magma chambers. *J. Petrology* **30,** 885-923.

Carslaw, H.S., & Jaeger, J.C., 1959. *Conduction of heat in solids.* Oxford University Press.

Huppert, H.E., & Sparks, R.S.J., 1980. The fluid dynamics of a basaltic magma chamber replenished by influx of hot, dense ultramafic magma. *Contr. Miner. Petrol.* **75,** 279-89.

Huppert, H.E., & Sparks, R.S.J., 1988. The generation of granitic magmas by intrusion of basalt into continental crust. *J. Petrology* **29,** 588-624.

Huppert, H.E., & Turner, J.S., 1981a. Double-diffusive convection. *J. Fluid Mech.* **106,** 299-329.

Huppert, H.E., & Turner, J.S., 1981b. A laboratory model of a replenished magma chamber. *Earth Planet. Sci. Lett.* **54,** 144-52.

Huppert, H.E., Turner, J.S., & Sparks, R.S.J., 1982. Replenished magma chambers: effects of compositional zonation and input rates. *Earth Planet. Sci. Lett.* **57,** 345-57.

Huppert, H.E., Sparks, R.S.J., Wilson, J.R., & Hallworth, M.A., 1986. Cooling and crystallization at an inclined plane. *Earth Planet. Sci. Lett.* **79,** 319-28.

Irvine, T.N., 1970. Heat transfer during solidification of layered intrusions. I. Sheets and sills. *Can. J. Earth Sci.* **7,** 1031-61.

Kerr, R.C., 1994. Melting driven by vigorous compositional convection. *J. Fluid Mech.* **280**, 255-85.

Kerr, R.C., Woods, A.W., Worster, M.G., & Huppert, H.E., 1989. Disequilibrium and macrosegregation during solidification of a binary melt. *Nature* **340,** 357-62.

Kerr, R.C., Woods, A.W., Worster, M.G., & Huppert, H.E., 1990a. Solidification of an alloy cooled from above. Part 1. Equilibrium growth. *J. Fluid Mech.* **216,** 323-42.

Kerr, R.C., Woods, A.W., Worster, M.G., & Huppert, H.E., 1990b. Solidification of an alloy cooled from above. Part 2. Non-equilibrium interfacial kinetics. *J. Fluid Mech.* **217,** 331-48.

Kerr, R.C., Woods, A.W., Worster, M.G., & Huppert, H.E., 1990c. Solidification of an alloy cooled from above. Part 3. Compositional stratification within the solid. *J. Fluid Mech.* **218,** 337-54.

Kress, V.C., & Ghiorso, M.S., 1993. Multicomponent diffusion in basaltic melts. *Geochim. Cosmochim. Acta* **57,** 4453-66.

Martin, D., 1990. Crystal settling and in situ crystallization in aqueous solutions and magma chambers. *Earth Planet. Sci. Lett.* **96,** 336-48.

Martin, D. & Campbell, I.H., 1988. Laboratory modelling of convection in magma chambers: crystallization against sloping floors. *J. Geophys. Res.* **93** (B7), 7974-88.

Martin, D. & Nokes, R., 1989. A fluid-dynamical study of crystal settling in convecting magmas. *J. Petrology* **30,** 1471-500.

Martin, D., Griffiths, R.W., & Campbell, I.H., 1987. Compositional and thermal convection in magma chambers. *Contr. Miner. Petrol.* **96,** 465-75.

McBirney, A.R., 1980. Mixing and unmixing of magmas. *J. Volcanol. Geotherm. Res.* **7,** 357-71.

McBirney, A.R., Baker, B.N., & Nilson, R.H., 1985. Liquid fractionation. Part 1: basic principles and experimental simulations. *J. Volcanol. Geotherm. Res.* **24,** 1-24.

Morse, S.A., 1986. Thermal structure of crystallizing magma with two-phase convection. *Geol. Mag.* **123,** 205-14.

Nilson, R.H., McBirney, A.R., & Baker, B.H., 1985. Liquid fractionation, Part II. Fluid dynamics and quantitative implications for magmatic systems. *J. Volcanol. Geotherm. Res.* **24,** 25-54.

Sparks, R.S.J., & Huppert, H.E., 1984. Density changes during fractional crystallization of basaltic magmas: fluid dynamic implications. *Contr. Miner. Petrol.* **85,** 300-9.

Sparks, R.S.J., Meyer, P., & Sigurdsson, H., 1980. Density variation amongst mid-ocean ridge basalts: implications for magma mixing and the scarcity of primitive lavas. *Earth Planet. Sci. Lett.* **46,** 419-30.

Sparks, R.S.J., Huppert, H.E., & Turner, J.S., 1984. The fluid dynamics of evolving magma chambers. *Phil. Trans. Roy. Soc. Lond..* **A310,** 511-34.

Stewart, B.M., & DePaolo, D.J., 1990. Isotopic studies of processes in mafic magma chambers: II. The Skaergaard Intrusion, East Greenland. *Contr. Miner. Petrol.* **104,** 125-41.

Stewart, B.M., & DePaolo, D.J., 1992. Diffusive isotopic contamination of mafic magma by coexisting silicic liquid in the Muskox Intrusion, Northwest Territories, Canada. *Science,* **255,** 708-11.

Stewart, B.M., & DePaolo, D.J., 1996. Isotopic studies of processes in mafic magma chambers: III. The Muskox Intrusion, Northwest Territories, Canada. *J. Geophys. Res.* (in press).

Taylor, H.P. Jr., & Forester, R.W., 1979. An oxygen isotope study of the Skaergaard Intrusion and its country rocks: a description of a 55 My old fossil hydrothermal system. *J. Petrology* **20,** 355-419.

Turner, J.S., 1973. *Buoyancy effects in fluids.* London: Cambridge University Press, 367 pp.

Turner, J.S., 1980. A fluid-dynamic model of differntiation and layering in magma chambers. *Nature* **285,** 213-5.

Turner, J.S., 1985. Multicomponent convection. *Ann. Rev. Fluid Mech.* **17,** 11-44.

Turner, J.S., 1986. Turbulent entrainment: the development of the entrainment assumption, and its application to geophysical flows. *J. Fluid Mech.* **173,** 431-71.

Turner, J.S., & Campbell, I.H., 1986. Convection and mixing in magma chambers. *Earth Sci. Rev.* **23,** 255-352.

Turner, J. S., & Gustafson, L.B., 1978. The flow of hot saline solutions from vents in the sea floor – some implications for exhalative massive sulfides and other ore deposits. *Econ. Geol.* **73,** 1081-100.

Wager, L.R., & Brown, G.M., 1968. *Layered Igneous Rocks.* Edinburgh and London: Oliver and Boyd, 588 pp.

Wager, L.R., & Deer, W.A., 1939. Geological investigations in East Greenland, Pt. III. The petrology of the Skaergaard Intrusion, Kangerdlugssuaq, East Greenland. *Medd. Grœnl.* **105,** 1-352.

Wilson, A.H., 1982. The geology of the Great "Dyke", Zimbabwe: the ultramafic rocks. *J. Petrology* **23**, 240-92.

Worster, M.G., & Leitch, A.M., 1985. Laminar free convection in confined regions. *J. Fluid Mech.* **156,** 301-19.

Worster, M.G., Huppert, H.E., & Sparks, R.S.J., 1990. Convection and crystallization in magma cooled from above. *Earth Planet. Sci. Lett.* **101**, 78-89.

Zhang, Y., Walker, D., & Lesher, C.E., 1989. Diffusive crystal dissolution. *Contr. Miner. Petrol.* **102,** 492–513.

LAYERED INTRUSIONS
R.G. Cawthorn (editor)

Texture Development in Cumulate Rocks

R.H. Hunter

Department of Earth Sciences, University of Liverpool, Brownlow Street, Liverpool, L69 3BX, U.K.

Abstract

In the past three decades, the cumulus terminology developed by Wager and co-workers has provided the framework for understanding texture development in crystal mushes. Much of the debate has concerned the conditions necessary for development of adcumulate rocks and has involved discussion of mechanisms of heat and mass transfer within mushes. In this article the historical development of ideas is reviewed and aspects of the nomenclature are discussed. The development of primary and secondary textures in mushes are then discussed, principally with respect to the relative roles of crystal overgrowth, compaction, and cementation.

Most crystal accumulation in moderate- to large-sized layered intrusions occurs on the floor, where crystal mushes develop by either *in situ* crystallization or crystal sedimentation. Except where a preferred crystal shape orientation occurs as a result of directional growth from a substrate, there are no definitive textural criteria for distinguishing *in situ* crystallization from crystal sedimentation in the accumulation of mushes. Mushes inherit primary textural characteristics that influence the subsequent texture development within the crystal pile. Primary porosity and permeability are influenced by initial packing and clustering characteristics of crystals which are a function of the way in which crystals accumulate and any subsequent mechanical reorganization. Crystal growth, solution/replacement, cementation, compaction, and recrystallization are competing processes involved in the secondary texture development of the crystal pile. The densification of a crystal mush involves the reduction of primary porosity of the cumulus grains. This may be by overgrowth on the grains or compaction. Either process will be restricted by the nucleation and growth of poikilitic grains which cement the granular crystal framework. These processes are analogous to syntaxial overgrowth, compaction and cementation involved in sediment diagenesis.

Whether crystals grow under near-isothermal conditions or during cooling depends upon whether the mush is open or closed to melt percolation but is independent of the mechanism of heat and mass transfer within the mush. Compaction, necessarily an open-system process, involves deformation (dislocation creep) or solution/reprecipitation of grains (diffusion creep) and usually results in an increase in the degree of local textural equilibration. However, recrystallization (*e.g.* by thermal annealing) also results in textural equilibration. Growth, compaction, and recrystallization are all competing processes and it is commonly not possible to isolate their contribution to any given texture; all produce rocks with the textural characteristics of adcumulates.

The extent of densification of a mush of cumulus grains depends critically on the timing of nucleation and growth of poikilitic cementing phases. In any given magma composition this is a function of the local phase relationships. A cyclicity will develop in the texture in a crystal mush that is a function of the balance of densification and poikilitic cementation. Repeated replenishment of a magma chamber may result in suppression of the cementation cycle and allow mushes to become highly densified. On the scale of an intrusion the texture which develops depends upon the interaction of fronts of densification and cementation and hence is dependent on intrusion geometry.

1. INTRODUCTION

Fractional crystallization remains central to ideas of magmatic evolution and to the understanding of magma chamber processes. Crystal settling was thought, for many years, to be the principal process involved. As the importance of boundary layer processes at the floor, roof, and walls of magma chambers became recognized, the idea that much primary crystallization occurs *in situ* within these boundary layers has become increasingly popular. The problem then is to understand how magma chambers evolve chemically and the interaction of boundary layer and magma reservoir processes has been a focus of attention. There is still considerable debate regarding whether crystals accumulate by sedimentation or grow *in situ* on the floor, walls, or roof of magma chambers. Whatever the mechanism however, there is little doubt that in disected basic and ultrabasic intrusions, most of the crystals appear to have accumulated on the floor. Roof and marginal border series are developed in sills and in some, generally small, intrusions but volumetrically these are minor in comparison to floor accumulations. This is the realm of the crystal mush. Although bulk-rock and mineral chemistry, experimentation, and theory have been applied to the understanding of the physiochemical evolution of crystal mushes, it is the interpretation of textures in slowly cooled layered intrusions that has remained central to the development of ideas.

In this contribution, I will first review the history of development of ideas, starting with the scheme of cumulus nomenclature devised by Wager *et al.* (1960), which has been a major contribution to the understanding of igneous textures and that, in various forms, has remained in general use. The review will also discuss the various additions and modifications to the scheme and highlight alternative approaches, in particular, the recognition that textural equilibration has played an important role in texture development. Attention has focussed on various physiochemical processes involved in movement of magma and fluid through porous crystal piles and the formation of adcumuluate texture, principally convection and compaction, and the development of ideas in these areas will be reviewed. The next sections will summarize the processes that influence the primary texture of accumulations of crystals and the factors which will be important in their subsequent development; principally, growth, compaction, reaction/replacement, cementation, and recrystallization. Although of some importance, I do not discuss exsolution, inversion or hydrothermal modifications. The emphasis in this section is on processes of densification, the rheology of crystal mushes, and the various creep mechanisms involved in compaction. Then, I will comment on the systematic development of textures in relation to typical phase relationships and discuss the timing of their development, in particular, the balance between various densification processes and cementation of the cumulus framework. These various aspects will then provide a framework for a critique of the cumulate model which will highlight some aspects of the textural interpretation of cumulus rocks that remain problematical.

2. HISTORICAL PERSPECTIVE AND DEVELOPMENT OF IDEAS

2.1. Cumulus nomenclature

The *cumulate* terminology developed by Wager *et al.* (1960) and subsequently amplified by Wager (1963) and Wager and Brown (1968), to describe the textures of igneous rocks in slowly cooled layered intrusions has had a profound influence on subsequent thinking about the way in which crystal mushes and magma chambers evolve. In the scheme, primary precipitate crystals that accumulated on the floor of a magma chamber, before any modification of the liquid in the pore spaces, were termed cumulus crystals and the interstitial liquid was called the intercumulus liquid. Rocks that formed from accumulation of one or more cumulus minerals, in which the unmodified intercumulus liquid crystallized to intercumulus material, were called orthocumulates. It was recognized that true orthocmulates were likely to be rare because a number of processes operated to modify the composition of the intercumulus liquid. Conditions for the formation of orthocumulates were promoted by fast bottom accumulation of crystals.

Hess (1939, 1960) postulated that during slow accumulation of crystals, material could diffuse from the magma above the crystal pile into and out of the interstitial liquid in the mush on the floor, to promote crystallization of minerals of constant composition. Wager *et al.* (1960) recognized that such a mechanism of enlargement of cumulus crystals at constant temperature could only take place at or near the top of any pile of crystals. They called this style of growth adcumulus growth. The adcumulus process gradually reduced the volume of intercumulus liquid by mechanically pushing it out of the pile and could result in vanishingly small quantities being preserved. Any intercumulus liquid remaining as a result of continued accumulation of crystals was termed trapped liquid which crystallized to the pore material. This liquid, and the subsequently crystallized pore material, necessarily had the composition of the contemporary magma. Rocks with small amounts of pore material were termed adcumulates.

Adcumulates and orthocumulates represented end-members of a continuum of rock types with increasing amounts of trapped liquid preserved as pore material. It was suggested however that the terms be used for rocks in which pore material is inconspicuous or absent (adcumulates) or those in which adcumulus growth was inconspicuous (orthocumulates). Rocks of intermediate character, showing moderate amounts of pore material, were called mesocumulates. Orthocumulates were characterized by zoned cumulus minerals and a variety of postcumulus material, representing crystallization during cooling from original intercumulus liquid trapped in interstitial pore spaces. Adcumulates had unzoned cumulus crystals with little or no interstitial pore material.

In orthocumulates, new minerals commonly grow as poikilitic or subpoikilitic crystals surrounding the cumulus crystals. These would show compositional zoning, reflecting cooling of the trapped intercumulus liquid. However, a class of rocks was recognized as having unzoned cumulus crystals surrounded by similarly unzoned poikilitic crystals. These oikocrysts must have nucleated within the pore liquid but have grown by enlargement from material in the main body of magma by the adcumulus process. They were thus recognized as a subclass of adcumulates and termed heteradcumulates.

The scheme of cumulus nomenclature focussed on the proportion of the rock representing crystallization from trapped interstitial liquid, the pore material. However, a rock composed of several cumulus minerals might not show much interstitial pore material, since most of it would overgrow the cumulus minerals. It is thus the presence or absence of zoning of the cumulus minerals that is the most important manifestation of trapped liquid. Wager and co-workers explored and developed further these ideas in relation to rocks of the Skaergaard and Rum layered intrusions (Wager, 1963; Wadsworth, 1961; Wager and Brown, 1968).

2.2. Jackson's (1961) contribution

At about the same time as the cumulus theory was being developed, Jackson (1961) was investigating similar problems in the Ultramafic Zone of the Stillwater layered intrusion. An important aspect of this work, which distinguishes it from that of Wager *et al.*, is the emphasis on shapes of crystals and mutual grain relationships. Like Wager and co-workers, Jackson regarded sedimentation of crystals as the principal mechanism involved in accumulation of the crystal pile. Jackson made the distinction between the primary precipitate minerals and those which crystallized from the pore space surrounding the crystal accumulate, drawing an analogy with the similar distinction made in clastic sedimentary rocks between detrital grains and cement. He recognized that two processes operated to obscure the relationships between euhedral settled crystals and space-filling interprecipitate material; secondary enlargement and reaction replacement. Secondary enlargement was the equivalent of adcumulus growth. Jackson noted that euhedral grains were associated with rocks with relatively large amounts of interstitial material. With increased secondary enlargement, the settled crystals developed polygonal mutual interference boundaries (mosaic texture). Grains developed mutual interference boundaries against adjacent grains but retained crystal faces when growing into pore spaces; also, in rocks with moderate amounts of overgrowth, the interstices retained their shape with decreasing volume. These observations are consistent with the operation of textural equilibration during crystallization (see section 5.3) although Jackson did not recognize this, *per se*. Jackson (1961) was also the first to undertake any systematic grain size and shape fabric analysis of crystals in cumulate rocks, an important area of study which has received remarkbly little attention until more recently (*e.g.* Benn and Allard, 1989; Conrad and Naslund 1989; Higgins, 1991; Wilson, 1992).

2.3. Crystal settling and other modifications

As noted earlier, Wager and co-workers and Jackson regarded the settling of crystals from the magma reservoir as the dominant mechanism of crystal accumulation; indeed, this was implicit to the cumulus theory of Wager *et al.* (1960). Wager and Brown (1968) also recognized that crystals could accrete against the walls and roof of an intrusion (congelation cumulates) or grow *in situ* inwards from the walls or upwards from the floor (crescumulates). Later workers have challenged the concept of crystal settling and favoured *in situ* growth as the principal mechanism of crystal accumulation, especially for feldspar-rich rocks (Campbell, 1978; Morse, 1979a; McBirney and Noyes, 1979). In the light of these ideas, Irvine (1982) suggested modifications to the nomenclature aimed at removing any genetic connotation as to the mechanism of crystal accumulation and formalizing the ranges of intercumulus modes appropriate for ortho-, meso-, and adcumulate, respectively. Wadsworth (1985) challenged the validity of these ranges but endorsed the use of poikilitic adcumulate as an alternative to heteradcumulate, as proposed by Irvine (1982). Morse (1979b) introduced the term residual porosity as a means of quantifying the amount of trapped liquid (pore material) based on the bulk chemistry of the rock.

At the present time, most workers use some form of the cumulus nomenclature modified to suit the specific problems of the intrusions being investigated (*e.g.* Wilson, 1992). However, there is still active debate about the relative roles of crystal settling versus *in situ* growth in the formation of cumulate rocks. In the past fifteen years, attention has centred on mechanisms and processes of heat and compositional transfer associated with adcumulus growth. However, for the most part, these studies have not addressed the development of textures themselves. In particular, the role of convection has been highlighted as a means of removing latent heat and excluded solute from the growing crystals and the crystal pile itself.

2.4. Convection

The original mechanism of adcumulus growth involved diffusive transfer of components between the intercumulus liquid and the magma overlying the crystal mush. Effectively, this restricted formation of adcumulates to a zone close to the magma-mush interface and implied slow crystal accumulation. This is because the effective diffusion length scale is only of the order of cm-dm on the time scale over which solidification would occur by conductive cooling. Morse (1986) proposed that convection within the magma reservoir was an efficient way of removing latent heat and solute from the magma in contact with the top of the crystal pile; this would enhance the rate of adcumulus growth promoting more-or-less complete solidification close to the top of the pile.

A significant development was the realization that convection of magma *within* the crystal pile could be important in promoting adcumulus growth at deeper levels in the mush (*e.g.* Tait *et al.*, 1984). This followed from laboratory tank experiments using aqueous solutions as analogues to model crystallization in mushes (see Sparks and Huppert, 1987). This convection (called compositional convection) is driven by the release of bouyant solute during crystallization, which rises through the pile and is replaced by undepleted melt from the reservoir above the mush. The process can work in reverse if released solute is more dense than the ambient intercumulus liquid and indeed would inhibit adcumulus growth if such a situation arose in mushes on the floor of the chamber. The efficacy and scale of this process in promoting adcumulus growth in crystal mushes in magma chambers is difficult to evaluate. It could enable adcumulates to grow at deeper levels within the mush than possible for diffusive exchange. On the other hand, it should promote rapid adcumulus growth near the top of the pile, since crystals there are the first to come into contact with undepleted magma. Thus, conversely, it may be effective in trapping melt at deeper levels and promoting orthocumulus growth (*e.g.* Campbell, 1987). It may also result in concentration of flow into vertical channels of high permeability (Tait and Jaupart, 1992). Calculations by Kerr and Tait (1986) suggest that porosity could be reduced to 10-20% by coupled compositional convection and isothermal crystal growth. However, their ability to produce such residual porosities in reality is not clear; it may be important only during the early stages of crystallization within a mush (Campbell, 1987).

2.5. Textural equilibration and compaction

An important aspect of the textural evolution of crystal mushes that was not recognized by Wager *et al.* (1960) or Jackson (1961) was the role of textural equilibration involving solution and reprecipitation during crystallization, leading to lower-energy grain-boundary geometries. This aspect was explored in detail by Hunter (1987) who recognized that many adcumulate rocks, including heteradcumulates, showed an approach to local textural equilibrium, a feature also identified by Campbell (1987). Hunter (1987) highlighted the fact that 'adcumulate' textures can be the result of a variety of processes, including compaction, and that textural equilibration is an important aspect of the compaction process. Grain coalescence, coarsening, Ostwald ripening, and solution/reprecipitation are all involved in the creep of crystals which facilitate the compaction process. Further, sub-solidus annealing of textures was also recognized as important in modifying earlier-formed textures. Aspects of annealing of cumulates had been investigated previously by Voll (1960), Vernon (1970), Ulmer and Gould (1982), Hulbert and von Gruenewaldt (1985), Reynolds (1985), and Mathison (1987).

Many of the observations of Jackson (1961) on the shapes of crystals and their mutual relationships are consistent with the operation of textural equilibration during crystallization. Although Jackson did not recognize this, he did note that the compaction involving both mechanical reorganization and deformation of crystals was an important aspect of the development of crystal mushes. Kink-banding in olivine was believed to be caused by 'deformational filter-pressing' prior to final crystallization of the interstitial material. In addition to secondary enlargement, the amount of compaction prior to cementation was considered to be important in defining porosity variations within the mush. Wager *et al.* (1960) also anticipated that compaction was a likely process: "in very thick piles of rapidly accumulated primary precipitate ... there would be an excess weight in the column of crystals over that in the column of liquid, giving a tendency for the crushing down of the lower part of the crystal column". During the mid 1980s, compaction was recognized as an important general process involved in melt migration and expulsion in the crust and mantle (*e.g.* McKenzie, 1985) and also in layered intrusions (Sparks *et al.*, 1985; Shirley, 1986; McKenzie, 1987). Textural equilibration is an important element of the microscopic creep of crystals involved in the compaction process and the work of Hunter (1987) provided the textural framework for understanding compaction in cumulate rocks.

Much of the work on textural equilibration in the presence of melt had involved laboratory melting experiments (*e.g.* Bulau *et al.*, 1979; Cooper and Kohlstedt, 1986; Toramaru and Fujii, 1986; von Bargen and Waff, 1986) and relied on earlier observations, and experiments in the materials sciences (*e.g.* Smith, 1948, 1964; Beere, 1975; Park and Yoon, 1985). This subject is now considerably more advanced (see review by Kohlstedt, 1992). Experiments have also been undertaken on thermally-driven compaction of olivines (Walker *et al.*, 1988; Lesher and Walker, 1988) and the results applied to the formation of adcumulate rocks. Another relevant laboratory experimental development has involved simulation of texture development during crystallization and associated deformation using aqueous solutions (Means and Park, 1994).

3. TEXTURE DESCRIPTION AND INTERPRETATION

There is an important distinction to be made between descriptive and interpretative approaches to the understanding of texture development in cumulate rocks. The description of a texture involves a quantification of its various elements. From such data, we can build models of how the texture developed, providing we have a knowledge of how various interacting processes influence the textural elements, the physical and chemical mechanisms involved, and the limiting length and time scales on which these processes occur. Thus, texture models depend upon observational, experimental, and theoretical considerations.

At some scale, cumulates have homogeneous textural characteristics; this scale may be up to many tens of metres but is typically on the centimetre to metre scale. Layering in cumulates is usually defined on the basis of differences in texture elements (*e.g.* mode) and individual layers may show considerable internal variation. Such variations form part of the lithofacies characterization of cumulates which must include the nature and scale of textural associations.

3.1. Texture description

The simple description of a texture involves a visual perception of shape factors (shape of grains, shape of grain boundaries and fabric), together with a visual estimate of readily quantifiable factors, such as cumulus *versus* postcumulus modes, relative cumulus modes, and presence or absence and extent of mineral zoning. A full description of texture involves point-counting (modal analysis), quantification of grain size and size distribution, characterization of grain shape and grain boundary geometry (curvature and contact angle distribution), fabric characterization (grain shape-preferred orientation (SPO) and lattice-preferred orientation (LPO)), and determination of packing/clustering characteristics of cumulus grains. It would normally also involve analysis of mineral compositions and zoning (*e.g.* by electron microprobe). At present, most workers undertake only a partial quantification of the texture, because the techniques for quantifying geometrical and clustering characteristics are not yet fully developed. All of the shape factors are inter-dependent but it is important to point out that quantification of texture elements does not involve interpretation, other than deciding what is a grain and what is a grain boundary.

Primarily, the shape of cumulus grains depends upon whether or not they impinge upon one another and hence is a function of volume fraction and packing. Isolated cumulus grains have a *form* which is usually visually estimated as euhedral, subhedral, or anhedral, with various modifiers (*e.g.* tabular). However, form can be quantified in terms of roundness or axial dimensions. For cumulus grains which impinge upon one another, terms such as euhedral, are inappropriate; their shape is defined by their grain boundary geometry. Visually, boundaries may be straight, smoothly or irregularly curved, or irregular. Particularly where cumulus grains are clustered, it is not uncommon for them to have both straight mutual boundaries with one another *and* euhedral or curved crystal faces preserved by cementing oikocrysts. The *curvature* of grain boundaries and their geometry of intersection (the general grain boundary *topology*) are important elements of the texture and usually relate to the extent of local textural equilibration.

Mode, grain size/shape distribution, packing/clustering characteristics, and fabric may all, or in part, be a result of either primary accumulation processes and/or of postcumulus modification. Hence, no single criterion will serve to distinguish one possible mode of formation or process from another. Only a full characterization of any given texture and a knowledge of its textural associations and context will allow us to understand cumulate rocks as part of dynamic magmatic systems.

3.2. Texture interpretation

Interpretation can occur at various levels. It is possible to discuss some general processes which we know must be operating and which have a first-order effect on the development of textures. Principally, these are growth and/or solution/replacement of cumulus grains, compaction (*s.l.*) of cumulus grains, cementation of cumulus grains, and recrystallization (annealing). It is clear from our present understanding of cumulate rocks, that a variety of open-system percolative processes *may* be operating on a variety of *scales* and that at some stage any given system locally becomes closed. From the point of view of understanding the texture, it is useful to be able to discuss the general processes of texture development independently of any reservoir which may be involved in any given open-system behaviour *and* of the mechanism of transport, driving forces, and implicit length scales involved in the percolative movement of magma/fluid. These are the realm of the textural model and usually involve input other than simple textural observations. Within the framework of these general processes, it should then be possible to develop specific models for the development of cumulate rocks which must be tailored to individual intrusions depending on particular boundary conditions.

4. PRIMARY TEXTURE DEVELOPMENT

An important aspect of the texture development of crystal mushes is whether the pore system is open to percolative movement of magma/fluid and the nature and extent of open-system behaviour. Thus, the porosity structure and permeability of mushes are important parameters which, at least initially, are determined by the way in which crystals accumulate and by the nature and extent of early-stage mechanical reorganization. Subsequent evolution of the texture is superimposed on any inherited depositional texture and fabric. It is appropriate, therefore, to outline the important factors involved in the early accumulation and development of crystal mushes insofar as they influence the later stages of texture development.

4.1. Mechanism of deposition

4.1.1. In situ growth

Crystals may accumulate by sedimentation or grow *in situ*. Crystals may nucleate and grow on an existing substrate or nucleate homogeneously and impinge to form a framework. They may form in isolation, in chains or in clusters. It is obviously inappropriate to define an 'initial' porosity during *in situ* growth. The 'packing' characteristics and developmental morphology will depend upon nucleation density and growth rate (degree of undercooling). A shape-preferred orientation may result from *in situ* growth; crescumulates (*e.g.* harrisites; Wager *et al.*, 1960) provide a prime example. Commonly, *in situ* growth also produces a lattice-preferred orientation as a result of preferential growth on certain crystal faces.

4.1.2. Sedimentation

Crystals may be periodically deposited as a 'rain' from a column of magma or from a magmatic current. An array of transport regimes may be involved, ranging from dilute suspensions to high-concentration (crystal-rich) gravity currents. Crystals may be maintained in suspension by a variety of mechanisms and grain interactions and be deposited individually or as chains or clusters. Deposition from dilute suspensions or waning flows commonly results in sorting with respect to grain size, shape, and density. Deposition from high-concentration currents involves progressive aggradation; *en masse* deposition is prevented by upward percolation of displaced interstitial melt (hindered settling). Crystals of contrasted hydraulic properties commonly are deposited together. The initial packing density of crystals may be varied and heterogeneous within layers (Hunter and Kokelaar, 1994). The maximum packing density of spherical grains deposited in a close-packed arrangement is ~73%. Typically, well-sorted, rounded sand grains have initial packing densities of 40-50% (Atkins and McBride, 1992). If grains are clustered, or have non-uniform grain-size distributions, initial packing densities may range between values of 20-60%.

A depositional fabric may result from settling of non-spherical (*e.g.* tabular or prismatic) grains; this will usually be a planar lamination. Deposition from a flow may result in a linear fabric (Benn and Allard, 1989; Higgins, 1991). The presence of a linear SPO may be the only way to distinguish simple settling from deposition from a current.

4.2. Mechanical reorganization of crystals

Crystal mushes which accumulate with a high initial porosity may increase their packing by mechanical compaction of grains and/or re-alignment by flow of magma or mush. The ability of grains to move relative to one another during mechanical deformation depends upon the nature and strength of grain boundaries and contacts. On a larger scale, this determines the intrinsic strength (rigidity) of the crystal mush and its ability to resist moderate to high strain-rate shear-deformation associated with sliding and slumping. Fluidization and liquefaction of mushes can initiate sliding and slumping but may only result in localized mechanical rearrangement and internal sorting of mushes.

4.3. Porosity structure and permeability

It will be clear that the mechanism of accumulation and early history of mush development strongly influence initial porosity structure and hence permeability, both on the microscopic (cm-dm) and mesoscopic (dm-m) scales. At the mesoscopic scale, the permeability of a mush is an average property depending on the average porosity, grain size and shape. However, on a small scale the pore-system microgeometry (*i.e.* distribution and interconnectivity) is important and hence, clustering, sorting, and packing of grains become important to the way in which textures develop on a microscopic scale. The distinction of these two scales is important because the former relates to the scale of individual layers, at which facies and textural associations are described, and the latter to the thin-section scale, at which textures themselves are described. Both scales are important in defining textural models.

5. SECONDARY TEXTURE DEVELOPMENT

Growth, solution/replacement, cementation, compaction (*s.l.*) of cumulus grains are all important secondary processes in the development of crystal mushes. Recrystallization (*e.g.* by annealing) is also important in their texture development. It is useful to describe each of these general processes in isolation from one another although, in general, they operate simultaneously. A specific texture will be a complex function of the interplay between these processes and the importance, locally, of any one in relation to the others.

5.1. Crystal growth: Replacement of pore volume

Cooling of magma results in growth of crystals and the kinetics are generally well understood. The growth of cumulus crystals into the pore spaces or nucleation and growth of a new phase or phases, in the pore spaces essentially involves the replacement of magma-filled pore space by crystal growth. This growth can be near-isothermal or can occur during cooling; either case could be a result of open-system percolation. Closed-system crystallization from pore magma will produce normally zoned cumulus crystals. Overgrowth of cumulus grains could, in theory, continue until no melt-filled pore space remained. In practice, as in clastic sediments, porosity is likely to be occluded when it approaches ~10%, resulting in at least some closed-system crystallization from 'trapped melt'.

Cementation of cumulus grains occurs by nucleation and growth of a new phase or phases in the pore spaces. This style of replacement of pore space results in a different pattern of texture development and can occur as a result of open-system or closed-system crystallization. If, for example, new phases nucleate early in the postcumulus evolution of a mush, their growth will restrict substantial overgrowth of some, but not all, cumulus grains, since the cumulus phase continues to precipitate. Poikilitic or subpoikilitic enclosure of cumulus grains will prevent their further direct growth from the melt, whilst incompletely enclosed cumulus grains will continue to enlarge through overgrowth. The resulting texture will be heterogeneous on the scale of the oikocryst dimensions; domains of enlarged cumulus grains, with little poikilitic cementing phase, pass into domains with successively smaller, poikilitically cemented cumulus grains. The spatial distribution of the oikocryst domains will be controlled by factors which influence their nucleation density and subsequent growth and will be some function of the diffusion/transport length scale within the mush. The volume fraction of syntaxial overgrowth *versus* interstitial (poikilitic) cementation is a critical function of the temperature of the mush in relation to the temperature of saturation of any poikilitic cementing phase, *i.e.* the temperature interval during which either overgrowth (or compaction) of the cumulus grains may occur before they are cemented by nucleation and growth of oikocrysts. Thus it is the timing of growth of oikocrysts which becomes important and this will be discussed further in section 6.1.

5.2. Reaction and replacement

Reaction and/or resorption of cumulus grains with pore melt/fluid can occur during open- or closed-system regimes. It may be a thermal or compositional effect produced by percolation of magma (dissolution of primocrysts) or volatile-enriched magma/fluid (reaction/replacement of primocrysts) or may involve a peritectic reaction with replacement of primocrysts by poikilitic cement (*e.g.* cumulus olivine-melt reaction-relationship producing poikilitic orthopyroxene). The possibility exists for complete reaction/replacement of cumulus grains by melt/fluid resulting in a metasomatic or replacement cumulate.

The reaction involving chemical equilibration of cumulus phases with melt in the pore spaces can occur whether the pore system is open to percolation or closed. It is particularly important in Fe-Mg exchange involving ferromagnesian silicates and oxides, and is recognized as the 'trapped-liquid shift effect'. Its magnitude depends critically upon the buffering effect of the mode (*e.g.* Barnes, 1984).

5.3. Textural equilibration

Both compaction and recrystallization result in grain-shape changes. Before describing the textural consequences of either process, it will be useful to discuss the general issue of textural equilibrium since they usually result in a lower-energy grain-boundary configuration. Aspects of textural equilibration of cumulates have been illustrated and considered in detail by Hunter (1987). Textural equilibration involves changes in the topology of a system of phases in such a way as to reduce the total (surface) energy of the system. Since any system of phases (crystals, fluid, or vapour) consists of regions of relatively homogeneous properties separated from one another by interfaces (*e.g.* grain boundaries), the total interface energy of a system will be a function of the fraction of the system represented by interfaces. The principal result of textural equilibration, therefore, is a change in geometry and area of the grain boundaries which minimizes their local surface energy. The driving force for textural equilibration is differences in local grain boundary curvature. Texturally equilibrated rocks have constant mean grain boundary curvature which results in constant grain boundary contact (dihedral) angles between like phases or combinations of phases. A specific equilibrium texture is a function of both the relative volume fraction of phases and the magnitude of the surface energy differences between phases. Since large grains have a lower mean grain-boundary curvature than smaller grains, textural equilibration will also result in an increase in the average grain size; this reduces the total area of grain boundary per unit volume and hence the total surface energy of the system. It is important to appreciate that textural equilibration is only achieved on a local scale, the magnitude of which is a function of the characteristic diffusion/transport length scales.

Textural equilibration can occur both in the presence of melt and in the sub-solidus, but the rate of equilibration is significantly enhanced in the presence of melt. Hunter (1987) noted that equilibrium dihedral angles between cumulus and post cumulus grains commonly are in the range 40-60°, mimicking likely cumulus grain-melt dihedral angles, and cited this as evidence for equilibration in the presence of melt. The aqueous fluid experiments of Means and Park (1994) demonstrated that textural equilibration involving solution-precipitation, grain growth, and Ostwald ripening (see below) could occur during super-solidus crystallization. These processes, coupled with grain boundary sliding, could modify textures during growth from liquid.

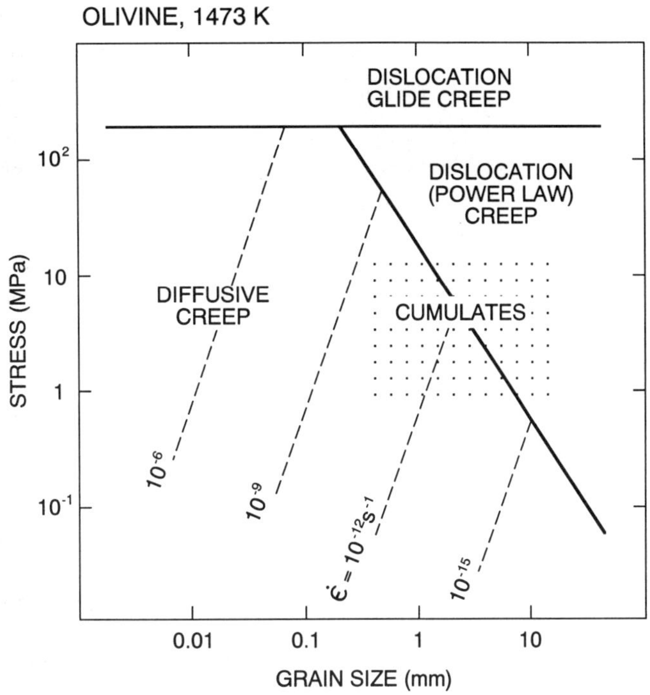

Figure 1. *Deformation map for olivine (at 1473 K) showing the dominant creep mechanisms for different grain sizes, stresses and strain rates (ε is strain rate). The field labelled cumulates shows the likely conditions appropriate for compacting crystal mushes in magma chambers (modified from Cooper and Kohlstedt, 1986).*

5.4. Compaction: Reduction of pore volume

Reorganization of grains during early stages of mush development can result in an increase in packing density of crystals and expulsion of magma from a mush; this can be termed mechanical compaction. If, however, pore volume is to be reduced further than maximum mechanical packing density, then compaction must involve viscous deformation of grains. There are three principal potentials involved in the microscopic deformation of polycrystalline materials: deviatoric stress (which may be buoyancy, *i.e.* gravity-driven, or applied), surface energy, and temperature. Materials can deform in the solid state or when fluid/melt or vapour are present. The kinetics of compaction are a function of temperature, melt fraction, grain size, poten-

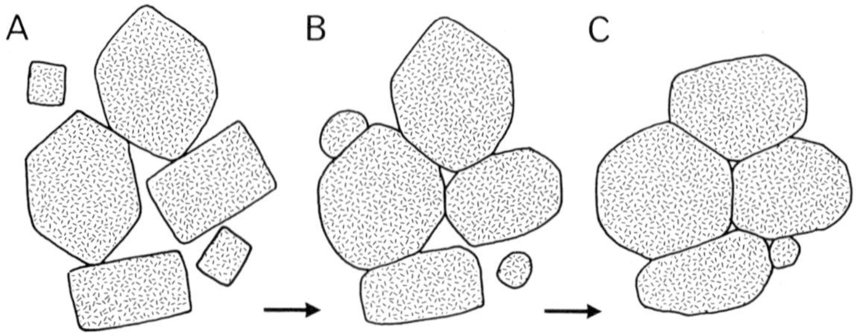

Figure 2. *Progressive changes of crystal shape and packing during compaction accommodated by textural equilibration. Pressure solution occurs at regions of high curvature (grain corners) and reprecipitation on regions of low curvature (crystal faces). Note how grains coalesce and grain sizes change as a result of grain growth and/or Ostwald ripening. The changes in grain shape and movement of grain boundaries lead to a lower-energy, texturally equilibrated, densified cumulate. Compare the changes with photomicrographs and drawings in Figures 5 and 6.*

tial gradient, and the rheological properties of the phases present. Irrespective of the driving forces involved, and providing the dominant creep mechanism is the same, the textural effects are predictable. The creep mechanism is a function of deviatoric stress, strain-rate, temperature, and grain size, and differs for different materials. However, for most crystalline phases in cumulates, and for the likely range of grain sizes, temperatures and low strain-rates involved in compaction, the dominant grain-scale deformation mechanisms will involve *diffusive creep* or *dislocation creep* (Figure 1). Although both creep mechanisms can operate simultaneously, the textural responses to diffusive creep and dislocation creep are different and will be summarized separately.

Deformation within the diffusive creep regime takes place by

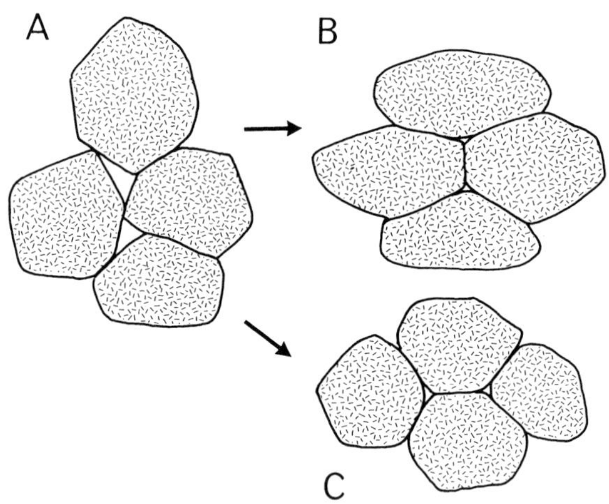

Figure 3. *Spatial changes associated with diffusive creep involving diffusion pathways: A-B through the grains (Nabarro-Herring creep); A-C along grain boundaries (Coble creep). Diffusion of material can also take place through intergranular melt channels (melt-enhanced diffusive creep). Spatial changes in the grains are accommodated by grain-boundary sliding.*

the transfer of material from regions of high potential to low potential (Figure 2). Although the potential may be stress, surface energy, or temperature, it is the deviatoric stress that has the greatest magnitude at the grain scale (Wheeler, 1991). In general, point contacts between grains focus grain-scale stresses and dissolution will preferentially occur at these points. Surface energies of grains are also highest where grains are strongly curved, *i.e.* at apices and edges. Material will preferentially dissolve from these regions and be deposited on regions of low curvature, *i.e.* flat faces. Re-deposition of material occurs in such a way as to minimize surface energies. Thus, during diffusive creep, textures generally mature to a lower-energy, texturally equilibrated topology (Figure 2).

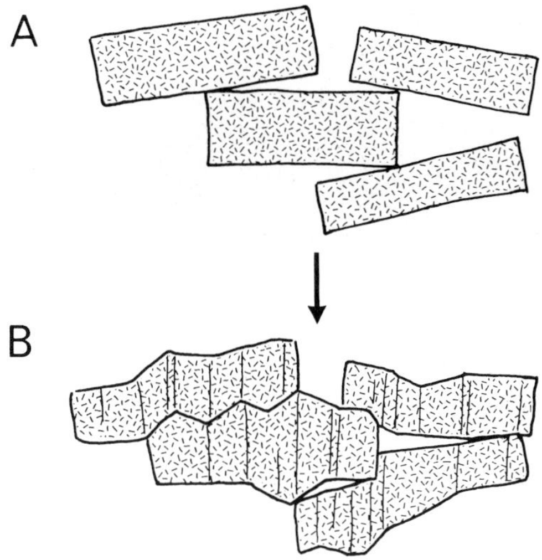

Figure 4. *Changes of grain shape produced by dislocation creep. Creep involves movement of dislocations both by glide and climb along slip systems (e.g. (010)[100] in olivine). Both shape and orientation changes are involved. Low-angle dislocation walls divide subgrains whose lattices are rotated by only a few degrees. Discrete new grains may form if substantial angular rotation occurs. This creep mechanism does not result in a minimum energy configuration, but local textural equilibration may occur through recovery.*

Diffusive transfer of material can take place through the crystals themselves (volume diffusion), along grain boundaries, or, if melt is present, through the melt-filled pore spaces. These three mechanisms have different activation energies and rates; they are called Nabarro-Herring creep, Coble creep, and melt-enhanced diffusive (MED) creep, respectively (Figure 3). At high melt fractions, MED creep is the dominant mechanism. As melt fraction (*i.e.* pore volume) decreases during compaction, grain boundary contact area increases and grain-boundary diffusion becomes the rate-limiting process (Cooper and Kohlstedt, 1986).

Dislocation creep involves glide and climb of dislocations within crystals, with movement occuring along specific slip systems (Figure 4). Dislocation walls separate subgrains with lattices re-oriented by a few degrees (Figures 5A and 5B) and these are low-energy boundaries. Textures produced during dislocation creep are un-equilibrated; serrated crystal boundaries are common with subgrain walls, when present, forming perpendicular to the uniaxial compaction direction. Evidence of 'bending' of crystals also may be present (Figure 5D). However, local textural equilibrium usually is achieved through recovery involving grain-boundary migration, particularly at smaller grain sizes.

For a given mineral, and all other factors being constant, an increase in grain size results in a change from diffusion-dominated to dislocation-dominated creep. This effect is offset at higher melt fractions because of the enhanced kinetics of diffusive transfer. However, a decrease in temperature results in expansion of the dislocation creep field, shown in Figure 1, at the ex-

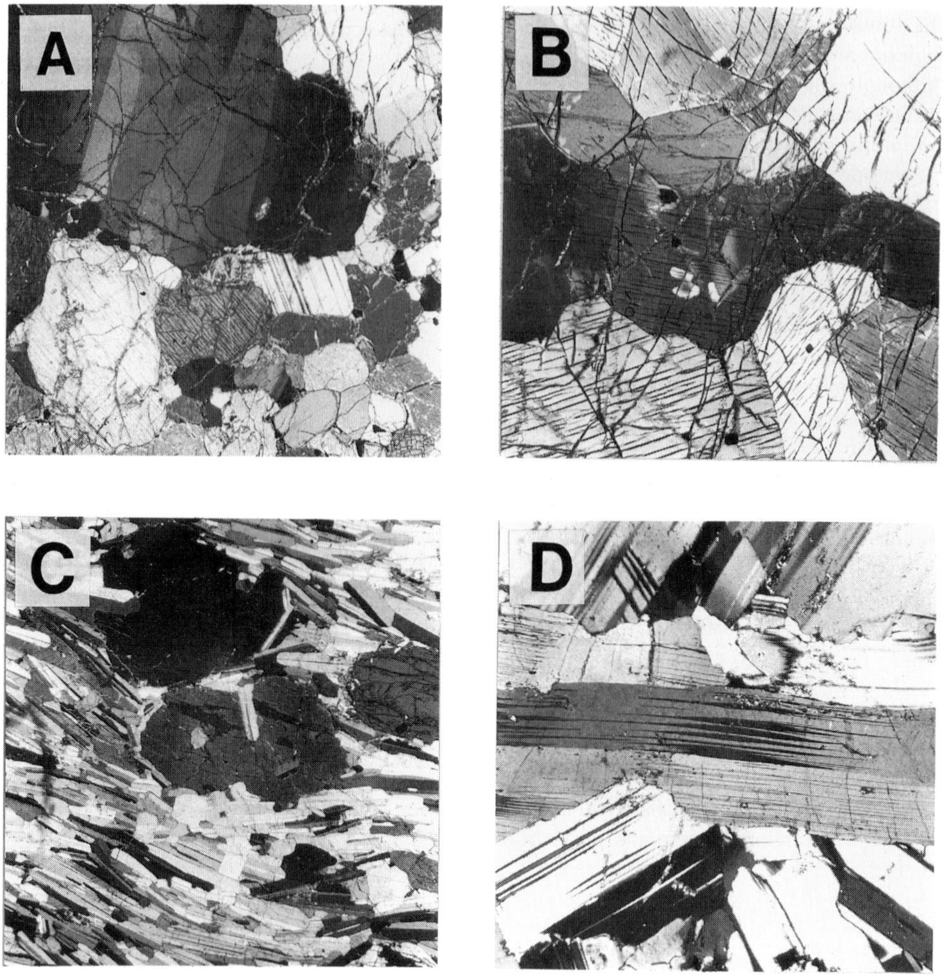

Figure 5. *Deformation during compaction involving dislocation creep. A. Olivine showing subgrain walls which form normal to the compaction direction. Mafic troctolite, Rum Intrusion, northwest Scotland. B. Orthopyroxene showing subgrain walls and irregular grain boundaries. Smaller orthopyroxene grains have equilibrated by diffusion creep. Great Dyke, Zimbabwe. C. Draping of plagioclase around olivines in troctolite from the Rum Intrusion. Away from the olivines, the feldspar shows a strong planar lamination and equilibrated grain geometry. D. Bent plagioclase crystals in gabbro of UZa, Skaergaard intrusion, east Greenland (section courtesy of A.R. McBirney). Note the equilibrated grain boundaries of the smaller grains. (Width of field in A = 5 mm; B = 8 mm; C = 1 cm; D = 8 mm.)*

pense of the diffusive creep field. In general, different minerals behave differently under the same boundary conditions. For example, olivine may deform dominantly by solution/reprecipitation, whereas plagioclase crystals of the same size might deform by dislocation creep. However, it should be emphasized that the behaviour of mixtures of phases during deformation is not fully understood and may either decrease strain rate (through grain-boundary pinning) or increase the strain rate (through superplasticity).

Grain boundaries in polycrystalline material can also accommodate strain. Diffusion-accommodated *grain-boundary sliding* can facilitate spatial changes of grains during compaction and, in conjunction with other creep mechanisms, can increase strain rates significantly. Planar lamination may be produced during compaction of cumulates. Much of this may be accommodated by grain-boundary sliding and laminations commonly are draped around enclosing oikocrysts (Figure 5C).

An increase in packing of grains results from both diffusive and dislocation creep mechanisms. Nabarro-Herring creep produces changes in grain shape but no spatial change in the relative position of grains. Coble creep and MED creep produce both shape and spatial changes (Figure 3). Dislocation creep produces a change in grain orientation as well as a shape change. Angular rotation of subgrains may ultimately result in formation of discrete new grains. The *scale* of pore volume loss during compaction depends upon the specific compaction process (driving force). The *extent* of compaction depends upon the timing and nature of cementation (see below).

5.5. Recrystallization/Annealing

Thermal annealing (static recrystallization) and strain recovery also constitute driving forces for grain-shape changes and such effects will operate in conjunction with the changes accommodating compaction. Static recrystallization of a polycrystalline aggregate involves an increase in grain size and, therefore, a reduction in total energy. Highly strained rocks also reduce their internal strain energy (recover) by increasing their grain size. The recrystallization involves migration of grain boundaries or nucleation and growth of strain-free grains. Large grains have lower relative grain-boundary curvature in comparison to smaller grains, so large grains will grow by grain-boundary migration and small grains will decrease in size and ultimately be consumed. This general process of coarsening is called *grain growth*; the coarsening of a dispersed phase by a similar process is called *Ostwald ripening*. Sometimes, a single large grain in an aggregate of smaller grains will undergo rapid growth, consuming adjacent smaller grains; this process is called *secondary* grain growth. The mobility of grain boundaries is restricted by the presence of dispersed phases. Very small volume fractions of a second phase can *pin* grain boundaries of the principal phase restricting grain-boundary mobility and hence grain growth. As a result, single phase aggregates will usually be coarser than polyphase aggregates at the same temperature.

It is important to appreciate that recrystallization can take place above or below the solidus. Because diffusion is involved, rates of annealing will be higher at higher temperatures or where high heat-flow is maintained. Ultrabasic and basic cumulates and systems open to repeated replenishment of magma will be more prone to such recrystallization than more evolved or lower temperature systems such as syenitic cumulates and granites.

6. DENSIFICATION AND CEMENTATION

In materials processing, a desired aim is often the reduction or elimination of porosity, which is usually achieved through compaction involving volume loss of pore space. The reduction of porosity, irrespective of how it is achieved, is termed *densification*. Densification commonly results in an increase in specific gravity but does not have to as, for example, in the densification of ice. The term *cementation* has common usage in the materials industry and in sedimentology. Within sedimentology, it refers to the replacement of primary porosity either by overgrowth on the detrital grains or nucleation and growth of new minerals in the pore spaces. Cementation and compaction are competing processes, replacing or reducing porosity, respectively. Cementation imparts rigidity to a granular framework and can limit the amount of porosity reduction by subsequent compaction. Although the mechanisms involved may differ in detail, physically, the processes of in-fill (replacement of pore volume) and compaction (reduction of pore volume) in crystal mushes are essentially analogous to those of sediments.

Both in the materials sciences and in sedimentology, the terms densification and cementation are used without reference to specific processes and, with some modification, can usefully be applied to cumulates. Cumulates represent aggregates of discrete grains or clusters of grains with magma-filled porosity. The porosity can be replaced by overgrowth on the cumulus grains or by cementation or be reduced by compaction. The replacement of porosity by overgrowth on the cumulus phases by classical adcumulus growth, finds analogy with authigenic, syntaxial overgrowth (*e.g.* in quartz-cemented sandstone). Both in sediments and in cumulates, it is often not a straightforward matter to distinguish such overgrowths from the primary grain morphology. This is particularly so if chemical re-equilibration with pore magma has taken place. Strictly speaking, we should term replacement of porosity by overgrowth on cumulus phases as cementation and reduction of pore volume as compaction. However, in practice, it may not be possible to distinguish the effects of each process from examination of the texture alone particularly when growth and compaction are competing processes.

The term densification can be used for all processes which increase the volume fraction of cumulus phases. This includes *growth* densification which results in replacement of pore volume and *compaction* densification which results in reduction of pore volume. When only one or two cumulus phases are present, cementation usually involves the nucleation and growth of new, usually poikilitic or subpoikilitic phases within the pore spaces. Within such rocks, the terms 'granular' and 'poikilitic' become usefully synonymous with densified and cemented, respectively. A fully or highly densified cumulate would have a granular texture, a partly densified, cemented cumulate would have a poikilitic texture; compare the two orthopyroxenites shown in Figure 6 and the sequence of sketches of olivine cumulates in Figure 7. In multiply saturated rocks, most overgrowth occurs on cumulus grains and poikilitic cementation may be limited or absent. Such rocks generally have a granular texture and would thus be densified; the presence or absence of compositional zoning would then define whether densification had been open or closed with respect to percolation of melt.

An adcumulate is a highly or fully densified cumulate which has developed during open-system percolation of melt. A heteradcumulate is a partly to highly densified cumulate with poikilitic cement developed during mostly open-system percolation. Orthocumulates are the closed-system equivalents (granular or poikilitic) and are distinguished by zoned phases. The advantage of the term densification is that it is not process specific; it can be used to describe

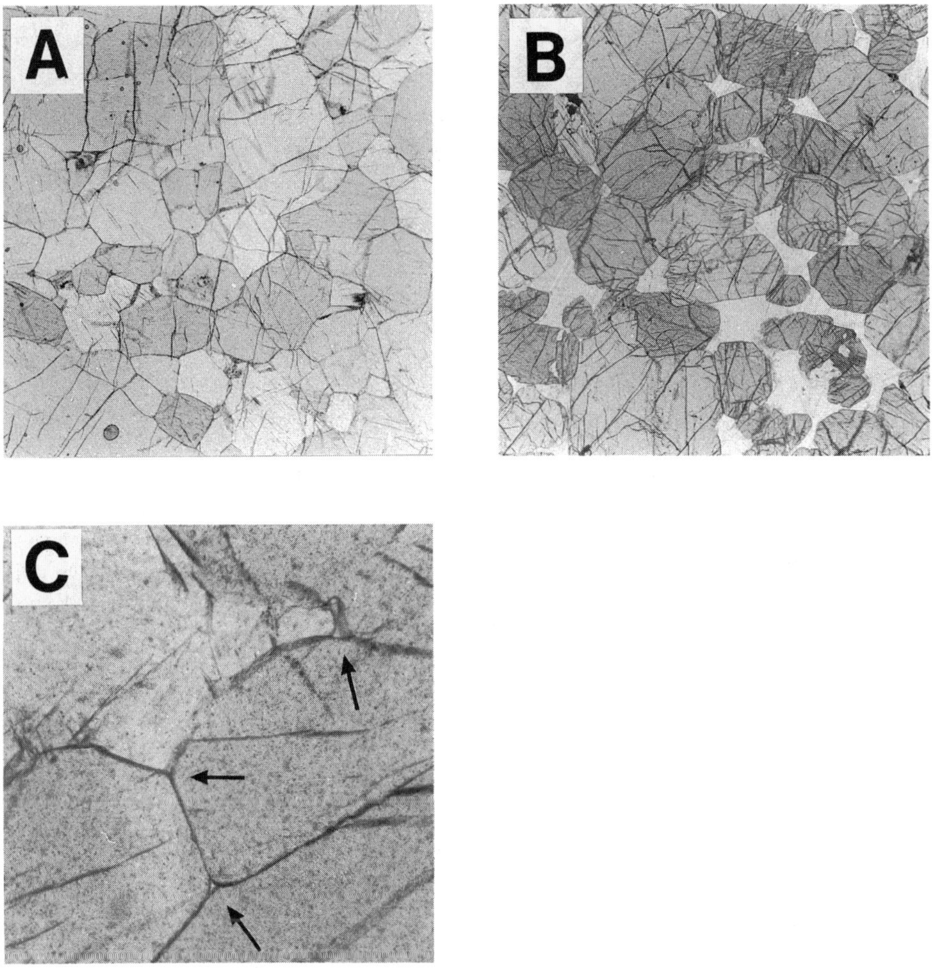

Figure 6. *Textures of orthopyroxene cumulates from the P1 pyroxenite of the Great Dyke, Zimbabwe (sections courtesy of A.H. Wilson). A is from the axis of the intrusion and shows a fully-densified, texturally equilibrated mosaic of granular orthopyroxene. B is from nearer to the margin of the intrusion. Partly equilibrated grains of orthopyroxene are cemented by poikilitic plagioclase feldspar. C shows tiny plagioclase grains (arrowed) at the triple junctions of orthopyroxene in the densified rock shown in A. These fill the interconnected porosity after most of the melt has been compacted out of the mush. (Width of field in A and B is 1 cm; in C is 1.5 mm.)*

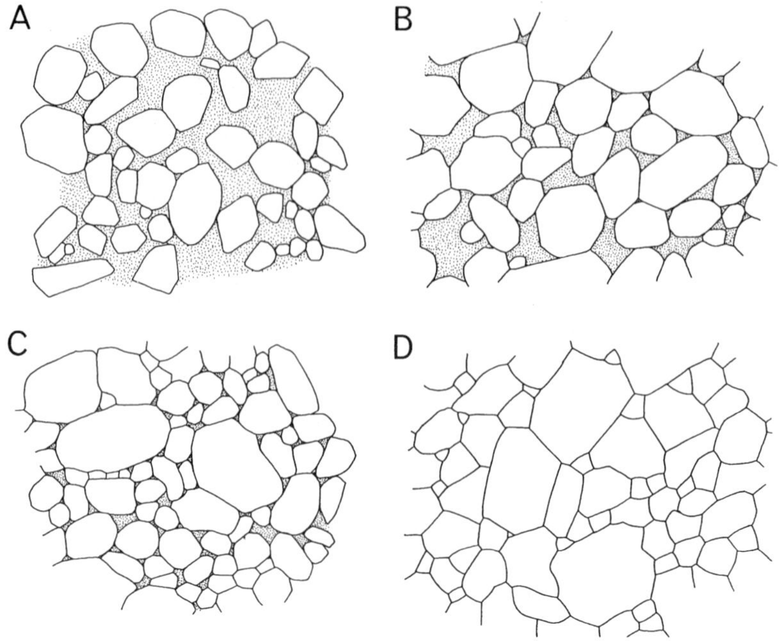

Figure 7. *Sketches showing changes in crystal shape and packing for differently densified olivine cumulates from the Rum intrusion, northwest Scotland. The cementing phase is plagioclase (stippled) in each case. Although the grains sizes and size distributions are slightly different in each case, the progression from A to D shows how the grains equilibrate during densification, resulting in a granular mosaic. Each sketch represents an area of ~6x4 mm.*

overgrowth on cumulus phases and/or compaction of cumulus phases. Evidence of creep involving grain-boundary equilibration or grain deformation would be required to distinguish the two processes. Unfortunately, a distinction cannot be made between a cumulate originally densified by overgrowth and subsequently thermally annealed and a cumulate densifed by compaction involving diffusive creep; their textures and chemistry would be identical.

6.1. Timing of densification and cementation

It has been emphasized that overgrowth, reaction/replacement, compaction and cementation are competing processes; any texture must ultimately reflect the timing and relative importance of each. As a first approach, it is instructive to consider the textural implications of the timing of pore-fill cementation in relation to syntaxial overgrowth, in the absence of compaction. This relates to the factors which influence the nucleation and growth of poikilitic crystals in mushes of cumulus grains. From observation and experiment, when grains of a particular phase exceed a critical volume fraction, ~50%, any second, or subsequent, phase nucleating in the pore spaces tends to grow poikilitically or subpoikilitically. When magmas are approaching multiple saturation, there is a relatively rapid transition to cotectic crystallization of the phases as discrete cumulus grains. The texture which develops is thus a function of the details of the

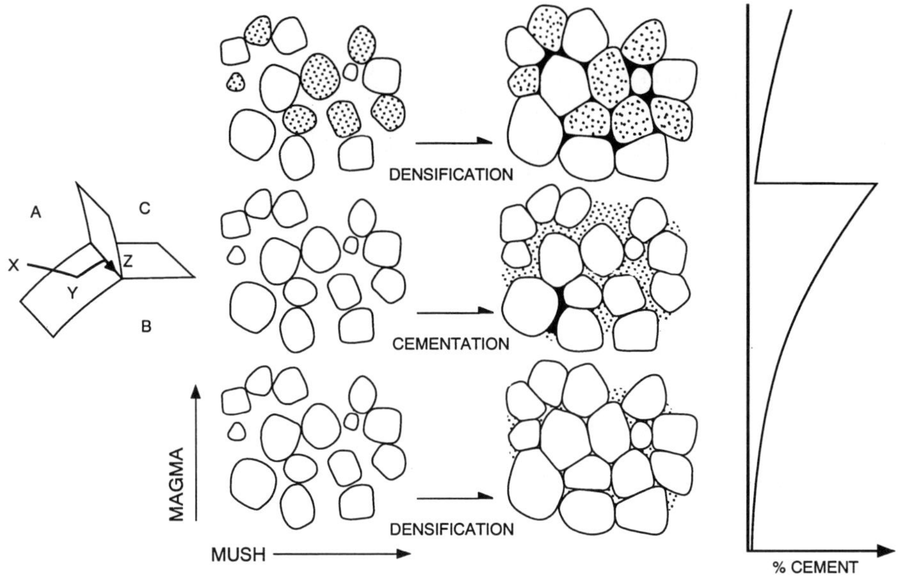

Figure 8. *Texture development during progressive accumulation of cumulus grains from a magma evolving from composition X through Y to Z in a hypothetical ternary system (shown at left) of phases A (white), B (stippled), and C (black). The initial packing geometry is shown for magmas at X (bottom left) and Y (top left). Each is able to densify almost completely by overgrowth before cementation, i.e.,* ΔT_{cem} *is large (densification could also be by compaction producing a similar texture). Middle left shows packing geometry for a magma close to cotectic saturation by B;* ΔT_{cem} *approaching zero. The mush becomes cemented by poikilitic crystals of B. The percentage of cement is shown schematically at the right. The cyclicity of texture development would continue with saturation of C. Replenishment of magma of composition X prior to saturation of B would produce a similar textural cyclicity with grains of A showing an increase in cementation (decrease in densification) upwards in each cycle.*

phase relationships and is best considered with reference to a hypothetical ternary system (Figure 8). Such diagrams have been used commonly in the past to describe sequences of appearance and disappearance of phases but not their texture development.

Consider the static crystallization of a magma of composition X in Figure 8, forming a mush of cumulus grains of A. Assume that the crystals at any given depth in the mush grow near-isothermally (adcumulus growth), but that the reservoir magma is progressively cooling and proceeding to cotectic saturation of phase B. Prior to saturation of B in the magma, cumulus grains of phase A continue to grow by overgrowth, resulting in densification of the mush. At the time of saturation of phase B in the magma (composition Y), the porosity of the mush will increase upwards (more densified downwards). At the base of the mush, B will nucleate and grow as poikilitic crystals locally enclosing and isolating cumulus grains of A from further overgrowth; grains of A not poikilitically enclosed by phase B continuing to enlarge by overgrowth. The volume fraction of poikilitic, pore-cementing B will increase upwards through the mush and at some point will form discrete cumulus grains. This will be somewhere below the point at which B forms directly from the magma. At the top of the mush, phase B will form cumulus grains in cotectic proportion with A. Thereafter, a mush of cumulus A and B will develop until saturation of the magma with C (composition Z). Subsequent texture development will follow the cycle of granular-poikilitic-granular with the proportion of pore-fill cement increasing in each cycle. Replenishment of the magma chamber with composition X at any stage will obviously arrest the rhythm but also result in cyclicity in the texture development.

The above scenario is a simplification. It takes no account of volume of magma in the reservoir or of the magnitude, mechanisms, and length scales of open-system behaviour. Nor does it take account of re-equilibration of cumulus phases with pore melt or possible reaction relationships. It assumes that the mush grows progressively (*e.g.* by *in situ* growth or steady sedimentation) but makes no statement regarding the thickness (depth) of mush. It does, however, provide a framework for interpreting textural associations and also for consideration of the implications of timing of cementation for densification of the mush either by overgrowth or by compaction.

6.2. Implications for development of crystal mushes

The nucleation and growth of a poikilitic cementing phase represents a thermal event and must correspond to the location of an isotherm within the mush at any given instant. The movement of this isotherm (particularly if locally planar) corresponds to a *cementation front.* If the cementing phase grows near-isothermally (*e.g.* as a result of compositional convection), the cementation front may correspond to the solidification front; in reality, it probably contributes significantly to the restriction of porosity and thus corresponds to a front of 'trapped' porosity.

Clearly the time available for densification of the mush by overgrowth on cumulus grains or by compaction is a function of the magma composition, more specifically, the temperature interval before saturation of a poikilitic cement phase (ΔT_{cem}). Significant poikilitic cementation prevents substantial unrestricted growth of cumulus grains. It also restricts compaction. Granular aggregates of cumulus grains will readily compact; the compaction is accommodated by the various creep mechanisms described earlier. Magma compositions precipitating one or more cumulus phases, but which are some way from saturation of a poikilitic cement phase (*i.e.* large ΔT_{cem}), will produce granular mushes that can densify by overgrowth or compaction. If the mush remains permeable, then theoretically the mush could become fully densified before $\Delta T_{cem} = 0$ and hence no poikilitic cementation occurs. In deep mushes formed from repeatedly replenished magma, or formed during sustained sedimentation, gravity-driven compaction may result in full densification. If the mush accumulates slowly, then thermal compaction or overgrowth may result in a high degree of densification close to the mush/magma interface (equivalent to a 'hard ground' or rapidly cemented hiatal surface in sedimentology). Clearly, therefore, the magnitude of ΔT_{cem} is critical to the style of texture development.

At the scale of a whole intrusion, the texture which develops in a mush will be a function of the interaction of cementation fronts and densification fronts. The former will relate to intrusion geometry, the latter will typically be sub-horizontal. In general, it would be expected that rocks close to the margins of intrusions would be more highly cemented than equivalent rocks in the axes or centres of intrusions which will be more densified. In the Great Dyke, this is the case (*e.g.* Wilson, 1992; see Figure 6) but more detailed and systematic textural studies are required in other intrusions.

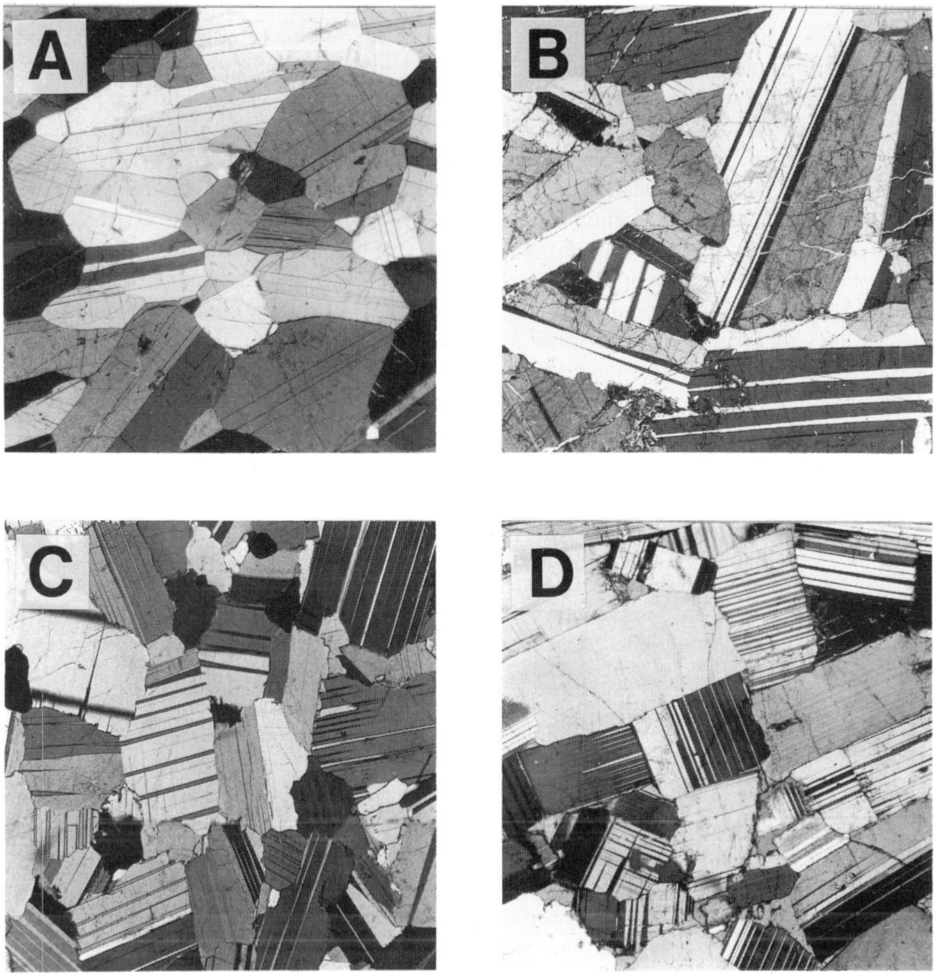

Figure 9. Textures in highly densified anorthositic cumulates. A. Texturally equilibrated geometry (Eastern Layered Series, Rum Intrusion). B. Partly equilibrated geometry (Middle Banded Zone, Stillwater Intrusion). C. Partly equilibrated geometry with serrated grain boundaries (Upper Critical Zone, Bushveld Intrusion). D. Densely packed tablets of feldspar with unequilibrated geometry (UZa, Skaergaard Intrusion). (Width of field in each is 1 cm.)

7. CUMULUS NOMENCLATURE

Throughout the preceding discussions, I have largely refrained from using the terms adcumulate, heteradcumulate, mesocumulate, and orthocumulate. It will be apparent that the discussion of densification and cementation largely revolves around rocks that, traditionally, would be refered to as adcumulates and heteradcumulates, respectively. In the original sense, orthocumulates would represent closed-system cementation and heteradcumulates represent open-system cementation. Adcumulates are indeed highly densified, usually granular and commonly texturally equilibrated. In the original scheme, they are a result of isothermal overgrowth of cumulus grains sustained either by diffusion or compositional convection. However, texturally, it may not be possible to distinguish cumulates formed in this way from cumulates densified by compaction. The latter will be texturally equilibrated, the former may be. Only if evidence of dislocation creep (*i.e.* subgrain or disclocation wall structures) or crystal drapes are preserved (Figure 5), would one be able to positively identify gravity-driven compaction as the cause. Compaction results in a reduction of porosity and, strictly speaking, the rocks are subtraction cumulates (*e.g.* Irvine, 1980). The term compactite might be an appropriate term for such a rock. However, since densification will usually be a result of both growth and compaction, use of this term is as ambiguous as adcumulate.

In large magmatic systems, where heat flux is maintained by repeated replenishment, thermal annealing is to be expected within mushes of crystals or in recently solidified rocks. With any form of textural equilibration (*via* compaction or annealing) maturation of the texture involves diffusive transport, and is usually accompanied by chemical equilibration. Thus, texturally equilibrated cumulates usually satisfy the adcumulate criterion. In theory, a thermally annealed orthocumulate could develop the texture of an adcumulate.

The problems associated with use of the term adcumulate are highlighted in Figure 9. Four anorthositic cumulates are shown, all of which would classically be described as adcumulates. Clearly, there are important differences between each texture, which would be hidden by use of the blanket term adcumulate. Any one could have been produced by several different processes.

Increasingly, open-system percolation is being proposed to account for features of cumulate rocks. Any magma coming to rest in pore spaces during open-system percolation of melt may ultimately cool, crystallize, and produce an orthocumulate texture. Orthocumulates, in the original definition, are rocks in which the *original* melt trapped in the pore spaces crystallizes as a closed system. Compaction is, by definition, an open-system process. In some cases, open-system behaviour could involve complete reaction and replacement of an existing texture. Such metasomatic rocks commonly display textures that are indistinguishable from rocks formed by primary accumulation of cumulus phases. Clearly, it is inappropriate to refer to such rocks even as cumulates. However, their metasomatic origin is usually evident from other criteria, such as field relations.

I would advocate that it is useful to retain the term cumulate (and hence cumulus and post-cumulus). The terms adcumulate and orthocumulate (plus meso- and heterad-) involve considerable ambiguity and in any case are model, not descriptive, terms. Textures should be described by terms which carry no model dependence. Densification and cementation are useful terms, as are granular and poikilitic. Interpretation and construction of texture models can then be built upon the basic texture description. Finally, it should be noted that postcumulus processes are superimposed on any initial textures which may be inherited from a wide variety of possible accumulation regimes and conditions. It should be clear, therefore, that the systematic study of textural *associations* in sequences of cumulates is at least as important as the study of the textures themselves.

8. ACKNOWLEDGEMENTS

Many people have been influential in the development of ideas presented in this contribution. I would like to single out only two; Sir Malcolm Brown, who introduced me to igneous rocks, and Dan McKenzie, who stimulated my interest in their textures. My thanks go to Mike Atherton, Mike Cheadle, and Henry Emeleus who provided helpful comments and Chip Lesher and David Shelley who provided constructive reviews. The Natural Environment Research Council (U.K.) and the University of Liverpool have provided financial support.

9. REFERENCES

Atkins, J.E., & McBride, E.F., 1992. Porosity and packing of Holocene river, dune and beach sands. *Bull. Amer. Assoc. Petroleum. Geol.* **76**, 339-55.

Barnes, S.J. 1984. The effect of trapped liquid crystallisation on cumulus mineral compositions in layered intrusions. *Contr. Miner. Petrol.* **93**, 524-31.

Beere, W. 1975. A unifying theory of the stability of penetrating liquid phases and sintering pores. *Acta. Metall.* **23**, 131-8.

Benn, K., & Allard, B., 1989. Preferred mineral orientations related to magmatic flow in ophiolite layered gabbros. *J. Petrology* **30**, 925-46.

Bulau, J.R., Waff, H.S., & Tyburczy, J.A., 1979. Mechanical and thermodynamic constraints on fluid distribution in partial melts. *J. Geophys. Res.* **84**, 6102-8.

Campbell, I.H., 1978. Some problems with the cumulus theory. *Lithos* **11**, 311-21.

Campbell, I.H., 1987. Distribution of orthocumulate textures in the Jimberlana Intrusion. *J. Geol.* **95**, 35-54.

Conrad, M.E., & Naslund, H.R., 1989. Modally-graded rhythmic layering in the Skaergaard Intrusion. *J. Petrology* **30**, 251-69.

Cooper, R.F., & Kohlstedt, D.L., 1986. Rheology and structure of olivine basalt partial melts. *J. Geophys. Res.* **91**, 9315-23.

Hess, H.H., 1939. Extreme fractional crystallization of a basaltic magma: the Stillwater igneous complex. *Trans. Amer. Geophys. Union, Reports & Papers, Volcanology* **3**, 430-2.

Hess, H.H., 1960. The Stillwater Igneous Complex, Montana: A quantitative mineralogic study. *Mem. Geol. Soc. Amer.* **80**, 230 pp.

Higgins, M.D., 1991. The origin of laminated and massive anorthosite, Sept Iles layered intrusion, Quebec, Canada. *Contr. Miner. Petrol.* **106**, 340-54.

Hulbert, L.J., & von Gruenewaldt, G., 1985. Textural and compositional features of chromite in the Lower and Critical Zones of the Bushveld Complex, south of Potgietersrus. *Econ. Geol.* **80**, 872-95.

Hunter R.H., 1987. Textural equilibrium in layered igneous rocks. In: Parsons, I. (ed.) *Origins of Igneous Layering.* Dordrecht: Reidel, 473-503.

Hunter, R.H., & Kokelaar, B.P., 1994. Igneous cumulates in sedimentological perspective. *Geoscientist* **4** (No. 3), 15-7.

Irvine, T.N., 1980. Magmatic infiltration metasomatism, double diffusive fractional crystallization and adcumulus growth in the Muskox and other layered intrusions. In: Hargreaves, R.B. (ed.) *Physics of Magmatic Processes.* Princeton: Princeton University Press, 325-83.

Irvine, T.N., 1982. Terminology for layered intrusions. *J. Petrology* **23**, 127-62.

Jackson, E.D., 1961. Primary textures and mineral associations in the ultramafic zone of the Stillwater complex, Montana. *U.S. Geol. Surv. Prof. Paper* **358**, 1-106.

Kerr, R.C., & Tait, S.R., 1986. Crystallisation and compositional convection in a porous medium with application to layered igneous intrusions. *J. Geophys. Res.* **91**, 3591-608.

Kohlstedt, D., 1992. Structure, rheology and permeability of partially molten rocks at low melt fractions. In: *Mantle Flow and Melt Generation at Mid-Ocean Ridges. American Geophysical Union Geophysical Monograph* **71**, 103-21.

Lesher, C.E., & Walker, D., 1988. Cumulate maturation in a temperature gradient. *J. Geophys. Res.* **93**, 10295-311.

Mathison, C.I., 1987. Pyroxene oikocrysts in troctolitic cumulates - evidence for supercooled crystallisation and postcumulus modification. *Contr. Miner. Petrol.* **97**, 228-36.

McBirney, A.R., & Hunter, R.H., 1995. The cumulate paradigm reconsidered. *J. Geol.* **103**, 114-22.

McBirney, A.R., & Noyes, R.M., 1979. Crystallization and layering of the Skaergaard Intrusion. *J. Petrology* **20**, 487-564.

McKenzie, D.P., 1985. The extraction of magma from the crust and mantle. *Earth Planet. Sci. Lett.* **74**, 81-91.

McKenzie D.P., 1987. The compaction of igneous and sedimentary rocks. *J. Geol. Soc. Lond.* **144**, 299-307.

Means, W.D., & Park, Y., 1994. New experimental approach to understanding igneous texture. *Geology* **22**, 323-6.

Morse, S.A., 1979a. Kiglapait Geochemistry I: Systematics, sampling and density. *J. Petrology* **20**, 555-90.

Morse, S.A., 1979b. Kiglapait Geochemistry II: Petrography. *J. Petrology* **20**, 591-624.

Morse, S.A., 1986. Convection in aid of adcumulus growth. *J. Petrology* **27**, 1183-214.

Park, H-H., & Yoon, D.N., 1985. Effect of dihedral angle on the morphology of grains in a matrix phase. *Metall. Trans.* **16**, 923-8.

Reynolds, I.M., 1985. The nature and origin of titaniferous magnetite-rich layers in the Upper Zone of the Bushveld Complex. *Econ. Geol.* **80**, 1089-108.

Shirley, D.N., 1986. Compaction of igneous cumulates. *J. Geol.* **94**, 795-809.

Smith, C.S., 1948. Grains, phases and interfaces: An interpretation of microstructure. *Trans. A.I.M.E.* **197**, 15-51.

Smith, C.S., 1964. Some elementary principles of polycrystalline microstructure. *Metall. Rev.* **9**, 1-48.

Sparks, R.S.J., & Huppert, H.E., 1987. Laboratory experiments with aqueous solutions modelling magma chamber processes. I: Discussion of their validity and geological application. In:Parsons, I. (ed.) *Origins of Igneous Layering.* Dordrecht: Reidel, 527-38.

Sparks, R.S.J., Huppert, H.E., Kerr, R.C., McKenzie, D.P., & Tait, S.R., 1985. Post-cumulus processes in layered intrusions. *Geol. Mag.* **122**, 555-68.

Tait, S.R., & Jaupart, C., 1992. Compositional convection in a reactive crystalline mush and melt differentiation. *J. Geophys. Res.* **97**, 6735-56.

Tait, S.R., Huppert, H.E., & Sparks, R.S.J., 1984. The role of compositional convection in the formation of adcumulus rocks. *Lithos* **17**, 139-46.

Toramaru, A., & Fujii, N., 1986. Connectivity of a melt phase in a partially molten peridotite. *J. Geophys. Res.* **91**, 9239-52.

Ulmer, G.C., & Gould, D.P., 1982. Monomineralicity and oikocrysts: keys to cumulus cooling rates? *Lunar Planet. Inst. Tech. Rept.* **80-01**, 154.

Vernon, R.H., 1970. Comparative grain boundary studies of some basic and ultrabasic granulites, nodules and cumulates. *Scott. J. Geol.* **6**, 337-51.

Voll, G., 1960. New work on petrofabrics. *Liverpool Manchester Geol. J.* **1**, 73-85.

von Bargen, N., & Waff, H.S., 1986. Permeabilities, interfacial areas, and curvatures of partially molten systems. Results of numerical computations of equilibrium microstructures. *J. Geophys. Res.* **91**, 9261-76.

Wadsworth, W.J., 1961. The layered ultrabasic rocks of south-west Rhum, Inner Hebrides. *Phil. Trans. Roy. Soc. Lond.* **244B**, 21-64.

Wadsworth, W.J., 1985. Terminology of postcumulus processes and products in the Rhum layered intrusion. *Geol. Mag.* **122**, 549-54.

Walker, D., Jurewicz, S.R., & Watson, E.B., 1988. Adcumulus dunite growth in a laboratory thermal gradient. *Contrib. Mineral. Petrol.* **99**, 306-19.

Wager, L.R., 1963. The mechanism of adcumulus growth in the Layered Series of the Skaergaard Intrusion. *Spec. Pap. Mineral. Soc. Amer.* **1**, 1-19.

Wager, L.R., & Brown, G.M., 1968. *Layered Igneous Rocks*. Edinburgh: Oliver and Boyd, 558 pp.

Wager, L.R., Brown, G.M., & Wadsworth, W.J., 1960. Types of igneous cumulate. *J. Petrology* **1**, 73-85.

Wheeler, J., 1991. A view of texture dynamics. *Terra Nova* **3**, 123-36.

Wilson, A.H., 1992. The geology of the Great Dyke, Zimbabwe. Crystallisation, layering and cumulate formation in the P1 Pyroxenite of Cyclic Unit 1 of the Darwendale subchamber. *J. Petrology* **33**, 611-63.

LAYERED INTRUSIONS
R.G. Cawthorn (editor)

A Review of Mineralization in the Bushveld Complex and some other Layered Intrusions

C.A. Lee

Geology Department, Anglo American Platinum Corporation Limited, P.O. Box 62179, Marshalltown, 2107, South Africa.

Abstract

Layered mafic intrusions are significant sources of the platinum-group elements, base metal sulphides, chromite, magnetite, and ilmenite. The distribution of these ores is reviewed, with special attention to the economic deposits and subeconomic occurrences. The geological setting, composition, mineralogy, and textures of the ores are described for the Bushveld and Stillwater Complexes, the Great Dyke, the Munni Munni Intrusion, complexes in Finland, and some smaller intrusions. Both the platinum-group element (PGE) mineralization and the often associated base metal sulphides have characteristic geochemical and mineralogical styles; these are variable in even a single layered intrusion, and are even more so when different intrusions are compared. The distinction between constant and variable metal contents in relation to thickness variations of the PGE sequences is emphasized. Oxide ore deposits are less variable but the compositions, especially for chromite, are specific to the layered intrusion in question. Subsolidus re-equilibration and ore-mineral alteration are usually present as variable processes in all the mineralized sequences. Mineralization models are briefly addressed in the light of these variations. The primary geochemical character of PGE ores, and the occurrence and character of the oxide ores, probably reflect the influence of the magma source region at depth rather than processes in the magma chamber at the site of emplacement.

1. INTRODUCTION

This paper reviews the primary metalliferous, economically exploitable, ores in the Bushveld Complex and certain other layered mafic intrusions; the commodities are the platinum-group elements (collectively or individually referred to as PGE), including Au and the associated base metal sulphide, chromite, magnetite, and ilmenite. Many layered intrusions have some form of metalliferous mineralization, at the scale of an occurrence in outcrop or in drill core, or with a potential to be exploited. In addition, the host rocks can under suitable circumstances, be exploited for non-metallic, industrial minerals, *e.g.* andalusite in metamorphosed country rocks and the layered rocks for dimension stone, but are not discussed here.

The dominance of the Bushveld Complex in world-wide production of minerals related to mafic layered intrusions, depicted in Figure 1, gives this intrusion an archetypal status in exploration and resource models for mafic intrusions and hence is emphasized in this review. Several other mafic intrusions are mineralized and produce one of the listed commodities, or have done so, but none have or are able to produce the range of commodities which comes from the Bushveld Complex. These are considered in somewhat less detail.

2. PLATINUM-GROUP ELEMENTS AND BASE METAL SULPHIDES

2.1. Bushveld Complex

The Bushveld Complex, South Africa (Eales and Cawthorn, this volume) (Figure 2), has two stratiform PGE sequences (the Merensky Reef and the Upper Group 2 chromitite), and

LAYERED COMPLEX or GEOGRAPHIC AREA	Pt >Pd	Pt ≈ Pd	Pt <Pd	Au	Ag	Cu	Ni	Co	S	Cr	Fe	V	Ti
Bushveld	■✓	■✓	◊	*	*	*	*	*	*	■✓	✓	■✓	+
Stillwater			■✓			✓	✓			✗			
Great Dyke		~				* ~	* ~			■✓			
Windimurra			◊									■~	
Munni Munni			◊			◊	◊						
Rooiwater											■	◊	◊
Ushushwana											■	◊	◊
Muskox		◊				◊	◊			◊	◊		
Skaergaard			◊	◊									
Finland	+	+				■✓	■✓			■		✗	
Norway			◊			■✓	■✓		✓				■✓
Australia		◊	◊				+			◊			
S.America			◊							◊		◊	
Canada		◊	◊			◊				◊			
CIS, Kola		◊	◊			◊	◊	◊		◊			
OTHER COMPLEXES FOR COMPARISON													
Sudbury		◊	*	*			■✓	✓	✓				
Norilsk			■			■✓	■✓		✓				
Kemi		◊								■✓			
Jinjuan			*			■✓	■✓						

■ Major component ✓ Producing ✗ Dormant * By product
~ Planned + Possible ◊ Occurs

Figure 1. *Chart summarizing mineralization in a variety of mafic intrusions, The information is obtained from a wide number of mineral industry-related journals (Mining Journal, London; Engineering and Mining Journal; The Northern Miner; Canadian Institute of Mining Bulletin; Canadian Mining Journal; Metals Bulletin Monthly; Minerals Industry International; Australian Journal of Mining, and similar publications). The ranking of a commodity is the author's interpretation, based on the reports available up to 1995. The Bushveld Complex has a wide variety of commodities compared with other mafic complexes. Notable is the range of PGE in respect of the Pt/Pd ratio, and the dominance of Pt in the Complex. By far the majority of other PGE occurrences are Pd-dominant.*

Figure 2. *(facing page) Regional geological map of the Bushveld Complex, with producing platinum mines, and the dunite pipes.*

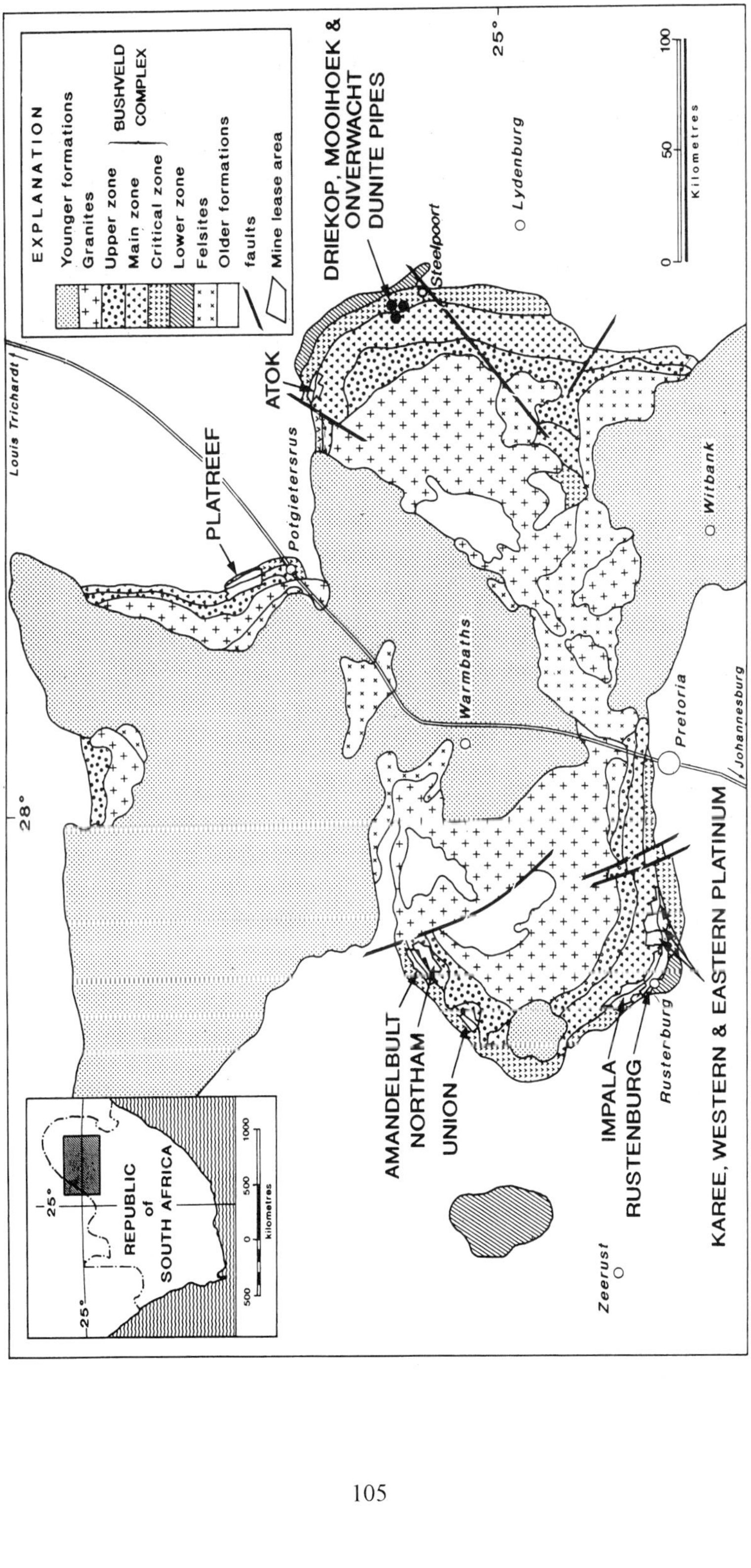
EXPLANATION
Younger formations
Granites
Upper zone
Main zone
Critical zone
Lower zone
BUSHVELD COMPLEX
Felsites
Older formations
faults
Mine lease area
DRIEKOP, MOOIHOEK & ONVERWACHT DUNITE PIPES
Steelpoort
Lydenburg
25°
0 50 100
Kilometres
ATOK
Louis Trichardt
PLATREEF
Potgietersrus
Witbank
Warmbaths
Pretoria
Johannesburg
28°
AMANDELBULT
NORTHAM
UNION
IMPALA
RUSTENBURG
Rusterburg
KAREE, WESTERN & EASTERN PLATINUM
Zeerust
REPUBLIC of SOUTH AFRICA
25°
25°
500 0 500 1000
kilometres

PGE-enriched base metal sulphide sequence (the Platreef) at the base of the northern sector of Complex. Minor PGE occur in the Lower, Main, and the Upper Zones.

2.1.1. Merensky Reef

Geology and mineralization. The Merensky Reef is the major source of PGE. It can be traced throughout the strike of the Complex. The dip ranges from 9° to 27°, with small sectors as steep as 65°. The thickness ranges from 4 cm to 4 m. Seismic surveys show reflectors correlated with the position of the Merensky Reef that can be traced as far as 50 km down-dip of outcrop, and as deep as 6 km below surface (du Plessis and Kleywecht, 1987). Lithological and stratigraphic variations of this sequence are well documented *e.g.* Vermaak (1976), and Wagner (1929) has the best summary of the different styles of Merensky Reef and the relative positions of the higher PGE values and base metal sulphides. In general, the reef consists of a texturally heterogeneous pegmatoidal feldspathic pyroxenite, partially pegmatoidal feldspathic pyroxenite, or feldspathic pyroxenite. The rock is an orthocumulate consisting of a framework of very coarse-grained subhedral to euhedral orthopyroxene constituting 70-90%, and up to 30% plagioclase as an intercumulus phase. Clinopyroxene oikocrysts up to 3 cm long occur throughout the rock. Mica is a common accessory. Two to four thin chromitite layers (1-2 cm) define the upper and lower limits of the main economic minerlization. The footwall is either plagioclase cumulate or, less common, feldspathic pyroxenite or harzburgite. A centimetres-thick anorthosite usually occurs below the lower chromitite when plagioclase cumulate is the footwall lithology. Olivine occurs sporadically in the reef at Rustenburg Section. In the north-west at Union and Amandelbult mines the reef generally contains olivine, and olivine-rich rocks occur in the footwall sequences. At the Atok mine, Merensky Reef footwall is generally feldspathic pyroxenite, but gabbronorite occurs in parts of the mine. The overlying rocks in all these geographic areas grade upwards through feldspathic pyroxenite, and norite to anorthosite, which in turn are followed by the pyroxenite - norite - anorthosite sequence of the Bastard unit. Base metal sulphide and PGE occur throughout the overlying rocks, which is reflected in the whole rock chemistry of the Merensky unit (Lee, 1983; Brown, 1994).

Some 3% base metal sulphides, and the associated platinum-group minerals (PGM), are interstitial to the silicates. The base metal sulphides range from disseminated to coarse-grained zoned aggregates dominated by pyrrhotite, pentlandite, chalcopyrite, pyrite, and cubanite and rare sulpharsenides, galena, and sphalerite. Ballhaus and Stumpfl (1986) note a common association of sulphides with hydrous silicates, and emphasize the role of hydrous fluids in these textural associations. Phlogopite is usually associated with sulphides, and often contains zircon and is intergrown with late-stage quartz. Amphiboles and talc locally have ragged contacts or are intergrown with sulphides where there is local alteration of pyroxene. Base metal sulphides and the PGE are highly correlated in the mineralized portion of the reef (Lee, 1983). Economic mineralization is concentrated in the pegmatoidal fraction of the sequence with the highest values associated with the chromitite layers. Mineralization with economic quantities of base metal sulphide and PGE are frequently dispersed into the hanging wall and footwall rocks (Kinloch, 1982), particularly in the thinner reef variants. The extent and the relative amount of the PGE and base metal sulphide mineralization in the Merensky Reef in the surrounding rocks appears to be a function of the reef thickness. Sub-economic values occur in the footwall of thick reef facies; higher, often economic, values occur in the footwall, and frequently the hanging-wall, of thin reef facies. The sulphide content of the hanging-wall rocks is usually greater than the footwall rocks. This sulphide distribution is most likely a consequence of the fixed metal and sulphide content of the Merensky Reef package. Sulphide migrates from the reef into the adjacent rocks in cases where the space volume to contain the sulphide melt exceeds the interstitial volume of the silicates, and the reef becomes over-endowed with sulphide relative to silicate in thinner facies.

The Merensky Reef is paraconformable with the underlying cumulates (Irvine, 1982; Kruger, 1990), the scale of the discontinuity ranging from the regional (kilometres) to the local (metres to centimetres). Where the Merensky Reef abruptly transgresses the footwall at the local scale the phenomenon is referred to as a "pothole". Several varieties of pothole reef are documented (Ballhaus, 1988; Kinloch and Peyerl, 1990; Viljoen and Hieber, 1986; Viljoen *et al.*, 1986a, 1986b). Pothole reef is where the Merensky Reef occurs below the normal footwall elevation for the geographic area, cutting down stratigraphically in a step-like transgression before again becoming conformable with the new footwall sequence. If the transgression is deep (>20 m) into the footwall, cross-cutting iron-rich replacement pegmatite occurs in places, but this is an uncommon and overemphasized relationship. Large numbers of potholes without cross-cutting pegmatite are known from mining and the oft-proposed association of potholes with pegmatite is not proven. Iron-rich pegmatite frequently occurs at normal reef elevation as pegmatite-replaced reef (Kinloch and Peyerl, 1990; Scoon and Mitchell, 1994), and transgressive iron-rich pegmatite also occurs. At Rustenburg the pothole reef rests on plagioclase cumulates, generally igneous-laminated norite. A contrast is found at Union Section where the shallow potholes have plagioclase-rich footwall rocks and the deeper, more common, pothole reef rests on harzburgite of the pseudoreef. This is located approximately 20 m below the Merensky Reef and is frequently base metal sulphide and PGE mineralized (Viljoen *et al.*, 1986a). At depth on this mine the harzburgite is the common footwall to the reef. This is referred to as regional pothole reef. At Western Platinum mine mapping has delineated narrow strike-parallel zones of steeper (25-30°) dipping Merensky Reef, where potholes appear to be more abundant per unit area than in normal (15-20°) dipping reef. These zones may be genetically related to monoclinal structural features sub-parallel to the edge of the Complex, which in turn control the location of syn-magmatic zones of extension and fracturing in the cumulate pile along which the potholes developed (Carr *et al.*, 1994).

An average grade of the Merensky Reef in the Rustenburg area is 8.1 ppm PGE + Au (Buchanan, 1988). The proportions of the precious metals are 4.82 ppm Pt, 2.04 ppm Pd, 0.66 ppm Ru, 0.24 ppm Rh, 0.08 ppm Ir, 0.26 ppm Au; the Cu/Ni ratio is 0.61. Proven ore reserves for the Rustenburg, Union, and Amandelbult mines combined are 204 million tonnes with a grade of 7.26 ppm Pt+Pd+Rh+Au. Probable reserves for the same operations amount to 390 million tonnes at 5.60 ppm. The smaller Atok mine has proven reserves of 69.6 million tonnes at 6.11 ppm, and probable reserves of 55.9 million tonnes at 5.24 ppm. These values are for a mining width and not for the geologically defined reef, which will be narrower or wider depending on the geographic location. These data thus represent minimum PGE values if used in geological modelling. Viljoen (1994) notes a general lateral consistency of PGE values over relatively large areas of Rustenburg Section. Using PGE sampling data for a mining width of 76 cm for the "best value zone" Viljoen shows over 70% of the reef mined deviates by less than one standard deviation (arithmetic) from a mean (no values are quoted, only patterns are discussed). However, this analysis concerns the grade of an economic mining width and ignores the geological definitions. The upper and lower chromitite layers, which commonly define the lithological boundaries of the reef, are crossed and the hanging-wall and footwall rocks are included in the analysis. The Merensky Reef as a geological entity has a constant metal content and thus the grade (value, in grams per tonne, relative to width) varies with the width of the reef, because of changes in the volume of silicate relative to the constant sulphide content.

A range of the Merensky Reef types has been documented by Kinloch and Peyerl (1990), based on the reef thickness, whether potholed or pegmatite-replaced, the composition of the footwall rocks, and the PGM assemblages. For the Rustenburg area the Merensky Reef thickness varies over a ten-fold range of 4 cm to 4 m; an average thickness is typically 30-80 cm thick. PGM are dominated by sulphide phases. In parts of Union Section the Merensky Reef is lithologically similar but thicker and has harzburgite at the base, and rests on plagioclase cumulates; the mineralization is confined to the upper portion of the pegmatoidal reef below the upper chromitite layer. Elsewhere on this mine the reef is thinner, in particular in areas where harzburgite is the footwall instead of anorthosite (Viljoen *et al.*, 1986a). In this environment the PGM are consistently of Pt-Fe alloy associated with base metal sulphides, and minor PGM sulphides; very high (1000 ppm) Pt occurs as solid solution in pyrrhotite and troillite. Wide reef is up to 1.5 m thick at Rustenburg and has Pt-Pd bismuth-telluride-arsenide semi-metal alloys in addition to PGE sulphide phases. The reef thickens eastward from Rustenburg, and generally becomes feldspathic pyroxenite largely devoid of pegmatoid; the PGE tend to concentrate towards the upper part of the reef as is the case at Western Platinum mine (Davey, 1992; Viljoen, 1994). The Merensky Reef at Atok Section, the only active platinum mine in the eastern Bushveld, is different. The PGE, dominantly sulphide PGM, and base metal sulphide mineralization, with higher pyrite than elsewhere, are located in cumulate textured feldspathic pyroxenite in a zone about 50 cm thick, bounded top and bottom by thin chromitite layers. Pegmatoidal feldspathic pyroxenite occurs below the lower chromitite layer, and this rarely has PGE values (Mossom, 1986). Despite these contrasted mineralization distributions the strontium isotope values of the Merensky Reef at Rustenburg and Atok are identical (Kruger, 1990; Lee and Butcher, 1990). In an alternative analysis of variations in the Merensky Reef, four reef facies have been recognized in the Rustenburg Section area and two facies in the Union Section area (Viljoen, 1994). These divisions are based on the thickness of the Merensky Reef and the abundance, size, and type of pothole structures.

Platinum-Group Minerals. An important aspect of PGE mineralization in general is the composition, texture, and size of the PGM, the relationship these have with the base metal sulphides and gangue minerals, and the impact these factors have on the potential to exploit a deposit (Cabri, 1988, 1994; Prendergast, 1990). The mineralogy has to be considered in any comparisons made between deposits, and in the evaluation of deposits for potential worth.

The composition and texture of the Merensky Reef PGM vary regionally around the Complex (Brynard *et al.*, 1976; Kinloch, 1982; Mostert *et al.*, 1982; Vermaak and Hendriks, 1976). Kinloch and Peyerl (1990) recognize fifteen types of Merensky Reef on geological and mineralogical criteria at Rustenburg mine, seven at Union, eleven at Amandelbult, and two at Atok. The PGM are dominantly Pt-Pd sulphides, and lesser and approximately equal amounts of PGE-arsenides, tellurides and other semi-metal phases. Ru sulphides and alloys, dominated by laurite, are associated with the reef chromitite layers. A variety of PGE-alloys, dominantly iron-rich phases, are regionally dominant, such as at the Union, Amandelbult and Northam mines. Electrum occurs in small quantities throughout. These PGM compositional variations can be ascribed to differences, some small, in the lithology of the footwall rocks to the Merensky Reef, either at normal or at pothole elevation. Kinloch (1982) noted that for any particular area the PGM of the UG2 chromitite and the overlying Merensky Reef show close compositional similarities. There is a regional pattern to the distribution of Pt-Pd sulphide phases and Pt-Fe alloys, whereas this is not the case for the semi-metal alloys. Rh-bearing PGM are rare in the Merensky Reef and the bulk of the Rh fraction of the PGE is located as solid solution in base metal sulphides.

The PGM of the Merensky Reef occur in three textural associations: PGM enclosed in or attached to base metal sulphides (38-97% of occurrences), PGM enclosed in silicates (3-62% of occurences), and to a lesser extent PGM enclosed in or attached to chromite or Fe-oxide. Trace quantities of graphite associated with PGM are frequently observed in undisturbed reef. Kinloch (1982) notes a correlation between high Pt-Fe alloy content and enclosure in silicates. Disturbed or potholed Merensky Reef and UG2 chromitite are generally sulphur-poor relative to undisturbed reef and the PGM are dominated by Pt-Fe alloys, with semi-metal PGM phases locally abundant. PGM grain size has two ranges in the Merensky Reef, 50-350 μm and 10-31 μm; overall the PGM are coarser in the Merensky Reef than in the UG2 chromitite.

The extreme lithological variability of the Merensky Reef has hindered a definition. Based on detailed work at the Rustenburg Mine, a definition of the reef for this review is: *A plagioclase-bearing (feldspathic) orthopyroxenite, olivine orthopyroxenite, or harzburgite layer, located at the base of the Merensky unit, and enriched in economic amounts of base metal sulphide and platinum-group elements. The texture is coarse-grained pegmatoidal, partly pegmatoidal, or medium-grained. Thin chromitite layers (two to four) are associated with the upper and lower limits of the economic mineralization. The Merensky Reef is conformably overlain by medium- to coarse-grained poikilitic feldspathic pyroxenite, constant in thickness. The Merensky Reef is paraconformable to the uppermost units of the Critical Zone. In the case where these units are plagioclase cumulates, the Merensky Reef may be directly underlain by an anorthosite, conformable with the Merensky Reef, and variable in thickness.*

The Merensky Reef is enriched in PGE, S, C, Ni, radiogenic elements, REE, P, and other incompatible elements, the Mg# (Mg/(Mg+Fe)) of the original unequilibrated orthopyroxene is less than the footwall orthopyroxene, and evolves upwards into the pyroxenite. The Merensky Reef appears to be a unique event possibly synchronous with the onset of Main Zone magmatism in which enriched residual Critical Zone source material was mobilized and added to the complex (Wilson *et al.*, 1995).

Platreef. The Platreef occurs in the northern sector of the Bushveld Complex (Buchanan, 1988; White, 1994). The succession within this sector differs from those of the east and west Bushveld in that the Critical Zone is not developed, and the Lower Zone is of limited extent in the south of the sector. The layered mafic rocks transgressively intrude metamorphosed sedimentary rocks of the Transvaal Supergroup (dolomite, shale, ironstone) in the south and Archean granite in the north. The Platreef occurs along some 30 km at the contact of the mafic rocks with either the sediment or granite floor rocks.

The Platreef strikes northwest and dips 40° southwest; the sequence varies in thickness and has an irregular footwall contact and an undulating upper contact with the overlying gabbronorite, which is equated with the Main Zone. The Platreef consists of feldspathic pyroxenite with three subdivisions based on texture and mode. The top of the Platreef ("C") is fine-grained poikilitic feldspathic pyroxenite containing up to 70% clinopyroxene in places. This is underlain by coarse-grained feldspathic pyroxenite ("B") with between 50 and 90% orthopyroxene. Base metal sulphides are common to abundant and there is sporadic chromite. The "B" pyroxenite is the main ore zone. The lowermost pyroxenite ("A") is heterogeneously textured feldspathic pyroxenite of variable grain size and with sporadic base metal sulphide mineralization.

Xenoliths of metadolomite and calc-silicate, ranging from 1-100 m across, are scattered through the Platreef (Gain and Mostert, 1982). These are frequently rimmed by or contain concentrations of sulphide, often with high Cu, Ni, and PGE values.

The "B" pyroxenite ore sequence has a broad zonal structure based on the distribution of base metal sulphides and the PGE content; Cu and Ni range between 0.1-0.25% and 0.15-0.35% respectively. PGE range from <2.5 ppm to 15 ppm and sometimes up to 25 ppm; the Pt/Pd ratio is approximately unity. Four zones, based on the lateral continuity of the base metal sulphides and PGE contents greater than 3 ppm, have been delineated from drilling and evaluation for the present open-pit mine. The uppermost zone is the thickest (2-39 m) and the most consistent with respect to the PGE grade limit of 3 ppm. There is a general correlation of Cu+Ni with the PGE where the base metals total less than 0.1-0.2%. With Cu and Ni values greater than this the correlation is poor, and high base metal abundances are not necessarily indicative of high PGE content (Gain and Mostert, 1982). There is a correlation between the Cu+Ni content and alteration where serpentinization of pyroxenite is noted (Kennedy, 1994). The PGE content in the Platreef does not vary with the thickness of the mineralized pyroxenite. In the present operations reserves are 23.5 million tonnes at a grade of 5.73 ppm Pt+Pd+Rh+Au. Further ore reserves along strike and to a depth of 250 m are estimated to be 151 million tonnes at 4.98 ppm.

The PGM are isoferroplatinum, sperrylite, cooperite, merenskyite, and a wide range of semi-metal alloys. Au occurs as electrum and in places ranges from 5-11 vol% of the total PGM, significantly higher than in the Merensky Reef. The PGM are associated with base metal sulphides, and a higher proportion are encapsulated in silicates; the mineral composition and the textural relationship vary along strike and appear to be related to floor rock lithology (Kinloch, 1982). Contamination of mineralized sequences by floor rock metasediments and the xenoliths has occurred in places and this has affected the PGM composition (Kinloch, 1982; Kinloch and Peyerl, 1990). Platreef resting on sediments has sulphide and semi-metal PGM phases associated with the base metal sulphides, whereas Platreef with a granite footwall is dominated by Pt-Fe alloys as well as the semi-metal phases. The PGM are on average coarser grained than those of the Merensky Reef and UG2; a mode value is 52 μm with a range from 2-350 μm.

The geochemical and silicate mineral characteristics of the Platreef have been used as a model for the origin of sulphide concentrations at the margins of certain layered complexes (*e.g.* in Finland (Alapieti and Piirainen, 1984; Alapieti *et al.*, 1989)). Silicic contamination and sulphur addition from floor rocks is an inferred mechanism (Buchanan, 1988; Buchanan and Rouse, 1984). The evidence presented for silicic contamination in the Platreef is the "reverse fractionation" iron-enrichment trend in the orthopyroxene down through the “A” pyroxenite of the Platreef towards the lower contact; this trend occurs in all intrusions regardless of whether sulphides occur. The reverse fractionation trend exists in the “A” pyroxenite at the base of the Platreef (Buchanan, 1988). Barton *et al.* (1986) and Cawthorn *et al.* (1985), however, demonstrated that contamination in the Platreef is a post-emplacement event and was not a likely mechanism for liquating sulphides, and Lee *et al.*, (1989) concluded that the sulphide mineralization is primary. The tenor of Cu and Ni in the sulphides and the concentration of PGE in the Platreef is difficult to reconcile with a local floor-derived sulphur source. Ripley (1986) has recognized this problem in the Duluth Complex, where sulphide deposits are due to interaction between floor rocks and magma. The Cu and Ni are possibly magmatic in origin, concentrated into small amounts of sulphide from magma at depth in a secondary chamber, followed by fractional crystallization and sulphur addition from the floor rocks after intrusion at the present site. This may apply to the Platreef, and also to similar models invoking contamination as an overriding mineralization process. Contamination by material from mobilized floor rocks can best be considered an ore-deposit modifying process rather than a primary trigger process. The result is local redistribution and decoupling of base metals, sulphides, and the PGE, and development of the lower temperature semi-metal alloy PGM assemblages. That the PGE tenors appear high relative to the Cu and Ni contents for sulphide fractionation in a secondary magma chamber may be due to the influence of co-formed PGM. Barnes (1993) has observed that sulphide melt may not be the sole collector and host of all PGE and has suggested that the coexisting sulphide and silicate melts are saturated with respect to PGM, brought about through oversaturation of the sulphide melt in PGE. PGM nucleating from sulphide, in addition to PGE held in sulphide solid solution, would give an apparently strongly PGE-enriched sulphide fraction.

PGE in the Lower Zone, Main Zone and Upper Zone. Ultramafic cumulates in the Lower Zone south of Potgietersrus contain a number of minor sulphide-rich layers with elevated Pt and Pd. A pyroxenite zone about 20 m thick has up to 0.24% Ni, 0.11% Cu, 1.39 ppm Pt and 1.73 ppm Pd (Hulbert and von Gruenewaldt, 1982). The highest values are located in the lower 6 m of the sequence. This and the other layers have a systematic up-dip decrease in the Pt/Pd ratio, attributed to fractionation. The Pt/Pd ratio is similar to the Platreef.

PGE-bearing sulphides occur sporadically in the Main Zone (Wagner, 1929) of the eastern Bushveld Complex. Lenses of medium-grained gabbronorite with interstitial sulphides (pyrrhotite, pyrite, pentlandite, and chalcopyrite) about 3 m by 0.5 m conform to the layering. PGE grade is 0.1 to 2 ppm, and the PGM is sperrylite. One occurrence was large enough to be explored by underground workings.

Sulphides occur in small quantities through much of the Upper Zone of the Bushveld Complex together with trace (ppb range) PGE. A sulphide-enriched anorthosite is located below one of the lower magnetite layers; the sulphides are chalcopyrite, pyrrhotite, pentlandite, and pyrite, in decreasing abundance. Total sulphide varies, but on average is 3% (Harney and Merkle, 1990). Small (1-40 μm) PGM grains are bismuth-telluride group and sperrylite. The majority are enclosed in silicates, with the remaining 15% being associated with the base metal sulphides. Harney and Merkle (1990) suggest hydrothermal emplacement for these assemblages.

2.2. Stillwater Complex

2.2.1. JM Reef

The JM reef of the Stillwater Complex, Montana, USA, (Todd *et al.*, 1982; Bow *et al.*, 1982) is a coarse-grained olivine-bearing sequence 1-3 m thick, dipping at 65° north, within the plagioclase-dominant Banded Series (McCallum, this volume). The reef consists of troctolite, anorthosite and norite with 0.5-2% disseminated sulphide; chromite is rare. Plagioclase-rich fragments scattered in the reef may represent blocks of footwall rock or foundered hanging-wall rock. In analogy with the Merensky Reef, the JM reef is located above the stratigraphic level where plagioclase becomes a major cumulus phase.

The reef is discordant to the footwall and, similar to the Merensky Reef, a thin anorthosite marks this lower contact. Pothole-like structures have been recorded where, over a diameter of

6-10 m or more, the reef and the hanging-wall rocks transgress the footwall stratigraphy. The reef in places grades into an olivine-poor facies, where olivine is replaced by orthopyroxene oikocysts with relict olivine. Locally there are coarse-grained plagioclase-dominant lithologies with intercumulus orthopyroxene, clinopyroxene, and coarse anhedral olivine. Rocks below the JM reef contain 500-2000 ppm Cr, whereas the rock above the reef have low (<200 ppm) Cr (Todd *et al.*, 1982).

The footwall norite and gabbronorite are patchily mineralized, in particular the more plagioclase-rich lithologies. This footwall mineralization extends four metres below the olivine-bearing JM reef. The sulphides occur as finely disseminated pentlandite and pyrrhotite within the plagioclase rocks and in the gabbroic sequences as coarse-grained but isolated segregations surrounded by haloes of disseminated sulphide.

Major sulphides in the JM reef are chalcopyrite, pyrrhotite, and pentlandite, generally interstitial to the silicates, and locally rare <3 μm sulphide spheres are included in silicates. Other sulphides are pyrite, millerite, cubanite and violarite. Millerite, occurring as a replacement of pyrrhotite, is attributed to Ni-release during serpentinisation of olivine, and strong oxidation has formed magnetite-replaced sulphides (Volbarth *et al.*, 1986). An average composition of 13 JM reef samples is: S, 0.43%; Cu, 0.14%; Ni, 0.24%; Pt, 6.26 ppm; Pd, 24.04 ppm. A 5.5 km long portion of the reef gave an average grade of 22.3 ppm Pt+Pd over a thickness of 2.1 m. The average Pt/Pd ratio is 0.35, and this is constant throughout the reef, despite variations in PGE mineralogical control. The distribution of sulphide within the reef is erratic, but the thickness of the mineralized intervals is less variable. The sulphides aggregate into large clots in the lower parts of the reef, and grade upwards through net-textured to disseminated at the top (Todd *et al.*, 1982). Sulphide abundance tends to be top-loaded in the olivine-rich reef, and more evenly distributed within olivine-poor reef (Bow *et al.*, 1982).

A generally non-uniform distribution of the economic mineralization in the JM reef is shown by a spatial analysis of the of the sulphides and PGE (Raedeke and Vian, 1986). This analysis of 575 underground drill core intersections at 15 m intervals along the reef shows variations in grade, thickness, stratigraphic position of the mineralization, and the distribution of the mineralization in the plane of layering. The drillhole data covered about 1000 m of strike and delineated four zones relative to the contact of the footwall gabbronorite with the olivine-pyroxenite zone of the JM reef. The footwall contact surface is defined by a series of peaks and depressions (some of which may be potholes) with a total relief of some 45 m. Mineralization occurs in the footwall gabbronorite, with local high grades (values are not published by Raedeke and Vian) in the footwall anorthosite and norite as local enrichments of finely disseminated sulphide. Basal zone mineralization straddles the lower contact of the JM reef. A main zone of mineralization contains most of the 'higher grade' ore; subzones within the main zone define elongate to irregular shapes, spaced at 60-90 m intervals. All ore above the main zone is classified as upper zone ore; this is sporadically distributed over lateral intervals of 50-150 m. This spatial distribution study has been largely ignored in the genetic modelling of PGE sequences; however the data reflect the real situation in the JM reef and the distributions recorded by Radeke and Vian cannot be ignored in the modelling of the PGE reefs in general. These authors invoke turbulent magma mixing to account for the JM reef mineralization patterns.

Strong geochemical interrelationships between the PGE and the base metal sulphide elements are evident from detailed studies on the more highly mineralized lithologies of the JM reef (Barnes and Naldrett, 1985). As with the Merensky Reef there is a high intercorrelation amongst the six PGE, and Cu, Ni, and S, indicating the close association of these elements with the sulphide fraction. Correlation of Au with all these elements is poor. Of interest is the abundance of Ru, Os, and Ir and the fair to good correlation of these elements with the base metal sulphides and the PGE. Generally Ru, Os, Ir are controlled by laurite, especially associated with chromite; abundant chromite is not reported in the JM reef. Whole-rock Cr values reported by Todd *et al.* (1982) show scattered peak values of 2000 ppm Cr, correlating with higher Ni and Cu; this suggests Cr and Ru are controlled as solid solution in base metal sulphide; Cr partitions into sulphides in amounts which could account for the Cr values in the JM reef (de Villiers and Kleyenstuber, 1993). In spite of the significant correlation for the data set as a whole, there is scatter in the data, which increases with increasing sulphur and metal values. This scatter is least in the olivine-bearing rocks and greatest in the feldspathic pyroxenite, and appears to indicate lithology-specific PGE abundances related to variations in the metal content of the rocks. For each of the host rocks scatter is large and interelement correlation is poor when sulphur is greater than 0.6%. This pattern suggests local element decoupling, which probably most affected Au. Element redistribution is indicated in the mineralogical studies (Volbarth *et al.*, 1986) and is a possible reason for the poor correlation of Au with the PGE, Cu, Ni, and S noted by Barnes and Naldrett (1985).

Variation in the PGM assemblages was noted by Volbarth *et al.* (1986) in a mineralogical study of the 35 km of strike of the JM reef. The nature of the mineralization is influenced by the extent of the alteration rather than by the rock type. From fresh olivine-pyroxene rock the grades of alteration range from complete olivine serpentinization and partial orthopyroxene alteration with fresh plagioclase to completely serpentinized with late magnetite and pyrite with only ghosts and relics of the original rock. Pt occurs mainly as PGM, and most Pd occurs as solid solution in pentlandite, where an average of 0.77% has been detected (Todd *et al.*, 1982). Moncheite is a common PGM, attached to chalcopyrite at the border with silicates, and also pentlandite and sporadically pyrrhotite. Sperrylite is the next most abundant mineral mostly as small grains in silicates, and larger grains within base metal sulphide aggregates. The Pt-bearing PGM are cooperite, braggite, and vysotskite which as a group are the most abundant mineral phases. Vysotskite exhibits some affinity for oxidised and magnetite-replaced sulphides. Electrum is the most abundant of the alloys, apparently associated with the altered and serpentinized silicates, cobaltite, and magnetite-associated Fe, Ni sulphides; this relationship probably accounts for the poor geochemical relationship of Au with the PGE, Cu, and Ni. Pt-Fe alloys which, although not common, occur in serpentinized rocks, and are also associated with later alteration. Similarities with disturbed or potholed Merensky Reef are evident in this textural feature of the PGM. Graphite-rich coarse-grained pyroxene pegmatoid occurs beneath the JM reef; the pegmatoid contains a variety of rarer ore minerals, including rhenium sulphide. The variation in the PGM assemblages in the JM reef is only local, in contrast to the regional variations in PGM composition for Merensky Reef and UG2, which are related to footwall lithology (Kinloch and Peyerl, 1990).

2.2.1. Picket Pin mineralization

The Stillwater Complex has several stratabound sulphide sequences in addition to the JM reef, such as the Picket Pin towards the top of the Complex (Boudreau and McCallum, 1986). This deposit is located in anorthosite, overlain by troctolite. The mineralization has been traced along the 22 km exposed, and occurs as podiform or lenticular sulphide concentrations with 1-5% sulphide interstitial to plagioclase. Medium-grained anorthosite immediately below the troctolite is dominated by PGE-poor sulphide and the underlying coarse-grained anorthosite is characterized by PGE-bearing sulphide. Whole-rock Pt+Pd ranges up to 3.0 ppm, Cu is up to 0.8%, and Ni up to 0.4%. The PGE show a broad correlation with whole-rock sulphur, and the individual PGE, except for Pd, are significantly intercorrelated. Ore minerals are pyrrhotite, chalcopyrite and pentlandite; PGM are sperrylite and Pd semi-metal alloys. A hydrothermal origin for this occurrence is suggested (Boudreau and McCallum, 1986). Transgressive and podiform "pipes" cutting through as much as 50 m of the layered sequence occur in the vicinity of the Picket Pin mineralization: these can contain 1-5% PGE-bearing sulphide (Boudreau and McCallum, 1985).

2.2.2. Basal massive and disseminated sulphides

Sulphur in the Stillwater Basal Series pyoxenites ranges from 0.1 to 21%, with values up to 3.5% being most frequent (Page *et al.*, 1985a; Zientek *et al.*, 1986). The Pt/Pd ratio decreases from 1.13 in rocks with <0.1% sulphur to 0.19 in rocks with up to 21% sulphur. Pt ranges from 1 to 37 ppb and Pd from 0.5 to 210 ppb. Rh is also reported and this ranges from 0.5 to 11 ppb. Sulphides in the metasedimentary floor rocks contain PGE similar to the igneous rocks and appear to be related to the Basal Series. There is a general lack of correlation between PGE and S, except at the low S content; this is ascribed to varying equilibration of sulphide liquid with the silicate melt. It is pertinent to note that the Pt/Pd ratio of these Basal Series sulphides is similar to the ratio in the JM reef. This would suggest that the magma source influences the overall geochemical character of a mineralized zone of a layered intrusion and overrides the effects of any *in situ* magmatic fractionation. Later modification at sub-magmatic temperatures is a superimposed process affecting PGE ratios.

2.3. Great Dyke

The Great Dyke in Zimbabwe is described in detail (Wilson, this volume). The sulphide-PGE mineralization is located within websterite and pyroxenite close to the top of the uppermost pyroxenite layer of the ultramafic sequence. Plagioclase crystallization did not occur at or below the level of mineralization (Wilson and Tredoux, 1990) and in this respect the Great Dyke contrasts with many other layered intrusions. Gabbro overlies the websterite but the contact is a few metres above the mineralization threshold. The mineralized zone is 1-5 m thick and contains up to 7% sulphide. PGE and base metal sulphide mineralization extend down from the websterite into the pyroxenite of the unit below. There are no chromitite layers at this stratigraphic level. The characteristics of the Great Dyke sulphide and PGE mineralization in the Darwendale subchamber (Wilson and Tredoux, 1990) and the Wedza-Mimosa subchamber further south (Prendergast, 1990, 1991) are very similar.

In the Wedza-Mimosa area the main mineralization is a 2.5 m thick sulphide-pyroxenite defined on the bulk sulphide Cu + Ni content. The thickness is constant throughout most of the 30 square kilometre extent of the deposit delineated by drilling. There are wide lateral variations in the bulk metal content, inter-element ratios, and PGE contents per unit sulphide. From the axis of the intrusion to the margins bulk sulphide content, Cu/Ni and Pd/Pt ratios increase and Pd+Pt and PGE contents per unit sulphide decrease towards the margins. In addition there is higher variability of metal contents along the margins relative to the axis. Throughout the Wedza-Mimosa subchamber there is an apparent discordant relationship between the modal sulphide and the major silicate phase layering; the sulphide zone straddles the base of the websterite and the host-rock lithology varies from medium-grained websterite at the top to pyroxenite at the base. Pegmatoid is found in the mineralized zone towards the margin of the intrusion but, in contrast to the Merensky Reef, the association of mineralization and pegmatoid is rare. Evidence for magma currents possibly related to new magma addition during crystallization of the mafic sequence is cross-lamination in the pyroxenite and channels eroded into the underlying cumulates.

In the Darwendale subchamber the economically important sulphide zone is a laterally continuous layer over 200 km of strike and at the same stratigraphic level. Sulphides occur throughout the uppermost pyroxenite layer with the narrow interval of maximum sulphide enrichment located at the top of this pyroxenite layer, 1-5 m below or straddling the contact with the overlying websterite. The zone is variable in thickness with respect to the longitudinal axis of the intrusion. Further sulphide mineralization occurs within pyroxenite below this zone.

The abundance of sulphide is variable and is also dependent on the position in the subchamber. Within the main layer sulphide increases rapidly upwards from <0.3% to as much as 8% over less than 2 m. Thereafter sulphide decreases to less than 1% over a 2-5 m interval. Within the sulphide zone there is poor correlation of the highest Pt and Pd values; higher Pt is recorded in the sequence above the higher Pd values. There is no coincidence of base metal sulphide concentration peaks and the higher PGE concentrations. The underlying sulphide zones vary greatly in width and contain trace to about 1% sulphide.

The spread of exploration drillholes in the Great Dyke has permitted a regional two-dimensional evaluation of the economic mineralization in at least part of the Darwendale sub-chamber (Wilson and Tredoux, 1990). The maximum values for Pt in the mineralization profile increase southwards along strike, and with increasing distance from the longitudinal axis. For Pt values exceeding a cut-off grade of 0.1 ppm, the amount of Pt in the entire thickness of the mineralized zone in a 1.5 m^2 column increases towards the axis of the intrusion. The thickness of the PGE mineralization also increases from 1.5 m at the margin to more than 12 m in the axial facies. Indications from the borehole data are that the PGE peak values decrease and the total metal content becomes increasingly dispersed over the mineralized zone from margin to axis of the intrusion. The mineralization is not lithologically constrained, so these variations are not related to the thickness of a single rock layer; which is in contrast to the Merensky Reef.

Ore reserves for the Darwendale subchamber are estimated at 2,500 million tonnes. The Pt/Pd ratio is approximately 1.65 but reduces to unity in the axial facies. The PGE proportions are 52% Pt, 35% Pd, 9.5% Au for a 2.8 m thick section, but mining widths may be narrower than this and the proportions will vary accordingly. The Cu/Ni ratio is 0.83.

The silicate phases of the mineralized zone have, at the margin of the Dyke, been altered to an intergrown assemblage of biotite, tremolite, talc, chlorite and hornblende, and minor quartz, epidote, and magnetite. The sulphides are intimately associated with this alteration. Along the longitudinal axis of the intrusion primary cumulus silicate mineralogy is preserved. The PGM are dominated by sperrylite and semi-metal Pt, Pd alloys, and electrum (Coghill and Wilson, 1993; Prendergast, 1990). Laurite and other PGE sulphides are scarce; the rarity of laurite probably correlates with the lack of chromite. Pt-rich PGM phases are the most abundant. The majority of the PGM are located on base metal sulphide-silicate boundaries, 34% are enclosed in silicates, and less than 10% are enclosed entirely by sulphide. The PGM range from <4 μm to 10 μm in size, and are thus very fine-grained compared to the Bushveld Complex PGM. The principal base metal sulphide is pyrrhotite, with minor pentlandite, chalcopyrite, and pyrite. Generally the PGM are associated with any of these sulphides, except for pyrite. Individual sulphide grains range from 1-3 mm in composite aggregates to finely disseminated grains less than 0.1 mm in size. Sulphides are enclosed in pyroxene crystals and as interstitial phases to the silicates. No graphite has been observed in any part of the sulphide zone (Prendergast, 1990)

An important contrast between the Great Dyke main sulphide zone and the Merensky Reef is found in the patterns of variation for the total metal content. Wilson and Tredoux (1990) observed the PGE and the base metal sulphide concentrations decrease from the margin to the axis of the Dyke, as does the metal content for a unit area. In the Merensky Reef the metal content of the reef, as a geological entity, is constant, a fact established from mine-wide sampling over many years of operation. The Great Dyke main mineralized zone also has different geochemical characteristics with respect to the PGE interrelationships. The Merensky Reef has a generally high interelement correlation for Cu, Ni and the six PGE, and Au to a lesser extent, with little variation around the Bushveld. This is not the case for the main mineralized zone in the Great Dyke.

2.4. Munni Munni Complex

The Munni Munni Complex, Western Australia, consists of a cyclically layered ultramafic sequence overlain by a gabbroic layered sequence. Despite the gross analogies drawn between the Munni Munni and the Great Dyke, chromite is rare in the lower part of the Munni Munni ultramafic sequence, and the uppermost rocks are magnetite gabbros. Sulphide mineralization is located within the ultramafic layered sequence, where it is PGE-bearing, and also at the base of the intrusion.

The PGE-enriched base metal sulphide zone is a 1-5 m thick layer which has been traced over 12 km of strike. It is located in feldspathic websterite situated close to the top the ultramafic sequence rocks in a zone about 20 m below the overlying gabbro. The major sulphide mineralization in this Complex is, similar to the Great Dyke, below the level at which cumulus plagioclase enters the sequence, although postcumulus plagioclase is present. PGE (Pt+Pd+Au) average 2.9 ppm. The Pt/Pd ratio is around 0.5, with the lower parts of the mineralized sequence higher in Pd than the upper parts; Cu dominates the base metal sulphides, the average Cu/Ni ratio being 1.5. The metal content does not vary with thickness. The PGM are sulphide phases, semi-metal alloys, and native Pt and Pd as 2-20 μm grains closely associated with the base metal sulphides (Hoatson and Glaser, 1989; Hoatson and Keays, 1989). The semi-metal alloys and native metals are the dominant PGM. The majority of the PGM are enclosed in silicates or occur at the silicate grain boundaries, and the remaining 20% of the PGM are enclosed in or attached to base metal sulphides. The PGM vary in composition from Pt-Pd tellurides at the margin to arsenides, sulphides and other complex phases at the central part of the intrusion (Hoatson and Keays, 1989).

Barnes *et al.* (1990, 1992) recognize two patterns to the cumulus sulphide mineralization. In the first there is coincidence of the Cu-rich base metal sulphides (chalcopyrite-pentlandite-pyrrhotite) and PGE, whereas in the second, overlying mineralization, the PGE maxima are offset below the maximum base metal sulphide peak. The offset mineralization is further subdivided into a PGE-rich upper offset and a PGE-poor lower offset. In contrast to the Merensky Reef and the JM reef there is positive correlation between Au and Pt and Pd, but the peak values of Au and Pt+Pd do not correlate. Au abundance is highest at the top of the geochemical distribution profiles of the offset sulphide intersections (Barnes, 1993), where the concentration peak coincides with the start of the increase in Cu and the decline in Pt and Pd abundances. Ir, Ru, and Rh tend to follow Pd rather than Pt in the geochemical profile; the highest concentrations of these elements are below the Pt-Pd peak and these fall sharply to low levels before Pd decreases.

The stratigraphic position of the mineralization varies across the intrusion. The contact of the ultramafic and gabbroic series is straddled by the offset sulphide layer at the margin and is at the contact in the central part of the intrusion. The underlying coincident PGE-sulphide mineralization is displaced increasingly below the contact across the intrusion. The coincident mineralization is located below the level in the sequence where there is a reduction in the Cr content in clinopyroxene. This is similar to the Finland layered intrusions and the Stillwater Complex, where sulphide mineralization is also located where Cr abundances decline. The change in mineralization style is possibly related to an abrupt change in parent magma composition (Barnes *et al.*, 1992). Magma addition and mixing together with crystal fractionation, resulting in a high silicate to sulphide mass ratio to partition PGE into sulphide, are proposed to account for the mineralization in the websterite. Decoupling of the various precious metals and the base metal sulphides, and the apparently high PGE content of the sulphide liquid, is ascribed to saturation of the sulphide melt in PGE to give primary PGM (Barnes, 1993). As both the unmineralized and the mineralized rocks have the same pattern of precious metal abundance, and all are Pd-dominant (Hoatson and Barnes, 1989), the primary geochemical character of the intrusion is probably governed by the magma source.

Disseminated sulphides occur in gabbronorite and olivine microgabbro at the base of the intrusion. There is up to 6% sulphide consisting of pyrrhotite, chalcopyrite, and pentlandite, with sulphur isotope ratios typical of Archean magmatic sulphides. The origin is ascribed to rapid chilling against granite country rock, with subsequent local contamination (Hoatson and Keays, 1989).

2.5. Intrusions in Finland

Some 20 layered mafic intrusions occur in northern Finland along the Tornio-Näränkävaara trend, about 300 km in length (Alapieti *et al.*, 1990). The intrusions are similar in structure and lithology. The intrusions consist of a variable number of megacyclic units grading upwards from olivine or pyroxene cumulates at the base, through norite and gabbronorite to anorthosite at the top. Disseminated chromite and chromitite layers occur sporadically. All the PGE mineralization in the layered sequences is located in the plagioclase-dominant rocks and, similar to the Stillwater and Munni Munni Complexes, the sequences below the mineralized zones are Cr-rich and those above are Cr-poor.

The Penikat Complex has been tectonically disrupted and metamorphosed to greenschist facies. It consists of five west-dipping blocks. The internal structure consists of a marginal series and the layered series, with five lithologically similar and repetitive cyclic units. Seven PGE-bearing zones, or reefs, have been described (Alapieti and Lahtinen, 1986; Halkoako, 1994), with a variety of mineralization styles. PGE are associated with base metal sulphide mineralization, with or without chromite, or disseminated in silicates.

The Sompujärvi (SJ) reef (Halkoaho *et al.*, 1990a), the lowermost of the PGE sequences, occurs over the entire 23 km length of the Penikat Complex. The main rock types are norite to gabbronorite with chromite disseminations, lenses, and non-continuous layers. The SJ reef varies from tens of centimetres to several metres wide, with an average width of one metre. Halkoaka *et al.* (1990b) recognize two broad SJ reef types: one generally associated with chromite and the other with base metal sulphides. PGM also occur in rocks adjacent to the mineralized sequences and represent a contrasting mineralization style. The chromite-rich SJ reef has the higher PGE concentrations of all these reef types. A further subdivision exists to give a four-fold classification of the SJ reef (Halkoaho, 1994): a chromite type, a chromite-base metal sulphide type, a base metal sulphide type, and a PGE only type. The chromite type is richer in PGE, has a Pt/Pd ratio of 0.3, and is depleted in Ru. This is notable as Ru is invariably contained in laurite which consistently associates with chromite in layered intrusions. The SJ chromites are highly corroded and PGE alloys are closely associated with these corroded grains. The paucity of laurite in this reef may support an interpretation in which fluids moving along thermal gradients associated with the regional metamorphism redistributed sulphide and PGE to the sites now observed. Fluid enriched in Pd and Pt but depleted in Ru may have been trapped beneath the impermeable anorthosite layers; this fluid reacting with the chromite grains, which served as nuclei for PGM deposition, could account for the unusual mineral associations in this reef (*e.g.* Stumpfl, 1993).

The API and APII Ala-Penika PGE reefs are 250-450 m above the SJ reef. They are 20-40 m thick and have been traced over the entire 23 km of the Complex (Halkoaho *et al.*, 1990b). The API and APII reefs are located in the uppermost part of the gabbronorite megacyclic unit and the mineralization extends into the overlying poikilitic anorthosite of both reefs. Mineralization is similar in both reefs. Sulphide contents are 1-5 vol%, and occur interstitial to silicates. Pyrrhotite, chalcopyrite, pentlandite, as well as a range of Pb, Zn, and Co sulphides occur as minor phases. There is no chalcopyrite in the APII reef. Pd in pentlandite ranges from 0.05-0.08%. The topmost PGE reef of the Penikat intrusion is the Paarsivaara reef, also located at the top of a megacyclic unit (Huhtelin *et al.*, 1990). The mineralogy is similar to the Ala-Penikka reefs, the mineralization is erratic and frequently dispersed deep into the footwall.

The Portimo Complex consists of two intrusions (Narkaus and Suhanko-Konttijärvi) and the Portimo Dykes. Cyclic units lithologically similar to Penikat characterize the layered sequence of the Complex, but there is a more extensively developed marginal package. The Portimo Complex is host to a variety of PGE mineralization styles, more so than any other mafic complex so far described. Iljina *et al.* (1992) and Iljina (1994) document six types:

1) PGE-enriched base metal sulphide (pyrrhotite-chalcopyrite-pentlandite) disseminated in the Suhanko-Konttijärvii marginal rocks.
2) Massive pyrrhotite close to the basal contacts of both the Portimo intrusions.
3) Two reef-like PGE sequences within the layered series of the Suhanko body almost devoid of base metal sulphide (0.1%) and 30 cm to 10 m thick (the Rytikangas reef) and in the Narkus body also sulphide deficient, one to several metres thick with barren interlayers (the Siika Kämä reef).
4) Cu-PGE mineralization offset below the Narkus intrusion, of limited extent and the most PGE enriched.
5) Sulphide bearing, chalcopyrite-dominant, mineralized "dykes" within the granite country rocks.
6) Scattered local PGE enrichments associated with mafic pegmatite, pyroxenite pegmatite pipes, and massive to disseminated chromitite.

In all these occurrences and reefs the PGM are dominated by Pd-arsenides, Pd-bismuth-tellurides, and similar semi-metal minerals. Sulphide and metal alloy phases are rare, except for Sn and Pb alloys. Mineral grain size is small, 5 -15 μm, and generally the grains are ragged to anhedral in habit. All the sequences are Pd rich and the Pt/Pd ratio ranges from 0.3 to 0.13 with Pd ranging from 50 ppb to 2 ppm and locally as high as 20 ppm. The metal content increases with increasing width in most of the mineralized zones.

Cu-Ni-PGE mineralization is located in the marginal series of the Koillismaar Complex, northeast Finland (Alapieti and Piirainen, 1984). Cu and Ni are 0.4 to 1.0%, with Cu/Ni ratio of 5 to as high as 30. Pt, Pd, and Au are each less than 1 ppm, and the Pt/Pd ratio is approximately 0.3. The marginal rocks have the characteristic Fe-enrichment reverse fractionation trend in the orthopyroxene, and more Na-rich plagioclase, than the overlying layered rocks. A contamination model, similar to the Platreef of the Bushveld Complex, has been invoked to account for this mineralization.

2.6. PGE and base metal sulphides in other intrusions

Several examples of mineralization in smaller layered mafic intrusions are found in the Skaergaard Intrusion, Greenland, the Muskox Intrusion in northern Canada, the Dovirensky Layered Complex in Siberia, and the intrusions of the Kola Peninsula. The Sudbury intrusion is an example of a mafic intrusion with large reserves of massive and disseminated sulphides with generally low PGE contents (Pye *et al.*, 1984), although locally high concentrations of Pd and Pt occur as PGM semi-metal alloys (Li and Naldrett, 1993). The massive and disseminated sulphides of the Noril'sk intrusion have significant PGE contents (Czamanske *et al.*, 1992; Lightfoot and Naldrett, 1994). Reviews of other intrusions with emphasis on the PGE are given by Hulbert *et al.* (1988).

2.6.1. Skaergaard Intrusion

The Skaergaard Intrusion accumulated Cu-rich immiscible sulphide during the final stages of fractionation (Wager *et al.*, 1957). A number of Pd-Au enriched sequences are located over a 20-45 m thick interval, below the Triple Group (Bird *et al.*, 1991). The mineralized zone is continuous throughout the intrusion, as deduced from chip sampling and diamond drilling. The sequence is Au-dominant in the uppermost parts and Pd-dominant in the lower part. Au exceeds 1 ppm over a 2-3 m thick interval, with sporadic values of 20 ppm. The Pt/Pd ratio averages about 0.1. A Pd dominant zone beneath the Au zone has a Au/Pd ratio of around 0.5 and the Pt/Pd ratio is 0.1. Two Pd-rich zones occur below these sequences, the lowermost being the 3 to 6 m thick Pd rich sequence with up to 2 ppm Pd, and some values up to 6 ppm.

Au and Pd occur as anhedral and subhedral Au-Cu, Au-Pd, and Pd-Cu alloys, with small amounts of Fe and Ag, 1-75 μm in diameter, poikilitically enclosed at the rims of plagioclase and pyroxene and in association with a variety of sulphides such as bornite, digenite, and chalcopyrite (Bird *et al.*, 1991). The Au/Cu ratio is very variable.

The mineralized layers do not coincide with any distinct lithological layers. There appear to be systematic variations in the Pd-Au concentrations relative to the assumed centre of the intrusion; the numeric value of the Au grade multiplied by the width decreases with increasing distance from the centre of the intrusion (Bird *et al.*, 1991). This variation in metal content with the geometry of an intrusion has also been noted for the Great Dyke (Wilson and Tredoux, 1990), and it may well be a characteristic feature of smaller or linear-shaped intrusions, not evident in a large complex such as the Bushveld.

2.6.2. Muskox Intrusion

The Muskox Intrusion in the North West Territory of Canada has PGE-bearing chromite and base metal sulphides in the marginal rocks (Irvine, 1988; Barnes and Francis, 1993). Two thin chromitite layers with about 50% chromite and 3-6% sulphides contain 300-1000 ppb Pt+Pd, with a Pt/Pd ratio of 0.28. These are associated with S (1.62%), Cu (0.32%), and Ni (0.25%) values, comparable to the higher sulphur UG2 chromitite of the Bushveld Complex.

Cu (and S) occurs erratically throughout the intrusion, generally in the range 50-1200 ppm. Pd and Pt (Pd>Pt) are correlated with Cu in the ultramafic rocks and the chromitites. Marginal rocks contain 1-3% sulphide with 0.2-0.5% Cu+Ni and 100-300 ppb Pt+Pd. High Cu (10%) and PGE (100 ppm Pt+Pd) are recorded in the massive to semi-massive sulphide concentrations at the footwall contact of the intrusion. High Cu and low PGE occur in the uppermost gabbros. Extensive exploration has not delineated economic concentrations of PGE or base metal sulphides in any of these sequences.

2.6.3. Dovirensky Layered Complex, Siberia

The Dovirensky Complex is one of a number of mafic-ultramafic intrusions in the 1500 km belt of supracrustal sequences between the Archean Aldan Shield and the Siberian Platform (Papunen *et al.*, 1992). This complex is about the size of the Stillwater Complex, 26 km long and 3.5 km wide with a subvertical dip. It consists of dunite grading upwards into troctolite and gabbro. Nickeliferous (0.3%) base metal sulphides occur in portions of the lower marginal sequences; pyrrhotite and pentlandite and rare chalcopyrite are the main phases. The PGE are Pd-rich with estimated Pt/Pd ratios of 0.09 to 0.05; highest Pd correlates with high Cu in the marginal and brecciated ores.

2.6.4. Kola Peninsula

More than 40 layered mafic intrusions occur in two zones in the Kola Peninsula of Russia which are significant geographic extensions to the Finland domains. The mineralization is base metal sulphide-PGE in the older intrusions, whereas the younger intrusions contain chromitite and V-Ti deposits and are low in sulphide.

The Fedorovo-Pansky intrusion is a metamorphosed layered gabbroic and ultrabasic sequence up to 3000 m thick. Four mineralized zones are recorded (Balabonin *et al.*, 1994). These are a lower marginal zone, a lower layered sequence which is the most significant, in the gabbronorite, and in the upper part of the intrusion. The PGE and base metal sulphide mineralization in the lower layered sequence occurs in several lithologies, but dominantly at the contacts between plagioclase cumulates. The sequence is Pd-dominant (Pt/Pd ratio around 0.2) and the Cu/Ni ratio is 1.2. Average Pt is 0.33 ppm and Pd is 2.31 ppm. Cu+Ni average is 0.3%. The base metal sulphide are pentlandite, pyrrhotite, chalcopyrite, and locally Cu-rich chalcopyrite, millerite, and bornite.

The Imandra Complex (Zhangurov *et al.*, 1994) is a 3000 m thick layered intrusion with Cu, Ni, and PGE mineralization associated with chromitite. Sulphur is low. Pd (0.29 ppm) exceeds Pt (0.19 ppm), and, consistent with the chromite association, Ru exceeds these values and averages 1.5 ppm.

3. PLATINUM-GROUP ELEMENTS AND CHROMITITES

3.1. UG2 chromitite, Bushveld Complex

The UG2 (Upper Group 2) chromitite layer in the Bushveld Complex is probably the largest PGE resource on Earth (Vermaak, 1985). It occurs 15-400 m below the Merensky Reef; this vertical separation is smallest in the western and greatest in the eastern Bushveld. Potholes are a common feature of the UG2 chromitite, and are described from all mines on this layer (Hahn and Ovendale, 1994). These features are generally circular in shape and range from less than 10 m to more than 50-100 m in diameter; in many respects they are similar to the Merensky Reef potholes. There is no spatial correlation of potholes in the Merensky Reef to those in the

UG2 (Cawthorn and Barry, 1992), but large scale disruptions to the layering often affect both sequences, particularly at Union mine where the stratigraphic separation is around 15 m.

The UG2 sequence consists of 0.5-1.0 m thick chromitite with a pegmatoidal feldspathic pyroxenite footwall in most cases, and more rarely anorthosite. Two to four minor (<1 cm) chromitites occur in the hanging-wall feldspathic pyroxenite. Chromite content is 60-90%; the average Cr/Fe ratio ranges between 1.26 -1.4 with 43.5% Cr_2O_3. PGE contents (Gain, 1985) are up to 10 ppm PGE+Au (3.6 ppm Pt; 3.81 ppm Pd; 0.33 ppm Rh; others 2.26 ppm). Cu and Ni are low, generally less than 0.05%, and the amount of base metal sulphide is low, but these are present throughout the layer. Locally sulphides can be more abundant without affecting the overall PGE tenor. Proven reserves of the UG2 chromitite for the Rustenburg, Union, and Amandelbult mines combined are 390.53 million tonnes at a grade of 5.60 ppm Pt+Pd+Rh+Au, and probable reserves are 406.3 million tonnes at 5.49 ppm. The Atok mine in the eastern Bushveld Complex has UG2 reserves of 65.5 million tonnes at 6.76 ppm and probable reserves of 151.5 million tonnes at 6.28 ppm. As with the Merensky Reef these values are for a mining width and not for the geological entity, and are not applicable in modelling.

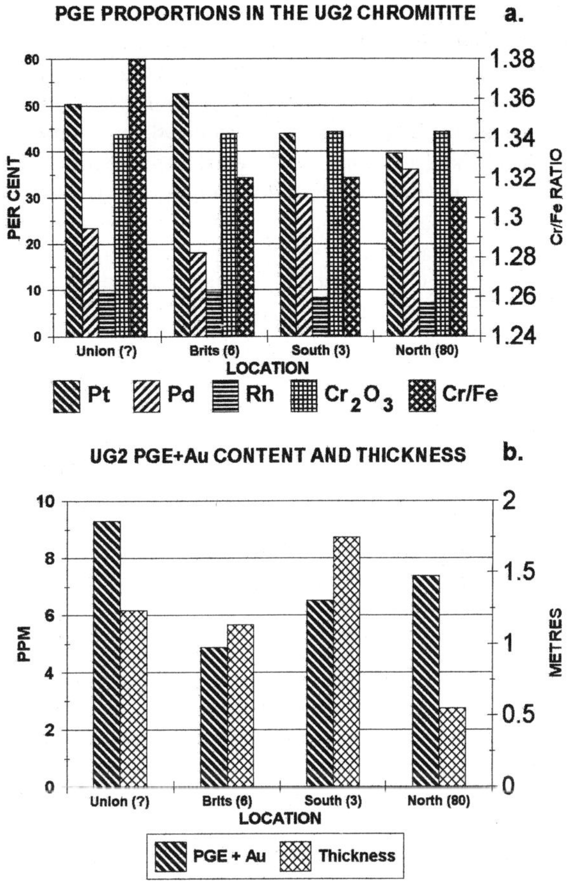

Figure 3. Diagrams illustrating regional composition variations of the UG2 chromitite, for the number of data points shown, in the Union and Brits area of the western Bushveld, and in the areas north and south of Steelpoort in the east Bushveld: (a) the proportions of PGE in the UG2 chromitite and the chromite composition; (b) PGE + Au content and thickness of the UG2. Data from Vermaak (1985).

There are frequently two peaks in the PGE distribution of the UG2 chromitite layer (McLaren and de Villiers, 1982; Gain, 1985, 1986; Hiemstra, 1985, 1986). The precious metals are concentrated towards the base of the layer, and also at the middle or towards the upper contact. This break in the PGE distribution appears to correlate with a textural change in the chromite-silicate gangue from a granular silicate-chromite to a poikilitic silicate-chromite texture, best displayed on slightly weathered exposures. The Pt/Pd ratio commonly changes vertically up the layer together with PGE abundance and also with changes in the PGM composition (McLaren and de Villiers, 1982); the ratio can be low in the lower half and high in the upper half, or it can be constant throughout. It could be speculated from the textural zoning and the bimodal PGE distributions that the UG2 is a composite chromitite layer made up of two successive chromitites without the development of intervening silicate layers.

The Pt/Pd ratio varies with geographic location: an average Pt/Pd ratio for the layer in the eastern Bushveld is 0.94 (Gain, 1985) and in the western Bushveld is 2.5. Pt and Pd for two western Bushveld UG2l chromitite intersections (Hiemstra, 1985, 1986) show high Pt/Pd ratios (5-15) towards the base of the layer, due to the low Pd. The Pt/Pd ratio progressively declines to 1 or 2 up-layer as Pd increases in abundance. For Pt/Pd ratios greater than 5 there is a negative relationship between Cu content and the Pt/Pd ratio, with the higher Pt/Pd ratios correlating with lower Cu abundances. These relationships suggest substantial upward redistribution of Pd-bearing phases or, more probably, of Pd in solid solution in base metal sulphides, while Pt remains as PGM in the low sulphide part of the UG2 chromitite.

The proportion and distribution of PGE in the UG2 are illustrated in Figures 3a and 3b. In the four, geographic areas depicted the Pt/Pd ratio and the Rh content are higher in the west of the complex than in the east. There is no correlation between PGE proportions or the total metal content and the composition of chromite at the ore grades of the UG2. The total PGE+Au and the thickness of the layer show the grade is higher in the thinner than in the thicker facies, in both the east and the west sectors (Figure 3b). There is thus some evidence that the precious metal content in the UG2 is constant, similar to the Merensky Reef.

The PGM are invariably interstitial to the chromite grains, and the only PGM commonly enclosed by chromite is laurite. PGM grain sizes are the smallest of the Bushveld mineralized sequences, with an average diameter of 9.3 µm. In spite of the low sulphide content, most PGM are attached to small base metal sulphide grains so the PGE may still be considered sulphide-associated; PGM also occur within the orthopyroxene gangue.

The association of Ru and chromite is noted as a consistent pattern in the Bushveld chromitites, and laurite is recorded throughout the UG2 (Kinloch, 1982). Regionally the UG2 PGM assemblage is dominated by Pt-Pd sulphides and laurite. The Rh content is proportionally and significantly higher in the UG2 compared to the Merensky Reef. Rh is contained in a widely occurring but unnamed PGM with a variable composition approximating Pt-Cu-S(Rh,Ir,Pd,Ni) which, with other unnamed Rh-PGM, can comprise 6-10% of the total PGM assemblage in the UG2. This mineral is a major host for and source of Rh in the Bushveld Complex. Isoferroplatinum and semi-metal PGM alloys occur where the UG2 is potholed and in areas with ultramafic replacement pegmatite. Peyerl (1982) has documented a progressive change in the PGM assemblage from sulphide phases some distance from the Driekop dunite pipe to PGM alloys close to the pipe. The associated base metal sulphides are replaced by bornite and troillite. This zonal alteration of PGM is ascribed to the influence of pipe-related volatiles, deduced by Schiffries (1982) to be chlorine-rich.

A sulphide-rich facies of the UG2 is recorded from the northwestern sector of the eastern Bushveld Complex (von Gruenewaldt *et al.*, 1990), in which the base metal sulphides are pyrite, pentlandite, chalcopyrite, and pyrrhotite. Cu + Ni attain 2000-5000 ppm and the PGM assemblages are typical of the UG2 elsewhere. The PGE content is not affected by the increased sulphide abundance.

3.2. PGE in other chromitites

Chromitites in layered mafic complexes are PGE-enriched compared with the adjacent silicate host rocks. A detailed study by Lee and Parry (1988) of the Bushveld Complex chromitites demonstrated that the PGE group Ru-Os-Ir is consistently present in the range 1-3 ppm in the Lower Group (LG), Middle Group (MG), and UG1 chromitite layers. The PGE group Pt-Pd-Rh is low in the LG and the lower MG chromitites in pyroxenite, whereas the MG chromitites within plagioclase cumulates are enriched in Pt-Pd-Rh in the 3-5 ppm range. Lee and Parry also noted that there is a relationship between modal chromite and the PGE content. Above 75 vol% chromite the Pt-Pd-Rh suite decline relative to Ru-Os-Ir, indicating contrasting textural controls on the overall PGE abundance. Similar geochemical trends are recorded for the LG and MG chromitite layers of the western Bushveld Complex (Teigler, 1990; von Gruenewaldt *et al.*, 1986).

A genetically controlled correlation between chromite composition and PGE distribution, in which the PGE increases in the chromitites are related to the upward changes in chromite composition, is suggested by Teigler and Eales (1993) for chromitites of the western Bushveld Complex. In this study no attention was paid to texture, silicate gangue proportions or the PGM phases and mineral association, all of which influence PGE proportions and abundance levels (Lee and Parry, 1988; Kinloch, 1982).

Chromitite layers (LG, MG, UG1) from two boreholes in the western Bushveld Complex, contain elevated amounts of PGE (Scoon and Teigler, 1994), and are poor in base metal sulphides. The LG chromitites have trace S and the MG and the UG1 have 0.01-0.02% S. PGE concentrations are higher (up to 2200 ppb) in the MG and the UG1 compared to the LG chromitites (up to 1050 ppb). The Pt/Pd ratio of these chromitites ranges from 1.5 to exceptional values of 17. This ratio is significant as it reflects the overall Pt-rich character of the Bushveld Complex mineralized sequences and silicate rocks. There is no suggestion of Pt and Pd fractionation with stratigraphic height in these chromitites. The very high Pt/Pd ratios in some of the chromitite samples is indicative of subsolidus decoupling in a process whereby Pd in solid solution with base metal sulphide is displaced from Pt-bearing PGM during annealing and densification of the chromitites. The MG chromitites and the UG1 are proportionally higher in Pt+Pd+Rh and the LG chromitites proportionally higher in Ru+Ir+Os. This pattern has been discussed by Lee and Parry (1988) for similar chromitites with respect to the modal volume of chromite and reduction of intergranular space, the effect this has on the amount and type of sulphide-related PGM, and thus the PGE metal ratios.

The composition and textural association of the PGM is a significant factor in interpreting the geochemical behaviour of the PGE in chromitites. The PGM of the MG chromitites with more than 3 ppm PGE+Au is Pd-rich braggite, cooperite, unnamed Pt-Cu-S(Rh,Ir,Pd,Ni), PtFe alloy, and Pt(or Pd)RhSAs (Lee and Parry, 1988). These are enclosed by, or are attached to, the base metal sulphides chalcopyrite, pentlandite, pyrite, and millerite. The sulphides are interstitial to the chromite grains and trace amounts occur within chromite or the interstitial silicates. Laurite comprises 30-50% of the PGM, and typically occurs enclosed in chromite, and is thus relatively insensitive to the reductions in intergranular space which occur on annealing. Merkle (1992) documented the PGM, chromite and base metal sulphide associations in the MG chromitites of the western Bushveld Complex, in sequences similar to those analyzed by Scoon and Teigler (1994). Laurite is enclosed in chromite, and laurite with other PGM sulphide phases are located at chromite-silicate interfaces. PGM sulphides and semi-metal PGM phases are associated with intercumulus base metal sulphides. The range and composition of the sulphide and semi-metal PGM phases increases in complexity up the stratigraphic sequence. Laurite enclosed in chromite is the dominant association in all the chromitite layers. These PGM variations are reflected in the bulk geochemistry of the chromitites described earlier. The PGM are 3-5 μm in size, and Merkle (1992) noted the problem associated with optical identification of very small PGM grains, as well as the difficulty of obtaining quantitative electron-microprobe analyses. Base metal sulphides are chalcopyrite and pentlandite; Cu in the MG chromitites ranges from 20-40 ppm, indicating the generally low sulphide content of these rocks. Merkle (1992) concluded that the sulphide mineralogy (dominance of chalcopyrite and pentlandite, replacement of pentlandite, presence of pyrite and lack of pyrrhotite) cannot represent a primary magmatic assemblage, and that S has been lost from the system or was hydrothermally modified, and Cu and Ni combined with PGE, in particular Rh, to form the un-named Rh-rich PGM. The PGM semi-metal phases probably formed in a similar way. These modifying events complicate the interpretation of whole rock PGE geochemistry, especially in chromitites.

A PGE-bearing chromitite layer rich in sulphide occurs in harzburgites of the southern part of the of the Potgietersrus sector of the Bushveld Complex, at the top of 1600 m thick sequence of ultramafic cumulates. The mineralized rock is about 40 cm thick and consists of 3-4% sulphide with 3.5-6 ppm PGE+Au, 0.24-0.27% Ni and 0.11-0.19% Cu (Hulbert and von Gruenewaldt, 1982).

Stillwater chromitite layers contain variable but low concentrations (generally 50 ppb, exceptionally up to 16 ppm) of Pt+Pd+Rh+Ir +Ru, with the highest values in the "A" chromitite layer, close to the base of the ultramafic sequence (Talkington and Lipin, 1986). The Pt/Pd ratio is 0.6 in the "A" layer and in all layers Pd>Pt>Rh>Ru >Ir. Significantly this Pd-dominance is characteristic of the complex as a whole, despite the low abundances. Laurite is the only

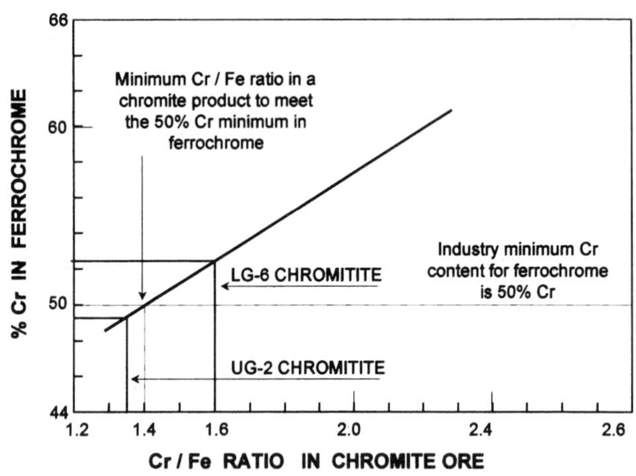

Figure 4. *Using the Bushveld Complex chromium ores as an example, this diagram illustrates the interrelationship of the Cr/Fe ratio in a processed chromite ore as feedstock and a ferrochromium product. On the 1994 industry standards the LG6 chromitite provides an acceptable ore for a 52% ferrochromium product. The UG2 chromite, produced as a by-product of PGE mining, is below the 1995 industry minimum for ferrochromium production without being used directly as a feed to stainless steel manufacture.*

PGM inclusion identified in the chromites, generally less than 20 μm in size; it also occurs at chromite grain boundaries. No relationship between the Ru content and the stratigraphic position of the chromitite layer has been found. Despite the PGE abundance reported and the systematic variation in PGE ratios, very few other PGM are recorded. Laurite is the dominant phase and the other PGM are PGE arsenides; traces of base metal sulphide probably contain Pd, Pt, and Rh in solid solution, reflected in the PGE ratios.

The Niquelandia Complex, Brazil, is a layered intrusion consisting of chromite-bearing pyroxenite and harzburgite cyclic units overlain by gabbro. The chromitites are a series of thin layers in about 1 m thickness of silicates. Pt+Pd+Rh (up to 3.4 ppm) and Ru+Pt+Pd (up to 170 ppb) associated with Au and Ag have been recorded (Ferrario and Garuti, 1988; Millioti and Stumpfl, 1993). Laurite, and the Os-rich phases erlichmanite and iridosmine, and Pt-Fe alloys are small inclusions (<10 μm) within the chromitite grains, and also in the surrounding dunite. Pt and Pd phases are bismuth-tellurides associated with minor base metal sulphides, or enclosed in silicates. Secondary processes such as serpentinization are considered agents for PGE redistribution.

Sulphide-bearing chromitite (the C1d layer) in the Great Dyke (Evans *et al.*, 1994) has low but detectable levels of Pt, Pd, and Rh coincident with sulphides in the silicate rocks beneath the chromitite. Pd exceeds Pt throughout the profile. Ru, Ir, and Os are highest in the chromitite layer, and correlate with the Cr content; this indicates control by a PGM such as laurite, included in chromite grains. PGM in the sulphide-rich silicates are Pt-Bi-Te alloys associated with the interstitial sulphides.

4. PLATINUM-GROUP ELEMENTS IN MAFIC AND IRON-RICH PEGMATITE

While discordant iron-rich and mafic to ultramafic pegmatites are widely distributed in the Bushveld Complex (Viljoen and Scoon, 1985; Scoon and Mitchell, 1994), and similar magnesium-rich and iron-rich bodies are recorded in the Stillwater Complex (Boudreau and McCallum, 1985) and the Finnish complexes, only three PGE-bearing occurrences with significant PGE enrichment are known and were mined. The Onverwacht, Mooihoek, and Driekop dunite pipes are discordant bodies orthogonal to the plane of igneous layering in the Critical Zone of the eastern Bushveld Complex (Figure 2). Close to the pipes the igneous layering of the country rocks dips steeply downwards towards the pipe. PGE are restricted to the iron-rich olivine cores of the concentrically zoned pipes. PGE grades were around 5-6 ppm with irregular patches up to 80-100 ppm. Xenoliths of Lower Group chromitites, mined from the Onverwacht and Mooihoek pipes, contained high concentrations of PGE compared to the chromitites in the adjacent layered rocks (Wagner, 1929). The platinum-group mineralogy of the Driekop pipe is described by Tarkian and Stumpfl (1975), Stumpfl and Rucklidge (1982), and Rudashevsky *et al.* (1992). The Pt/Pd ratio is very high and PGM concentrates contain dominantly iron-bearing Pt, sperrylite, hollingworthite and irasite, and a variety of rare PGM, all considered to represent post-magmatic low-temperature assemblages.

Iron-rich Pt alluvial grains concentrated in streams draining the Mooihoek pipe led to the discovery of the pipes and recognition of the PGE potential, and subsequently focused the attention of prospectors onto the Merensky Reef, discovered in the vicinity in 1924.

5. OXIDE DEPOSITS

5.1. Chromite

Early Precambrian major mafic layered intrusions are unique in terms of size and the amount of chromite mineralization (Roberts *et al.*, 1990; Stowe, 1994). About 90% of the world's proven economic chromite reserves occur in the two Southern African intrusions, the Bushveld Complex and the Great Dyke. The only limiting factors concerning reserves are ore quality and amenability to mining.

Figure 4 illustrates the effect chromite composition has on potential for exploitation; in the case of the Bushveld Complex chromitites in which Cr/Fe ratio, Cr content of a ferrochromium product, and the quality of the processed chromite ore are compared. A Cr/Fe ratio of 1.4 (in lumpy ore and in fine-grained chromitite cleaned of the interstitial silicate) is the minimum acceptable to the industry at present: this would give a "Grade C" ferrochromium product. The LG6, MG1, and MG2 chromitite layers are sufficiently thick and have Cr/Fe 1.4, and are therefore mined on a large scale. Regional variations in chromitite chemistry (Figures 5 and 6) and layer thickness are additional important economic considerations in the Bushveld and any other layered complex. By-product chromite is produced in substantial amounts from the UG2 chromitite, mined for the PGE content, and presently no market has been developed for the low-Cr-grade chromite from

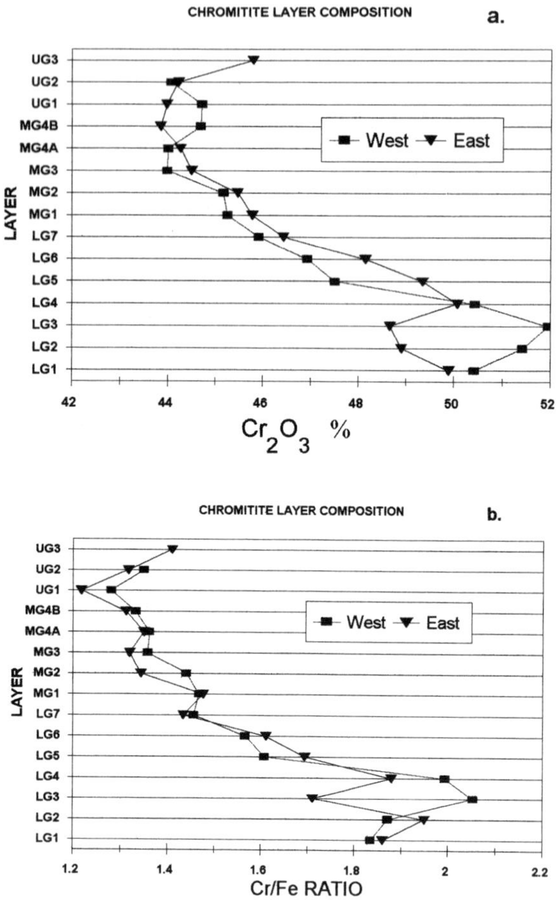

Figure 5. *The subdivision of the chromitite layers of the Bushveld Complex. Note the regional contrasts in (a) Cr_2O_3 content and (b) the Cr/Fe ratio, the upward three-fold grouping of the variation trends, and the occurrence of an additional chromitite layer (UG3) in the eastern Bushveld Complex (data from Cameron, 1977; Vermaak, 1985). The regional variations affect the economics of mining in different parts of the Complex. The same considerations would apply to any mafic complex.*

this ore. The UG2 has an average of 44% Cr_2O_3 and a Cr/Fe ratio of 1.35 which gives rise to a 48% Cr ferrochrome alloy, below the lower limit (50%) the market will presently accept for charge ferrochrome alloy. There is potential however for a UG2-derived ferrochrome to be used as direct feed into, for example, a stainless steel manufacturing plant, which is the case at Kemi in Finland, where chromite with 26% Cr_2O_3 is utilized (Duke, 1983).

5.1.1. Bushveld Complex

Chromitite layers in the Bushveld Complex occur in the pyroxenite-harzburgite sequences of the Lower Zone in the Potgietersrus area and in the far western Bushveld Complex. The chromitites of the lower and upper Critical Zone (Cameron, 1977, 1980) are the most extensive and occur as three stratigraphically delineated clusters of layers (Hatton and von Gruenewaldt, 1987, 1990). The Lower Group (LG) comprises seven chromitites, all within feldspathic pyroxenite of the lower Critical Zone; the thickest layer, the LG6, 0.8 -1.2 m thick, is mined in several parts of the Complex. This and almost all other layers can be traced through the entire complex. Four major and several minor (1-5 cm) layers make up the Middle Group (MG), which straddles the boundary of the lower and upper Critical Zone, defined by the first appearance of cumulus plagioclase. The two (in the western Bushveld) or three (in the eastern Bushveld) layers of the Upper Group (UG) occur in upper Critical Zone norite- and anorthosite-dominated sequences. Chromite in the plagioclase cumulates is a departure from the pattern found in the Great Dyke and the Stillwater Complex.

Figures 5a and b illustrate Cr_2O_3 and Cr/Fe variations for Bushveld Complex LG6 chromitite. The LG6 has a shallow dip of 8-12° and is thus amenable to trackless and conventional mining.; trackless mining techniques are widely used. MG chromitites are also exploited, in particular in the western Bushveld Complex where these chromitites are thicker and have higher Cr/Fe ratios compared to the eastern Bushveld sequences. In the LG6 chromitite layer the Cr_2O_3 content of 46-48% and the Cr/Fe ratio of 1.56-1.6 are both higher in the east of the Complex, which is the major contributor to the total chromite reserves of the complex. The highest Cr/Fe ratios (2.13-2.83) occur in the Lower Zone hosted chromitites in the Grasvally area south of Potgietersrus and far western parts of the Bushveld Complex at Groot Marico (Figure 6).

Data for 19 successive chromitite layers of the Critical Zone in the western Bushveld show Ti, Fe^{2+} and Al increase and Cr and Mg decrease with stratigraphic height (Teigler and Eales, 1993). Three geochemical groups of Critical Zone chromitite layers, as distinct from the stratigraphically defined grouping, have been described (Teigler and Eales, 1993): type 1a (LG1-LG4, Cr/Fe > 1.75; $Fe^{3+} < 1.45$; Ti < 0.1), type 1b (LG5-MG1, Cr/Fe < 1.72; $Fe^{3+} > 1.48$; Ti > 0.1), and type 2 (MG2-UG2; Cr/Al < 1.73; Ti > 0.13). This grouping is broadly shown by the variation of Cr_2O_3 and the Cr/Fe ratio depicted in Figure 6. These parameters increase in value up to the level of the LG3; from the LG4 to the MG1 they fall steadily up through the pyroxenite sequence. At the MG2, where cumulus plagioclase becomes abundant, there is a sharp decrease and the subsequent upward composition change is small and erratic. An inflection in the Sr-isotope profile at the MG chromitites suggests a change in magma composition and magma addition at this level in the intrusion (Hatton *et al.*, 1986).

Chromium spinel compositions are sensitive to several geochemical effects and thus generalizations based on composition are acceptable only in the broadest sense (Eales, 1987). Factors such as the composition of the magma, the modal ratio of silicate to chromite, the composition of the associated silicate liquidus phase, and subsolidus reaction affect chromite composition.

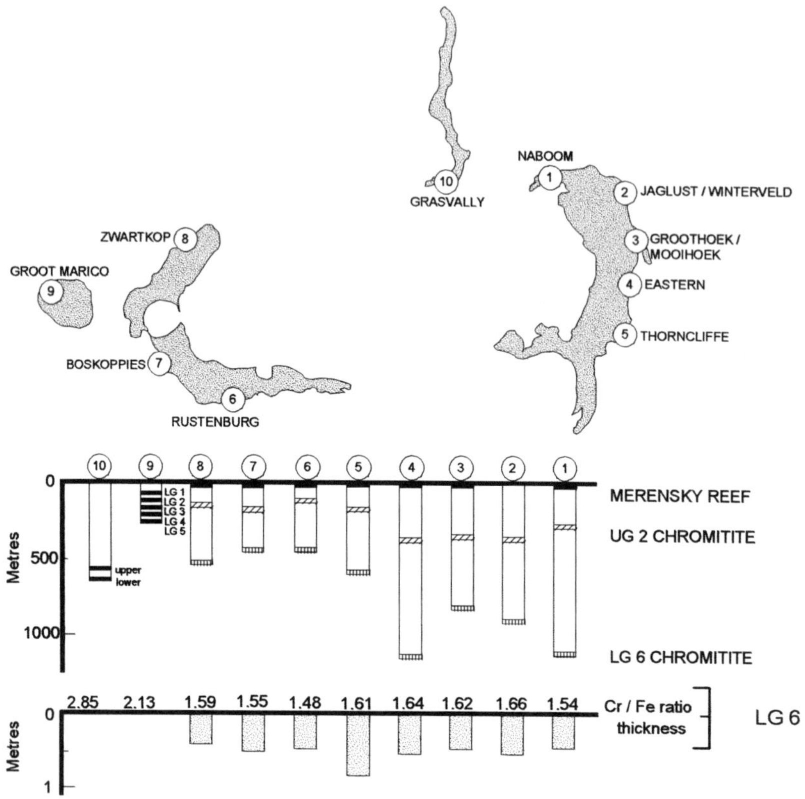

__Figure 6__. The depth of the LG6 and UG2 chromitites below the Merenksy reef in different sectors of the Bushveld Complex. There are regional variations in the Cr/Fe ratio of the LG6 chromitite, and smaller variations in the thickness of the layer. Sections 9 and 10 are for lower Critical Zone and Lower Zone hosted chromitites respectively with higher Cr/Fe ratios. Regional variations in the stratigraphic position of layers such as these are to a large extent due to "missing" or greatly thinned sequences in the intervening silicate cumulates, a feature typical of mafic layered intrusions. The LG6 chromitite is entirely within pyroxenite cumulate in all the sections depicted, and the reason for the variation in the Cr/Fe ratio around the Complex is presently not evident; this may be related to local variations in bulk magma composition due to the large size of the Complex, with subsolidus re-equilibration as a superimposed process.

Systematic vertical chemical variation in the LG6 is related to the modal % of chromite through the layer; as the proportion of chromite increases, the Cr/Fe, Mg#, and Al_2O_3 content of chromite also increase. Cameron (1975) concluded that this cryptic variation is primary, rather than related to postcumulus re-equilibration. However, postcumulus changes in chemistry and texture are documented for the LG chromitites (Hatton and von Gruenewaldt, 1985) and also in those of the Lower Zone (Hulbert and von Gruenewaldt, 1985).

The mineral chemistry of the UG1 and UG2 chromitites and the associated minor layers has been studied in detail by Eales and Reynolds (1986) and Eales (1987). A range of textures from densely packed chromite in a 100% chromite rock to 50% to trace amounts of chromite occur in the UG1 and UG2 sequences. Disseminated chromite grades upwards into massive chromitite, and sharp contacts between chromitite and silicate layers are also frequent. Within the chromitite layers chromite grains surrounded by silicate are smaller than grains in mutual contact. Touching grains anneal and enlarge, and result in a variety of textural forms. Eales (1987) classified these textures as corpuscular (discrete grains) through coalescent to interstitial, the latter texture consisting of continuous masses of completely annealed chromite grains. Ferric iron increases with increasing textural complexity of these assemblages, reflecting the chemical change that accompanies chromite grain re-equilibration. Textures similar to those documented by Eales occur in the lower chromitite layer of the Merensky Reef (Kinloch and Peyerl, 1990). Chromite in the main UG1 and UG2 layers is higher in Al, Fe^{3+}, and Cr than accessory chromite in the footwall rocks. Within the UG2 chromitite there is an upward increase in Fe^{2+}, Fe^{3+}, Cr, and a decrease in Mg and Al (McLaren and de Villiers, 1982). However, laterally Mg varies sympathetically with Fe^{3+} and Cr.

The UG1 chromitite has an anorthosite footwall characterized by an abundance of minor layers and lenses of chromitite, as well as disrupted. disseminated chromite-anorthosite layers. These features are ascribed to post-deposition mobilization and downward injection of chromite crystal mush (Lee, 1981). The PGE mineralized UG2 chromitite commonly lies on a coarse-grained pegmatoidal feldspathic pyroxenite, possibly derived through *in situ* recrystallization of feldspathic pyroxenite footwall (Cawthorn and Barry, 1992). Locally, such as at Western Platinum Mine, the UG2 lies on anorthosite.

The chromitite-anorthosite association is rare. The Fisskenaesset Complex is another example of this association, where chromitite occurs in multiple layers 0.5 to 3.0 m thick and as elongate lenses up to 120 m long in the anorthosite Complex (Ghisler and Windley, 1967).

5.1.2. Stillwater Complex

The chromitite layers in the Stillwater Complex are situated in the lowermost ultramafic zone, at the base a series of dunite-harzburgite-pyroxenite cyclic units. Ten layers (A at the base to K at the top of the ultramafic zone) and a number of minor layers have been recorded. Average Cr_2O_3 ranges from 44.2% (A layer) to 41.6% (K layer) and Cr/Fe is 1.42 to 1.90. Highest Cr contents and Cr/Fe ratios are in the G and H chromitites (Jackson, 1968). The G and H layers are 60-140 m apart and contain the equivalent amount of chromite to form a single layer one metre thick, although this is dispersed through a series of rhythmic sequences of thin layers over 2-5 m. Cr in the chromite is highest at the base of a unit (Campbell and Murck, 1993). The Cr_2O_3 content of massive ore in the G and H layers is 2-3% higher than in the disseminated ore; Cr/Fe ratio is also higher in the massive ore. The average composition of the ore at the mill was 40% Cr_2O_3 with a Cr/Fe ratio of 1.08. The steep dip (60°+) of the Stillwater rocks aided mining, but the layers are disrupted by faulting, and also contain a higher proportion of silicate gangue compared to Bushveld and Great Dyke chromitites (Page *et al.*, 1985b).

5.1.3. Great Dyke

Chromitite layers in the Great Dyke occur at the base of mafic rhythmically layered units, similar to the Stillwater Complex. Great Dyke chromitites (Prendergast 1987) occur within a 1880 m thick sequence of harzburgite and pyroxenite and all the chromitite layers have variations in bulk composition and thickness. Up to 11 layers are known; a broad subdivision into an upper group (the topmost three layers) and a lower group (the remaining layers) is noted by Prendergast (1987). The eight chromitite layers in the lower 1300 m, dominantly harzburgite, are metallurgical grade (Cr/Fe ratio 2.6-3.5). The upper chromitites are chemical grade (Cr/Fe ratio 1.9-2.5), vary from 10-50 cm wide, and yield a lumpy to semi-friable ore. The lower layers are thinner (10-15 cm) and, being in serpentinized zones, yield a friable ore. The chromite compositions have been modified through subsolidus reaction (Wilson 1982).

There is a morphological contrast between the upper and lower layers of the Great Dyke. Prendergast (1987) observed that the upper layers have notable lateral variation over short distances in stopes, from massive chromitite to disseminated chromite-olivine rock. These variations in the ratio of olivine to chromite and the number and thickness of the mineral layers are often associated with trough structures, interpreted to be of magmatic current origin. The lower group of chromitites is less variable in thickness and composition than the upper chromitites, except where later alteration and serpentinization are recorded. Prendergast concluded that the broad subdivision is largely the effect of primary magmatic differentiation, with repeated injections of basic magma and blending with resident magma. Regional variations are ascribed to the effects of the Great Dyke magma chamber geometry. The elongate form (Wilson, this volume) resulted in contrasted rates of heat loss from the margins relative to the axis and hence different crystallization sequences. Magma would mix in different amounts along the length of the chamber, relative to the feeder, and this could affect the f_{O_2}, the magma convection patterns, and the magma composition resulting from the variable blending (Prendergast, 1987).

5.1.4. Comparison of chromite domains

An analysis of the chemistry of chromites in the Great Dyke, the Bushveld and Stillwater Complexes reveals regional composition contrasts; the lower Cr content of the Stillwater chromitites compared with the Great Dyke chromitites suggests host rock lithology is not a specific controlling factor on chromite composition (Roberts *et al.*, 1990). However, a pseudofractionation trend within chromites of the Great Dyke and Bushveld is evident in comparisons of Cr^{3+} against Fe^{3+} or Ti. The Ti variation, which increases from 0.01 cations in Great Dyke dunites to 0.21 cations in Bushveld feldspathic pyroxenites, illustrates the impact of host rocks on aspects of chromite composition (Roberts *et al.*, 1990). As with the Great Dyke, the Stillwater chromitite sequences occur in the ultramafic series, at the base of olivine to pyroxenite cyclic units, yet more evolved compositions of the chromitites are evident, even though the composition of the silicate phases is comparable.

Stratiform chromite deposits occurring in layered intrusions are confined to stable continental shields. Economic grades are constrained to a time interval of 2900-2000 Ma, and coincide with the peak in continental growth rates (Stowe, 1994). Stowe notes that the Cr/Fe ratio, a significant parameter in the economics of chromite, declines with younger ages, perhaps indicating that the younger mafic intrusions crystallized from chromium-depleted magmas. Since the major stratiform base metal sulphide and PGE deposits also occur in chromite-bearing intrusions in this time interval, the interpretation that decreasing radiogenic heat lowered continental thermal gradients and thus retained high Cr levels in magma as it ascended (Stowe, 1994) may equally apply to the capacity of a magma to retain dissolved sulphur, and PGE ligands.

The great lateral persistence of the thicker chromitite layers in these layered intrusions governs thoughts concerning the origins of these layers. However lateral variations in the stratigraphy of the chromitite layers exist and these have to be considered in the genesis of the chromitite layers, in particular the Bushveld Complex (Hatton and von Gruenewaldt 1987). From the range of theories in the literature (*e.g.* pressure change, immiscibility, differential gravity settling, assimilation of siliceous country rocks, f_{O_2} changes), Eales (1987), Hatton and von Gruenewaldt (1987), Roberts *et al.* (1990), Campbell and Murck (1993), opted for magma addition and blending (Irvine and Sharpe, 1986) as the most plausible mechanism for chromitite layer formation. However, Lipin (1993) has argued in favour of pressure changes as a mechanism, particularly with respect to the Stillwater Complex. In the case of minor chromitite layers (generally less than 1 cm thick), often at lithologically contrasted contacts, Lee *et al.* (1983) deduced an *in situ* reaction mechanism for the origin. *In situ* growth through volatile flux and metasomatism has been proposed for the minor chromitite layers of the Merensky Reef (Nicholson and Mathez, 1991).

5.2. Magnetite and ilmenite

Through fractionation late-stage liquids in layered intrusions become iron enriched, manifest as magnetitite layers and also locally by iron-rich pegmatite (Scoon and Mitchell 1994). As a general statement few layered intrusions contain an entire suite of potentially economic oxide-rich layers or sequences; a complex is either tectonically disrupted (*e.g.* the Stillwater and Penikat Complexes presumably had an iron-rich "upper zone"), or it may have been derived from an iron-rich tholeiite magma (*e.g.* the Ushushwana and Rooiwater Complexes in South Africa, without chromitite but with thick ilmenite-rich vanadiferous magnetitite layers (Winter, 1965; Reynolds, 1978)) and thus have only magnetite-rich layers. The Bushveld Complex (Reynolds, 1985) and the Windimurra Complex in Western Australia (Parks and Hill, 1986) are examples of iron-rich trends leading to the development of magnetite layers in the upper parts of the complex. In the Muskox Intrusion 5-10% magnetite and ilmenite occur in the uppermost gabbro (Irvine, 1988)

Magnetitite layers in the Bushveld Complex occur in the 1300-1500 m thick Upper Zone; the layers range in thickness from 3 cm-10 m. Disseminated magnetite is consistently present in the Upper Zone silicate cumulates as an accessory phase. As with the chromitite layers, the magnetitites persist for many tens of kilometres of strike around the complex. Some 30 layers, concordant with the layering of the silicate cumulates, have been mapped (Cawthorn and Molyneux, 1986). The layers occur in four groups numbered consecutively upwards (lower layers 1-4; the main magnetite layer and layers 1-7; layers 8-14; layers 15-21) separated by sequences of gabbro, ferrogabbro, and anorthosite. Regular geochemical trends have been documented for these magnetitite layers: V decreases and Ti increases with stratigraphic height, from 10% TiO_2 and 2% V_2O_3 at the base of the Upper Zone to 16-20% TiO_2 and 0.2% V_2O_3 at the top of the zone. The V_2O_3 content of the main magnetitite layer is 1.32%, with 12% TiO_2. Only the main magnetitite layer (1.5-2 m thick) in the lower part of the Upper Zone is mined for the vanadium content. Vanadium also occurs in transgressive magnetite pipes in the Bushveld Complex, such as the Kennedy's Vale pipe (van Rensburg, 1962), presently producing V.

Magnetite and ilmenite grains in many oxide-rich sequences exhibit a range of grain boundary relationships. In the Bushveld ores curved grain boundaries and triple junctions to magnetite, with interstitial ilmenite, give a densely textured rock. Smaller sized magnetite aggregates with a similar texture and with cumulus plagioclase crystals between the aggregates also occur. The denser textures in the ore are the result of high temperature subsolidus annealing and enlargement of the Ti-magnetite grains (Reynolds, 1985), much the same as proposed for the chromitites.

Several patterns of cryptic variation in the Cr and V content have been noted in the main layer and other magnetitites of the Bushveld Complex. Single and multiple cycles of trace element depletion vertically through and laterally along a single layer have been ascribed to magma convection during the growth of the layers at the cumulate-magma interface (McCarthy *et al.*, 1985).

The Bjerkreim-Sokndal intrusion contains cryptically layered concentrations of Fe-Ti oxides in plagioclase-pyroxene-olivine cyclic units (Duchesne, 1972). Trace elements in the oxides change progressively up these sequences. V follows Fe^{3+} and decreases upwards in magnetite and ilmenite, whereas Mn and Zn in magnetite increase. Cr and Ni are enriched in magnetite near the bottom of a cyclic unit and decrease rapidly upwards, to increase again at the base of the next unit (Jensen *et al.*, 1993).

Granular ilmenite and apatite occur in the upper magnetitite layers of the Bushveld Complex, and von Gruenewaldt (1993) has suggested the layers may be considered a source of ore for these minerals. Granular ilmenite, as opposed to ilmenite-magnetite exsolution intergrowths, more typical of the lower magnetitite layers (Reynolds, 1985), may prove to be amenable to extraction and provide a source for Ti.

The Windimurra Complex contains V-bearing titaniferous magnetite layers, presently considered to be subeconomic (Parks and Hill, 1986).

Magnetite gabbros of the Koitelainen Complex in northern Finland (Mutanen, 1989) contain up to 0.25% V; an average of 0.5 ppm Pt+Pd+Au is also present, represented by discrete PGM within the magnetite grains. A correlation of PGE+Au and V is reported.

6. REVIEW

Platinum-group elements and the related base metal sulphides, chromite, magnetite, and ilmenite ore deposits associated with mafic layered intrusions have been described in this overview. Figure 7 summarizes the mineralization styles and hosts in layered mafic intrusions in a combination of genetically-based classifications (*e.g.* Macdonald, 1987) and mineralogically-based classifications (*e.g.* Cabri, 1994).

The oxide ores are the least complicated of the deposits: these occur in laterally persistent layers over many kilometres, there is little internal, structural or thickness variation in any one layer, but there may be lateral chemical variation. Cryptic variation occurs in many chromitite and magnetitite sequences. The oxide-mineral composition is specific to the intrusion and to the stratigraphic position, is frequently related to the host-rock composition and the gangue minerals, and varies only a little in a particular layer of an intrusion. The special problem in modelling the origin of these layers tends to focus on the mechanism which would concentrate what is a minor element in the magma into the amounts required to form an economic layer.

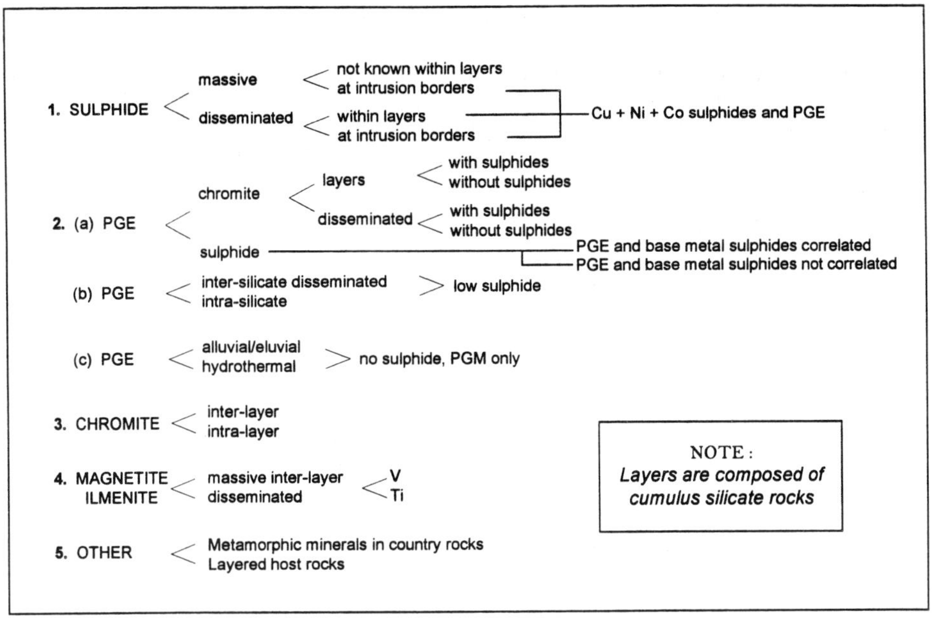

***Figure** 7. Diagram summarizing the mineralization styles and associations in layered mafic intrusions, and the position of the mineralization in an intrusion in relation to the overall layering and the mineral associations.*

Some mechanisms are summarized in Figure 8, and these may apply to the chromitites and magnetitites.

In contrast the PGE occur in a wide diversity of settings and are very variable in rock and mineral association, metal distribution, and metal ratios. The PGM assemblages, the range in composition, in grain size, and in the textural association of the PGM with the base metal sulphides and the gangue minerals, are significant considerations for PGE deposits. Two different styles of mineralization exist with respect to the metal content of a PGE-bearing sequence, and these have to be considered in the application of models. In the first style the total amounts of precious metal and sulphide are constant and are diluted by increasing volumes of silicate minerals. The effect of this dilution is that the thicker the unit becomes the lower the grade with respect to width or, as it thins, the grade increases. This is the situation for the Merensky Reef, may also be so for the UG2 chromitite, and appears to hold for the JM reef. In this first case the mineralization must be related to the source which gave rise to the PGE-host rocks. In the second widely occurring style the metal content increases with increase in the thickness of the unit in question; this implies mineralization must have been added to an existing assemblage, or was introduced into the system and trapped in some way, or was later modified. The compositional complexity and diversity of PGE occurrences in even a single mafic layered intrusion, down to the scale of a mineralized layer, have also to be accounted for in the search for an explanation of the origins and controls of this type of mineralization. At the

MAGMATIC & POST-MAGMATIC CHEMICAL MECHANISMS

PROCESS - gradual, internal, inter-relational to cumulate pile [Fe, Ti, Cr, PGE, S]
- Fractionation
- Saturation onset, fugacity shift, redox shift [S, and Fe, Ti in particular]
- Incompatible element transfer and accumulation at magma/cumulate interface [PGE, S]

PROCESS - trigger, external, non-relational to cumulate pile [Fe, Ti ,Cr, PGE ,S]
- Magma addition and blending
- Contamination

PROCESS - source, external
- New magma [PGE, S, Cr]

PROCESS - overprinting, origin from within cumulate pile [PGE]
- hydrothermal carrier
- volatile transfer

PROCESS - immiscibility [Fe, Ti]

MAGMA CHAMBER PHYSICAL MECHANISMS

- Pressure change (increase or decrease) [Cr, Fe]
- Tectonic imprint, mush deformation [PGE]
- Temperature change (decrease) [S]
- Heat flow, duration, rate of cooling

Figure 8. *A summary of the various processes and mechanisms, divided into chemical and physical, proposed to account for mineralization in layered complexes, in relation to the sequences described in this review. Five mechanisms related to the possible chemical processes are listed together with the mineralization. It is doubtful if any one of these processes can be singled out to account for mineralization, given the wide range in styles noted in Figure 8.*

regional scale there is some evidence that the mantle source must have an initial control on why the majority of mafic layered intrusions are Pd-dominant, such as the Stillwater, and why Pt-dominant intrusions are rare, as is the case for the Bushveld Complex. The invariable association of PGE with chromitite, but only in the case of the UG2 at economic amounts, the association of more than trace amounts of PGE with sulphides, which are universally abundant, in only certain settings, and PGE occurring as PGM only in yet other settings, render the processes listed in Figure 8 specific to an intrusion, if not to the deposit.

Metal ratio diagrams have been suggested as aids to PGE prospecting (Barnes, 1988). These diagrams tend to characterize the mineralized domains as related to the proposed effects of fractionation of chromite, olivine, or sulphides and/or PGE. The fields delineated in these plots would be self-evident from the geology and setting of the rocks in question, and this would be confirmed by use of metal ratios. Other metal-diagrams are derived by normalising the PGE to chondrite or mantle abundances. At best these plots reflect the PGM control on PGE abundance and distribution and thus the application in modelling is doubtful without due regard to the composition and amount of PGM or to the effects of alteration. Furthermore, a data distortion in the chondrite-normalized plot exists in the relative positions of the Pt and Pd points; for certain concentration ranges of Pt and Pd with Pt/Pd ratios between 1 and 1.5 the

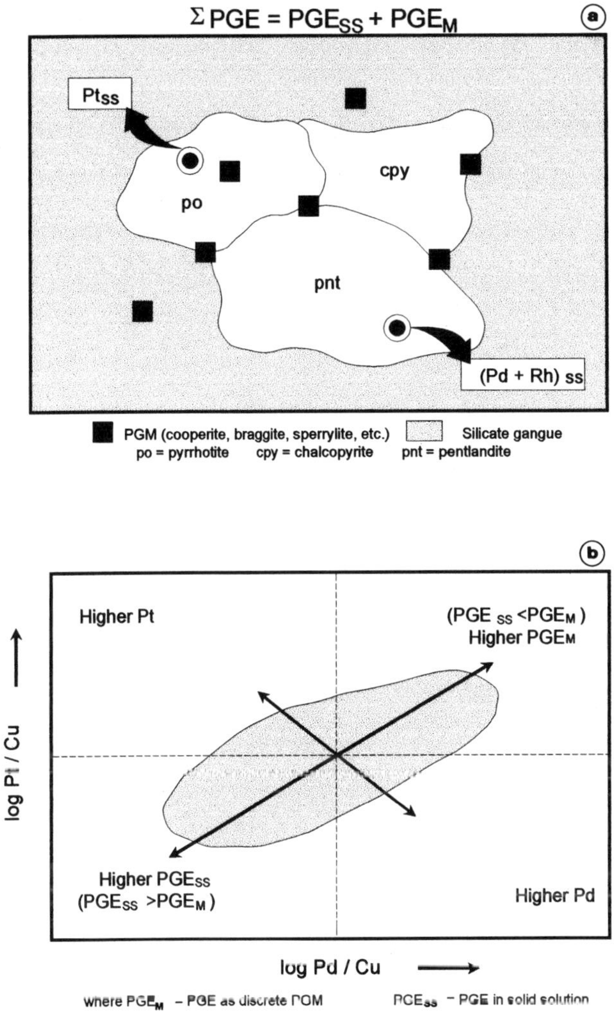

Figure 9. *(a) Graphical illustration of total PGE content controlled as PGM and as PGE in solid solution in base metal sulphides. In magmatic assemblages Pt tends to be higher in pyrrhotite and Pd and Rh are higher in pentlandite, whereas in chalcopyrite these elements are low to non-detectable (e.g. Czamanske et al., 1992). PGM occur within sulphide grains, at sulphide margins and triple junctions, at sulphide-gangue margins, and completely enclosed in chromite and in silicate gangue. (b) Use of Pt/Cu against Pd/Cu as a metal ratio plot to deduce the relative contributions of PGE as solid solution in sulphide (PGE_{SS}) and PGE as discrete Pt minerals (PGE_M), as well as the ratio of sulphide to PGM and the relative Pt/Pd ratio.*

Pd points plot above the Pt points on the normalized curve, and hence masks the economically significant Pt/ Pd ratio. Similarly, the technique of reducing and reporting raw PGE values to 100% sulphide tends to distort basic observation on the distribution of these elements and pre-supposes the distribution and abundance of the elements are fundamentally controlled by sulphide liquation (Barnes *et al.*, 1988).

Analytical data are routinely available during exploration of layered intrusions for PGE. Metal ratios from these data can be applied, prior to detailed mineralogical assessments, to characterizing the finer details of an intrusion or a mineralized sequence. The ratios assist in determining the relative contributions discrete PGM and solid-solution PGE make to the metal content of the sequence. Having established the ratio of PGM to PGE in solid solution, deductions concerning the PGE sequence can be made, compared with known examples, and perhaps be modelled and evaluated. Figure 9a depicts the base metal sulphide and PGM controls on PGE distribution and in Figure 9b these are shown as four fields empirically delineated in a logarithmic plot of Pt/Cu against Pd/Cu. Evaluation of ore-grade PGE deposits is a distinct issue from the evaluation of PGE in the widely occurring trace abundances, but the use of metal ratios in this fashion will assist in characterizing a deposit at an early stage in exploration.

There are several models for the PGE and related base metal sulphide occurrences (*e.g.* Macdonald, 1987), as well as for chromitites and magnetitites, based on the composition of magma, tectonic setting and age of the intrusion. Examples are the cratonic or inter-cratonic setting of the intrusion, whether the processes were orthomagmatic or hydrothermal or a combination, whether high-magnesium or alkaline magma types were involved. A number of chemical and physical processes within the magma chamber have been proposed to account for mineralization in layered intrusions, as summarized in Figure 8. Magma influx and mixing together with PGE scavenging by liquated sulphide is a widely applied model to account for stratiform PGE sequences. The frequent occurrence of PGM-alloy dominated mafic bodies in

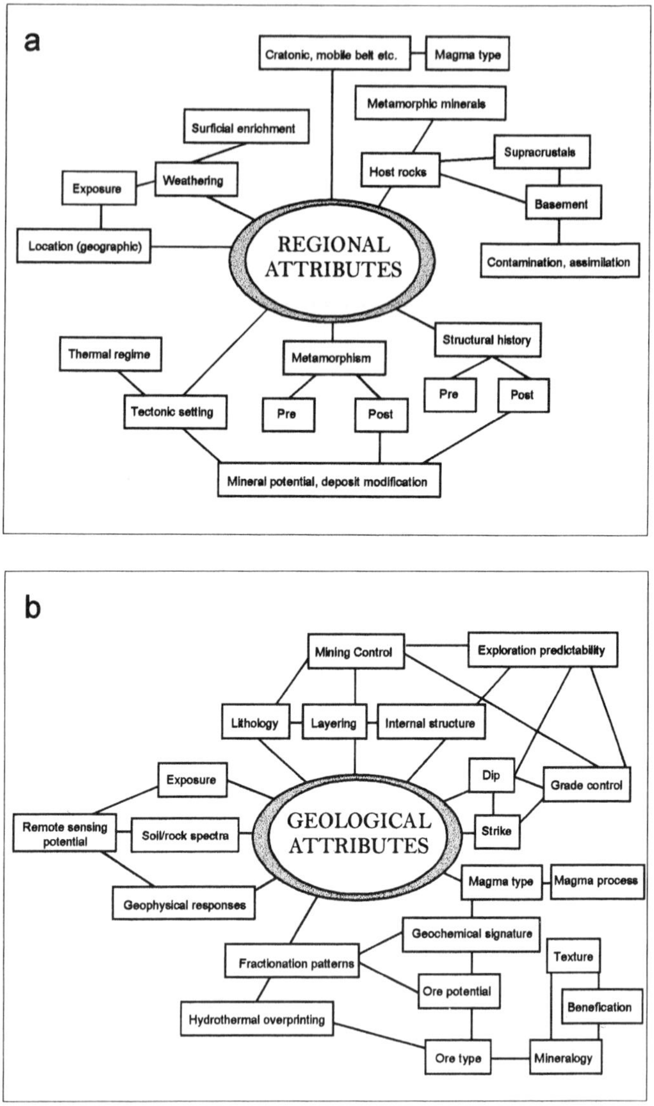

Figure 10. *Attributes of a layered mafic intrusion to be considered in descriptive models for exploration and evaluation (a): the regional attributes and the interactions. (b): the geological attributes and the interactions. The outer ring of attributes applies to the potential for exploiting a layered intrusion.*

the absence of sulphide is an apparent contradiction of this model, and serves as a caveat concerning rigorous application of models in exploring for and evaluating mafic intrusions for PGE potential. Hydrothermal overprinting, and redistribution or alteration of PGM and sulphides by contamination, appear to be inevitable down-temperature events and imprint a particular characteristic onto a deposit or mineral occurrence. Many of these aspects should be evident from the diverse styles of PGE mineralization described in this review.

Descriptive geological models of deposits, formulated for exploration targets or for the purposes of economic evaluation, often form the foundation for the genetic models. The genetic models become increasingly relevant as observations are added to the descriptive models, through the experience gained in applying the model. With time the genetic model replaces the descriptive model, depending on the deposit (Mosier and Bliss, 1992). Whatever the model preferred, it may be taken that mineralization in mafic intrusions is formed during the growth and the development of the intrusion. The syngenetic nature of mineralization in layered mafic intrusions emphasizes the interactive attributes depicted in Figures 10a and b. These attributes are of greater or lesser significance depending on the stage reached in exploration, evaluation, exploitation, and beneficiation, and may be regional (Figure 10a) or morphological (Figure 10b), or both. No one feature can be singled out as significant in determining the potential worth of an intrusion. All the attributes are of significance, to a variable extent, and specific to each case. Emphasizing one or other genetic model in isolation is inappropriate given the complexity of mineralization in layered intrusions.

With regard to generally applicable exploration models based on petrological considerations alone, Eales *et al.* (1993) have concluded that this has yet to be realized. Eales *et al.* (1993) further conclude that processes active within the chamber, rather than intrinsic properties of the liquids emplaced within it, hold the key to successful exploration models, but this is debateable. For example, arguments for a PGE-enriched source, in particular in the Kaapvaal craton, suggest that as early as Archean times a proto-craton enriched in PGE may have developed (Tredoux *et al.*, 1989). A potential PGE accumulation could thus have existed to contribute to, and imprint at source, on early Proterozoic magmatic events such as the Bushveld Complex. The extent to which subsequent mantle or crustal reservoir fractionation exerts control on the Pt/Pd ratio, amongst other metal ratios, and thus gives a layered complex a particular geochemical characteristic, has yet to determined. The close association of complexes with both chromite and PGE in space and time bears consideration in concluding a mineralization model.

A future enhancement to petrological modelling, which is essentially a two-dimensional exercise, is provided by the application of seismic reflection surveys to determine the down-dip extent and the internal structure of a layered intrusion, as a complement to gravity and magnetic surveys and the compilation of detailed geological sections from mapping and drill-core. Seismic reflection surveys of the Sudbury intrusion and the eastern and western Bushveld Complex have the shown extent of layering and the variations in thickness of seismically defined layers (du Plessis and Kleywegt, 1987; Odgers and du Plessis, 1993). In the case of the Sudbury intrusion the application of geophysical techniques has led to modifications of genetic interpretations based on petrology alone (in Pye *et al.*, 1984). Structural models, derived from mapping and well-constrained geophysical data, should be incorporated into discussion on the origin and controls of mineralization in layered intrusions in general.

7. ACKNOWLEDGEMENTS

The Management of Anglo American Platinum Corporation is thanked for permission to compile and publish this review. Critical comments by R.T. Brown, K. Lomberg, R.W. Hieber, and A.H. Wilson on early drafts were helpful and all are thanked.

8. REFERENCES

Alapieti, T.T., & Lahtinen, J.J., 1986. Stratigraphy, petrology, and platinum group element mineralisation of the early Proterozoic Penikat layered intrusion, northern Finland. *Econ. Geol.* **81**, 1126-36.

Alapieti, T.T., & Piirainen, T., 1984. Cu-Ni-PGE mineralisation in the marginal series of early Proterozoic Koillisma layered igneous complex, northeast Finland. In: Buchanan, D.L., & Jones, M.J. (eds.) *Sulphide deposits in mafic and ultramafic rocks.* Inst. Mining Metallurgy, London, 123-31.

Alapieti, T.T., Lahtinen, J.J., Huhma, H., Hanninen, E., Piirainen, T., & Sivonen, S.J., 1989. Platinum-group element bearing Cu-Ni sulphide mineralisation in the marginal series of early Proterozoic Suhanko Konttijärvi layered intrusion, northern Finland. In: Prendergast, M.D. (ed.) *Magmatic sulphides - the Zimbabwe volume.* Inst. Mining Metallurgy, London, 177-88.

Alapieti, T.T., Filen, B.A., Lahtinen, J.J., Lavrov, M.M., Smolkin, V.F., & Voitsekhovsky, S.N., 1990. Early Proterozoic layered intrusions in the northeastern part of the Fennoscandian shield. *Miner. Petrol.* **42**, 1-22.

Balabonin, N.L., Korchagin, A.U., Latypov, R.M., & Subbotin, V.V., 1994. Ferodorovo-Pansky Intrusion. In: Mitrofanov, F., & Torkov, M. (eds.) *Field Guide to Kola PGE Geology.* Geological Institute, Apatity, 71-108.

Ballhaus, C.G., 1988. Potholes in the Merensky Reef at Brakspruit shaft, Rustenburg Platinum Mines: primary disturbance in the magmatic stratigraphy. *Econ. Geol.* **83**, 1140-58.

Ballhaus, C.G., & Sumpfl, E.F., 1986. Sulfide and platinum mineralisation in the Merensky Reef: evidence from hydrous silicates and fluid inclusions. *Contr. Miner. Petrol.* **94**, 193-204.

Barnes, S.J., 1988. The use of metal ratios in platinum-group element prospecting. *Explore* **64**, 8-10.

Barnes, S.J., 1993. Partitioning of the platinum group elements and gold between silicate and sulphide magmas in the Munni Munni Complex, western Australia. *Geochim. Cosmochim. Acta* **57**, 1277-90.

Barnes, S.J., & Francis, D., 1993. The distribution of platinum-group elements, gold, nickel and copper in the Muskox intrusion of the North West Territories, Canada. *Terra Abstracts* **3**, 41.

Barnes, S.J., & Naldrett, A.J., 1985. Geochemistry of the JM reef of the Stillwater Complex, Minneapolis adit area 1: sulphide chemistry and sulphide olivine equilibrium. *Econ. Geol.* **80**, 627-45.

Barnes, S.J., Boyd, R., Korneliussen, A., Nilsson, L.P., Often, M., Pedersen, R.B., & Robins, B., 1988. The use of mantle normalisation and metal ratios in discriminating between the effects of partial melting, crystal fractionation and sulphide segregation on platinum-group elements, gold, nickel and copper: examples from Norway. In: Prichard, H.M., Potts, P.J., Bowles, J.F.W., & Cribb, S.J. (eds.) *Geoplatinum 87.* Elsevier Appl. Sci., London and New York, 113-43.

Barnes, S.J., McIntyre, J.R., Nisbet, B.W., & Williams, C.R., 1990. Platinum-group element mineralisation in the Munni Munni Complex, western Australia. *Miner. Petrol.* **42**, 141-64.

Barnes, S.J., Keays, R.R., & Hoatson, D.M., 1992. Distribution of sulphides and PGE within the porphyritic websterite zone of the Munni Munni Complex, western Australia. *Austral. J. Earth Sci.* **39**, 289-302.

Barton, J.M., Jr., Cawthorn, R.G., & White, J.A., 1986. The role of contamination in the evolution of the Platreef of the Bushveld Complex. *Econ. Geol.* **81**, 1096-148.

Bird, D.K., Brooks, C.K., Gannicott, R.A., & Turner, P.A., 1991. A gold-bearing horizon in the Skaergaard intrusion, east Greenland. *Econ. Geol.* **86**, 1083-92.

Boudreau, A.E., & McCullum, I.S., 1985. Features of the Picket Pin deposit. In: Czamanske, G., & Zientek, M.L. (eds.) *The Stillwater Complex, Montana; geology and guide.* Special Publication, Montana Bureau of Mines and Geology **92**, 346-57.

Boudreau, A.E., & McCallum, I.S., 1986. Investigations of the Stillwater Complex: III. The Picket Pin Pt/Pd deposit. *Econ. Geol.* **81**, 1953-75.

Bow, C., Wolfgram, A., Turner. A,, Barnes, S., Evans, J., Zdapski, M., & Boudreau, A.E., 1982. Investigations of the Howland reef of the Stillwater Complex, Minneapolis adit area: stratigraphy , structure, and mineralisation. *Econ. Geol.* **77**, 1481-92.

Brown, R.T., 1994. *The petrology and geochemistry of the Merensky Reef in the Rustenburg area.* Unpubl. M.Sc. thesis, University of Natal, 101 pp.

Brynard, H.J., de Villiers, J.P.R., & Viljoen, E.A., 1976. A mineralogical investigation of the Merensky Reef at the western platinum mine, near Marikana, South Africa. *Econ. Geol.* **77**, 1299-307.

Buchanan, D.L., 1988. Platinum-group element exploration. *Developments in economic geology* **26**, Elsevier Science Publishers, 184 pp.

Buchanan, D.L., & Rouse, J.E., 1984. Role of contamination in the precipitation of sulphides in the Platreef of the Bushveld Complex. In: Buchanan, D.L., & Jones, M.J. (eds.) *Sulphide deposits in mafic and ultramafic rocks.* Inst. Mining Metallurgy, London, 141-6.

Cabri, L.J., 1988. Overview on application of platinum mineralogy to mineral exploration and processing. In: Carson, D.J.T., & Vassilou, A.H. (eds.) *Process Mineralogy VIII: Applications of mineralogy to mineral beneficiation technology, metallurgy, and mineral exploration and evaluation, with emphasis on precious metal ores.* The Minerals, Metals and Materials Society, 23-31.

Cabri, L.J., 1994. Current status of determination of mineralogical balances for platinum-group element-bearing ores. *Trans. Inst. Min. Metall.* (Sect. B: Appl. Earth Sci.) **103**, 3-9.

Cameron, E.N., 1975. Postcumulus and subsolidus equilibration of chromite and coexisting silicates in the eastern Bushveld Complex. *Geochim. Cosmochim. Acta* **39**, 1021-33.

Cameron, E.N., 1977. Chromite in the central sector of the eastern Bushveld Complex, South Africa. *Am. Miner.* **62**, 1082-96.

Cameron, E.N., 1980. Evolution of the lower critical zone, central sector, eastern Bushveld Complex, and its chromite deposits. *Econ. Geol.* **75**, 845-71.

Campbell, I.H., & Murck, B.W., 1993. Petrology of the G and H chromitite layers in the Mountain View area of the Stillwater Complex, Montana. *J. Petrology* **34**, 291-316.

Carr, H.W., Groves, D.I., & Cawthorn, R.G., 1994. The importance of synmagmatic deformation in the formation of Merensky Reef potholes in the Bushveld Complex. *Econ. Geol.* **89**, 1398-410.

Cawthorn, R.G., & Barry, S.D., 1992. The role of intercumulus residua in the formation of pegmatoid associated with the UG2 chromitite, Bushveld Complex. *Austral. J. Earth Sci.* **39**, 263-76.

Cawthorn, R.G., & Molyneux, T.G., 1986. Vanadiferous magnetite deposits of the Bushveld Complex. In: Anhaeusser, C.R., & Maske, S. (eds.) *Mineral Deposits of Southern Africa,* **2**. Johannesburg: Geol. Soc. S. Afr., 1251-66.

Cawthorn, R.G., Barton, J.M., Jr., & Viljoen, M.J., 1985. Interaction of floor rocks with the Platreef on Overysel, Potgietersrus, northern Transvaal. *Econ. Geol.* **80**, 988-1006.

Coghill, B.M., & Wilson, A.H., 1993. Platinum-group minerals in the Selukwe subchamber, Great Dyke, Zimbabwe - implications for PGE collection mechanisms and post-formation redistribution. *Miner. Mag.* **57**, 613-33.

Czamanske, G.K., Kunilov, V.E., Zientek, M.L., Cabri, L.J., Likhachev, A.P., Calk, L.C., & Oscarson, R.L., 1992. A proton-probe study of magmatic sulphide ores from the Noril'sk-Talnakh district, Siberia. *Canad. Miner.* **30**, 249-87.

Davey, S.R., 1992. Lateral variations within the upper Critical zone of the Bushveld Complex on the farm Rooikoppies 297JQ Marikana, South Africa. *S. Afr. J. Geol.* **95**, 141-9.

de Villiers, J.P.R., & Kleyenstüber, A., 1993. The partitioning of chromium between sulphide and silicate melts at controlled partial pressures of oxygen and sulphur. Mintek Mineralogy Division (Johannesburg), Report No. M139D.

Duchesne, J.C., 1972. Iron-titanium oxide minerals in the Bjerkreim-Sogndal massif, South-western Norway. *J. Petrology* **13**, 57-81.

Duke, J.M., 1983. Ore deposit models #7. Magmatic segregation deposits of chromite. *Geoscience Canada* **10**, 1524.

du Plessis, A., & Kleywecht, R.J., 1987. A dipping sheet model for the mafic lobes of the Bushveld Complex. *S. Afr. J. Geol.* **90**, 1-6.

Eales, H.V., 1987. Upper critical zone chromitite layers at RPM Union section mine, western Bushveld Complex. In: Stowe, C.W. (ed.) *Evolution of the chromium ore fields.* Evolution of ore fields series, van Nostrand Reinhold, New York. 144-68.

Eales, H.V., & Reynolds, I.M., 1986. Cryptic variations within chromitites of the upper critical zone, northwestern Bushveld Complex. *Econ. Geol.* **81**, 1056-66.

Eales, H.V., Botha, W.J., Hattingh, P.J., de Klerk, W.J., Maier, W.D., & Odgers, A.T.R., 1993. The mafic rocks of the Bushveld Complex: a review of emplacement and crystallization history, and mineralisation, in the light of recent data. *J. Afr. Earth Sci.* **16**, 121-42.

Evans, D.M., Buchanan, D.L., & Parry, S.J., 1994. The Bohmke reef: platinum mineralisation associated with the C1d chromitite of the Great Dyke, Zimbabwe. Abstracts, 8th International Platinum Symposium, Moscow, 25.

Ferrario, A., & Garuti, G., 1988. Platinum-group minerals in chromite-rich horizons of the Niquelandia Complex (central Goias, Brazil). In: Prichard, H.M., Potts, P.J., Bowles, J.F.W., & Cribb, S.J. (eds.) *Geoplatinum 87.* London and New York, Elsevier Appl. Sci., 261-72.

Gain, S.B., 1985. The geologic setting of the platiniferous UG2 chromitite layer on the farm Maandagshoek, eastern Bushveld Complex. *Econ. Geol.* **80**, 925-43.

Gain, S.B., 1986. The upper group chromitite layers at Maandagshoek, eastern Bushveld Complex. In: Anhaeusser, C.R., & Maske, S. (eds.) *Mineral Deposits of Southern Africa,* **2**, Johannesburg: Geol. Soc. S. Afr., 1197-208.

Gain, S.B., & Mostert, A.B., 1982. The geological setting of the platinoid and base metal sulphide mineralisation in the Platreef of the Bushveld Complex in Drenthe, north of Potgietersrus. *Econ. Geol.* **77**, 1395-404.

Ghisler, M., & Windley, B.F., 1967. The chromite deposits of the Fisskenaeset region, west Greenland. *Greenland Geol. Survey Report* **12**, 139 pp.

Hahn, U.F., & Ovendale, B., 1994. UG2 chromitite layer potholes at Wildebeesfontein north mine, Impala Platinum Limited. Proceedings 15th CMMI Congress. South African Institute of Mining and Metallurgy, 195-200.

Halkoaho, T.A.A., 1994. The Sompujarvi and Ala-Penika PGE reefs in the Penikat intrusion, northern Finland. *Acta Univ. Oul.* **A 249**, 122 pp.

Halkoaha, T.A.A., Alapieti, T.T., & Lahtinen, J.J., 1990a. The Sompujarvi PGE reef in the Penikat layered intrusion, northern Finland. *Miner. Petrol.* **42**, 39-56.

Halkoaho, T.A.A., Alapieti, T.T., Lahtinen, J.J., & Lerssi, J.M., 1990b. The Ala-Penika PGE reefs in the Penikat layered intrusion, northern Finland. *Miner. Petrol.* **42**, 23-38.

Harney, D.M.W., & Merkle, R.K.W., 1990. Pt Pd minerals from the upper zone of the eastern Bushveld Complex, South Africa. *Canad. Miner.* **28**, 619-28.

Hatton, C.J., & von Gruenewaldt, G., 1985. Chromite from the Swartkop chrome mine: an estimate of the effects of subsolidus reequilibration. *Econ. Geol.* **80**, 911-24.

Hatton, C.J., & von Gruenewaldt, G., 1987. The geological setting and the petrogenesis of the Bushveld chromitite layers. In: Stowe, C.W. (ed.) *Evolution of the chromium ore fields.* Evolution of ore fields series, van Nostrand Reinhold, New York, 109-43.

Hatton, C.J., & von Gruenewaldt, G., 1990. Early Precambrian layered intrusions. In: Hall, R.P., & Hughes, D.J. (eds.) *Early Precambrian basic magmatism.* Blackie, Glasgow, 56-82.

Hatton, C.J., Harmer, R.E., & Sharpe, M.R., 1986. Petrogenesis of the middle group chromitite layers, Doornvlei, eastern Bushveld Complex. In: Gallagher, M.J., Ixer, R., Neary, C.R., & Prichard, H.M. (eds.) *Metallogeny of basic and ultrabasic rocks.* Institution of Mining and Metallurgy, London, 241-7.

Hiemstra, S.A., 1985. The distribution of some platinum group elements in the UG2 chromitite layer of the Bushveld Complex. *Econ. Geol.* **80**, 944-57.

Hiemstra, S.A., 1986. The distribution of chalcophile and platinum-group elements in the UG2 chromitite layer of the Bushveld Complex. *Econ. Geol.* **81**, 1080-6.

Hoatson, D.M., & Glaser, L.M., 1989. Geology and economics of the platinum-group metals in Australia. Bureau of Mineral Resources, Geology and Geophysics, Resource Report 5.

Hoatson, D.M., & Keays, R.R., 1989. Formation of platiniferous sulphide horizons by crystal fractionation and magma mixing in the Munni Munni layered intrusion, west Pilbara block, Western Australia. *Econ. Geol.* **84**, 1775-804.

Huhtelin, T.A., Alapieti, T.T., & Lahtinen, J.J., 1990. The Paasivaara PGE reef in the Penikat layered intrusion, northern Finland. *Miner. Petrol.* **42**, 57-70.

Hulbert, L.J., & von Gruenewaldt, G., 1982. Nickel, copper, and platinum mineralisation in the lower zone of the Bushveld Complex, south of Potgietersrus. *Econ. Geol.* **77**, 1296-306.

Hulbert, L.J., & von Gruenewaldt, G., 1985. Textural and compositional features of chromite in the lower and critical zones of the Bushveld Complex south of Potgietersrus. *Econ. Geol.* **80**, 872-95.

Hulbert, L.J., Duke, J.M., Eckstrand, O.R., Lydon, J.W., Scoates, R.F.J., & Cabri, L.J., 1988. Geological environments of the platinum-group elements. In: Hulbert, L.J., Duke, J.M., Eckstrand, O.R., Lydon, J.W., Scoates, R.F.J., & Cabri, L.J. (eds.) *Geological environments of the platinum-group elements.* Geological Survey of Canada, Open File 1440.

Iljina, M.J., 1994. The Portimo layered Complex. *Acta Univ. Oul.* **258**, 158 pp.

Iljina, M.J., Alapieti, T.T., & McElduff, B.M., 1992. Platinum-group element mineralisation in the Suhanko-Konttijarvi intrusion, Finland. *Austral. J. Earth Sci.* **39**, 303-13.

Irvine, T.N., 1982. Terminology for layered intrusions. *J. Petrology* **23**, 127-62.

Irvine, T.N., 1988. Muskox Intrusion, Northwest Territories. In: Hulbert, L.J., Duke, J.M., Eckstrand, O.R., Lydon, J.W., Scoates, R.F.J., & Cabri, L.J. (eds.) *Geological environments of the platinum-group elements.* Geological Survey of Canada, Open File 1440, 25-39.

Irvine, T.N., & Sharpe, M.R., 1986. Magma mixing and the origin of stratiform oxide ore layers in the Bushveld and Stillwater Complexes. In: Gallagher, M.J., Ixer, R., Neary, C.R., & Prichard, H.M. (eds.) *Metallogeny of basic and ultrabasic rocks.* Institution of Mining and Metallurgy, London. 183-98.

Jackson, E.D., 1968. The chromite deposits of the Stillwater Complex, Montana. Amer. Inst. Min. Metall. and Petroleum Engineers **2**, 1495-510.

Jensen, J.C., Nielsen, F.M., Duchesne, J.C., Demaiffe, D., & Wilson, J.R., 1993. Magma influx and mixing in the Bjerkreim-Sokndal layered intrusion, south Norway; evidence from the boundary between two megacyclic units at Storeknuten. *Lithos* **29**, 311-25.

Kennedy, D.C., 1994. Datamine modelling as a tool for ore reserve evaluation and mine design and the need for standardisation in data generation. Proceedings 15th CMMI Congress. South African Institute Mining & Metallurgy, Johannesburg, 143-52.

Kinloch, E.D., 1982. Regional trends in the platinum-group mineralogy of the critical zone of the Bushveld Complex, South Africa. *Econ. Geol.* **77**, 1328-47.

Kinloch, E.D., & Peyerl, W., 1990. Platinum-group minerals in various rock types of the Merensky Reef; genetic implications. *Econ. Geol.* **85**, 537-55.

Kruger, F.J., 1990. The stratigraphy of the Bushveld Complex: a reappraisal and the relocation of the Main Zone boundaries. *S. Afr. J. Geol.* **94**, 376-81.

Lee, C.A., 1981. Post deposition structures of the Bushveld Complex mafic sequence. *J. Geol. Soc. Lond.* **138**, 327-41.

Lee, C.A., 1983. Trace and platinum group element geochemistry and the development of the Merensky Reef of the western Bushveld Complex. *Miner. Deposita* **18**, 173-90.

Lee, C.A., & Butcher, A.R., 1990. Cyclicity in the Sr-isotope stratigraphy through the Merensky and Bastard reefs, Atok section, eastern Bushveld Complex. *Econ. Geol.* **85**, 877-83.

Lee, C.A., & Parry, S.J., 1988. Platinum-group element geochemistry of the lower and middle group chromitites of the eastern Bushveld Complex. *Econ. Geol.* **83**, 1127-39.

Lee, C.A., Cawthorn, R.G., & Barton, J.M., Jr., 1989. Further Sr-isotope and trace element studies on the Platreef, Bushveld Complex. *Bull. Geol. Soc. Finland* **61**, 21.

Lee, C.A., Sharpe, M.R., & Viljoen, E.A., 1983. The chemistry of minor chromitite layers, with special reference to chromite plagioclase equilibration. Programme with Abstracts, Symposium on the Bushveld Complex, University Pretoria Institute for Geological Research on the Bushveld Complex, 60-62.

Li, C., & Naldrett, A.J., 1993. Platinum-group minerals from the deep copper zone of the Strathcona deposit, Sudbury, Ontario. *Canad. Miner.* **31**, 31-44.

Lightfoot, P.C., & Naldrett, A.J. (eds.) 1994. Proceedings of the Sudbury - Noril'sk Symposium. Ontario Geological Survey, Special Volume 5.

Lipin, B.R., 1993. Pressure increases in the formation of chromite seams and the development of the ultramafic series in the Stillwater Complex, Montana. *J. Petrology* **34**, 955-76.

Macdonald, A.J., 1987. Ore deposit models #12. The platinum group element deposits: classification and genesis. Geoscience Canada, **14**, 155-66.

McCarthy, T.S., Cawthorn, R.G., Wright, C.J., & McIver, J.R., 1985. Mineral layering in the Bushveld Complex: implications of Cr abundances in magnetite from closely spaced magnetitite and intervening silicate-rich layers. *Econ. Geol.* **80**, 1062-74.

McLaren, C.H., & de Villiers, J.P.R., 1982. The platinum-group chemistry and mineralogy of the UG2 chromitite layer of the Bushveld Complex. *Econ. Geol.* **77**, 1348-66.

Merkle, R.K.W., 1992. Platinum group minerals in the middle group of chromitite layers at Marikana, western Bushveld Complex; indications for collection mechanisms and post magmatic modification. *Canad. J. Earth Sci.* **29**, 209-21.

Milliotti, C.A., & Stumpfl, E.F., 1993. Platinum-group mineral inclusions, textures and distribution in the chromitites of the Niquelandia Complex, Brazil. In: Jost, H., & Macdo, D. (eds.) *Brazilian meeting on Platinum-Group Elements, Extended Abstracts.* Brasilia, 33-5.

Mosier, D.L., & Bliss, J.D., 1992. Introduction and overview of mineral deposit modelling. In: Bliss, J.D. (ed.) *Developments in mineral deposit modelling.* U. S. Geol. Surv. Bull. **2004**, 1-5.

Mossom, R.J., 1986. The Atok platinum mine. In: Anhaeusser, C.R., & Maske, S. (eds.) *Mineral Deposits of Southern Africa,* **2,** Johannesburg: Geol. Soc. S. Afr., 1143-54.

Mostert, A.B., Hofmeyer, P.K., & Potgieter, G.A., 1982. The platinum-group mineralogy of the Merensky Reef at the Impala platinum mines. *Econ. Geol.* **77**, 1385-94.

Mutanen, T., 1989. Koitelainen intrusion and Keivitsa-Satovaara Complex. Geol. Survey, Finland, excursion guide **28**, 49 pp.

Nell, J., 1985. The Bushveld metamorphic aureole in the Potgietersrus area: evidence for a two stage metamorphic event. *Econ. Geol.* **80**, 1129-52.

Nicholson, D.M., & Mathez, E.D., 1991. Petrogenesis of the Merensky Reef in the Rustenburg section of the Bushveld Complex. *Contr. Miner. Petrol.* **107**, 293-309.

Odgers, A.T.R., & du Plessis, A., 1993. Interpretation of a regional reflection seismic survey in the northeastern Bushveld Complex. S. Afr. Geophys. Assoc., 3rd Technical Meeting, Cape Town, 125-9.

Page, N.J., Zientek, M.L., Czamanske, G., & Foose, M.P., 1985a. Sulphide mineralisation in the Stillwater Complex and the underlying rocks. In: Czamanske, G., & Zientek, M.L. (eds.) *The Stillwater Complex, Montana; geology and guide.* Special Publication, Montana Bureau of Mines and Geology, **92**, 93-6.

Page, N.J., Zientek, M.L., Lipin, B.R., Raedeke, L.D., Wooden, J.L., Turner, A.R., Loferski, P.J., Foose, M.P., Moring, B.C., & Ryan, M.P., 1985b. In: Czamanske, G., & Zientek, M.L. (eds.) *The Stillwater Complex, Montana; geology and guide.* Special Publication, Montana Bureau of Mines and Geology, **92**, 147-209.

Papunen, H., Distler. V., & Sokolov, A., 1992. PGE in the upper Proterozoic Dovirensky layered Complex, north Baikal area, Siberia. *Austral. J. Earth Sci.* **39**, 327-34.

Parks, J., & Hill, R.E.T., 1986. Phase compositions and cryptic variation in a 2.2 km section of the Windimurra layered gabbroic intrusion, Yilgarn block, Western Australia - a comparison with the Stillwater Complex. *Econ. Geol.* **81**, 1196-202.

Peyerl, W., 1982. The influence of the Driekop dunite pipe on the platinum-group mineralogy of the UG2 chromitite in its vicinity. *Econ. Geol.* **77**, 1432-8.

Prendergast, M.D., 1987. The chromite ore field of the Great Dyke, Zimbabwe. In: Stowe, C.W. (ed.) *Evolution of the chromium ore fields.* Evolution of ore fields series, van Nostrand Reinhold, New York, 89-108.

Prendergast, M.D., 1990. Platinum-group minerals and hydrosilicate alteration in the Wedza Mimosa platinum deposit, Great Dyke, Zimbabwe - genetic and metallurgical implications. *Trans. Inst. Mining Metallurgy Sect. B* **99**, 91-105.

Prendergast, M.D., 1991. The Mimosa Wedza platinum deposit, Great Dyke, Zimbabwe: layering and stratiform PGE mineralisation in a narrow mafic magma chamber. *Geol. Mag.* **128**, 235-49.

Pye, E.G., Naldrett, A.J., & Giblin, P.E. (eds.) 1984. *The geology and ore deposits of the Sudbury structure.* Ontario Geological Survey, Special Volume 1.

Raedeke, L.D., & Vian, R.W., 1986. A three dimensional view of mineralisation in the Stillwater JM reef. *Econ. Geol.* **81**, 1187-95.

Reynolds, I.M., 1978. Mineralogical studies of South African titaniferous iron ores: their applications to extractive metallurgy. *Trans. Geol. Soc. S. Afr.* **81**, 233-40.

Reynolds, I.M., 1985. The nature and origin of titaniferous magnetite-rich layers in the upper zone of the Bushveld Complex: a review and synthesis. *Econ. Geol.* **80**, 1089-108.

Ripley, E.M., 1986. Origin and concentration mechanisms of copper and nickel in the Duluth Complex sulphide zones - a dilemma. *Econ. Geol.* **81**, 974-8.

Roberts, S., Foster, R.P., & Nesbitt, R.W., 1990. Mineralisation associated with early Precambrian basic magmatism. In: Hall, R.P., & Hughes, D.J. (eds.) Early Precambrian basic magmatism. Blackie, Glasgow, 157-90.

Rudashevsky, N.S., Avontsev, S.N., & Dneprovskaya, M.B., 1992. Evolution of PGE mineralisation in hortonolitic dunites of the Mooihoek and Onverwacht pipes, Bushveld Complex. *Miner. Petrol.* **47**, 37-72.

Schiffries, C.M., 1982. The petrogenesis of a platiniferous dunite pipe in the Bushveld Complex; infiltration metasomatism by a chloride solution. *Econ. Geol.* **77**, 1439-53.

Scoon, R.N., & Mitchell, A.A., 1994. Discordant iron-rich ultramafic pegmatites in the Bushveld Complex and their relationship to iron-rich intercumulus and residual liquids. *J. Petrology* **35**, 881-917.

Scoon, R.N., & Teigler, B., 1994. Platinum-group element mineralisation in the critical zone of the western Bushveld Complex: 1. Sulphide poor chromitites below the UG2. *Econ. Geol.* **89**, 1094-121.

Stowe, C.W., 1994. Compositions and tectonic settings of chromite deposits through time. *Econ. Geol.* **89**, 528-46.

Stumpfl, E.F., 1993. Fluids: a prerequisite for platinum metals mineralisation. Current research in Geology Applied to Ore Deposits, 15-21.

Stumpfl, E.F., & Rucklidge, J.C., 1982. The platiniferous dunite pipes of the eastern Bushveld Complex. *Econ. Geol.* **77**, 1419-31.

Talkington, R.W., & Lipin, B.R., 1986. Platinum-group minerals in chromite seams of the Stillwater Complex, Montana. *Econ. Geol.* **81**, 1179-86.

Tarkian, M., & Stumpfl, E.F., 1975. Platinum mineralogy of the Driekop mine, South Africa. *Miner. Deposita* **10**, 71-85.

Teigler, B., 1990. Platinum group element distribution in the lower and middle group chromitites in the western Bushveld Complex. *Miner. Petrol.* **42**, 165-79.

Teigler, B., & Eales, H.V., 1993. Correlation between chromite composition and PGE mineralisation in the critical zone of the western Bushveld Complex. *Miner. Deposita* **28**, 291-302.

Todd, S.G., Kieth, D.W., LeRoy, L.W., Schissel, D.J., Mann, E.L., & Irvine, T.N., 1982. The JM platinum palladium reef of the Stillwater Complex, Montana:1 stratigraphy and petrology. *Econ. Geol.* **77**, 1454-80.

Tredoux, M., de Wit, M.J., Hart, R.J., Armstrong, R.A., Lindsay, N.M., & Sellschop, J.P.F., 1989. Platinum-group elements in a 3.5 Ga nickel-iron occurence: possible evidence of a deep mantle origin. *J. Geophys. Res.* **94**, 795-813.

van Rensburg, W.C.J., 1962. The geology of the Dwars river fragment and the ore minerals of the magnetite deposit on Kennedy's Vale 361KT, eastern Transvaal. Unpubl. M.Sc. thesis, Univ. Pretoria.

Vermaak, C.F., 1976. The Merensky Reef - thoughts on its environment and genesis. *Econ. Geol.* **71**, 1270-98.

Vermaak, C.F., 1985. The UG2 layer - South Africa's slumbering chromitite giant. *Chromium Review* **5**, 9-23.

Vermaak, C.F., & Hendriks, L.P., 1976. A review of the mineralogy of the Merensky Reef, with specific reference to new data on the precious metal mineralogy. *Econ. Geol.* **77**, 1244-69.

Viljoen, M.J., 1994. A review of regional variations in facies and grade distribution of the Merensky Reef, western Bushveld Complex, with some mining implications. Proceedings 15th CMMI Congress. South African Institute of Mining and Metallurgy, 183-94.

Viljoen, M.J., & Hieber, R.W., 1986. The Rustenburg section of Rustenburg Platinum Mines Limited, with reference to the Merensky Reef. In: Anhaeusser, C.R., & Maske, S. (eds.) *Mineral Deposits of Southern Africa,* **2,** Johannesburg: Geol. Soc. S. Afr., 1107-34.

Viljoen, M.J., & Scoon, R.N., 1985. The distribution and the main geologic features of discordant bodies of iron-rich ultramafic pegmatite in the Bushveld Complex. *Econ. Geol.* **80**, 1109-28.

Viljoen, M.J., de Klerk, W.J., Coetzer, P.M., Hatch, N.P., Kinloch, E.D., & Peyerl, W., 1986a. The Union section of Rustenburg Platinum Mines Limited, with reference to the Merensky Reef. In: Anhaeusser, C.R., & Maske, S. (eds.) *Mineral Deposits of Southern Africa,* **2,** Johannesburg: Geol. Soc. S. Afr., 1061-90.

Viljoen, M.J., Theron, J., Underwood, B., Walters, B.M., Weaver, J., & Peyerl, W., 1986b. The Amandulbult section of Rustenburg Platinum Mines Limited, with reference to the Merensky Reef. In: Anhaeusser, C.R., & Maske, S. (eds.) *Mineral Deposits of Southern Africa,* **2,** Johannesburg: Geol. Soc. S. Afr., 1041-60.

Volbarth, A., Tarkian, M., Stumpfl, E.F., & Housely, R.M., 1986. A survey of the Pd-Pt mineralisation along the 35km strike of the JM reef, Stillwater Complex, Montana. *Canad. Miner.* **24**, 329-46.

von Gruenewaldt, G., 1993. Ilmenite-apatite enrichments in the upper zone of the Bushveld Complex: a major titanium-rich phosphate resource. *Internat. Geol. Rev.* **35**, 987-1000.

von Gruenewaldt, G., Hatton, C.J., Merkle, R.K.W., & Gain, S.B., 1986. Platinum group element - chromitite associations in the Bushveld Complex. *Econ. Geol.* **81**, 1067-79.

von Gruenewaldt, G., Dicks, D., de Wet, J., & Horsch, H., 1990. PGE mineralisation in the western sector of the eastern Bushveld Complex. *Miner. Petrol.* **42**, 71-95.

Wager, L.R., Vincent, E.A., & Smales, A.A., 1957. Sulphides in the Skaergaard intrusion, east Greenland. *Econ. Geol.* **52**, 855-903.

Wagner, P.A., 1929. *The platinum deposits and mines of South Africa.* Edinburgh: Oliver and Boyd, 326 pp.

White, J.A., 1994. The Potgietersrus project geology and exploration history. Proceedings 15th CMMI Congress. South African Institute of Mining and Metallurgy, 173-82.

Wilson, A.H., 1982. The geology of the "Great Dyke", Zimbabwe: the ultramafic rocks. *J. Petrology* **23**, 240-92.

Wilson, A.H., & Tredoux, M., 1990. Lateral and vertical distribution of platinum-group elements and the petrogenetic controls on the sulphide mineralisation in the P1 pyroxenite layer of the Darwendale subchamber of the Great Dyke, Zimbabwe. *Econ. Geol.* **85**, 556-84.

Wilson, A.H., Lee, C.A., & Brown, R.T., 1995. Characterisation and reassessment of PGE, base metal and silicate compositional variation in the Merensky Reef, Rustenburg area. 1995 IGCP Project 336, Proceedings, Duluth, 209-10.

Winter, P.E., 1965. The Ushushwana Igneous Complex. *Swaziland Geol. Surv. Bulletin* **5**, 29 pp.

Zhangurov, T.B., 1994. Imandra Layered Intrusion. In: Mitrofanov, F., & Torkov, M. (eds.) *Field Guide to Kola PGE Geology.* Geological Institute, Apatity, 42-70.

Zientek, M.L., Foose, M.P., & Mei, L., 1986. Palladium, platinum, and rhodium contents of rocks near the lower margin of the Stillwater Complex, Montana. *Econ. Geol.* **81**, 1169-78.

LAYERED INTRUSIONS
R.G. Cawthorn (editor)

The Skaergaard Intrusion

A.R. McBirney

Department of Geological Sciences, University of Oregon, Eugene, Oregon 97403, U.S.A.

Abstract

Thanks to its magnificent exposures and extraordinarily complete sequence of strongly differentiated rocks, the Skaergaard Intrusion has long served as a prime example of shallow magmatic differentiation and as a testing ground for a wide range of petrologic concepts. During the magmatic episode accompanying the opening of the North Atlantic about 55 Ma ago, a moderately evolved tholeiitic magma was intruded, apparently in a single, prolonged pulse, into Archean gneisses and Tertiary basalts close to the eastern edge of Greenland. Over a period of about 10,000 years, the Layered Series crystallized on the floor, while similar sequences crystallized on the walls and under the roof to form the Marginal and Upper Border Series. In all three places, the minerals follow parallel trends with plagioclase progressing steadily from basic labradorite to sodic oligoclase and olivine and Ca-rich pyroxene evolving to pure fayalite and hedenbergite. The original magma, which was unusually rich in phosphorus and titanium, followed a trend of differentiation characterized by exceptionally strong iron enrichment and relatively little increase of silica until the very latest stages of differentiation when the magma split into two liquids, one very rich in iron and the other in silica. The original compositions and textures of the rocks have been altered by late-stage metasomatism which, in extreme cases, resulted in closely associated anorthosites and pyroxenites. The principal mechanism of crystal-liquid fractionation during formation of the Layered Series was compaction, but convective fractionation seems to have become important in the late stages of evolution.

1. INTRODUCTION

The descriptions of the Skaergaard Intrusion given by Wager and his co-workers (Wager and Deer, 1939; Wager and Brown, 1968) have stood the test of time remarkably well. Together with the subsequent work they inspired, they have made the intrusion a prime example of igneous differentiation and a testing ground for many of the theoretical advances of the last half century. As can be seen from the geological map (in the back envelope), increasingly detailed studies over the last twenty-five years have not changed the basic characterization of the geology, but a wealth of new geochemical and petrologic information has greatly clarified the processes by which the intrusion crystallized and differentiated.

2. GEOLOGIC SETTING AND STRUCTURE

The Skaergaard Intrusion (Figure 1) is one of a chain of gabbroic and syenitic bodies intruded near the eastern coast of Greenland during an intense magmatic episode associated with the opening of the North Atlantic about 55 Ma ago (Deer, 1976; Brooks and Nielsen, 1982). Swarms of dykes near the continental margin fed fissure eruptions of flood basalts that reached thicknesses of at least 3 or 4 km (Wager and Deer, 1938; Nielsen, 1978). This Eocene volcanic sequence, which forms the upper walls on the southern and most of the eastern side of the in-

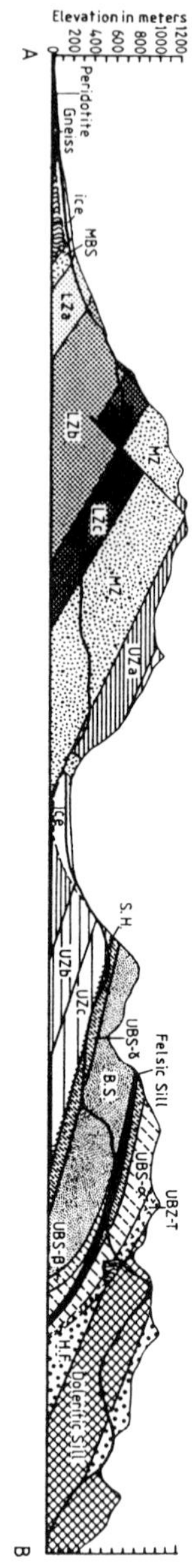

Figure 1. *Simplified geologic map (facing page) and north-south section (adjacent) through the Skaergaard Intrusion. The Basistoppen Sheet (BS) is a later intrusion that was fed by the large dyke cutting the east wall of the main intrusion. The solid line in the section represents the topography along AB onto which higher topography has been projected. A full-colour, 1:20,000 scale map is in the envelope at the back of the book.*

trusion, rests on 10 to 100 metres of Late Cretaceous to Paleocene arkoses and siltstones and a much thicker complex of Archean gneisses and amphibolites, at least 3000 Ma old (Kays *et al.*, 1989). The Precambrian rocks make up the walls of the lower two thirds of the intrusion.

The Skaergaard gabbros cut a set of slightly earlier north-trending basaltic dykes and are themselves intruded by later dykes and sills. Most of the latter strike more or less east-west, and, whereas the early dykes are chiefly tholeiitic, later ones are more varied and tend to be alkaline or transitional between alkaline and tholeiitic basalts (Nielsen, 1978). An important member of the later group is the Vandfaldsdalen Macrodyke (White *et al.*, 1989), which cuts the eastern margin of the Skaergaard Intrusion and appears to have been the feeder from which the Basistoppen Sill was intruded into the upper part of the Skaergaard Intrusion shortly after the latter had solidified (Hughes, 1956; Douglas, 1961; Naslund, 1989). In addition to these doleritic and gabbroic bodies, a swarm of relatively small granophyric dykes and sills intruded the central and upper parts of the intrusion (Hirschmann, 1992). The largest of these felsic bodies, the Tinden Sill, intruded at a level only a few hundred metres below the roof.

The entire coastal region was tilted southward by about 20° less than 250,000 years after the Skaergaard magma had crystallized but before it had cooled enough to acquire remnant magnetization at a temperature of about 500°C (Schwarz *et al.*. 1979). The tilting was not a simple flexure as originally thought; Nielsen (1978) has shown that it was due to antithetic rotation of several fault blocks.

3. AGE

Owing to their low potassium contents and extensive late-stage recrystallization, the gabbros making up the main mass of the intrusion have yet to yield reliable age determinations, but estimates have been obtained from fission-track measurements (Brooks and Gleadow, 1977) and from Rb/Sr (Hirschmann, 1992). Both methods give ages close to 55 Ma. This makes the Skaergaard ten to fifteen million years younger than any of the other large intrusions in this part of eastern Greenland.

4. FORM AND PRINCIPAL UNITS

The intrusion rose along a north-trending fissure to the level of the unconformity between the metamorphic basement and overlying volcanic rocks, then formed a wedge between the two

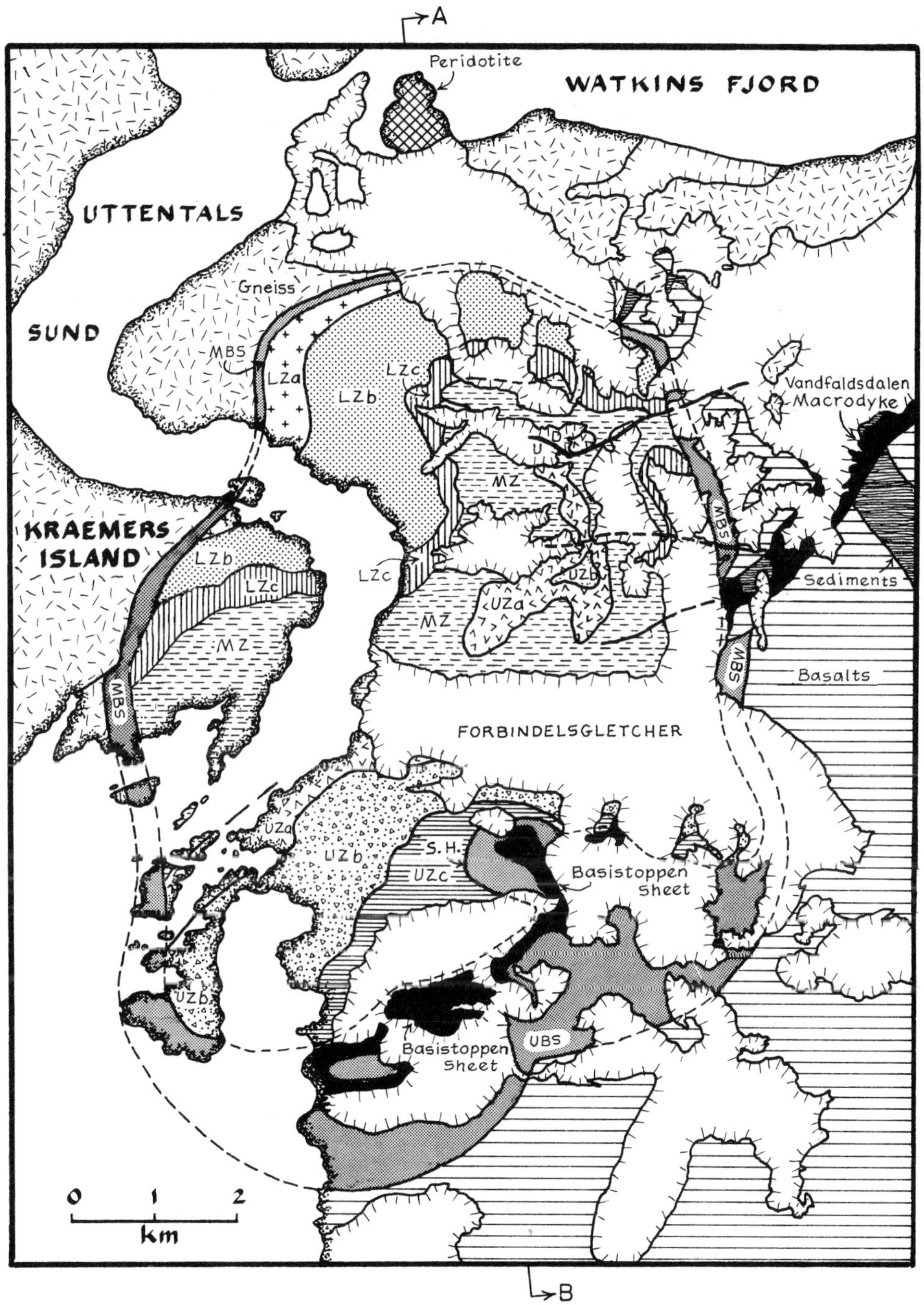
A
Peridotite
WATKINS FJORD
UTTENTALS
SUND
Gneiss
MBS
LZa
LZb
LZc
Vandfaldsdalen
Macrodyke
MZ
KRAEMERS
ISLAND
UZb
UZa
Sediments
Basalts
FORBINDELSGLETCHER
S. H.
UZc
Basistoppen
Sheet
UBS
0 1 2
km
B

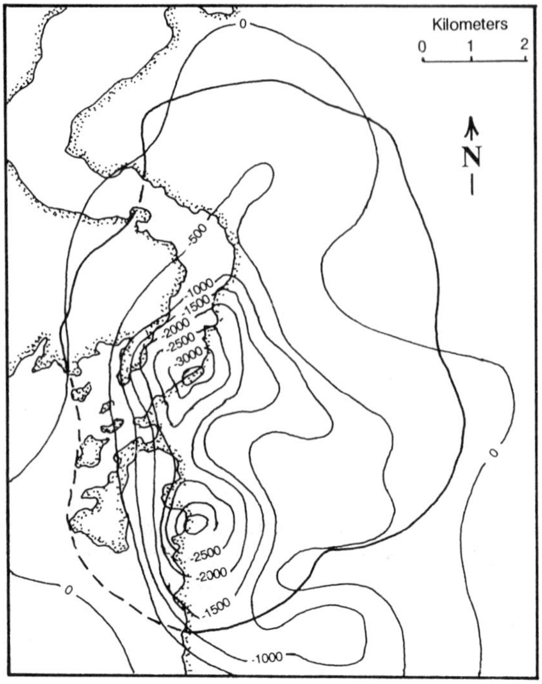

Figure 2. *A map view shows the subsurface form of the intrusion as defined by contours at 500 metre intervals down to a depth of 3.5 km. Subsurface contours are from Blank and Gettings (1973). The heavy line denotes the boundary of the intrusion at its present level of exposure.*

two series extending an unknown distance toward the north. It was bounded on the south by steep faults related to the coastal flexure. Owing to high relief on the pre-volcanic surface, the unconformity is approximately 800 metres lower on the east side.

At its present level of exposure, the Skaergaard Intrusion measures about 6 by 11 km. By virtue of its topographic relief of more than 1200 m and a southward tilt of 15 to 25°, an almost unbroken vertical section of more than 3.5 km is accessible to direct examination. Gravity and magnetic surveys (Blank and Gettings, 1973) have shown that the subsurface part of the body narrows sharply to a small feeder under the south-central part of the intrusion (Figure 2). As a result, the proportion of unexposed rock is substantially less than originally thought. This "Hidden Layered Series" was probably of the order of 10 to 15% of the original intrusion.

Because little of the roof is preserved, the original form and depth of the upper part of the body can be estimated only from indirect evidence. The relations of polymorphs of silica and Fe-rich pyroxenes (Lindsley *et al*,, 1969) indicate that the Sandwich Horizon crystallized at a pressure of about 500 bars. Judging from the thickness and density of overlying rocks, the roof could not have been much deeper than a kilometre below the Eocene surface.

The intrusion is asymmetrical in both a north-south and an east-west direction. The southern and western walls are nearly vertical, but the contact on the eastern side dips inward at about 45 to 60° and decreases to as little as 25 to 30° at the northern margin. These differences are due in part to the southeastward tilt, but most of the asymmetry resulted from shallow structural features that influenced the original form of the intrusion.

At least three steep, north-dipping normal faults cut the steep walls of the glacial valleys near the eastern contact; they appear to end before reaching the western half of the Layered Series. This faulting seems to have been associated with the intrusion of the Vandfaldsdalen Macrodyke that cuts the eastern margin and with a similar body, the Mikis Fjord Macrodyke about four kilometres to the southeast (White *et al.*, 1989).

The intrusion has been divided into three major units (Figure 3). A Layered Series that formed on the floor of the intrusion is lithologically and structurally distinct from the Marginal Border Series that crystallized on the walls and from the Upper Border Series that crystallized under the roof. The boundary between the Layered and Marginal Border Series on the western side is marked by a sharp angular discordance, whereas the division between these two units on the eastern side is more difficult to define, because the layering steepens gradually before becoming parallel to the wall. The upper boundary of the Layered Series is defined by its contact with the Upper Border Series along a level to which Wager and Deer (1939) gave the name Sandwich Horizon. This is not a separate unit but rather the demarcation between rocks that crystallized from the floor up and from the roof down. Where the two series are easily distinguished, as they are on the eastern side of Basistoppen Peak, the Sandwich Horizon is recognizable by the contrast between very mafic ferrodiorites below and more plagioclase-rich rocks above. Elsewhere, coarse melanogranophyres make up a large proportion of the uppermost Layered Series and lowermost Upper Border Series, and the boundary between the two is a transitional zone, 10 to 20 metres thick.

The compositions of the two Border Series broadly parallel those of the Layered Series. Part of the outermost, unbanded "Tranquil Division" of the Marginal Border Series is equivalent to the unexposed rocks beneath the Layered Series and to the uppermost member of the Upper Border Series. Blocks of wehrlite and olivine gabbro, the "gabbro picrite" of Wager and Deer (1939), were carried up in the magma at the time of intrusion and accumulated near the base, mainly along the northern edge of the intrusion. Once thought to have been picked up by the intruding magma from the unexposed lowermost part of the "Hidden Zone", the rocks are now known to have come from an older body immediately to the north and presumably below the Skaergaard Intrusion (Kays and McBirney, 1982). Thus, they have no petrologic relation to the Skaergaard magma and played no role in its evolution.

The exposed areas of the Layered, Upper Border, and Marginal Border Series have a ratio of about 20:7:1, but if the three-dimensional form of the intrusion is taken into account, the combined volume of the two border series was comparable to that of the Layered Series and may even have been greater.

5. COOLING HISTORY, HYDROTHERMAL ALTERATION, AND MINERALIZATION

Cooling of the intrusion was enhanced by a large-scale hydrothermal system that was especially vigorous in the highly permeable basalts of the roof and upper walls. Circulation through the less permeable metamorphic rocks was limited mainly to widely spaced fractures (Taylor and Forester, 1979). Hydrothermal fluids altered the oxygen isotopic ratios of the rocks to differing degrees depending on the temperatures and amounts of water that moved through different parts of the intrusion. Calculations based on these effects indicate that the magma solidified completely after about 130,000 years, but the hydrothermal system continued to circulate for at least another 370,000 years (Norton and Taylor, 1979; Bird *et al.*, 1988; Manning and Bird, 1991).

Copper and iron sulphides are particularly prominent near the top of Middle Zone. Gold associated with these sulphides (Nielsen and Schønwandt, 1990) has not proved to be of minable grade, and no other minerals of economic importance have been found in the intrusion.

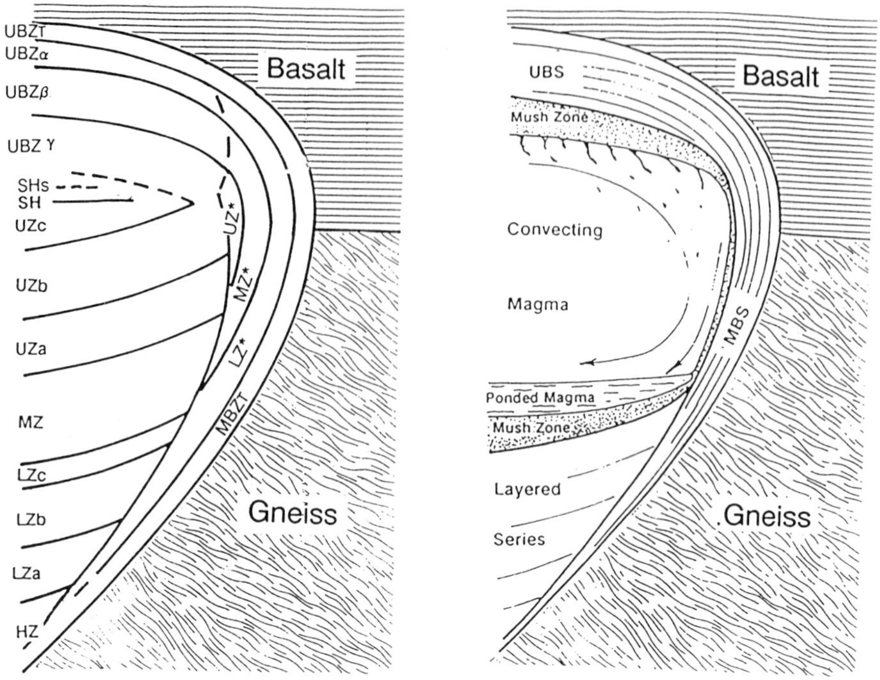

Figure 3 *The three main petrologic units, the Layered Series, Upper Border Series, and Marginal Border Series, crystallized concurrently along essentially parallel trends of differentiation. The Marginal Border Series has zones (LZ*, MZ*, etc.) corresponding to all the units of the Layered Series except UZc. The "Tranquil Zone" (MBZ_T) of the Marginal Border Series has an equivalent in the Upper Border Series (UBZ_T) and corresponds to the Hidden Zone (HZ) below the Layered Series. The Layered Series crystallized on the floor, while the Upper Border Series formed under the roof and the Marginal Border Series at the wall. The dense liquid produced by crystallization at the wall is believed to have descended to pond on the floor, while some of the heavy liquid and crystals produced under the roof probably sank into the main magma. The two fronts of crystallization converged at the Sandwich Horizon (SH), but the interstitial liquid continued to differentiate and migrate upward. As a result, incompatible elements reach maximum concentrations at a secondary Sandwich Horizon (SHs) in the lower part of the Upper Border Series. Diagrams are not to scale.*

6. LITHOLOGIC DIVISIONS

By projecting the surface outcrop shown in Figure 1 to a plane perpendicular to the original horizontal, the principal lithologic units are shown in the spatial relations they would have had in a vertical section through the intrusion (Figure 5).

6.1. Layered Series

The Layered Series is divided into Lower, Middle, and Upper Zones by the disappearance of abundant primary olivine at the base of Middle Zone and by its reappearance at the base of Upper Zone (Figure 4). Olivine is present in Middle Zone only as rare grains and as thin reaction products between pyroxene and Fe-Ti oxides. Lower Zone is further divided into three subzones, a, b, and c (LZa, LZb, and LZc) by the distinctive poikilitic texture of pyroxene in Lza and by the appearance of abundant Fe-Ti oxide minerals at the base of LZc. In a similar way, Upper Zone is divided into three subzones (UZa, UZb, and UZc) by the appearance of abundant, coarse apatite at the base of UZb and by the mosaic form of inverted ferrobustamite (formerly ferro-wollastonite) in UZc. Inverted pigeonite is found in all rocks up to the middle of UZa but is rarely abundant. Interstitial granophyre is common in the upper part of Upper Zone.

All the gabbros of the Layered Series are medium- to coarse-grained with local pegmatitic facies. Although, on average, the rocks become progressively more mafic upward, the modal proportions of plagioclase and mafic minerals range between those of anorthosite and pyroxenite. Even seemingly homogeneous rocks vary both vertically and along strike. Layering is prevalent in all units up to the lower part of UZb. Apart from the poikilitic pyroxene and olivine in LZa,

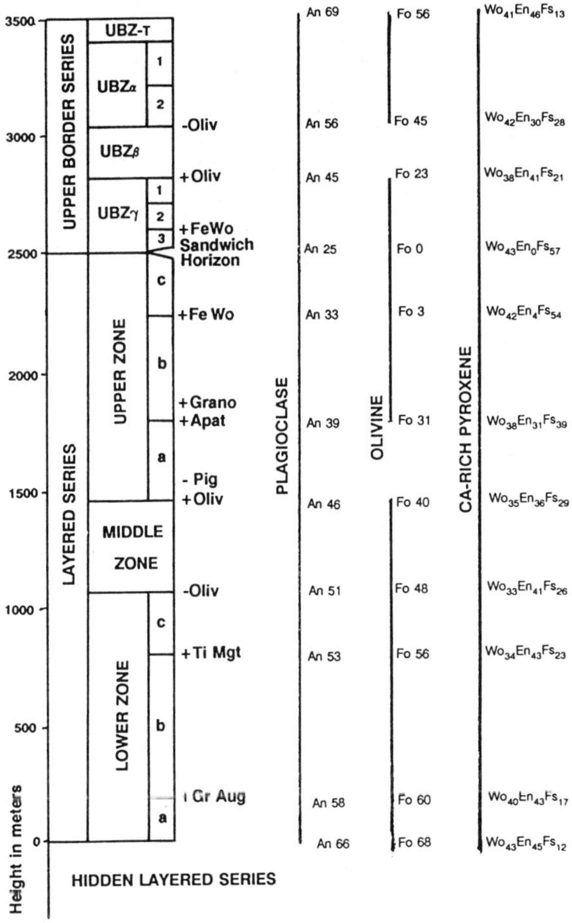

Figure 4. *The units of the Layered Series are defined by the appearance or disappearance of primary phases or, in the case of Lower Zone A, by the poikilitic form of the pyroxene (denoted Gr Aug). The Upper Border Series is divided into three main zones, α, β, and γ, equivalent to the Lower Middle and Upper Zones of the Layered Series. Upper Border Zone γ is subdivided to correspond to the three subzones of Upper Zone; Upper Border Zone α has upper and lower parts, 1 and 2, that are not directly equivalent to the three subzone of Lower Zone. The compositions given for the main silicate minerals at zone boundaries are approximate and may differ along the horizon extent of the boundary.*

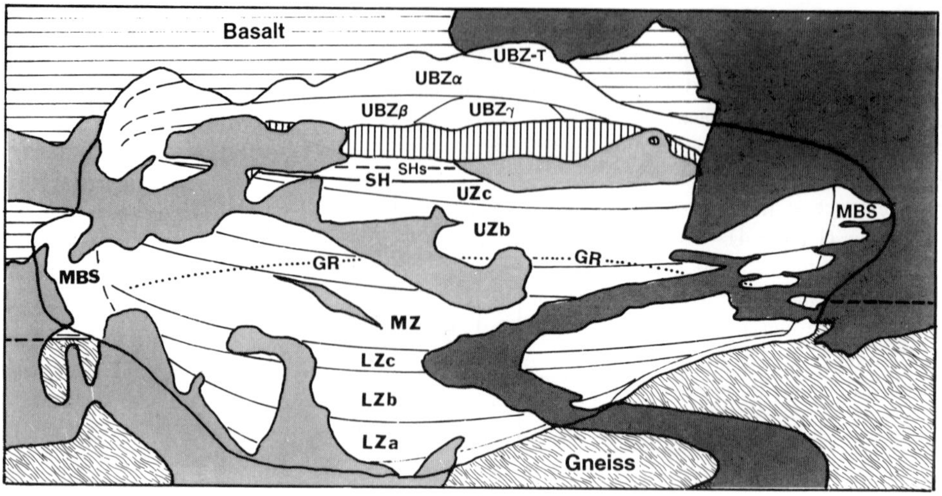

Figure 5. *Outcrop areas and zone boundaries are shown here as they would appear to an observer looking down at the angle the intrusion has been tilted toward the south-southeast. Such a view is not a true vertical section but a projection of the irregular erosion surface. Dark shading is water, light shading ice. Vertical ruling indicates the post-Skaergaard Basistoppen Sill. The dotted line labeled GR indicates the lower limit of rocks containing abundant interstitial granophyre. Other abbreviations are explained in the text and Figure 3.*

most mafic minerals have equigranular textures, the main exception being inverted pigeonite, which normally surrounds partly resorbed olivine and augite. Plagioclase commonly has a tabular form and a preferred orientation in the plane of layering (Brothers, 1964). Hydrothermal alteration is weak and largely confined to veins (Bird *et al.*, 1988); weathering is limited almost entirely to the mechanical effects of freezing and thawing.

6.2. Upper Border Series

Wager (1960) and Douglas (1961) divided the UBS into zones α, β, and γ, mainly on the basis of plagioclase compositions, corresponding to Lower, Middle, and Upper Zones of the Layered Series. These zones were modified by Naslund (1984a) on the basis of more detailed knowledge of their mineral compositions, but their equivalence to the Layered Series has been maintained. An uppermost unit, previously referred to as Upper Border Zone ultra-α, has been redefined as Upper Border Zone T. It is equivalent to the outer, unlayered "tranquil" zone of the Marginal Border Series and, presumably, to part of the Hidden Layered Series. It is a fine- to medium-grained rock containing primocrysts of olivine in a framework of plagioclase laths. UBZα is a coarse-grained gabbro with abundant plagioclase and smaller amounts of primary olivine, ilmenite, magnetite, and apatite but few if any pyroxene primocrysts. UBZβ contains primocrysts of Ca-rich pyroxene but none of olivine, whereas the assemblage of UBZγ includes olivine. Large crystals of apatite are scattered sporadically throughout the sequence, and quartz is a minor constituent at all levels. Ca-poor pyroxene is abundant in UBZ-T but rare in the other zones.

UBZα has been divided into two subzones, UBZα_1 and UBZα_2, and UBZγ into three subzones, UBZγ_1, UBZγ_1, and UBZγ_3, on the basis of their equivalence to corresponding parts of the Layered Series, but mineralogical distinctions are less pronounced than in the Layered Series, and the rocks are more heterogeneous. The zones differ widely in thickness from one section to the next, mainly because large sections broke away and sank into the underlying Layered Series. The thickness of the exposed section is about a third of that of the Layered Series. It seems to have maintained this ratio throughout the entire sequence of crystallization.

On average the rocks are more felsic and coarser grained than those of the Layered Series. Skeletal and dendritic crystals are very common, particularly among iron oxides, and some of the apatite crystals are hopper-shaped (Naslund, 1984b).

6.3. Marginal Border Series

The sequence of gabbros that crystallized on the walls (Hoover, 1989a) was divided by Wager and Brown (1968) into two main units, an outer "Tranquil Division" and an inner "Banded Division". Because the former is only weakly layered, it was thought to have crystallized before the onset of convection. It now appears that the change from unlayered to layered rocks was related in some way to the rates of cooling and *in situ* nucleation.

A chilled margin, a metre or so thick, is easily distinguished wherever the contact is exposed. Most of the fine-grained rocks chilled against the walls have hornfelsic textures and appear to have been altered to differing degrees from their initial compositions and textures, but a small section of doleritic rocks at the western margin seems to have preserved its original composition reasonably intact (Hoover, 1989b). The sample EG4507, considered by Wager and Deer (1939) to be most representative of the original magma, also has a well-preserved doleritic texture, but later attempts to locate the source of this critical rock have led to the conclusion that it was taken from a dyke.

Two notable members of the Tranquil Division are the perpendicular-feldspar and wavy-pyroxene rocks, first described by Wager and Deer (1939). The former is characterized by elongated plagioclase crystals oriented normal to the contact. The latter contains lenticular dark clots of pyroxene 2 to 5 cm long and a centimetre or so thick, crudely oriented parallel to the contact.

The units of the Banded Division, designated LZa*, LZb*, etc., correspond to equivalent units of the Layered Series. No rocks corresponding to UZc are exposed. Though much thinner, the units have a closer petrographic resemblance to the Layered Series than do those of the Upper Border Series. They parallel the walls in regular inward order, but each unit starts a few tens of metres above the base of its equivalent in the Layered Series.

Generally coarser than most gabbros of the Layered Series, the Banded Division has large proportions of pegmatitic rocks of both mafic and felsic compositions. Rounded clots of felsic pegmatite, normally with mafic rinds on the side towards the wall, are scattered throughout the Marginal Border Series but tend to be less common and more indistinct toward the interior.

Although its outcrop area accounts for only 5% of the entire intrusion, the total volume of the Marginal Border Series must have been proportionately much greater, perhaps as much as 15 to 20% of the original body.

6.4. Granophyres and felsic pegmatites

Felsic differentiates of the Skaergaard magma are limited almost exclusively to local segregations of pegmatite and melanogranophyre. The former are mostly zoned pods and slender diapirs in the lower part of the Layered Series but form dykes and sills at higher levels (Larsen

Figure 6. *(A) Modal layering in UZa. The dark zones are composed of ferroaugite, hortonalite, and iron oxides; light layers are mostly plagioclase. (B) cross bedding in Lza close to the Marginal Border Series.*

and Brooks, 1994). The latter are derived from immiscible liquids that separated to form pods and schlieren in the uppermost levels of the Layered Series and equivalent parts of the Upper Border Series (McBirney and Nakamura, 1973; McBirney, 1975; Naslund, 1983a). Granophyric dykes intruded the Layered Series in a crudely radial pattern about a centre overlying the northern root of the intrusion. Their large concentrations of lithophile elements and radiogenic strontium show that they are largely products of crustal anatexis with differing proportions of a mafic magmatic component not unlike the early Skaergaard liquids (Hirschmann, 1992). The large Tinden Sill which intruded the Upper Border Series is almost pure granophyre with a composition close to that of the ternary minimum in the system quartz-albite-potassium feldspar.

Many of the clots of granophyre and felsic pegmatite that are so common in much of the Marginal Border Series have compositions inherited from the xenolithic blocks of gneiss that were caught up in the margins of the intrusion (Kays *et al.*, 1981).

7. INTERNAL STRUCTURE

7.1. Layering

Although layering of some sort can be found in almost all parts of the intrusion, it varies

***Figure** 7. View of the central part of the Layered Series. The three prominent layers near the top, known as the Triple Group, are a useful stratigraphic horizon marking the top of Middle Zone. Gold mineralization has been found near the base of these three layers. Some of the large blocks that fell from the roof can be seen in the wall below the Triple Group. Note that layering in the adjacent gabbro has been propagated through a large anorthositic block.*

greatly in form, frequency, and spacing. Modal and grain-size layering is conspicuous in all but the upper three or four hundred metres of the Layered Series, but it tends to become more distinct upwards before ending abruptly near the middle of UZb. Layers are especially numerous in the vicinity of blocks that have fallen from the Upper Border Series. Much of the modal layering in the banded division of the Marginal Border Series is strong, but it tends to be wavy with large variations in grain size; in the tranquil division, it is much less pronounced and confined almost entirely to textural rather than modal variations. Modal and textural layering is seen in the Upper Border Series but it is much more diffuse than in the Layered Series.

In its most common form, layering is manifested in the modal proportions of dark and light minerals, but in places, particularly in the Marginal Border Series, it also results from variations of grain size and crystal orientation. Although graded layers are common, in many places dark and light layers form distinct pairs or single isolated layers (Figure 6A). In paired or graded layers of the Layered Series, the dark minerals are concentrated below the plagioclase-rich zone; in the Marginal Border Series they are on the side nearer the wall.

Much of the layering near the margins of the Layered Series is cut by shearing, crossbedding, and other conspicuous forms of deformation (Figure 6B). It is characterized by sharp modal contrasts, strong foliation, and linear orientation of elongated minerals (McBirney and Nicolas, in review). Most of the sharp layering in the interior of the Layered Series is weakly foliated but lacks lineation and is thought to have been formed mainly by compaction (McBirney and Boudreau, in prep.). Another less common variety of layering in the Layered Series is thought to be due to non-dynamic processes, such as variations in rates of nucleation and crystal growth. It tends to be more diffuse than most other types. The most notable example is the set of three prominent layers known as the "Triple Group" near the top of Middle Zone (Figure 7). Conspicuous when seen from a distance, the layers are hardly noticeable when viewed at close range.

Thin, dark layers regularly spaced at intervals of about 2 cm are locally well developed, especially in UZa and the upper part of MZ. This type of layering is limited to local exposures a metre or two thick and at most a few tens of metres in lateral extent. It is probably the result of cyclic recrystallization and Ostwald ripening (Boudreau, 1987).

7.2. Trough structures

A distinctive type of layering, referred to by Wager and Deer (1939) as "trough bands", occurs along the central and western parts of the boundary between UZa and UZb. Twenty or so of these synformal structures are oriented more or less radially with respect to the north-central part of the Layered Series. Strongly bimodal mafic and felsic layers, each a few centimetres thick, are separated by more homogeneous gabbro with about the same thickness as the paired layers (Figure 8). The curvature tends to become more pronounced toward the margins and upward in the sequence. Although the stacks of layers deviate little from the vertical axis, they rarely if ever have cross-cutting relations.

Originally thought be the products of turbidity currents flowing across the floor (Wager and Deer, 1939; Irvine, 1980), the remarkable regularity, lack of cross-bedding, and anomalous compositional features of the troughs have raised doubt about this interpretation (Taylor and

---

Figure 8. *(facing page) Trough structures in Upper Zone. (A) Trough "G". (B) Closer view of trough layering. Pocket knife in crack indicates scale.*

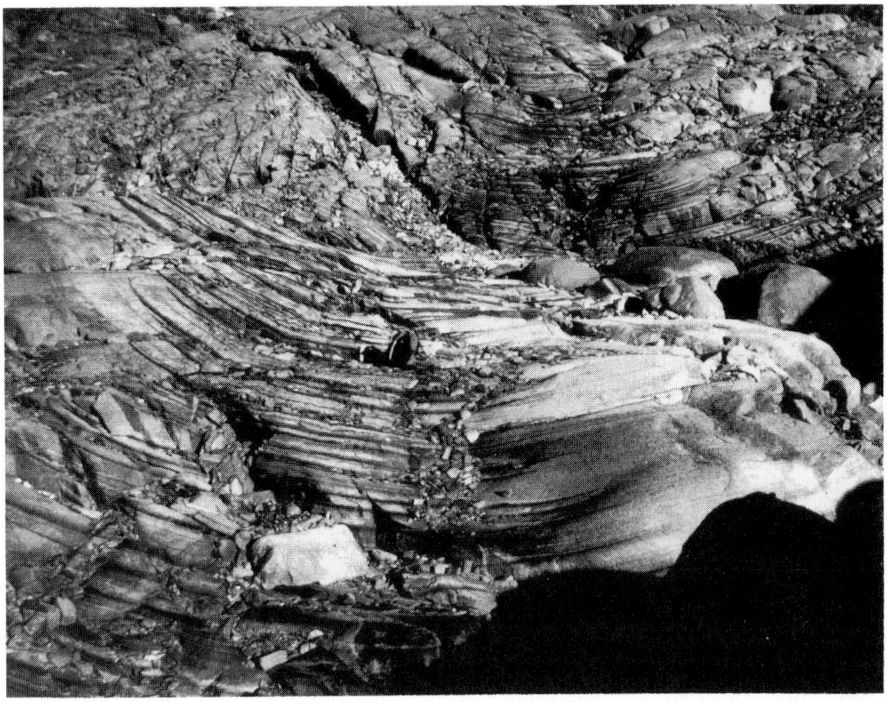

Forester, 1979; Sonnenthal, 1992; McBirney and Naslund, this volume).

7.3. Current erosion and slumping

In many places near the base of the intrusion and adjacent to the Marginal Border Series in the northern and western parts of the Layered Series the layering has been scoured by currents that eroded and redistributed the weakly consolidated crystals. Though much less common, erosional features can also be found at higher levels in the interior of the Layered Series, mostly in the vicinity of swarms of blocks that fell from the roof.

Near the margins, especially near the base of LZa, layers are offset by steeply dipping shears with displacement downwards towards the interior. The fabric of the rocks, particularly the orientation of platy plagioclase, shows that they were disturbed by slumping while still partly liquid (McBirney and Nicolas, in review).

Little if any evidence of slumping is seen in the Marginal Border Series. The Upper Border Series, however, has stratigraphic unconformities due to large missing sections where parts of the roof series became detached and fell away.

7.4. Inclusions

A variety of xenoliths and autoliths are found throughout the Layered and Marginal Border Series (Figure 9). In addition to the blocks of wehrlite and gabbro picrite mentioned earlier, the Border Series contains inclusions of gneiss and amphibolite in all stages of digestion (Kays *et al.* 1981). Basaltic xenoliths are much less common, having been noted at only three localities in the Layered Series (McBirney, 1979). Far more numerous are the blocks of leucogabbro and anorthosite that have fallen from the Upper Border Series into Lower and Middle Zones of the Layered Series. Although the compositions of these blocks are now more felsic than those of the Upper Border Series (McBirney and Sonnenthal, 1990), there is little doubt that the blocks fell from the roof. Major unconformities in UBZα and β can be related to swarms of blocks now found in parts of the Layered Series that were forming at the time sections of the roof became dislodged.

Small rafts of what appear to have been silty shales from the base of the Tertiary section have been found near the outer part of the Marginal Border Series. They have been converted to hercynite-rich cordierite hornfels (M. A. Kays, pers. comm.).

7.5. Metasomatic alteration

The importance of late-stage "post-cumulus" processes has become increasingly apparent as detailed studies have revealed that many of the rocks no longer retain their original textures and compositions (McBirney, 1987; McBirney and Sonnenthal, 1990; McBirney and Hunter, 1995; McBirney, 1995). These effects are most conspicuous in localized parts of the Lower Zone and the western Marginal Border Series (Figure 10). Irregular masses of anorthosite and olivine pyroxenite cut across the original layered gabbros in what appears to have been a volume-for-volume replacement. Mafic and felsic rocks of this kind are also found in the contact aureole of the Basistoppen Sill where they appear to be products of partial melting (Naslund, 1986), but in other places it is difficult to find evidence of remelting. Despite their total recrystallization, the rocks have textures that would normally be interpreted as igneous "cumulates" (McBirney and Hunter, 1995).

Figure 9. *(A) Blocks of wehrlite and gabbro picrite in the Marginal Border Series at the lowest exposed levels of the intrusion. Width of view ca. 10 m. (B) Blocks of anorthositic gabbro in Middle Zone.*

Figure 10. *(A) Anorthosites and olivine pyroxenites produced by metasomatic replacement of layered gabbros of the Marginal Border Series, Ivnarmiut Island. Width of view ca. 15 m. (B) Irregular segregations of anorthosite and pyroxenite in LZa.*

8. MINERAL ASSEMBLAGES

The principal minerals making up the three major series are essentially the same; they differ only in their average modal proportions. The most notable difference is the greater amount of plagioclase in the Upper Border Series. The order of appearance of liquidus phases is the same in the Layered and Marginal Border Series, but the Upper Border Series differs in that apatite, Fe-Ti oxides, and rare Ca-poor pyroxene appeared before augite. All three series become more mafic with differentiation, but the increase is most pronounced in the Layered Series.

As Figure 11 shows, the compositions of plagioclase, olivine, and pyroxene become steadily more evolved inwards from the floor, roof, and walls. When equivalent zones of the early and

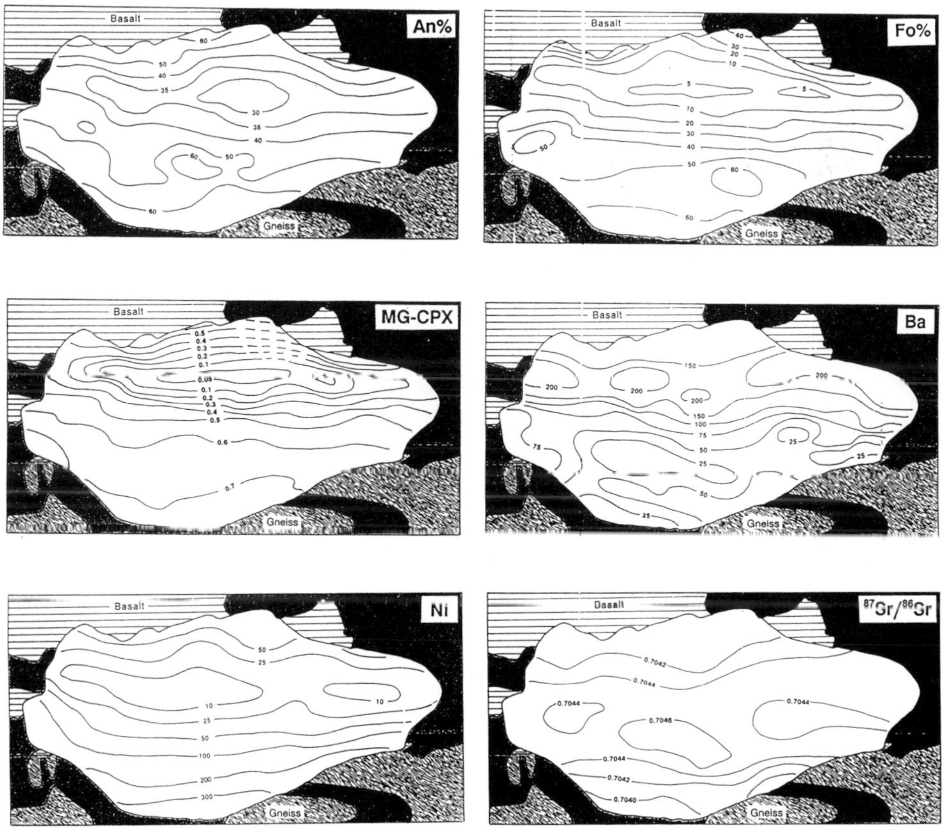

Figure 11. Spatial variations of various components are shown on the same projection as Figure 5: (A) anorthite content of plagioclase, (B) forsterite content of olivine, (C) mg ratio (Mg/Mg+Fe) of Ca-rich pyroxene. Bulk-rock variations of Ba (D), Ni (E), and initial $^{87}Sr/^{86}Sr$ (F). See McBirney (1995) for further details.

middle stages of crystallization are compared, the plagioclase and pyroxene of the Marginal Border Series tend to be less evolved than those of the Layered Series. The reverse seems to be true of the Upper Border Series, but in the late stages of differentiation, these differences became less pronounced.

The large numbers of new mineral analyses obtained by electron microprobe have revealed no basic difference in the trends of evolution of the principal minerals, but they have shown that the rocks are less homogeneous than earlier bulk analyses of mineral separates seemed to indicate. The compositions of plagioclase and pyroxene cores can differ by as much as 10 mole % An or En in a single thin section, but the normal range is closer to 2 or 3 %. The rims of plagioclase and pyroxene tend to be more strongly zoned in rocks near the margins of the Layered Series. Because these same rocks are richer in incompatible elements, the zoning of minerals must be due to greater amounts of trapped interstitial liquid.

A conspicuous anomaly is seen near the centre of the Layered Series where plagioclase is more An-rich and olivine is somewhat more forsteritic. The rocks in this area are greatly enriched in iron oxides and have anomalously small concentrations of incompatible elements, but their strontium isotopic ratios are the most radiogenic in the entire intrusion. The anomaly has been attributed to the metasomatic effects of a late, iron-rich fluid that rose from the roots of the intrusion and permeated the overlying solidification zone.

8.1. Feldspar

The Skaergaard feldspars have been studied by Gay and Muir (1962), and Maaløe (1976). The most anorthitic plagioclase of the Layered Series, found in a bore hole that penetrated about 150 m below the lowest exposed level of LZa, has a composition of An_{68} (Maaløe, 1976), whereas the most calcic plagioclase of the Marginal and Upper Border Series is An_{72}. The most albitic plagioclase cores, about An_{25}, are found in the central part of the Sandwich Horizon. Though common in the plagioclase of all three series, zoning rarely exceeds 5 mole %. Corroded cores are found in the plagioclase of many parts of the intrusion, but they are most conspicuous in the lower part of the Layered Series, the Hidden Zone, and equivalent parts of the two Border Series. They are especially common in anorthositic blocks that have fallen from the roof and in other rocks that seem to have suffered some degree of metasomatic recrystallization.

Potassium feldspar is a minor component of plagioclase; it reaches maximum proportions of about Or_4 in the most albitic plagioclase. Although the bulk-rock composition of the Upper Border Series is richer in K_2O than the Layered Series, the reverse is true of the plagioclase, probably because of the greater fugacity of water and lower temperature of the magma crystallizing under the roof (Naslund, 1984b). Shimizu (1978) has noted that the concentrations of K in the Skaergaard plagioclase are much greater than would be expected from the normal distribition coefficient. Alkali feldspar is seen only in granophyric intergrowths with quartz and inverted tridymite.

8.2. Pyroxenes

The Skaergaard pyroxenes have been studied by Brown (1957), Brown and Vincent (1963), Nwe (1976), and Coleman (1978). The most magnesian Ca-rich and Ca-poor pyroxenes have *mg* numbers (En/(En+Fs)) of 52.7 and 68.8 respectively. The most iron-rich compositions, found in and near the Sandwich Horizon, are virtually pure hedenbergite. The ferroaugites of UZc crystallized initially with the crystal structure of ferrobustamite but inverted on cooling to clinopyroxene. In the very latest rocks, however, hedenbergite crystallized directly without undergoing a polymorphic inversion.

Inverted pigeonite occurs both as separate crystals with well-developed exsolution lamellae on both (100) and (001) and as rims on resorbed olivine and augite. Although it does not seem to have been a liquidus phase during the early stages of crystallization of most of the Layered Series, at later stages it must have followed closely or even accompanied the precipitation of olivine. Apart from primary hypersthene found in a few of the earliest rocks of the two Border Series, all Ca-poor pyroxene has the form of inverted pigeonite.

8.3. Olivine

The most magnesian olivine in the Marginal Border Series is Fo_{74}, whereas that of the Layered Series is Fo_{68}. More forsteritic olivine in xenolithic blocks of wehrlite and gabbro picrite is not directly related to the Skaergaard magma. The olivine of the Sandwich Horizon is essentially pure fayalite. Throughout the intrusion, zoning is weak if not undetectable. Because olivine has re-equilibrated more than plagioclase and pyroxene, its composition is more sensitive to the amounts of interstitial liquid and tends to show wide variations that are crudely related to the bulk-rock concentrations of incompatible elements.

For a given composition of plagioclase, olivine is slightly more iron rich in the Upper Border Series than in the other two series.

In addition to normal primocrysts of olivine, most gabbros of LZc, as well as many of MZ and UZa, have small amounts of secondary olivine that is a product of the reaction between pyroxene and magnetite. During cooling and re-equilibration, iron and magnesium from the oxide minerals combined with Ca-poor pyroxene exsolved from augite to produce olivine with a composition close to that of primocrysts in the same rocks.

8.4. Fe-Ti oxides

Iron-titanium oxides are very abundant in all parts of the intrusion that crystallized after LZb. Ilmenite was the dominant mineral, even before the oxides exsolved and recrystallized. Compositions of the two oxide minerals indicate that they continued to equilibrate down to submagmatic temperatures of less than 800°C and oxygen fugacities as low as -16.4 (Buddington and Lindsley, 1964).

8.5. Apatite

Apatite is a primary mineral in all rocks later than UZa. It also occurs as an accessory mineral in earlier rocks of the Layered and Marginal Border Series, but it may have been a primary phase throughout crystallization of most of the Upper Border Series (Naslund, 1984a). The F-Cl ratios of apatites increase abruptly near the level of the trough structures. At the same time, the concentrations of rare-earth elements in the apatite increase sharply (Sonnenthal, 1992).

8.6. Accessory minerals

Quartz is present in interstitial granophyric intergrowths with alkali feldspar in most of the Layered Series above UZa. It does not appear to have been a primary mineral, even after Ca-poor pyroxene ceased to crystallize and olivine reached a high degree of iron enrichment. Inverted tridymite is common in much of the Upper Border Series, especially UBZγ.

Biotite is a widespread accessory mineral in LZa. Zircon and sphene are seen in a few rocks of the Sandwich Horizon and late members of the Upper Border Series. Lamellae of baddeleyite have been found in ilmenite of the Basistoppen Sill (Naslund, 1987).

Sulphides of Cu and Fe, principally bornite, chalcopyrite, and pyrite are sporadically present in most parts of the intrusion but are particularly abundant in UZc and in the vicinity of the Triple Group layers near the top of MZ.

Wager and Brown (1968) reported spinel and orthopyroxene in symplectite-like intergrowths at the edges of Fe-Ti oxides in LZc and MZ, but analyses by electron microprobe show that the minerals are chiefly olivine and plagioclase with traces of biotite (D. Russell, pers. comm., 1986). Naslund (1983b) has identified ilvaite as a product of subsolidus alteration of olivine in the Upper Border Series.

9. BULK-ROCK COMPOSITIONS

9.1. Major elements

Average bulk-rock compositions for the principal units of the Layered, Marginal, and Upper Border Series are summarized in Table 1. Compositional variations of key components are shown in Figure 12. The most conspicuous feature of the intrusion as a whole is its uninterrupted trend of iron enrichment. Silica declines in the early stages and does not increase notably until UZc; it reaches a maximum of about 49 % at the Sandwich Horizon.

Within the Layered Series, compositions vary horizontally as well as vertically. As noted in an earlier section, gabbros near the margins tend to be richer in iron and alkalies than those in the centre, mainly because of stronger zoning of the principal minerals. The Upper Border Series is uniformly less mafic but more silica-rich than the Layered Series. These differences are a reflection of the greater proportion of plagioclase and smaller amounts of

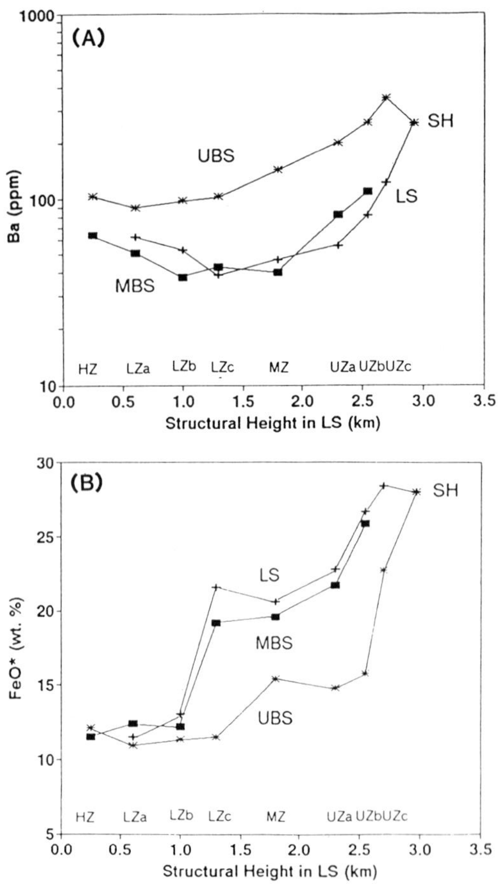

Figure 12. *The trend of differentiation of the Layered Series, illustrated here by the variations of Ba (A) and total iron as FeO* (B) as functions of structural height, closely parallels those in equivalent units of the Marginal and Upper Border Series, but the concentrations of incompatible elements in rocks under the roof are uniformly greater than those in the Layered Series. The two border series are shown at the structural height of the Layered Series units to which they correspond petrologically.*

olivine, pyroxene, and oxide minerals in rocks that crystallized under the roof. The Marginal Border Series more closely resembles the Layered Series but, on average, has less iron and titania.

9.2. Trace elements

Early investigations of trace-element variations (Wager and Mitchell, 1951; Haskin and Haskin, 1969; Paster *et al.*, 1974) have been supplemented by more detailed sampling and analyses designed to determine the spatial distribution of components in all accessible parts of the intrusion.

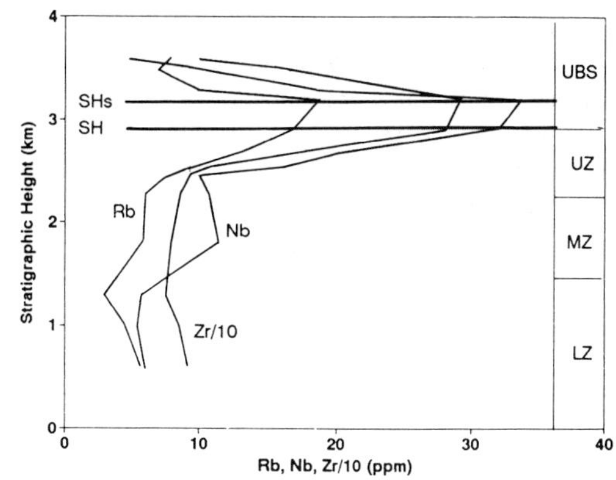

Figure 13. Average bulk-rock concentrations of Zr, Rb, and Nb with height in the Layered and Upper Border Series. Note that the abundances initially decline and then increase to reach maximum concentrations above the Sandwich Horizon (SH). Nb begins to increase at LZc, because it is weakly included in the abundant iron oxides of that zone. The level of maximum concentrations of incompatible elements constitutes a secondary Sandwich Horizon (SHs), in which the last low-density liquid was concentrated.

The abundances of included trace elements, such as Ni, Cr, Co, and V decline exponentially with evolution of all three series until they reach the limit of analytical detection near the Sandwich Horizon. The behaviour of incompatible elements is more complex. Ba, Zr, Nb, Rb, and the rare-earths decline from LZa up to Middle Zone and then reverse their trend and increase sharply through Upper Zone (Figure 13). These elements do not return to the bulk-rock concentrations they have in the lowest part of the Layered Series until high in the Upper Zone where about 90 % of the intrusion has crystallized. Most incompatible elements reach maximum concentrations in the lower part of the UBSγ, a hundred metres or so above the Sandwich Horizon. The entire Upper Border Series is richer in incompatible elements than the corresponding units of both the Layered and Marginal Border Series.

Concentrations of trace elements in minerals follow the bulk-rock trends but have second-order variations that seem to reflect the crystallization history of individual rocks. Included elements are weakly correlated with the modal abundance of the minerals in which they are concentrated. Concentrations of Ni in the olivine of olivine-rich rocks, are less than in olivine-poor rocks, as would be expected if the crystals grew from a limited mass of liquid.

In theory the trace-element compositions of minerals should provide a means of estimating the concentrations of these elements in the liquid with which the crystals last equilibrated. Unfortunately, large discrepancies are found between the calculated liquid concentrations obtained for coexisting minerals. For example, the potassium content of plagioclase is greater than what would be predicted from the distribution coefficient and potassium content of the magma (Shimizu, 1978). Moreover, the Sr contents of plagioclase and augite, when divided by their distribution coefficients, do not indicate the same liquid concentrations; the values

Table 1
Average bulk-rock compositions of the principal units of the Layered Series and equivalent rocks of the Marginal and Upper Border Series.

Layered Series								
	LZa	LZb	LZc	MZ	UZa	UZb	UZc	SH
SiO_2	48.12	48.84	41.10	42.79	43.07	41.78	46.00	49.09
TiO_2	1.35	1.44	6.92	6.79	5.67	4.06	2.63	2.22
Al_2O_3	16.81	12.55	11.02	11.53	11.17	9.51	7.86	7.88
FeO*	11.13	12.84	21.10	20.00	22.52	26.64	28.67	27.68
MnO	0.16	0.21	0.26	0.26	0.31	0.41	0.65	0.25
MgO	9.42	10.13	7.61	6.24	5.62	3.41	0.38	0.09
CaO	10.11	11.57	9.77	9.87	8.62	9.36	10.14	8.17
Na_2O	2.52	2.13	1.97	2.23	2.55	2.59	2.42	2.70
K_2O	0.27	0.20	0.20	0.21	0.26	0.36	0.41	0.72
P_2O_5	0.11	0.09	0.05	0.08	0.22	1.88	0.84	0.53
Ba	62	53	38	48	56	81	110	257
Li	6	6	5	6	6	8	8	9
Rb	6	6	4	6	6	7	8	15
Sr	285	219	199	218	233	244	263	450
Co	71	71	95	79	89	66	19	14
Cr	93	162	106	29	17	16	12	7
Sc	18	47	47	52	36	36	30	38
Ni	212	161	118	57	29	13	6	nd
Cu	118	105	101	375	1005	794	450	133
Zn	87	95	154	139	156	189	273	442
Zr	93	81	72	80	95	97	135	324
La	5.62	4.68	3.84	3.09	3.49	14.57	22.68	57.8
Sm	2.47	3.08	2.34	2.10	2.10	10.00	15.27	34.5
Lu	0.20	0.41	0.26	0.26	0.22	0.44	1.73	2.8

calculated for plagioclase being about an order of magnitude greater than those for augite. The explanation for these discrepancies has yet to be found.

9.3. Isotopic compositions

The rise of Tertiary basaltic magma through Rb-rich Archean gneisses afforded the ideal conditions for detecting crustal contamination before and after emplacement of the intrusion. Variations of strontium and neodymium isotopes have been examined by several investigators (Hamilton, 1963; Leeman and Dasch, 1978; Stewart and DePaolo, 1990), but a comprehensive study of the entire body remains to be completed (Creaser *et al.*, work in progress).

Results obtained thus far have revealed what seems to be a systematic pattern of vertical and horizontal variations. The Upper Border Series is isotopically different from contemporary rocks of the Layered Series. Even though equivalent units of the two series are thought to have crystallized from the same liquids, the former has distinctly less radiogenic Sr. Despite the

Table 1 (continued)

Marginal Border Series

	MBST	LZa*	LZb*	LZc*	MZ*	UZa*	UZb*	UZb*2
SiO_2	48.12	49.43	50.30	44.86	43.46	44.33	47.03	45.35
TiO_2	0.75	1.00	1.08	4.91	5.44	6.24	3.42	3.61
Al_2O_3	15.49	13.71	14.13	11.35	12.48	10.05	8.11	10.52
FeO*	10.29	11.87	11.83	19.02	20.51	21.47	25.59	25.89
MnO	0.16	0.19	0.18	0.26	0.23	0.29	0.39	0.33
MgO	10.62	10.77	9.42	6.60	5.64	5.57	2.07	1.76
CaO	10.83	10.76	10.89	10.59	9.54	9.39	6.63	8.87
Na_2O	1.99	1.93	1.88	2.03	2.24	2.08	2.07	2.65
K_2O	0.15	0.25	0.22	0.26	0.35	0.34	1.10	0.84
P_2O_5	0.04	0.08	0.07	0.13	0.12	0.24	1.59	0.18
Ba	21	51	34	45	40	82	109	184
Rb	11	8	7	7	8	11	12	19
Sr	277	234	181	201	191	259	264	280
Co	67	57	70	69	72	62	53	56
Cr	222	219	113	71	13	13	13	nd
Sc	32	35	66	48	56	46	47	35
Ni	334	138	140	82	66	22	3	nd
Cu	88	116	91	143	126	370	921	412
Zn	69	173	104	133	138	161	191	183
Zr	13	71	61	73	77	112	188	250
La	1.55	5.25	3.63	5.44	5.01	9.60	22.3	32.3
Sm	0.58	2.40	2.41	2.60	2.68	4.63	12.8	15.9
Lu	0.06	0.16	0.23	0.21	0.23	0.31	0.5	0.7

abundant xenoliths of gneiss in both border series and a well-documented exchange of Sr across the contact, the magma assimilated surprisingly small amounts of Sr from the metamorphic basement complex. The greater concentrations of incompatible elements in the Upper Border Series were long thought to reflect greater contamination by metamorphic rocks dislodged at the time of intrusion, but this explanation cannot be correct, for the Sr in these rocks is actually less radiogenic than that of the Layered Series (Figure 11F).

Neodymium isotopes vary much less than strontium and have no apparent irregularities in their distribution in the intrusion as a whole. The late-stage processes that seem to have affected the Sr had little effect on the Nd.

Although oxygen isotopes show that much of the body was affected by a hydrothermal system that circulated through the country rock and gabbro (Taylor and Forester, 1979; Norton and Taylor, 1979), no correlation has been found between the isotopes of oxygen and those of either Sr or Nd. The oxygen isotopes have been reset mainly in the upper and eastern parts of the intrusion where the wall rocks are permeable basaltic lavas and near faults that cut the eastern side of the Layered Series. Because the Sr isotopes are not notably affected in these

Table 1 (continued)

Upper Border Series

	UBZ-T	UBZα_1	UBZα	UBZα_2	UBZβ	UBZγ_1	UBZγ_2	UBZγ_3
SiO_2	49.24	49.47	49.58	49.68	46.99	51.96	53.86	51.55
TiO_2	1.97	2.33	2.41	2.48	4.47	2.51	1.98	1.50
Al_2O_3	18.27	16.81	16.20	15.69	12.98	13.12	11.96	8.92
FeO*	10.09	10.76	11.11	11.45	15.62	14.90	16.49	23.00
MnO	0.13	0.17	0.17	0.17	0.20	0.21	0.23	0.29
MgO	3.77	4.58	4.68	4.78	4.89	1.87	0.95	0.14
CaO	11.78	10.49	10.36	10.23	9.48	7.94	6.58	8.51
Na_2O	2.83	3.00	3.00	3.00	2.69	3.54	3.85	2.87
K_2O	0.30	0.41	0.40	0.38	0.57	0.98	1.30	0.82
P_2O_5	0.19	0.18	0.21	0.23	0.21	0.84	0.69	0.29
Ba	104	51	34	45	40	82	109	257
Rb	6	8	7	7	8	11	12	15
Sr	313	234	181	201	191	-	-	450
Co	39	57	70	69	72	62	53	14
Cr	108	219	113	71	13	13	13	7
Sc	31	31	35	29	34	27	25	28
Ni	62	138	140	82	66	22	3	nd
Cu	152	116	91	143	126	370	921	133
Zn	79	173	104	133	138	161	191	442
Zr	101	182	118	187	235	190	401	613
La	8.46	16.06	9.92	14.45	19.55	18.95	38.33	48.6
Sm	3.52	6.58	4.30	5.86	7.61	7.75	15.68	21.6
Lu	0.25	0.42	0.30	0.39	0.54	0.52	.	1.7

Compiled from data of McBirney (1989), Hoover (1989a), and Naslund (1984a) with additions and corrections based on new analyses. Because an equivalent of UZc is not exposed in the Marginal Border Series, the composition of the innermost exposed rocks of UZb* (Uzb*2) is given as the most differentiated composition of that series.

areas, any redistribution of Sr under subsolidus conditions must have been largely independent of the hydrothermal system. The noble gases in rocks of the Layered Series (Smith, 1984) have an isotopic character reflecting both atmospheric and juvenile components with the former being most prominent in regions that have been most strongly affected by the circulation hydrothermal solutions.

10. CRYSTALLIZATION HISTORY AND DIFFERENTIATION

10.1. Physical conditions

The temperatures and pressures at which the intrusion crystallized have been estimated by means of several geothermometres. The most important of these is based on the polymorphic forms of SiO_2 and $CaFeSi_2O_6$. Inversions of both the silica minerals and iron-rich pyroxenes occurred close to the Sandwich Horizon (Figure 14). Because the experimentally determined temperatures at which ferrobustamite inverts to hedenbergite are relatively insensitive to pressure (Lindsley *et al.*, 1969), the boundaries of the stability fields intersect the calculated tridymite-β quartz inversion curve at a high angle. This fortuitous combination provides an excellent grid of temperature and pressure to which the observed mineral assemblages can be related. The rocks near the Sandwich Horizon are narrowly confined to a liquidus temperature of about 940°C and a pressure close to 500 bars. Judging from the densities of the overlying gabbros and basalts, this corresponds to a depth of about 2 km for the Sandwich Horizon and places the roof less than a kilometre below the surface.

As mentioned in an earlier section, Fe-Ti oxides no longer preserve their original magmatic compositions and cannot be used to estimate temperatures and oxygen fugacities for the Layered Series. Nevertheless, the equilibrium relations of olivine, pyroxene, silica, and magnetite yield results (Morse *et al.*, 1980; Hoover, 1989b; McBirney, 1993) (Figure 15) in fairly good agreement with those obtained from independent, experimental studies of the stability relations of the natural rocks (McBirney and Naslund, 1990).

10.2. Composition of the initial magma

The composition of the initial magma favored by Wager (1960) and, until recently, used in nearly all calculations involving compositional variations, was based on that of a sample, reportedly taken from the chilled margin at the southwestern contact of the intrusion against basalts. As mentioned in an earlier section, the validity of this sample is open to question. A thorough study of the chilled margin by Hoover (1989b) has provided a more reliable composition based on samples judged to be closest to the original unaltered magma. An average of multiple analyses of 3 samples from a locality on the western side of the intrusion is given in Table 2.

This composition is notably richer in TiO_2, FeO*, SiO_2, P_2O_5, and alkalies and poorer in MgO and Al_2O_3 than the previously accepted one. It represents a magma that had already evolved from a primary mantle-derived liquid, presumably by differentiation prior to intrusion at the present level. These differences greatly reduce the amount of crystallization needed in the Hidden Zone from that required with the starting liquid assumed in earlier calculations. The

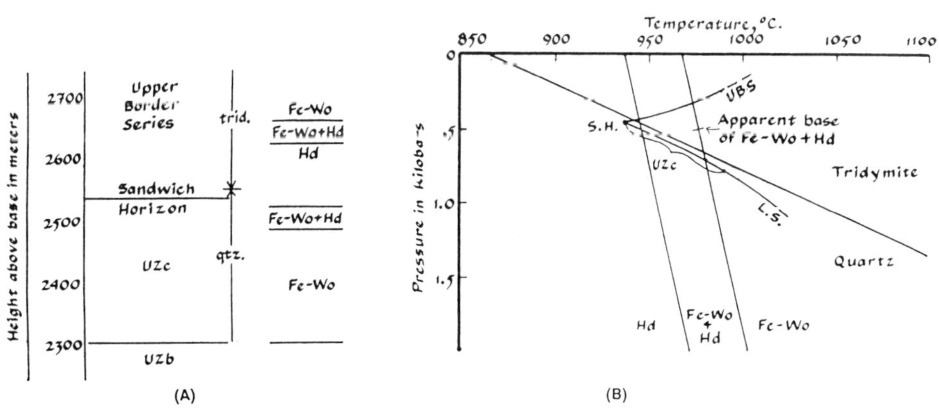

Figure 14 Temperature and pressure relations in the vicinity of the Sandwich Horizon (A) have been defined by the stability limits of the polymorphic forms of SiO_2 and $CaFeSi_2O_6$ in UZc and $UBZ\gamma$ (B).

Table 2
Compositions of the chilled margin.

	EG-4507	KT-39		EG-4507	KT-39		EG-4507	KT-39
SiO_2	48.1	49.62	Ba	48	96	Ni	120	141
TiO_2	1.17	2.61	Li	-	11	Cu	118	105
Al_2O_3	17.2	13.25	Rb	6	15	Zn	69	191
FeO*	9.6	13.03	U	-	0.43	Zr	93	159
MnO	0.16	0.22	Th	0.65	1.43	Hf	2.6	4.4
MgO	8.6	7.23	Ta	0.46	1.05	La	5.6	6.8
CaO	11.4	10.13	Sr	285	241	Ce	12.1	17.1
Na_2O	2.37	2.39	Co	54	57	Nd	9.2	13.3
K_2O	0.25	0.45	Cr	330	199	Sm	2.5	2.9
P_2O_5	0.10	0.22	Sc	23	35	Eu	1.04	1.13
						Tb	0.44	0.50
						Yb	1.20	1.47
						Lu	0.19	0.20

EG-4507 is Wager's (1960) preferred sample. KT-39 is an average composition of 3 samples from Hoover's (1989b) chilled margin. The average values shown for trace elements in KT-39 includes additional analyses not available to Hoover.

titanium- and phosphorus-rich composition resembles that of basalts found in certain parts of the continental interiors, such as the Great Basin of the western United States.

The intrusion may not have been a single impulse of uniform magma. The variations of Sr isotopic ratios, together with the declining abundances of incompatible elements, can be taken as evidence that the initial influx of magma continued, possibly until as late as Middle Zone time (McBirney, 1995), but the smooth, exponential decline of included elements, such as Ni and Cr, shows no evidence of a sudden injection of more primitive magma during crystallization of the Layered Series. A prolonged influx of slowly changing composition cannot be ruled out. Because of the unusual spatial variations of Sr isotopes and their poor correlation with trace elements or Nd isotopes, the question of additions to the initial intrusion has not yet been resolved.

10.3. Order of crystallization

Petrographic and experimental criteria indicate that the magma contained phenocrysts of plagioclase and subordinate amounts of olivine at the time it was emplaced. Ca-rich pyroxene appeared slightly later, first in the "wavy pyroxene rock" of the outer Marginal Border Series, then in poikilitic form in LZa, and finally as allotriomorphic granular crystals in LZb. The textural character of LZa was originally interpreted as the result of "orthocumulate" crystallization, but is now thought to be due chiefly to crystallization from relatively small numbers of pyroxene nuclei (McBirney and Noyes, 1979; McBirney and Hunter, 1995), much in the manner proposed by Wager (1961) for similar textures in the chilled margin.

This sequence of crystallization is consistent with experimental studies of the chilled margin composition by Hoover (1989b). On the magnetite-wustite and quartz-fayalite-magnetite buffers, plagioclase was found to have a liquidus temperature of 1170° and olivine 1165 to 1160°C, but if conditions had been closer to those of the nickel-nickel oxide buffer, these temperatures would have been about 1163 and 1154°C respectively. The temperature of 1135°C at which pyroxene was found to appear in experiments does not seem to have been measurably affected by oxygen fugacity.

Fe-Ti oxides began to precipitate in abundance at the base of LZc. Because of their extensive recrystallization, it is difficult to say whether both ilmenite and magnetite appeared together at this time. The fact that ilmenite is by far the more abundant oxide, and TiO_2 declines in subsequent rocks that continued to become more iron rich, indicates that ilmenite had a greater effect than magnetite on the evolution of liquid compositions. The appearance of Fe-Ti oxides was strongly influenced by the phosphorus and titanium contents of the liquid. Even small amounts of P_2O_5 suppress the field of magnetite in favour of ilmenite (Toplis and Libourel, 1992). The observed variations in the Layered Series are consistent with the dominance of ilmenite over magnetite from LZc to the Sandwich Horizon.

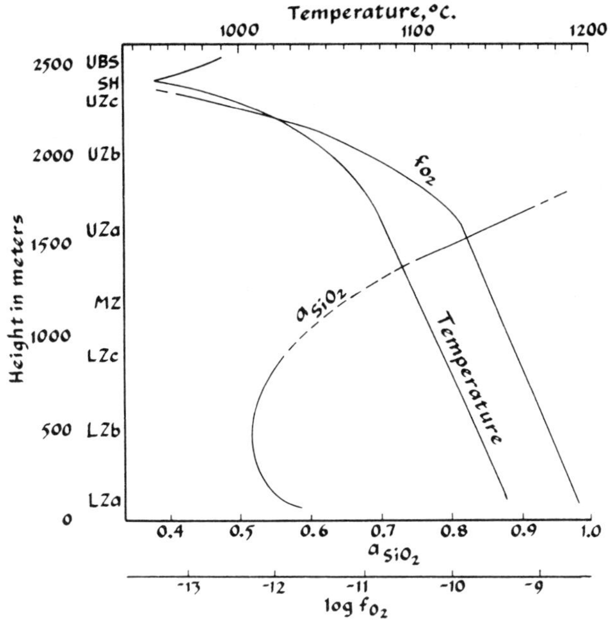

Figure 15. *Calculated temperatures, oxygen fugacities, and silica activities as functions of height in the intrusion. Temperatures and oxygen fugacities are based on experiments of McBirney and Nakamura (1973) and McBirney and Naslund (1990). Activity of SiO_2 has been calculated from the relations of olivine, Ca-poor pyroxene, and silica (McBirney, 1993).*

Olivine was a liquidus phase in the chilled margin, but it ceased to crystallize at the base of Middle Zone and reappeared at the base of Upper Zone. The mineral could not have been far from the liquidus during this hiatus, for occasional crystals of olivine are found in many of the rocks of Middle Zone. Moreover, throughout this interval, thin rims of olivine were produced by reaction between pyroxene and Fe-Ti oxides. The reappearance of olivine in UZa was followed closely by the disappearance of Ca-poor pyroxene and the appearance of interstitial granophyre.

10.4. Compositions of differentiated liquids

Several techniques have been employed to deduce the composition of the Skaergaard magma at successive stages of differentiation, but a fully reliable method has yet to be found. The early attempts by Wager (1960) were based primarily on mass-balance relations between an initial composition taken from what was erroneously thought to be a representative chilled

Table 3
Experimentally determined compositions of liquids in equilibrium with the principal units of the Layered Series.

	LZa	LZb	LZc	MZ	UZa	UZb	UZc
T (°C)	1150	1135	1115	1097	1082	1056	1002
log f_{O_2}	-8.7	-8.9	-9.2	-9.5	-10.0	-11.1	-12.4
SiO_2	48.8	48.3	47.3	46.3	46.9	46.3	44.5
TiO_2	2.5	3.4	5.0	5.5	3.6	3.4	3.2
Al_2O_3	14.1	13.3	11.1	10.3	9.5	8.2	7.9
FeO*	13.6	15.2	17.5	19.7	22.8	26.0	30.5
MnO	0.2	0.2	0.3	0.3	0.4	0.5	0.6
MgO	6.6	5.5	4.8	4.3	3.2	1.9	0.3
CaO	11.3	11.1	11.1	10.6	10.0	10.0	10.0
Na_2O	2.4	2.4	2.4	2.5	2.4	2.4	2.3
K_2O	0.3	0.3	0.2	0.2	0.5	0.5	0.5
P_2O_5	0.2	0.2	0.3	0.3	1.2	1.3	0.9

After McBirney and Nakamura (1973) and McBirney and Naslund (1990).

margin, and averaged compositions of successive zones the volumes of which were taken as proportional to their thickness. Various refined versions of this method have proved useful for specific purposes. For example, mass-balance relations can be used to estimate the relative amount of crystallization needed to change the composition of the initial magma to one that would crystallize the rocks of LZa. The value of about 25 % obtained in this way is consistent with the estimated volume of the Hidden Layered Series and equivalent rocks of the Marginal and Upper Border Series. But calculations of this kind cannot be carried beyond this initial step without introducing large cumulative errors, for the problem of determining the relative volumes of successive units seems insurmountable (McBirney and Naslund, 1990).

Experiments designed to restore the "trapped liquid" retained in rocks of the Layered Series (Table 3) indicate the general character of liquids that could be in equilibrium with the observed mineral assemblages, but because all the compositions were probably affected by late-stage processes, partial melting may not reproduce the exact composition of the main magma from which these minerals crystallized. Calculations based on crystal-liquid equilibria are theoretically feasible but entail multiple assumptions as to the mechanisms of differentiation and require a more complete knowledge of the history of the rocks than is now possible. A more pragmatic approach uses the compositions of dykes found in close association with the intrusion (Brooks and Nielsen, 1978; 1990). Although this method is based on observed rocks that clearly crystallized from liquids, it rests on the assumption that the dykes differentiated under the same conditions as the larger intrusion. It is also limited by the restricted range of compositions seen in dykes.

Despite their individual shortcomings, the various methods tend to converge on a general course of differentiation that differs little from that deduced by Wager (1960). The trend of iron enrichment with little enrichment of SiO_2 indicated in Table 3 is in good accord with several well-established petrologic observations, such as the unusually large abundances of apatite and ilmenite and the immiscible relations between the melanogranophyres and ferrogabbros. In the late stages of differentiation when the magma reached a high degree of iron enrichment it began to separate a second liquid, richer in silica and poorer in iron than the main magma (McBirney and Nakamura, 1973; McBirney, 1975). This immiscible relation accounts for pods of melanogranophyre that make up as much as 15% of UZc and the Sandwich Horizon.

The trend of strong iron enrichment was originally attributed to exceptionally low fugacities of oxygen, but as Ping and Libourel (1991) have shown, it is more likely due to the composition of the initial liquid and its position with respect to a thermal divide separating the trend of iron enrichment from that of silica. This effect was magnified by the unusually large amounts of TiO_2 and P_2O_5 in the magma. Kushiro (1975) has shown that both of these components have the effect of suppressing silica enrichment, and Toplis and Libourel (1992) found that under conditions near the Ni-NiO buffer, phosphorus increases the proportion of melt to crystals at any given temperature. The proportions of pigeonite and ilmenite increase at the expense of olivine and magnetite, and the SiO_2 content of the liquid is reduced.

10.5. Mechanisms of differentiation

Most of the evolution of the Skaergaard magma can be explained as the result of processes operating within the zones of crystallization at the roof, floor, and walls (McBirney, 1995). Assimilation seems to have played only a minor role. The distribution of trace elements, particularly incompatible elements, in rocks that crystallized where the solidification front had different orientations with respect to gravity provides a way of assessing the relative importance of different mechanisms of crystal-liquid fractionation. This is less true of strongly included elements, such as Ni, V, and Cr. Locked into the crystalline fraction of the solidification zone, they remained relatively immobile and largely unaffected by removal or replacement of their coexisting liquids.

Although diffusive exchange at the solidification front was originally considered the basic mechanism of cumulate processes (Hess, 1939; Wager, 1963), it does not seem to have played a major role in the Skaergaard case, for no obvious difference is seen in the behaviour of components of widely differing diffusivities (McBirney, 1995). The rates of enrichment of high-field-strength and large-ion-lithophile elements, for example, are virtually indistinguishable (Figure 13). The effects of diffusive exchange seem to have been overwhelmed by those of mass-transfer in porous flow.

Compaction must have played an important role in segregation of residual liquids from the rocks forming on the floor, for the Layered Series retained a much smaller proportion of late interstitial liquids than did corresponding rocks formed under the roof. Thus, it is the absence of compaction rather than contamination with basement rocks that best explains the greater concentrations of incompatible elements in the Upper Border Series. The upward-declining concentrations of incompatible components in the lower part of the Layered Series is readily explained as the result of an increased efficiency of compaction as the rate of cooling declined, the thickness of the crystal mush increased, and the forces tending to expel interstitial liquids became greater. At the same time, however, progressive enrichment of these components in the remaining liquid eventually caused this trend to be reversed. From the Middle Zone on, concentrations of incompatible elements increase exponentially, even though the slower rate of cooling would have led to smaller amounts of trapped liquids. Compaction continued to be effective until the fronts of crystallization finally converged at the Sandwich Horizon. These inferences based on the distribution of trace elements are supported by petrographic evidence, such as bent laths of plagioclase and equilibrated grain boundaries.

Convective exchange was driven by differences between the compositional densities of the main magma and interstitial liquids. Increasingly iron-rich liquids of the main stages of differentiation tended to migrate downwards, whereas buoyant, late-stage liquids rich in incompatible elements and volatile components infiltrated upwards, much as they have in Hawaiian lava lakes (Helz, 1987; Helz *et al.*, 1989). The consequences at the floor were enrichment of iron and depletion of incompatible elements; under the roof they were the reverse. Convection seems to have been most effective at the steep walls, where a dense boundary layer descended to pond on the floor. Under the roof, iron-rich liquids and a certain proportion of the mafic minerals separated from the zones of crystallization and returned to the main magma leaving the Upper Border Series somewhat more felsic than equivalent rocks in other parts of the intrusion. At more advanced stages of crystallization when density relations were reversed, a buoyant liquid, rich in volatiles and incompatible elements, rose from the Layered Series and permeated the upper part of the intrusion. The level at which this occurred is marked by an abrupt increase in the F-Cl ratio in apatites, probably reflecting exsolution of a vapour phase (Sonnenthal, 1992). The line defining the lower limit of gabbros with abundant interstitial granophyre and greater concentrations of incompatible trace elements (Figure 5) probably marks the lower boundary of rocks that were saturated in this way.

The late-stage liquid continued to evolve after the lower and upper crystallization fronts met at the Sandwich Horizon. As a result, most incompatible elements reach maximum concentrations at a secondary Sandwich Horizon located in the lower part of the Upper Border Series (Figure 13).

11. SUMMARY AND CONCLUSIONS

The descriptions of basic geologic features and petrologic units by earlier workers have not been materially altered by more recent studies. The main trends of evolution of rocks and minerals also remain essentially valid. A geophysical survey has shown that the subsurface form of the body is much smaller than previously believed, and the "gabbro picrite", once thought to come from the Hidden Layered Series, is now known to be exogenous. The relatively small strontium-isotope ratios of the Upper Border Series show that the greater concentrations of incompatible elements in gabbros crystallized under the roof are not the result of contamination with basement gneiss.

Detailed sampling and new analytical data show that the compositional and petrographic variations are more complex than originally thought. Much of the intrusion has been affected by late-stage, metasomatic processes that, in places, have extensively altered the bulk compositions and mineral assemblages of the rocks to anorthosites and pyroxenites. Owing to these late-stage affects, it is difficult to explain the present compositional variations in terms of a simple mechanism of crystal-liquid fractionation based on crystal settling. A combination of processes must have operated through a long period of cooling and re-equilibration. The chief mechanism responsible for differentiation was compaction, but convective fractionation was also important, especially in the upper levels of the intrusion.

12. ACKNOWLEDGEMENTS

Recent studies of the Skaergaard Intrusion have been carried out by the co-operative efforts of the University of Oregon and several other institutions, chiefly the California Institute of Technology, the State University of New York at Binghamton, the University of Alberta, and

Stanford University. In addition, we have benefitted from the co-operation and assistance of countless individuals, including Peter Brown, L. C. Coleman, Julius E. Dasch, W. P. Leeman, Charles Lesher, Troels Nielsen, and Adolphe Nicolas. Most of the work reported here was carried out with the support of the National Science Foundation. The Danish Geodetic Institute prepared a topographic base map using ground control and aerial photography provided by the University of Oregon.

13. REFERENCES

Bird, D.K., Manning, C.E., & Rose, N.M., 1988. Hydrothermal alteration of Tertiary layered gabbros, East Greenland. *Am. J. Sci.* **288**, 405-57.

Blank, H.R. & Gettings, M.E., 1973. Subsurface form and extent of the Skaergaard intrusion, East Greenland. *Trans. Am. Geophys. Un.* **54**, 507.

Boudreau, A.E., 1987. Pattern formation during crystallization and the formation of fine-scale layering. In: Parsons, L. (ed.) *Origins of Igneous Layering.* NATO ASI Ser. C, **196**, 453-71.

Brooks, C.K., & Gleadow, A.J.W., 1977. A fission-track age for the Skaergaard Intrusion and the age of the East Greenland basalts. *Geology* **5**, 539-40.

Brooks, C.K., & Nielsen, T.F.D., 1978. Early stages in the differentiation of the Skaergaard magma as revealed by a closely related suite of dyke rocks. *Lithos* **11**, 1-14.

Brooks, C.K., & Nielsen, T.F.D., 1982. The Phanerozoic development of the Kangerdlugssuaq area, East Greenland. *Meddr. Gronland, Geosci.* **9**, 1-30.

Brooks, C.K., & Nielsen, T.F.D., 1990. A discussion of Hunter and Sparks (Contrib Mineral Petrol **95**: 451-461). *Contr. Miner. Petrol.* **104**, 244-7.

Brothers, R.N., 1964. Petrofabric analyses of Rhum and Skaergaard layered rocks. *J. Petrology* **5**, 255-74.

Brown, G.M., 1957. Pyroxenes from the early and middle stages of the Skaergaard intrusion, East Greenland. *Miner. Mag.* **31**, 511-43.

Brown, G.M., & Vincent, A.E., 1963. Pyroxenes from the late stages of fractionation of the Skaergaard Intrusion, East Greenland. *J. Petrology* **4**, 175-97.

Buddington, A.F., & Lindsley, D.H., 1964. Iron-titanium oxide minerals and synthetic equivalents. *J. Petrology* **5**, 309-57.

Coleman, L.C., 1978. Solidus and subsolidus compositional relationships of some coexisting Skaergaard pyroxenes. *Contr. Miner. Petrol.* **66**, 221-7.

Deer, W.A., 1976. Tertiary igneous rocks between Scoresby Sund and Kap Gustov Holm, East Greenland. In: Escher, A., & Watt, W.S., (ed.) *Geology of Greenland.* Grønlands Geologiske Undersøgelse, 405-29.

Douglas, J.A.V., 1961. Geologic investigations in East Greenland: Part VII. The Basistoppen sheet, a differentiated basic intrusion into the upper part of the Skaergaard Complex, East Greenland. *Meddr. Gronland* **164**, 1-66.

Gay, P., & Muir, I.D., 1962. Investigation of the feldspars of the Skaergaard Intrusion. *J. Geol.* **70**, 565-81.

Hamilton, E.I., 1963. The isotopic composition of strontium in the Skaergaard Intrusion, East Greenland. *J. Petrology* **4**, 383-91.

Haskin, L.A., & Haskin, M.A., 1969. Rare-earth elements in the Skaergaard intrusion. *Geochim. Cosmochim. Acta* **32**, 433-47.

Helz, R.T., 1987. Differentiation behavior of Kilauea Iki lava lake, Kilauea Volcano, Hawaii: An overview of past and current work. Mysen, B. (ed.) *Magmatic Processes: Physicochemical Principles,* Geochemical Society, Spec. Publ. **1**, 241-58.

Helz, R.T., Kirschenbaum, H., & Marinenko, J.W., 1989. Diapiric transfer of melt in Kilauea Iki lava lake, Hawaii: A quick, efficient process of igneous differentiation. *Geol. Soc. Am. Bull.* **101**, 578-94.

Hess, H.H., 1939. Extreme fractional crystallization of a basaltic magma: The Stillwater igneous complex. *Trans. Am. Geophy. Un.* pt. 3, 430-32.

Hirschmann, M., 1992. Origin of the transgressive granophyres from the Layered Series of the Skaergaard Intrusion, East Greenland. *J. Volc. Geoth. Res.* **52**, 185-207.

Hoover, J.D., 1989a. Petrology of the Marginal Border Series of the Skaergaard Intrusion. *J. Petrology* **30**, 399-439.

Hoover, J.D., 1989b. The contact gabbro and initial liquid composition of the Skaergaard Intrusion. *J. Petrology* **30**, 441-76.

Hughes, C.J., 1956. Geological investigations in East Greenland, Part VI. A differentiated basic sill enclosed in the Skaergaard Intrusion, East Greenland, and related sills injecting the lavas. *Meddr. Gron.* **137**, 1-27.

Irvine, T.N., 1980. Magmatic density currents and cumulus processes. *Am. J. Sci.* **280-A**, 1-58.

Kays, M.A., Goles, G.G., & Grover, T. W., 1989. Precambrian sequence bordering the Skaergaard Intrusion. *J. Petrology* **30**, 321-61.

Kays, M.A., & McBirney, A.R., 1982. Origin of the picrite blocks in the Marginal Border Group of the Skaergaard Intrusion, East Greenland. *Geochim. Cosmochim. Acta* **46**, 23-30.

Kays, M.A., McBirney, A.R., & Goles, G.G., 1981. Xenoliths of gneiss and the conformable clot-like granophyres in the Marginal Border Group of the Skaergaard Intrusion, East Greenland. *Contr. Miner. Petrol.* **76**, 265-84.

Kushiro, I., 1975. On the nature of silicate melt and its significance in magma genesis: regularities in the shift of the liquidus boundaries involving olivine, pyroxene, and silica minerals. *Am. J. Sci.* **275**, 411-31.

Larsen, R.B., & Brooks, C.K., 1994. Origin and evolution of gabbroic pegmatites in the Skaergaard Intrusion, East Greenland. *J. Petrology* **35**, 1651-79.

Leeman, W.P., & Dasch, E.J., 1978. Strontium, lead, and oxygen isotopic investigation of the Skaergaard Intrusion, East Greenland. *Earth Planet. Sci. Lett.* **41**, 47-59.

Lindsley, D.H., Brown, G.M., & Muir, I.D., 1969. Conditions of the ferrowollastonite-ferrohedenbergite inversion in the Skaergaard Intrusion, East Greenland. *Miner. Soc. Am. Spec. Publ.* **2**, 193-201.

Maaløe, S., 1976. Zoned plagioclase of the Skaergaard Intrusion, East Greenland. *J. Petrology* **17**, 398-419.

Manning, C.E., & Bird, D.K., 1991. Porosity evolution and fluid flow in the basalts of the Skaergaard magma-hydrothermal system, East Greenland. *Am. J. Sci.* **291**, 201-57.

McBirney, A.R., 1975. Differentiation of the Skaergaard Intrusion. *Nature* **253**, 691-4.

McBirney, A.R., 1979. Effects of assimilation. In: Yoder, H.S. (ed.) *The Evolution of the Igneous Rocks*, Princeton: Princeton University Press, 307-38.

McBirney, A.R., 1987. Constitutional zone refining of layered intrusions. In: Parsons, I. (ed.) *Origins of Igneous Layering*, Dordrecht: Reidel Publ. Co., 437-52.

McBirney, A.R., 1989. The Skaergaard Layered Series: I. Structure and average compositions. *J. Petrology* **30**, 363-97.

McBirney, A.R., 1993. *Igneous Petrology*, 2nd Ed., Boston: Jones & Bartlett, 508 pp.

McBirney, A.R., 1995. Mechanisms of differentiation of layered intrusions: Evidence from the Skaergaard Intrusion. *J. Geol. Soc. Lond.* **152**, 421-35.

McBirney, A.R., & Hunter, R.H., 1995. The cumulate paradigm reconsidered. *J. Geology* **103**, 114-22.

McBirney, A.R., & Nakamura, Y., 1973. Immiscibility in late-stage magmas of the Skaergaard Intrusion. *Carnegie Inst. Wash. Geophy. Lab. Rpt.* 348-52.

McBirney, A.R., & Naslund, H.R., 1990. The differentiation of the Skaergaard Intrusion: a discussion of Hunter and Sparks (Contr. Miner. Petrol., **95**, 451-61). *Contr. Miner. Petrol.* **104**, 235-40.
McBirney, A.R., & Noyes, R.M., 1979. Crystallization and layering of the Skaergaard Intrusion. *J. Petrology* **20**, 487-564.
McBirney, A.R., & Sonnenthal, E.L., 1990. Metasomatic replacement in the Skaergaard Intrusion, East Greenland: Preliminary observations. *Chem. Geol.* **88**, 245-60.
Morse, S.A., Lindsley, D.H., & Williams, R.J., 1980. Concerning intensive parameters in the Skaergaard Intrusion. *Am. J. Sci.* **280-A**, 159-70.
Naslund, H.R., 1983a. The effect of oxygen fugacity on liquid immiscibility in iron-bearing silicate melts. *Am. J. Sci.* **283**, 1034-59.
Naslund, H.R., 1983b. Ilvaite, an alteration product replacing olivine in the Skaergaard Intrusion. *Am. Miner.* 68, 1004-8.
Naslund, H.R., 1984a. Petrology of the Upper Border Series of the Skaergaard Intrusion. *J. Petrology* **24**, 185-212.
Naslund, H.R., 1984b. Supersaturation and crystal growth in the roof-zone of the Skaergaard magma chamber. *Contr. Miner. Petrol.* **86**, 89-93.
Naslund, H.R., 1986. Disequilibrium partial melting and rheomorphic layer formation in the contact aureole of the Basistoppen Sill, East Greenland. *Contr. Miner. Petrology* **93**, 359-67.
Naslund, H.R., 1987. Lamellae of baddeleyite and Fe-Cr spinel in ilmenite form the Basistoppen Sill, East Greenland. *Can. Miner.* **25**, 91-6.
Naslund, H.R., 1989. Petrology of the Basistoppen Sill, East Greenland: A calculated magma differentiation trend. *J. Petrology* **30** , 299-319.
Nielsen, T.F.D., 1978. The Tertiary dyke swarms of the Kangerdlugssuaq area, East Greenland. An example of magmatic development during continental break-up. *Contr. Miner. Petrol.* **67**, 63-78.
Nielsen, T.F.D., & Schønwandt, H.K., 1990. Gold and platinum group metal mineralization in the Skaergaard Intrusion, southern East Greenland. Rapp. *Grøn. Geol. Unders.* **148**, 101-3.
Norton, D., & Taylor, H.P. Jr, 1979. Quantitative simulation of the hydrothermal systems of crystallizing magmas on the basis of transport theory and oxygen isotope data: an analysis of the Skaergaard Intrusion. *J. Petrology* **20**, 421-86.
Nwe, Y.Y., 1976. Electron-probe studies of the earlier pyroxenes and olivines of the Skaergaard Intrusion, East Greenland. *Contr. Miner. Petrol.* **55**, 285-300.
Paster, T.P., Schauwecker, D.S., & Haskin, L.A., 1974. The behavior of some trace elements during solidification of the Skaergaard layered series. *Geochim. Cosmochim. Acta* **38**, 1549-77.
Ping, S., & Libourel, G., 1991. The effect of FeO on the system CMAS at low pressure and implications for basaltic crystallization processes. *Contr. Miner. Petrol.* **108**, 129-45.
Schwarz, E.J., Coleman, L.C., & Cattroll, H.M., 1979. Paleomagnetic results from the Skaergaard Intrusion, East Greenland. *Earth Planet. Sci. Lett.* **42**, 437-44.
Shimizu, N., 1978. Analysis of the zoned plagioclase of different magmatic environments: a preliminary ion-microprobe study. *Earth Planet Sci. Lett.* **39**, 398-406.
Smith, S.P., 1984. Atmospheric and juvenile noble gases in the Skaergaard layered igneous intrusion. *Geochim. Cosmochim. Acta* **48**, 1033-41.
Sonnenthal, E.L., 1992. Geochemistry of dendritic anorthosites and associated pegmatites in the Skaergaard Intrusion, East Greenland: Evidence for metasomatism by a chlorine-rich fluid. *J. Volc. Geoth. Res.* **52**, 209-30.
Stewart, B.M., & DePaolo, D.J., 1990. Isotopic studies of processes in mafic magma chambers: II. The Skaergaard Intrusion, East Greenland. *Contr. Miner. Petrol.* **104**, 125-41.

Taylor, H.P., Jr., & Forester, R.W., 1979. An oxygen and hydrogen isotope study of the Skaergaard Intrusion and its country rocks: a description of a 55-MY-old fossil hydrothermal system. *J. Petrology* **20**, 355-419.

Toplis, M., & Libourel, G., 1992. Phosphorus in basalts and ferro-basalts: an experimental study. *EOS, Trans. Amer. Geoph. Un.* **73**, 615.

Wager, L.R., 1960. The major element variation of the Layered Series of the Skaergaard Intrusion and a re-estimation of the average composition of the Hidden Layered Series and of the successive residual magmas. *J. Petrology* **1**, 364-98.

Wager, L.R., 1961. A note on the origin of ophitic texture in the chilled margin of the Skaergaard Intrusion. *Geol. Mag.* **98**, 353-69.

Wager, L.R., 1963. The mechanism of adcumulate growth in the layered series of the Skaergaard Intrusion. *Spec. Paper Miner. Soc. Am.* **1**, 1-9.

Wager, L.R., & Brown, G. M., 1968. *Layered Igneous Rocks*, Edinburgh: Oliver & Boyd, 588 pp.

Wager, L.R., & Deer, W.A., 1938. A dyke swarm and coastal flexure in East Greenland. *Geol. Mag.* **75,** 39-46.

Wager, L.R., & Deer, W.A., 1939. Geological investigations in East Greenland, Part III. The petrology of the Skaergaard Intrusion, Kangerdlugssuaq, East Greenland. *Meddr. Gronland.* **105**, 352 pp.

Wager, L.R., & Mitchell, R.L., 1951. The distribution of trace elements during strong fractionation of basic magma - a further study of the Skaergaard Intrusion, East Greenland. *Geochim. Cosmochim Acta* **1**, 129-208.

White, C.M., Geist, D.J., Frost, C.D., & Verwoerd, W.J., 1989. Petrology of the Vandfaldsdalen Macrodike, Skaergaard region, East Greenland. *J. Petrology* **30**, 271-98.

LAYERED INTRUSIONS
R.G. Cawthorn (editor)

The Bushveld Complex

H.V. Eales[a] and R.G. Cawthorn[b]

[a]Department of Geology, Rhodes University, PO Box 94, Grahamstown, 6140, South Africa.

[b]Department of Geology, University of the Witwatersrand, Private Bag 3, Wits, 2050, South Africa.

Abstract

The layered mafic rocks of the Bushveld Complex are derived from three magmatic lineages - a lower part comprising the Lower and Critical Zones which crystallized from high-Mg and -Si parent liquids, an isotopically distinct Lower Main Zone derived from more evolved aluminous tholeiitic liquids, and a succession above the Pyroxenite Marker, which includes the entire Upper Zone, stemming from the mixing of residua of these earlier liquids with a final major injection of tholeiitic liquid. The varied mineralogy and thickness of the Marginal Zone indicate that it is not a quenched magma, but represents variable cumulus enrichment into several chemically different magmas, only some of which may be representative of magma producing the layered rocks.

The complex was emplaced by repeated injections of magma ranging in volume from small to large. Evidence supporting the concept of multiple injections lies in the presence of distinct breaks in initial Sr-isotopic ratio, textures indicating partial resorption of earlier crystallizing phases, abrupt reversals of fractionation trends defining saw-tooth profiles through successive units, and protracted reversals in mineral compositions through hundreds of metres of section that culminate in primitive olivine-rich cumulates in the Lower and Critical Zones. The small influxes yielded localized partial cyclic units <1 m thick. The size of the larger influxes may be gauged from the thickness and areal extent of chromitite layers which require hundreds to thousands of times their own thickness of liquid to satisfy the Cr budget for their formation.

Studies of the Critical Zone in the Western limb, for which more comprehensive data are available than elsewhere, reveal a gradation along *ca.* 200 km of strike from a more primitive proximal facies in the northwest to a more evolved distal facies in the southeast. This concept helps to explain major problems in the Cr budget of the entire limb, the greater proportion of chromite- and olivine-rich cumulates in the proximal facies, and the more feldspathic character of the distal sequence. In the Eastern limb changes in relative spacing between, and thicknesses of, chromitite layers from north to south suggest a similar process, but the transition is abrupt rather than gradational.

No single process can account for all forms of layering. Despite certain limitations, large-scale magma mixing is the most plausible mechanism to bring the liquid composition into the primary phase volume of chromite to produce chromitite layers. By contrast, the trace-element chemistry of magnetitite layers suggests that they were derived from relatively thin liquid layers, and that magma addition and mixing did not occur. The origin of the incomplete cycles which may range from pyroxene- and/or olivine-rich to feldspar-rich layers of the Lower, Critical, and Main Zones appears to be intimately bound up with the emplacement of fresh magma batches. These mixed with partially crystallized residual liquids to produce cumulates bearing incompletely resorbed plagioclase inclusions, highly variable Sr-isotopic ratios, and non-cotectic proportions of phases. Chemical fingerprinting of pyroxenes in Critical Zone rocks shows that gravitational sorting must have played some part in the generation of pyroxenites and in yielding non-cotectic norites.

Prominent layering near the top of the Main Zone in the Eastern limb cannot be explained by injection of fresh magmas, oscillatory nucleation or crystal settling, leaving the action of density currents as the most probable mechanism. Whereas layering is well developed in parts of the Upper Zone, the cyclicity of the Critical Zone is absent. The existence of an extremely thick liquid column in the chamber during the later stages is indicated by the 2500 m-thick, isotopically homogeneous sequence above the Pyroxenite Marker. Layering and reversals in mineral composition within this interval may have resulted from the breakdown of stratified liquid layers or convective overturn, rather than addition of magma.

1. INTRODUCTION

Much has been learned about the Bushveld Complex since the synthesis by Wager and Brown (1968). This chapter only briefly reviews the well-established features, and highlights new information stemming from exploration and mining operations, and from recent mineralogical and geochemical studies on the layered mafic and ultramafic rocks.

Within an immense areal extent of *ca.* 65 000 km^2 and thickness of 7-9 km of layered mafic rocks, lithologies range between the extremes of dunite and pyroxenite to anorthosite and pure oxide layers, with almost all conceivable intermediate varieties. Layering is present at all vertical scales from millimetres to hundreds of metres and is very continuous laterally. There are significant cryptic and modal variations, not only vertically, but also along strike, and significant variations of ratios exist in the various isotopic systems. Despite an age of 2.06 Ga (Walraven *et al.*, 1990), the mafic rocks have experienced only mild or local alteration and no metamorphism.

2. GEOLOGICAL SETTING

2.1. Major limbs

The complex outcrops in four apparently discrete regions (Figure 1), with a fifth hidden below younger sediment. Whether they were once joined at depth is unknown, but drilling has revealed extensions of the Western limb at its northern end beneath Bushveld Granite, east of the Crocodile River Fault, and of the Eastern limb beneath Karoo sedimentary cover west of the Wonderkop Fault. The Western limb (Figure 2a) extends 200 km along an arc from near Thabazimbi to north of Pretoria. Outcrop is generally very poor. The 200 km-long Eastern limb (Figure 2b), extending from Stoffberg to Chuniespoort, underlies rugged terrain where surface exposure is better, though still far from complete. The other three limbs are less well known. The Far Western limb is an eroded remnant extending to the Botswana border. The Northern or Potgietersrus limb is partially hidden beneath younger rocks, with exposures confined to its eastern edge and near Villa Nora. The Southeastern or Bethal limb was identified on the basis of a gravity high and is known only from bore-core information.

***Figure 1.** (facing page) Simplified geological map of the entire Bushveld Complex.*

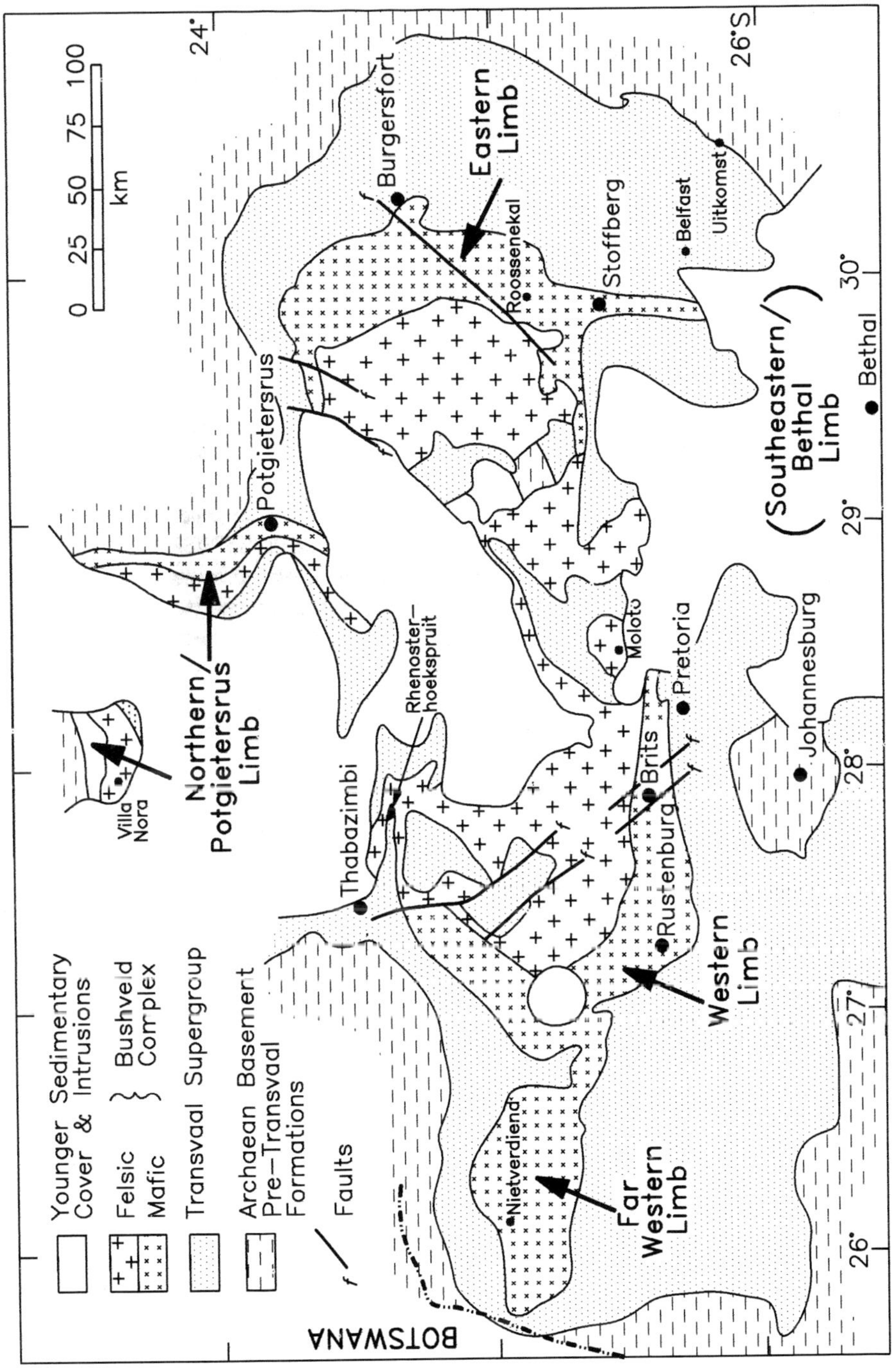

Younger Sedimentary Cover & Intrusions
Felsic
Mafic
Bushveld Complex
Transvaal Supergroup
Archaean Basement Pre-Transvaal Formations
Faults
0 25 50 75 100
km
24°
26°S
30°
29°
28°
27°
26°
BOTSWANA
Villa Nora
Northern/ Potgietersrus Limb
Potgietersrus
Burgersfort
Eastern Limb
Roossenekal
Stoffberg
Belfast
Uitkomst
(Southeastern/ Bethal Limb)
Bethal
Moloto
Pretoria
Johannesburg
Rhenoster-hoekspruit
Thabazimbi
Brits
Rustenburg
Western Limb
Nietverdiend
Far Western Limb

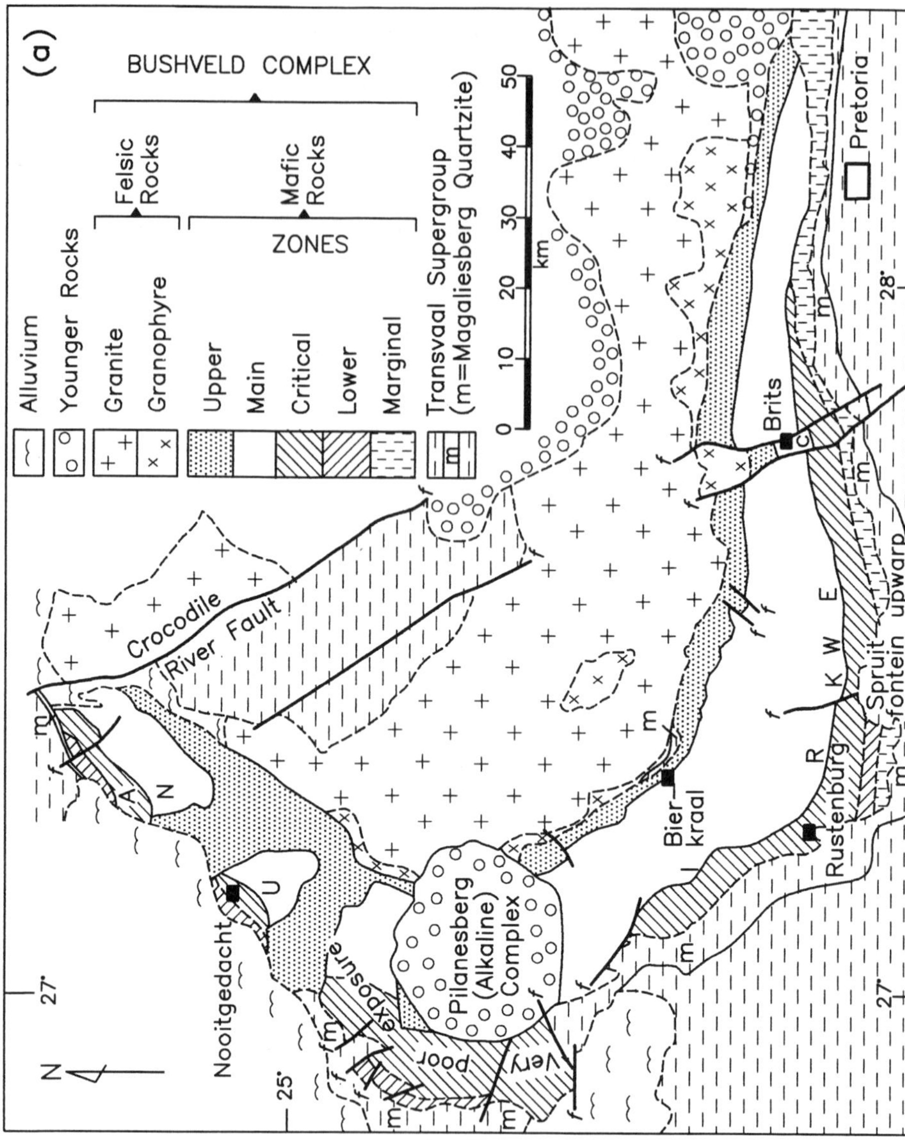

***Figure 2**. (above and facing page) Geological maps of the Western (a) and Eastern (b) limbs of the Bushveld Complex. The abbreviations in (a) refer to the platinum mines: A - Amandelbult Section, N - Northam, U - Union Section, I - Impala, R - Rustenburg Section, K - Karee, W - Western, and E - Eastern Platinum Mines. The only active platinum mine in the Eastern limb (b) is denoted A - Atok. Generally, the Marginal Zone is too narrow to include, but is almost always present.*

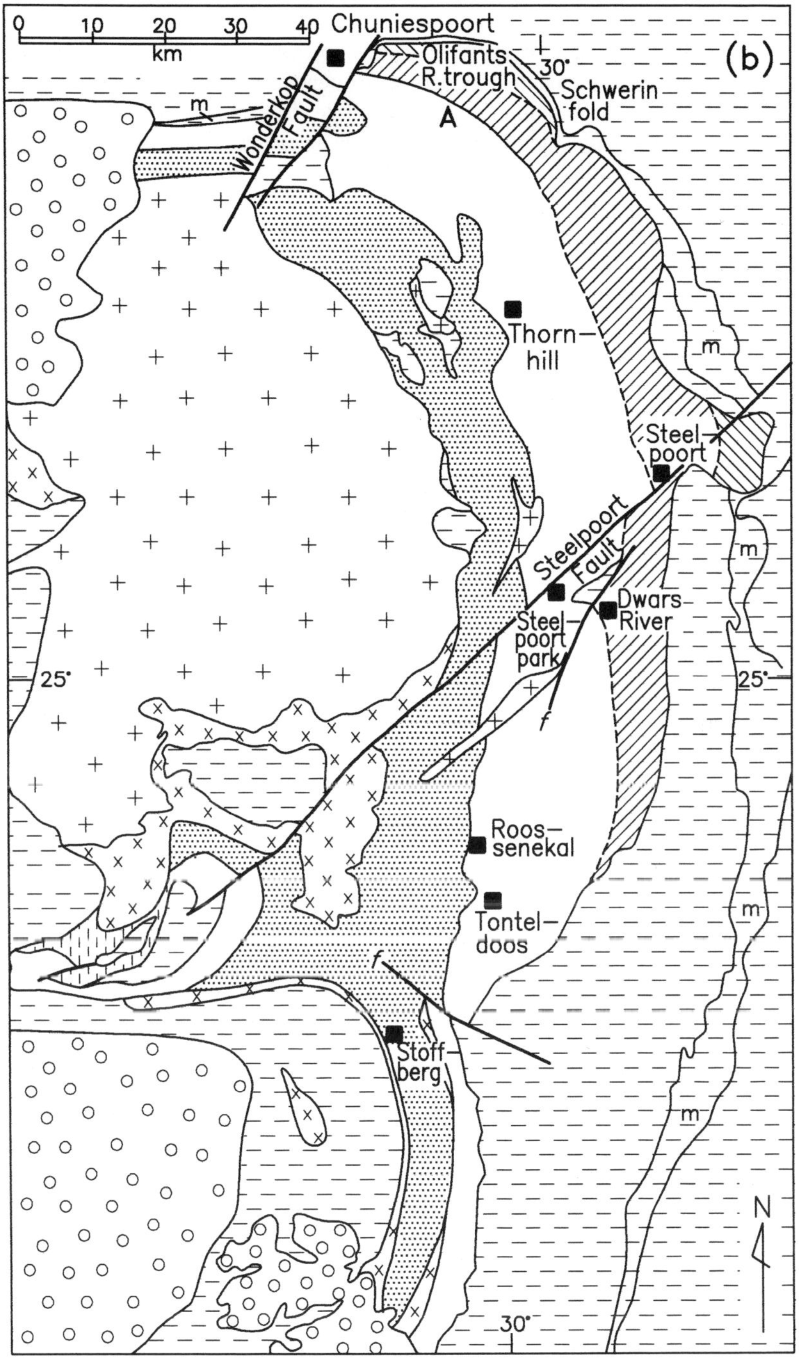
0 10 20 30 40
km
Chuniespoort
(b)
Olifants R.trough
30°
Schwerin fold
A
m
Wonderkop Fault
Thorn-hill
m
Steel-poort
Steelpoort Fault
m
Dwars River
Steel-poort park
25°
25°
f
Roos-senekal
Tontel-doos
m
f
Stoff-berg
m
N
30°

2.2. Floor and roof relations

The complex intruded the Transvaal Supergroup, which consists of a basal quartzite, followed by a dolomitic and banded ironstone sequence, an alternating quartzite and shale package, and an upper basaltic and acid volcanic phase (Button, 1976). Ages of the Transvaal Supergroup are uncertain, but the basal rocks may be as old as 2.5 Ga, whereas the uppermost Rooiberg volcanics are 2.06 Ga (Walraven *et al.*, 1990), making them almost identical in age to the Complex.

In the Western and Far Western limbs the complex intruded at the level of the Magaliesberg Quartzite. In the Eastern limb, north of Steelpoort, emplacement occurred at this same level, but to the south it transgressed upwards through more than 2 km of sediments, so that near Stoffberg the Dullstroom basaltic volcanics are preserved in the floor (Button, 1976). In the Northern limb intrusion occurred at the level of the Magaliesberg Quartzite in the south, but transgressed downwards towards the north, until the mafic rocks abut Archaean granitic gneiss (van der Merwe, 1976; Cawthorn *et al.*, 1985).

Relations in the roof are harder to decipher because of intense metamorphism of the roof rocks, and emplacement of the Bushveld Granite. The Rooiberg rhyolites, and granophyres, comprise the roof rocks. The granophyres may be metamorphosed and remelted rhyolite or quartzo-feldspathic sediments (Walraven, 1985). However, the apparent absence of recognizable Dullstroom basaltic volcanics and other sedimentary rocks in the roof to the north of Stoffberg led Cheney and Twist (1991) to suggest that the mafic intrusion occurred along a regional unconformity. The Dullstroom basalts and Rooiberg rhyolites discordantly overlay this proposed unconformity.

The Southeastern limb was emplaced at the level of the Malmani Dolomite (Buchanan, 1977). Its uppermost sequence was eroded and unconformably covered by Karoo sediments. The Far Western limb has no preserved roof.

3. LITHOLOGY AND STRATIGRAPHIC SUBDIVISION

Traditionally, the mafic sequence has been divided into Marginal, Lower (LZ), Critical (CZ), Main (MZ) and Upper (UZ) Zones, although their exact boundaries have been the subject of debate (Kruger, 1990). Generalized sections through the two major limbs are shown in Figure 3. Maximum thicknesses are shown in Figure 3, but lateral variations, especially of the LZ and CZ, are significant. The rocks of the complex are mineralogically simple, with the general order of appearance, and disappearance (shown by negative sign) of cumulus minerals being orthopyroxene (in the Eastern limb), chromite, olivine + orthopyroxene (in the Western limb), –olivine, plagioclase, –chromite, clinopyroxene, pigeonite (now inverted to orthopyroxene), –orthopyroxene, magnetite, ilmenite, ferrian olivine and apatite. However, because of lateral facies variations this sequence is not always followed. Most rocks contain only minor post-cumulus minerals and low incompatible element abundances, and so are close to adcumulates in the sense of Wager and Brown (1968).

3.1. Marginal Zone

Medium-grained, unlayered rocks underlie most of the complex. Unlike the Skaergaard intrusion, there are no upper marginal or border facies. Sequences of up to 800 m of heterogeneous noritic rocks forming the base of the complex, assigned by Vermaak (1976) and Engelbrecht (1985, 1990) to the Marginal Zone, cannot represent parental magmas chilled on the edge of the main chamber, but may represent composite sills, or the distal facies of evolved magmas from within the chamber. Even where thick sequences of ultramafic rock are developed at the base of the complex (*e.g.* in the Olifants River trough of the Eastern limb), a marginal norite is present (Cameron, 1978), suggesting that this norite is not genetically related to the overlying cumulates. The grain size, texture, and chemical composition further suggest

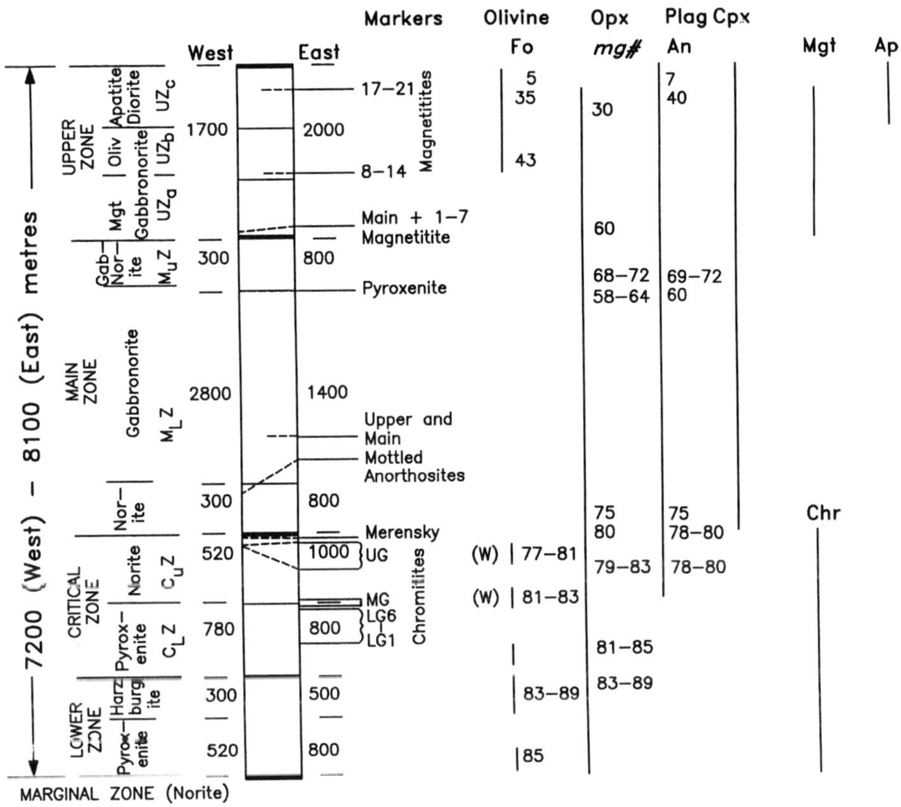

*Figure 3. Generalized stratigraphic sections through the Western and Eastern limbs, showing the appearance of cumulus phases. Maximum thicknesses of the different portions of zones in each limb are shown. West and East refer to Western and Eastern limbs. Some of the major marker horizons are shown, however, these may not occur in both Eastern and Western limbs, and so are shown as incomplete dashed lines in the log. Numbers 1-7, 8-14, and 17-21 refer to clustered magnetitite layers. Ranges of cumulus mineral compositions are indicated, but these can vary along strike as well as vertically (mg# refers to the 100*Mg/(Mg+Fe(total)) atomic ratio). The figure is compiled from numerous sources, but primarily Cameron (1978, 1980), von Gruenewaldt (1973) and Molyneux (1974) for the Eastern limb, and Teigler and Eales (in press), Mitchell (1990) and unpublished data for the Western limb.*

that they are cumulates (Cawthorn *et al.*, 1981). The extremely complex, multi-intrusive nature of this zone is evident in the Eastern limb, where different generations of mafic rocks interdigitate (Sharpe, 1981).

Two areas do not contain marginal norite. North of Potgietersrus the contact rocks are extremely coarse-grained melanorites and pyroxenites constituting the sulphide- and PGE-

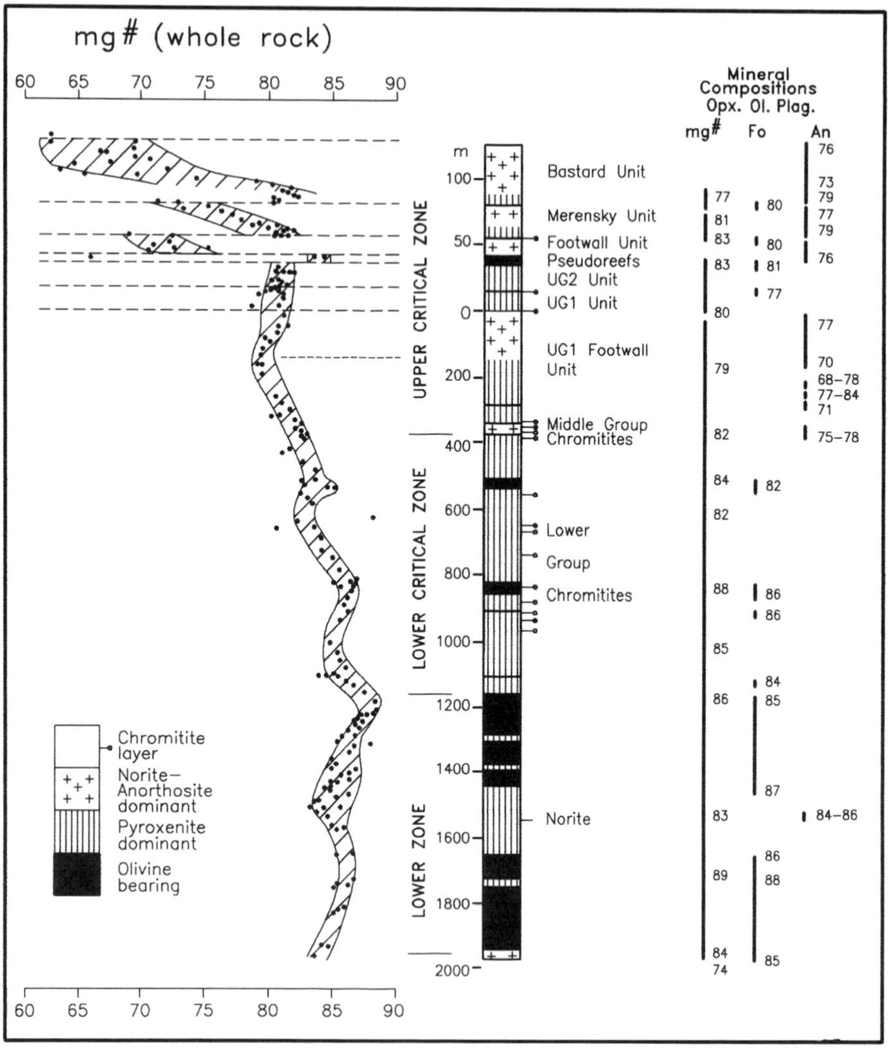

Figure 4. Section of LZ and CZ in the Union Section area, illustrating cryptic variations. Whole-rock mg# calculated assuming a Fe_2O_3:FeO ratio of 0.1. All mineral data are by electron microprobe. Note change of scale at the base of the UG1 Unit. The norite layer at 1490 m depth is too thin to show in the log.

bearing Platreef (White, 1994). Near Burgersfort, in the Eastern limb, coarse harzburgite is in contact with metasediments. Here, only a distinct increase in phlogopite content indicates proximity to the footwall contact.

3.2. Lower Zone

Poor exposure and a dearth of economic mineral deposits contribute towards the LZ being relatively poorly studied, and major discrepancies are found in maps purporting to show the Marginal, Lower, and Lower Critical (C_LZ) Zones, especially in the Western limb. Floor topography and structure have undoubtedly controlled the distribution and thickness of the LZ. The Spruitfontein upwarp of the sedimentary floor (Scoon and Teigler, 1994) is a site of attenuated thickness of the LZ and C_LZ in the Western limb (Figure 2a), as is the Schwerin fold (Cameron, 1978) in the Eastern limb (Figure 2b). Facies changes across these upwarps show that they imposed a degree of compartmentalization during the early stages of crystallization of the complex (Scoon and Teigler, 1994).

The four thickest sections through the LZ are to be found in the Olifants River trough in the Eastern limb, in drill-core from near the Union Section Platinum Mine in the Western limb, in the Far Western limb, and south of Potgietersrus in the Northern limb.

Core from the Nooitgedacht boreholes (Figure 2a) at Union Section provide a complete section from the Marginal to the Upper Critical (C_UZ) Zones (Teigler, 1990; Teigler and Eales, in press). Here the 800 m-thick LZ is an olivine- and orthopyroxene-rich sequence, with accessory chromite and variable amounts of intercumulus plagioclase, biotite, and clinopyroxene (Figure 4). The lowermost 345 m comprises countless thin layers and wisps, commonly no more than a few mm thick, of dunite, poikilitic and granular harzburgite, olivine pyroxenite and feldspathic pyroxenite. Rarely, individual layers reach up to 40 m. Logging on a scale of mm shows that 23% of this interval is pyroxenite and the remainder dunite or harzburgite. There follows a 175 m interval within which pyroxenite dominates (87%). The succeeding 290 m is again dominated by olivine-rich cumulates, with the proportion of pyroxenite dropping to 32%. Layering within this interval is commonly on the scale of 1-5 m, with rare individual layers reaching 40 m. Zonal subdivision cannot therefore be based on characteristic rock types so much as the overall ratio of olivine-rich to olivine-poor cumulates.

Cameron (1978) employed a four-fold division of the LZ in the Eastern limb into basal feldspathic pyroxenite with minor harzburgite, lower pyroxenite, harzburgite, and upper pyroxenite. Maximum thicknesses are 400, 400, 500, and 250 m respectively. However, it is only in the Olifants River trough (Figure 2b) that all units are well developed. Cyclicity, consisting of dunite, harzburgite, and pyroxenite, occurs in the Harzburgite Subzone, on scales ranging from cm to tens of m (Figure 5a). Cameron (1978) defined the base of the CZ as the level where interstitial plagioclase in pyroxenite increases abruptly from 2% to 6%. As a matching increase is not found in the Western limb, Teigler *et al.* (1992) proposed that the top of the LZ be defined by the termination of significant olivine in the sequence. In terms of this definition, the uppermost pyroxenite of Cameron's LZ and the feldspathic pyroxenite at the base of the CZ (absent at Union Section) now fall into the C_LZ.

A thin norite layer occurs in the lower pyroxenite of the LZ in both the Olifants River trough and the Western limb. This is not a break defining any zonal boundary, but merely a marker horizon.

In the Far Western limb the LZ contains nine cyclic units of dunite-harzburgite-pyroxenite some 1050 m thick (Engelbrecht, 1985). The LZ in the Northern limb is 1700 m thick, and

a
d
b
e
c
f

Figure 5. (facing page) (a). Cyclic layering of dunite, harzburgite and pyroxenite in the Olifants River trough. (b). Contact between C_LZ and C_UZ in Olifants River trough. The upper part of the photograph shows anorthosite (the lowest appearance of cumulus plagioclase) overlying a very thin chromitite layer (MG2). Below the MG2 is a feldspathic pyroxenite, with a thin chromitite (MG1) towards the base. The MG chromitites are extremely thin in this section. (c). Magnetitite layer 1 (2 m above the Main Magnetitite layer), with sharp basal and gradational upper contacts with anorthosite, Magnet Heights, Eastern limb (Molyneux, 1974). (d). Layering between leucocratic and melanocratic gabbronorites in the Main Zone, with non-cotectic proportions of plagioclase and pyroxenes (Quadling and Cawthorn, 1994). (e) Oikocrysts of pyroxene in anorthosite displaying variation in diameter with height, and hence producing post-cumulus layering, near Tweefontein Chrome Mine, Eastern limb. (f). Photomicrograph of abundant, partially resorbed relicts of plagioclase (white, some showing twinning) enclosed within the outer zone of an orthopyroxene grain, with the core being free of inclusions, pyroxenite orthocumulate, UG1 Footwall Unit, C_UZ, at Union Section. Field width is 1mm.

consists of 37 cyclic units of olivine-, chromite- and orthopyroxene-bearing cumulates (Hulbert and von Gruenewaldt, 1985). This limb is anomalous in that it is the only limb where the LZ contains significant PGE occurrences and chromitite layers.

In the Southeastern limb, drilling has revealed only some tens of metres of LZ harzburgite, which are directly overlain by evolved magnetite-rich rocks of the UZ (Buchanan, 1975, 1977).

3.3. Critical Zone

The CZ carries two of the world's largest platinum-bearing ore bodies, the Upper Group 2 Chromitite (UG2) and the Merensky Reef, and huge reserves of chromite within the Lower (LG), Middle (MG), and Upper (UG) Group chromitite layers (Hatton and von Gruenewaldt, 1987), as shown in Figure 4. Individual layers are labelled from the base upwards LG1-7, MG1-4 and UG1-2 (and UG3 only in the east). The thickest exceed 1 m. The C_LZ comprises a thick succession of pyroxenitic cumulates overlying the Harzburgite Subzone of the LZ. At Union Section Mine this is 780 m thick (Figure 4), of which 86% is pyroxenite, 11% olivine-rich, 0.5% chromitite, and the remainder pyroxenite which bears an average of 28% chromite (Teigler, 1990). A distinctive stratigraphic marker extends from 220 to 330 m above the base of the subzone in the west (Figure 4), comprising 30 m of olivine-rich cumulates, and then 45 m of pyroxenite capped by a further 35 m of olivine-rich cumulates bearing a robust chromitite layer (LG4). Cameron (1980) identified a similar distinctive sequence in the Eastern limb. A second olivine-rich interval occurs 300 m above this package at Union Section Mine, but has no counterpart in the east.

The transition from C_LZ to C_UZ occurs between MG2 and 3 chromitite layers, where the first cumulus plagioclase-bearing rocks appear in the Eastern limb and most of the Western limb (Figure 5b). Eight cyclic units are recognized in the C_UZ, consisting of partial or complete sequences from a base of ultramafic cumulates (chromitite, harzburgite, pyroxenite) through norite to anorthosite. The base of each unit is sharp, but internal contacts within the cycles may be sharp or gradational. Relative thicknesses in metres are:

Cyclic Unit	Union	Rustenburg	Maandagshoek
Bastard Unit	65	90	38
Merensky Unit	20	9	42
Merensky Footwall Unit and Pseudoreefs	25	130	250 (as UG3)
UG2 Unit	15	20	100
UG1 Unit	17	70	90
MG4 Unit	290	160	235
MG3 Unit	50	30	140
MG2 Unit	5	9	20

(Data are from de Klerk, 1982, 1991; Viljoen *et al.*, 1986a, 1986b; Viljoen and Hieber, 1986; Gain, 1985).

Whole-rock geochemical parameters may define either different narrow ranges (*e.g.* Sr/Al ratios), or saw-tooth trends through successive units *(e.g. mg#* [100*Mg/(Fe^{2+}+Mg)] or Ni/Sc ratios; Eales *et al.*, 1986). Averaged compositions of the more common rocks are given in Table 1. Significant lithological and chemical variation prevails along strike within the Western limb:- (i) Whereas the C_LZ is exclusively ultramafic at Union Mine, cumulus plagioclase is prominent below the MG2 chromitite layer in the Brits area to the east (Teigler *et al.*, 1992). (ii) The LG chromitites degenerate eastwards along the southern arc of the Western limb (Hatton and von Gruenewaldt, 1987; Teigler *et al.*, 1992). (iii) The UG1 and UG2 Units at Union Mine comprise chromitite and pyroxenite only, but become capped by norite and anorthosite further south (Viljoen *et al.*, 1986a, 1986b; Viljoen and Hieber, 1986). (iv) The harzburgitic Pseudoreefs between the UG2 and Merensky Reef at Union Mine degenerate to olivine norite towards the northeast at Amandelbult (Viljoen *et al.*, 1986b) and along the southern arc of the Western limb (Maier, 1991). (v) Well-defined ultramafic layers, such as the UG2 and Bastard pyroxenites, become somewhat less magnesian, and the interval between the UG2 pyroxenite and Merensky Reef becomes progressively thicker and more feldspathic both north and south of Union Mine.

These, and related features, have prompted the definition of proximal and distal facies in the Western limb, with a feeder zone located in the proximal area (Eales *et al.*, 1988, 1993a; Maier and Eales, 1994a).

The degeneration of the LG chromitite layers from northwest to southeast in the Western limb is contrasted by an increasing thickness of the MG layers in the same direction (Hatton and von Gruenewaldt, 1987; Teigler *et al.*, 1992; Scoon and Teigler, 1994). A similar geometry is recognized in the Eastern limb, where close to the Steelpoort Fault there is a relatively abrupt change from a northern sector with thick LG and thin MG chromitites to a southern sector where the LG package degenerates rapidly and the MG layers become robust (Scoon and Teigler, 1995). However, unlike the sequence in the Western limb, there are no observed changes in accompanying rock type, but a lack of mineral compositional data inhibits further comparisons.

In the Far Western limb five chromitite layers within the CZ are regarded as the equivalent of the LG1-5 layers (Engelbrecht, 1985) as they occur within dominantly pyroxenitic cumulates.

Table 1
Average compositions of the dominant rocks of the Upper Critical Zone.

	Harzburgite	Pyroxenite	Norite	Anorthosite
SiO_2	43.75	52.93	50.67	49.20
TiO_2	0.15	0.25	0.09	0.06
Al_2O_3	4.08	4.78	22.24	30.63
Fe_2O_3	15.76	13.04	4.75	1.53
MnO	0.22	0.25	0.07	0.02
MgO	33.45	24.33	9.13	1.31
CaO	3.27	3.86	11.22	14.77
Na_2O	0.38	0.61	1.85	2.43
K_2O	0.13	0.13	0.13	0.14
P_2O_5	0.03	0.01	0.01	0.01
Total	99.89	99.01	99.77	99.98
Rb	d-8	d-20	d-10	d-10
Sr	70	30-70	100-400	400-500
Y	d-7	d-15	d-8	d-10
Zr	d-10	d-50	d-15	d-15
Zn	80	60-160	d-60	d-20
Cu	20-50	20-80	d-50	d-40
Ni	1500	400-1200	50-500	d-100
Co	160	80-200	10-100	d-20
Cr	1000	2000-4000	100-2000	d-100
V	50	90-150	3-30	d-30
Sc	10	20-50	3-30	d-7

Data from Maier (1991), representing 275 samples collected along *ca.* 170 km of strike of the Western limb of the Bushveld Complex. Ranges in trace-element levels from d (lower limit of detection) to highest levels are given in ppm.

No obvious equivalent of the C_LZ is seen in the Potgietersrus limb (van der Merwe, 1976), but it is conceivable that facies variations have rendered the C_LZ indistinguishable from the LZ. Relatively fine-grained gabbronorites immediately overlie the LZ (Hulbert and von Gruenewaldt, 1985). The entire CZ wedges out to the north (van der Merwe, 1976).

Merensky Cyclic Unit: At the top of the CZ are two distinctive cyclic units, the Merensky and Bastard Units. The Merensky Unit contains basal chromitite, pyroxenite (sometimes with olivine), norite and anorthosite (see Lee, this volume). The mined interval is olivine-bearing in the proximal facies, but more generally it is a pyroxenite, which is pegmatitic at Rustenburg and Union Platinum Mines (Vermaak, 1976), but only locally so elsewhere (Leeb-du Toit, 1986; Viljoen *et al.*, 1986a , 1986b; Viljoen and Hieber, 1986). Several different facies have been designated by Viljoen (1994). Most of the platinum-group element (PGE) mineralization is concentrated near the chromitite layers, especially the upper one (Viljoen, 1994). The Bastard Unit is similar to the Merensky, except for its sub-economic platinum content. Locally, it may be devoid of chromitite, and display a basal norite rather than pyroxenite. A significant break occurs in the initial Sr-isotope ratio (Sr_i) at the

Marsh, 1982) in the Western limb, a feature which is repeated at the Bastard Reef in the east (Lee and Butcher, 1990).

3.4 Main Zone

A satisfactory boundary between CZ and MZ is difficult to define on mineralogical criteria. It is currently taken at the top of the Giant Mottled Anorthosite, a distinctive marker bearing large oikocrysts of pyroxene (up to 10 cm across), capping the Bastard Cyclic Unit. Overlying rocks are mineralogically more homogeneous norites with few texturally identifiable markers (von Gruenewaldt, 1973; Mitchell, 1990).

The MZ is a thick succession of cumulates devoid of olivine and chromian spinel, in general lacking both the fine-scale layering of the CZ, and its extremes of lithological diversity. Augite (at some 300 m above the base of the MZ) and then pigeonite (now inverted to orthopyroxene) attain cumulus status within it, but pyroxenites are rare and anorthosites confined mainly within two short intervals, the Main Mottled and Upper Mottled Anorthosites (Figure 3). The Upper Mottled Anorthosite has not been identified in the Western limb. Furthermore, the Main Mottled Anorthosite occurs in noritic rocks in the west, but in gabbronoritic rocks in the east (Figure 4), thus complicating detailed correlation between the limbs.

In the Eastern limb (von Gruenewaldt, 1973; Molyneux, 1974) there is a basal norite followed by gabbronorite from 1-1.5 km thick. This is overlain by a 1 km-thick interval where both primary orthopyroxene and pigeonite occur together, in the Eastern limb (von Gruenewaldt, 1973; Molyneux, 1974), before pigeonite becomes the sole primary Ca-poor pyroxene. A further 1 km higher in the sequence primary pigeonite is replaced by primary orthopyroxene at an horizon referred to as the Pyroxenite Marker. A reversal in mineral compositions and a break in Sr-isotopic ratio occur close to this level (Sharpe, 1985), and are also noted in the Western limb (Mitchell, 1990; Cawthorn *et al.*, 1991). Von Gruenewaldt (1973) divided the MZ into three subzones, the central one containing pigeonite, and the upper and lower containing primary orthopyroxene. In contrast, Mitchell (1990) subdivided the MZ in the Western limb into a Lower Main Zone (M_LZ) comprising Norite Units I-II, overlain by Gabbronorite Units I-IV, separated from an Upper Main Zone (M_UZ) by the Pyroxenite Marker. Gabbronorite Unit I is sandwiched between the Porphyritic Gabbro Marker (or Porphyritic Cluster Norite) and the Main Mottled Anorthosite. The former is a distinctive interval of some 60 m bearing orthopyroxene grains up to 10 mm in size (enclosing lath-shaped plagioclase inclusions) in a finer-grained gabbroic groundmass of plagioclase and augite. The Main Mottled Anorthosite carries a strong magnetic signature by which it is traceable, in both

---

Figure 6. *(facing page) Plot of distribution of magnetitite layers (shown by solid triangles on height axis) in the Western limb of the UZ, based on borehole intersections BK1, BK2 and BK3, from Bierkraal (Figure 2a), north of Rustenburg (Walraven and Wolmarans, 1979; Kruger et al., 1987). The compositions of plagioclase (a) and V_2O_3 contents of magnetite (b) were determined by XRF spectrometry on mineral separates (unpublished data of RGC). Samples from the upper borehole (BK1) and lower borehole (BK2) are designated with open symbols, to distinguish them from those in the middle (from BK3). The correlation between boreholes is based on the presence of magnetitite layers and the appearance of apatite and olivine. Samples from magnetitite layers are designated with square symbols. The inset (c) shows detail from the main section for magnetite compositions.*

Eastern and Western limbs (Viljoen and Scoon, 1985).

One section of the MZ shows fine-scale layering in the Eastern limb (Molyneux, 1974; Quadling and Cawthorn, 1994). Some 80 m below the Pyroxenite Marker, sharply defined layers of leucogabbronorite alternate with melagabbronorite over a 10 m interval (Figure 5d). Individual layers are typically a few cms thick, with over 100 layers recorded in one profile. This package has not been reported in the west.

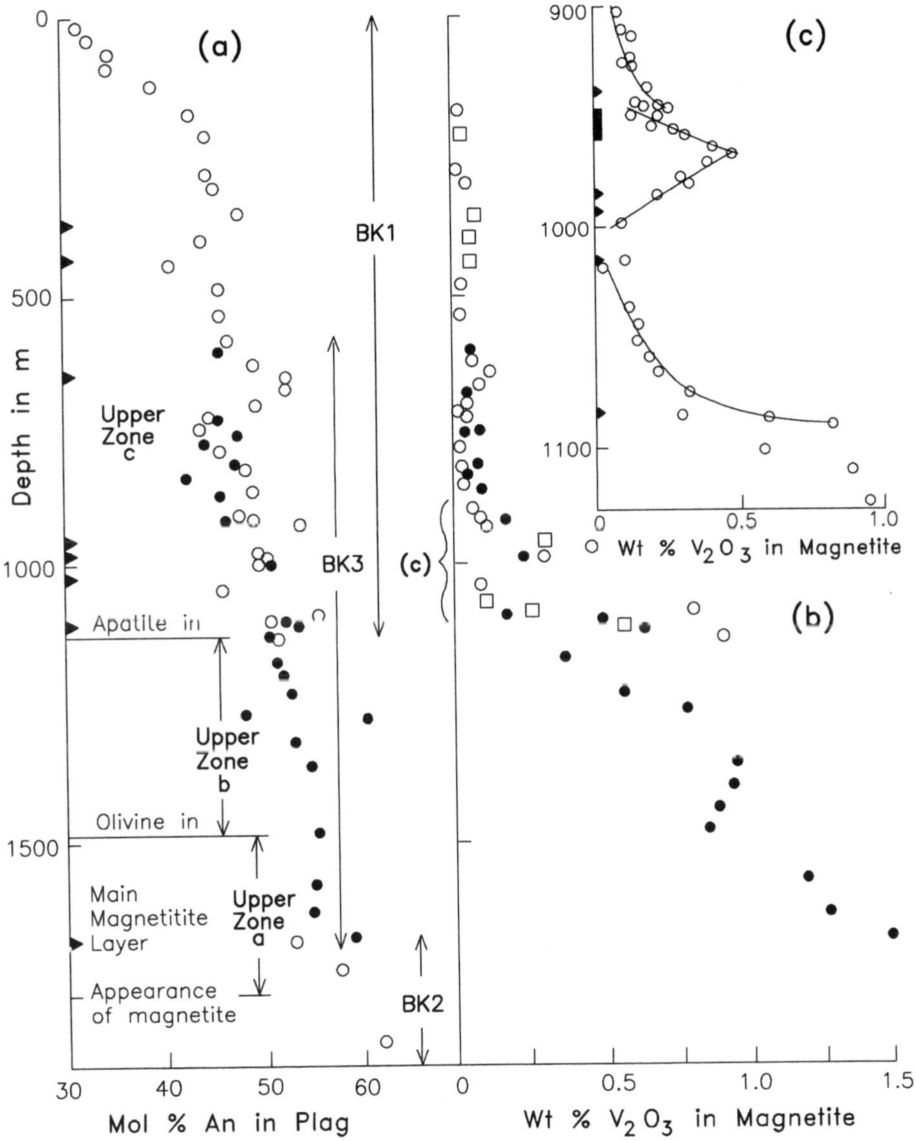

3.5. Upper Zone

The base of the UZ is defined by the appearance of cumulus magnetite. There are isotopic grounds for redefining the base of the UZ at the Pyroxenite Marker (Kruger, 1990), but the more easily applied criterion of the presence of cumulus magnetite is retained here. This occurs only a few metres below the lowest magnetitite layer (Molyneux, 1974), although large, isolated oikocrysts of magnetite may appear in the uppermost MZ. The UZ is some 2000 m thick and is intermittently well-layered. Its most striking feature is the presence of some 25 magnetitite layers in the Eastern limb (Molyneux, 1974), which cluster into four groups, each with up to seven layers (Figure 3). Magnetitite layers typically have sharp bases, but gradational tops (Figure 5c), and in the eastern limb are bounded by anorthosites. The thickest is 6 m, while the Main Magnetitite Layer, near the base of the UZ, is 2 m thick and mined for its 1.3% V_2O_3 content. Poor outcrop on the Western limb makes it impossible to determine exactly how many layers are present, but eight major layers in three groups have been identified in drill-core (Figure 6). Most rocks within the UZ contain >50% plagioclase, and hence are dominated by anorthosite and leucogabbro with variable amounts of cumulus ferrian pyroxenes and olivine, magnetite, and apatite. Primary orthopyroxene gives way to pigeonite near the base of the UZ, but may disappear near the top (von Gruenewaldt, 1973, Molyneux, 1974). Towards the top of the zone ilmenite may exceed magnetite in abundance (Reynolds, 1985). Interstitial biotite, hornblende, and, especially over the uppermost 200 m, quartz and alkali feldspar are also present.

The UZ has been divided into three subzones by SACS (1980), as shown in Figure 3. Subzone *a* comprises some 700 m of anorthosite and magnetite ferrogabbro. Near the base are three thin magnetitite layers. The Main Magnetitite Layer occurs 130 m above the base, and is closely overlain by a further seven magnetite layers. The incoming of ferrian olivine defines the base of the 580 m-thick Subzone *b*, where anorthosite, troctolite, and olivine and magnetite ferrogabbro are host to seven more magnetitite layers. Cumulus apatite marks the base of Subzone *c*, close to which the plagioclase composition becomes more sodic than An_{50}. This 1000 m-thick sequence is composed of olivine diorite, with anorthosite, magnetite-rich diorite, and another seven magnetitite layers. A similar mineralogical evolution exists in the west, but poorer exposure inhibits accurate subdivision and detailed lateral correlation.

The UZ in the Northern limb shows an overall resemblance to other exposures, but appears to be compressed through Subzones *a* and *b* to about half the normal thickness, and has 20 magnetitite layers (van der Merwe, 1976).

The 1900 m interval of UZ rocks intersected by drilling into the Southeastern limb (Buchanan, 1975) contains a succession of anorthosites, norites and gabbros, with ferrian olivine and apatite appearing in dioritic rocks towards the top. At least 17 magnetitite layers are present.

3.6. Discordant bodies

A number of extremely coarse-grained discordant bodies, usually pipe- or carrot-like and perpendicular to layering, cut the complex. They range from magnesian dunite (some containing platinum), through iron-rich ultramafic pegmatites to gabbroic anorthosites and nickel sulphide-rich plugs (Viljoen and Scoon, 1985). The two commonest types are iron-rich dunites and wehrlites, and oxide dominated (Scoon and Mitchell, 1994). The former are found from the C_LZ to M_LZ, and the latter in the M_UZ and UZ, although some pipes may be composite. The largest reaches 1.5 km in diameter. Evidence, such as high Cl contents of fluid inclusions, suggests that the dunites may be hydrothermal or fluid-dominated metasomatic in origin (Schiffries, 1982). However, Viljoen and Scoon (1985) and Scoon and Mitchell (1994) suggested that they are magmatic. The magnesian dunites are considered to be injections of new magma, whereas the iron-rich dunites and wehrlites are thought to be residual liquid from anorthosites, and the oxide-dominated rocks the products of liquid immiscibility, both of which passively replaced the existing layered rocks, especially plagioclase-rich rocks, by downward percolation. Although the field relations are consistent with these suggestions, certain petrological problems pertain to these iron-rich ultramafic pegmatitic rocks. There are no wehrlitic layered rocks which might be expected if the pipe wehrlites are simply the products of normal differentiation. Differentiation and liquid immiscibility should produce liquids saturated in plagioclase, which is inconsistent with the preferential replacement of plagioclase in the layered sequence. Finally, the initial $^{87}Sr/^{86}Sr$ ratios for the pegmatitic wehrlites differ from those of the immediately adjacent layered rocks (Scoon and Mitchell, 1994).

3.7. Lateral extent of Zones

All zones are not equally laterally extensive (Figure 2). The LZ occurs in three basins in the northern part of the Eastern limb, only from Thabazimbi to Rustenburg in the Western limb, and only south of Potgietersrus in the Northern limb. The Critical Zone is identified around most of the Western limb, but is absent south of Roossenekal in the Eastern limb (Figure 2b), and again north of Potgietersrus in the Northern limb (van der Merwe, 1976). The MZ is similarly truncated in the Northern limb (van der Merwe, 1976), and partially so south of Stoffberg in the Eastern limb. The UZ is therefore the most laterally extensive in all limbs. This geometrical relation is attributed to periodic influx of magma which inflated the chamber vertically and laterally. Unlike these gradual on-lap relations, the Upper Zone cuts steeply into older cumulates until it comes into direct contact with the sedimentary floor in so-called "gap areas" north of the Pilanesberg Intrusion (Figure 2a). They are considered the result of tectonic redistribution of magma within the chamber (Wilson *et al.*, 1994).

3.8. Satellite bodies

Several disparate bodies, considered to be coeval with the Bushveld Complex, occur over a considerable area, and considerably increase the extent of the Bushveld magmatic province. The largest is the Molopo Farms Complex, which is totally hidden beneath Karoo sediments and Kalahari sand in southwestern Botswana and the Northern Cape Province in South Africa, and so is known only from drilling. Gravity data suggest it covers an area in excess of 1300 km^2. It consists largely of olivine and orthopyroxene cumulates, with no chromitite layers, and only a thin noritic component (Reichhardt, 1994). The Uitkomst Intrusion (Kenyon *et al.*, 1986; Gauert *et al.*, 1995) outcrops 60 km from the Eastern limb, due east of Belfast (Figure 1). It is a long, 1 km-wide northwest-trending trough, containing up to 400 m of harzburgite and pyroxenite with a chromitic zone some 20 to 60 m thick, overlain by a more widespread gabbro less than 200 m thick. It contains abundant nickel-copper sulphide mineralization at its base. The Losberg Intrusion occurs 105 km south of Rustenburg at the same stratigraphic horizon as the Bushveld Complex. Despite being only 120 m thick, it has a well-developed basal harzburgite 20 m thick, with similar mineral compositions to the Lower Zone of the Bushveld Complex, overlain by a granophyric gabbro (Abbott and Ferguson, 1965). The Moloto intrusion occurs between the Western and Eastern limbs, 50 km northeast of Pretoria (Figure 1). There is no outcrop, but it was identified by a gravity anomaly in the Bushveld Granite. Over 300 m of unlayered olivine-apatite-magnetite gabbro was intersected by drilling

(Walraven, 1987). The Rhenosterhoekspruit body occurs 50 km east of the northeastern limit of the Western limb (Figure 1). Its outcrop is only 5 by 1 km, but it contains at least six substantial magnetitite layers in 1250 m of Upper Zone rocks.

4. CRYPTIC VARIATIONS IN MINERAL COMPOSITIONS

Variations in mineral composition record subtle changes in magma composition far more effectively than the presence or absence of specific phases. However, a caveat applies in that primary compositions may be changed by late-stage processes such as reaction with interstitial liquid and sub-solidus equilibration. Determined mineral compositions may thus, in part, be a function of modal proportions (Barnes, 1986a). Sub-solidus equilibration between chromite, orthopyroxene and olivine results, for example, in *mg#* increase in the silicate and decrease in the spinel phases (Eales and Reynolds, 1986), with the greater compositional shift being shown by the minor phase. Where modal orthopyroxene drops below about 30% in noritic rocks, reaction with interstitial liquid may shift its initial composition to apparently more evolved compositions (Scoon and Mitchell, 1994; Cawthorn, 1996), leading to possibly erroneous conclusions about fractionation trends.

The most commonly used indices of cryptic variation involve the major elements, such as *mg#* of mafic phases, and An content of plagioclase, but minor and compatible trace-elements may also yield valuable information. Such would be Al, Ti and Mn in pyroxenes, Ni in olivine, Ti in chromite, and V and Cr in magnetite. As nearly all Sr resides in plagioclase, whole-rock Sr isotope data really represent a plagioclase cryptic variation profile.

General trends are shown in Figure 3, although it is becoming increasingly apparent that while vertical variations dominate in modelling fractionation and magma rejuvenation processes, systematic lateral variations exist within the complex on a regional scale.

4.1. Olivine

Lower Zone: The rare harzburgites in the Pyroxenite Subzone in the Olifants River trough contain olivine *ca.* Fo_{85} (Cameron, 1978). Within the Harzburgite Subzone, some 850 m higher, there is a muted reversal through 500 m to Fo_{87}. In the equivalent sequence at Union Section (Figure 4), olivine in the lowermost dunites is Fo_{85-88}, and the reversal extending through the underlying pyroxenites to a peak value near the top of the Harzburgite Subzone is more pronounced, from *ca.* Fo_{84} to Fo_{88} over 500 m (Teigler, 1990; Eales *et al.*, 1993a, 1994). In the Kroondal area, 8 km east of Rustenburg, the olivine composition declines to Fo_{83}, indicating lateral facies variations (Teigler, 1990). In the Potgietersrus limb olivine ranges from Fo_{86-90} (Hulbert and von Gruenewaldt, 1985), but the sample spacing is here too wide to discern trends.

Critical Zone: At Union Section, most olivine in the C_LZ is Fo_{84-86}; compositions more magnesian than this are attributable to equilibration with chromite. Equivalent values in the more distal facies at Brits decline to Fo_{81-83} (Teigler, 1990). Towards the top of the C_LZ, the olivine-rich interval between the LG and MG chromitites represents a local peak of reversal in the pattern of cyclic variation of *mg#* (Figure 4) but at relatively low values of Fo_{81-83}. Further decline to Fo_{77-82} is evident through the C_UZ in olivine-bearing rocks of the UG2, Pseudoreef, Merensky and Bastard units at Union Mine, but olivine is rare in the C_UZ east of Rustenburg (Maier and Eales, 1994a).

Upper Zone: In the northern sector of the Eastern limb olivine changes from Fo_{63} to Fo_{35} over 900 m and then to Fo_5 over the next and uppermost 200 m (Molyneux, 1974). Further south, von Gruenewaldt (1973) documented a similar pattern, but with compositions up to 10% more Fe-rich, suggesting a regional change from north to south. In the western Bushveld, Hoyle (1993) showed a decline from Fo_{35} to nearly pure fayalite, with several prominent reversals, through the uppermost 1500 m of the UZ.

4.2. Orthopyroxene

As orthopyroxene is an almost ubiquitous mineral, and shows both major- and minor-element variations, it is the most useful and extensively studied phase.

Lower Zone: The orthopyroxene composition is nearly constant throughout the Olifants River trough (Cameron, 1978), its *mg#* increasing from 83-89 in the olivine-rich units. In the northern sector of the Western limb, *mg#* values of 84-81 characterize the lowermost pyroxenites. Above this, cyclic variations (Figure 4) are within the range 89-83, defining a close correlation between olivine and orthopyroxene compositions in which $mg\#_{opx} = 0.87\ mg\#_{ol} + 12.5$ (Teigler, 1990). Highest values occur in olivine-rich rocks. In the Potgietersrus limb the composition ranges from *mg#* 89-93 (Hulbert and von Gruenewaldt, 1985).

Critical Zone: Upwards through the C_LZ of the Olifants River trough, there is a regular trend from *mg#* 83 to 85 and back to 81, and through the C_UZ a general decline to 77 at the level of the Merensky Reef (Cameron, 1980, 1982). No sharp breaks are recorded. At Union Section there is an overall decline of *mg#* through the C_LZ, from typical LZ values to 82, but with two prolonged reversals (Figure 4) culminating in olivine-rich cumulates (Teigler, 1990). In the C_UZ *mg#* values are typically 75-83 with the lowest values within the Bastard Unit (de Klerk, 1991). Intercumulus grains within anorthosites decline further to *mg#* values of 54 (Eales *et al.*, 1993b).

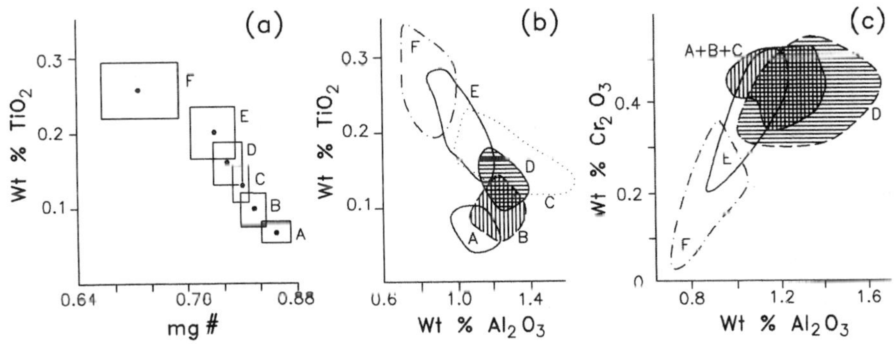

***Figure** 7. Variation in minor-element abundances in orthopyroxene (from 306 samples) in the LZ and CZ (from Eales et al., 1993b). Samples have been drawn from stratigraphic units as follows: A - samples from uppermost 300 m of LZ and lowermost 500 m of CZ (Figure 4); B - C_LZ up to the base of the MG Chromitites; C - samples straddling the MG Chromitites; D and E - noritic samples from the MG Chromitites to the top of the CZ (D from the proximal facies at Union Section; E from the distal facies at Brits); F - anorthositic samples with intercumulus orthopyroxene from the C_UZ. (a) Variation in TiO_2 versus mg#. Box size represents one standard deviation. (b) Variation in TiO_2 versus Al_2O3. (c) Variation in Cr_2O_3 versus Al_2O_3. (Reproduced with permission of Mineralogical Magazine.)*

There is a steady, near-threefold increase in Ti and Mn in orthopyroxene (Figure 7) through the entire CZ (Eales *et al.*, 1993b). Al_2O_3 first increases through the C_LZ from 1.1% to 1.3% (Groups A-C, Figure 7), and then declines to below 1% in the C_UZ where plagioclase is a cumulus phase. In contrast, Cr_2O_3 remains virtually constant throughout the entire LZ and CZ at 0.4-0.5%, even adjacent to thick chromitite layers. Only within intercumulus orthopyroxene do levels fall to *ca.* 0.2% (Group F, Figure 7).

Pyroxene compositions in individual cyclic units of the C_UZ show an upward decline of >10% in *mg#*, from pyroxenite to anorthosite, followed by an abrupt increase at the pyroxenitic base of the next cyclic unit (Kruger and Marsh, 1985; Naldrett *et al.*, 1986; Field, 1987). A systematic lateral decline of *ca.* 3% in *mg#* is evident in orthopyroxene of nearly all layers when traced over 100 km from the proximal to the distal facies of the Western limb

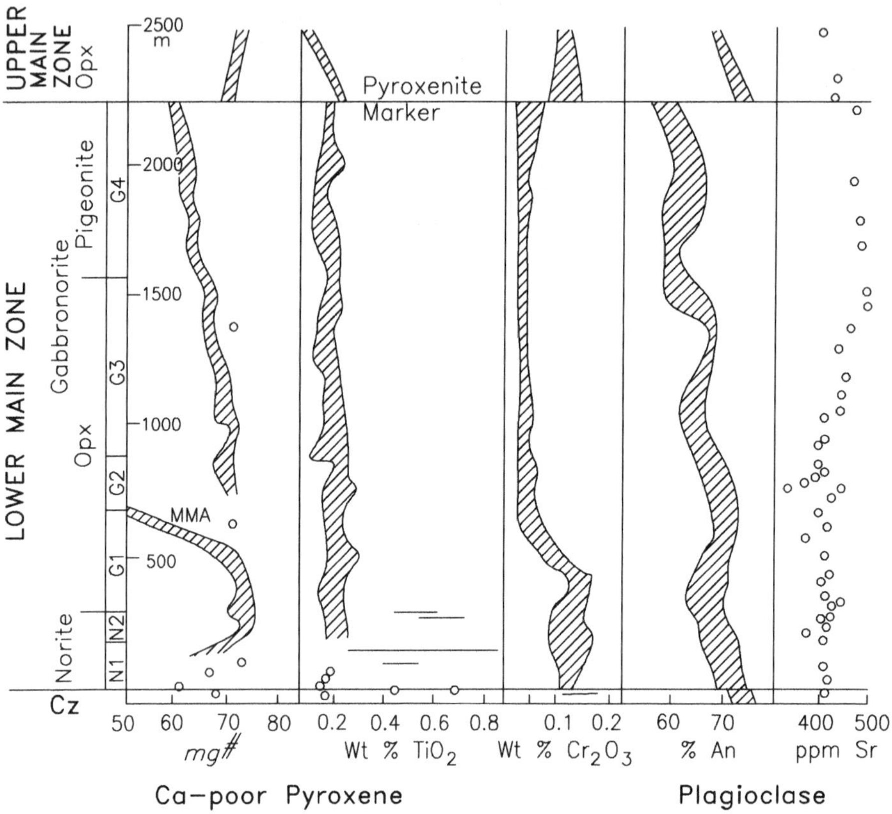

Figure 8. *Plot of mg#, wt % TiO_2 and wt % Cr_2O_3 in Ca-poor pyroxene, and An and Sr contents in plagioclase in the M_LZ after Mitchell (1990). N1-N2 and G1-G4 refer to norite and gabbronorite units identified by Mitchell. MMA is Main Mottled Anorthosite where pyroxene is intercumulus. (Reproduced with permission of the South African Journal of Geology.)*

(Eales *et al.*, 1994). However, an unique trait of the basal metre of the Bastard pyroxenite is a constant *mg#* of 80 along a strike-length >170 km. The Cr content of pyroxene drops rapidly over a short vertical interval in this unit.

Main Zone: There are fewer chemical analyses through this and the UZ, but comparisons show good agreement between trends determined by older optical and subsequent electron microprobe methods. Above the Bastard Unit in the Western limb there is a continuation of fractionation trends to *mg#* 65, then a reversal to *mg#* 75 through the 100 m Norite Unit 2 (Figure 8) (Mitchell, 1990). Cr_2O_3 is typically 0.1-0.15% in the lowest norites, but drops to 0.05%, as does TiO_2 from 0.4 to 0.2%, once clinopyroxene attains cumulus status. Throughout the overlying 2000 m of M_LZ the *mg#* declines slowly to 60 except where the pyroxene has intercumulus status (Mitchell, 1990).

Both pyroxenes show a reversal of 10% in *mg#* over an interval of 100-200 m spanning the Pyroxenite Marker in the Eastern and Western limbs (Groeneveld, 1970; von Gruenewaldt, 1973; Molyneux, 1974; Sharpe, 1985; Mitchell, 1990; Cawthorn *et al.*, 1991), but the absolute values at both the base and top of the interval decline when traced southwards in the eastern limb (Klemm *et al.*, 1985a). Wilson *et al.* (1994) and Hoyle (in prep.) reported an abrupt *decrease* in the order of 10% in bore-core drilled in the Northam area. The Ca-poor pyroxene is inverted pigeonite above the Marker in these cases.

Upper Zone: Optical data from the northern sector of the Eastern limb show a change from *mg#* 60 at the Main Magnetitite Layer to *mg#* 30 some 1300 m above, close to the top of the UZ (Molyneux, 1974). Near Roossenekal, where orthopyroxene is less abundant, and virtually disappears some distance below the roof, von Gruenewaldt (1973) reported a composition of *mg#* 57 at the Main Magnetitite Layer. Further south, near Stoffberg, Groeneveld (1970) reported a composition *mg#* 53 at the same horizon, suggesting a southwards decrease in *mg#* in the liquid at the same stratigraphic level. In the Western limb microprobe analyses at this level give *mg#* 55 (Cawthorn *et al.*, 1991). In the Northam area Hoyle (in prep.) showed an overall decline from *mg#* 55 above the Pyroxenite Marker to *mg#* 50 in Subzone B, with Ca-poor pyroxene being virtually absent in Subzone C.

4.3. Clinopyroxene

There are relatively fewer analyses of clinopyroxene in the intrusion, but the *mg#* always mirrors the trends for Ca-poor pyroxene (Atkins, 1969; Teigler, 1990; Cawthorn *et al.*, 1991).

4.4. Plagioclase

Electron microprobe analyses of cumulus plagioclase grains show marked zoning, which introduces ambivalence in defining precise trends.

Lower and Lower Critical Zones: The plagioclase is interstitial, and strongly zoned (An_{77-50}), and analyses cannot be used to infer fractionation trends. A single, thin norite layer in the LZ at Union Section bears plagioclase An_{86-84}, whereas in C_LZ norites of the distal facies near Brits values are An_{81-77} (Teigler, 1990).

Upper Critical Zone: Cameron (1982) showed that in the Olifants River trough the lowermost cumulus plagioclase is An_{78}, declining to An_{73} at the Merensky Reef. Detailed studies through each cyclic unit in the Western limb yield values closer to An_{80} (Kruger and Marsh, 1985; Naldrett *et al.*, 1986). Between the proximal and distal facies averaged values for cumulus plagioclase in the UG2-Merensky Reef interval decline from $An_{76.4}$ to $An_{74.6}$ (Maier and Eales, 1994a), with lower values in ultramafic rocks where plagioclase has intercumulus status. Above the Merensky Reef in the Western sector, there is a decline to An_{70} in some profiles, but in most a near-constant value close to An_{75} is maintained through the succession of anorthosites capping the C_UZ (de Klerk, 1991).

Main Zone: In the Western limb (Figure 8) there is no break in plagioclase composition at the base of the MZ, nor is there a reversal in composition in the lowest rocks of the MZ as is found for orthopyroxene. In the Eastern limb a reversal is reported for both pyroxene and plagioclase (von Gruenewaldt, 1973). Mitchell (1990) traced a gradual decline, with small oscillations, from An_{70} to An_{60} at the Pyroxenite Marker at Union Section (Figure 8), similar to that determined by von Gruenewaldt (1973) and Molyneux (1974) in the Eastern limb. All these studies record a *ca.* 10% increase in An content across the Marker, with the peak being reached some distance above it. In the southern portion of the Eastern limb the plagioclase is more evolved (Klemm *et al.*, 1985a), and the magnitude of the reversal is less. In the anomalous Northam profiles where abrupt shifts towards lower *mg#* occur, Wilson *et al.* (1994) and Hoyle (in prep.) show a comparable shift in plagioclase composition.

Upper Zone: Von Gruenewaldt (1973) and Molyneux (1974) suggested that fractionation continued without interruption above the Pyroxenite Marker, based on continuous trends displayed by pyroxene and plagioclase, the latter declining to An_{40} 30 m from the roof contact. Detailed microprobe studies across some of the magnetitite layers (Harney *et al.*, 1990) showed oscillations, but a wide range of core compositions makes precise trends difficult to identify.

Plagioclase separates extracted from the Bierkraal borehole core from the Western limb (Figure 2a) and analyzed by XRF spectrometry display a rather diffuse trend (Figure 6). Isolated samples show anomalously high and low values, but the only sustained breaks of significance occur 400 and 650 m below the roof. Microprobe data of Hoyle (in prep.) based on a number of profiles in the Northam area show an overall decline through the UZ to An_{44} some 150 m below the roof, with a prominent reversal from An_{51} to An_{62} at the base of Upper Subzone B. In this, and several lesser reversals, the trends are matched by abrupt shifts in the *mg#* of clinopyroxene. Within the diorites underlying the granite roof, the An content declines further to $An_{23\text{-}07}$.

Sr-isotope ratios: Numerous determinations of initial $^{87}Sr/^{86}Sr$ ratio in whole-rock and mineral separates from the Western limb (Figure 9), summarized by Kruger (1994), define a plagioclase cryptic variation trend. Through the LZ the ratio increases erratically from 0.705-0.707, and then declines progressively to 0.705 in the lowermost CZ. Through the chromitite-bearing interval of the C_LZ the ratio oscillates around this value. A slow and irregular increase then begins below the first appearance of cumulus plagioclase, reaching 0.7065 immediately below the Merensky Reef. Here there is a very distinct break to 0.7075, and a rapid climb to a peak value of 0.709 in the M_LZ, before becoming essentially constant at 0.7085 through the gabbronorites. A steady decrease begins 200 m below the Pyroxenite Marker, and then the entire sequence from the Marker to the roof in both lobes has a constant Sr_i of 0.7073 (Sharpe, 1985; Kruger *et al.*, 1987). The constant Sr_i ratio of 0.7073±1 up to within 20 m of the roof contact in both the Eastern and Western lobes indicates minimal assimilation of roof rocks by the UZ magma.

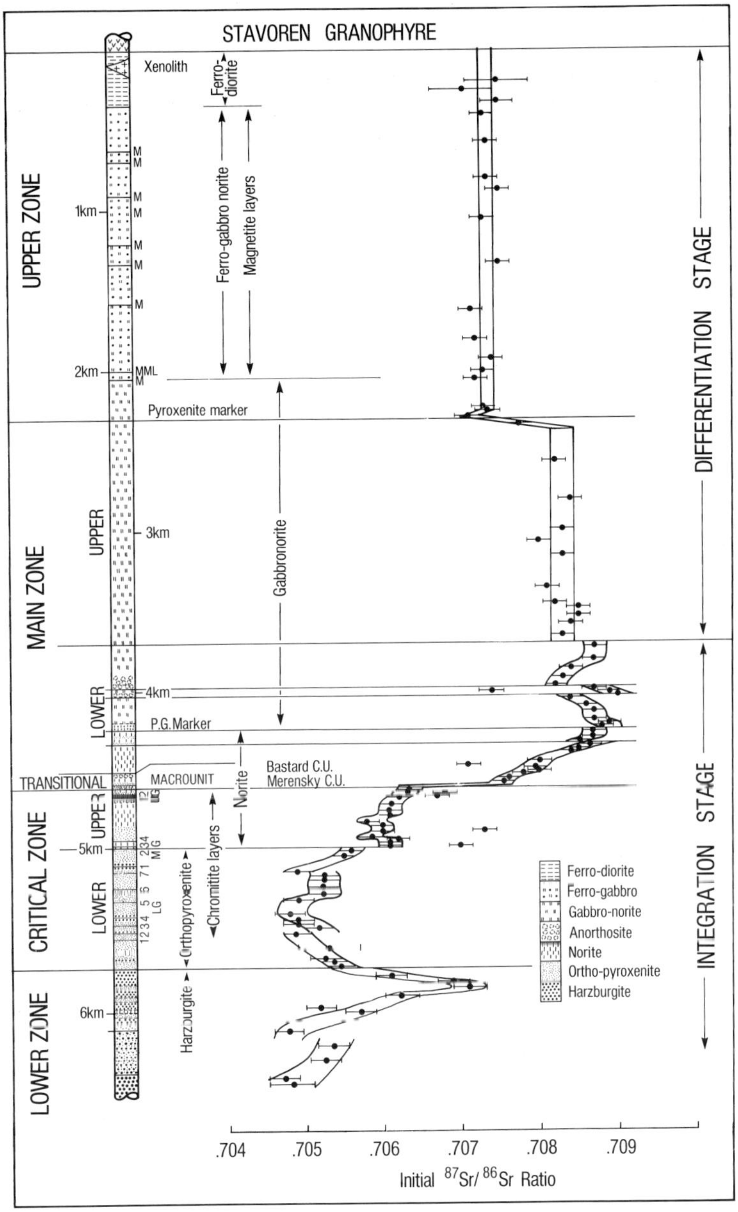

***Figure 9**. Plot of initial $^{87}Sr/^{86}Sr$ ratio versus height for samples from the Western limb (from Kruger, 1994, with permission of South African Journal of Geology).*

4.5. Chromite

Only where there is a high modal proportion of chromite in massive layers, which creates a buffer to chemical change, is a close approach to true magmatic chromite compositions preserved. Such data, however, offer an intermittent rather than continuous record of magma evolution. Trends at Union Section (proximal facies) and Brits (distal) in the Western limb

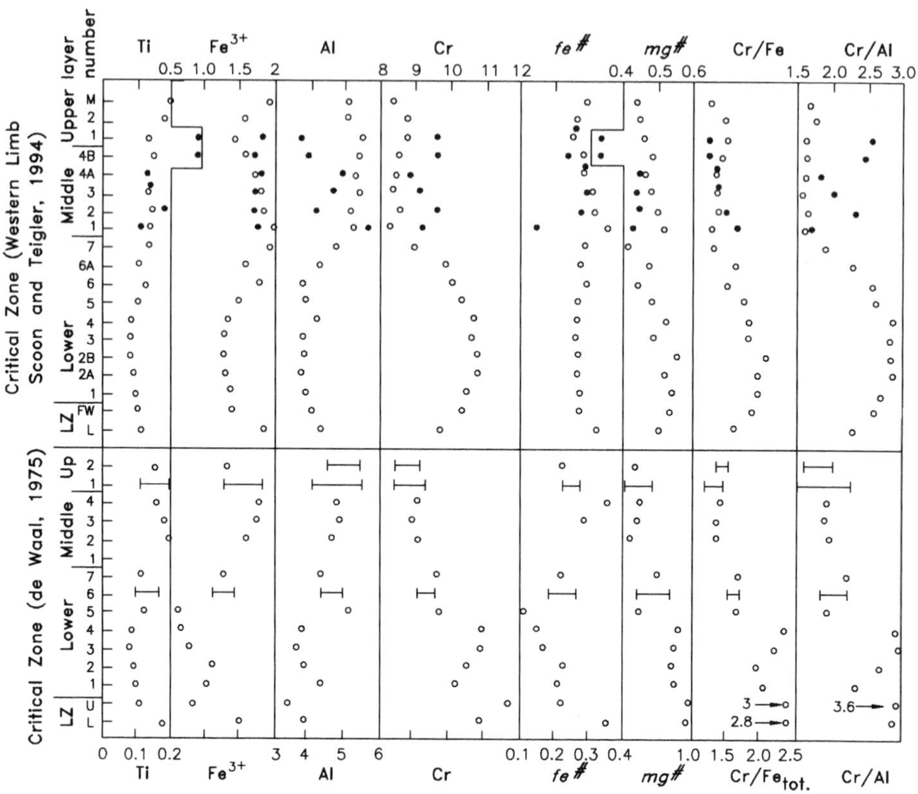

Figure 10. Chemical trends for chromite from the Western limb are shown in the upper part of the diagram (from Scoon and Teigler, 1994), with open circles representing Union Section (proximal facies) and solid circles representing Brits (distal facies). Lower part of diagram from de Waal (1975) is a composite representing both limbs. Values for Ti, Fe^{3+}, Al, and Cr are in cations. fe# is $Fe^{3+}/(Fe^{2+}+Fe^{3+})$. Analyses by Scoon and Teigler (1994) are by electron microprobe (with ferric iron content determined by stoichiometry), and those by de Waal (1975) are by wet chemistry on chromite separates taken from the entire layer. Horizontal bars indicate the range when more than three analyses of different sections of a layer are reported. In the upper diagram L refers to a lens of chromite in the Lower Zone, FW refers to analyses of chromite from the Far West limb, and M to Merensky chromitite. In the lower diagram L and U refer to the lower and upper chromitite layers from the Grasvally mine (Hulbert and von Gruenewaldt, 1985).

(Scoon and Teigler, 1995), and a composite of samples from both limbs (de Waal, 1975) are shown in Figure 10.

Successive chromitite layers reveal a broad trend (Figure 10) of upward decline in Cr/Fe ratios, from 2.2 to 1.3, although in different sections through the intrusion absolute values may differ. At Union Section Cr declines between the lowest and highest layers, while Al initially increases through the C_LZ and then remains constant once cumulus plagioclase appears (Teigler and Eales, 1993). The *mg#* shows an initial increase up to the LG4 layer, matching the broad reversal displayed by olivine and orthopyroxene through the same interval. An overall decline is then traceable through the remaining layers up to the LG7 layer. An abrupt reversal to higher values in the Middle Group is then succeeded by further decline through the C_UZ, with the lowest *mg#* in the thin chromitites of the Merensky and Bastard units. Ti levels increase almost three-fold between the C_LZ and Merensky Reef chromitites.

4.6. Magnetite

Magnetite shows several solid-solution substitutions which yield relatively systematic changes in composition through the UZ. Magnetite-ulvospinel solid solution at magmatic temperatures results in high TiO_2 contents, which increase regularly from 12%-20% with increasing height (Klemm *et al.*, 1985b). However, because of extensive oxidation-exsolution, electron microprobe analyses of low-temperature magnetite yield TiO_2 contents of only *ca.* 4%.

Many publications cite V_2O_5 contents of magnetite. However, V^{3+} is present in magnetite, not V^{5+}. V_2O_5 is the commercially extracted end-product. Also, some report whole-rock contents in magnetitite layers, whereas others quote analyses on mineral separates. V_2O_3 contents in magnetite are highest in the lowest magnetitite layers with a general upward decrease both in massive layers and intervening disseminated samples from about 1.7 to 0.2% in the Eastern, Western and Northern limbs (see Figure 6; Molyneux, 1970; Klemm *et al.*, 1985b; van der Merwe, 1976). Mg and Al contents are higher in massive than in disseminated layers (Klemm *et al.*, 1985b), which probably reflects re-equilibration of magnetite with associated silicates. Lateral variations in composition have not been studied in detail, but Willemse (1969) and Cawthorn and Molyneux (1986) reported that the Main Magnetitite Layer shows a decrease in V when traced northwards in the Northern limb and southwards in the Eastern limb. In both cases this decrease occurs as the layer approaches the floor contact as a result of the MZ/UZ transgression. No change has been recorded in the composition of the Main Magnetitite Layer in the Western limb.

Cr is highly compatible in magnetite, and consequently Cr in magnetite decreases rapidly upwards (Cawthorn and McCarthy, 1980), even within a single layer. The highest Cr content typically found at the base of the Main Magnetite Layer is 8000 ppm, but this may drop by a factor of ten within one metre, and major reversals occur within this layer. Similar depletion profiles, and usually reversals, are found in all layers studied (McCarthy *et al.*, 1985).

5. PARENTAL MAGMAS

Mapping of the marginal zone rocks and sills in the immediate floor of the Western and Eastern limbs has led to the recognition of several different magma types, although relationships to the layered sequence itself are still incompletely resolved. The Marginal Zone contains such a complexity of rock types, with variable cumulus enrichment, that their compositions cannot be taken as representative of magmas. Furthermore, Sharpe (1981) compositions cannot be taken as representative of magmas. Furthermore, Sharpe (1981) regarded this zone as a precursor rather than parental to the main intrusion. The correlation of magma types with sections of the layered sequence has been attempted in two different ways. Cawthorn *et al.* (1981) used mineralogical and chemical attributes of sills in the floor of the Western limb to suggest correlations, whereas Sharpe (1981) used geographic proximity of marginal rocks to different sections of the layered sequence.

Magnesian sills with spinifex-textured orthopyroxene and hopper olivine crystals were suggested by Davies *et al.* (1980) to represent magma parental to the LZ (analysis 1, Table 2) but Cawthorn *et al.* (1981) concluded that there had been considerable assimilation by these sills. Subsequent analysis of rocks with similar major-element composition by Harmer and Sharpe (1985) and Sharpe and Hulbert (1985) yielded lower Rb and K_2O contents (analysis 2, Table 2) that may represent more appropriate values. A remarkable attribute of this and every magma within the intrusion is the high Sr_i ratio, as first established by Hamilton (1977). The origin of such high values, and the debate as to links with pyroxenitic komatiite (Cawthorn and Davies, 1985), siliceous high-magnesia basalt (Barnes, 1989) or boninite (Hatton and Sharpe, 1989) are clouded by uncertainty about, and large variations in, the trace-element chemistry of

Table 2
Analyses of proposed magmas parental to various sequences in the Bushveld Complex.

	1	2	3	4	5
SiO_2	55.70	53.17	48.50	50.70	49.32
TiO_2	0.36	0.36	0.75	0.41	0.81
Al_2O_3	12.74	11.36	16.49	16.03	15.67
FeO(T)	8.80	10.72	12.41	9.14	12.77
MnO	0.09	0.20	0.19	0.17	0.19
MgO	12.44	14.93	7.57	9.21	6.08
CaO	6.96	7.47	11.15	11.14	10.83
Na_2O	2.02	1.57	2.17	2.52	2.94
K_2O	1.03	0.17	0.14	0.23	0.25
P_2O_5	0.10	0.07	0.18	0.08	0.07
mg#	71.5	71.4	52.1	64.2	45.9
Ni	292	337	93	162	77
Cr	970	1240	226	205	111
Rb	37	4	3	7	4
Sr	195	183	359	324	350
Ba	439				176
Zr	70	47	34	20	31
Y		15	21	12	25

1 - Proposed magma parental to Lower Zone (Davies *et al.*, 1980). 2 - Average B1 magma of Sharpe (1981), originally proposed as being related to Marginal Zone only, but considered by Sharpe and Hulbert (1985) as parental to Lower Zone. 3 - Average B2 magma of Sharpe (1981), proposed as parental to Critical Zone. 4 - Average B3 magma of Sharpe (1981), proposed as parental to Main Zone. 5 - Proposed parental magma to Upper Zone, intruded at the level of the Pyroxenite Marker (Davies and Cawthorn, 1984).

Wager and Brown (1968) accepted that there had been repeated injection of magma during an Integration Stage (the combined LZ and CZ), but it was implicit that each replenishment was of the same composition. Differences in Sr_i ratio led Hamilton (1977) to conclude that each zone was derived from an isotopically discrete magma, but the major-element composition of these magmas was not addressed. Identification of successive magma compositions, and relating these to the layered sequence, is fraught with problems such as the extent of magma mixing within the chamber. Sharpe (1981) suggested that a discrete magma (analysis 3, Table 2) was parental to the CZ, but its low Cr content of 200 ppm appears inconsistent with the Cr budget of the intrusion, as discussed below.

The major break in isotopic ratio (Kruger and Marsh, 1982) and rapid transition from noritic to gabbronoritic cumulates above the Merensky Reef demands a fundamentally different magma for the MZ. Based on geographic criteria, Sharpe (1981) suggested a MZ magma composition not very different from that of the CZ (analysis 4, Table 2). Diagnostic parameters to confirm this are not easily defined. One criterion is that this proposed MZ magma has half the Ti content of the CZ magma, consistent with the sudden decrease in the Ti content of pyroxenes at this level. The similar Cr contents of the proposed CZ and MZ magmas do not, however, equate with the abrupt decrease of Cr in pyroxenes at the base of the MZ. Further, the higher *mg#* of the MZ liquid relative to the CZ liquid (analyses 3 and 4, Table 2) is hard to reconcile with the observed trends in *mg#* of the mafic phases (Figure 3).

Addition of magma close to the Pyroxenite Marker is indicated by cryptic reversals. An appropriate magma type has been identified in a discordant microgabbronorite body intruding the Marginal and Critical Zones in the Western Bushveld (Davies and Cawthorn, 1984). Melting experiments on this composition (analysis 5, Table 2) yield the appropriate mineralogy and mineral composition (Cawthorn and Davies, 1983), and it has the appropriate Sr_i ratio (Kruger *et al.*, 1987). Intriguingly, this composition is very similar to that of Sharpe's (1981) CZ magma (analysis 3, Table 2). Clearly, the significance of these magma types needs to be re-assessed.

Significantly, compositions 3, 4, and 5 have *mg#* values that are far from those of primary magmas. The inference is that considerable fractionation must have taken place at depth, and hence that there may be a paired, more primitive intrusion beneath the Bushveld Complex.

6. THE NATURE AND ORIGIN OF LAYERING IN DIFFERENT ZONES OF THE COMPLEX

Layering of different types exists within the complex. **Phase layering** refers to the appearance or disappearance of specific phases generally related to fractionation or rejuvenation of the resident magma. Examples of this would be the restriction of chromite and magnesian olivine to the LZ-CZ sequence, the change from orthopyroxene to pigeonite in the MZ, and the restriction of Fe-Ti-oxides, ferrian olivine and apatite to the UZ.

Modal layering is not ubiquitous in the complex. Many thick, vertical sections of the MZ and UZ display homogeneous lithologies. Modal layering does occur in the Harzburgite Subzone of the LZ, but is most prominent in the CZ, where orderly, though variably incomplete, chromitite-harzburgite-pyroxenite-norite-anorthosite sequences give rise to **cyclic units**. Individual layers within such cyclic units may be traceable for hundreds of km. The absence of recognisable patterns prohibits similar identification of cycles in the UZ.

Layer contacts may be sharp or diffuse. Chromitite and magnetitite layers generally have sharp bases but more diffuse upper contacts and most are internally not totally homogeneous. Noritic layers commonly bear pyroxene- or feldspar-rich laminae, and anorthosites planar concentrations of pyroxene oikocrysts. Eales *et al.* (1994) have shown that in the C_UZ gradation between pyroxenite and norite is rare or confined to very thin intervals, whereas continuous gradation exists between norite and anorthosite.

Textural layering may be manifested by distinct changes in grain size, crystal habit or fabric. Plagioclase-rich rocks frequently display a layer-parallel orientation of bladed grains, as may pyroxenites, such as in the LZ and the UG2 Unit. Oikocryst size may also produce layering within an otherwise homogeneous rock (Figure 5e), and may vary systematically upwards through anorthosites.

Cryptic layering refers to changes in whole-rock or mineral composition, and is generally not recognizable without analytical data on closely spaced samples.

Layering may be disturbed by post-depositional events. Localized slumping is well illustrated by sagging of the immediate footwall beneath detached pods of chromitite (Viljoen *et al.*, 1986a; their figures 7 and 8). The review by Lee (1981) describes a variety of post-depositional features in the C_UZ and lower MZ, including liquid-escape structures, load casting, flame structures and detached inclusions of one rock type within another.

6.1. Dunite-pyroxenite layering in the Lower and Lower Critical Zones

Cyclic layering and cryptic variations provide some constraints on the processes that operated during the formation of the LZ and C_LZ. No sharp breaks in mineral composition separate the LZ and C_LZ, or C_LZ and C_UZ, at either Union Section or in the Olifants River trough (Cameron, 1978; Eales *et al.*, 1993a, 1994). Through the LZ and C_LZ at Union Section, there are five sections, from 60 to >300 m thick, defining reversals of normal fractionation trends, preceding comparable intervals tracing normal trends (Figure 4). Integration of these cycles with the lithology yields a clearly defined pattern. Intervals of normal fractionation begin either within or near the tops of thick successions of olivine-rich rocks, and continue into thick, overlying pyroxenitic intervals. Reversals of trend are then initiated within the middle or upper reaches of the pyroxenitic intervals, and propagated upwards until they terminate within the next olivine-rich interval. Peaks in the curve depicting *mg#* values are thus coincident with the five olivine-rich intervals in Figure 4. A similar pattern was reported by Bristow (1989) in the C_LZ in the Eastern limb. These trends imply that olivine-rich intervals of the LZ and C_LZ do not reflect sudden events within the chamber so much as the end-products of protracted periods of magma rejuvenation. The cyclicity is attributable to alternation of periods during which either crystal fractionation, or relatively primitive magma addition, was dominant.

The lowest sequence in the Olifants River trough is a pyroxenite, 400 m thick in troughs, but decreasing to some 250 m over the Schwerin fold (Cameron, 1978). This geometry implies that the troughs were not separate basins of magma, but interconnected. The pyroxenite averages >98% orthopyroxene. The formation of such a thick, virtually monomineralic sequence, of near-constant composition, implies the presence of a very large volume of magma.

Overlying this, the Harzburgite Subzone contains up to 350 m of dunite-harzburgite-pyroxenite cyclic units (Figure 5a). Chromite is present, but never exceeds 0.3%. On the Schwerin fold upwarp, the number of cycles decreases, but individual cycles have comparable thickness to those within the trough. This Harzburgite Subzone of the LZ and the Lower

Pyroxenite Subzone of the CZ collectively thin from 600 to 200 m over the Schwerin fold (Cameron, 1978), suggesting that these upwarps were not static, primary structural features, but continued to develop during accumulation. The greater number of cycles present in the trough, and thinning of the entire sequence over the arch would be consistent with added, dense magma ponding between upwarps. Conversely, the presence of dunite on the upwarps implies that on occasions the thickness of the magma layer crystallizing olivine must have exceeded the vertical relief between trough and upwarp. Reconciliation of these two inferences must await a greater understanding of the early-stage geometry of the intrusion.

Each cycle within the Harzburgite Subzone shows a vertical textural change as the proportion of orthopyroxene increases, from poikilitic, to both olivine and orthopyroxene being polygonal, to olivine becoming interstitial to orthopyroxene. One explanation for this latter feature might be that the composition of the basal layer of magma lay in the olivine primary phase field, while the overlying layer held orthopyroxene crystals in suspension during initial turbulence (Huppert and Sparks, 1980). Orthopyroxene may then have sank into the basal layer to become enclosed by intercumulus olivine.

In the olivine-bearing interval of the C_LZ there is layering of harzburgite and pyroxenite on a scale down to cms, but this is not systematic enough to be termed cyclic (Cameron, 1980). If the harzburgite layers reflect magma additions, they had to be frequent and of low volume, and to have compositions very close to the olivine-orthopyroxene peritectic. This spacing might also suggest the concept of oscillatory nucleation (Maaløe, 1978). However, where two phases show a reaction relationship, rather than one of co-precipitation, this process cannot operate, as the crystallization of pyroxene will not drive the liquid back towards olivine saturation. It is also unlikely that a process of crystal ageing, used to model inch-scale layering in the Stillwater intrusion (Boudreau, 1994), could apply where the two minerals display a reaction relationship.

Cycles ranging from dunite to pyroxenite are predictable from the crystallization of appropriate parental magmas (Cawthorn and Davies, 1983). However, a major question relates to the paucity of olivine in the lowermost part of the LZ, especially in the east. The most forsteritic olivine in the Olifants River trough is Fo_{85}, but at Union Section it is Fo_{89} and at Potgietersrus Fo_{90}. This may reflect either injection of more evolved magmas in the east, or progressive lateral change. The concept of proximal and distal facies within the Western limb rests upon substantial regional variations in lithology, and progressive evolution of cumulates along strike. If this also occurred in the east, there may be more primitive dunites, as yet unidentified, in the Eastern limb.

Within the ultramafic LZ at both Union Section and in the Olifants River trough there occurs a 3 m norite layer. At Union Section it occurs 470 m above the base (*ca.* 1490 m in Figure 4) and caps a 210 m sequence through which olivine-rich cumulates give way to pyroxenitic cumulates. Concomitantly, the orthopyroxene composition declines steadily from *mg#* of 89.5 to 83.6, establishing it as a fractionating sequence (Teigler, 1990). The Sr_i value of 0.7052 for the norite layer is unremarkable within a batch of six samples spaced from 325 m beneath to 86 m above the norite layer. These latter values range from 0.7048 to 0.7062, averaging 0.7054 (analytical data of F.J. Kruger) and point to an autochthonous origin for the norite layer. This demonstrates the capacity of LZ liquids to reach the cotectic with plagioclase within a limited degree of fractionation (*ca.* 25%, from the experimental data of Cawthorn and Biggar, 1993). As cumulus plagioclase did not reappear before more than 1000 m of ultramafic cumulates had been laid down, the system must repeatedly have been rejuvenated by further additions of fresh magma, which restrained it from reaching the plagioclase cotectic.

6.2. Chromitite layers

The abundance and thickness of chromitite layers present major problems in geochemical modelling. A summation of total Cr in the chromitite layers and pyroxenes of the CZ in the east yields a value of between 6000 and 13000 ppm (Cameron, 1980, 1982). At Union Section the cumulative thickness of all chromitite layers is 8.25 m. A calculation based on modal and microprobe data of Teigler (1990), Maier (1991), and de Klerk (1991), allowing for Cr in disseminated chromite and in orthopyroxene, yields an average of 8250 ppm Cr in the CZ, and 6585 ppm for the combined LZ and CZ. As experimental data of Barnes (1986b) indicate a maximum solubility of *ca.* 1000 ppm in feasible parental liquids (13% MgO), a great volume of magma must have been processed.

The implications of this are far-reaching. The thickness of the layered suite in the proximal facies of the Western limb is *ca.* 7.7 km, of which *ca.* 2 km are the Cr-rich rocks of the LZ and CZ. The *ca.* 2.2 km of Cr-poor M_LZ rocks are identified by isotopic evidence as a discrete magma injection post-dating at least the major part of the CZ. Thus, only a few hundred metres of cumulates at the base of the MZ could conceivably be identified as the *in situ* residua to the LZ-CZ cumulates. There is a massive discrepancy in the Cr budget here, and it is necessary to envisage lateral flow and subsequent crystallization of the Cr-impoverished residual liquids elsewhere within the original chamber, the boundaries of which are not preserved.

Hypotheses for forming chromitite layers include gravitational sorting, increases in oxygen fugacity, pressure changes, and magma mixing. The sharp contacts and remarkable lateral continuity of layers demand that, whatever the processes, they must have operated at the same time over the entire chamber. A consideration of cotectic proportions rules out simple gravitative sorting for these layers. The cotectic proportions of chromite:olivine are approximately 0.3:99.7. To form sufficient chromite to make a layer 1 m thick would require that the equivalent of over 300 m of olivine remain suspended in the magma. Density contrasts and convection forces suggest that this is implausible (Sparks *et al.*, 1993). It appears inescapable that chromitite layers result from events which bring the magma into the chromite primary phase volume.

Oxidation of magma on a chamber-wide scale does not seem feasible, especially when it is considered that in the C_UZ the chromitite layers define the bases of cycles, and oxygen fugacity would have little effect on the relative stability of plagioclase and pyroxene. The idea of pressure change driving the magma into the chromite field (Cameron, 1980; Lipin, 1993) is appealing, as a pressure increase would also increase the stability of pyroxene relative to plagioclase. However, in a chamber as large as the Bushveld Complex the roof could not have been rigid, but merely floating on the magma. Hence, mechanisms which could increase the pressure at the base of the chamber are difficult to envisage.

The magma-mixing hypothesis has two variants, one being the addition of primitive magma (Irvine, 1977; Murck and Campbell, 1986) and the other addition of plagioclase-saturated magma (Irvine *et al.*, 1983). The composition of chromite can be used to test the latter model. Dick and Bullen (1984) showed that the Cr/Al ratio of chromite is extremely sensitive to the Si/Al ratio of the magma. Two different magmas with orthopyroxene and plagioclase on their liquidi, respectively, would have very different Si/Al ratios. Mixing in different proportions would produce chromite with different compositions. The systematic vertical variation in chromite composition and lateral uniformity (Figure 10) would be difficult to explain by this model.

The Harzburgite Subzone of the LZ contains cycles of dunite to pyroxenite. If this is attributed to addition of undifferentiated magma, magma mixing might be expected to have initiated chromite precipitation. There is, however, only a small and fairly constant proportion of <0.3% chromite throughout this subzone in both Eastern and Western limbs, which is inimical to the mixing model involving primitive magma. It is also significant that in the C_LZ all but one of nine major chromitite layers are enclosed in pyroxenite, not harzburgite.

Experimental work by Murck and Campbell (1986) offers a possible explanation for the paucity of chromite in the LZ. The solubility of Cr in basaltic liquids, plotted against temperature, yields a strongly curved line. The mixing of evolved and primitive liquids could be represented by a chord to this curve. The greater the contrast between evolved and primitive magma components during mixing, the greater would be the penetration of the hybrid liquid (represented by a point on the chord) into the field of chromite. Within the LZ, contrasts between resident and newly-emplaced liquid would have been much less, and chromite-oversaturation more short-lived, than in the more evolved parts of the CZ where the robust chromitites appear.

The model invoking introduction of primitive magma could be questioned in the light of the absence of cryptic variations of orthopyroxenes above and below chromitite layers. For example, pyroxenes immediately above and below the LG6 chromitite layers in the Eastern limb have comparable *mg#* and Cr contents (Cameron, 1980). This is also true of the Grasvally chromitites (Hulbert and von Gruenewaldt, 1985), the LG1-LG4 chromitites of the Far Western limb (Engelbrecht, 1985), and the UG2 chromitite at Union (Cawthorn and Barry, 1992; Eales *et al.*, 1993b) and Amandelbult Sections (Eales *et al.*, 1988). A further problem for the mixing model concerns the volumes of magmas involved. Deposition of a 1 m layer of chromitite without significant depletion of the Cr content of the magma requires the processing of great volumes of liquid. Mixing on such a scale is implausible if the added magma is introduced as a dense, basal flow.

Clearly, no genetic model for the origin of chromitite layers fully meets all geochemical, mineralogical or physical constraints. Despite the problems listed above, the magma-mixing model probably still enjoys wider support than others.

6.3. Layering in the Upper Critical Zone

Here layering makes a spectacular display in repetition of complete or incomplete chromitite-harzburgite-pyroxenite-norite-anorthosite cycles, ranging in thickness from <1 to 300 m. Phase layering is illustrated by the absence of olivine in the lower part of the C_UZ at Union Section and its reappearance in the UG2 unit, the harzburgitic Pseudoreefs, and the Merensky Reef. Rhythmic layering is developed on a scale of cms in the Footwall Unit beneath the Merensky Reef (Viljoen *et al.*, 1986a, their figure 13) and on a scale of mm in the noritic footwall to the UG1 chromitite (*op. cit.*, their figure 8). Parameters such as *mg#* and trace-element ratios define saw-tooth patterns of cryptic cyclicity repeated from one unit to the next (Eales *et al.*, 1986, 1988, 1993a).

There is little support for the notion that such cycles represent simple, progressive crystallization of discrete batches of liquid. This conclusion is justified by various criteria:

1) Significant variations of Sr_i ratio exist within successive layers of individual units, and even between individual feldspar grains of single hand specimens (Eales *et al.*, 1990a, 1990b).

2) Orthopyroxenes evolve towards more ferrian varieties from base to top of "complete" cyclic units such as the Merensky Unit, but this is not matched by parallel evolution of plagioclase towards more sodic varieties (Naldrett *et al.*, 1986; Eales *et al.*, 1993a).
3) The proportion of leuconorite plus anorthosite in the Merensky Unit ranges from 33% to as much as 85% in different sectors of the complex (Vermaak, 1976). It seems improbable that such variations in total feldspar content could stem from the crystallization of a single parent liquid without the intervention of some additional process.

The model that best fits the data is one in which cyclic units stem from hybrid magmas generated by the mixing of more primitive liquids with resident residual magmas crystallizing plagioclase and orthopyroxene. The proportions in which these mixed was variable, as indicated by the variable proportions of ultramafic to felsic lithologies within individual cyclic units, from place to place.

The mechanism, whereby new magma batches were emplaced and then mixed with residua, remains debatable at present. In the "plume hypothesis" (*inter alia*, Campbell *et al.*, 1983; Naldrett *et al.*, 1987) newly-emplaced liquids are deemed to have risen through the residua in the chamber, until they spread laterally as extensive liquid layers once a level of neutral buoyancy was reached. Finger-mixing with the underlying liquid layer, and then downward plunging of crystal-laden density currents, would then spread cumulates along the floor. Alternatively, direct deposition of cumulates could occur from batches of hybrid magma moving along the crystalline floor from proximal to distal facies (Eales *et al.*, 1986, 1988, 1990a). Persuasive evidence for the latter is the entry of cumulus plagioclase into the distal facies of the C_LZ at Brits, and the cryptic evolution of pyroxenes, along strike, in the same direction (Teigler *et al.*, 1992).

An approach that may accommodate both concepts (Maier and Eales, 1994a) reasons that regionally extensive units showing only limited compositional and thickness variations (*e.g.*, the UG2 chromitite - pyroxenite couplet) may stem from the pluming process. This would be consistent with generation of their robust chromitite layers from large volumes of mixed liquids. By contrast, layers such as the harzburgitic Pseudoreefs are seen as cumulates deposited from bottom-flows. This would explain their restriction to the proximal facies in the Western limb, their degradation along strike to olivine norites, and the trivial thicknesses of associated chromitite, which is the consequence of less efficient mixing.

The concept of magma rejuvenation is supported by distinctive textural features. Partially resorbed, embayed inclusions of plagioclase are trapped within the orthopyroxene and olivine grains of many CZ cumulates (Figure 5f), which are in turn enclosed by later-generation intercumulus plagioclase. This paradoxical texture appears some 20 m below the first cumulus plagioclase in the C_UZ at Union Section, and then persists through the UG1 Footwall, Merensky and Bastard units. It is attributed to addition of fresh magma batches to resident, partially crystallized liquids that had reached the orthopyroxene-plagioclase cotectic (Eales *et al.*, 1991; Maier and Eales, 1994b). Partial resorption of the feldspar was effected during mixing, with the resultant hybrids remaining in the primary phase volume of orthopyroxene or olivine. The association of the texture with reversals of normal fractionation trends (increase of *mg#*) is consistent with this interpretation.

Levels of minor elements Ti and Mn in orthopyroxene increase with stratigraphic height through the LZ and C_LZ, but Al increases at first and then declines again in the C_UZ (Figure 7). This conforms with the finding of Grove and Bence (1977) and Shearer *et al.* (1989) that Ti/Al ratios increase once plagioclase nucleates, because of the partitioning of Al into plagioclase.

This chemical fingerprint indicates that a significant proportion of the pyroxene population in C_UZ pyroxenites crystallized on the cotectic with plagioclase, but was then incorporated into basal layers of pyroxenite by crystal settling, or by crystal-liquid density flows (Eales *et al.*, 1994).

The origin of anorthositic layers in the C_UZ raises further issues. The Bushveld anorthosites are not to be compared with the thick sequences in the Stillwater intrusion (see McCallum, this volume). A maximum of only 53 m is attained in the Giant Mottled Anorthosite capping the Bastard Unit. Their low pyroxene content does not allow the C_UZ anorthosites to be interpreted simply as end-products of fractionation of mafic magmas. Explanations for their origin currently fall into two categories. (i) They crystallized from feldspathic melts in which plagioclase was the sole crystallizing phase before the cotectic was reached, or (ii) physical processes, such as gravitational sorting, acted upon partially crystallized melts. The concept of periodic injection of feldspathic (aluminous or A-type) magmas during the accumulation of the C_UZ (Irvine *et al.*, 1983; Schurmann, 1993) has evoked controversy. In support of the concept, Hatton (1986) cited an inverse relationship between Sr_i ratios and *mg#* found across the C_LZ/C_UZ transition in the Eastern limb. This, he argued, was evidence for the blending of magma with a low *mg#* and high Sr_i ratio (Main Zone, A-type) with resident magma with high *mg#* and low Sr_i ratio (U-type). This antipathetic relationship is not, however, generally prevalent, as a sympathetic relationship has been proved at Union Section. Moreover, anorthosites and leuconorites of the C_UZ do not display Sr_i ratios significantly higher than associated norites and feldspathic pyroxenites, as would be expected if they were early precursors to emplacement of the Main Zone (Eales *et al.*, 1990a, 1990b).

If, as is concluded here, the anorthositic layers are derivatives of the same magmas that gave rise to the pyroxenites of the C_UZ, some physical process must have acted to yield cumulates devoid of cumulus ferromagnesian phases. Partial remelting of plagioclase-orthopyroxene-liquid mushes, initially on the cotectic but with the bulk composition shifted into the plagioclase field by settling of the more dense orthopyroxene phase, would, during remelting, be capable of generating limited volumes of melt from which plagioclase would have crystallized alone during subsequent cooling. Evidence supporting such settling exists in the wide modal variations in C_UZ cyclic units: analysis of 284 C_UZ samples has indicated no clustering around putative cotectic proportions of plagioclase:pyroxene (Eales *et al.*, 1990a, 1994).

The mineable concentrations of PGE in the Merensky Reef are discussed by Lee (this volume), and here are only discussed in terms of the relationship to layering in the C_UZ. All chromitites at the bases of cyclic units of the C_UZ hold anomalous concentrations of PGE, usually with little associated sulphide (Scoon and Teigler, 1994). In terms of mineral paragenesis, the Merensky Unit is comparable to one overlying and several underlying cycles, and the compositions of its plagioclase and pyroxene are not notably different. One major feature is a pronounced break in Sr_i ratios at its base (Kruger and Marsh, 1982), but even this is not unique, in view of a similar Sr_i profile at the base of the Bastard, as well as Merensky, Units at Atok Mine (Lee and Butcher, 1990).

Naldrett *et al.* (1986) have suggested that the Merensky Reef resulted from a mixing of resident magma with more primitive magma comparable with that of the LZ. To produce the high levels of PGE, very large volumes of magma are required. If the ratio of silicate to sulphide (the R-factor of Campbell *et al.*, 1983) is $>10^5$, the 2-3 cm of sulphide in the Reef demands that some 2000-3000 m of magma mixed rapidly and became saturated with sulphide. There is a chemical imbalance in this argument. The Sr_i ratio increases from 0.7063 to 0.7075 over this interval and, unless the new magma had an anomalously high initial ratio, the proportion of new magma must have been large. However, the composition of orthopyroxene in the Reef, compared to that in its footwall, is not consistent with a source relatively enriched in Mg (Kruger, 1992; Cawthorn, 1996). The model of Campbell *et al.* (1983) also requires that the magma producing the footwall and the hanging-wall sequence be sulphide-saturated, yielding rocks with about 0.1-0.2% sulphide. The extremely low levels of S in the order of 100-200 ppm in these rocks at Impala Mine (Schurmann, 1993), suggest that this was not the case. The same problem arises for the UG2 chromitite, an ore body with a greater PGE reserve than the Merensky Reef, but with a sulphur content averaging <1000 ppm (Gain, 1985). A totally different model for the PGE mineralization, involving vertical transport of PGE by hydrothermal fluid or residual liquid has been proposed by Mathez (1989) and Boudreau and McCallum (1992). The extensive mineralization in the Platreef along the floor of the Potgietersrus limb (White, 1994) is difficult to explain by this model. Further, Viljoen *et al.* (1986b) report that where replacement bodies cut the Merensky and associated rocks, the PGE mineralization can still be traced at its original horizon, and has suffered minimal remobilization. In conclusion, the origin of the PGE mineralization in the UG2 and Merensky Reefs has still to be satisfactorily accounted for.

The uppermost cyclic unit in the C_UZ is the Bastard Unit, which appears identical to the Merensky Reef except for uneconomic levels of PGE. Thus the processes producing cycles in the C_UZ continued after the formation of the Merensky Unit.

6.4. Layering in the Main Zone

The relative paucity of layering in the MZ is remarkable, and requires that, with few exceptions, plagioclase and pyroxene accumulated in their cotectic proportions. The Main and Upper Mottled Anorthosites represent exceptions to this conclusion. Their origin has not been the subject of any detailed study, but two other examples of modal layering have been investigated.

Layered gabbronorites: Some 50-100 m below the Pyroxenite Marker is a sequence of strongly layered gabbronorites (Quadling and Cawthorn, 1994). These are best developed in the northern part of the Eastern limb, become intermittent near Roossenekal, and are absent further south. This layered package ranges up to 10 m in total thickness, with over 100 layers being recorded at one locality. The cumulus phases within it are the same as in the enclosing gabbronorites, but in widely variable, non-cotectic proportions. The melanocratic layers pinch out both down dip and along strike, but only rarely do the leucocratic layers. All contacts are sharp, internal modal variation within individual layers is lacking, and there is no correlation between thickness of adjacent leuco- and melanocratic layers. Mineral compositions show no variation through adjacent leuco- and melanocratic layers, Sr_i ratios remain constant throughout.

As plagioclase and the two pyroxenes remain cumulus phases in almost every layer,

---

Figure 11. *(facing page) Plot of mineral compositions in gabbronorites from four traverses across the Pyroxenite Marker interval in the Eastern limb shown in Figure 2b (X-ray fluorescence analyses by Marais (1977) of mineral separates). (a) Plot of mg# in Ca-poor pyroxene and An in plagioclase versus height relative to the Pyroxenite Marker. (b) Plot of mg# in orthopyroxene minus An in plagioclase versus height.*

rhythmic nucleation (Maaløe, 1978) is ruled out. The absence of grain-size variation argues against post-cumulus maturation processes (Boudreau, 1994). The possible influence of different magmas is excluded on the grounds of constant phase and Sr-isotopic compositions in successive layers. The likely density of a magma crystallizing two pyroxenes and plagioclase is such that plagioclase should not be capable of sinking through it. This feature, the uniformity of mode within individual layers, and the sharp contacts between them, imply that here gravitational sorting of crystals offers no explanation. A more fruitful approach is suggested by the behaviour of crystal-liquid slurries (Irvine, 1980), where the bulk densities of different

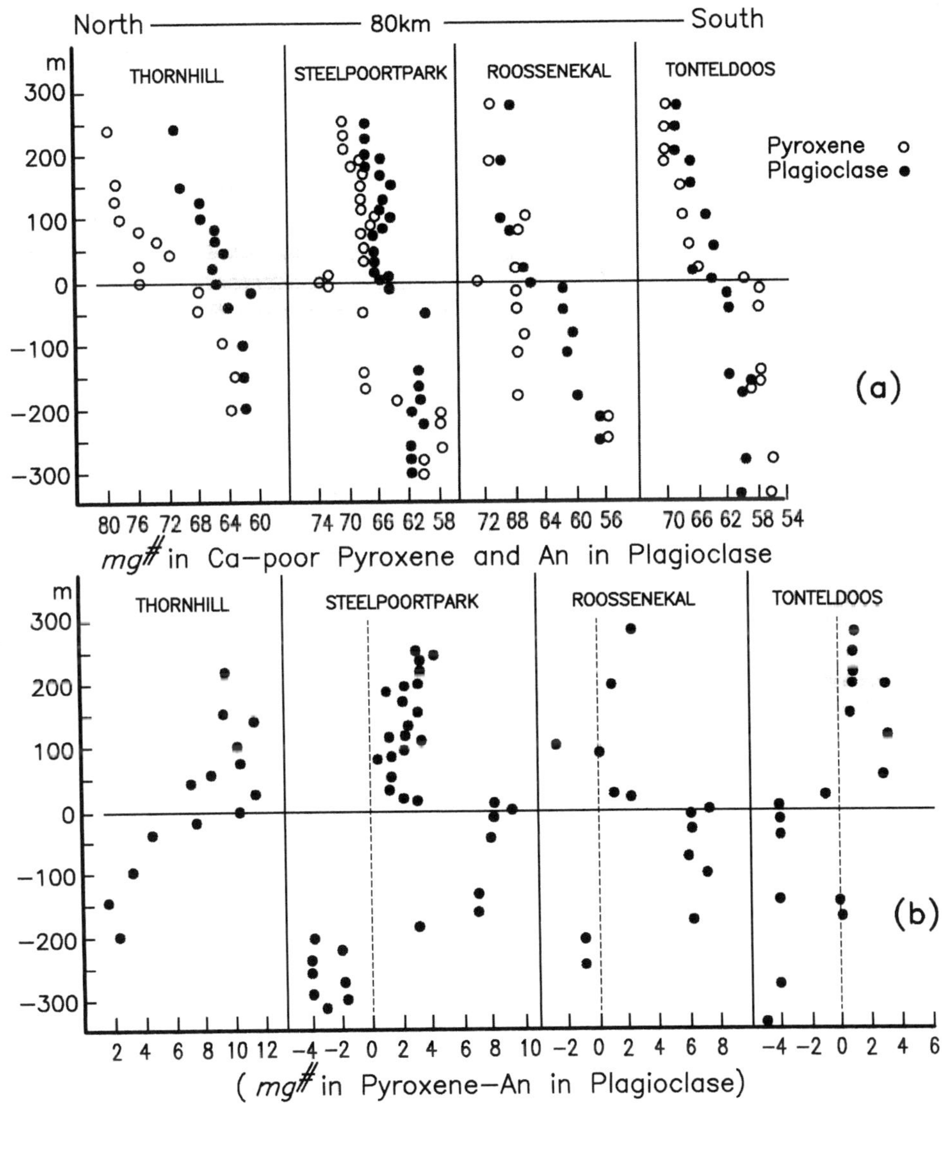

slurries is more a function of crystal content than of composition of the liquid phase (Quadling and Cawthorn, 1994; Rice and Eales, 1995).

Pyroxenite Marker interval: The break in Sr_i across this interval clearly indicates addition of magma. However, the geometry and timing of this intrusion have been modelled in different ways. Sharpe (1985) concluded that the MZ magma was injected between the CZ floor and its overlying residual magma. The M_LZ then crystallized from this new magma, while the residual magma from the CZ remained suspended and did not commence crystallizing until the level of the Pyroxenite Marker. A consideration of the mineral compositions in the M_LZ and M_UZ shows that this hypothesis is implausible as it would require a cooler, more evolved magma to continue crystallizing beneath a hotter, more primitive magma.

All samples above the Marker define an isochron, which requires that there was efficient mixing of all magmas in the chamber, and no further additions thereafter (Kruger *et al.*, 1987). It is remarkable that the Sr_i ratio is identical at 0.7073 in the Eastern and Western limbs (Hamilton, 1977; Sharpe, 1985; Kruger *et al.*, 1987) which demands mixing of identical proportions of new and residual magmas in both compartments. An alternative postulate might be that the residual magma was flushed out of the chamber by new magma with a Sr_i ratio of 0.7073, but the lower ratios at the level of the Marker (Kruger *et al.*, 1987) are more supportive of a mixing model.

The lateral variations which occur across the Pyroxenite Marker in the Eastern limb from north to south are shown in Figure 11a. North of Roossenekal the most evolved plagioclase composition in the gabbronorite below the Marker is relatively constant at An_{60}, whereas the orthopyroxene composition changes from *mg#* of 64 to 58 from north to south. South of Roossenekal both plagioclase and pyroxene below the Marker become more evolved, and minor magnetite appears in the assemblage (Klemm *et al.*, 1985a).

The geometry and change in phase compositions superficially resemble those in the Fongen-Hyllingen intrusion described by Wilson *et al.* (1987), who suggested that such lateral variation in mineral composition against a sloping edge could result from a stratified magma. There are two problems in applying this model here. Firstly, in the northern sector the pyroxene changes composition without concomitant change in plagioclase composition, suggesting that some process other than differentiation controlled the compositions of the phases. Secondly, the density relationships are the reverse of those required. In the case of the Fongen-Hyllingen intrusion, the fractionating magma decreased in density because magnetite was a major crystallizing phase. Residual magma could thus have moved both upwards and laterally. By contrast, fractionation of gabbronorite here would produce a more dense residual liquid, which could not migrate upwards. Consequently, the application of this type of model to the sequence below the Pyroxenite Marker is questionable.

There is normally a close positive correlation between the *mg#* in the orthopyroxene and An content of plagioclase in the complex, with the numeric value ($mg\#_{opx}$ - An_{plag}) being close to zero. Differentiation cannot cause a sudden change in this value. At 200 m below the Marker, the value ($mg\#_{opx}$ - An_{plag}) increases abruptly from *ca.* -3 to +8 in two profiles (Steelpoortpark and Roossenekal) and decreases abruptly again to *ca.* +2 at the Marker (Figure 11b). The value of +8 suggests disequilibrium between the phases. This might be explained by injection of a more primitive magma at some higher level in the resident magma column when crystallization had reached within 200 m of the present position of the Marker. The sinking of pyroxene crystals from the upper liquid layer, and subsequent equilibration with more ferrian pyroxenes crystallizing in the lower liquid layer, could have yielded pyroxenes with higher *mg#*

than would be expected to coexist with the evolved plagioclase. An objection to this model would be that it requires the mode of sequences with anomalous ($mg\#_{opx}$ - An_{plag}) values to be more pyroxene-rich than normal. This is not the case. What the data in Figure 11 do illustrate is that there is a dichotomy between the macroscopically observable features (the Pyroxenite Marker) and cryptic variation (whereby a break some 200 m below the Marker is indicated).

A further unresolved problem related to the Pyroxenite Marker is the observation in the Northam area that there is a forward jump, rather than a reversal in *mg#*, in some profiles (Wilson *et al.*, 1994; Hoyle, in prep.). One proposal, by Wilson *et al.* (1994), was that downward emplacement of dense magma, crystallizing the Upper Zone, displaced footwall rocks.

6.5. Magnetitite layers

Von Gruenewaldt (1973) and Molyneux (1974) considered the entire section from the Pyroxenite Marker to the roof to be the product of closed-system fractionation. They cited as evidence the regular decrease in the V content of magnetite. The partition coefficient for V in magnetite is about 25 (Irving, 1978). As the UZ contains an average of 8% magnetite by weight (Cawthorn and Molyneux, 1986), the bulk partition coefficient is *ca.* 2. Modelling the crystallization of the entire UZ as a single magma body, using these data, gives a trend essentially consistent with that observed in Figure 6b and reported by Klemm *et al.*, (1985b; their figure 4). However, there is a slow reversal in V in magnetite at approximately 1000 m below the top of the UZ (Figure 6c). Changes in the partition coefficient for V in magnetite can result from changes in f_{O_2}, but the increase in V in magnetite is coincident with a slow increase in the An content of the plagioclase (Figure 6a), negating the possibility of f_{O_2} fluctuations being the sole process involved.

The formation of magnetitite layers has been attributed to magma-mixing processes analogous to those for chromitite layers. However, the remarkable constancy of the Sr_i throughout the entire UZ presents a problem to this model. If the UZ magma is the end-product of mixing of several residual and new magma injections, it is unlikely that new magma could have the same Sr_i as the blended magma already in the chamber.

Klemm *et al.* (1985b) reported that magnetite in the massive layers has lower V than adjacent disseminated magnetite. As the partition coefficient for V in magnetite decreases with increasing f_{O_2}, these authors argued that the layers formed in response to an increase in f_{O_2}. However, their data also show that Ti contents are higher in massive than in disseminated magnetite. Application of the magnetite-ilmenite buffer indicates that increasing f_{O_2} should cause a decrease, not increase of Ti in magnetite. Thus, the data for V and Ti in magnetitite layers yield contradictory conclusions about the possible sense of changes in f_{O_2}.

In contrast to the gradual trends for V and Ti, the Cr content of magnetite shows extreme variations. Rapid vertical depletion over tens of cms in massive layers, with abrupt reversals, are reported by Cawthorn and McCarthy (1980). This decline has been modelled in terms of both *in situ* crystallization, and crystallization from a single basal layer in a stratified liquid column. In the model of Cawthorn and McCarthy (1980) several tens of cm of magnetite crystallized, liberating a low-density liquid, which eventually broke away from its boundary layer, formed an upward-migrating plume, and permitted replenishment by Cr-undepleted magma in the zone of crystallization. Kruger and Smart (1987) suggested that the lowest stratified layer crystallized magnetite while convecting rapidly, and decreased in density until it mixed with the overlying layer. Both models indicate that only a thin layer of liquid was chemically processed during formation of a magnetitite layer. In the former model a boundary layer, *ca* 30 m thick, developed a chemical gradient controlled by the rate of downward diffusion of Cr. In the latter model, a homogeneous, rapidly convecting layer some 50 m thick was invoked. Both of these processes would yield saw-tooth patterns of compatible-element abundance, without requiring intrusion of new magma to produce such reversals.

The crystallization of one metre of magnetitite containing less than 1% Cr causes massive depletion of Cr in the residual liquid. By contrast, pyroxenes directly below and above metre-thick chromitite layers contain identical Cr contents. Thus, despite apparent similarities, chemical differences between chromitite and magnetitite layers warn against assuming that they formed by comparable processes.

Harney *et al.* (1990) found an increase of 4% in the Sr content, but no difference in An content or Sr_i ratio, of plagioclase across the Main Magnetitite Layer. They attributed this to addition of new magma, but the data are entirely predictable in terms of crystallization from a single magma. The separation of 2 m of magnetite from a basal layer of magma would increase the level of incompatible elements, including Sr, without changing the Ca/Na ratio in the liquid.

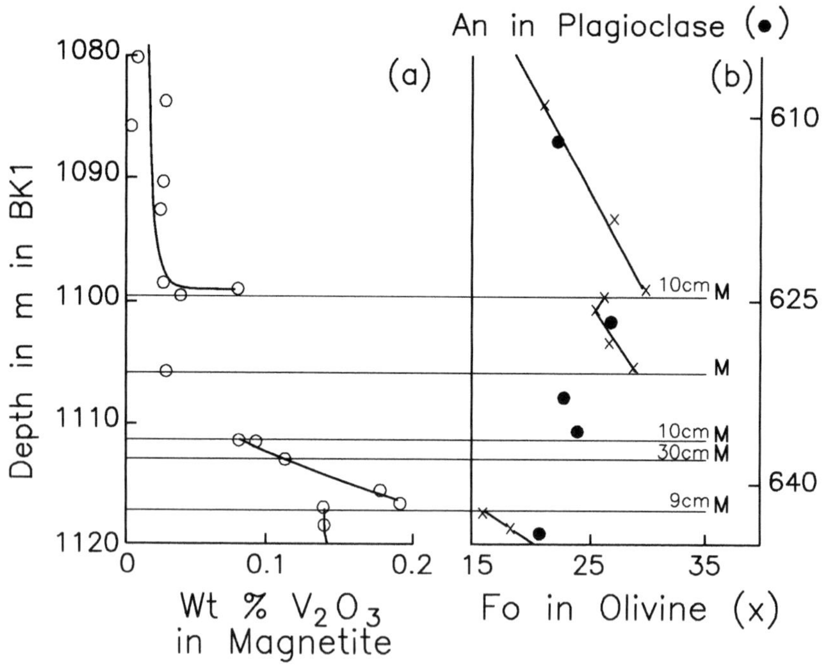

Figure 12. *Plot of V_2O_3 in magnetite and An in plagioclase determined by X-ray fluorescence on mineral separates, and Fo in olivine determined by electron microprobe in a short section of borehole BK1 in the Upper Zone of the Western limb (see Merkle and von Gruenewaldt, 1986; Kruger et al., 1987). Positions of magnetitite layers and their thicknesses are denoted M. (Unpublished data of RGC.)*

Thus once plagioclase resumed crystallization, an increase in its Sr content without a change in An content is to be expected.

A reversal in olivine composition associated with one magnetitite layer in the Western limb (Merkle and von Gruenewaldt, 1986) was attributed to the breakdown of stratified liquid layers in the chamber, rather than magma addition. This succession, bearing five layers of magnetitite, has now been studied in greater detail (Figure 12). Beneath the lowest layer, the olivine is Fo_{17}. A reversal to Fo_{29} occurs at the upper two layers. The Fo content then decreases rapidly from Fo_{29} to Fo_{21} within 15 m above the last reversal. Plagioclase compositions show a smaller variation from An_{49-53} throughout this interval. The ratio of oxide to olivine and pyroxene in this interval is high, and Reynolds (1985) favoured a process of subsolidus re-equilibration between olivine and magnetite to explain the more magnesian nature of the olivine. Alternatively, as magnetite is present in greater than cotectic proportions in this section, excess Fe depletion from the liquid must have occurred with removal of relatively little Mg. This led to an increase in the Mg/Fe ratio in the boundary layer of magma, and in the Fo content of resulting olivine. The decrease in Fo above the fifth layer, to a composition comparable to that observed below the magnetitite package, is far more rapid than could be accounted for by fractional crystallization of a large magma reservoir, but consistent with excessive magnetite crystallization in a thin boundary layer. Plagioclase compositions would have remained constant during this process.

Small-scale reversals in Cr in magnetitite and Fo in olivine in magnetite-rich packages can be interpreted as evidence for boundary-layer crystallization within the chamber. This latter model provides a mechanism for mineral reversals without magma addition and so may provide a compromise between the evidence of Sr_i isotope ratios in the Upper Zone (which suggest that there was no magma addition), and reversals in plagioclase composition and V content of magnetite which suggest otherwise.

The proposal that magnetitite layers result from the separation of dense, immiscible liquids (Reynolds, 1985) does not clarify how this would yield monomineralic layers. The experimental data of Philpotts and Doyle (1983) show that these Fe-rich liquids contain a maximum of only 28% total FeO, compared to the conjugate silicate-dominated liquid with 10-15% FeO. To produce a magnetitite layer from this Fe-rich liquid still requires separation of the magnetite from a greater proportion of silicate phases. The nature of the basal contact of magnetitite layers also militates against this model. Basal contacts with anorthosite are sharp and planar, whereas the low viscosity and high density of iron-rich liquid would cause it to penetrate downwards into underlying crystal mushes, producing irregular basal contacts. On the other hand, the liquid immiscibility model does explain the form of some of the oxide-dominated discordant pipes (Scoon and Mitchell, 1994).

6.6. Apatite in the Upper Zone

Unlike the sharp modal layering of chromite and magnetite, apatite displays a different pattern. It never produces monomineralic layers. The highest modal abundances of apatite are typically close to 10% (von Gruenewaldt, 1973; Molyneux, 1974; Grobler and Whitfield, 1970; Cawthorn and Walsh, 1988). The only exception is a 20 m section, identified in the Bierkraal boreholes near Rustenburg, with a basal abundance of 40% apatite, decreasing to 10% at the top (von Gruenewaldt, 1993). Figure 13 shows the data for one typical interval of 150 m near the top of the UZ, yielded by a study of a borehole drilled in the Brits area, from which continuous core has been analyzed every 1 m. Except for variations in the abundance of apatite, this is homogeneous magnetite diorite. Sections 30-40 m thick have a near-constant apatite content of 6.5-10.5 wt.% (2.8-4.5% P_2O_5). Interspersed are thicker sections with <0.5% apatite (0.03-0.2% P_2O_5). The decline (at 20-40 m in Figure 13) and increase (at 95-115 m) in apatite occur smoothly over less than 20 m.

The proportion of apatite recorded in the Upper Zone (Cawthorn and Walsh, 1988; Figure 13) can be discussed in terms of its expected cotectic proportions. At this level, the intrusion is dioritic, and temperatures of crystallization would have been in the order of 1000°C, for which experimental data yield P_2O_5 solubilities of *ca.* 1% (Green and Watson, 1982). Thus the cotectic proportion of apatite is *ca.* 2.5%. However, it frequently is uniformly concentrated in continuous sections by a factor of three to four relative to this true cotectic proportion (Figure 13). In contrast, intervening sections, in the order of 50 m thick, contain <0.2% P_2O_5 (<0.5% apatite) which indicates little or no cumulus apatite in these rocks. The reason for the

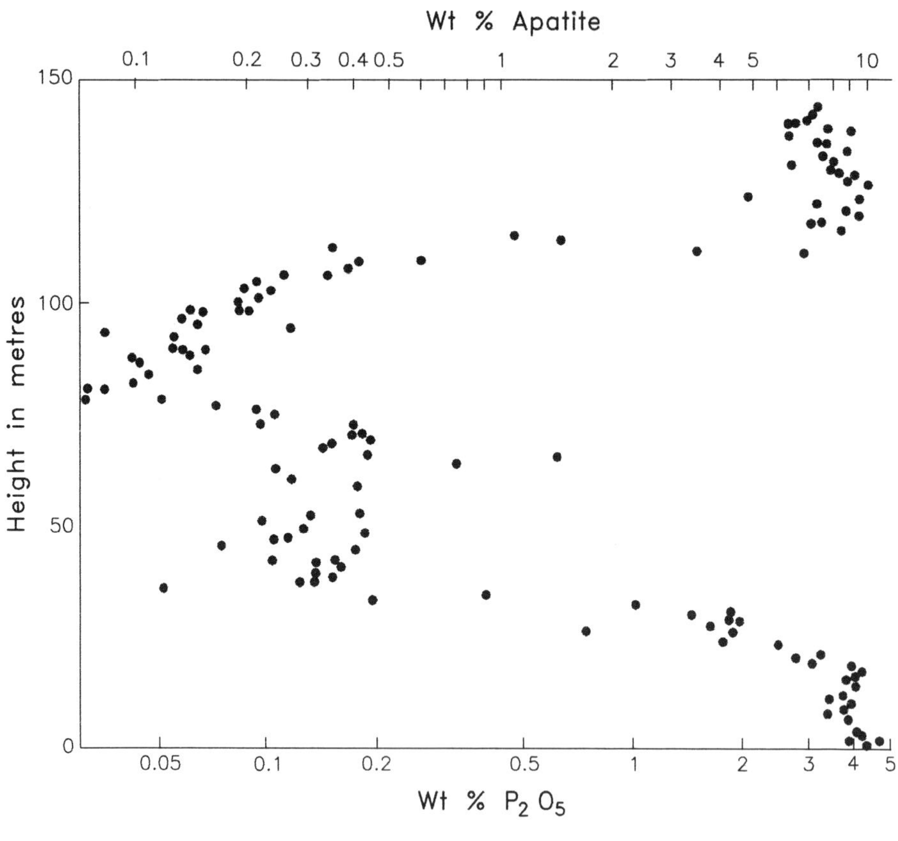

Figure 13. *Plot of whole-rock P_2O_5 and calculated apatite abundance (on a logarithmic scale) versus height through a short section near the middle of the Upper Zone from bore-core taken near Brits (unpublished data of Gencor (SA) Ltd.). The data show a bimodal distribution of apatite in the rocks, with smooth, rather than abrupt transitions.*

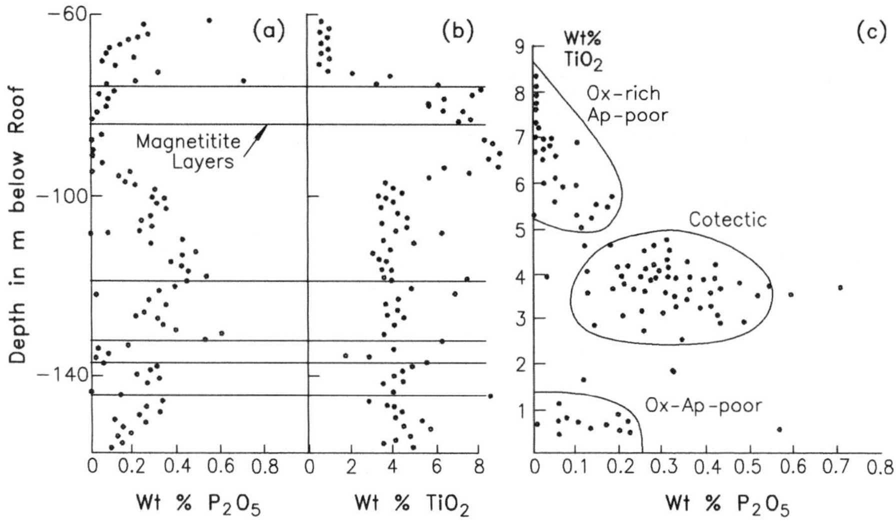

Figure 14. *Plot of whole-rock analyses for TiO_2 and P_2O_5 for a section immediately below the roof, from bore-core taken near Brits (unpublished data of Gencor (SA Ltd.). (a) Plot of P_2O_5 and (b) TiO_2 abundance versus height below the roof, showing the positions of numerous magnetitite layers. (c) Plot of TiO_2 versus P_2O_5 showing the subdivision of all samples into three groups with differing modal abundances of apatite and oxide.*

alternation of apatite-rich and -poor intervals on a scale of tens of metres, the relatively constant apatite abundance between *ca.* 6-10% in the former, and the absence of rocks with cotectic proportions in this particular section, remain unexplained.

In some instances oxide-rich layers are also enriched in apatite (von Gruenewaldt, 1993). However, there are other examples (above the level of apatite saturation) where this is not observed. Figure 14, which covers a vertical section intersecting seven oxide-rich layers close to the roof in the Brits area, illustrates the relationship between oxides and apatite. This plot of TiO_2 versus P_2O_5 shows that the whole-rock samples fall into three groups: one with neither cumulus oxide nor apatite, a second with high oxide abundance but little or no apatite, and a third containing 2.5-4.5% TiO_2 and 0.2-0.4% P_2O_5. At temperatures <1000°C, and with increasing SiO_2 content, the solubility of apatite decreases markedly (Green and Watson, 1982). The clustering of analyses in the third group probably reflects the true cotectic proportions in an evolved dioritic liquid. However, again there are rocks which contain no cumulus apatite or oxide within this interval. As experimental studies show that phosphorus would concentrate in such liquids (*e.g.* Kolker, 1982), the absence of high apatite abundances in the oxide-rich horizons argues against the concept of the involvement of iron-rich immiscible liquids producing the oxide-rich layers.

7. SUMMARY

Despite the great age of the Bushveld Complex, it is still remarkably unmetamorphosed and undeformed. It is characterized by the immense lateral continuity of even thin layers, especially of the uppermost rocks.

The most plausible parental magma for the Lower Zone is a Mg- and Si-rich basalt. Evidence for a plagioclase-rich parent for the Critical Zone is lacking. The Main and Upper Zones resulted from addition of more typical continental tholeiitic magma. However, all magmas had variably high initial $^{87}Sr/^{86}Sr$ ratios. Extremely slow evolution of mineral compositions, with frequent reversals in the Lower and Critical Zones indicate frequent magma replenishment, as suggested by the term *Integration Stage*, coined by Wager and Brown (1968). The origin of the chromitite layers is still poorly understood, not least from mass balance considerations, which require far larger volumes of magma to be processed to produce such thick layers, than is exposed in vertical sections through this intrusion. The lateral flow of such magma to areas now eroded not only explains the volume paradox, but also the facies changes observed from Union Section to Brits in the Western limb.

The Main Zone is remarkable for its absence of layering and for the extremely slow rate of evolution of mineral compositions. By far the largest reversal in mineral composition occurs near the Pyroxenite Marker. This level defines the last major addition of magma, but remarkably, the mixing of the assorted magmas was so thorough as to produce a single isochron and initial Sr_i in both the Eastern and Western limbs of the intrusion.

The Upper Zone contains thick magnetitite layers, but a consideration of the Cr budget for these contrasts strongly with the implications of chromitite layers, and comparable modes of origin cannot be assumed. Monomineralic layers of apatite are never formed, but apatite modal abundance oscillates from four times cotectic proportions to almost zero on scales of tens of metres.

The processes which generated this enormous volume of magma and then created layering on all scales are possibly even more elusive than when Wager and Brown (1968) summarized their findings, but there is now a very impressive data base of information with which to test any hypothesis.

8. ACKNOWLEDGMENTS

Much of the authors' research has been supported by the Foundation for Research Development (Pretoria) and their respective Universities, but would not have been possible without the logistical support of many mining companies, especially Rustenburg Platinum Mines of Johannesburg Consolidated Investments, Impala Platinum of Gencor (SA) Ltd, Gold Fields of S.A. Ltd., and Rand Mines, Ltd. We thank Roger Scoon, Richard Wilson, Jean-Clair Duchesne, and Johan Kruger for constructive comments on the manuscript, and Lyn Whitfield and Di du Toit for drafting the diagrams.

9. REFERENCES

Abbott, D., & Ferguson, J., 1965. The Losberg intrusion, Fochville, Transvaal. *Trans. geol. Soc. S. Afr.* **68**, 31-52.

Atkins, F.B., 1969. Pyroxenes of the Bushveld Intrusion, South Africa. *J. Petrology* **10**, 222-49.

Barnes, S.J., 1986a. The distribution of chromium among orthopyroxene, spinel and silicate liquid at atmospheric pressure. *Geochim. Cosmochim. Acta* **50**, 1889-909.

Barnes, S.J., 1986b. The effect of trapped liquid crystallization on cumulus mineral compositions in layered intrusions. *Contr. Miner. Petrol.* **93**, 524-31.

Barnes, S.J., 1989. Are Bushveld U-type parent magmas boninites or contaminated komatiites? *Contr. Miner. Petrol.* **101**, 447-57.

Boudreau, A.E., 1994. Crystal aging in two-crystal, two-component systems and the formation of fine-scale layering. *S. Afr. J. Geol.* **97**, 473-85.

Boudreau, A.E., & McCallum, I.S., 1992. Concentration of platinum-group elements by magmatic fluids in layered intrusions. *Econ. Geol.* **87**, 1830-48.

Bristow, D.M., 1989. The petrology and geochemistry of the Lower Critical Zone, Winterveld Chrome Mine. Unpubl. Ph.D. thesis, Univ. Natal, 356 pp.

Buchanan, D.L., 1975. The petrography of the Bushveld Complex intersected by boreholes in the Bethal area. *Trans. geol. Soc. S. Afr.* **78**, 335-48.

Buchanan, D.L., 1977. Cryptic variation in minerals from the Bushveld Complex of the Bethal area. *Trans. geol. Soc. S. Afr.* **80**, 49-52.

Button, A., 1976. Stratigraphy and relations of the Bushveld floor in the Eastern Transvaal. *Trans. geol. Soc. S. Afr.* **79**, 3-12.

Cameron, E.N., 1978. The Lower Zone of the Eastern Bushveld Complex in the Olifants River trough. *J. Petrology* **19**, 437-62.

Cameron, E.N., 1980. Evolution of the lower Critical Zone, central sector, eastern Bushveld Complex. *Econ. Geol.* **75**, 845-71.

Cameron, E.N., 1982. The upper Critical Zone of the eastern Bushveld Complex. *Econ. Geol.* **77**, 1307-27.

Campbell, I.H., Naldrett, A.J., & Barnes, S.J., 1983. A model for the origin of the platinum-rich sulfide horizons in the Bushveld and Stillwater complexes. *J. Petrology* **24**, 133-65.

Cawthorn, R.G., 1996. Re-evaluation of magma compositions and processes in the uppermost Critical Zone of the Bushveld Complex. *Miner. Mag.* **60**, 131-48.

Cawthorn, R.G., & Barry, S.D., 1992. The role of intercumulus residua in the formation of pegmatoid associated with the UG2 chromitite, Bushveld Complex. *Austral. J. Earth Sci.* **39**, 263-76.

Cawthorn, R.G., & Biggar, G.M., 1993. Crystallization of titaniferous chromite, magnesian ilmenite and armalcolite in tholeiitic suites in the Karoo Igneous Province. *Contr. Miner. Petrol.* **114**, 221-35.

Cawthorn, R.G., & Davies, G., 1983. Experimental data at 3 Kbars pressure on parental magma to the Bushveld Complex. *Contr. Miner. Petrol.* **83**, 128-35.

Cawthorn, R.G., & Davies, G., 1985. Possible komatiitic affinity of the Bushveld Complex, South Africa. In: Arndt, N.T. & Nisbet, E.G. (eds.) *Komatiites*. London: George Allen and Unwin, 91-6.

Cawthorn, R.G., & McCarthy, T.S., 1980. Variations in Cr content of magnetite from the Upper Zone of the Bushveld Complex - evidence for heterogeneity and convection currents in magma chambers. *Earth Planet. Sci. Lett.* **46**, 335-43.

Cawthorn, R.G., & Molyneux, T.G., 1986. The vanadiferous magnetite deposits of the Bushveld Complex. In Anhaeusser, C.R. & Maske, S. (eds.) *Mineral Deposits of Southern Africa.* Johannesburg: Geol. Soc. S. Afr., 1251-66.

Cawthorn, R.G., & Walsh, K.L., 1988. The use of phosphorus contents in yielding estimates of the proportion of trapped liquid in cumulates of the Upper Zone of the Bushveld Complex. *Miner. Mag.* **52**, 81-9.

Cawthorn, R.G., Barton, J.M., & Viljoen, M.J., 1985. Interaction of floor rocks with the Platreef on Overysel, Potgietersrus, northern Transvaal. *Econ. Geol.* **80**, 988-1006.

Cawthorn, R.G., Davies, G., Clubley-Armstrong, A., & McCarthy, T.S., 1981. Sills associated with the Bushveld Complex, South Africa: an estimate of the parental magma composition. *Lithos* **14**, 1-15.

Cawthorn, R.G., Meyer, S.P., & Kruger, F.J., 1991. Major addition of magma at the Pyroxenite Marker in the western Bushveld Complex, South Africa. *J. Petrology* **32**, 739-63.

Cheney, E.S., & Twist, D., 1991. The conformable emplacement of the Bushveld mafic rocks along a regional unconformity in the Transvaal succession of South Africa. *Precamb. Res.* **52**, 115-32.

Davies, G., & Cawthorn, R.G., 1984. Mineralogical data on a multiple intrusion in the Rustenburg Layered Suite of the Bushveld Complex. *Miner. Mag.* **48**, 469-80.

Davies, G., Cawthorn, R.G., Barton, J.M., & Morton, M., 1980. Parental magma to the Bushveld Complex. *Nature* **287**, 33-5.

de Klerk, W.J., 1982. Geology, geochemistry and silicate mineralogy of the Upper Critical Zone of the North-Western Bushveld complex. Unpubl. M.Sc. thesis, Rhodes Univ., Grahamstown, 200 pp.

de Klerk, W.J., 1991. Petrogenesis of the Upper Critical Zone in the western Bushveld Complex. Unpubl. Ph.D. thesis, Rhodes Univ. 294 pp.

de Waal, S.A., 1975. The mineralogy, chemistry and certain aspects of reactivity of chromite from the Bushveld Igneous Complex. *National Institute for Metallurgy, Johannesburg, S. Africa*. Report No. **1709**, 80 pp.

Dick, H.J.B., & Bullen, T., 1984. Chromian spinel as a petrogenetic indicator in abyssal and alpine-type peridotites and spatially associated lavas. *Contr. Miner. Petrol.* **86**, 54-76.

Eales, H.V., & Reynolds, I.M., 1986. Cryptic variations within chromites of the Upper Critical Zone, Northwestern Bushveld Complex. *Econ. Geol.* **81**, 1056-66.

Eales, H.V., Marsh, J.S., Mitchell, A.A., De Klerk, W.J., Kruger, F.J., & Field, M., 1986. Some geochemical constraints upon models for the crystallization of the upper Critical Zone - Main Zone interval, Northwestern Bushveld Complex. *Miner. Mag.* **50**, 567-82.

Eales, H.V., Field, M., De Klerk, W.J., & Scoon, R.N., 1988. Regional trends of chemical variation and thermal erosion in the Upper Critical Zone, Western Bushveld Complex. *Miner. Mag.* **52**, 63-79.

Eales, H.V., De Klerk, W.J., Butcher, A.R., & Kruger, F.J. 1990a. The cyclic unit beneath the UG1 chromitite (UG1FW unit) at RPM Union Section Platinum Mine. *Miner. Mag.* **54**, 23-43.

Eales, H.V., De Klerk, W.J., & Teigler, B., 1990b. Evidence for magma mixing processes within the Critical and Lower Zones of the Northwestern Bushveld Complex. *Chem. Geol.* **88**, 261-78.

Eales, H.V., Maier, W.D., & Teigler, B., 1991. Corroded plagioclase feldspar inclusions in orthopyroxene and olivine of the Lower and Critical Zones, Western Bushveld Complex. *Miner. Mag.* **55**, 479-86.

Eales, H.V., Botha, W.J., Hattingh, P.J., De Klerk, W.J., Maier, W.D., & Odgers, A.T.R., 1993a. The mafic rocks of the Bushveld Complex: a review. *J. Afr. Earth. Sci.* **16**, 121-42.

Eales, H.V., Teigler, B., & Maier, W.D., 1993b. Cryptic variations of minor elements Al, Cr, Ti and Mn in Lower and Critical Zone orthopyroxenes of the Western Bushveld Complex. *Miner. Mag.* **57**, 257-64.

Eales, H.V., De Klerk, W.J., Teigler, B., & Maier, W.D., 1994. Nature and origin of orthopyroxenites in the Western Bushveld Complex, in the light of compositional data. *S. Afr. J. Geol.* **97**, 399-407.

Engelbrecht, J.P., 1985. The chromites of the Bushveld Complex in the Nietverdiend area. *Econ. Geol.* **80**, 896-910.

Engelbrecht, J.P., 1990. The Marico hypabyssal suite, and the marginal zone of the Bushveld Complex in the Marico district. *S. Afr. J. Geol.* **93**, 318-28.

Field, M., 1987. The petrology and geochemistry of the Upper Critical Zone of the Bushveld Complex, at the Amandelbult Section of Rustenburg Platinum Mines, Ltd. Unpubl. M. Sc. thesis, Rhodes Univ., Grahamstown, 129pp.

Gain, S.B., 1985. The geologic setting of the platiniferous UG2 chromitite layer on the farm Maandagshoek, Eastern Transvaal. *Econ. Geol.* **80**, 925-43.

Gauert, C.D.K., de Waal, S.A., & Wallmach, T., 1995. The geology of the Uitkomst Complex, eastern Transvaal - evidence for a magma conduit. *J. Afr. Earth Sci.* **21**, 553-70.

Green, T.H., & Watson, E.B., 1982. Crystallization of natural magmas under high pressure, hydrous conditions with particular reference to 'orogenic' rock series. *Contr. Miner. Petrol.* **79**, 96-105.

Grobler, N.J., & Whitfield, G.G., 1970. The olivine-apatite magnetitites and related rocks in the Villa Nora occurrence of the Bushveld Igneous Complex. *Geol. Soc. S. Afr. Spec. Publn.* 1, 208-27.

Groeneveld, D., 1970. The structural features and the petrography of the Bushveld Complex in the vicinity of Stoffberg, Eastern Transvaal. *Geol. Soc. S. Afr., Spec. Publ.* **1**, 36-45.

Grove, T.L., & Bence, A.E., 1977. Experimental study of pyroxene-liquid interaction in quartz-normative basalts. *Proc. Eighth Lunar Sci. Conf.* 1549-79.

Hamilton, P.J., 1977. Sr isotope and trace element studies of the Great Dyke and Bushveld mafic phase and their relation to early Proterozoic magma genesis in Southern Africa. *J. Petrology* **18**, 24-52.

Harmer, R.E., & Sharpe, M.R., 1985. Field relations and strontium isotope systematics of the marginal rocks of the Eastern Bushveld Complex. *Econ. Geol.* **80**, 813-37.

Harney, D.M.W., Merkle, R.K.W., & von Gruenewaldt, G., 1990. Plagioclase composition in the Upper zone, eastern Bushveld Complex - support for magma mixing at the Main Magnetite Layer. *Inst. Geol. Res. Bushveld Complex,* Univ. Pretoria **87**, 19 pp.

Hatton, C.J., 1986. Regional variations in the Middle Group of chromitite layers, Bushveld Complex. *Extended Abstracts Geocongress '86,* Geol. Soc. S. Afr., Johannesburg, 595-98.

Hatton, C.J., & Sharpe, M.R., 1989. Significance and origin of boninite-like rocks associated with the Bushveld Complex. In: Crawford, A.J. (ed.) *Boninites.* London: Unwin Hyman, 174-203.

Hatton, C.J., & von Gruenewaldt, G., 1987. The geological setting and petrogenesis of the Bushveld chromitite layers. In: Stowe, C.W. (ed.) *Evolution of chromium ore fields.* New York: Van Nostrand Reinhold Co., 109-43.

Hoyle, P., 1993. Chemical and modal variations at the top of the Upper Zone in the Northwestern Bushveld Complex. *Symposium on Layering in Igneous Complexes, Abstracts Vol.,* Univ. Witwatersrand, Johannesburg.

Hoyle, P., (in prep). Petrology and geochemistry of the Upper Zone, Western Bushveld Complex. Unpubl. Ph.D. thesis, Rhodes Univ., Grahamstown.

Hulbert, L.J., & von Gruenewaldt, G., 1982. Nickel, copper and platinum mineralization in the Lower Zone of the Bushveld Complex, south of Potgietersrus. *Econ Geol.* **77**, 1296-306.

Hulbert, L.J., & von Gruenewaldt, G., 1985. Textural and compositional features of chromite in the Lower and Critical Zones of the Bushveld Complex south of Potgietersrus. *Econ. Geol.* **80**, 872-95.

Huppert, H.E., & Sparks, R.S.J. 1980. The fluid dynamics of a basaltic magma chamber replenished by influxes of hot, dense ultrabasic magma. *Contr. Miner. Petrol.* **75**, 279-89.

Irvine, T.N., 1977. Origin of chromitite layers in the Muskox intrusion and other stratiform intrusions. *Geology* **5**, 273-7.

Irvine, T.N., 1980. Magmatic density currents and cumulus processes. *Amer. J. Sci.* 280-A, 1-58.

Irvine, T.N., Keith, D.W., & Todd, S.G., 1983. The J-M platinum-palladium reef of the Stillwater Complex, Montana. II. Origin by double diffusive convective magma mixing and implications for the Bushveld Complex. *Econ. Geol.* **78**, 1287-334.

Irving, A.J., 1978. A review of experimental studies of crystal/liquid trace element partitioning. *Geochim. Cosmochim. Acta* **42**, 743-70.

Kenyon, A.K., Attridge, R.L., & Coetzee, G.L., 1986. The Uitkomst nickel-copper deposit, Eastern Transvaal. In: Anhaeusser, C.R., & Maske, S. (eds.) *Mineral Deposits of Southern Africa,* **1**, Johannesburg: Geol. Soc. S. Afr., 1009-17.

Klemm, D.D., Ketterer, S., Reichhardt, F., Steindl, J.. & Weber-Diefenbach, K., 1985a. Implications of vertical and lateral compositional variations across the Pyroxenite Marker and its associated rocks in the upper part of the Main Zone in the eastern Bushveld Complex. *Econ. Geol.* **80**, 1007-15.

Klemm, D.D., von Gruenewaldt, G., Henckel, J., & Dehm, R., 1985b. The geochemistry of titanomagnetite in magnetite layers and their host rocks. *Econ. Geol.* **80**, 1075-88.

Kolker, A., 1982. Mineralogy and geochemistry of Fe-Ti oxide and apatite (nelsonite) deposits and evaluation of the liquid immiscibility hypothesis. *Econ. Geol.* **77**, 1146-58.

Kruger, F.J., 1990. The stratigraphy of the Bushveld Complex: a reappraisal of the Main Zone boundaries. *S. Afr. J. Geol.* **93**, 376-81.

Kruger, F.J., 1992. The origin of the Merensky Cyclic Unit: Sr-isotopic and mineralogical evidence for an alternative orthomagmatic model. *Austral. J. Earth Sci.* **39**, 255-61.

Kruger, F.J., 1994. The Sr-isotopic stratigraphy of the western Bushveld Complex. *S. Afr. J. Geol.* **97**, 393-8.

Kruger, F.J., & Marsh, J.S., 1982. Significance of $^{87}Sr/^{87}Sr$ ratios in the Merensky Cyclic Unit of the Bushveld Complex. *Nature* **298**, 53-5.

Kruger, F.J., & Marsh, J.S., 1985. The mineralogy, petrology and origin of the Merensky Cyclic Unit in the western Bushveld Complex. *Econ. Geol.* **80**, 958-74.

Kruger, F.J., & Smart, R., 1987. Diffusion of trace elements during bottom crystallization of double-diffusive convection systems: the magnetitite layers of the Bushveld Complex. *J. Volcanol. Geotherm. Res.* **34**, 133-42.

Kruger, F.J., Cawthorn, R.G., & Walsh, K.L., 1987. Strontium isotope evidence against magma addition in the Upper Zone of the Bushveld Complex. *Earth. Planet. Sci. Lett.* **84**, 51-8.

Lee, C.A., 1981. Post-deposition structures in the Bushveld Complex mafic sequence. *J. Geol. Soc. Lond.* **138**, 327-41.

Lee, C.A., & Butcher, A.R., 1990. Cyclicity in the Sr-isotopic stratigraphy through the Merensky and Bastard Reef Units, Atok Section, Eastern Bushveld Complex. *Econ. Geol.* **85**, 877-83.

Leeb-du Toit, A., 1986. The Impala Platinum Mines. In: Anhaeusser, C.R., & Maske, S. (eds.) *Mineral Deposits of Southern Africa,* **2**, Johannesburg: Geol. Soc. S. Afr., 1091-106.

Lipin, B.R., 1993. Pressure increases, the formation of chromite seams, and the development of the Ultramafic Series in the Stillwater Complex, Montana. *J. Petrology* **34**, 955-76.

Maaløe, S., 1978. The origin of rhythmic layering. *Miner. Mag.* **42**, 337-345.

Maier, W.D., 1991. Geochemical and petrological trends in the UG2-Merensky Unit interval of the Upper Critical Zone. Unpubl. Ph. D. thesis, Rhodes Univ., Grahamstown, 241 pp.

Maier, W.D., & Eales, H.V., 1994a. A facies model for the interval between the UG2 and Merensky Reef, Western Bushveld Complex, South Africa. *Trans. Inst. Min. Metall. (Sect. B: Appl. Earth Sci.)* **103**, B22-30.

Maier, W.D., & Eales, H.V., 1994b. Plagioclase inclusions in orthopyroxene and olivine of the UG2-Merensky Reef interval: regional trends in the western Bushveld Complex. *S. Afr. J. Geol.* **97**, 408-14.

Marais, C.L., 1977. An investigation of the Pyroxenite Marker and the associated rocks in the Main Zone of the eastern Bushveld Complex. Unpubl. M.Sc. thesis, Univ. Pretoria, 87 pp.

Mathez, E.A., 1989. Interactions involving fluids in the Stillwater and Bushveld Complexes: observations from the rocks. *Rev. Econ. Geol.* **4**, 167-79.

McCarthy, T.S., Cawthorn, R.G., Wright, C.J., & McIver, J.R., 1985. Mineral layering in the Bushveld Complex: implications of Cr abundances in magnetite from closely-spaced magnetitite and intervening silicate-rich layers. *Econ. Geol.* **80**, 1062-74.

Merkle, R.K.W., & von Gruenewaldt, G., 1986. Compositional variation of Co-pentlandite: relation to the evolution of the Upper Zone of the western Bushveld Complex. *Can. Miner.* **24**, 529-46.

Mitchell, A.A., 1990. The stratigraphy, petrography and mineralogy of the Main Zone of the Northwestern Bushveld. *S. Afr. J. Geol.* **93**, 818-31.

Molyneux, T.G., 1970. The geology of the area in the vicinity of Magnet Heights, Eastern Transvaal, with special reference to magnetic iron ore. *Geol. Soc. S. Afr., Spec. Publ.* **1**, 228-41.

Molyneux, T.G., 1974. A geological investigation of the Bushveld Complex in Sekhukhuneland and part of the Steelpoort valley. *Trans. geol. Soc. S. Afr.* **77**, 329-38.

Murck, B.W., & Campbell, I.H., 1986. The effects of temperature, oxygen fugacity and melt composition on the behaviour of chromium in basic and ultrabasic melts. *Geochim. Cosmochim. Acta* **50**, 1871-87.

Naldrett, A.J., Gasparrini, E.C., Barnes, S.J., von Gruenewaldt, G., & Sharpe, M.R., 1986. The Upper Critical Zone of the Bushveld Complex and the origin of Merensky-type ores. *Econ. Geol.* **81**, 1105-17.

Naldrett, A.J., Cameron, G., von Gruenewaldt, G., & Sharpe, M.R., 1987. The formation of stratiform PGE deposits in layered intrusions. In: Parsons, I. (ed.) *Origins of igneous layering*. Dordrecht: Reidel Publ. Co., 313-98.

Philpotts, A.R., & Doyle, C.D., 1983. Effect of magma oxidation state on the extent of silicate liquid immiscibility in a tholeiitic basalt. *Amer. J. Sci.* **283**, 967-86.

Quadling, K., & Cawthorn, R.G., 1994. The layered melagabbronorite-leucogabbronorite succession from the Main Zone of the Bushveld Complex. *S. Afr. J. Geol.* **97**, 442-54.

Reichhardt, F.J., 1994. The Molopo Farms Complex, Botswana: history, stratigraphy, petrography, petrochemistry and Ni-Cu-PGE mineralization. *Explor. Mining Geol.* **3**, 263-84.

Reynolds, I.M., 1985. Contrasted mineralogy and textural relations in the uppermost Ti-magnetite layers of the Bushveld Complex in the Bierkraal area north of Rustenburg. *Econ. Geol.* **80**, 1027-48.

Rice, A.R., & Eales, H.V., 1995. The densities of Bushveld melts: textural and hydrodynamic criteria. *Miner. Petrol.* **54**, 45-53.

Robins, B., 1982. Finger structures in the Lille Kufjord layered intrusion, Finmark, northern Norway. *Contr. Miner. Petrol.* **81**, 290-5.

SACS (South African Committee for Stratigraphy), 1980. Kent, L.E. (compiler) *Stratigraphy of South Africa*. Geol. Surv. S. Afr., Pretoria, Handbk., **8**, 690 pp.

Schiffries, C.M. 1982. The petrogenesis of a platiniferous dunite pipe in the Bushveld Complex: infiltration metasomatism by a chloride solution. *Econ. Geol.* **77**, 1439-53.

Schurmann, L.W., 1993. The geochemistry and petrology of the Upper Critical Zone of the Boshoek section of the Western Bushveld Complex. *Geol. Surv. S. Afr. Bull.* **113**, 88 pp.

Scoon, R.N., & de Klerk, W.J., 1987. The relationship of olivine cumulates and mineralization to cyclic units in part of the Upper Critical Zone of the Western Bushveld Complex. *Can. Miner.* **25**, 51-77.

Scoon, R.N., & Mitchell, A.A., 1994. Discordant iron-rich pegmatites in the Bushveld Complex and their relationship to iron-rich intercumulus and residual liquids. *J. Petrology* **35**, 881-917.

Scoon, R.N., & Teigler, B., 1994. Platinum-group element mineralization in the Critical Zone of the western Bushveld Complex. *Econ. Geol.* **89**, 1094-121.

Scoon, R.N., & Teigler, B., 1995. A new LG6 chromite reserve at Eerste Geluk in the boundary zone between the central and southern sectors of the Eastern Bushveld Complex. *Econ. Geol.* **90**, 969-82.

Sharpe, M.R., 1981. The chronology of magma influxes to the eastern compartment of the Bushveld Complex as exemplified by its marginal border groups. *J. Geol. Soc. Lond.* **138**, 307-26.

Sharpe, M.R., 1985. Strontium isotope evidence for preserved density stratification in the Main Zone of the Bushveld Complex. *Nature* **316**, 119-26.

Sharpe, M.R., & Hulbert, L.J., 1985. Ultramafic sills beneath the eastern Bushveld Complex. *Econ. Geol.* **80**, 849-71.

Shearer, C.K., Papike, J.J., Simon, S.B. & Shimizu, N., 1989. An ion microprobe study of the intra-crystalline behaviour of REE and selected trace elements in pyroxene. *Geochim. Cosmochim. Acta* **53**, 1041-54.

Sparks, R.S., Huppert, H.E., Koyaguchi, T., & Hallworth, M.A., 1993. Origin of modal and rhythmic igneous layering by sedimentation in a convecting magma chamber. *Nature* **361**, 246-9.

Teigler, B., 1990. Mineralogy, petrology and geochemistry of the Lower and Lower Critical Zones, Northwestern Bushveld Complex. Unpubl. Ph.D. thesis, Rhodes Univ., Grahamstown, 247 pp.

Teigler, B., & Eales, H.V., 1993. Correlation between chromite composition and PGE mineralization in the Critical Zone of the Western Bushveld Complex. *Mineral. Deposita* **28**, 291-302.

Teigler, B., & Eales, H.V., (in press). The Lower and Critical Zones of the western limb of the Bushveld Complex as intersected by the Nooitgedagt boreholes. *Geol. Surv. S. Afr., Bull.*, Pretoria.

Teigler, B., Eales, H.V., & Scoon, R.N., 1992. The cumulate succession in the Critical Zone of the Rustenburg Layered Suite at Brits, western Bushveld Complex. *S. Afr. J. Geol.* **95**, 17-28.

van der Merwe, M.J., 1976. The layered sequence of the Potgietersrus limb of the Bushveld Complex. *Econ. Geol.* **71**, 1337-51.

Vermaak, C.F., 1976. The Merensky Reef - thoughts on its environment and genesis. *Econ. Geol.* **71**, 1270-98.

Viljoen, M.J., 1994. A review of regional variations in facies and grade distribution of the Merensky Reef, Western Bushveld Complex, with some mining implications. In: Anhaeusser, C.R. (ed.) *Proc. XVth CMMI Congress.* S. Afr. Inst. Mining Metal., Johannesburg, **3**, 183-94.

Viljoen, M.J., & Hieber, R., 1986. The Rustenburg Section of Rustenburg Platinum Mines Limited. In: Anhaeusser, C.R., & Maske, S. (eds.) *Mineral Deposits of Southern Africa*, **2** Johannesburg: Geol. Soc. S. Afr., 1107-34.

Viljoen, M.J., & Scoon, R.N., 1985. The distribution and main geologic features of discordant bodies of iron-rich ultramafic pegmatoid in the Bushveld Complex. *Econ. Geol.* **80**, 1109-28.

Viljoen, M.J., De Klerk, W.J., Coetzer, P.M., Hatch, N.P., Kinloch, E., & Peyerl, W., 1986a. The Union Section of Rustenburg Platinum Mines Limited. In: Anhaeusser, C.R., & Maske, S. (eds.) *Mineral Deposits of Southern Africa,* **2**, Johannesburg: Geol. Soc. S. Afr., 1061-90.

Viljoen, M.J., Theron, J., Underwood, B., Walters, B.M., Weaver, J., & Peyerl, W., 1986b. The Amandelbult Section of Rustenburg Platinum Mines Limited. In: Anhaeusser, C.R., & Maske, S. (eds.) *Mineral Deposits of Southern Africa,* **2**, Johannesburg: Geol. Soc. S. Afr., 1041-60.

von Gruenewaldt, G., 1973. The Main and Upper Zones of the Bushveld Complex in the Roossenekal area, eastern Transvaal. *Trans. geol. Soc. S. Afr.* **76**, 207-27.

von Gruenewaldt, G., 1993. Ilmenite-apatite enrichments in the Upper Zone of the Bushveld Complex: a major titanium-rock phosphate resource. *Internat. Geol. Rev.* **35**, 987-1000.

Wager, L.R., & Brown, G.M., 1968. *Layered Igneous Rocks.* Edinburgh: Oliver and Boyd, 588 pp.

Walraven, F., 1985. Genetic aspects of the granophyric rocks of the Bushveld Complex. *Econ. Geol.* **80**, 1166-80.

Walraven, F., 1987. Geochronological and isotopic studies of Bushveld Complex rocks from the Fairfield borehole at Moloto, northeast of Pretoria. *S. Afr. J. Geol.* **90**, 352-60.

Walraven, F., & Wolmarans, L.G., 1979. Stratigraphy of the upper part of the Rustenburg Layered Suite, Bushveld Complex, in western Transvaal. *Geol. Surv. S. Afr. Annals,* **13**, 109-14.

Walraven, F., Armstrong, R.A., & Kruger, F.J., 1990. A chronostratigraphic framework for the north-central Kaapvaal Craton, the Bushveld Complex and Vredefort structure. *Tectonophysics* **171**, 23-48.

White, J.A., 1994. The Potgietersrus prospect - geology and exploration history. In: Anhaeusser, C.R. (ed.) *Proc. XVth CMMI Congress.* S. Afr. Inst. Mining Metal., Johannesburg, **3**, 173-81.

Willemse, J., 1969. The vanadiferous magnetic iron ore of the Bushveld Igneous Complex. *Econ. Geol. Monogr,* **4**, 187-208.

Wilson, J.R., Cawthorn, R.G., Kruger, F.J., & Grundvig, S., 1994. A major unconformity in the western Bushveld Complex: the northern "gap". *S. Afr. J. Geol.* **97**, 462-72.

Wilson, J.R., Menuge, J.F., Pedersen, S., & Engell-Sorensen, O., 1987. The southern part of the Fongen-Hyllingen layered mafic complex, Norway. In: Parsons, I. (ed.) *Origins of Igneous Layering.* Dordrecht: Reidel Publ. Co., 145-84.

LAYERED INTRUSIONS
R.G. Cawthorn (editor)

The Bjerkreim-Sokndal Layered Intrusion, Southwest Norway

J.R. Wilson[a], B. Robins[b], F.M. Nielsen[a], J.C. Duchesne[c] and J. Vander Auwera[c].

[a]Department of Geology, University of Aarhus, 8000 Aarhus C, Denmark.

[b]Department of Geology, Allégt. 41, University of Bergen, 5007 Bergen, Norway.

[c]Laboratoires associés de Géologie, Pétrologie et Géochimie, Université de Liège, B-4000 Sart Tilman, Belgium.

Abstract

The Bjerkreim-Sokndal Layered Intrusion is a large (~230 km^2), discordant, Late Proterozoic, post-orogenic pluton in the Egersund-Farsund Igneous Province. The intrusion was emplaced shortly after massif-type anorthosite plutons and is cut by jotunite dykes. It contains a >7000 m thick Layered Series consisting of rocks belonging to the anorthosite kindred: andesine anorthosite, leuconorite, troctolite, norite, gabbronorite, mangerite, and quartz mangerite. Cumulates in the Layered Series are organized in 6 megacyclic units (MCU 0 to IV), individually up to 1800 m thick, but varying considerably in thickness and development along strike. The highest-temperature cumulates are troctolites containing plagioclase of ~An_{54} and olivine of ~Fo_{77}. Phase contacts in the macrocyclic units reflect crystallization of the silicate minerals in the order plagioclase (± olivine), orthopyroxene, Ca-rich pyroxene, pigeonite. Ilmenite crystallized early and apatite appeared as a cumulus mineral at about the same time as Ca-rich pyroxene. Cumulus magnetite followed orthopyroxene and preceded Ca-rich pyroxene in MCU III and IV, but crystallized after Ca-rich pyroxene in MCU IB. MCUs 0, IA and II do not contain cumulates with cumulus magnetite or Ca-rich pyroxene. Olivine (~Fo_{50}) reappears in the uppermost part of the Layered Series where there is a rapid stratigraphic transition to mangerite and quartz mangerite. The basal parts of MCUs III and IV are characterized by thin sequences of plagioclase, plagioclase-orthopyroxene-ilmenite and orthopyroxene-ilmenite cumulates in which there are systematic upward decreases in initial Sr isotope ratios. They are overlain by troctolite (plagioclase-olivine cumulate) and are believed to have crystallized from hybrid magmas. The MCUs, the discordant geometry of phase contacts, the stratigraphic variations in initial $^{87}Sr/^{86}Sr$ ratio (0.7049-0.7085), and the abundance of xenoliths suggest crystallization of the cumulates at the base of a periodically-replenished, compositionally-zoned magma chamber that was continually assimilating country rocks. The parent, as indicated by medium-grained jotunite along country-rock contacts, appears to have been an evolved, Ti-rich magma similar to ferrobasalt, but poor in diopside components. Systematic stratigraphic variations in initial $^{87}Sr/^{86}Sr$ ratio at the base of MCU III and MCU IV suggest that influx of magma into the chamber was accompanied by mixing with resident, contaminated magma.

1. INTRODUCTION

The Bjerkreim-Sokndal Intrusion (BKSK) (Michot, 1960, 1965; Duchesne, 1987) is a large (40 km long and up to 15 km wide) layered intrusion that occupies an area of about 230 km^2 (Figure 1). Lithologically the intrusion consists of rock types belonging to the anorthosite kindred; *i.e.* andesine anorthosite, troctolite, leuconorite, norite, gabbronorite, jotunite (monzonorite), mangerite (hypersthene monzonite) and quartz mangerite. The BKSK is emplaced in granulite-facies quartzo-feldspathic and mafic gneisses as well as anorthosite and leuconorite belonging to the Egersund-Ogna (Michot and Michot, 1970; Duchesne and Maquil, 1987) Håland-Helleren (Michot, 1961) and Åna-Sira (Krause *et al.*, 1985; Duchesne and Michot, 1987) massifs, and xenoliths of all these host rocks are common within the intrusion itself (Duchesne, 1970). Zircons within an orthopyroxene megacryst from the Egersund-Ogna anorthosite have recently yielded an U-Pb age of 929±2 Ma (Schärer *et al.*, 1992).

The BKSK and the various anorthosite massifs are cut by members of a suite of small plutons and wide, laterally-persistent dykes of jotunite, some of which are differentiated (Duchesne *et al.*, 1989; Wilmart *et al.*, 1989). The most voluminous of the jotunite dykes that cut the northern part of the BKSK is the Lomland dyke (Duchesne *et al.*, 1989) (Figure 1). Zircon and baddeleyite from the Tellenes ilmenite norite, belonging to the same suite of intrusions, have yielded U-Pb ages of 920±2 Ma (Schärer *et al.*, 1992; Duchesne *et al.*, 1993). The BKSK is also cut by members of the Egersund swarm of basaltic dykes (Figure 1), that were intruded at 630-650 Ma (Sundvoll, 1987).

Several workers have suggested that the BKSK was emplaced in two quite separate stages: crystallization of the lower part of the pluton, the Layered Series (generally referred to as the "leuconoritic phase"), taking place long before the overlying mangerite and quartz mangerite (Wielens *et al.*, 1980; Rietmeijer, 1984). The present authors do not ascribe to this view but regard the BKSK as a single, differentiated intrusion, emplaced in the course of a relatively short-lived (~10 Ma), post-Sveconorwegian (= post-Grenvillian) magmatic event, together with anorthosite massifs and jotunite intrusions (Duchesne *et al.*, 1993).

This paper gives a review of present knowledge of the BKSK and presents evidence for development of its cumulates in a repeatedly-replenished, continuously-fractionating magma chamber in which country rocks were assimilated. Evidence is also presented for the development of compositional stratification of the magma, and the nature of the parental magma is discussed. The description deals principally with the Bjerkreim lobe of the intrusion where the Layered Series is most completely developed.

2. PLUTONIC SETTING OF THE BJERKREIM-SOKNDAL INTRUSION

The BKSK is part of the autochthon that appears south of the Caledonian nappe complex of the Stavanger area and forms part of the Sveconorwegian (= Grenvillian) Orogen. The autochthon in southwest Norway is composed of migmatitic ortho- and paragneisses, together with isolated occurrences of supracrustal rocks and synorogenic syenites and granitoids (Hermans *et al.*, 1975). The gneisses and metasediments are intruded by post-Sveconorwegian plutons, including the BKSK, that constitute the Egersund-Farsund Igneous Province. Individual intrusions in the province vary in composition from andesine anorthosite and leuconorite (which dominate in terms of area), to norite, jotunite, syenite, charnockite, and granite. The structural and metamorphic history of the gneissic envelope to the BKSK was complex and protracted. Intense Sveconorwegian deformation was coeval with upper amphibolite to granulite facies metamorphism (M1) at pressures of 5-7 kb (Jansen *et al.*, 1985; Maijer, 1987). M2 intermediate pressure (4-6 kb) granulite-facies metamorphism increased in grade towards the Egersund-Farsund Igneous Province as marked by isograds that follow its margin (Tobi *et al.*, 1985; Jansen *et al.*, 1985). M2 metamorphic assemblages were variably overprinted

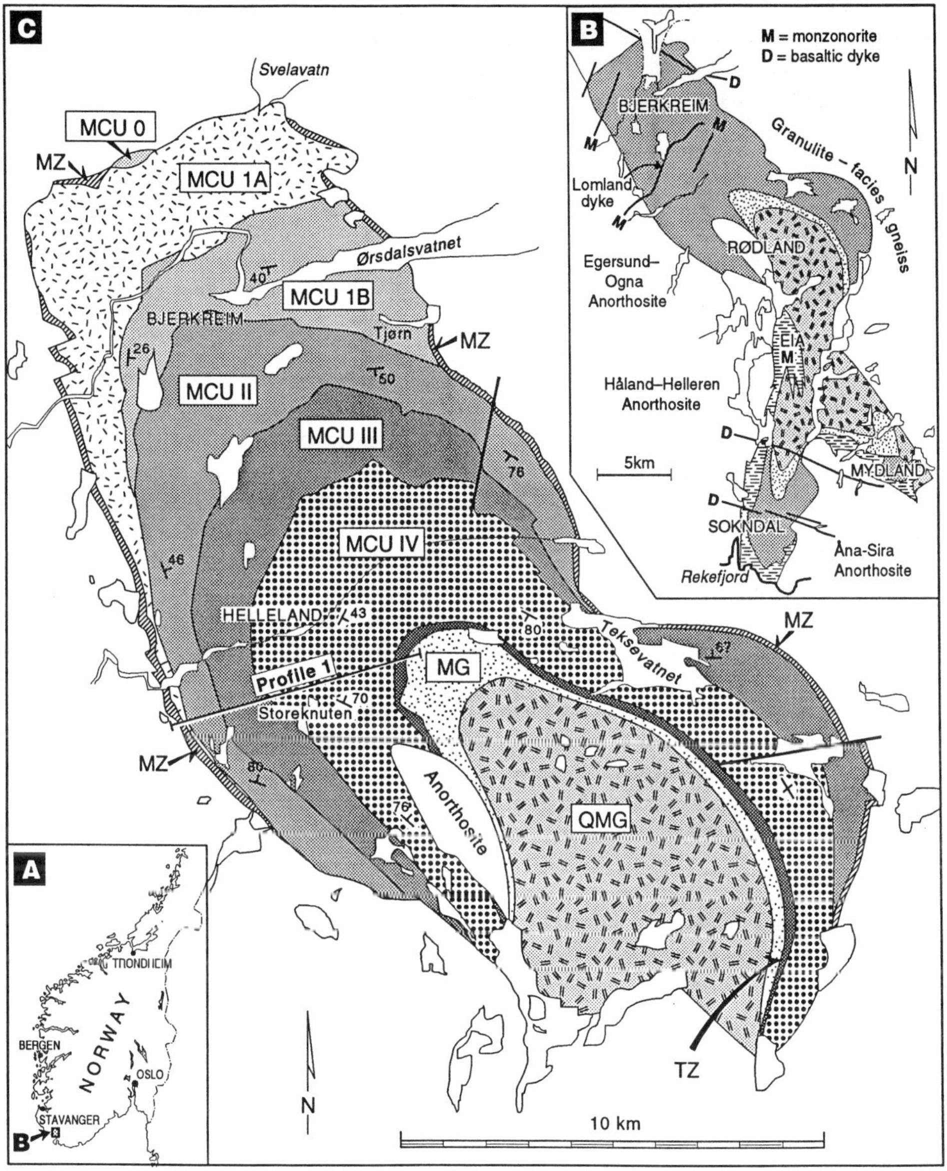

Figure 1. Location (A), simplified geological map of the Bjerkreim-Sokndal Layered Intrusion (B), and the distribution of megacyclic units 0 to IV in the Bjerkreim lobe of the intrusion (C). In Figure 1B, intrusions marked as M are jotunite and those marked as D are basaltic dykes belonging to the Egersund Swarm. In Fig 1C, MG = mangerite, QMG = quartz mangerite, TZ = Transition Zone, MZ = Marginal Zone (jotunite, norite, and leuconorite).

during the retrogressive M3 metamorphic event (at 3-5 kb, Jansen *et al.*, 1985).

The depth at which the BKSK crystallized has been estimated from the mineral assemblages present in mangerite and quartz mangerite. The coexistence of Fo_6 + quartz and evidence of primary orthopyroxene with a composition of $Wo_2En_{15}Fs_{83}$ (Rietmeijer, 1979) bracket pressures to 5.7-7.5 kb using the Bohlen and Boettcher (1981) geobarometer (Wilmart and Duchesne, 1987). The low Al_2O_3 contents of orthopyroxenes in the BKSK suggest, however, that crystallization took place at pressures less than 5 kb (Vander Auwera and Longhi, 1994).

3. THE FORM OF THE BJERKREIM-SOKNDAL INTRUSION

The Bjerkreim-Sokndal Intrusion has generally been described as a lopolith, but recent detailed mapping shows it to be a trough-like, discordant intrusion. Modelling of the associated +10-30 mgal gravity anomaly gives a minimum depth to the base of the intrusion of about 4 km (Smithson and Ramberg, 1979).

Layering within the intrusion is deformed into a deep, doubly-plunging syncline that branches in the south around a dome cored by the Ana-Sira anorthosite massif (Figure 1). The core of the syncline is occupied by massive quartz mangerite, which does not exhibit modal or textural layering, and this is separated in places from the underlying mangerite by a zone rich in wall-rock xenoliths. The size of the gravity lows over the quartz mangerite suggest a maximum thickness of about 2 km (Smithson and Ramberg, 1979). There is no evidence that the roof of the intrusion is preserved anywhere within the confines of the present outcrop.

The BKSK consists of three lobes; the Bjerkreim lobe in the northwest, and the smaller Sokndal and Mydland lobes to the south and south-east respectively (Figure 1). Modal layering and phase contacts in the Bjerkreim lobe are disposed in a syncline that plunges southeast at 20-40°. In the steep limbs of the syncline the cumulates are generally foliated, essentially in the plane of the modal layering. In places, cumulus minerals form augen in a foliated matrix, small shear zones are developed, and there is a strong mineral lineation. Linear fabrics dominate around the hinge of the syncline (Paludan *et al.*, 1994). Cumulus plagioclase grains are strained or recrystallized to shape-oriented polygonal aggregates, while prismatic Ca-poor pyroxenes are commonly kinked or bent. Uniform paleomagnetic vectors in different parts of the intrusion (Poorter, 1972) suggest that the deformation and development of the synformal disposition of the layering took place at temperatures in excess of the Curie point (550-650°C). Michot and Michot (1970) and others have attempted to relate the deformation to a late expression of the Sveconorwegian Orogeny. Since the age of emplacement appears to be ~70 Ma later than the last recorded regional deformation, the crystallization and deformation of the BKSK must have taken place entirely in a anorogenic environment. The foliation, mineral lineation and the synformal disposition of the Layered Series seem, therefore, to be due to high-temperature, gravitational foundering (Glazner, 1994) consequent on the crustal-scale density inversion (demonstrated by the present gravity anomaly) that resulted from the location of the BKSK within lower-density gneiss and anorthosite.

4. SUBDIVISION OF THE LAYERED SERIES

The Layered Series has a thickness of >7000 m in the axial region of the syncline and can be divided into 6 megacyclic units (MCU 0 to IV) which repeat characteristic sequences of cumulates (Figures 1 and 2). The numbering of Michot (1960, 1965) has been retained as far as pos-

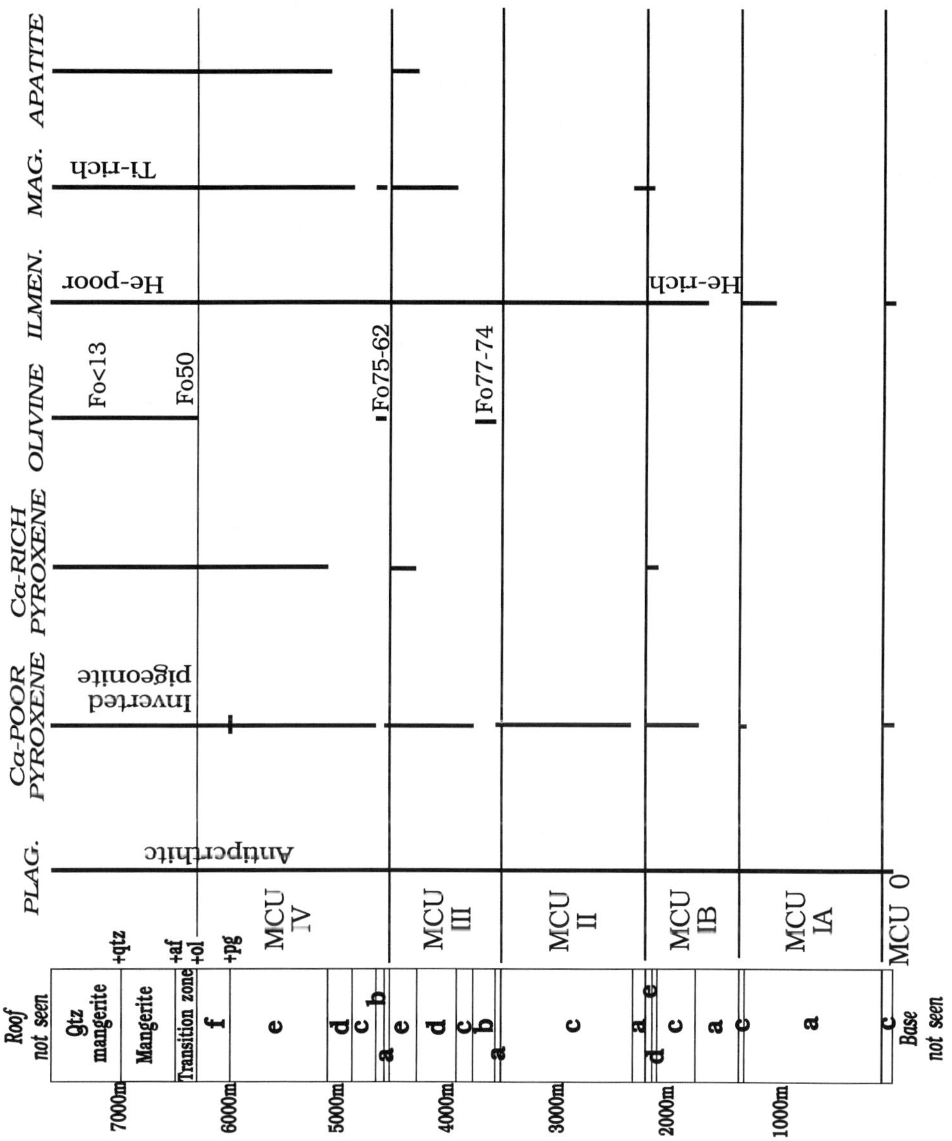

Figure 2. Generalized stratigraphy of the Bjerkreim-Sokndal Layered Series as developed in the northern part of the intrusion along the axial trace of the syncline defined by the layering. The lower case letters (a, b, c, d, e, and f) refer to stratigraphic zones that are characterized by different assemblages of cumulus minerals.

sible in that his units 2, 3, and 4 are referred to as MCU II, III, and IV, in agreement with papers by Nielsen and Wilson (1991) and Jensen *et al.* (1993). We have not been able to confirm the existence of Michot's unit 5 and consider that MCU IV grades into the Transition Zone (Figure 3). A consequence of the numbering system is that we refer to the lowest three units as MCU O, IA and IB. The megacyclic units can be further subdivided into zones a-f, based on assemblages of cumulus minerals (Figure 2).

The megacyclic units vary in stratigraphic thickness, lateral persistence and in the nature of the layer sequences they exhibit. The lower three megacyclic units, exposed only in the northernmost part of the intrusion, are individually as much as 1300 m thick but show a pronounced southward thinning in the western limb of the syncline and are not developed in the southern parts of the Bjerkreim lobe.

The lowermost cumulates are exposed in the northwestern part of the Bjerkreim lobe and consist of plagioclase-hypersthene-ilmenite cumulates (phiC). They are regarded as the top of MCU 0, the rest of which, together with an unknown thickness of cumulates, is hidden. These cumulates are overlain successively by pC, piC and phC belonging to MCU IA (~1300 m thick in a profile along the axial trace of the syncline). This sequence is repeated in MCU IB (~875 m thick) which also contains more evolved lithologies with the entry of cumulus Ca-rich pyroxene, followed by magnetite. MCUs 0-IB are characterized by the presence of plagioclase megacrysts (up to 10 cm long) in all rocks with the exception of the most evolved cumulates at the top of MCU IB.

MCU II (reaching a thickness of 1300 m) consists of a thin layer of magnetite-bearing piC overlain by phiC. The appearance of cumulus magnetite in the leuconorites at the base of MCU II and its absence in the overlying cumulates suggests affinities with the olivine-bearing zones near the bases of succeeding MCUs.

MCU III (maximum ~1300 m thick) has, in places, a thin zone consisting mainly of pC, but with interlayered iC, phiC, and hiC at the base (Zone IIIa in Figure 2), overlain by leucotroctolite that contains cumulus magnetite in addition to plagioclase, olivine and ilmenite. The leucotroctolite is in turn overlain by phiC (in places with an intervening thin piC), followed by magnetite norite and gabbronorite with the successive (re-)entry of cumulus magnetite and then apatite together with Ca-rich pyroxene.

MCU IV (maximum thickness ~1800 m) repeats the same sequence as in MCU III. MCU IV contains, however, additional, more-evolved cumulates. Michot (1960) recognized the prominent olivine-bearing zone near the base of MCU IV and referred to it as the "Svalestad horizon". It has a thickness of about 100 m and is laterally persistent along strike for about 24 km. Olivines in the olivine-bearing zones near the bases of MCU III and IV are partially or completely replaced by orthopyroxene-Fe-Ti oxide symplectites, but the zones are texturally distinctive even where olivine is absent. Small amounts of biotite and hornblende also occur in the olivine-bearing zones. Ca-poor pyroxene is inverted pigeonite in the upper part of MCUIV which grades into overlying mangerite through a jotunitic Transition Zone (TZ) whose base is defined by the entry of Fe-rich olivine, which more or less coincides with the appearance of interstitial alkali feldspar (Duchesne *et al.*, 1987). With the appearance of cumulus mesoperthite the rocks grade upwards from jotunite to mangerite, which in turn passes into massive quartz mangerite. Even in these highly-evolved rocks, hydrous phases are not abundant: calcic amphibole is generally a minor mineral (except in the uppermost part of the quartz mangerite where it may occur as large oikocrysts) and biotite is generally an accessory mineral. The aggregate thickness of the mangerite and quartz mangerite is >350 m (Rietmeijer, 1979) and may be as much as 2.5 km. Viewed on a broad scale (Figure 2), the lower part of the Layered Series is dominated by plagioclase cumulates, the middle part by plagioclase-hypersthene-ilmenite cumulates and the upper part by plagioclase-hypersthene/pigeonite-augite-ilmenite-magnetite-apatite cumulates. Combined with the reversals to relatively primitive mineral assemblages at the bases of the MCUs, this is strong, first-order evidence that the Layered Series crystallized in a continuously-fractionating, periodically-replenished magma chamber. The base of MCU IV seems to reflect the last major influx of magma into the BKSK magma chamber. After this event fractional crystallization was relatively uninterrupted, though there is some mineralogical evidence of a minor regressive discontinuity in the cryptic layering at the base of the TZ.

5. THE GEOMETRY OF PHASE CONTACTS

A prominent feature of the Layered Series is the discordant geometry of phase contacts within individual megacyclic units. The angles which phase contacts subtend with boundaries between MCUs (2-15°) have the result that zones characterized by particular cumulus assemblages may wedge out laterally. This is particularly prominent in MCUs IB and II where the basal plagioclase cumulates have maximum thicknesses of around 800 and 170m respectively where they abut the northeastern margin of the Bjerkreim lobe, but they thin to the southwest and eventually disappear. The most evolved cumulates in MCU IB also pinch out in a southwesterly direction, so that the unit cannot be recognized further along strike (Figure 1), but they extend beyond the termination of the associated basal piC.

The two basal zones of MCU III (MCU IIIa and the olivine-bearing MCU IIIb) can be traced for about 19 km from the southwestern flank of the Bjerkreim lobe where they abut onto anorthosites of the Håland-Helleren massif, round the hinge zone of the syncline, to the northern flank. Here these two zones thin gradually towards the south-east and wedge out completely approaching the Teksevatnet area where phiC of MCU IIIc overlies apparently identical phiC of MCU IIc. The country-rock gneisses form a protrusion in the Teksevatnet

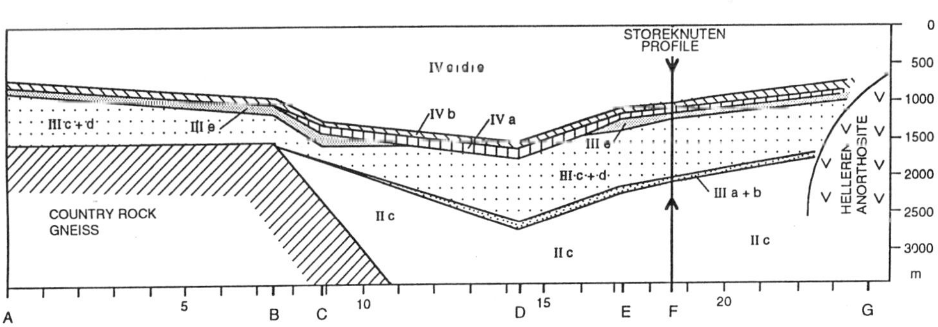

Figure 3. *The geometrical arrangement of zones during crystallization of the upper part of the Layered Series in the Bjerkreim lobe (based on Nielsen and Wilson, 1991). The vertical exaggeration is x2. The thicknesses of individual zones is measured from the entry of cumulus pigeonite (inverted) which is shown as a reference line the top of the figure. Faults are not shown. Recent detailed mapping has shown that zone IIIe terminates abruptly towards the centre of the intrusion, rather than wedging out.*

area, to the east of which only MCUs III and IV appear to be developed. These relationships are illustrated in Figure 3 which is based on seven profiles studied in detail by Nielsen and Wilson (1991). The base of MCU IV is in contact with modally layered cumulates of MCU IIIe in all the profiles except D (in Figure 3) where it overlies MCU IIId. The olivine-bearing zone (MCU IVb) has a fairly constant thickness (75-100 m) along its entire strike length of ~24 km, whereas the underlying MCU IVa is absent on the northern flank, reaches a maximum thickness of ~90 m in profile D in the central, axial region where MCU IIIe is absent, and thins towards the southern flank where it has a thickness of about 10 m. MCU IVc varies in thickness from about 40-100 m, but this variation does not appear to be related to the location of the profiles.

These geometrical relationships imply that contrasting cumulus assemblages crystallized simultaneously on different parts of the magma-chamber floor. At certain times, higher temperature cumulus assemblages were forming on the central, deeper regions of the floor while lower temperature cumulates were precipitating towards the margins of the chamber. We believe that the geometry of phase boundaries in the BKSK is a result of an interaction between vertical compositional gradients in the magma column and a basinal or half-basinal floor to the chamber, as will be discussed later.

6. MODAL LAYERING

The BKSK displays small-scale modal and textural layering in zones characterized by two

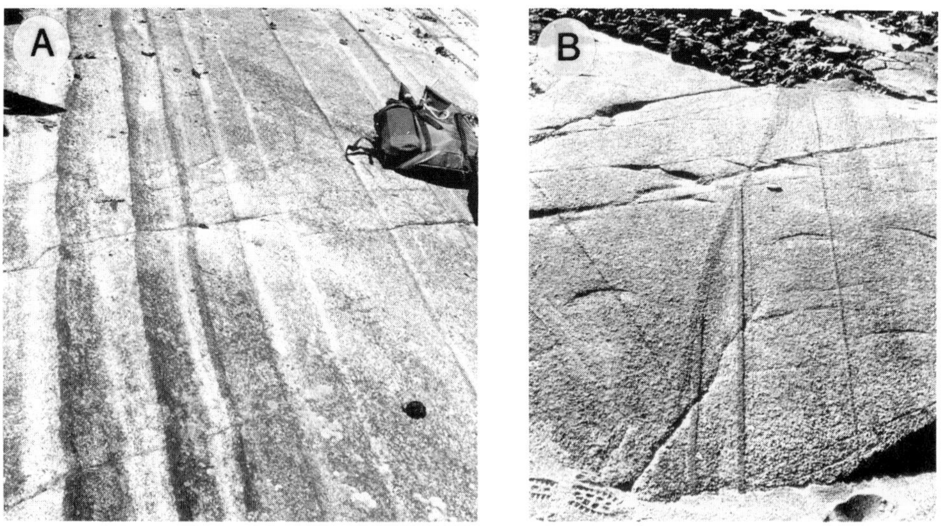

Figure 4. *A) Modally-graded layers alternating with more homogeneous ilmenite-magnetite gabbronorite (phaimC) within MCU IIIe (Figure 2) near Storeknuten (Figure 1C). Stratigraphic top is to the right (NE). B) Minor erosional unconformity in steeply dipping, modally layered ilmenite-magnetite norite (phimC) within MCU IIId (Figure 2) on the northern shore of Teksevatnet (Figure 1C). Stratigraphic top is to the left (SW). Boot print for scale.*

or more cumulus minerals, and modal layering is usually particularly well developed in the upper, more mafic parts of the MCUs. Igneous lamination is almost ubiquitous, but commonly is overprinted by a deformational fabric.

Layers commonly exhibit modal grading from pyroxene, ilmenite or magnetite-rich bases to plagioclase-rich tops (Figure 4A). Modal grading in the lower part of the Layered Series, where plagioclase megacrysts are present, is accompanied by reverse size grading, the megacrysts being concentrated towards the tops of modally-graded layers. Thicker isomodal layers generally separate modally-graded layers (Figure 4A). Isomodal layers of ilmenite pyroxenite and ilmenitite are common in the basal zones of MCUs III and IV. A unique thin, discontinuous layer of pyrrhotite pyroxenite is found just beneath the base of MCU III where it is associated with disseminated sulphides in both the underlying and overlying phiC. The TZ in the Sokndal lobe is characterized by the occurrence of isomodal ultramafic layers up to 4 m thick consisting of cumulus Fe-Ti oxides, pyroxene and Fe-rich olivine (Duchesne *et al.*, 1987).

Modally-layered intervals sometimes display slump structures. Trough structures and minor unconformities are developed locally (Figure 4B).

7. INCLUSIONS

Large rounded to slab-like inclusions (up to 600 x 400 m) of massive or foliated anorthosite and leuconorite are common in the lower part of the Layered Series and in the south west part of the Bjerkreim lobe. These are believed to be derived from the adjacent Egersund-Ogna and Håland-Helleren anorthosite massifs. Based on the work of Michot (1961), the Håland part of the latter massif was previously thought to intrude the southern flank of the Bjerkreim lobe of the BKSK in the Rødland area; our detailed mapping has shown that the BKSK cuts anorthosite and leuconorite which in this area form abundant, large inclusions. This implies not only that the BKSK is younger than all of the anorthosite plutons, but also that the inclusion of foliated anorthositic rocks indicates that emplacement took place after the deformation of at least parts of the anorthosite massifs.

Xenoliths of anorthosite and leuconorite are particularly frequent in the plagioclase cumulates in MCUs IA and IB. In these megacyclic units there appears to be an upward decrease in both the number and size of the leucocratic xenoliths. Blocks of quartzo-feldspathic gneiss and mafic granulite are especially abundant in the upper part of the Layered Series and are believed to have been derived from the roof. Large gneissic xenoliths are numerous in the upper part of MCU IV and extend into the TZ and mangerite. The presence of a zone rich in gneissic xenoliths, the "septum xenolithique" of Michot (1960), in the vicinity of the contact between the mangerite and quartz mangerite has long been recognized.

8. MINERAL CHEMISTRY

The compositional variations in plagioclase and Ca-poor pyroxene with stratigraphic height as displayed along the profile marked in Figure 1 are presented in Figure 5.

Plagioclase is the dominant cumulus mineral in the BKSK and is unusually sodic. It shows surprisingly little systematic, stratigraphic or lateral variation in composition in the lower part of the Layered Series. There also seems to be little systematic correlation between the composition of plagioclase and the cumulus paragenesis. Plagioclase in thin sections typically exhibit no compositional zoning but there are significant compositional differences between individual grains. This is believed to be the result of the recrystallization of cumulus crystals that originally were compositionally zoned. In MCU IA and IB plagioclases are in the range An_{49-36}. At the base of MCU II there is a distinct regression to slightly more calcic plagioclase (An_{52-43}).

In Figure 5 plagioclase has a composition of An_{50} in the basal, marginal leuconorites, and becomes slightly more sodic upwards, reaching An_{45} near the top of MCU II. There is a reversal to about An_{51} at the base of MCU III, followed by a trend to slightly more evolved compositions upwards, reaching An_{46} in MCU IIIe. There is an abrupt reversal to the most calcic plagioclase compositions in the entire BKSK intrusion (An_{54}) at the base of MCU IV (shown in more detail in Figure 6). This is followed by a systematic trend to more evolved plagioclase compositions through the ~2000 m-thick MCU IV, reaching An_{37} near the top. Plagioclase in the TZ changes rapidly from An_{37} to about An_{24} in the mangerite. According to Rietmeijer (1979) plagioclase in mangerite and quartz mangerite varies between An_{43} and An_{14}. Plagioclase in the upper part of MCU IV is typically antiperthitic. Antiperthite also occurs sporadically in the lower megacyclic units. Sr concentrations in plagioclase vary between 1500 and 850 ppm and are not simply correlated with the anorthite content. Sr decreases with falling anorthite content until Ca-rich pyroxene and apatite appear as cumulus phases, after which Sr increases as plagioclase becomes more albitic, reaches a maximum on the appearance of cumulus alkali feldspar, and then decreases (Duchesne, 1978).

Ca-poor pyroxene in MCU IA and IB has 100Mg/Mg+Fe (Mg#) between 75 and 63. Ca-poor pyroxene oikocrysts at the base of the profile in Figure 5 have Mg# 72, above which it becomes a cumulus phase with about Mg# 74. This composition persists into the lower part of MCU III, above which the Ca-poor pyroxene becomes gradually more evolved, reaching Mg# 70 in the upper part of MCU III. There is a reversal to Mg# 74 at the base of MCU IV, followed by a trend to more iron-rich compositions, reaching Mg# 55 in inverted pigeonite at the top of MCU IV. There is an abrupt decrease to Mg# 28 in inverted pigeonite oikocrysts across the MCU IV - TZ boundary in Figure 5, followed by a slight increase to Mg# 32 in the

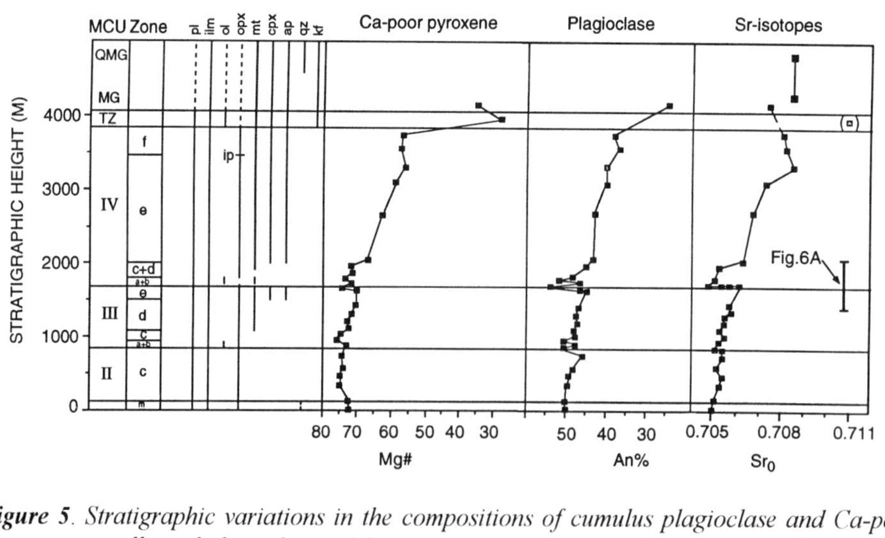

***Figure 5**. Stratigraphic variations in the compositions of cumulus plagioclase and Ca-poor pyroxene as well as whole-rock initial Sr isotope ratios along profile 1 (Figure 1C). Based on Nielsen et al. (1996).*

mangerite. Elsewhere in the Bjerkreim lobe there is a gradual regression in the cryptic variation beneath the MCU IV - TZ boundary and the TZ contains inverted pigeonite with Mg# of 57-43 while the mangerite contains more Fe-rich compositions (*e.g.* Mg# 32). The TZ in the Sokndal lobe appears to be anomalous in that its lower part reportedly contains cumulus Fe-rich orthopyroxene with Mg# from 34 to 24 (Rietmeijer, 1979), rather than inverted pigeonite. The upper part of the TZ and the overlying mangerite, however, contain inverted pigeonite with Mg# between 31 and 25 (Rietmeijer, 1979), partly as large oikocrysts.

Ca-poor pyroxene in the BKSK is much poorer in Al_2O_3 than pyroxene megacrysts of comparable Mg# in the adjacent anorthosite massifs. The BKSK pyroxenes contain 0.9-2.6 wt % Al_2O_3 compared to 7-9 % in the anorthosites (Duchesne and Michot, 1987).

Ca-rich pyroxene has a more restricted distribution than Ca-poor pyroxene: It is only found as a cumulus mineral in the upper parts of MCUs IB, III, and IV, as well as in the TZ, mangerite and quartz mangerite. Ca-rich pyroxene in MCUs IB and III has the composition $Ca_{45\text{-}42}Mg_{43\text{-}40}Fe_{16\text{-}14}$ (Mg# 71-75). Through MCU IV the Mg# in Ca-rich pyroxene decreases upwards from 75 to about 62. The TZ is characterized by Ca-rich pyroxenes with Mg# between 69 and 45 (Duchesne *et al.*, 1987). In the mangerite and quartz mangerite Ca-rich pyroxenes have Mg# in the range 42-15 (Rietmeijer, 1979). Ca-rich pyroxene is the only pyroxene present in the upper part of the quartz mangerite.

Olivine is restricted in occurrence in the lower part of the Layered Series, as noted above, principally to two stratigraphic intervals near the bases of MCUs III and IV. It has a composition of $Fo_{77\text{-}74}$ in MCU IIIb, while MCU IVb is characterized by olivine of $Fo_{75\text{-}62}$. Iron-rich olivine re-appears in the Transition Zone and changes rapidly in composition from Fo_{50} to Fo_{19} over a stratigraphic thickness of 50-80 m. Mangerite and quartz mangerite contain $Fo_{13\text{-}4}$ (Rietmeijer, 1979).

Ilmenite exhibits a general decrease in hematite content through the Layered Series, from 16-20 % in the lower part of the series where it is the only cumulus Fe-Ti oxide, to about 2 % in the mangerite and quartz mangerite (Duchesne, 1972). This pattern is repeated on a smaller scale in the individual MCUs. Ilmenite contains about 0.3 % V_2O_3 in the lower part of the Layered Series and vanadium shows an identical variation to haematite content. Manganese in ilmenite, however, increases almost continuously through the Layered Series, from about 0.3 % in leuconorites near the base to 1.0 % MnO in mangerite; breaks in the trend at contacts between MCUs are slight. Nickel and chromium are enriched in ilmenite at the bases of MCUs where concentrations can be as high as 1000 ppm Ni and 1.4 % Cr_2O_3.

Magnetite is a cumulus phase in the upper parts of MCUs IB, III, and IV, in the TZ where it occurs in oxide-rich layers, and in the leuconorites and leucotroctolites near the bases of MCUs II, III, and IV. Its TiO_2 content increases systematically from <2 % at its reappearance in the upper part of MCU III to as much as 19 % (corresponding to $Usp_{58}Mt_{42}$) in the TZ. As with ilmenite, the manganese concentrations increase (to ~0.25 % at the top of MCU IV) and vanadium decreases upwards through the Layered Series (from ~1.3 % to 0.02 % V_2O_3 at the top of MCU IV) (Duchesne, 1972). Magnetite in leucotroctolite at the base of MCU IV contains lower V concentrations (1.0-0.75 % V_2O_3) than the magnetite in the uppermost part of MCU III and on its reappearance in the upper part of MCU IV, possibly due to the presence in the leucotroctolite of small amounts of amphibole that has a high partition coefficient for V (Jensen *et al.*, 1993). Ni concentrations are generally low (<40 ppm), except in the leucotroctolites at the bases of MCUs III and IV where concentrations are 900-600 ppm. Nickel concentrations decrease with stratigraphic height in the leucotroctolite at the base of MCU IV. Chromium exhibits the same behaviour as Ni, concentrations in the leucotroctolite at the base of MCU IV varying from 1.4-0.4 % Cr_2O_3. Chromium contents in magnetite elsewhere in the BKSK are very low.

Apatite is a cumulus mineral in the uppermost parts of MCUs III and IV and in mangerite. It exhibits light-element enriched chondrite-normalized REE patterns with pronounced negative Eu-anomalies (Roelandts and Duchesne, 1979; Holdam, 1990). The concentration of REE in cumulus apatite increases significantly from MCU III to the mangerite, but there is at present insufficient data to document any cryptic variation which may be present in individual megacyclic units.

9. Sr ISOTOPE RATIOS

Initial $^{87}Sr/^{86}Sr$ ratios (Sr_0) (recalculated for an age of 930 Ma) for rocks and minerals from the BKSK lie between 0.7049-0.7085, with the highest values being found in mangerite and quartz mangerite (Versteeve, 1975; Demaiffe *et al.*, 1979; Wielens *et al.*, 1980; Jensen *et al.*, 1993). A Sr isotope study of the Bjerkreim lobe of BKSK has recently been carried out by Nielsen *et al.* (1996), whose results are shown in Figures 5 and 6. The variation in Sr_0 in Figure 5 is broadly parallel to the variation in An% in plagioclase and Mg# in Ca-poor pyroxene, showing that there is a close relationship between fractionation and contamination in BKSK. Sr_0 shows an overall increase through the first ~3400 m in Figure 5 from about 0.7050 to 0.7085. Superimposed on this trend is a major reversal to lower Sr_0 values (reaching 0.7049) at the base of MCU IV, and a minor, indistinct reversal at the base of MCU III. Above 3400 m the correlation with mineral composition breaks down, with Sr_0 decreasing slightly and reaching constant values of 0.7085 in the quartz mangerite.

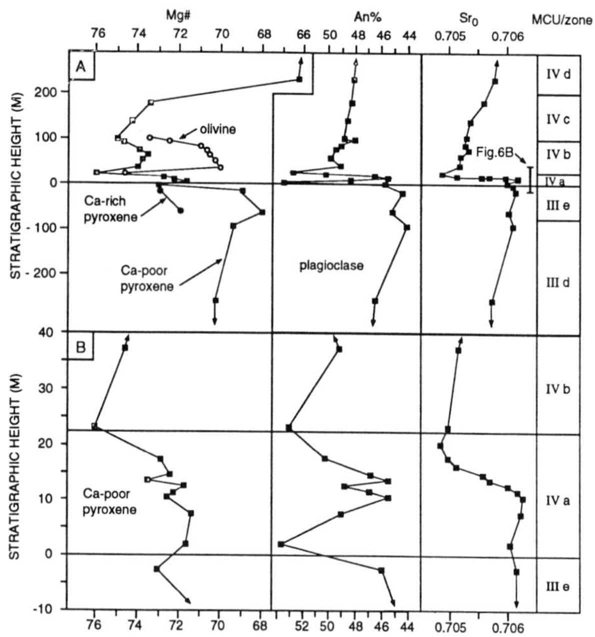

Figure 6. A) Stratigraphic variations in mineral compositions and Sr-isotopes (both whole-rock and in separated plagioclase) across the MCU III - MCU IV boundary at Storeknuten (Figure 1C). Ca-poor pyroxene in MCU IVb replaced olivine and is not a cumulus phase. B) The detailed compositional variations at the base of MCU IV at Storeknuten. Based on Nielsen et al. (1996).

The fact that the lowermost sample at the base of the profile in Figure 5, collected from close to the country rock contact, has a relatively low Sr_0 value (0.7050) implies that there is little local contamination from the gneissic country rock (Sr_0 = 0.7196; Versteeve, 1975) which forms the floor in this part of the intrusion.

Previous workers (Demaiffe *et al.*, 1979) considered that the Layered Series had Sr_0 values around 0.705-0.706, considerably lower than the mangerite and quartz mangerite (0.7085). This apparently significant difference in isotopic ratios was used as a major argument against a direct genetic relationship between these two major parts of the complex (Wiebe, 1984). The gradual increase in Sr_0 through MCU IV, ending with fairly constant values in the mangerite and quartz mangerite, combined with the field, mineralogical and cryptic variations, implies that the two parts of the complex are related.

10. VARIATIONS ACROSS THE MCU III/IV BOUNDARY

The most primitive rocks in the entire Bjerkreim-Sokndal intrusion occur near the base of MCU IV in MCU IVb (Figures 5 and 6). A detailed section across this boundary (at Storeknuten, on profile 1 in Figure 1) was investigated by Jensen *et al.* (1993). The Storeknuten profile is part of profile F in Figure 3 which is based on Nielsen and Wilson (1991). Further details of the Storeknuten section have been studied by Nielsen *et al.* (1996). Some of the results of these studies are included here.

Figure 6A shows the lithological variation in a ~600 m thick section, together with compositional variations of the major silicates and variations in Sr_0. The modally layered phimC of MCU IIId about 250 m below the base of MCU IV contains plagioclase and Ca-poor pyroxenes with An_{47} and Mg# 70 respectively, together with an Sr_0 of 0.7058 in separated plagioclase. Slightly more evolved mineral compositions, and systematically higher Sr_0 values occur at the top of MCU IIId and in the phimcaC of MCU IIIe. A few meters below the base of MCU IV (Figure 6B) plagioclase and Ca-poor pyroxene show compositional reversals from An_{44} to An_{46} and from Mg# 69 to Mg# 74 respectively. The compositions of these two phases, especially plagioclase, vary erratically in MCU IVa and are not in equilibrium with each other. Sr_0 remains constant at about 0.7061 in the upper part of MCU IIIe and lowest 12 m of MCU IVa, after which it shows a very systematic decrease to 0.7048 at the top of MCU IVa.

The leucotroctolite at the base of MCU IVb contains olivine of Fo_{74} and plagioclase of $An_{52.5}$. The olivine composition reaches Fo_{70} in the next sample, after which there is a regression to $Fo_{73.5}$ through the remainder of MCU IVb. The Ca-poor pyroxene in MCU IVb replaces olivine and their compositions are clearly in equilibrium with each other. Above MCU IVb, cumulus Ca-poor pyroxenes become increasingly evolved with stratigraphic height through MCU IV (Figures 5 and 6A). Plagioclase shows a slightly erratic trend from An_{53} at the base of MCU IVb to more sodic compositions in the overlying zones (Figures 5 and 6A). After the systematic upward trend to lower values in the upper part of MCU IVa, Sr_0 shows a gradual increase through MCU IVb, c, and d (Figures 5 and 6A).

As will be discussed later, the compositional regression across the MCU III - IV boundary reflects a major addition of magma to the chamber. MCU IVa is interpreted as having formed by crystallization during the mixing of new, primitive magma and resident, evolved, contaminated magma.

Table 1
Major- and trace-element analyses of selected fine- to medium-grained jotunites from the discordant northern margin of the Bjerkreim-Sokndal Intrusion.

Major-element data			
Sample #	B90	B93	B95
SiO_2	49.8	50.6	48.8
TiO_2	4.2	4.1	4.2
Al_2O_3	15.0	14.8	14.7
Fe_2O_3	1.1	1.5	1.9
FeO	11.6	10.7	10.8
MnO	0.2	0.2	0.2
MgO	5.0	5.0	4.5
CaO	6.7	6.3	6.8
Na_2O	2.4	2.5	2.5
K_2O	0.9	1.3	1.2
P_2O_5	0.8	0.7	1.1
LOI	1.2	1.1	2.9
Total	98.9	98.8	99.6
Trace-element (INAA, XRF and MS) data			
ppm (XRF)			
S	1060	1390	1290
V	237	245	248
Cr	50	48	48
Co	33	30	33
Ni	56	56	53
Cu	23	21	22
Zn	133	114	133
Y	36	28	36
Zr	201	155	169
ppm (INAA)			
La	16.0	19.4	24.9
Ce	39.5	46.9	61.0
Nd	25.8	26.9	34.8
Sm	8.0	7.0	9.7
Eu	3.1	2.2	2.8
Tb	1.5	1.1	1.1
Ho		1.0	1.1
Tm	0.7	0.4	n.d.
Yb	3.4	2.2	2.6
Lu	0.5	0.4	0.5
ppm (MS)			
Rb	2.90	13.45	12.82
Sr	446.67	439.95	432.70
Nd	29.474	25.200	50.237
Sm	7.204	5.326	11.232
$^{87}Sr/^{86}Sr_{930}$	0.703021	0.708907	0.713724
$^{143}Nd/^{144}Nd_{930}$	0.511355	0.511377	0.511454

XRF = X-ray fluorescence analysis. INAA = Instrumental neutron activation analysis. MS = Mass spectrometry.

11. MARGINAL JOTUNITES

Fine- to medium-grained, granular rocks are present at several places along the steep, discordant northern margin of the BKSK and occur along the outer margin of a Marginal Zone of layered jotunite and norite that separate cumulates belonging to MCU IA and IB and quartzo-feldspathic gneiss of the envelope (Figure 1). The width of the Marginal Zone varies considerably. In places cumulates are in sharp contact with gneiss, while elsewhere there are up to 100 m of marginal rocks.

The outermost rocks of the Marginal Zone are generally sparsely to markedly porphyritic. Plagioclase phenocrysts are in places up to 2 cm in length and some exhibit skeletal overgrowths. The marginal rocks are generally massive, but where they have their thickest development their grain size increases inwards from the contact with the gneiss, plagioclase phenocrysts become more abundant and a layering defined by textural and modal variations appears. The layering is sub-parallel with the contact with the gneiss, but may be tortuous or corrugated on flat exposures. Along some contacts, the inner contact of the marginal rocks is gradational over ~20 m into coarse-grained ilmenite leuconorite. These features suggest that the marginal rocks belong to a discontinuous Marginal Series and that the outermost granular rocks may be chilled representatives of the magma that was parental to the oldest part of the Layered Series.

The compositions of 3 samples of sparsely-porphyritic, fresh, marginal jotunite analysed by a variety of techniques, are reported in Table 1. These samples, from three different localities, have been selected as the most probable representatives of a parental magma whose composition has been least modified by assimilation of country rock gneiss. They have compositions similar to Ti-rich ferrobasalt and are characterized by high FeO_{total} (11.1-12.9 wt %), MgO in a narrow range between 3.8 and 5.0 and low CaO (5.4-6.7 wt %). They exhibit chondrite-normalized, rare-earth patterns with either a small positive (B90) or no Eu anomaly (B93 and B95), suggesting that previous fractionation or accumulation of plagioclase phenocrysts was very limited. The light REE-enriched patterns of B93 and B95 are almost identical, but sample B90 exhibits a lesser degree of light-REE enrichment, and generally lower REE concentrations. Their compositions are similar to jotunite between anorthosite xenoliths in the plagioclase cumulates of MCU IB (at Tjørn) that likewise has been claimed to be representative of a parental magma (Duchesne and Hertogen, 1988), and also to marginal jotunites of the Hidra Leuconorite (Duchesne *et al.*, 1974).

A total of 7 samples of marginal jotunite have been analysed. They show a considerable variation in $^{87}Sr/^{86}Sr_{930}$ from 0.70302 to 0.71372, corresponding to εSr_{930} of -5.6 to +146.4 respectively. εNd_{930} shows a more limited variation, from +0.3 to -2.7. These isotopic compositions are similar to those of the jotunite dykes present in the region that exhibit Sr_0 of 0.704-0.710 and εNd_{930} in the range +5.4 to -1 (Duchesne, 1990).

12. DISCUSSION

12.1. Fractionation trend, conditions of crystallization, and parental magma composition

The Layered Series of the BKSK reflects crystallization in a periodically-replenished, continuously-fractionating magma chamber into which country rocks were being progressively assimilated. The sequence of crystallization represented by the megacyclic units in the BKSK suggests that they crystallized from a magma which initially had plagioclase ± olivine on the liquidus, was sufficiently enriched in TiO_2 for ilmenite to be an early crystallizing phase, and had a low Ca content in comparison with common basalts, as reflected by the delayed appearance of augite as a cumulus mineral and the relatively sodic cumulus plagioclase. The marginal jotunite has a composition consistent with these features of the Layered Series, supporting the hypothesis that they are representative of a parent magma that was compositionally similar to a Ti-rich ferrobasalt, but poor in augite components.

Anhydrous melting experiments using a fine-grained jotunite from the BKSK shows that such melts have plagioclase as the sole liquidus phase up to ~7 kb at temperatures of 1150-1165°C and oxygen fugacities of between FMQ-2 and FMQ-4 (Vander Auwera and Longhi, 1994). Olivine, ilmenite and Ca-poor pyroxene (which crystallizes together with olivine) appear successively at lower temperatures. At higher pressure (7-13 kb) plagioclase and Ca-poor pyroxene are the liquidus phases. These are joined by ilmenite ~20°C below the liquidus and either olivine (~Fo_{60}) or Ca-rich pyroxene at lower temperatures. Allowing for the low oxygen fugacity in the melting experiments, which stabilizes olivine relative to Ca-poor pyroxene and suppresses magnetite saturation, the crystallization sequence of the jotunite at moderate pressure is similar to that in the BKSK Layered Series.

During much of the early evolution of the chamber the magma which flowed into the chamber generated thick sequences of plagioclase cumulates (leuconorite and minor anorthosite) found in MCUs IA and IB. Plagioclase was successively joined on the liquidus at lower temperatures by ilmenite, orthopyroxene, Ca-rich pyroxene, and magnetite.

Magmas fed into the chamber during the later stages of its evolution, as represented by cumulates towards the bases of MCUs III and IV, crystallized plagioclase, olivine, ilmenite, and magnetite. At slightly lower temperature olivine was replaced by orthopyroxene and at the same time magnetite ceased crystallization. After a further temperature decrease magnetite resumed crystallization, and later, Ca-rich pyroxene and apatite became cumulus phases. MCU II appears to represent an intermediate stage in the evolution of the magma chamber in that the basal rocks are generally plagioclase-ilmenite-magnetite cumulates that are succeeded by plagioclase-orthopyroxene-ilmenite cumulates. Olivine is present only locally at the base of this unit.

The differences in the order of crystallization of cumulus minerals in the lower and upper MCUs may be due to changes in silica activity or oxygen fugacity in the magmas introduced, changes in pressure, or variations in the extent of mixing of inflowing and resident magmas.

Morse (1979a) has shown that silica activity and oxygen fugacity in magmas usually vary in concert, decreases in either would have increased the stability of olivine relative to orthopyroxene. Thus magmas which were emplaced into the chamber during the later stages of its evolution may have had slightly lower a_{SiO2} or f_{O2} than those that intruded earlier. Crystallization of introduced magmas under lower pressure would also favour olivine rather than orthopyroxene as a liquidus or near-liquidus phase. It can be envisaged that the pressure gradually decreased during the evolution of the BKSK magma chamber, due for instance to tectonic exhumation during crustal extension, causing the precipitation of olivine from magmas emplaced during the later stages.

The anomalous appearance of cumulus magnetite in the basal leucotroctolitic cumulates of the upper MCUs, its subsequent absence over significant stratigraphic intervals and later reappearance appears to be the result of ephemeral conditions related to the emplacement of new magma. It is noteworthy that cumulus magnetite exhibits this anomalous behaviour only in MCUs II, III, and IV, where the basal cumulates rest on lower-temperature, magnetite-bearing cumulates. The experiments on hybrids conducted by Sharpe and Irvine (1983), although involving the mixing of olivine boninite (with chromite alone as the liquidus phase) and aluminous tholeiite (with plagioclase as the liquidus phase) compositionally removed from the magmas involved in the BKSK, show the crystallization sequence plagioclase + olivine + spinel (presumed to be enriched in FeO, Fe_2O_3, and TiO_2), plagioclase + orthopyroxene + olivine, plagioclase + orthopyroxene, similar to that observed in the lower parts of MCUs III and IV. Magnetite in the leucotroctolites of the BKSK is believed to have been stabilized by the Cr content of new magma that flowed into the chamber and mixed with relatively small amounts of magnetite-saturated resident magma.

The scarcity of hydrous minerals in the BKSK, except in the most evolved rocks, suggests crystallization from relatively anhydrous magmas. This is supported by the occurrence of hypersolvus alkali feldspar and geothermometry based on coexisting igneous pyroxenes in the mangerite and quartz mangerite which gives 908±70°C and 839±70°C by the Wells (1977) and Wood and Banno (1973) methods respectively (Rietmeijer, 1979). Crystallization temperatures and oxygen fugacities calculated from the estimated compositions of isolated hemo-ilmenite and Ti-magnetite grains that have preserved their magmatic compositions vary from 900-975°C and 10^{-11} atm. (slightly above the Ni-NiO buffer) for cumulates in the upper magnetite-bearing part of MCU IV to 750-800°C and 10^{-16} atm. (slightly below FMQ) for the mangerite (Duchesne, 1972). The ubiquitous occurrence of hemo-ilmenite in the less-evolved cumulates suggests oxygen fugacities of at least 10^{-11} atm. during the early crystallization history of the BKSK. According to Snyder *et al.* (1993), the crystallization of ilmenite prior to magnetite from basaltic magmas is favoured by low oxygen fugacities and that fractionation of ilmenite (and silicates) results in an increase in oxygen fugacities (due to extraction of FeO). They suggest that this may be a general feature of tholeiitic magmas. The experiments on which these conclusions are based were conducted at varying oxygen fugacities less than FMQ, but have little relevance for the T-fO_2 evolution of the BKSK magmas since the synthesized ilmenite had a low hematite content.

Morse (1979b) has discussed how plagioclase fractionation is affected by augite saturation and showed that fractional crystallization of plagioclase alone leads to a slow rate of change in the anorthite content of residual liquids. He has suggested that the augite effect may be responsible for the limited range in feldspar compositions in augite-poor, massif-type anorthosites (Morse, 1982). The BKSK seems to be a natural example of these principles. Both the phase layering and the composition of marginal rocks suggest derivation of the cumulates from augite-poor magmas and cryptic variation in plagioclase composition is inconspicuous in the BKSK Layered Series until Ca-rich pyroxene joins the fractionating assemblage. In addition, the occurrence of fayalite (Fo_4) with plagioclase of around An_{12} shows that residual liquids were relatively Ca-rich due to a low effective distribution coefficient for An_{plag}/An_{liq} (Morse, 1979). Recent experiments on a jotunite from the BKSK (near Tjørn, Duchesne and Hertogen, 1988) show that the distribution coefficient for liquidus plagioclase is inversely related to pressure (Vander Auwera and Longhi, 1992; Longhi *et al.*, 1993). Crystallization of the BKSK at elevated pressure will therefore have promoted fractionation of relatively sodic plagioclase and enhanced the augite effect.

12.2. Evidence for magma stratification and the mechanism of magma replenishment

The unfolding of the synclinal structure of the BKSK intrusion to bring the layering to an approximately horizontal disposition results in a broad, saucer-shaped magma chamber. The breadth/thickness ratio of the Bjerkreim lobe of the magma chamber at the present erosion level is about 3:1 (lateral extent ~24 km, thickness ~8 km), but for each individual magma influx it would have been much greater. For example, when the final batch of magma began to crystallize MCU IVb, this ratio must have been about 8:1.

The magma in the BKSK chamber is believed to have been strongly compositionally stratified, at least by the time that MCU III was crystallizing. The repeated influx of relatively dense, hot, primitive magma along the floor successively elevated the more buoyant, cooler, evolved resident magma, giving rise to compositional stratification from the base up. Extensive roof melts (see section 12.3) were buoyant and remained at the top of the chamber, leading to stratification downwards from the roof.

The jotunitic magmas that periodically discharged into the BKSK chamber had densities estimated from the partial molar volumes of Lange and Carmichael (1987) and Toplis *et al.* (1994) of ~2.74-2.77 g cm^{-3} at 1150°C and 5 kb, assuming 0.0-0.5 % H_2O (Vander Auwera and Longhi, 1994). Fractional crystallization of plagioclase from such magmas must have increased the densities of residual liquids. Modelling of the density evolution shows, however, that extraction of plagioclase, ilmenite and orthopyroxene or olivine in reasonable proportions could eventually have led to a decrease in density.

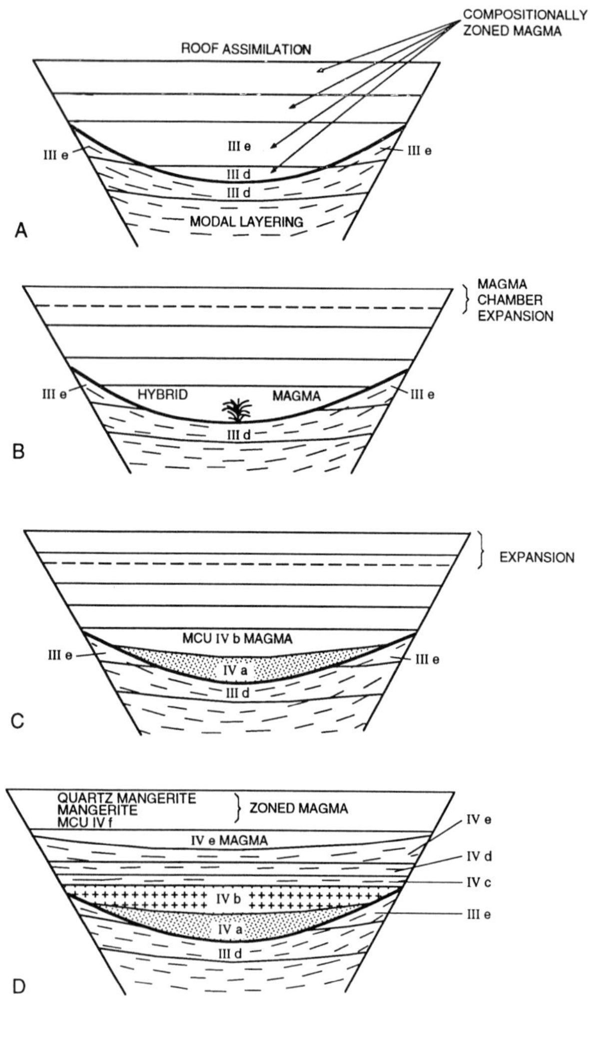

The geometry of phase contacts between MCUs III and IV (Figure 3) was explained by Nielsen and Wilson (1991) by crystallization of compositionally-stratified magma against a saucer-shaped floor (Figure 7). Zone d of MCU III was crystallizing from the basal layer of the compositionally-stratified magma in the deepest, central parts of the chamber while at the same time MCU IIIe was crystallizing on the flanks from a less dense, more evolved magma layer (Figure 7A). The absence of MCU IIIe in the central part of the magma chamber is therefore explained by non-deposition; this model also accounts for the symmetrical disposition of this unit on both flanks of the saucer-shaped chamber. It was at the stage of magma chamber evolution illustrated in Figure 7A that new magma entered the chamber to form MCU IV (Figure 7B).

The compositional reversal through MCU IVa (Figures 5 and 6) reflects crystallization during magma influx and mixing between new, primitive magma and residual, contaminated magma. At the lowest point of the saucer-shaped chamber, reversals in mineral chemistry and Sr-isotopes would take place simultaneously when the new magma entered the chamber. In Figure 6B, however, it is apparent that there is a reversal in mineral chemistry at the top of MCU IIIe, below a change in cumulus assemblage (from plagioclase-orthopyroxene-clinopyroxene-ilmenite-magnetite-apatite cumulates to plagioclase ± ilmenite ± orthopyroxene cumulates) that defines the base of MCU Iva. The systematic isotopic regression sets in about half way through zone IVa. This can be explained by the location of the Storeknuten section of Figure 6 on the flank of the saucer-shaped magma chamber floor (Figures 3 and 7). Space for the new magma was made available by the roof being pushed upwards and the magma chamber expanding laterally. While the new, dense magma started to mix with the basal layer of the resident magma, the overlying magma layers were elevated.

In the Storeknuten profile on the flank of the saucer-shaped chamber, hybrid magma first reached the crystallization front after the residual magma column had passed by and produced the compositional reversal in mineral chemistry at the top of MCU IIIe. The first hybrid magma that reached the Storeknuten profile produced a change in cumulus assemblage but did not change the Sr-isotope ratio, indicating that the hybrid was dominated by residual magma. Incomplete mixing and/or kinetic effects resulted in the crystallization of plagioclase and orthopyroxene that were not in equilibrium with each other in the lower part of MCU IVa (Figure 6B). The start of the gradual decrease in Sr isotope ratios in Figure 6B reflects crystallization from hybrid magma with a systematically increasing proportion of new magma. Eventually the hybrid reached a composition with the lowest Sr isotope ratio (0.7049) in the entire Layered Series at the top of MCU IVa. This marks the cessation of magma influx after which the basal magma layer in the chamber crystallized the most primitive cumulates in BKSK - the leucotroctolites in MCU IVb.

***Figure** 7. (facing page). Schematic illustration of the evolution of the Bjerkreim-Sokndal magma chamber during the time interval represented by zones IIId-IVf (Figure 2). The vertical scale is considerably exaggerated in these sketches. Based on Nielsen and Wilson (1991). A) Crystallization of the upper part of MCU III from a compositionally-zoned magma column. Zone IIId is crystallizing from the central, lowest part of the basinal floor while zone IIIe crystallizes towards the margins from more evolved and less dense magma. B) Emplacement of dense, more primitive magma as a fountain in which mixing produces hybrid magma. The magma column is elevated and higher-temperature magma migrates up the inwardly-sloping floor, producing the gradual compositional regression in zone IIIe. The decrease in initial Sr isotope ratio seen in zone IVa reflects an increasing proportion of new magma in the basal hybrid. The roof of the chamber is being lifted to accommodate the new magma. C) Hybrid magma nearly fills the basin on the floor of the magma chamber and crystallizes zone IVa, producing an almost horizontal floor. D) Zone IVb and subsequent zones crystallize over the entire floor of the chamber.*

The fact that MCU IVa is thickest in the central, axial region of the magma chamber and wedges out towards the northern flank (Figures 3 and 7) is another result of the saucer-shaped form of the magma chamber floor. Magma influx and mixing took place in the lowest part of the chamber; the northern flank did not become flooded until the final stages of hybridization had been reached. The fact that the olivine-bearing unit MCU IVb has fairly constant thickness (~100 m) in the intrusion implies that the magma chamber floor was very close to horizontal during its crystallization; the hybrid magma producing MCU IVa must have filled in the central part of the saucer-form. If the floor was horizontal when MCU IVb started to crystallize, the lateral variation in thickness of MCU IVa can be used to estimate the slope of the floor. MCU IVa gradually wedges out from 90 m in profile D to between profiles B and C in Figure 3, a lateral distance of about 6 km. This implies an apparent slope of about 2° in the exposed section; the true slope must have been somewhat greater than this. Calculations based on geochemical modelling, the thickness of cumulate stratigraphy repeated, and Sr isotope ratios indicate that the new magma influx at the base of MCU IV had a thickness of 350-500 m in the Storeknuten section, and that MCU IVb represents about 20-30 % crystallization of this influx (Jensen *et al.*, 1993).

Mixing of new and resident magmas probably took place in a turbulent fountain (Campbell and Turner, 1986) such that the proportion of new to residual magma in the hybrid gradually increased.

12.3. Role of assimilation

The remarkable sympathetic variation in the composition of cumulus minerals and Sr_0 present in the BKSK cumulates (Figure 5) suggests that assimilation of country rocks was continuous during crystallization of the Layered Series. Furthermore, this relationship implies coupling of fractional crystallization and assimilation. In the BKSK the rate of fractional crystallization may have been governed by assimilation. The abundance of country-rock xenoliths preserved in parts of the Layered Series suggests that assimilation occurred as a result of the melting of blocks of gneiss derived from the roof of the chamber while they were being transported to the floor.

The role of assimilation in the BKSK magma chamber has been assessed quantitatively by Nielsen *et al.* (1996). They assumed that the parental magma had an Sr_0 of 0.7049 (the lowest ratio measured in the Layered Series) and 600 ppm Sr, whereas the contaminant had an Sr_0 of 0.7196 and 160 ppm Sr (based on a charnockitic migmatite at 947±26 Ma, Versteeve, 1975). These end-member compositions give about 28 % contamination by gneiss just before the final major magma influx, and 56 % during the later stages of the life of the chamber as represented by the upper part of MCU IV the TZ, mangerite, and quartz mangerite. Assimilation of >50 % gneiss is improbable on a large scale and this result may cast doubt on the choice of the contaminant, which could have been more radiogenic than the average charnockitic migmatite. In a saucer-shaped magma chamber most of the heat will, however, be released through the roof. Buoyant, relatively low-temperature country rock melts can remain at the roof and be chemically decoupled from crystallization at the floor. Mixing of roof melts and liquids produced largely by fractional crystallization can have resulted in the high initial Sr ratios observed in rocks in the uppermost part of the BKSK.

12.4. Origin of the Bjerkreim-Sokndal parental magmas

The ultimate origin of the magmas parental to the BKSK, and the Rogaland jotunitic intrusions generally, is still unclear. It is apparent that they could not have been derived directly from the mantle, since they were saturated in plagioclase when emplaced into the crust. An alternative hypothesis is removal of olivine, aluminous pyroxene and possibly garnet from mantle-derived basalt at or close to the base of thickened continental crust. This is sufficient to explain the differentiated composition and the poverty in diopside and anorthite components. One of us favours the view that the parental magmas were generated by batch melting of noritic rocks in the lower crust (Duchesne, 1990), a model that is supported by the results of melting experiments on high-Al basalts (Baker and Eggler, 1987). In both of these hypotheses the composition of the ferrobasaltic liquids would have essentially been controlled by equilibrium with aluminous pyroxenes and possibly garnet, but not with plagioclase.

A major unresolved problem in modelling the origin of the parental magmas is the variable isotopic composition of the marginal jotunites of the BKSK. Interestingly, the Holocene Snake River Plain ferrobasalts that are compositionally similar to the Rogaland jotunites, have high $^{87}Sr/^{86}Sr$ and similar variations (Leeman *et al.*, 1976), though they seem to be more systematically related to the degree of differentiation. Leeman *et al.* (1976) suggested that this may be due to selective contamination by small amounts of very radiogenic crustal materials, an *ad hoc* explanation that could also be applied to the BKSK, but that finds no support in the isotopic composition of the cumulates.

The nature of any relationship which may exist between the ferrobasalts parental to the BKSK, the monzonoritic minor intrusions and the anorthosite massifs also remains obscure. The lack of negative Eu anomalies in jotunite indicates that the ferrobasalts cannot be magmas residual after extensive plagioclase fractionation. One of us (J.C. Duchesne) has suggested that the ferrobasalt magmas may have resulted from partial melting of crustal rocks due to the thermal effects of the diapiric emplacement of the anorthosites. However, a more direct relationship is indicated by the jotunite chilled margin of the Hidra leuconorite and the thickness of leuconoritic cumulates present in the BKSK. The very aluminous pyroxene megacrysts suggest that the initial crystallization of the anorthosite massifs occurred at elevated pressures. However, the Cr contents (600-950 ppm) of the pyroxenes are difficult to reconcile with crystallization from ferrobasaltic magmas as represented by the jotunite, which generally contains <50 ppm Cr (Duchesne *et al.*, 1985 and Table 1), using generally accepted values for the partition coefficient. However, the distribution coefficient is very dependant on temperature and oxygen fugacity (Barnes, 1986). Recent experiments on a jotunite from the BKSK have shown a change in $D_{Cr(opx/melt)}$ with pressure from <3 at 1 atm to 14 at 10 kb (Vander Auwera *et al.*, 1993).

13. CONCLUSIONS

The BKSK shows that members of the anorthosite kindred (andesine anorthosite, leuconorite, troctolite, norite, gabbronorite, mangerite, and quartz mangerite) can be generated by differentiation of Ti-rich, diopside-poor basic magmas with low Mg# within crustal magma chambers. These rock types in the BKSK Layered Series were generated principally by fractional crystallization, accompanied by assimilation of roof gneisses, in a periodically-replenished magma chamber.

There is a record in the Layered Series of the BKSK of 6 major episodes of magma emplacement, each accompanied by lateral and upward enlargement of the chamber. At least two of these episodes were relatively persistent and were associated with mixing of new and lower-temperature resident magmas. Mixing resulted in hybrid magmas and is represented by mappable sequences of cumulates that exhibit regressive cryptic variations.

The generalized order of appearance of the main minerals during uninterrupted fractional crystallization at a late stage in the evolution of the magma chamber was plagioclase + olivine + he-rich ilmenite + Cr-bearing magnetite, orthopyroxene (- magnetite), Ti-poor magnetite, Ca-rich pyroxene + apatite, pigeonite, alkali feldspar, Fe-rich olivine, quartz. The earliest stages were characterized by the sequence plagioclase, he-rich ilmenite, orthopyroxene, Ca-rich pyroxene, Ti-poor magnetite. The latter order of crystallization is only fully developed in MCU IB; MCUs IA and II illustrate only the early part of the sequence. The differences are believed to be due to several causes. The earlier appearance of cumulus apatite and magnetite in the upper MCUs was probably due to entrainment into the inflowing magmas of the more-evolved magmas residing in the chamber that were already enriched in P_2O_5 and Fe_2O_3 due to previous fractional crystallization. The absence of cumulus olivine in the early history of the chamber may have been due to the emplacement of magmas with higher silica activities due to a higher degree of previous fractional crystallization at higher pressure, or an increased degree of mixing of inflowing and resident magmas.

The discordant geometry of phase contacts in the MCUs that results in the lateral wedging out of certain cumulus assemblages, is interpreted as due to the persistence of a vertical chemical zonation in the column of magma which existed in the chamber and the development of a basinal or half-basinal chamber floor.

The remarkable sympathetic variation of mineral chemistry and initial Sr ratios that exists in the upper part of the BKSK Layered Series suggests that assimilation of country rocks was continuous and that there also was a fairly constant relationship between rates of assimilation and fractional crystallization. Assimilation was probably promoted by the incorporation into the chamber of large numbers of xenoliths of gneiss from the roof.

14. REFERENCES

Baker, D.R., & Eggler, D.H., 1987. Compositions of anhydrous and hydrous melts coexisting with plagioclase, augite and olivine or low-Ca pyroxene from 1 atm to 8 kbar: Application to the Aleutian volcanic centre of Atka. *Am. Miner.* **72**, 12-28.

Barnes, S.J., 1986. The distribution of chromium among orthopyroxene, spinel and silicate liquid at atmospheric pressure. *Geochim.Cosmochim. Acta* **50**, 1889-909.

Campbell, I.H., & Turner, J.S., 1986. The influence of viscosity on fountains in magma chambers. *J. Petrology* **27**, 1-30.

Demaiffe, D., Duchesne, J.C., & Hertogen, J., 1979. Trace element variations and isotopic composition of charnockitic acidic rocks related to anorthosites (Rogaland, SW Norway). In Ahrens, L.H. (ed.) *Origin and Distribution of the Elements.* Pergamon Press, 417-29.

Demaiffe, D., & Michot, J., 1985. Isotope geochronology of the Proterozoic crustal segment of southern Norway. In Tobi, A.C., & Touret, J.L.R. (eds.) *The Deep Proterozoic Crust in the North Atlantic Provinces.* NATO ASI Ser. C 158, D. Reidel Publishing Co.: Dordrecht, 411-33.

Duchesne, J.C., 1970. Sur la provenance de xénolithes anorthositiques dans le massif de Bjerkreim-Sogndal (Norvège). *Ann. Soc. Géol. Belg.* **93**, 643-56.

Duchesne, J.C., 1972. Iron-titanium oxide minerals in the Bjerkreim-Sogndal massif, South-western Norway. *J. Petrology* **13**, 57-81.

Duchesne, J.C., 1978. Quantitative modelling of Sr, Ca, Rb, and K in the Bjerkreim-Sogndal Layered Lopolith (S.W. Norway). *Contr. Miner. Petrol.* **66**, 175-84.

Duchesne, J.C., 1987. The Bjerkreim-Sokndal massif. In Maijer, C., & Padget, P. (eds.) *The geology of Southernmost Norway.* Norges geol. unders. Spec. Pub. **1**, 56-9.

Duchesne, J.C., 1990. Origin and evolution of monzonorites related to anorthosites. *Schweiz. Mineral. Petrogr. Mitt.* **70**, 189-98.

Duchesne, J.C., & Hertogen, J., 1988. Le magma parental du lopolithe de Bjerkreim-Sokndal (Norvège méridionale). *C.R. Acad. Sci. Paris* **90**,45-48.

Duchesne, J.C., & Maquil, R., 1987. The Egersund-Oga massif. In Maijer C, & Padget P (eds.) The geology of Southernmost Norway. *Norges geol. unders. Spec. Pub.* **1**, 50-6.

Duchesne, J.C., & Michot, J., 1987. The Rogaland intrusive masses: Introduction. In Maijer C, & Padget P (eds.) *The geology of Southernmost Norway.* Norges geol. unders. Spec. Pub. **1**, 48-50.

Duchesne, J.C., Denoiseux, B., & Hertogen, J., 1987. The norite-mangerite relationships in the Bjerkreim-Sokndal layered lopolith (S.W. Norway). *Lithos* **20**, 1-17.

Duchesne, J.C., Roelandts, I., Demaiffe, D., Hertogen, J., Gijbels, R., & De Winter, J., 1974. Rare-earth data on monzonoritic rocks related to anorthosites and their bearing on the nature of the parental magma of the anorthositic series. *Earth Planet Sci. Lett.* **24**, 325-35.

Duchesne, J.C., Roelandts, I., Demaiffe, D., & Weis, D., 1985. Petrogenesis of monzonoritic dykes in the Egersund-Ogna anorthosite (Rogaland, S.W. Norway): trace elements and isotopic (Sr, Pb) constraints. *Contr. Miner. Petrol.* **90**, 214-25.

Duchesne, J.C., Wilmart, E., Demaiffe, D., & Hertogen, J., 1989. Monzonorites from Rogaland (Southwest Norway): a series of rocks coeval but not comagmatic with massif-type anorthosites. *Precamb. Res.* **7**, 111-28.

Duchesne, J.C., Schärer, U., & Wilmart, E. 1993. A 10 Ma period of emplacement for the Rogaland anorthosites, Norway: evidence from U-Pb ages. *Terra Nova* **5**, 64.

Glazner, A.F., 1994. Foundering of mafic plutons and density stratification of continental crust. *Geology* **22**, 435-8.

Hermans, G.A.E., Tobi, A.C., Poorter, P.E., & Maijer, C., 1975. The high-grade metamorphic Precambrian of the Sirdal-Ørsdal area, Rogaland/Vest-Agder, SW Norway. *Norges geol. unders.* **318**, 51-74.

Holdam, H.K., 1990. *Bjerkreim-Sokndal intrusjonen, Sydnorge: En undersøgelse af den nordøstlige flankes petrologi og mineralkemi.* Unpubl. Cand. scient. thesis, Univ. of Aarhus, Denmark.

Jansen, J.B.H., Blok, R.J.P., Bos, A., & Scheelings, M., 1985. Geothermometry and geobarometry in Rogaland and preliminary results from the Bamble area, S. Norway. In Tobi, A.C., & Touret, J.L.R. (eds.) *The Deep Proterozoic Crust in the North Atlantic Provinces.* NATO ASI Ser. C 158, D. Reidel Publishing Co.: Dordrecht, 499-516.

Jensen, J.C., Nielsen, F.M., Duchesne, J.C., Demaiffe, D., & Wilson, J.R., 1993. Magma influx and mixing in the Bjerkreim-Sokndal layered intrusion, South Norway: evidence from the boundary between two macrocyclic units at Storeknuten. *Lithos* **29**, 311-25.

Krause, H., Gierth, E., & Schott, W., 1985. Ti-Fe deposits in the South Rogaland igneous complex with special reference to the Åna-Sira anorthosite massif. *Norges geol. unders. Bull.* **402**, 25-37.

Lange, R.A., & Carmichael, I.S.E., 1987. Densities of Na_2O-K_2O-CaO-MgO-FeO-Fe_2O_3-Al_2O_3-SiO_2 liquids: New measurements and derived partial molar volumes. *Geochim. Cosmochim. Acta* **51**, 2931-46.

Leeman, W.P., Vitaliano, C.J., & Prinz, M., 1976. Evolved lavas from the Snake River Plain: Craters of the Moon National Momument, Idaho. *Contr. Miner. Petrol.* **56**, 35-60.

Longhi, J., Fram, M.S., Vander Auwera, J., & Montieth, J.N., 1993. Pressure effects, kinetics, and reology of anorthositic and related magmas. *Am. Miner.* **78**, 1016-30.

Maijer, C., 1987. The metamorphic envelope of the Rogaland intrusive complex. In Maijer, C., & Padget, P. (eds.) *The geology of Southernmost Norway.* Norges geol. unders. Spec. Pub. **1**, 68-73.

Michot, J., 1961. The anorthositic complex of Haaland-Helleren. *Norsk geol. Tidsskrift* **41**, 157-72.

Michot, J., & Michot, P., 1970. The problem of the anorthosites. The South Rogaland igneous complex (South Western Norway). In: Isachsen, Y.W. (ed.) *Origin of anorthosites and related rocks*. New York State Mus. Sci. Ser. Mem. **18**, 399-410.

Michot, P., 1960. La géologie de la catazone: le problème des anorthosites, la palingenèse basique et la tectonique catazonale dans le Rogaland méridionale (Norvège méridionale). *Norges geol. unders.* **212**, 1-54.

Michot, P., 1965. Le magma plagioclasique. *Geol. Rundschau* **54**, 956-76.

Morse, S.A., 1979a. Reaction constants for En-Fo-Sil equilibria: An adjustment and some applications. *Am. J. Sci.* **279**, 1060-9.

Morse, S.A., 1979b. Influence of augite on plagioclase fractionation. *J. Geol.* **87**, 202-8.

Nielsen, F.M., & Wilson, J.R., 1991. Crystallization processes in the Bjerkreim-Sokndal layered intrusion, south Norway: evidence from the boundary between two macrcyclic units. *Contr. Miner. Petrol.* **107**, 403-14.

Nielsen, F.M., Campbell, I.H., McCulloch, M., & Wilson, J.R. (1996). A strontium isotopic investigation of the Bjerkreim-Sokndal layered intrusion, southwest Norway. *J. Petrology* **37, 171-94**.

Paludan, J., Hansen, U.B., & Olesen, N.Ø. (1994). Structural evolution of the Precambrian Bjerkreim-Sokndal Intrusion, South Norway. *Norsk geol. Tidsskrift.* **74**, 185-98.

Pasteels, P., Michot, J., & Lavreau, J., 1970. Le complexe éruptiv du Rogaland méridional (Norvège). Signification pétrogénétique de la farsundite et de la mangérite quartzique des unités orientales: arguments géochronologiques et isotopiques. *Ann. Soc. Géol. Belg.* **93**, 453-76.

Pasteels, P., Demaiffe, D., & Michot, J., 1979. U-Pb and Rb-Sr geochronology of the eastern part of the South Rogaland igneous complex. *Lithos* **12**, 199-208.

Poorter, R.P.E., 1972. Palaeomagnetism of the Rogaland Precambrian (Southwestern Norway). *Phys. Earth Planet. Interiors* **5**, 167-76.

Rietmeijer, F.J.M., 1979. Pyroxenes from iron-rich igneous rocks in Rogaland, S.W. Norway. *Geol. Ultraiectina* **21**, 341 pp.

Rietmeijer, F.J.M., 1984. Pyroxene (re-)equilibration in the Precambrian terrain of SW Norway between 1030-990 and reinterpretation of events during regional cooling. *Norsk geol. Tidsskrift* **64**, 7-20.

Roelandts, I., & Duchesne, J.C., 1979. Rare-earth elements in apatite from layered norites and iron-titanium oxide ore bodies related to anorthosites (Rogaland, S.W. Norway). In Ahrens, L.H. (ed.) *Origin and Distribution of the Elements*. Pergamon Press, 199-212.

Schiellerup, H., 1991. *Bjerkreim-Sokndal intrusjonen. En petrologisk, mineral- og geokjemisk undersøgelse af Bjerkreimloben i området omkring Helleland.* Unpubl. Cand. scient. thesis, Univ. of Aarhus, Denmark.

Schärer, U., Wilmart, E., & Duchesne, J.C., 1992. U-Pb age constraints on the emplacement of Rogaland anorthosites. *Abstract, IGCP 290 1992 meeting, Rogaland, Norway.*

Sharpe, M.R., & Irvine, T.N., 1983. Melting realtions of two Bushveld chilled margin rocks and implications for the origin of chromitite. *Carnegie Inst. Yrbk.* **82**, 295-300.

Smithson, S.B., & Ramberg, I.B., 1979. Gravity interpretation of the Egersund anorthosite complex, Norway: Its petrological and geothermal significance. *Geol. Soc. Am. Bull.* **90**, 199-204.

Snyder, D., Carmichael, I.S.E., & Wiebe, R.A., 1993. Experimental study of liquid evolution in an Fe-rich, layered mafic intrusion: constraints of Fe-Ti oxide precipitation on the T-fO_2 and T-ρ paths of tholeiitic magmas. *Contr. Miner. Petrol.* **113**, 73-86.

Sundvoll, B., 1987. The age of the Egersund dyke swarm SW Norway: Some tectonic implications. *Terra Cognita* **7**, 180.

Tobi, A.C., Hermans, G.A.E.M., Maijer, C., & Jansen, J.B.H., 1985. Metamorphic zoning in the high-grade Proterozoic of Rogaland - Vest Agder, SW Norway. In Tobi, A.C., & Touret, J.L.R. (eds.)

The Deep Proterozoic Crust in the North Atlantic Provinces. NATO ASI Ser. C 158, D. Reidel Publishing Co., Dordrecht, 477-97.

Toplis, M.J., Libourel, G., & Carroll, M.R., 1994. The role of phosphorus in crystallisation processes of basalts: An experimental study. *Geochim. Cosmochim. Acta* **58**, 797-810.

Vander Auwera, J., & Longhi, J., 1992. Phase equilibria and fractionation paths from 1 atm up to 16 Kb of the monzonoritic Tjørn chilled facies: Implications for the Bjerkreim-Sokndal Lopolith (Southern Norway). *Abstract IGCP 290 Origin of anorthosites, Moi meeting 1992.*

Vander Auwera, J., Longhi, J., & Duchesne, J.C., 1993. Jotunites from the Rogaland Province (Norway): constraints from experimental data and the partitioning of Sr (plag/melt) and Cr (opx/melt). *EOS* **74**, 659.

Vander Auwera, J., & Longhi, J. (1994). Experimental study of a jotunite (hypersthene monzodiorite): constraints on the parent magma composition and crystallisation conditions (P, T, fO_2) of the Bjerkreim-Sokndal layered intrusion (Norway). *Contr. Miner. Petrol.* **118**, 60-78.

Versteeve, A.J., 1975. Isotope geochronology in the high-grade metamorphic Precambrian of south-western Norway. *Norges geol. unders.* **318**, 1-50.

Wells, P.R.A., 1977. Pyroxene thermometry in simple and complex systems. *Contr. Miner. Petrol.* **62**, 129-39.

Wiebe, R.A., 1984. Co-mingling of magmas in the Bjerkreim-Sogndal lopolith (S.W. Norway): evidence for the composition of residual liquids. *Lithos* **17**, 171-88.

Wielens, J.B.W., Andriessen, P.A.M., Boelrijk, N.A.I.M., Hebeda, E.H., Priem, H.N.A., Verdurmen, E.A.T., & Verschure, R.H., 1980. Isotope geochronology in the high-grade metamorphic Precambrian of southwestern Norway: New data and interpretations. *Norges geol. unders.* **359**, 1-30.

Wilmart, E., & Duchesne, J.C., 1987. Geothermobarometry of igneous and metamorphic rocks around the Åna-Sira anorthosite massif: Implications for the depth of emplacement of the South Norwegian anorthosites. *Norsk geologisk Tidsskrift* **67**, 185-96.

Wilmart, E., Demaiffe, D., & Duchesne, J.C., 1989. Geochemical constraints on the genesis of the Tellnes ilmenite deposit, Southwest Norway. *Econ. Geol.* **84**, 1047-56.

Wood, B., & Banno, S., 1973. Garnet-orthopyroxene and orthopyroxene-clinopyroxene relationships in simple and complex systems. *Contr. Miner. Petrol.* **42**, 109-24.

LAYERED INTRUSIONS
R.G. Cawthorn (editor)

Layered Intrusions of the Duluth Complex, Minnesota, USA

J.D. Miller, Jr.[a] and E.M. Ripley[b]

[a]Minnesota Geological Survey, University of Minnesota, 2642 University Ave., St. Paul, Minnesota 55114, U.S.A.

[b]Department of Geological Sciences, Indiana University, Bloomington, Indiana, 47405, U.S.A.

Abstract

The Duluth Complex and associated subvolcanic intrusions comprise a large (5,000 km^2) intrusive complex in northeastern Minnesota that was emplaced into comagmatic volcanics during the development of the 1.1 Ga Midcontinent rift in North America. In addition to anorthositic and felsic intrusions, the Duluth Complex is composed of many individual mafic layered intrusions of tholeiitic affinity. The cumulate stratigraphies and cryptic variations of six of the better exposed and better studied intrusions are described here to demonstrate the variability in their cumulus mineral paragenesis. Although the general paragenetic sequence is:

$$Ol(\pm Pl) \rightarrow Pl+Ol \rightarrow Pl+Cpx+Ox(\pm Ol\pm Opx) \rightarrow Pl+Cpx+Ox+Ap(\pm Ol)$$

considerable differences exist among the six intrusions in the relative order and timing of cumulus arrivals, most notably with regard to augite (Cpx) and Fe-Ti oxide (Ox).

The various cumulate stratigraphies and cryptic variations represented by the six intrusions described here largely reflect differences in the degree of open-system behaviour to recharge, eruption, and country-rock assimilation, but also may have been influenced by differences in parent magma composition, in the efficiency of fractional crystallization, and in the conditions of crystallization (*e.g.* P_{total}, f_{O_2}, P_{H_2O}). The Sonju Lake intrusion, wherein cumulus augite arrived before ilmenite, formed by essentially closed-system fractional crystallization. However, its compositional evolution may have been affected to an uncertain degree by assimilation of a granitic hanging-wall. The monotonous, thick troctolitic cumulate sequences of the lower parts of the South Kawishiwi and Partridge River Intrusions appear to represent the effects of frequent magma recharge coupled with *in situ* (or boundary layer) crystallization though other interpretations have been proposed. Extensive Cu-Ni sulphide mineralization at the base of these intrusions is attributed to country-rock contamination of the earliest intruded magmas. The cyclical progression of cumulates in the Layered Series at Duluth, wherein cumulus augite and oxide arrived nearly simultaneously, formed in a moderately open system characterized by periodic eruption and recharge. The Wilder Lake Intrusion is different from other intrusions in that cumulus ilmenite appears before augite and that olivine and augite composition define an inverted cryptic variation. The latter phenomenon may be related to a strong upward gradation toward lesser amounts of trapped liquid in the cumulates. Finally, the incomplete cumulate stratigraphy of the Bald Eagle Intrusion, which is composed of approximately equal thicknesses of troctolite (PO) and gabbro (PAO) adcumulates, is unique in that Fe-Ti oxide did not arrive as a cumulus phase despite prolonged crystallization of cumulus augite. These different cumulus parageneses probably produced a variety of derivative magmas, which may have contributed in part to the compositional diversity of the Midcontinent rift volcanics.

1. INTRODUCTION

The Duluth Complex (DC) is an immense igneous complex that includes over a dozen discrete mafic intrusions exposed over a 5,000 km^2 area in northeastern Minnesota (Figure 1). Whereas earliest studies (Grout, 1918a, 1918b) considered the DC to be one intrusive entity, subsequent mapping (Grout and Schwartz, 1939; Grout *et al.*, 1959; Taylor, 1964; Green *et al.*, 1966; Phinney, 1972*b*; Davidson, 1972; Bonnichsen, 1971, 1972) revealed its multiple intrusive nature as summarized by Weiblen and Morey (1980). Because over half the exposure area of the Duluth Complex is obscured by glacial drift (Figure 1), further insight into the geology of the Duluth Complex has been more recently provided by high-resolution aeromagnetic data (Chandler, 1990). Most recently, U-Pb dating of zircons from various intrusions of the Duluth Complex and related intrusions has determined (Paces and Miller, 1993) that episodic intrusive activity spanned an 11 Ma period (1107-1096 Ma), or close to half the 23 Ma period of magmatism associated with Midcontinent rift (1109-1086 Ma), of which the Duluth Complex is the major intrusive component.

Despite the recognition and general characterization of many individual intrusions comprising the Duluth Complex, petrologic studies delineating the emplacement and crystallization histories of these bodies have been conducted on only a few such intrusions and, with only a few exceptions, at a rudimentary level. Nevertheless, these studies show significant variability in the internal structure and stratigraphic changes in mode, texture, and composition within individual intrusions that in turn indicate (1) comparable variability in their parental magma compositions, (2) the physical conditions of their emplacement and crystallization, (3) the efficiency of their fractional crystallization, and (4) most signficantly, their openness to recharge, eruption, and assimilation. In this paper, we attempt to highlight this varibility by summarizing the distinct cumulate stratigraphies of six mafic layered intrusions that compose part of the Duluth Complex. Given the continued state of uncertainty as to how such mafic system physically and chemically evolve (Jaupart and Tait, 1995), our intention here is not so much to explain this variability as it is to emphasize that it is an inherent characteristic of the Duluth Complex.

Figure 1. *(facing page) Generalized geology of northeastern Minnesota. Units outlined in dashed lines indicate areas where the bedrock geology is poorly exposed due to glacial till cover and is largely interpreted from aeromagnetic data (Chandler, 1990; Meints* et al., *1993). The exposed extent of the Beaver Bay Complex is distinguished from other areas of subvolcanic intrusions by a darker shade. Locations of the layered mafic intrusions described in this chapter are indicated (SLI - Sonju Lake Intrusion, PRI - Partridge River Intrusion, SKI - South Kawishiwi Intrusion, BEI - Bald Eagle Intrusion, WLI - Wilder Lake Intrusion, DLS - layered series at Duluth). The general internal structure of the SKI and PRI bodies is denoted by strike and dip symbols; the internal structure of the other intrusions is shown in Figures 2, 9, 11, 13, and 15. NSVG - North Shore Volcanic Group. Inset shows the position of the Duluth Complex in relation to the igneous rocks of the Midcontinent rift.*

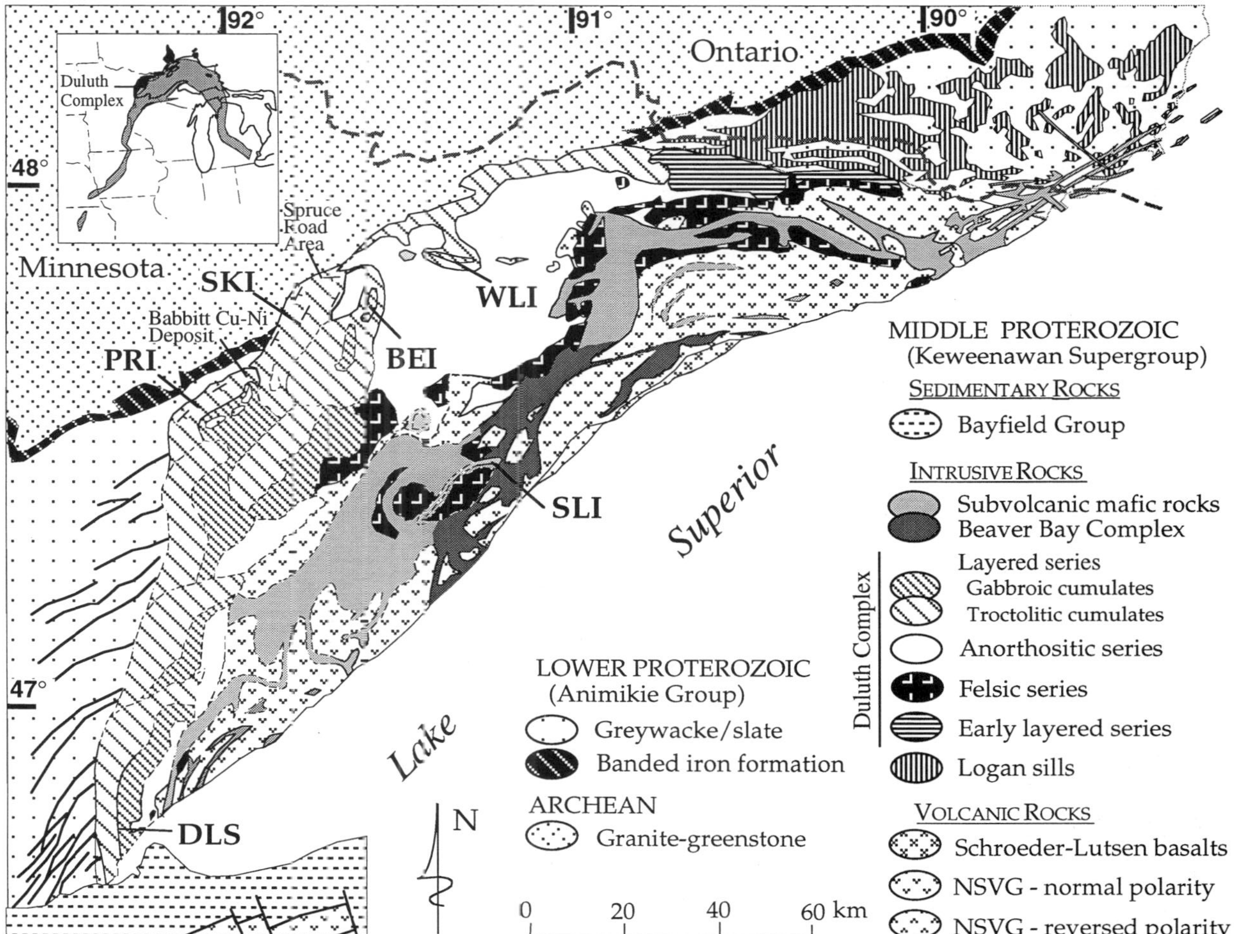
92°
91°
90°
Ontario
48°
47°
Duluth Complex
Minnesota
Spruce Road Area
SKI
Babbitt Cu-Ni Deposit
PRI
BEI
WLI
SLI
DLS
Lake
Superior
N
MIDDLE PROTEROZOIC
(Keweenawan Supergroup)
SEDIMENTARY ROCKS
Bayfield Group
INTRUSIVE ROCKS
Subvolcanic mafic rocks
Beaver Bay Complex
Duluth Complex
Layered series
Gabbroic cumulates
Troctolitic cumulates
Anorthositic series
Felsic series
Early layered series
Logan sills
VOLCANIC ROCKS
Schroeder-Lutsen basalts
NSVG - normal polarity
NSVG - reversed polarity
LOWER PROTEROZOIC
(Animikie Group)
Greywacke/slate
Banded iron formation
ARCHEAN
Granite-greenstone
0
20
40
60 km

2. GEOLOGIC SETTING

The mafic intrusive rocks and comagmatic flood basalts underlying most of northeastern Minnesota were emplaced during the development the Middle Proterozoic (Keweenawan) Midcontinent rift (MCR). The MCR can be traced by exposure in the Lake Superior region and by its geophysical signature elsewhere, along a 2,000 km-long segmented, arcuate path from Kansas in the southwest to lower Michigan to the southeast (Figure 1, inset). Geologic mapping and, more critically, geophysical modelling of aeromagnetic, gravity, and seismic data show the rift to be composed of a deep (≤30 km) asymmetric basin infilled with a lower sequence of volcanic rocks, locally as much as 20 km thick, and an upper sequence of fluvial sedimentary rocks (Cannon *et al.*, 1989; Allen *et al.*, 1994). Petrologic evidence, particularly radiogenic isotope data, suggests that the magmatic evolution of the MCR can be attributed to a mantle plume (Nicholson and Shirey, 1990; Hutchinson *et al.*, 1990; Shirey *et al.*, 1994). The increased influence of compressional forces associated with the Grenville orogeny is thought to have prevented the intracontinental rift from developing into an ocean basin and also to have caused the late reversed faulting which created horst structures throughout much of the rift (Cannon and Hinze, 1992; Cannon, 1994).

The Middle Proterozoic geology of northeastern Minnesota (Figure 1) is characterized by a thick (7-10 km) sequence of plateau lavas and minor interflow sedimentary rocks that was intruded at various stratigraphic levels by numerous mafic intrusions and fewer felsic intrusions. The volcanic rocks, termed the North Shore Volcanic Group (NSVG), were erupted onto a peneplained crust composed of early Proterozoic greywacke, slate, and iron-formation and Archean granite-greenstone terrane. The lava flows are predominantly tholeiitic basalts, ranging in composition from high-Al olivine tholeiite to basaltic andesite (Green, 1972). Intermediate compositions are rare, but felsic lavas (rhyolite, dacite, icelandite) comprise 10-25% of the NSVG section (Green and Fitz, 1993). Among the plutonic rocks, two major supersuites are distinguished, based on the general stratigraphic level of their emplacement into the volcanic edifice. The suite of mafic to felsic intrusions emplaced in the vicinity of the unconformity of the NSVG and older crust are collectively termed the Duluth Complex. Many intrusions were emplaced into medial and upper parts of the volcanic pile as well. A particularly high concentration of at least six subvolcanic intrusions occurs in the central part of the NSVG basin (generalized in Figure 1) and are referred to as the Beaver Bay Complex (Grout and Schwartz, 1939; Miller and Chandler, in press).

The Duluth Complex is typically subdivided into four major series based on lithology, internal structure, and intrusive relationships (Weiblen and Morey, 1980): (1) an early layered series (commonly referred to as Nathan's layered series), (2) a felsic series, (3) an anorthositic series, and (4) a layered (or troctolitic) series. The early layered series occurs along the northern tier of the Duluth Complex as a layered sequence of oxide-rich gabbroic, troctolitic, and gabbronoritic cumulates (Figure 1). Its older age, originally implied by intrusive relationships and its reversed magnetic polarity, was recently confirmed by a U-Pb zircon date of 1106.9±0.6 Ma (Paces and Miller, 1993), roughly 8 Ma older than the main phases of the Duluth Complex. This intrusive series has received minimal attention beyond Nathan's (1969) original field and petrographic study wherein he concluded that the layered sequence was formed by repeated sheet-like intrusions of variably evolved magmas. The felsic series refers to the collection of granitoid (commonly granophyre) intrusions occurring along the roof zone of the Duluth Complex. Intrusive relationships with adjacent gabbroic rocks are commonly equivocal. Along with the felsic series, the anorthositic series typically occupies the structually highest levels of the complex as a multiple-intrusive, structurally complicated suite of laminated, but unlayered, plagioclase cumulates. In one area (~91°00', 47°45', Figure 1), anorthositic rocks are clearly intrusive into granophyre. The erratic internal structure, feldspathic composition, porphyritic character, and lack of signs of internal differentiation have been interpreted as resulting from emplacement and crystallization of plagioclase crystal mushes (Grout, 1918b; Taylor, 1964; Miller and Weiblen, 1990). By contrast, the layered series is composed of many, discrete layered mafic intrusions (some of the larger-scale bodies are delineated in Figure 1). These layered series intrusions are commonly found intruded into anorthositic series rocks as well as into one another. However, nearly identical 1099 Ma U-Pb dates of two layered series and two anorthositic series samples from different parts of the complex (Paces and Miller, 1993) disprove a long-standing interpretation of these intrusive relationships, namely that the anorthositic series is significantly older than the layered series, and imply that the two series have a closer, but yet unexplained, petrogenetic relationship.

What follows is a summary of the geologic setting, cumulate stratigraphy, and cryptic variation of six intrusions of the layered series that exemplify the various styles of formation of layered mafic intrusions associated with the MCR in northeastern Minnesota. These intrusions were formed from generally similar tholeiitic basaltic magmas but differ fundamentally in their openness to magmatic recharge, eruption, and assimilation during crystallization. At one end of the scale is the Sonju Lake Intrusion of the Beaver Bay Complex, which displays features consistent with a nearly closed magmatic system. At the opposite end are the Partridge River and South Kawishiwi Intrusions of the northwestern Duluth Complex, which show evidence of frequent magmatic recharge and country-rock assimilation. Intermediate to these end-member styles of formation is the Layered Series at Duluth. In light of features indicative of strong crystallization differentiation frequently interrupted by recharge and eruption events, it is fitting that these complexly formed intrusive bodies be considered the type intrusions of the Duluth Complex layered series. Finally, we describe two relatively poorly studied layered intrusions, the Wilder Lake Intrusion and the Bald Eagle Intrusion, which display unique phase layering and cryptic variations. Before describing these bodies, however, the rock type and unit classification schemes are defined which we apply to Duluth Complex rocks.

3. TERMINOLOGY

A persistent problem among petrologic studies of the Duluth Complex has been the inconsistent use of rock nomenclature. Most studies have adopted variants of a modal classification scheme based on proportions of the major mafic mineral phases (*e.g.* Phinney, 1972a; Severson, 1994; Miller and Weiblen, 1990), and here we employ the scheme developed by Miller (1995). However, a modal classification alone gives no indication of how various gabbroic rocks may have crystallized and often leads to much ambiguity. The texture of individual minerals is very important to incorporate into the rock term. Therefore, the use of textural modifiers such as ophitic, subophitic, and intergranular, which generally indicates the texture of pyroxene, is used throughout the rock descriptions to follow.

In addition to a modal nomenclature, we also employ, where appropriate, a shorthand notation for cumulate rock types that is based on mineral habit and relative mode (Table 1). It is similar to cumulate classification schemes such as that developed for the Stillwater Complex (McCallum *et al.*, 1980; Zientek *et al.*, 1985) and a similar one applied to the Duluth Complex

Table 1
Terminology and abbreviations for cumulate rocks

Cumulus/Intercumulus Mineral Codes			
PP/P/p	plagioclase*	H/h	hypersthene
O/o	olivine	Hb/hb	hornblende
A/a	augite	B/b	biotite
F/f	Fe-Ti oxide	Ap/ap	apatite
IP/ip	inverted pigeonite	/gp	granophyre

*Use PP for anorthositic compositions

Cumulate	Conventional Rock Term
PPaof	ophitic olivine gabbroic anorthosite
OP	melatroctolite
PaOf	ophitic olivine gabbro
PAFO	intergranular olivine-oxide gabbro
PAFo	intergranular oxide gabbro with poikilitic olivine
PAIPf	intergranular oxide-bearing gabbronorite
PAFbgp	intergranular granophyric biotitic gabbro

by Foose and Cooper (1978), but differs in that it incorporates the modal and textural characteristics of the intercumulus mineral assemblage. It applies to rocks that show some igneous lamination or modal layering, thereby indicating segregation of mineral phases from a liquid or from each other. It denotes granular (cumulus) mineral phases with upper case letter abbreviations and interstitial (intercumulus) mineral phases with lower case abbreviations and lists all minerals composing greater than about 2 vol% in decreasing order of abundance regardless of their cumulus status.

Another nomenclature problem for the Duluth Complex is the inconsistent use of unit designations to internally subdivide layered intrusions. We have adopted the recommendations of Irvine (1992) for the use of terms such as *zone, unit,* and *layer.* An exception to Irvine's suggestions and common usage in other layered intrusions is our use of the term *series.* Whereas this term is commonly used to structurally subdivide layered intrusions (*e.g.* marginal series, layered series, border series; Wager and Brown, 1968), *series* in the Duluth Complex has been historically used for groups of rocks that are lithologically and structurally similar regardless of their intrusive-stratigraphic relationships (*e.g.* layered series, anorthosite series, felsic series). In this way, the term has a useful genetic connotation that broadly groups rocks that formed from genetically similar parent magmas and by similar processes.

4. SONJU LAKE INTRUSION

A mapping and petrologic study by Stevenson (1974) of layered gabbroic cumulates exposed in sporadic outcrop near Finland, Minnesota showed the rocks to belong to a strongly differentiated layered intrusion, which he termed the Sonju Lake Intrusion (SLI, Figure 2). He recognized that the cumulate paragenesis of the intrusion was very similar to the Skaergaard Intrusion, the classic example of closed-system differentiation of a tholeiitic mafic magma

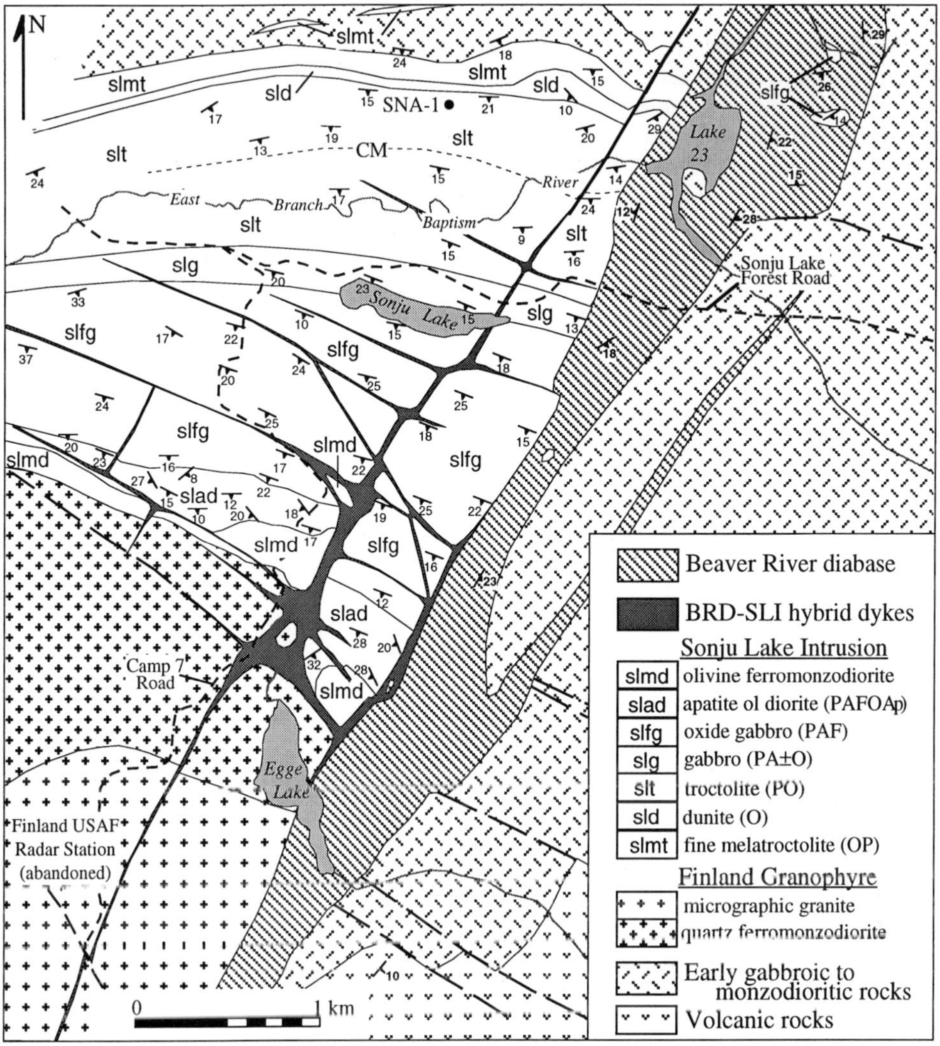

Figure 2. *Geology of the Sonju Lake Intrusion (modified from Miller et al., 1993a). Strike and dip symbols denote the orientation of igneous lamination and rare modal layering. The locations of drill hole SNA-1 and a horizon (CM) marking an abrupt change in mode and texture (Figure 3) are shown.*

(Wager and Brown, 1968). Recent detailed mapping (Miller *et al.,* 1993a) and petrologic studies (Miller and Chandler, in press) reconfirm Stevenson's conclusions and show that the SLI represents nearly closed-system differentiation of an evolved tholeiitic magma by moderately efficient, unidirectional (bottom-up) fractional crystallization.

4.1. Geologic Setting

The SLI is a moderately south-dipping, 1,200 m thick sequence of layered mafic cumulates exposed over a thickly forested, 9 km^2 area centered on Sonju Lake (Figure 2). The intrusion is part of the multiply intrusive Beaver Bay Complex (Miller and Chandler, in press) and has an emplacement age of 1096.1±0.8 Ma (Paces and Miller, 1993). Although its exposed strike-length is only about 3 km, its distinctive aeromagnetic signature indicates that it extends along a curved path to the west and southwest for at least 20 km beneath a thick cover of glacial drift (Miller *et al.*, 1993a). The nearly constant width of the anomalies suggests that the intrusive sheet retains a uniform thickness over most of this distance. The eastern margin of the SLI is abruptly truncated by a diabase dyke that is part of an extensive dyke and sill network extending over much of the Beaver Bay Complex and has a similar age (1095.8±1.2 Ma, Paces and Miller, 1993). Also apparently associated with the emplacement of the diabase dyke is an orthogonal set of narrow dykes that cut the SLI west of the truncating dyke (BRD-SLI hybrid dykes, Figure 2). These dykes are composed of hybrid rock types that appear to represent various degrees of mixing and assimilation of basaltic and felsic magma, the latter perhaps representing a residual melt in the upper part of the SLI.

The hanging-wall of the SLI is composed of granitic rocks covering an extensive area to the south and east that comprise the Finland granophyre (Miller *et al.*, 1993a). The granophyre is composed of two general lithologic units which have a map distribution which generally conforms to that of the underlying SLI cumulate units (Figure 2). Immediately overlying the upper SLI cumulates is an extensive area of mottled pink to green, medium- to coarse-grained, granophyric, quartz ferromonzodiorite. It contains granular Fe-oxides and prismatic Fe-silicates which grade in combined abundance from about 25% near the top of the SLI to 5-10% near its transition with the main granite phase of the Finland granophyre. This main phase is composed of a pinkish-orange, strongly micrographic leucogranite containing abundant miarolitic cavities and less than 5% Fe-silicates and oxides. The petrogenetic relationship of the Finland granophyre to the SLI poses one of the most interesting questions about the differentiation history of the SLI, as discussed below.

The base of the SLI is in intrusive contact with a complex mixture of gabbroic to dioritic rocks, granophyre, and volcanic hornfels. Although the actual contact is rarely visible in outcrop, it is evident in a drill core (SNA-1, Figure 2), where it is identified as a fine-grained, homogeneous moderately laminated melatroctolite in sharp contact with a coarse-grained, modally macrolayered troctolite to olivine gabbro. Although fine grained, the melatroctolite does not correspond to a liquid composition but rather is a primitive adcumulate of Fo_{79} olivine and An_{81} plagioclase. Up-section of the contact, the melatroctolite unit (*slmt*, Figure 2) becomes progressively more olivine rich, and more medium grained. By 70 m above the contact, a nearly pure olivine cumulate (*sld*, Figure 2) is developed that is taken to mark the base of the main cumulate sequence. The adcumulate nature of the *slmt* unit indicates that the SLI parent magma was emplaced into a warm country-rock. However, the footwall provided a sufficient heat sink to cause the magma, which initially had only olivine on the liquidus, to also crystallize progressively more plagioclase nearer the contact due to greater undercooling.

4.2. Cumulate Stratigraphy

Sandwiched between lower and upper marginal zones (*slmt* and *slmd*, respectively, Figure 2), five cumulate units are distinguished in the SLI based on the successive cumulus arrivals of olivine, plagioclase, augite, ilmenite, and apatite (Figure 3). Grading up from the lower melatroctolite unit, the base of the main cumulus section is marked by a 10 to 12 m thick interval of the olivine adcumulate (dunite) unit (*sld*, Figure 2). The reappearance of cumulus plagioclase, which is indicated by an abrupt change from poikilitic oikocrysts to lath-shaped, well-aligned crystals, defines the base of the overlying 400 m thick troctolite unit (*slt*, Figures 2 and 3). However, strong modal layering in the lower 25 to 40°m indicates that some time was required for cotectic modal equilibrium to be attained. Above this well-layered zone, ophitic augite troctolite (POaf cumulate) is well laminated but displays only intermittent schlieren and thin isomodal layering. Small granular (cumulus?) Cr-Mg-Al spinel grains compose about 2% of the middle section of the *slt* unit. In the upper 100 m, the troctolite locally becomes leucocratic, and a postcumulus assemblage of augite and ilmenite is typically more abundant than olivine and is commonly joined by as much as 5% subophitic inverted pigeonite (Figure 3). About halfway up the *slt* unit (CM line, Figures 2 and 3), a coarse-grained ophitic augite troctolite with up to 8% inverted pigeonite abruptly gives way to a medium-grained augite-poor troctolite. This abrupt textural and modal break also corresponds to a slight compositional shift toward more primitive compositions upsection (Figure 3). It is the only feature recognized in the SLI that likely represents a significant magmatic recharge event.

The transition from PO cumulate to PAO cumulates of the *slg* unit (Figures 2 and 3) occurs over a <3 m thick interval where augite habit changes from coarsely ophitic to anhedral granular. Olivine abundance varies nonsystematically between <2 to 20% through the well-laminated but unlayered *slg* unit, and anhedral granular to subophitic inverted pigeonite abundance is consistently in the range of 2 to 6%. The transition from PAO to PAF(±O) cumulates of the *slfg* unit (Figures 2 and 3) is marked by the abrupt increase in ilmenite mode from about 3% to 8-12% and its change in habit from subpoikilitic to anhedral granular. Corresponding to the cumulus appearance of ilmenite, is the nearly complete disappearance of olivine. Where olivine occurs, it typically composes less than 2% of the rock and is coarse-grained, ubiquitously altered, and irregularly shaped (corroded?). With the exception of rare intervals of modal layering, the 370-400 m thick *slfg* unit is remarkably homogeneous in texture and mode (Figure 3A). It does display, however, a smooth cryptic variation in the *mg#* of pyroxene and olivine (Figure 3C). Near the top of the *slfg* unit, the oxide gabbro locally is cut by subconformable lenses of coarse-grained, decussate, apatitic olivine ferromonzodiorite that are very similar to rocks of the *slmd* unit at the top of the cumulate section (Figures 2 and 3).

The uppermost cumulate unit of the SLI (*slad*, Figure 2) is characterized by the presence of cumulus apatite. Prismatic apatite composes an average of about 3-5 vol.% of the *slad* unit in comparison to only about 0.5% in the *slfg* unit. The arrival of cumulus apatite also roughly corresponds to the reappearance of olivine. Though typically anhedral granular, this Fe-rich olivine also occurs locally as ameboidal to subpoikilitic (*i.e.*, postcumulus) grains. In total, the apatite olivine ferrodiorite is a five-phase (PAFOAp) mesocumulate. Elongate crystals of ferroaugite, ilmenite, and plagioclase tend to be moderately to well laminated and modally layered in places. Entirely postcumulus phases include 2-5% hornblende, 0-3% biotite, and 1-12% orthoclase and quartz that are commonly micrographically intergrown. Deuteric alteration of primocryst phases is commonly extensive.

The upper contact zone between the cumulates of the SLI and the Finland granophyre is complex. Although part of this complexity in the area of exposure (Figure 2) is due to the deformation attending the emplacement of the BRD-SLI hybrid dykes, most of it is an inherent feature of the irregularly gradational contact between the SLI and the granophyre. Because of

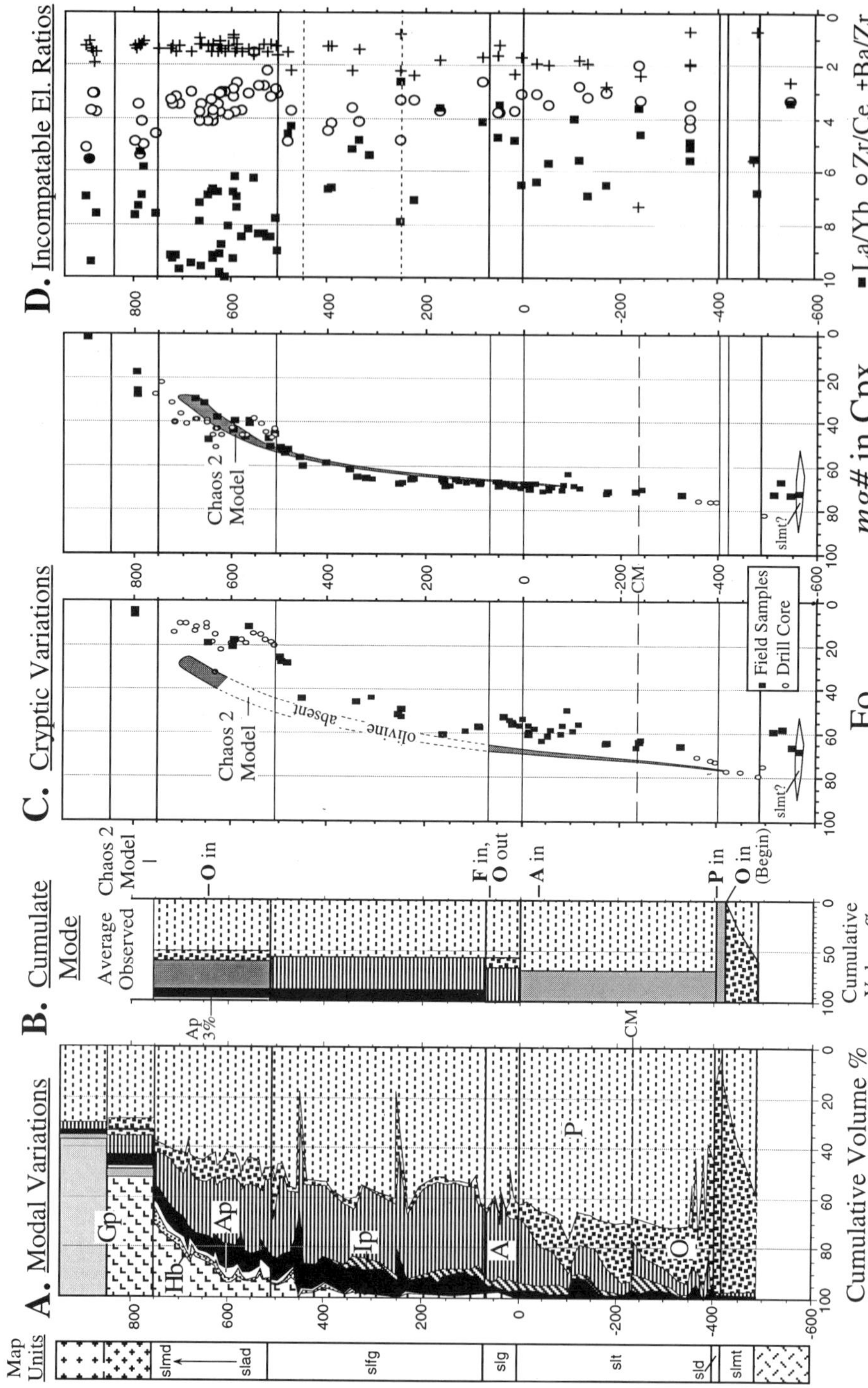

Map Units
slmd
slad
slfg
slg
slt
sld
slmt
A. Modal Variations
Gp
Hb
Ap
Ip
A
P
O
Cumulative Volume %
B. Cumulate Mode
Ap 3%
CM
Average Observed
Cumulative Volume%
Chaos 2 Model
O in
F in, O out
A in
P in
O in (Begin)
C. Cryptic Variations
Chaos 2 Model
olivine absent
Field Samples
Drill Core
slmt?
Fo
CM
Chaos 2 Model
mg# in Cpx
D. Incompatable El. Ratios
La/Yb
Zr/Ce
Ba/Zr

Figure 3. *(facing page) Stratigraphic variations of (A) measured modal mineralogy, (B) average cumulus mineral mode, (C) measured cryptic variations in Fo in olivine and* mg# *in augite, and (D) variations of La/Yb, Zr/Ce, and Ba/Zr determined from whole-rock analyses. Mineral abbreviations as in Table 1; map units as in Figure 2 and described in text. Stratigraphic height (in m) of field samples collected along five sampling profiles and from two drill cores are positioned relative to the horizon marking the cumulus arrival of augite. Horizon labeled CM marks an abrupt change in mode and grain size that may indicate a recharge event (see text). Also shown are the levels at which cumulus phases appear and disappear and the cryptic variation of primocrystic olivine and augite calculated by applying the Chaos 2 fractional crystallization model to the SLI bulk composition (Table 2) (Nielsen, 1990). Other input parameters presume an average trapped liquid proportion of ~30% and* f_{O_2} *= -1 to -2 log units below QFM.*

the difficulty in sorting through the various complexities of the upper contact zone from intermittent outcrop, a 250 m long drill core was obtained by the Minnesota Geological Survey in 1991 (Meints *et al.*, 1993). The drill site was located on Minnesota Highway 1, about 5 km to the west of the map area shown in Figure 2, and was positioned on the basis of aeromagnetic anomalies known to correlate with the upper contact zone in the area of exposure. The continous sequence displayed in drill core suggests that the complex relationship implied from outcrop between the laminated, apatite olivine ferrodiorite (*slad*) unit and a decussate, olivine ferromonzodiorite, mapped in Figure 2 as *slmd*, is cyclical in nature.

Through the lower 220 m of the core, five macrocycles are recognized which range in thickness from 40 to 70 m. In general, each cycle is characterized by a lower section of well-laminated, medium-grained, apatite olivine ferrodiorite (*slad*-like) that gradually gives way to a coarser grained, poorly laminated, commonly granophyric apatite olivine ferromonzodiorite (*slmd*-like) at the top (Figure 3). The base of each cycle tends to be marked by a relatively abrupt (cm-scale) to narrowly gradational (m-scale) change in rock type. Although all cycles have about the same scale of thickness, each succesively higher cycle contains a significantly greater proportion of the *slmd*-like ferromonzodiorite (Figure 3). In the drill core, the coarse-grained, decussate olivine ferromonzodiorite of the uppermost cycle grades over several metres into a pink, coarsely prismatic quartz ferromonzodiorite that closely resembles the outer (lower) phase of the Finland granophyre. Only 11 m of this quartz ferromonzodiorite was intersected at the top of the core, and so it is not known whether this rock type alone persists upsection or whether the cyclicity evident between the *slad-slmd* units carries through into the Finland granophyre.

4.3. Relationship to granophyre

The relationship of granophyre to mafic layered intrusions has spurred debate about even the best-studied layered intrusion—the Skaergaard (see Hunter and Sparks, 1990 and related discussions). The SLI and Finland granophyre present similar problems. Several characteristics of the transition between the two units suggest that the granophyre may be an upper differentiate of the mafic cumulate series. These include (1) the conformable zonation of granitic and quartz ferromonzodioritic units of the Finland granophyre to the internal structure of the mafic intrusion (Miller *et al.*, 1993a), (2) an irregularly gradational to cyclic transition in modal composition (Figure 3A), (3) a smooth compositional transition in cryptic mineral variations (Figure 3C), and (4) no clear change in incompatible element ratios (Figure 3D). A major problem with this interpretation, however, is that there is too much granophyre to have been generated from the volume of mafic cumulates exposed. The proportionality problem, which is very evident from the map areas of the two units (Figure 2; Miller *et al.*, 1993a), has been confirmed by unpublished geophysical modelling of gravity and aeromagnetic data. This modelling shows that the volume of felsic material is at least as great as the amount of mafic cumulates. Further evidence that the Finland granophyre was formed independently of the SLI has been provided by recently acquired radiogenic isotopic data that show the granitic rocks to have different ε_{Nd} and $^{87}Sr/^{86}Sr$ initial ratios from the mafic cumulates (A. Basu, pers. comm.). Lastly, the most forsteritic olivine composition found at the base of the intrusion (Fo_{79}) constrains the amount of iron-rich quartz ferromonzodiorite rock type (avg. wt% FeO^t = 12.8) that may compose the bulk composition of the SLI (see below).

It appears most likely that the Finland granophyre predates the SLI emplacement. It could have acted as a density barrier to the upward passage of mafic magma, thereby causing SLI magma to underplate the granophyric body. Underplating of SLI magma would have undoubtably caused melting of the lower part of the Finland granophyre. Such a viscous felsic cupola explains the lack of an upper border series to the SLI (Marsh, 1988), but poses other questions as to the extent of assimilation across such a contrasting felsic-mafic boundary. Was assimilation confined to and manifested by the cyclicity of the *slad-slmd* units (Figure 3)? Or does the quartz ferromonzodiorite represent melting of the Finland leucogranite and its subsequent contamination by partial assimilation of the underlying mafic magma? How much granophyre represents felsic melt generated by differentiation of SLI magma? Resolution of these problems will require evaluation of mass and thermal diffusion models and radiogenic and oxygen isotope data.

4.4. Composition and closed-system differentiation of the SLI parent magma

Understanding the details of the differentiation of the SLI system is complicated not only by the as yet unknown thermal and chemical effects of the granitic roof, but also by a continued uncertainty as to the pertinant chemical and dynamic process affecting all such large mafic systems (Jaupart and Tait, 1995). Despite this uncertainty, evermore elaborate "parameterized" geochemical equilibria-based models are being developed (*e.g.* Frenkel *et al.*, 1989; Langmuir, 1989; Nielsen, 1990; Nielsen and de Long, 1992; Ariskin *et al.*, 1993; Ghiorso and Sack, 1995) that attempt to account for various dynamic igneous processes which affect the cooling and crystallization of mafic intrusions (*e.g.* thermal and compositional convection, boundary layer formation, gravity settling, recharge, eruption, and assimilation) and which occur over a range of physical conditions. Work toward evaluating the applicability of these models to the SLI should give useful insight as to the processes responsible for its differentiation. However, at this preliminary stage of study, a simple fractional crystallization model is employed to demonstrate that the unidirectional cumulus phase layering and cryptic variation and the constancy of incompatible element ratios (or at least a lack of systematic variability) evident through the igneous stratigraphy of the SLI (Figure 3) are consistent with its formation by upward directed, moderately efficient fractional crystallization driving magmatic differentiation in a generally closed system.

The model used is Chaos 2 (variant of ERUPT, Nielsen, 1990) which simulates low-pressure differentiation of basaltic magma by homogeneous fractional (Rayleigh-type) or equilibrium crystallization and is based on experimentally determined mineral-melt phase equilibria and distribution coefficients. The model allows for variation in crystallization increment, oxygen fugacity within +2 to -4 log units of the QFM buffer, periodicity and volume of recharge, eruption and assimilation, and composition of an assimilant. Making reasonable estimates of these parameters and with input of a parent magma composition, it calculates the proportions and compositions of cumulus phases and derivative liquids as crystallization proceeds. The most critical input factor is an accurate estimate of the parent magma composition.

Chilled samples of the SLI parent magma have not been found. Although the fine-grained nature of the melatroctolite forming the lower margin of the SLI (*slmt*, Figure 2) indicates relatively rapid cooling, the laminated texture and incompatible element-poor composition of this rock indicates that it is a PO cumulate. In the gabbroic footwall of the SLI, several small sheet-like bodies of olivine diabase occur that may be offshoots of the SLI (Figure 2). However, least-squares mixing calculations for one such diabase sample indicate that it too is enriched in primocrysts of olivine and plagioclase. Given the absence of crystal-free chilled margins, an estimation of the parent magma composition is best accomplished by determining the bulk composition of the intrusion.

Because the shape of the intrusion is tabular, with a strike-to-thickness ratio of greater than 20:1, lateral variations in rock type are minimal relative to stratigraphic variations. Consequently, the cumulative composition of a stratigraphic profile through the intrusion should approximate the bulk composition. Because modal layering is generally not common and typically subtle except in the lower part of the troctolite unit (*slt*, Figure 3), a manageable number of representative analyses should provide a reasonably accurate bulk estimate. The greatest unknown in this calculation is not knowing the extent of assimilation between the SLI mafic magma and the Finland granophyre and therefore not knowing the geochemical top of the intrusion. Until radiogenic isotope data determine the degree of cross-contamination, the simplest assumption to make is that the granophyric *slmd* unit represents the end result of differentiation of the SLI magma. Assuming this, a total of 76 whole-rock analyses of outcrop and drill-core samples, selected from various parts of the 1100-m thickness of the SLI intrusion, were used to calculate the bulk composition of the SLI in Table 2. One indication of the plausibility of this composition to be the SLI parent magma is that it should produce primocrystic olivine with an Fo content of 79-80 (assuming 85-90% of total iron is as FeO and that the Fe-Mg $Kd^{(Ol\text{-}melt)} = 0.30$; Roeder and Emslie, 1970), which is exactly that observed in the basal melatroctolite ($Fo_{79.3\pm0.2}$). More convincing evidence that this calculated bulk composition is a good approximation of the SLI parent magma is demonstrated by the results generated by application of this composition to the Chaos 2 model.

Applying the calculated bulk SLI composition to the Chaos 2 model, with the parameters set to simulate closed-system fractional crystallization, immediately gives a good match to the observed phase paragenesis (O $\rightarrow$ PO $\rightarrow$ PAO $\rightarrow$ PAF $\rightarrow$ PAFO(+Ap)). The best scaling for the relative cumulus arrivals of augite and iron oxide is achieved with f_{O_2} set at between -1 and -2 log units below QFM. A less-than-ideal fit is achieved however by the too rapid rate at which differentiation progresses. This is demonstrated by cryptic layering being too extreme and cumulus arrivals of augite and Fe-Ti oxide being too early. These scaling problem can be accounted for in two ways: 1) by accumulating some trapped liquid along with the cumulus phases and 2) by periodically recharging the system with fresh magma. The eruption parameter of the Chaos 2 model can be used to extract an aliquot of liquid from the magma reservoir with each crystallization increment and thereby simulate the effect of trapping liquid with each

Table 2
Compositions related to Duluth Complex Intrusions

	1 Sonju Lake Intrusion bulk intrusion	2 S. Kawi-shiwi Intrusion sum Unit VII	3 Partridge River Intrusion parent magma	4 Duluth Layered Series UCZ avg. of 6	5 Wilder Lake Intrusion marginal gabbro	6 North Shore Volcanic Group Ol tholeiite
SiO_2	47.6	47.6	48.2	49.3	45.9	49.5
TiO_2	2.28	1.27	2.94	3.43	3.03	0.80
Al_2O_3	14.0	19.5	15.01	12.2	16.4	18.2
FeO_t	14.7	11.0	15.84	13.8	14.8	8.4
MnO	0.21	0.13	0.20	0.21	0.18	0.13
MgO	8.3	7.07	5.19	4.9	7.5	8.8
CaO	9.4	9.65	8.21	9.1	8.6	11.4
Na_2O	2.47	2.66	2.63	2.75	2.91	2.38
K_2O	0.55	0.50	1.32	0.80	0.24	0.19
P_2O_5	0.30	0.18	0.40	0.63	0.04	0.06
Volatiles	0.20	0.33	-	0.20	-	-
Total	100.0	100.0	99.9		99.6	99.8
mg#	50.2	53.4	36.9	38.7	47.4	65.3
Trace Elements (ppm)						
Sc	34	-	22.5	37	24	27
V	192	-	-	330	280	-
Cr	111	105	-	102	227	361
Co	75	-	57.5	39	70	49
Ni	185	50	53	71	165	277
Rb	20	-	-	30	<15	4
Sr	233	250	189	207	284	207
Ba	171	100	-	235	119	58
Y	20	15	-	62	2	13
Zr	114	93	-	317	-	60
Nb	17	3.9	-	25	-	-
Hf	3.1	1.7	-	7.8	<1	-
La	14.7	6.3	20.8	38.7	1.1	4.2
Ce	33.4	15.7	-	85.0	<3	10.3
Sm	4.1	2.3	-	10.2	0.5	1.9
Eu	1.6	1.0	-	2.8	1.0	0.83
Tb	0.8	0.5	-	2.0	<0.5	-
Yb	2.1	1.3	-	5.9	0.3	1.38
Lu	0.32	0.22	0.51	0.82	0.07	0.21

Table 2 continued
(1) Bulk composition based on weighted summation of 76 whole rock analyses through the Sonju Lake Intrusion (see text). (2) Bulk composition obtained by weighted summation of 13 whole rock analyses through Unit VII of the South Kawishiwi Intrusion (Lee and Ripley, in press). (3) Estimate of parent magma composition based on average trapped liquid composition through 525 m of the lower Partridge River layered Intrusion (Chalokwu *et al.*, 1993, Table 2). (4) Average of 6 analyses of very fine-grained diorite in "chilled" contact with anorthositic series rocks (locations in Figure 8). (5) Analysis of fine-grained subophitic olivine gabbro margin of Wilder Lake Intrusion (Sample 711, Figure 12). (6) Average composition of 14 unaltered, primitive olivine tholeiite basalts containing less than 3% glass (Brannon, 1984, Table 5.5).

---

crystal accumulation event. Assuming that about 30 wt% magma accumulates with the cumulus phases, a good fit of the model to observed phase and cryptic variations through the SLI can be achieved (Figure 3). This proportion of trapped liquid is in general agreement with the average abundance of intercumulus minerals (pyroxene, oxide, and overgrowth zones on plagioclase) observed in SLI troctolitic cumulates (Figure 3A); intercumulus mineral abundances in gabbroic cumulates being more difficult to estimate. Although this fit does not preclude that periodic recharge was responsible for the expanded scaling of cryptic and phase layering in the SLI, there is little empirical evidence, save the CM event (Figure 3), that magma recharge was an important process.

Although the calculated cryptic variations in *mg#* of primocrystic olivine and augite match the measured variations in form, only the pyroxene data match in fact (Figure 3C). The difference in Fo between measured olivine and the primocryst model curve can be attributed to a trapped liquid shift, a phenomenon caused by re-equilibration between primocrystic and more iron-rich postcumulus mafic phases and thought to routinely occur in mafic cumulates (Barnes, 1986; Chalokwu and Grant, 1987). Such a process is qualitatively consistent with the need to incorporate at least 30% intercumulus liquid in the Chaos 2 fractional crystallization modelling. If the mismatch of the model to the measured olivine compositions is due to the trapped liquid shift, however, the seemingly good fit of the clinopyroxene data to the primocryst model curve (Figure 3C) is in fact problematic because a similar trapped liquid shift should also affect pyroxene. Perhaps the experimental phase equilibrium data on which the pyroxene curve is based are inappropriate for magmas that are also crystallizing an iron oxide phase. An alternative explanation consistent with their low abundance of accessory mineral phases (Figure 3A) is that the oxide gabbro rocks are more adcumulate in nature. Greater stock should be taken of the fact that the measured data define trends that are consistent with model trends, indicating closed-system fractional crystallization.

In conclusion, the reasonable fit of the Chaos 2 model calculations to the observed cumulus phase layering and cryptic variation through the SLI provides supporting evidence that the intrusion formed via closed-system, moderately efficient fractional crystallization. These results support the earlier conclusion by Stevenson (1974) that the SLI evolved very much like the Skaergaard Intrusion. Moreover, it provides independent verification of the estimated bulk composition (Table 2) is a reasonable estimate of the parent magma composition. When the extent of felsic assimilation of the Finland granophyre is better understood, a more rigorous evaluation of the dynamic and chemical evolution of SLI can be attempted.

5. PARTRIDGE RIVER AND SOUTH KAWISHIWI LAYERED INTRUSIONS

Layered series rocks along the northwestern margin of the Complex, known as the Hoyt Lakes-Kawishiwi area (Weiblen and Morey, 1980), have received special attention due to the presence of Cu-Ni-PGE mineralization in the basal zone (Bonnichsen, 1974a; Hauck *et al.*, in press). Although exposure is limited, over 1000 cores have been drilled along a 50 km stretch of the basal contact by exploration companies over the past 45 years. Most of these drill cores, which are now in the public domain, penetrate the lower 1-1.5 km of the Duluth Complex to its contact with Archean granitic rocks and early Proterozoic iron-formation, graywacke, and slate (Figure 1). Two intrusive bodies are recognized in the area, the South Kawishiwi Intrusion (SKI) in the northeast (Phinney, 1969; Foose and Cooper, 1978) and the Partridge River Intrusion (PRI) to the southwest (Bonnichsen, 1974b; Severson, 1994).

Although both intrusions, where exposed and drilled, are predominantly composed of troctolitic (PO) cumulates, Bonnichsen (1972, 1974b) first recognized a difference in the igneous stratigraphy southeast and northwest of the Dunka River. The presence of more abundant magnetite in the SKI was noted by Pasteris (1985) as a major mineralogical difference between the two bodies. However, more recent work (*e.g.* Severson, 1994; Lee and Ripley, in press) shows that although titanomagnetite-rich layers occur in the SKI, titanomagnetite is not ubiquitous, and its abundance alone should not be taken as a key distinguishing feature. The best evidence that the two masses are indeed separate intrusions has come from detailed relogging of hundreds of drill cores in the Hoyt Lakes-Kawishiwi area by Severson and coworkers at the Natural Resources Research Institute of the University of Minnesota, Duluth (Severson, 1991; 1994; Severson and Hauck, 1990; Geerts, 1991; Zanko *et al.*, 1994). Their megascopic examination of over 490 drill cores from the PRI and over 165 cores from the SKI, along with some petrographic and geochemical studies, has documented distinct differences in the igneous stratigraphy of the lower parts of the two intrusions and established the nature of the contact between them (Figure 4).

Although no intrusive contact between the two intrusions has been mapped or is clearly recognizable in drill core, Severson (1994) concluded that the PRI is older than the SKI based on the development of a heterogeneous contact zone in the PRI where it is adjacent to the SKI. All stratiform units of the PRI become unrecognizable in the contact zone, and footwall inclusions and cross-cutting oxide ultramafic bodies are common. The same heterogeneity is not evident in the SKI which Severson (1994) explained as resulting from preheating of the country-rock by the earlier emplacement of the PRI. If the SKI were intrusive into the PRI, some type of marginal contact zone, wherein the stratiform SKI units lose their integrity near the contact with the PRI, would be expected, even if the PRI were still quite hot. Except for a steepened dip of some upper units, however, it is not clear from Severson's (1994) description that any type of marginal zone is developed in the SKI. Complicating the picture of the contact zone is its coincidence with a prominent fault zone, the Grano fault, which was active during and after magmatic activity.

It should be noted that Martineau (1989) interprets the stratigraphic relationships in the Hoyt Lakes – Kawishiwi area differently from Severson (1994). Martineau (1989) postulates that at least six separate intrusions occur along the western margin of the complex. The pinch-out of units and trangression of some units into others in the SKI and PRI were also recognized by Severson (1994). However, in the larger picture of the entire Duluth Complex, these pinch-outs and transgressions are minor in scale and contacts between these internal units are

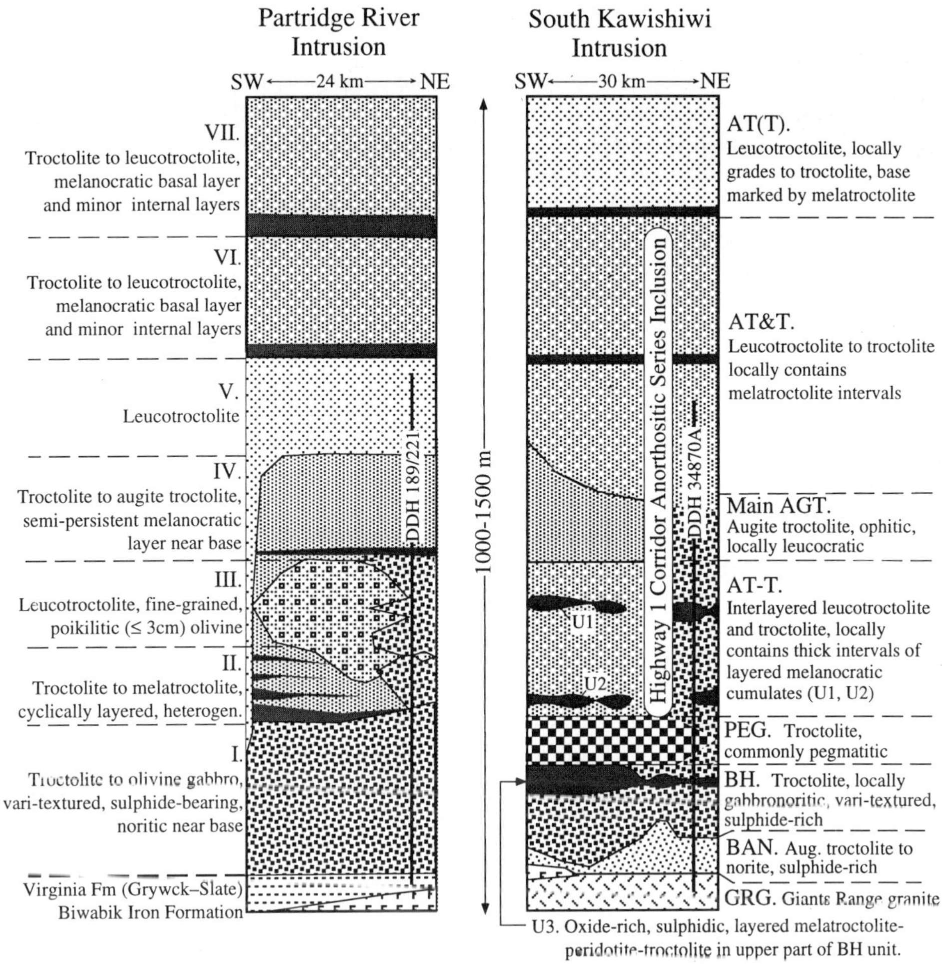

Figure 4. *Generalized igneous stratigraphy of the lower parts of the Partridge River and South Kawishiwi Intrusions determined from drill core logging (modified from Figures 8 and 10 of Severson, 1994). Unit labels and descriptions after Severson (1994). Heavy vertical lines denote sections intersected by drill core studied in detail: DDH 221 (Chalokwu and Grant, 1987, 1990; Grant and Chalokwu, 1992; Chalokwu et al., 1993), DDH 189 (Taib and Ripley, 1993; Severson, 1991; see Figures 7 and 8), DDH 34870-A (Lee and Ripley, in press; see Figure 5).*

generally gradational. Our work with drill core from both these intrusions leads us to agree with the interpretations of Severson (1994) that each intrusion is a single open magmatic system formed by repetitive magma inputs. We present data below from our studies of drill core from these intrusions that demonstrates the openness of each system.

5.1. Geologic setting and igneous stratigraphy of the South Kawishiwi Intrusion

From south to north, the basal contact of the SKI completely trangresses downward through a shallow dipping sequence of early Proterozoic sedimentary rocks (Virginia Formation and Biwabik Iron Formation) and comes into contact with Archean granitic rocks of the Giants Range batholith (Figure 1). Drill-core data indicate that this transgression of the contact also occurs downdip. The abundance of metasedimentary hornfels along the entire basal contact indicates that the early Proterozoic units were removed by emplacement of the SKI. The northeastern contact of the SKI is formed against gabbroic anorthosites of the anorthositic series (Figure 1), but its exact nature is unclear. The orientation of igneous lamination and locally developed modal layering in the SKI dip gently away from all exposed contacts to define a southward-deepening asymmetric basin whose axis is skewed to the east. Glacial till cover in the southern part of SKI masks its contact relationships to other intrusive rocks. Of particular interest is its relationship to the Bald Eagle Intrusion (see below). Weiblen and Morey (1980) suggested that the steep-walled Bald Eagle Intrusion may have acted as a feeder to the SKI. The occurrence of a troctolitic dyke connecting northern ends of the two intrusions (Figure 1) would indicate such a linkage. However, the troctolitic rocks of the SKI are much more evolved than those of the Bald Eagle (*e.g.* Fo_{63-55} in SKI vs. Fo_{74-70} in the Bald Eagle; Phinney, 1969). Moreover, aeromagnetic anomaly patterns suggest that the SKI is cut by the Bald Eagle Intrusion (Chandler, 1990; Meints *et al.*, 1993).

Bedrock mapping by several groups of workers (Green *et al.*, 1966; Bonnichsen, 1971; Morey and Cooper, 1977; and Foose and Cooper, 1978) has shown that the exposed part of the SKI is composed almost exclusively of a 2.5 km thick section of plagioclase-olivine cumulates. Based largely on the abundance and texture of intersitial augite and Fe-Ti oxide, four stratiform lithologic units have been generally delineated:

1) Basal contact zone: Within 50 to 250 m of the footwall contact is a vari-textured and lithologically heterogeneous unit composed of troctolite, olivine gabbbro, norite, ultramafic rocks, and anorthositic rocks that is host to many types of hornfels inclusions, as well as most of the sulphide mineralization (*scz* of Green *et al.*, 1966; *g* of Bonnichsen, 1971; and *tcz* of Morey and Cooper, 1977);
2) Augite troctolite: Subophitic to ophitic augite troctolite that is commonly biotitic and locally leucocratic (*sat* of Green *et al.*, 1966; *tam* of Bonnichsen, 1971);
3) Ophitic troctolite: Characterized by large (4-8 cm) augite oikocrysts but a lower total abundance of augite and oxide than unit below (*spt* of Green *et al.*, 1966; *ta* of Bonnichsen, 1971; POCx of Foose and Cooper, 1978);
4) Troctolite with anorthositic layers: The uppermost part of the intrusion is composed of troctolite containing only minor augite and oxide (POCu and POCl of Foose and Cooper, 1978; *t* of Bonnichsen, 1971). Detailed mapping in the eastern SKI by Foose and Cooper identified two 5 to 10 m thick conformable layers of plagioclase cumulates (*PC1* and *PC*) that, although broken up by small-scale faulting, could be traced about 12 km along strike.

All units are generally gradational into one another and, except for a general increase in grain size and reduction in the amount of biotite away from the contact (Weiblen and Morey, 1980), no regular modal or cryptic variation is evident. Aeromagnetic data do not indicate the presence of significant oxide-bearing gabbroic upper cumulates in the unexposed southern part of the SKI. Their absence probably indicates a deep level of erosion of the basin-shaped SKI.

Relogging of over 165 exploration drill core by Severson (1994) and Zanko *et al.* (1994) has revealed a much greater complexity to the igneous stratigraphy of the lower part of the

SKI. Although these drill holes penetrate only the lower two units recognized by field mapping (*e.g.* units *scz* and *sat* of Green *et al.,* 1966), Severson (1994) has delineated 17 different lithostratigraphic units over a strike length of 31 km. Individual units rarely persist along the entire strike length, but the stratigraphic relationships remain constant. In one area (the Highway 1 corridor), the general igneous stratigraphy evident elsewhere is complicated by the existence of a large (>1 km thick) mass of gabbroic anorthositic rocks that probably represents a large inclusion of the anorthositic series (Figure 4). Severson (1994) interpreted this compartmentalized nature of the SKI to indicate a complicated multiple-intrusive history.

5.2. The South Kawishiwi Intrusion in the Spruce Road area

Lee and Ripley (in press) have studied a 600 m thick portion of the lower SKI in the Spruce Road area (Figures 1 and 4). Their work is based on examination of four drill cores, and a detailed study of DDH 34870-A (Figure 5). They divided the igneous sequence into seven distinct units that broadly conform with the divisions defined by Severson (1994). Subdivision was based on a combination of macroscopic characteristics, petrologic features, major- and trace-element concentrations, and stable isotopic values. Briefly, Units I and II rocks are very heterogeneous in texture (fine- to coarse-grained, intergranular to ophitic) and rock type (gabbronorite, olivine gabbro, troctolite, melatroctolite, and troctolitic anorthosite are common). Randomly oriented (decussate) plagioclase and ophitic pyroxene impart a noncumulate character to the rocks. Unit III contains melatroctolite, troctolite, and olivine gabbro, with local oxide-rich layers containing primocryst titanomagnetite. Unit IV contains two to three cycles of basal melatroctolite overlain by troctolite and olivine gabbro. Unit V is a troctolite sequence with local pegmatitic horizons. Unit VI is strongly altered, but bears similarity to Units III and IV. Unit VII is similar to Unit V and is composed of homogeneous troctolite to leucotroctolite. All units above Unit II are PO cumulates though some in unit III are POF cumulates.

Oxygen isotopic values of Units III, IV, V, and VII are normally within the restricted range of 5.7 to 6.4‰, and show little evidence for crustal contamination. Unit II and Unit VI $\delta^{18}O$ values range from 6.5 to 7.1‰ and are consistent with sulphur isotopic evidence which suggests that crustal contamination has been important. Units I, II, and III, along with VI are sulphide-bearing. $\delta^{34}S$ values range from 4 to 10‰ and indicate that crustally derived sulphur has been involved in sulphide mineralization

Cryptic variations through the core 34870-A are cyclical and most correspond to lithologic breaks (Figure 5). Because incompatible elements (*e.g.* Zr, Y, P) are not concentrated in either plagioclase or olivine and their ratios remain essentially constant throughout the sequence; an increase in their absolute abundance records either variations in the amount or composition of trapped parental magma (now postcumulus minerals). Elevated incompatible element abundance is particularly common in most pegmatitic layers and suggests an enrichment during crystallization and possible compositional convection. The cyclical increase in incompatible element concentrations may indicate differentiation cycles or upward increases in the amount of trapped liquid (Figure 5). Abrupt reversals in both mineral compositions and incompatible element abundances may reflect new magma input.

Lee and Ripley (in press) estimated a possible parental magma composition for the South Kawishiwi Intrusion that is very similar to many MCR intermediate olivine tholeiite intrusions (*e.g.* Pigeon Point Sill, Weiblen and Morey, 1980) and volcanics (*e.g.* Brannon, 1984); and to the weighted average of Unit VII of the SKI (Table 2). This composition is significantly

Figure 5. *(facing page) Stratigraphic variation of modal rock type, sulphide concentration, Ni concentration and average Fo in olivine, most An-rich composition in plagioclase cores, and whole rock concentrations of* P_2O_5*, Zr and Y in DDH 34870-A from the Spruce Road area of the South Kawishiwi Intrusion (Figures 1 and 5). Data from Lee and Ripley (in press). Long dashes indicate units distinguished by lithologic and mineralization characteristics. Short dashes indicate compositional breaks that, along with the lithologic breaks, are interpreted to separate differentiation or mixing cycles.*

evolved relative to the primitive high-Al olivine tholeiites (Table 2), which are thought to be primary to most Midcontinent rift magmas (Green, 1982). Polybaric fractionation of such a primary magma at lower and mid-crustal levels is thought to be largely responsible for generating the various compositions now observed in the lavas and plutonic rocks of the MCR (Klewin, 1989; Miller and Weiblen, 1990; Jerde, 1991).

Evaluating several crystallization models, Lee and Ripley (in press) found that a model involving *in situ* (or boundary layer) equilibrium crystallization, convective fractionation of interstitial liquid after 60-70% crystallization and its mixing with the main magma reservoir, and periodic magma recharge was most consistent with observed chemical, as well as isotopic data from the lower part of the SKI (Figures 5 and 6). Many of the compositional variations within individual units (Figure 5) can be modelled by a predominantly *in situ* process of 60-70% equilibrium crystallization of olivine and plagioclase within narrow boundary layers. Calculations were made using a starting magma composition similar to the calculated bulk composition of Unit VII (Table 2). Phase equilibria-based liquid-line-of-descent computer algorithms of Weaver and Langmuir (1990) and Ariskin *et al.* (1993) were used in the evaluation. The model is similar to those of Langmuir (1989) and Nielsen and de Long (1992) where fractionated liquid from a bottom crystallization zone is expelled upwards and mixes with remaining liquid in the chamber. Lee and Ripley (in press) show that H_2O content of residual liquid is an important factor in the possible buoyant rise of interstitial liquid. With an initial H_2O content of 0.5 wt% (a reasonable value for continental basalts - *e.g.* Basaltic Volcanism Study Project, 1981; Ripley *et al.*, 1993), the computed density of a fractionated (iron-enriched) liquid derived from an intermediate olivine tholeiite becomes less than that of the initial melt (2.7 $g.cm^{-3}$) after about 60% crystallization. Therefore, convective fractionation and mixing with overlying liquid are not precluded by the iron-enrichment of the fractionated liquid. The prominent reversals in compositional trends and the enrichment of olivine at the base of some compositional cycles suggest recharge of parental magma. The enrichment of olivine may indicate early density segregation of olivine from plagioclase in the intruded magma pulse. Alternatively, it may indicate that the intruded magma was initially saturated only in olivine.

In summary, major units in the Spruce Road area may be explained by boundary layer equilibrium crystallization, some differential crystal settling, limited assimilation of crustal country-rocks, interstial melt migration and mixing, and magma recharge. Processes that affected crystallization of Duluth Complex magmas at their final sites may have differed from those that were important in deeper staging chambers. Differences such as addition of crustal sulphur in deep staging chambers or during early stages of rift evolution may have exerted a major control on the generation of potential sulphide ore-bearing magmas.

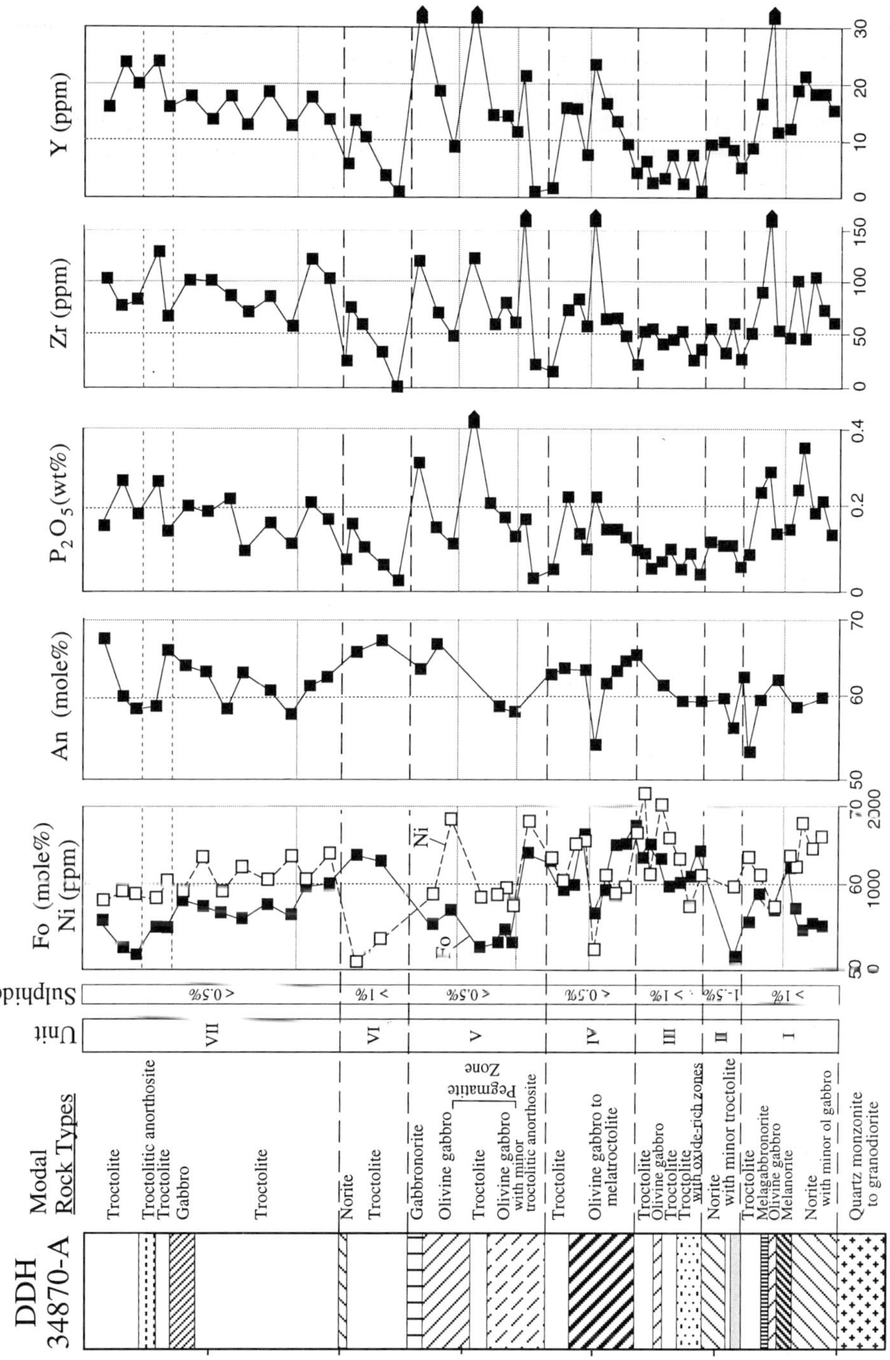

DDH 34870-A
Modal Rock Types
Unit
Sulphide
Fo (mole%) Ni (ppm)
An (mole%)
P2O5 (wt%)
Zr (ppm)
Y (ppm)
Troctolite
Troctolitic anorthosite
Troctolite
Gabbro
Troctolite
Norite
Troctolite
Gabbronorite
Olivine gabbro
Troctolite
Olivine gabbro with minor troctolitic anorthosite
Pegmatite Zone
Troctolite
Olivine gabbro to melatroctolite
Troctolite
Olivine gabbro
Troctolite
Troctolite with oxide-rich zones
Norite with minor troctolite
Troctolite
Melagabbronorite
Olivine gabbro
Melanorite
Norite with minor ol gabbro
Quartz monzonite to granodiorite
VII
VI
V
IV
III
II
I
< 0.5%
> 1%
< 0.5%
< 0.5%
> 1%
1-5%
> 1%
Ni
Fo
m
100
200
300
400
500
600

5.3. Geology and igneous stratigraphy of the Partridge River Intrusion

The lower, extensively drilled part of the PRI is like the SKI in terms of being composed of a heterogeneous, sulphide-bearing lower contact zone overlain by relatively massive troctolite cumulates. However, the detailed igneous stratigraphy of the PRI is distinctive. Severson and Hauck (1990), Geerts (1991) and Severson (1991, 1994) correlated seven generally troctolitic stratiform units over much of the 25 km strike distance of the PRI (Figure 1). The lower three units tend to be less continuous and more variable in thickness and lithology, whereas the upper units are generally composed of homogeneous troctolite separated by picritic to peridotitic layers (Figure 4). A unique feature of the PRI is the presence of coarse-grained oxide-rich dunitic to peridotitic bodies that crosscut the stratigraphy at different horizons. These bodies tend to occur along linear trends, suggesting structural control, and tend to be spatially (and probably genetically) associated with occurrences of iron-formation both in the footwall and as inclusions.

Moderately exposed, but incompletely mapped, bedrock outcrop of the PRI also reveals a very distinct igneous stratigraphy compared to the SKI. Reconnaissance mapping by Bonnichsen (1971 and unpublished field maps) delineated a variety of gabbroic, troctolitic, and anorthositic rock types above the lower sequence of homogeneous troctolite, which he speculated may represent the upper differentiates of the PRI (Bonnichsen, 1974b). More recent reconnaissance mapping in the area by Severson and Hauck (1990) led them to link this structurally and lithologically complex mix of generally evolved gabbroic cumulates to the PRI troctolites. They termed this upper sequence the "Partridge River gabbro complex" and the lower, well-characterized sequence the "Partridge River troctolitic series". We prefer the term "zone" in place of "complex" and "series". One of the more interesting components of the upper gabbro zone is a large mass of fine-grained, granoblastic gabbro, termed the Colvin Creek hornfels (Bonnichsen, 1972). These rocks have chemical compositions and relict structures (amydaloidal zones, flow contacts, cross-bedding) indicative of their origin as a sequence of lower NSVG lava flows and interflow sedimentary rocks. More detailed mapping is needed to better delineate the complex relationships

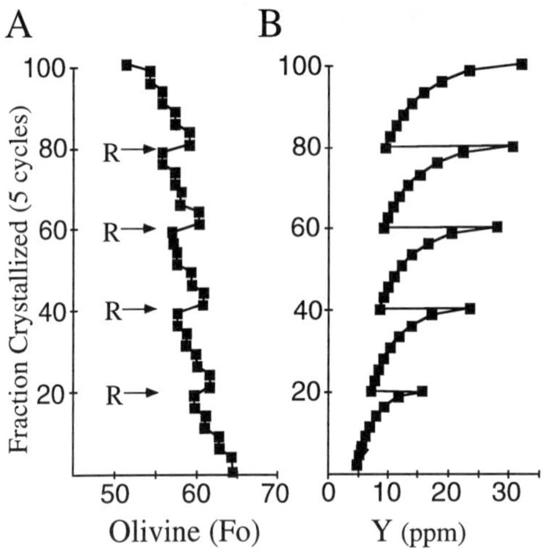

Figure 6. *Modelled variation in Fo content of olivine (A) and whole-rock concentration of Y (B) produced during in situ equilibrium crystallization of an intermediate olivine tholeiite, coupled with magma recharge (R). Within each crystallization cycle, trapped liquid is expelled upward after 70% crystallization of a mass of magma equivalent to 20% of the total magma mass (see Lee and Ripley (in press) for computational details).*

evident in the gabbro zone and its relationship to the lower troctolite zone.

5.4. Petrology of the Partridge River Intrusion in the Babbitt area

Of all Duluth Complex rocks, the greatest number of petrological studies have been conducted on drill core from the Partridge River Intrusion (*e.g.* Grant and Molling, 1981; Rao and Ripley, 1983; Tyson and Chang, 1984; Ripley and Alawi, 1986; Martineau, 1989; Chalokwu and Grant, 1987, 1990; Grant and Chalokwu, 1992; Chalokwu *et al.*, 1993; Taib and Ripley; 1993). While these studies are valid for the sections they portray, it should be understood that they do not profile the total stratigraphy of the PRI, nor are they entirely representative of lateral variations in the lower section of the PRI. All studies have focussed exclusively on detailed evaluation of drill core from the lower 500-600 m of what is probably a 2.5 km thick intrusion the estimated thickness of the PRI if the hornfels bodies and gabbroic anorthosites in the upper gabbro zone are taken to approximate the intrusion hanging-wall, and if internal structure is assumed to dip an average of 20° to the southeast (Bonnichsen, unpubl. maps; Severson and Hauck, 1990). Moreover, all of these studies focussed on drill core from the Babbitt Cu-Ni deposit at the northeastern end of the PRI (Figure 1). Severson's (1994) compilation of the igneous stratigraphy of the PRI troctolitic zone shows that some units (II and III) are not represented in the Babbitt area. Despite the common areal focus of the petrologic studies on the lower troctolitic rocks of the PRI in the vicinity of the Babbitt Cu-Ni deposit, they have led to decidedly different petrogenetic interpretations.

The detailed petrographic, geochemical, isotopic, and mineral chemical evaluation of drill core DDH 221 (Chalokwu and Grant, 1987, 1990; Grant and Chalokwu, 1992; Chalokwu *et al.*, 1993) led to the conclusion that the PRI formed by single-stage emplacement of a plagioclase-olivine mush. The authors further concluded that this mush crystallized *in situ* with limited differential movement of crystals from liquid, save some flotation of plagioclase. Chemical and textural discontinuities evident in the core were due primarily to variable nucleation and growth rates and subsolidus re-equilibration between trapped liquid and primocrysts. They based their conclusions on calculations of trapped liquid composition through the drill core that showed a remarkable uniformity of major- and trace-element abundance. Earlier estimates of this trapped liquid composition based on mass balance considerations (Grant and Chalokwu, 1992) was nearly identical to the ferrodioritic composition (FeO ~20 wt%) of the lower contact zone, which they interpreted to be a chill. However, this composition is much too evolved to be in equilibrium with the primary compositions calculated for cumulus olivine (avg. $Fo_{76\pm5.7}$) and measure cores of plagioclase ($An_{66.3\pm3}$) and led them to conclude that the parental magma was a disequilibrium crystal mush containing primitive xenocrysts in an evolved liquid (Chalokwu and Grant, 1990; Grant and Chalokwu, 1992). In a later paper, using the geochemical thermometry technique of Frenkel *et al.* (1989), Chalokwu *et al.* (1993) recalculated the trapped liquid compositions and found them to be less iron-rich than previously estimated. This revised composition (Table 2), which should produce primocrystic Fo_{67} olivine, gives a better match to earlier estimates of average primary olivine ($Fo_{69\pm2.7}$) composition based on least-squares regression (Chalokwu and Grant, 1987). They reinterpreted the basal ferrodiorite as representing a leading edge liquid from the evolved part of a zoned staging chamber. The main mass of troctolite was formed by emplacement and *in situ* equilibrium crystallization of a crystal mush initally composed of 17-30% phenocrysts in a magma of the composition listed in Table 2.

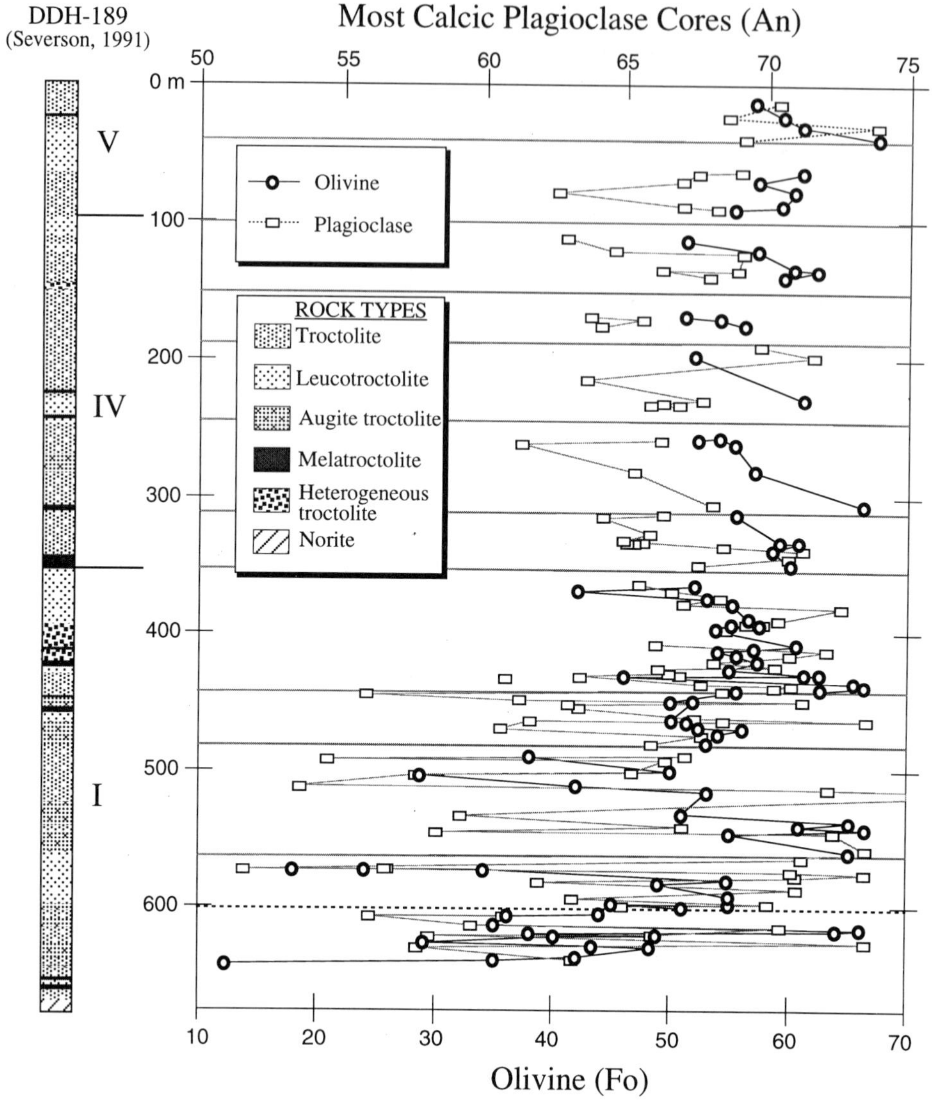

Figure 7. *Stratigraphic variation in rock type (modified from Severson, 1991) and olivine and plagioclase composition in DDH 189 from Babbitt Cu-Ni deposit in the Partridge River Intrusion (Figure 1). Lithostratigraphic units I, IV, and V are based on regional correlation of rock types identified in drill core from the lower part of Partridge River Intrusion (Severson and Hauck, 1990; Severson, 1994). Dashed lines divide the sequence into possible differentiation cycles characterized by upward decreases in Fo content of olivine, or possible mixing cycles marked by upward increases in Fo.*

Taib and Ripley (1993) have worked on similar drill core from the lower PRI and interpret the petrochemical data in a different fashion from Chalokwu *et al.* (1993). Figure 7 is a plot of olivine Fo and plagioclase An versus depth in drill core DDH 189, which is 500 m from DDH 221 and profiles a nearly identical stratigraphic section (Figure 4). Lower Fo and An compositions and enrichment in incompatible elements, such as P, Zr, and Y, in the lower portion of the core is evident and has been observed by many researchers (*e.g.*, Grant and Molling, 1981; Rao and Ripley, 1983; Ripley and Alawi, 1986; Martineau, 1989; Chalokwu and Grant, 1990). However, several other salient features of the core are noteworthy. First, less evolved rock types with more Fo-rich olivine are intercalated with the ferrogabbros of the basal unit. Second, several cycles characterized by upward increases in Fo are present in the overlying troctolitic material. Melatroctolite, containing the most Mg-rich olivine and commonly Ca-rich plagioclase, occurs at the base of many of these cycles (Figure 7). The occurrence of significant melatroctolite that is traceable over most of the lower PRI is the basis on which Severson and Hauck distinguished many of their units (II, IV, VI, and VII) that are otherwise composed predominantly of homogeneous troctolite to anorthositic troctolite. However, only Units I, IV and V of Severson and Hauck (1990) are represented in drill cores 221 and 189 (Figure 4), and significant melatroctolite is found at and near the base of unit IV in both cores. However, it should be noted that Chalokwu and Grant (1990, their figure 1) do not attribute any special signficance to this melatroctolite which occurs at about 270 m depth in DDH 221.

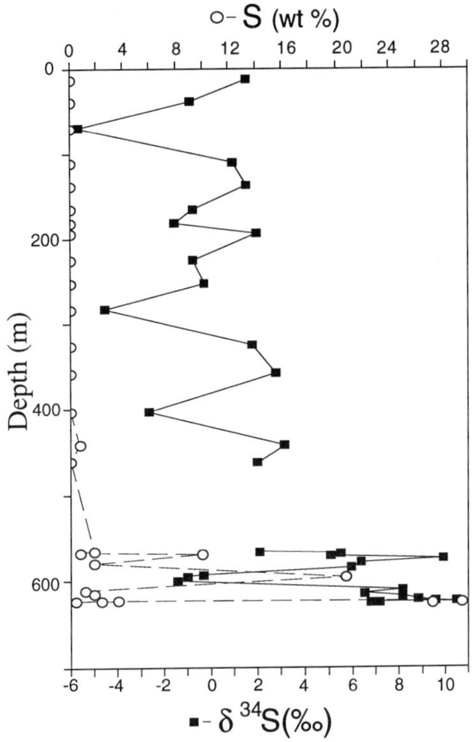

Figure 8. Sulfur concentration (values from 0 to 400 m = 250-500 ppm) and $\delta^{34}S$ values in DDH 189 from Babbitt Cu-Ni deposit in the Partridge River Intrusion (Figure 1).

Another important geochemical feature of DDH 189 is illustrated by S - $\delta^{34}S$ relationships (Figure 8). Sulphide mineralization occurs primarily in the ferrogabbroic layers of Unit I and is characterized by $\delta^{34}S$ values between 5 and 11‰ (characteristic of crustally derived sulphur). The overlying, more primitive rock types are characterized not only by very low sulphide abundance, but also $\delta^{34}S$ values near 0‰ (uncontaminated, mantle derivation). Taken together the data strongly indicate that the Fe-rich layers represent evolved magmas that were contaminated at depth by crustally-derived sulphur. Overlying units are primitive and less contaminated, which is also consistent with $\delta^{18}O$ data of Ripley and Al-Jassar (1987) and Taib and

Ripley (1993), and $^{87}Sr/^{86}Sr$ data of Grant and Chalokwu (1992). These data are consistent to the leading edge ferrodiorite model proposed by Chalokwu *et al.* (1993).

We contend that the recognition of cycles in both the regional lithostratigraphy of the lower PRI delineated by Severson and Hauck (1990) and Severson (1991) as well as in the detailed modal mineralogy and mineral chemistry of a portion of that stratigraphy evident in a single drill core (Figure 7) supports a petrogenetic model of open-system crystallization forming the lower PRI. Severson (1994) notes that the melatroctolitic to feldspathic peridotite layers at the base of most of his PRI units have sharp bases and gradational tops, which he attributes to mechanical segregation of olivine from a new magma pulse. The elevated Fo contents of olivine in such melatroctolites observed in DDH 189 (Figure 7) are consistent with such an interpretation. As demonstrated by the Sonju Lake Intrusion and the base of the Layered Series at Duluth (see below), most moderately evolved tholeiitic magmas emplaced into the Duluth Complex were initially oversaturated to some degree in olivine. The reason for this is unclear, but it is empirical evidence that magmatic recharge in Duluth Complex may be recognized by such enrichment in olivine. Perhaps the enrichment is due to a slight shift in the olivine-plagioclase cotectic toward plagioclase with adiabatic decompression of a moderately evolved tholeiitic magma rising from a deeper crustal staging chamber. Unfortunately, although experimental petrologic studies have detailed the phase equilibrium of primitive tholeiitic magmas at mantle and deep crustal pressures (*e.g.* Presnall *et al.*, 1978), such studies have not thoroughly investigated the phase equilibrium of iron and alkali enriched magmas at moderate to shallow crustal pressures.

The cyclical variations in Fo without olivine enrichment evident in DDH 189 (Figure 7) may also reflect small-scale recharge or perhaps eruption events, although the inconsistent variability of the most calcic plagioclase in each cyclical trend is problematic. Chalokwu and Grant (1990) observed similar variablity in DDH 221, but attributed it to varied proportions of trapped liquid to cumulus olivine. The difference in intrepretations is that where Taib and Ripley (1993) see cyclical varibility in the core, Chalokwu and Grant (1990) see random variation. Still the question arises, what would cause even random variation if the system was emplaced in a single event as a well-mixed crystal mush and solidified by *in situ* equilibrium crystallization as Chalokwu *et al.* (1993) concluded?

Throughout their series of papers, Chalokwu and Grant have argued that the variations observed in DDH 221 are inconsistent with fractional crystallization and instead must reflect *in situ* equilibrium crystallization of a single stage magma. Their main evidence, as recently restated in Chalokwu *et al.* (1993, p. 541), is the uniformity of mineral and calculated liquid compositions through the core. However, the authors fail to acknowledge that this 500 m thick core represents only a small percentage (~20%) of the total PRI stratigraphy. As model and empirical observations demonstrate, very little compositional variation is observed in the early stages of fractional crystallization of a mafic tholeiitic magma, even in a closed system such as the Sonju Lake Intrusion (Figure 3). Accepting that frequent magmatic recharge influenced the crystallization of at least the lower PRI, it is understandable why little compositional variability is evident in the drill core from the Babbitt deposit area.

In summary, although the different igneous stratigraphies of the PRI and SKI suggest that each system evolved independently of the other, they share many lithostratigraphic and petrochemical features. Both formed from similarly evolved tholeiitic magmas and their earliest magmas were variably contaminated, S-enriched ferrodioritic liquids that crystallized to form sulphide-bearing basal units in each body. Most significantly and controversially, we believe that mineralogic and cryptic homogeneity evident in the lower parts of both intrusions (assuming the upper part of the SKI was removed by erosion) is due in part to frequent magmatic recharge of a common parental magma and in part to inefficient fractional crystallization.

6. LAYERED SERIES AT DULUTH

With 120 years of field, structural and petrologic studies, the well-exposed gabbroic rocks forming the escarpment above the city of Duluth have long been recognized as the type section of the Duluth Complex. The pioneering studies of F.F. Grout, set forth in a series of papers published in 1918 (Grout, 1918a and b) stand as a major contribution to our understanding of the Duluth Complex and the petrology of the mafic intrusions, in general. He recognized the complex to be a multiply intrusive body predominantly composed of a suite of early gabbroic anorthosites, younger layered gabbros, and granophyric rocks of uncertain age and genesis. Taylor (1964) produced the first large-scale (1:24,000) geologic map of the complex in the Duluth area and defined the main series classification – anorthositic, layered (troctolitic), and felsic – that has since been adopted throughout the complex (Weiblen and Morey, 1980). Moreover, based on field, petrographic, and very limited geochemical data, Taylor recognized basic similarities between the layered series at Duluth and the Skaergaard Intrusion, which Wager and Brown (1968) had established as the classic example of fractional crystallization of a tholeiitic magma. Recent detailed mapping in the Duluth area has delineated much more about the structure and cumulate stratigraphy of the layered series (Miller *et al.*, 1993*b*), and ongoing petrologic studies are targeted towards unraveling the details of its crystallization history. What follows is a summary of the preliminary results of these ongoing studies.

6.1. Geologic setting

The Layered Series at Duluth (DLS)* forms the southernmost exposure of the Duluth Complex (Figure 1) and is well exposed along a 15 km escarpment rising 200 m above the harbour separating Duluth from Superior, Wisconsin. It occurs as a 3 to 5 km thick, well-differentiated, moderately east-dipping sequence of troctolitic to gabbroic cumulates that was emplaced near the base of the North Shore Volcanic Group (Figure 9). Layering in the DLS and modelling of aeromagnetic and gravity data indicate that its basal contact is not conformable with shallow-dipping (~15°) Keweenawan lava flows in the footwall, but instead dips at a steeper angle (>35°; Miller *et al.*, 1993b). Aeromagnetic data further suggest that the basalts are cut out by the DLS to the north and lower Proterozoic greywacke and slates of the Thomson Formation compose the footwall at the present level of exposure (Figure 9). Although the DLS is in contact with early Proterozoic pelitic sediments, which are thought to have contributed the sulphur for the extensive Cu-Ni sulphide deposits in the SKI/PRI bodies, contamination by footwall material appears to have been minimal in the DLS. Oxygen isotopic values of troctolitic to gabbroic rocks of the DLS range from 5.5 to 6.9‰ and indicate little or no crustal contamination. The monzodioritic and granophyric rocks have $\delta^{18}O$ values up to 8.7‰, however, suggestive of possible contamination along the upper margins of the intrusion.

---

* Although this body would qualify to be termed a layered intrusion, we retain the term "Layered Series at Duluth" because of its well-established usage (Taylor, 1964; Bonnichsen, 1972; Weiblen, 1982).

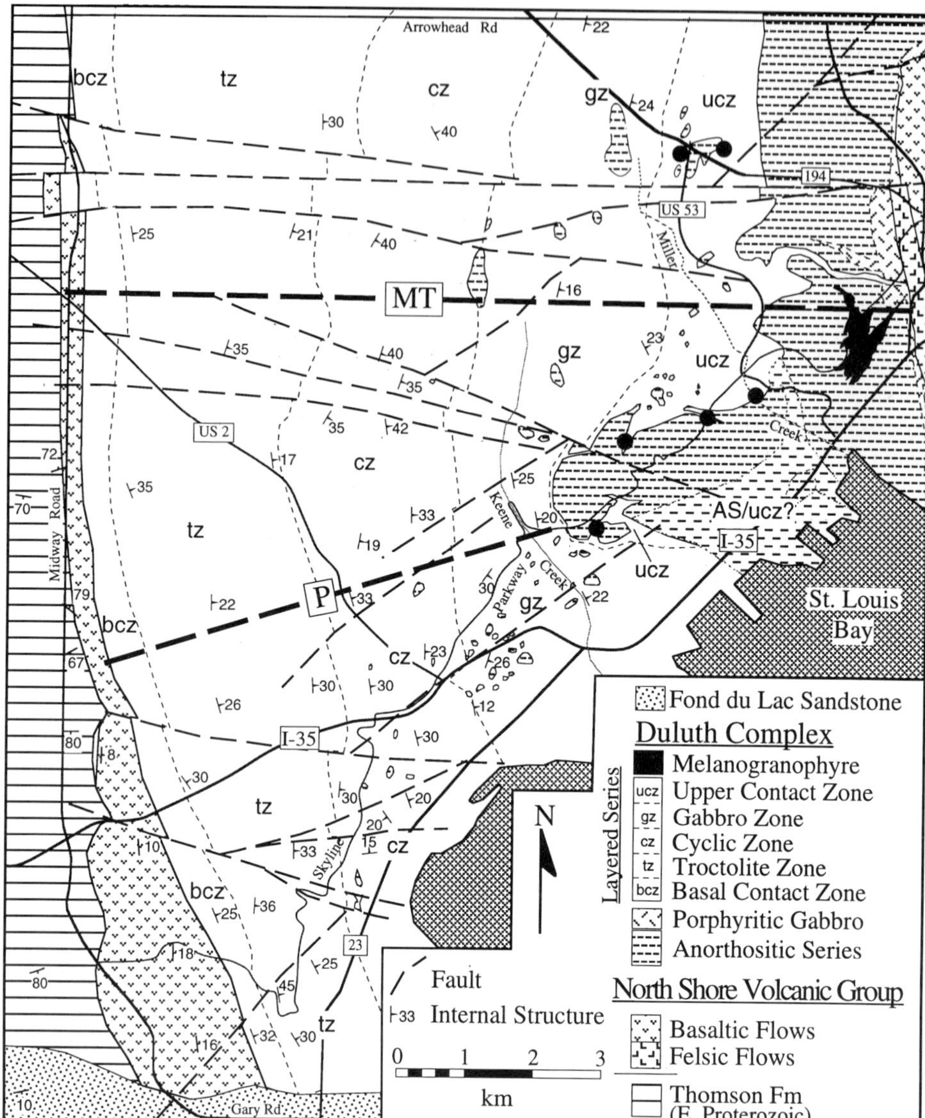

Figure 9. *Geology of the Duluth Complex in the Duluth area. Dashed lines represent the Proctor (P) and Morris Thomas (MT) profiles along which samples were analysed for their mineral chemistry and plotted in Figure 10. AS/ucz? represents an area of no outcrop which may be underlain by rocks either of the anorthositic series or DLS upper contact zone. Locations of six DLS "chill" samples (solid circles) are denoted and their average composition is listed in Table 2.*

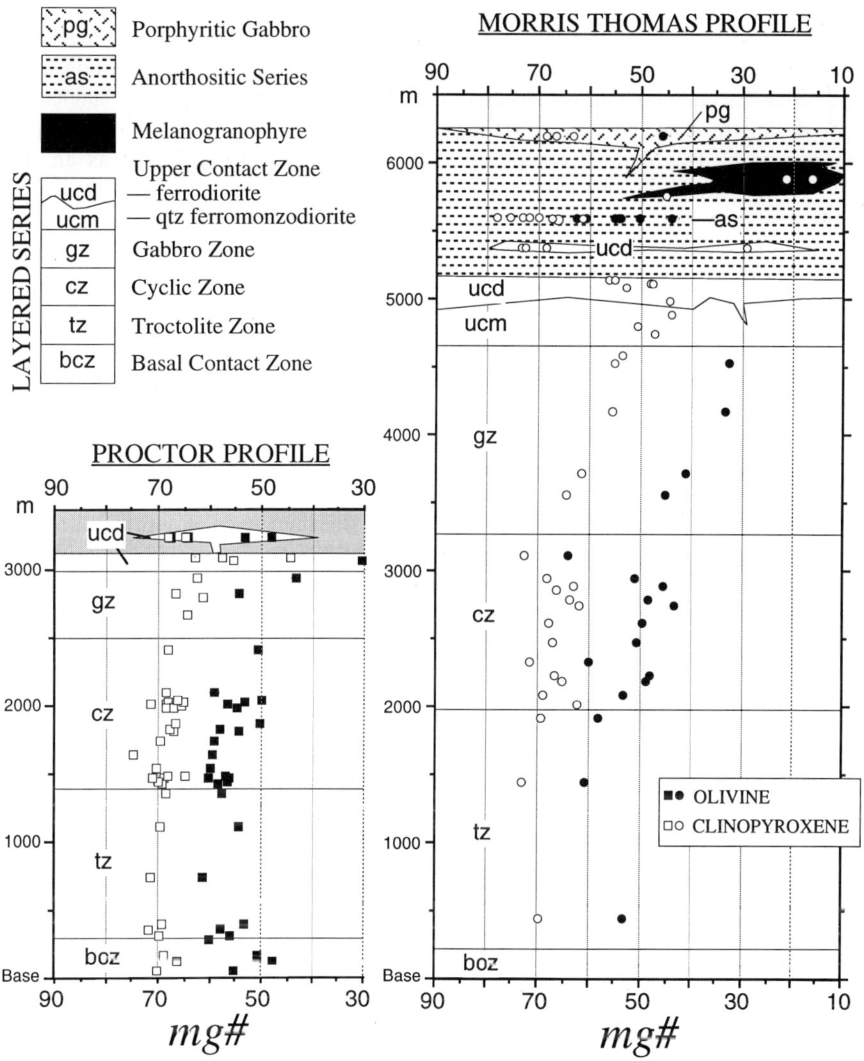

Figure 10. *Stratigraphic variations of average MgO/(MgO+FeO) (mol%) in olivine and clinopyroxene from samples from the Layered Series at Duluth taken along the Proctor and Morris Thomas profiles (Figure 9). In the Morris Thomas plot, data from anorthositic series rocks (as) and dykes of diorite (ucd) and melanogranophyre (mg) which cut the anorthositic series are collectively plotted at different, arbitrarily chosen stratigraphic positions for purposes of comparison to the DLS sequence.*

The hanging-wall of the DLS is composed of coarse-grained plagioclase cumulates of the anorthositic series which is structurally complex both internally and in its relationship to the layered series. Its complex internal structure is manifest by erratic plagioclase lamination on a metre to decimetre scale and the occurrence of anorthositic inclusions within other varieties of anorthositic rock. This internal complexity, which Taylor (1964) described as an "igneous breccia," is a ubiquitous characteristic of the anorthositic series throughout the Duluth Complex and is thought to indicate its formation from multiple emplacements of viscous, plagioclase crystal mush (Miller and Weiblen, 1990). The average rock type of the anorthositic series is an altered, coarse-grained, moderately laminated, ophitic olivine leucogabbro with about 80% plagioclase. With the exception of some occurrences of granular olivine, subhedral to euhedral plagioclase is the only cumulus phase in these rocks.

The complex shape of the contact between the anorthositic series and the layered series (Figure 9) is the consequence of multiple intrusions of layered series magmas into the anorthositic cupola and its resultant disaggregation. A variety of rock types ranging from diabase to layered gabbro to granophyre (larger bodies shown in Figure 9) cut the anorthositic cap as irregular bodies and dykes and indicate that DLS magmas were repeatedly intruded into the cap at various stages of differentiation. Moreover, the abundance of anorthositic inclusions in the upper half of the DLS (Figure 9) indicates that blocks periodically broke free of the cupola and settled to the upward-accumulating floor of the chamber. In fact, the western projection of anorthositic series at its southern extent (Figure 9) probably represents the detachment and foundering of a very large mass of the anorthositic cupola. More limited cryptic and lithologic variations in the DLS along Proctor profile, compared to more extensive variations along the Morris Thomas profile, supports this interpretation (Figures 9 and 10).

6.2. Igneous stratigraphy of the Layered Series at Duluth

The igneous stratigraphy of the DLS may be informally divided into five zones on the basis of their cumulus mineral assemblages and phase layering characterisitics:

Basal contact zone. The lower 200-300 m of the DLS is composed of macrolayers (each 50-150 m thick) that internally grade from lower intervals of medium-grained, well-laminated augite troctolite and melatroctolite to upper sequences of coarse-grained, decussate, ophitic olivine gabbro. The base of each macrolayer is typically marked by strongly layered melatroctolite in sharp contact with olivine gabbro below. Locally, small, irregular bodies of coarse-grained, biotitic ilmenite peridotite to dunite cut across layering in enclosing gabbro and troctolite (Ross, 1985). The bodies resemble the oxide ultramafic intrusions which cut the Partridge River intrusion (Severson and Hauck, 1990).

Troctolite zone. This 1 to 1.5 km thick unit is composed of a sequence of homogeneous, medium- to coarse-grained, moderately laminated, ophitic augite troctolite and troctolite (PO cumulates). Modal layering of cumulus olivine and plagioclase is locally developed, especially in the lower part of the zone, but is variable in frequency, scale, modal extremes, lateral continuity, and orientation. Some troctolite locally contains as much as 4% granular (cumulus) Cr-Al-Mg-bearing ulvospinel. Enrichment in augite oikocrysts (1-5 cm) and interstitial Fe-Ti oxide clots (<2 cm) to as much as 15 vol% is very nonsystematic through the zone. Cryptic variations in olivine and augite compositions through the zone are minor and also nonsystematic (Figure 10).

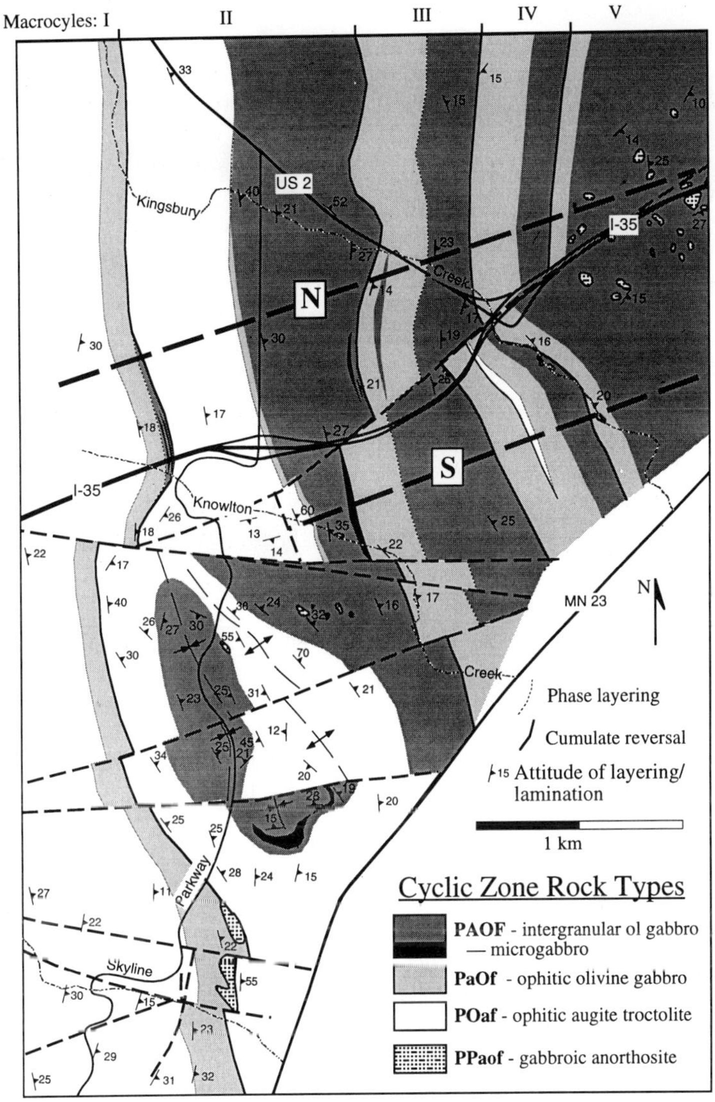

Figure 11. *Geology of the cyclic zone of the DLS near Interstate Highway 35 (I-35). Alternations between three general rock types define five macrocycles (I-V). PPaof cumulates are interpreted to be inclusions of anorthositic series rocks. Boundaries between macrocyles are defined by typically abrupt PAFO → PO cumulate regressions. Long dashed lines labelled N and S denote sample profiles north and south of I-35 (Figure 12). Cumulate rock abbreviations as described in Table 1.*

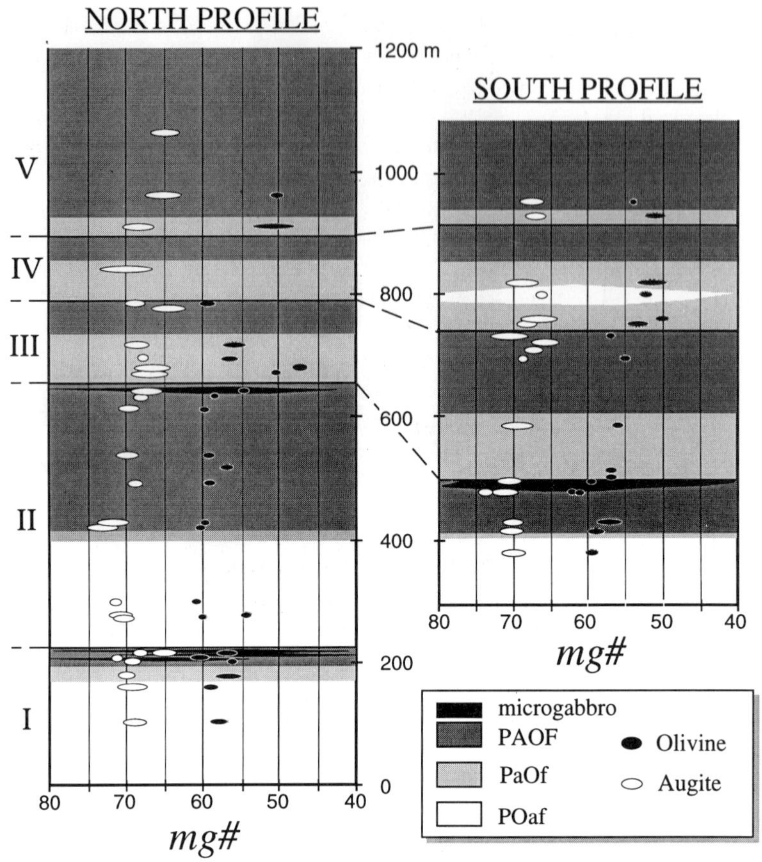

Figure 12. *Generalized cumulus phase layering and cryptic variations in* mg# *in olivine and augite in samples collected along profiles north and south of Interstate Highway 35 (Figure 11). Widths of elliptical data points reflect standard deviations of multiple microprobe analyses. Macrocycles I - V are defined by cyclical cumulus phase layering as described in the text (Figure 11). In general, cryptic variation shows little to no correlation with phase layering.*

Cyclic zone. Within the medial zone of the DLS, a 1-km-thick interval is characterized by cyclical variations in rock type that represent a transition from dominantly two-phase (PO) to three- or four-phase (PAF±O) cumulates (Figures 11 and 12). At least five, 50 to 200 m thick, major cycles compose the zone. In the lower two cycles, POaf cumulates grade into a narrow interval of coarse-grained PaOf cumulates and then aburptly into PAFO cumulates. In the upper three cycles, PaOf cumulates predominate in the lower parts and abruptly grade into PAF±O cumulates. The boundaries between all cycles are narrowly gradational to sharp and are locally marked by gabbroic anorthosite inclusions. Gabbroic anorthosite inclusions occur throughout the PAFO cumulate intervals as well (Figure 11). Fundamentally, these macrocycle boundaries mark a regression in the cumulus paragenesis of the DLS from PAFO back to PO. Curiously, cryptic variations in olivine and augite composition through the cyclic zone show no systematic correlation with the phase transitions. Another interesting feature is the common occurrence just below the macrocycle boundaries of centimetre- to metre-scale layers and lenses of fine-grained ilmenitic olivine gabbro (microgabbro) within the medium-grained PAFO cumulates (Figure 11). The modes and mineral chemistries of the microgabbro occurrences are generally similar to the enclosing PAFO cumulates (Figure 12).

Gabbro zone. This 700 to 1200 m thick upper interval is composed of iron-rich, well-laminated PAF ± O ± IP ± Ap cumulates which are commonly host to many anorthositic series inclusions. Cryptic layering through this zone varies by up 20% in *mg#* of olivine and augite (Figure 10). The contact between the gabbro zone and the overlying upper contact zone (Figure 9) is defined as the approximate level where obvious lamination in the gabbro is lost.

Upper contact zone. This zone hugs the irregular upper contact of DLS against the anorthositic series (Figure 9) and contains hybridized noncumulate mixtures of fine-grained, biotitic ilmenite ferrodiorite and medium-grained, apatitic quartz ferromonzodiorite. The fine-grained ferrodiorite is everywhere adjacent to the main Layered Series –Anorthosite Series contact and has been commonly interpreted to be a chill of the DLS parent magma against the anorthositic cap (Grout, 1918b; Taylor, 1964). At the westward extension of the anorthositic mass between Miller and Keene Creeks (Figure 9), the ferrodiorite alone forms the narrow (< 50 m) upper contact zone and abruptly grades into underlying gabbro cumulates. In the thicker parts of the upper contact zone (Figure 9), however, irregular bodies of medium-grained, apatitic, hornblende-pyroxene quartz monzodiorite intrude into and eventually predominate over the fine-grained ilmenite ferrodiorite. Although not displaying textural characteristics of a cumulate rock, the quartz monzodiorite has progressively more evolved compositions upwards through the upper contact zone that are continuous with compositions observed in the underlying gabbro zone (MT profile, Figure 10).

Although not directly connected to the DLS, an irregular composite intrusion of intermediate to felsic rocks cutting the anorthositic series (melanogranophyre, Figure 9) probably formed from magmas representing the most evolved stages of differentiation of the layered series (Figure 10). Over 90% of the body is composed of a salmon-red melanogranophyre (3-10% Fe-silicates and oxides) grading locally to a medium-grained, intergranular quartz ferromonzodiorite. Other rock types in the margin of the body include a medium-grained, layered olivine ferrogabbro, which is similar to rocks in the gabbro zone, and a coarse-grained, strongly altered, leucocratic hornblende monzogabbro. The textural and mineralogic variability of the monzogabbro and its compositional characteristics led Seifert and Kracher (1992) to conclude that it represents a heterogeneous hybrid of the granophyre/monzodiorite and the host gabbroic anorthosite. The gabbroic anorthosite in this area is strongly altered and tends to contain abundant granophyric mesostasis. These charateristics suggest that the anorthositic series remained partially molten during crystallization of the underlying DLS and was infiltrated by late felsic melts and hydrothermal fluids.

The succession of cumulate rock types in the DLS and their cryptic variations (Figure 10) are consistent with bottom-upward fractional crystallization of a tholeiitic basaltic magma under low-pressure conditions. Local reversals in cumulus mineral paragenesis and cryptic layering, especially evident in the cyclic zone, suggest that periodic perturbations caused minor setbacks to the overall differentiation of the DLS system (Miller, 1993). As discussed in the next section, these perturbations probably were caused by magma replenishment and eruption events.

6.3. Significance of the DLS "chill" and the cyclic zone

Observations of very fine-grained "chilled" mafic rock in sharp intrusive contact with coarse-grained gabbroic anorthosite, of gabbroic anorthosite inclusions in DLS cumulates, and of DLS-like ferrogabbro intrusions in the anorthositic series led Grout (1918b) and Taylor (1964) to the reasonable conclusion that the anorthositic series was emplaced much earlier than the layered series. Subsequent mapping in other parts of the Duluth Complex documented generally similar relationships between these two main rock series (Phinney, 1972b; Bonnichsen, 1972; Davidson, 1972). These intrusive relationships and the unique lithologic and structural attributes of the two series led ostensibly to the acceptance of a basic paradigm of the Duluth Complex that the two series are temporally and petrogenetically distinct. This fundamental tenet was further embellished by the development of two-stage petrogenetic models (Weiblen and Morey, 1980; Miller and Weiblen, 1990). However, the recent discovery of virtually identical U-Pb ages (~1099 Ma) of zircons from anorthositic and layered series rocks from Duluth and from the northwestern part of the complex (Paces and Miller, 1993) severely challenges this basic paradigm and has compelled a reassesment the field relations which led to its development, particularly the significance of the DLS "chill."

With a cursory investigation of the mineralogy and geochemistry of the fine-grained intergranular, biotitic olivine ferrodiorite that forms the DLS "chill" against coarse-grained anorthositic series rocks, it is immediately apparent that this composition is too evolved to be parental to the DLS. The average whole-rock composition of six samples taken from different parts of the DLS-AS contact (Figure 9) corresponds to an evolved high-Ti-P, low-Al tholeiitic ferrobasalt (Table 2). The Chaos 2 model (Nielsen, 1990) calculates that this compositions should initially crystallize a cumulus mineral assemblage of augite (*mg*#71), plagioclase (An_{56}), and Fe-Ti oxide. Clearly, this composition could not have generated the PO cumulates forming the troctolitic zone of the DLS. Rather, its composition is more appropriately comagmatic with the PAF±O cumulates of the cyclic and gabbro zones.

The evolved composition of the DLS "chill" and the similar ages of the DLS and anorthositic series suggest that, rather than the "chill" being a thermal quench of initial DLS magma against an older and colder anorthositic cap, it instead represents the decompression quenching of DLS magma at some intermediate stage of differentiation. Miller (1993, 1995) proposed the following model to explain the relationships of the DLS "chill," the cyclic zone, and the microgabbro intervals. At intermediate stages of fractional crystallization and under low-pressure conditions (1-2 kb), the DLS magma was saturated in olivine and plagioclase and was nearing saturation in augite and ilmenite. Although the transitions from PO to PAFO cumulates within the cyclic zone may have formed from normal crystallization differentiation, the abruptness of the transitions and the lack of systematic cryptic variation across the units (Figure 12) suggest that abrupt changes in physical conditions of crystallization also may have played a role. Such a change may have been an increase in pressure resulting from volatile (CO_2 and perhaps H_2O) saturation of the magma in the roof zone. Experimental data indicate that CO_2 will be saturated in tholeiitic magmas at shallow crustal depth and, because of its large molar volume at low pressure, will result in a significant overpressurization of the magma upon degassing (Bottinga and Javoy, 1990). The buildup of volatiles in the roof zone is consistent with the pervasive hydrothermal alteration of the anorthositic series and could have caused failure of the cupola leading to eruption from, and decompression of, the magma chamber. Because of the strong positive effect of pressure on water solubility in mafic magma at low total pressures (Burnham, 1980), decompression may have resulted in water-saturated conditions and a consequent dramatic rise in the solidus temperature of the magma, thereby quenching it to form the DLS "chill" zone. The lenticular microgabbro layers may represent semi-crystallized chill that detached from the roof zone and settled down to the crystal-liquid interface. Water-saturated conditions in the roof zone are indicated by the common occurrence of biotite phenocryts in the DLS "chill." Moreover, the hydrothermally altered character of gabbroic anorthosite inclusions in the unaltered PAFO cumulates indicates that theses xenoliths were altered prior to their detachment from the anorthositic series cupola. The return to troctolitic cumulate crystallization above the microgabbro layers indicates either recharge and re-inflation of the magma chamber by new magma or simply a return to lithostatic pressures following cessation of volcanic eruption and escape of volatiles from the roof zone. The lack of obvious shifts in *mg#* across cumulate regression boundaries (Figure 12) suggest that the latter process may be more likely.

In conclusion, a decompression-quench origin for the DLS "chill", as opposed to a thermal quench, better accounts for the similar ages of the DLS and the anorthositic series and for the evolved composition of the "chill." Moreover, with the age data indicating that the anorthositic and layered series formed in the same magmatic event, new models for the overall petrogenesis of the Duluth Complex must be considered. Regardless of how these two series are related, it is clear that the DLS formed by processes intermediate to the closed-system differentiation of the Sonju Lake Intrusion and the open-system evolution of the Partidge River and South Kawishiwi Intrusions. Despite frequent interruptions and periodic setbacks due to recharge and eruption, the DLS system evolved to a strongly differentiated state of iron-enrichment and late granophyre generation typical of tholeiitic magmatic systems (Wager and Brown, 1968). Given its intermediate style of formation, good exposure, and accessibililty, it is clearly appropriate that the DLS be considered the "type" layered series intrusion of the Duluth Complex. As such, more study of the DLS is needed to test some of the ideas put forth here.

7. WILDER LAKE INTRUSION

Reconnaissance mapping in the northwestern part of the Duluth Complex (Miller, 1986) has revealed a mafic layered intrusion, termed the Wilder Lake Intrusion (WLI), which is distinct in its structural setting, cumulate stratigraphy, and cryptic layering from other layered series intrusions of the complex. The little that is known about the petrology of this intrusion comes from petrographic and microprobe analyses of about 15 samples taken in the the North Wilder Lake area of the Boundary Waters Canoe Area Wilderness (Figure 13). Aeromagnetic anomaly images and reconnaissance mapping in the Lake Insula 7.5' quadrangle by W.C. Phinney (unpublished data) suggest that the WLI is a tabular-shaped body which extends at least 10 km to the east-southeast of the North Wilder Lake area. Although it is unknown whether the approximately 600 m thick stratigraphic section evident at the western end of the WLI (Figure 13) is representative of the entire, largely unmapped, intrusion, several features observed in the North Wilder Lake area nevertheless make the WLI unique among Duluth Complex intrusions.

One of the more obviously distinct characteristics of the WLI is its moderate (25-35°), northeast dip of igneous lamination and layering. Most troctolitic series intrusions of the

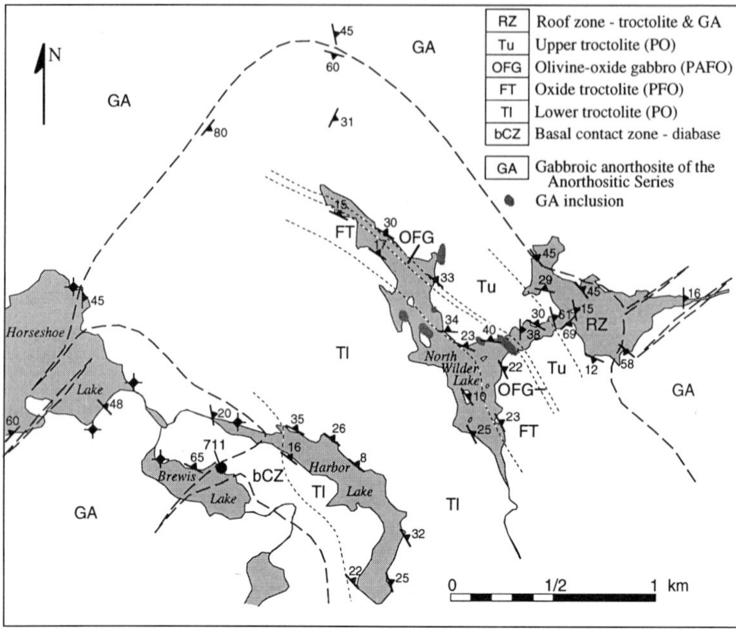

__Figure 13.__ Geology of the Wilder Lake Intrusion in the vicinity of North Wilder Lake (after Miller, 1986). Location of marginal gabbro sample 711 (Table 2) is shown. Strike and dip symbols denote the orientation of igneous lamination and rare modal layering. Abbreviations in parentheses indicate the cumulate rock types as defined in Table 1.

northwestern part of the Duluth Complex (*e.g.* the PRI) occur as thick sequences that dip monoclinally to the southeast (*i.e.*, toward the rift). The structural orientation of the South Kawishiwi and Partridge River Intrusions are examples (Figure 1). Rather than being sandwiched between Archean to Lower Proterozoic basement rocks and a cap of anorthositic rocks, the WLI is completely enveloped by the structurally complex plagioclase cumulates of the anorthositic series (Figure 1). The unique orientation of the WLI may reflect the control of a southeast-trending structural ridge of Archean rocks, which has been postulated by Miller and Chandler (in press) to separate the Duluth Complex and related hypabyssal intrusions into two intrusive "basins." The unique orientation, higher structural emplacement into the anorthositic series, and the presence of a chilled lower margin (see below) suggest that WLI is younger than the layered series intrusions along the base of the Duluth Complex.

Another unique characteristic of the WLI is its cumulus paragenesis. Gradationally upsection, above a 25 to 100 m thick basal contact zone of fine- to medium-grained, decussate to poorly laminated, subophitic olivine gabbro (bCZ), a moderately to well-laminated lower troctolite unit forms the lower part of the cumulate sequence (Figures 13 and 14). The troctolite, which comprises about half the thickness of the entire intrusion, grades from an intermittently layered ophitic augite troctolite in the lower part to a homogeneous augite-poor troctolite in the upper section. The middle section of this unit was not mapped. As observed in several localities around North Wilder Lake (Figure 13), the adcumulate troctolite narrowly grades into an ilmenite troctolite (POF cumulate) containing about 8% modal ilmenite. This unit (FT) is about 50 m thick and is massive, well-laminated, and modally homogeneous throughout.

In contrast to the Sonju Lake Intrusion, which heralds the imminent arrival of cumulus augite by becoming enriched in large, high-density oikocrysts, the FT unit contains only minor (<2%) interstitial augite right up to the arrival of cumulus augite. The overlying OFT unit is less than 25 m thick and is composed of well-laminated, intergranular olivine-oxide gabbro (PAFO cumulate). This unit abruptly gives way to well-laminated and locally modally layered, augite-poor troctolite (Tu) which persists upward through the rest of the section and into a structurally complex, anorthositic inclusion-rich roof zone (RZ). Gabbroic anorthosite inclusions are observed in all units exposed around North Wilder Lake (Figure 13), but are particularly abundant in the roof zone.

Field and geochemical data are insufficient to determine the significance of the cumulus reversal between the olivine-oxide gabbro and upper troctolite units. One possibility is that it represents a recharge event of more primitive liquid. Another possibility is that it represents the merger of an upper border series with the main layered series, akin to the sandwich horizon of

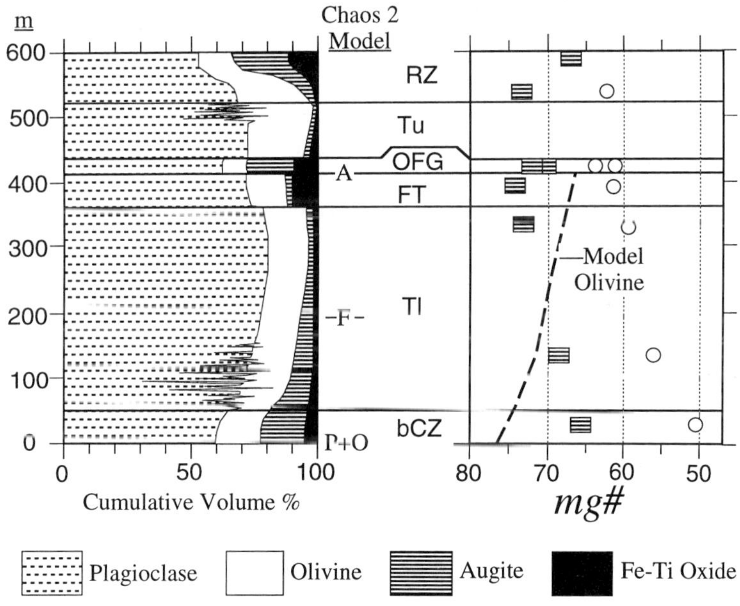

Figure 14. *Modal and cryptic variations through the Wilder Lake Intrusion. Intervals characterized by isomodal layering of olivine schematically denoted by interdigitation of plagioclase and olivine fields. A model curve for cumulus olivine composition and arrivals of cumulus phases generated by closed-system fractional crystallization of the marginal gabbro composition (711, Table 2) was calculated by the Chaos 2 progam (Nielsen, 1990). The model results were scaled to the observed arrival of augite (PAFO unit).*

the Skaergaard Intrusion (Wager and Brown, 1968). The lack of felsic differentiates at the contact and presence of more evolved cumulates in the upper troctolite may reflect the fact that this area is at the fringe of the intrusion. Perhaps more differentiated rock types occur to the east, toward the interior of the intrusion.

In contrast to the Sonju Lake Intrusion, where cumulus augite arrival preceeded Fe-Ti oxide, and the Layered Series at Duluth, where augite and oxide arrived approximately simultaneously, cumulus ilmenite abruptly appeared well before augite arrival in the WLI (Figure 14). The cause of this unique cumulus paragenesis is unclear. Although a uniquely Fe- and Ti-rich and Si-poor parent magma composition is suggested by the whole-rock composition of a fine-grained, intergranular to subophitic olivine diabase dyke intrusive into the coarse-grained gabbroic anorthosite footwall of the WLI (711, Figure 13, Table 2), the low incompatible element abundance of this diabase suggests that it is not a true liquid composition. Nevertheless, applying the Chaos 2 crystallization model to this diabase composition (Nielsen, 1990), assuming closed system fractional crystallization, correctly predicts the observed cumulus paragenetic sequence of PO $\rightarrow$ POF $\rightarrow$ PAFO, but not the relative timing of the arrivals (Figure 14).

A more anomalous feature of the WLI is the reversed cryptic variations of *mg#* in both olivine and augite upwards through the cumulate sequence (Figure 14). The result is that the most primitive olivine and augite compositions occur in the most mineralogically evolved cumulates. Olivine in the olivine-oxide gabbro cumulate has an Fo content almost 15 mol% greater than olivine from the basal contact zone (bCZ). The *mg#* of augite in the ilmenite troctolite cumulate is almost 10 mol% greater than augite at the upper and lower contacts. Suprisingly, where augite is cumulus in the gabbro, it is more iron-rich than in adjacent units where it is intercumulus and in minor abundance. Given the change from noncumulates and orthocumulates at the margins to the nearly pure adcumulates of the overlying units, a reasonable explanation for the reversed cryptic variations is that they reflect progressively less trapped liquid in the cumulates and consequently less of a trapped liquid shift. Figure 14 shows the variation in Fo of primocrystic olivine due to fractional crystallization that is calculated by applying the Chaos 2 modelling program to the diabase composition (711, Table 2) and scaling the results to the cumulus arrival of augite. Comparing the model to the observed olivine compositions, the trapped liquid shift is about 27% Fo in the basal contact zone and is less than 5% Fo in the four-phase cumulate. The 27% trapped liquid shift would imply that about 25% primocrystic olivine of Fo_{76} composition re-equilibrated with 75% postcumulus olivine with an average composition of Fo_{41}. The 5% shift in the four-phase cumulate corresponds to 88% primocrystic Fo_{67} olivine re-equilibrating with 12% postcumulus olivine of an average Fo_{32} composition. These cumulus to postcumulus proportions are consistent with the bCZ rocks being orthocumulates to porphyritic diabases and the layered sequence adcumulates.

8. BALD EAGLE INTRUSION

Finally, brief mention is given here to another layered intrusion of the Duluth Complex which displays an incomplete, but enigmatic cumulate stratigraphy, the Bald Eagle Intrusion (BEI; Wieblen, 1965; Weiblen and Morey, 1980). Like the Wilder Lake Intrusion, the BEI occurs in the northwestern part of the Duluth Complex (Figure 1) and was emplaced almost entirely within rocks of the anorthositic series. As described by Weiblen (1965), the BEI is an elliptical body, 3 x 10 km in plan view, but internal stuctures and gravity data indicate it has a funnel shape which is markedly steeper on its east side (Figure 15). Recent interpretations of aeromagnetic data (Chandler, 1990) suggest that the BEI extends considerably south of the limit of outcrop (compare dashed line in Figure 15 and outline shown in Figure 1). Unlike the Wilder Lake Intrusion, the margin of the BEI is unchilled and is locally observed to be coarse grained against gabbroic anorthosite. This suggests that the BEI was emplaced while the anorthositic series was still hot and therefore formed close to the time that the nearby PRI and SKI layered series intrusions were emplaced (~1099 Ma).

The BEI consists of two distinct rock types (Weiblen, 1965) an outer zone of troctolitic adcumulates and an inner core of intergranular olivine gabbro adcumulates. All rocks have excellent igneous lamination and in some cases, prismatic minerals show a down-dip lineation. Isomodal layering of olivine and plagioclase abundance is common in the troctolite especially where it is steeply dipping. Graded layering is rare. Interstitial Fe-Ti oxide, including minor chromian ulvospinel, occurs exclusively in troctolite, but composes less than 1% of either rock type. The transition from troctolite to olivine gabbro cumulates occurs over a 100 m wide zone wherein interstitial pyroxene increases in abundance, leucocratic compositions become common, and olivine gabbroic and troctolitic rock types become interlayered. Weiblen (1965) noted general cryptic differences in mafic silicates between these two cumulate rock types, but not in plagioclase. Microprobe analyses show that olivine in the troctolite zone varies between Fo_{77} and Fo_{70} and in the gabbro core between Fo_{69} to Fo_{62}. However, a systematic study of the cryptic variation through the intrusion was not conducted.

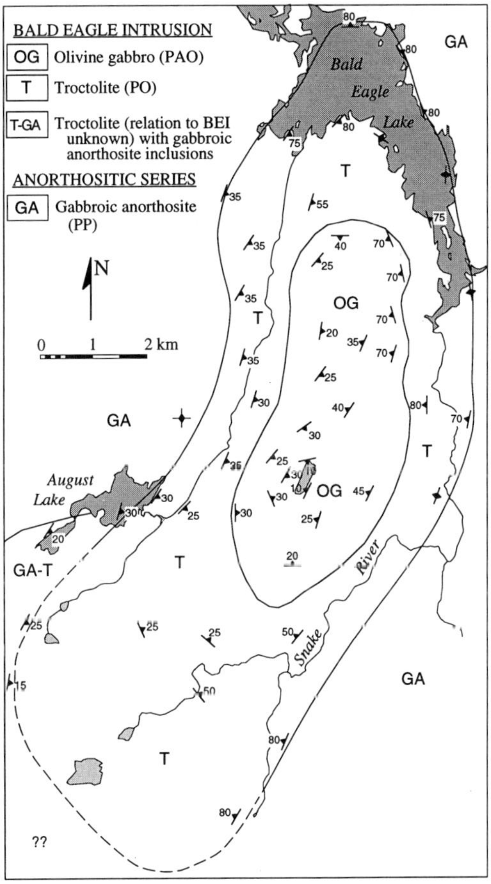

Figure 15. Geology of the Bald Eagle Intrusion (from Weiblen, 1965). Strike and dip symbols denote the orientation of well-developed igneous lamination and rare modal layering. Cumulate rock abbreviations as described in Table 1.

In contrast to all other Duluth Complex intrusions described above wherein the cumulus arrival of Fe-Ti oxide is close to that of augite, oxide never makes an appearance in the BEI despite about half the intrusion being composed of olivine gabbro. Because of its upright funnel shape and the depth of erosion, whatever upper differentiates may have existed in the intrusion have not been preserved. Weiblen (1965) suggested that exposures of olivine-oxide gabbro cumulates, observed some 14 km south of the assumed southern margin of the BEI, may represent upper differentiates of the BEI or an intrusion similar to it. However, recent intrepretations of aeromagnetic anomaly patterns (Meints *et al.*, 1993) indicate that these are part of a separate, largely unexposed layered intrusion, termed the Greenwood Lake Intrusion. Despite an incomplete cumulate sequence, it is nevertheless remarkable that about 50% of the approximately 2.5 km of stratigraphic section is composed of olivine gabbro.

The cause of the extended delay in the onset of oxide crystallization is unclear. The absence of a chilled margin and the thoroughly adcumulate nature of the intrusion preclude any speculation as to whether this delay is due to a unique parent magma composition. A possible explanation for the delay is that the intrusion was frequently recharged by a differentiated magma. Weiblen and Morey (1980) interpreted the limited cryptic variation, the steep dip of lamintion and layering, and adcumulate nature of the intrusion as indicating that the BEI served as an open conduit to higher level intrusions (*e.g.* the South Kawishiwi Intrusion) and perhaps volcanic flows. The present data base does not permit estimation of the volume and periodicity of magmatic recharge or the extent of internal crystal differentiation. Nevertheless, if recharge was an important process in the formation of the BEI, Weiblen (1965) properly noted that the recharging magma must have itself become progressively more differentiated with time in order to account for the rather unidirectional change in cumulate assemblage and cryptic variation between the troctolite and olivine gabbro units. Alternatively, recharge of a differentiated magma may not have occurred until crystallization of the inner gabbro. The interlayering of troctolite and olivine gabbro cumulates in the narrow transition zone between the two units pales in scale and contrast to the recharge/eruption-produced macrocyclic cumulus reversals observed in the Layered Series at Duluth. Clearly, a more detailed characterization of cryptic variations through the BEI stratigraphy is needed to evaluate the unique cumulus paragenesis of the BEI.

9. CONCLUDING REMARKS

With this summary, we hope to have conveyed a sense of the variety of cumulate stratigraphies, cumulus paragenetic sequences, and cryptic variations in six of the many mafic layered intrusions which make up the Duluth Complex. Although explanations for these variations involve a combination of factors including differences in parent magma composition, extent of contamination by country-rocks, physicochemical conditions of crystallization, efficiency of fractional crystallization, and most importantly openness to recharge and eruption, we stress that much more study of even the best known of these intrusions is needed. Moreover, with recent U-Pb dates indicating a contemporaneity between anorthositic series and layered series magmatism (Paces and Miller, 1993), thereby breaking down the long-held temporal paradigm of Duluth Complex, it is clear that these lithologically distinct series can no longer be studied in isolation.

A better understanding of the different, dynamic intrusive systems composing the Duluth Complex will also help toward a more complete understanding of the entire magmatic system of the Midcontinent rift. With high precision geochronologic data being able to establish the contemporaneity of intrusive and volcanic rocks of the rift and with evidence presented here and elsewhere (Miller and Chandler, in press) indicating that some intrusions contributed to surface eruptions, it seems reasonable to expect that some of the compositional diversity evident in the volcanic rocks (*e.g.* BVSP, 1981) is the result of magmatic differentiation of the Duluth Complex and related subvolcanic intrusions. However, the fact that every parental magma to Duluth Complex intrusions studied thus far is itself differentiated from what is commonly considered to be the primary composition of Midcontinent rift magmas (*e.g.* NSVG olivine tholeiite, Table 2) demonstrates that deeper-seated differentiation has also played a major role in creating that diversity. While parameterized models can be generally constructed to estimate magma compositions which could be generated by such deep-seated differentiation, the challenge and promise of petrologic studies of the Duluth Complex intrusions is the potential to more directly determine the magma compositions that were created by differentiation of these various shallow intrusions. Calculations of liquid lines of descent from cumulate rocks do not always produce precise, complete, or unique results, especially in incomplete systems like the Bald Eagle Intrusion or those complicated by frequent recharge such as the Partridge River Intrusion and the Layered Series at Duluth. Nevertheless, such calculations come much closer to the observations than do models of deep-seated differentiation which rely upon many simplifying assumptions. Models of deep-seated differentiation can actually benefit from an understanding of what volcanic compositional variations *cannot* be attributed to shallow differentiation of Duluth Complex and related intrusions.

10. ACKNOWLEDGEMENTS

The authors wish to thank the many researchers past and present who have added and continue to add to our understanding of this complex magmatic system. We hope we have portrayed their ideas accurately. Our own understanding of the complex has benefitted over the years from discussions with Paul Weiblen, Chris Chalokwu, Mark Severson, Bernie Saini-Eidukat, and John Green. JDM wishes to thank the Minnesota Geological Survey, the U.S. Geological Survey, the Minerals Coordinating Committee of the Minnesota State Legislature, and the Minnesota Department of Natural Resources for past financial support of mapping and research projects in the Duluth Complex. EMR acknowledges the support of the National Science Foundation for work dealing with sulphide mineralization and petrogenesis of the Duluth Complex.

11. REFERENCES

Allen, D.J., Hinze, W.J., Dickas, A.B., & Mudrey, M.G., Jr., 1994. Geophysical investigations of the Midcontinent rift system: A new model for western Lake Superior and northern Wisconsin. *40th Inst. Lake Superior Geol.*, Houghton, MI, 1-2.

Ariskin, A.A., Frenkel, M.Y., Barmina, G.S., & Nielsen, R.L., 1993. COMAGMAT: A FORTRAN program to model magma differentiation processes. *Comput. Geosci.* **19**, 1155-70.

Basaltic Volcanism Study Project (BVSP), 1981. *Basaltic Volcanism on the Terrestrial Planets.* New York: Pergamon Press, 1286 pp.

Barnes, S.J., 1986. The effect of trapped liquid crystallization on cumulus mineral compositions in layered intrusions. *Contr. Miner. Petrol.* **93**, 524-31.

Bonnichsen, B., 1971. Outcrop map of southern part of Duluth Complex and associated Keweenawan rocks, St. Louis and Lake Counties, Minnesota. *Minnesota Geol. Surv. Miscellaneous Map* **M-11**, scale 1:125,000.

Bonnichsen, B., 1972. Southern part of the Duluth Complex. In: Sims, P.K., & Morey, G.B. (eds.) *Geology of Minnesota: A Centennial Volume.* St. Paul: Minnesota Geol. Surv., 361-88.

Bonnichsen, B., 1974a. Copper and nickel resources in the Duluth Complex, northeastern Minnesota. *Minnesota Geol. Surv. Info. Circ.* **IC-10**, 24pp.

Bonnichsen, B., 1974b. Geology of the Ely-Hoyt Lakes district, northeastern Minnesota. *Minnesota Geol. Surv. Open-file Rep.*

Bottinga, Y., & Javoy, M., 1990. MORB degassing: bubble growth and ascent. *Chem. Geol.* **81**, 255-70.

Brannon, J.C., 1984. Geochemistry of successive lava flows of Keweenawan North Shore Volcanic Group. Unpubl. Ph.D. thesis, Washington University, St. Louis, 212 pp.

Burnham, C.W., 1980. The importance of volatile constituents. In: Yoder, H.S., Jr. (ed.) *The Evolution of the Igneous Rocks.* Princeton: Princeton University Press, 439-82.

Cannon, W.F., 1994. Closing of the Midcontinent rift — A far-field effect of Grenvillian compression. *Geology* **22**, 155-8.

Cannon, W.F., & Hinze, W.J., 1992. Speculations on the origin of the North American Midcontinent rift. *Tectonophysics* **213**, 49-55.

Cannon, W.F., Green, A.C., Hutchinson, D.R., Lee, M.W., Milkereit, B., Berhent, J.C., Halls, H.C., Green, J.C., Dickas, A.B., Morey, G.B., Sutcliffe, R.H., & Spencer, C., 1989. The North American Midcontinent rift beneath Lake Superior from GLIMPCE seismic reflection profiling. *Tectonics* **8**, 305-32.

Chandler, V.W., 1990. Geologic interpretation of gravity and magnetic data over the central part of the Duluth Complex, northeastern Minnesota. *Econ. Geol.* **85**, 816-29.

Chalokwu, C.I., & Grant, N.K., 1987. Reequilibration of olivine with trapped liquid in the Duluth Complex, Minnesota. *Geology* **15**, 71-4.

Chalokwu, C.I., & Grant, N.K., 1990. Petrology of the Partridge River intrusion, Duluth Complex, Minnesota: I. Relationships between mineral compositions, density, and trapped liquid abundance. *J. Petrology* **31**, 265-93.

Chalokwu, C.I., Grant, N.K., Ariskin, A.A., & Barmina, G.S., 1993., Simulation of primary phase relations and mineral compositions in the Partridge River intrusion, Duluth Complex, Minnesota: implications for the parent magma composition. *Contr. Miner. Petrol.* **114**, 539-49.

Davidson, D.M., Jr., 1972. Eastern part of Duluth Complex. In: Sims, P.K., & Morey, G.B. (eds.) *Geology of Minnesota: A Centenial Volume.* St. Paul: Minnesota Geol. Surv., 354-60.

Foose, M.P., & Cooper, R.W., 1978. Preliminary geologic report on the Harris Lake area, northeastern Minnesota. *U.S. Geol. Surv. Open-file Rep.* **78-385**, 24 pp.

Frenkel, M.Y., Yaroshevsky, A.A., Ariskin, A.A., Barmina, G.S., Koptev-Dvornikov, E.V., & Kireev, B.S., 1989. Convective-cumulative model simulating the formation processes of stratified intrusions. In: Bonin, B. (ed.) *Magma-crust Interactions and Evolution.* Athens: Theophrastus Publications S.A., 3-88.

Geerts, S.D., 1991. Geology, stratigraphy, and mineralization of the Dunka Road Cu-Ni prospect, northeastern Minnesota. *Natural Resources Res. Inst., Univ. Minnesota, Duluth, Tech. Rept.*, NRRI/TR-91-14, 63 pp.

Ghiorso, M.S., & Sack, R.O., 1995. Chemical mass transfer in magmatic processes IV. A revised and internally consistent thermodynamic model for interpolation and extrapolation of liquid-solid equilibria in magmatic systems at elevated temperatures and pressures. *Contr. Miner. Petrol.* **109**, 197-212.

Grant, N.K., & Chalokwu, C.I., 1992. Petrology of the Partridge River intrusion, Duluth Complex, Minnesota: II. Geochemistry and strontium isotope systematics in drill core DDH-221. *J. Petrology* **33**, 1007-38.

Grant, N.K., & Molling, P.A., 1981. A strontium isotope and trace element profile through the Partridge River troctolite, Duluth Complex, Minnesota. *Contr. Miner. Petrol.* 77, 296-305.

Green, J.C., 1972. North Shore Volcanic Group. In: Sims, P.K., & Morey, G.B. (eds.) *Geology of Minnesota: A Centennial Volume*. St. Paul: Minnesota Geol. Surv., 294-332.

Green, J.C., 1982. Geology of Keweenawan extrusive rocks. In: Wold, R.J., & Hinze, W.J. (eds.) *Geology and Tectonics of the Lake Superior Basin*. Geol. Soc. Am. Mem. **156**, 47-56.

Green, J.C., & Fitz, T.J., 1993. Extensive felsic lavas and rheoignimbrites in the Keweenawan Midcontinent rift plateau volcanics, Minnesota: petrographic and field recognition. *J. Volc. Geotherm. Res.* **54**, 177-96.

Green, J.C., Phinney, W.C., & Weiblen, P.W., 1966. Gabbro Lake quadrangle, Lake County, Minnesota. *Minnesota Geol. Surv. Misc. Map* **M-2**, scale 1:24,000.

Grout, F.F., 1918a. Internal structures of igneous rocks; their significance and origin with special reference to the Duluth Gabbro. *J. Geol.* **26**, 439-58.

Grout, F.F., 1918b. A type of igneous differentiation. *J. Geol.* **26**, 626-58.

Grout, F.F., & Schwartz, G.M, 1939. The geology of anorthosites of the Minnesota Coast of Lake Superior. *Minnesota Geol. Surv. Bull.* **28**, 119 pp.

Grout, F.F., Sharp, R.P., & Schwartz, G.M, 1959. The geology of Cook County, Minnesota. *Minnesota Geol. Surv. Bull.* **39**, 163 pp.

Hauck, S.A., Barnes, S.-J., Morton, P., Aliminas, H., Foord, E.E., & Dahlberg, E.H., in press. An overview of the geology and oxide, sulphide, and platinum group element mineralization along the western and northern contacts of the Duluth Complex. In: Ojakangas, R.W, Dickas, A.B., & Green, J.C. (eds.) *Mid-Proterozoic to Cambrian Rifting, Central North America*. Geol. Soc. Am. Spec. Paper.

Hunter, R.H., & Sparks, R.S.J., 1990. The differentiation of the Skaergaard Intrusion (Replies to Discussions by McBirney, Naslund, Morse, Brooks, and Nielsen). *Contr. Miner. Petrol.* **104**, 248-54.

Hutchinson, D.R., White, R.S., Cannon, W.F., & Shultz, K.F., 1990. Keweenawan hot spot: Geophysical evidence for a 1.1 Ga mantle plume beneath the Midcontinent rift system. *J. Geophys. Res.* **95**, 10 869-84.

Irvine, T.N., 1982. Terminology for layered intrusions. *J. Petrology* **23**, 127-62.

Jaupart, C., & Tait, S.R., 1995. Dynamics of differentiation in magma resevoirs. *J. Geophys. Res.* **100**, 17 615-36.

Jerde, E.A., 1991. Geochemistry and petrology of hypabyssal rocks associated with the Midcontinent rift, northeastern Minnesota. Unpubl. Ph.D. thesis, University of California, Los Angeles, 305 pp.

Klewin, K.W., 1989. Polybaric fractionation in an evolving continental rift: evidence from the Keweenawan Mid-continent rift. *J. Geol.* **97**, 65-76.

Langmuir, C.H., 1989. Geochemical consequences of in situ crystallization. *Nature* **340**, 199-205.

Lee, I., & Ripley, E.M., in press. Mineralogic and stable isotopic studies of the South Kawishiwi intrusion, Spruce Road Area, Duluth Complex, Minnesota. *J. Petrology.*

Marsh, B.D., 1988. Crystal capture, sorting, and retention in convecting magma. *Geol. Soc. Am. Bull.* **100**, 1720-37.

Martineau, M.P., 1989. Empirically derived controls on Cu-Ni mineralization: a comparison between fertile and barren gabbros in the Duluth Complex, Minnesota, U.S.A.. In: Prendergast, M.D., & Jones, M.J. (eds.) *Magmatic Sulphides - the Zimbabwe Volume,* 117-37.

McCallum, I.S., Radke, L.D., & Mathez, E.A., 1980. Investigations of the Stillwater Complex: Part 1. Stratigraphy and structure of the Banded zone. *Am. J. Sci.* **280-A**, 59-87.

Meints, J.P., Jirsa, M.A., Chandler, V.W., & Miller, J.D., Jr., 1993. Scientific core drilling in parts of Itasca, St. Louis, and Lake Counties, northeastern Minnesota. *Minnesota Geol. Surv. Info. Circ.* **37**, 159 pp.

Miller, J.D., Jr., 1986. The geology and petrology of anorthositic rocks in the Duluth Complex, Snowbank Lake quadrangle, northeastern Minnesota. Unpubl. Ph.D. thesis, University of Minnesota, Minneapolis, 280 pp.

Miller, J.D., Jr., 1993. Evidence of interruptions during fractional crystallization of the Duluth Complex Layered Series at Duluth. *39th Institute on Lake Superior Geology.* Eveleth, MN, 58-9.

Miller, J.D., Jr., 1995. The Duluth Complex at Duluth. In: Miller, J.D., Jr. (ed.) *Field Trip Guidebook for the Geology and Ore Deposits of the Midcontinent Rift in the Lake Superior Region. Minnesota Geol. Surv. Guidebook Series* **20**, 123-48.

Miller, J.D., Jr., & Chandler, V.W., in press, Geology, petrology, and tectonic significance of the Beaver Bay Complex, northeastern Minnesota. In: Ojakangas, R.W, Dickas, A.B., & Green, J.C. (eds.) *Mid-Proterozoic to Cambrian Rifting, Central North America.* Geol. Soc. Am. Spec. Paper.

Miller, J.D., Jr., & Weiblen, P.W., 1990. Anorthositic rocks of the Duluth Complex: Examples of rocks formed from plagioclase crystal mush. *J. Petrology* **31**, 295-339.

Miller, J.D., Jr., Green, J.C., Boerboom, T.B., & Chandler, V.W., 1993a, Geology of the Doyle Lake and Finland quadrangles, Lake County, Minnesota. *Minnesota Geol. Surv. Misc. Map Series* **M-73**, scale 1:24,000.

Miller, J.D., Jr., Green, J.C., & Chandler, V.W., 1993b, Preliminary geologic map of the Duluth area, St. Louis County, Minnesota. *Minnesota Geol. Surv. Open-file Rep.* **93-2**, scale 1:48,000.

Morey, G.B., & Cooper, R.W., 1977. Hoyt Lakes - Kawishiwi area, St. Louis and Lake Counties, northeastern Minnesota, bedrock geology. *Minnesota Geol. Surv. Open-file Rep.*, scale 1:48,000.

Nathan, H.D., 1969. The geology of a portion of the Duluth Complex, Cook County, Minnesota. Unpubl. Ph.D. thesis, University of Minnesota, Minneapolis, 198 pp.

Nicholson, S.W., & Shirey, S.B., 1990. Midcontinent rift volcanism in the Lake Superior region: Sr, Nd, and Pb isotopic evidence for a mantle plume origin. *J. Geophys. Res.* **95**, 10 851-68.

Nielsen, R.L., 1990. Theory and application of a model of open magmatic system processes. In: Nicholls, J., & Russell, J.K. (eds.) *Modern Methods of Igneous Petrology.* Miner. Soc. Am. Rev. Miner. **24**, 56-106.

Nielsen, R.L., & de Long, S.E., 1992. A numerical approach to boundary layer fractionation: application to differentiation in natural magma systems. *Contr. Miner. Petrol.* **110**, 355-69.

Paces, J.B., & Miller, J.D., Jr., 1993. Precise U-Pb ages of Duluth Complex and related mafic intrusions, northeastern, Minnesota: new insights for physical, petrogenetic, paleomagnetic and tectono-magmatic processes associated with 1.1 Ga Midcontinent rifting. *J. Geophys. Res.* **98**, 13 997-14 013.

Pasteris, J.D., 1985. Relationships between temperature and oxygen fugacity among Fe-Ti oxides in two regions of the Duluth Complex. *Can. Miner.* **22**, 39-53.

Phinney, W.C., 1969, The Duluth Complex in the Gabbro Lake quadrangle, Minnesota. *Minnesota Geol. Surv. Rept. Invest.* 9, 20 p.

Phinney, W.C., 1972a. Duluth Complex, history and nomenclature. In: Sims, P.K., & Morey, G.B. (eds.) *Geology of Minnesota: A Centenial Volume.* St. Paul: Minnesota Geol. Surv., 333-4.

Phinney, W.C., 1972b. Northwestern part of Duluth Complex. In: Sims, P.K., & Morey, G.B. (eds.) *Geology of Minnesota: A Centenial Volume.* St. Paul: Minnesota Geol. Surv., 335-45.

Presnall, D.C., Dixon, J.R., O'Donnell, T.H., Brenner, N.L., Shrock, R.L., & Dycus, D.W., 1978. Liquidus phase relations on the join diopside—fosterite—anorthite from 1 atm to 20 kbar: their bearing on the generation and crystallization of basaltic magma. *Contr. Miner. Petrol.* **66**, 203-20.

Rao, B.V., & Ripley, E.M., 1983. Petrochemical studies of the Dunka Road Cu-Ni deposit, Duluth Complex, Minnesota. *Econ. Geol.* **78**, 1222-38.

Ripley, E.M., & Alawi, J.A., 1986. Sulphide mineralogy and chemical evolution of the Babbitt Cu-Ni deposit, Duluth Complex, Minnesota. *Can. Miner.* **24**, 347-368.

Ripley, E.M., & Al-Jassar, T., 1987. Sulfur and oxygen isotope studies of melt-country-rock interaction, Babbitt Cu-Ni deposit, Duluth Complex, Minnesota. *Econ. Geol.* **82**, 87-101.

Ripley, E.M., Butler, B.K., Taib, N.I., & Lee, I., 1993. Hydrothermal alteration in the Babbitt Cu-Ni deposit, Duluth Complex: mineralogy and hydrogen isotopic systematics. *Econ. Geol.* **88**, 679-96.

Roeder, P.L., & Emslie, R.F., 1970. Olivine-liquid equilibrium. *Contr. Miner. Petrol.* **29**, 275-89.

Ross, B.A., 1985. A petrologic study of the Bardon Peak peridotite, Duluth Complex. Unpubl. M.S. thesis, University of Minnesota, Minneapolis, 140 pp.

Seifert, K.E., & Kracher, A., 1992. Rock-magma mixing in the Duluth Complex, Enger Hill, Duluth: Part 1 - bulk compositions. *EOS, Transactions, Am. Geophys. Union* **73-43**, 655.

Severson, M.J., 1991. Geology, mineralization, and geostatistics of the Minnamax/Babbitt Cu-Ni deposit (Local Boy area), Minnesota, Part I: geology. Natural Resources Res. Inst., Univ. Minnesota, Duluth, Tech. Rept., NRRI/TR-91/13a, 96 pp.

Severson, M.J., 1994. Igneous stratigraphy of the South Kawishiwi Intrusion, Duluth Complex, northeastern Minnesota. Natural Resources Res. Inst., Univ. Minnesota, Duluth, Tech. Rept., NRRI/TR-93/34, 210 pp.

Severson, M.J., & Hauck, S.A., 1990. Geology, geochemistry, and stratigraphy of a portion of the Partridge River intrusion. Natural Resources Res. Inst., Univ. Minnesota, Duluth, Tech. Rept., NRRI/GMIN-TR-89-11, 236 pp.

Shirey, S.B., Klewin, K.W., Berg, J.H., & Carlson, R.W., 1994. Temporal changes in the sources of flood basalts: Isotopic and trace element evidence from the 1100 Ma old Keweenawan Mamainse Point Formation, Ontario, Canada. *Geochim. Cosmochim. Acta* **58**, 4475-90.

Stevenson, R.J., 1974. A mafic layered intrusion of Keweenawan age near Finland, Minnesota. Unpubl. M.S. thesis, University of Minnesota, Duluth, 160 pp.

Taib, N.I., & Ripley, E.M., 1993. Distribution and genesis of Cu-Ni sulphides in a multiple intrusive sequence, Babbitt are, Duluth Complex, MN. (abstr.) *Geol. Soc. Am. Abstr. with Prog.* **25**, A400.

Taylor, R. B., 1964. Geology of the Duluth Gabro Complex near Duluth, Minnesota. *Minnesota Geol. Surv. Bull.* **44**, 63 pp.

Tyson, R.M., & Chang, L.L.Y., 1984. The petrology and sulphide mineralization of the Partridge River troctolite, Duluth Complex, Minnesota. *Can. Miner.* **22**, 23-38.

Wager, L.R., & Brown, G.M., 1968. *Layered Igneous Rocks.* San Francisco: W.H. Freeman, 588 pp.

Weaver, J.S., & Langmuir, C.H., 1990. Calculation of phase equilibrium in mineral-melt systems. *Comput. Geosci.* **16**, 1-19.

Weiblen, P.W., 1965. A funnel-shaped gabbro-troctolite intrusion in the Duluth Complex, Lake County, Minnesota. Unpubl. Ph.D. thesis, University of Minnesota, Minneapolis, 161 pp.

Weiblen, P.W., 1982. Keweenawan intrusive igneous rocks. In: Wold, R.J., & Hinze, W.J. (eds.) *Geology and tectonics of the Lake Superior Basin.* Geol. Soc. Am. Mem. **156**, 57-82.

Weiblen, P.W., & Morey, G.B., 1980. A summary of the stratigraphy, petrology, and structure of the Duluth Complex. *Am. J. Sci.* **280-A**, 88-133.

Zanko, L.M., Severson, M.J., & Ripley, E.M., 1994. Geology and mineralization of the Serpentine copper-nickel deposit, Duluth Complex, Minnesota. *Natural Resources Res. Inst., Univ. Minnesota, Duluth, Tech. Rept.*, NRRI/GMIN-TR-93-52, 90 pp.

Zientek, M.L., Czamanske, G.K., & Irvine, T.N., 1985. Stratigraphy and nomenclature for the Stillwater Complex. In: Czamanske, G.L., & Zientek, M.L. (eds.) *Stillwater Complex, Montana: Geology and Guide.* Montana Bur. Mines & Geol. Spec. Publ. **92**, 21-32.

LAYERED INTRUSIONS
R.G. Cawthorn (editor)

The Fongen-Hyllingen Layered Intrusive Complex, Norway

J.R. Wilson and H.S. Sørensen

Geologisk Institut, Aarhus Universitet, 8000 Aarhus C, Denmark.

Abstract

The 160 km^2 Fongen-Hyllingen Complex, situated 60 km southeast of Trondheim, Norway, is a synorogenic, layered mafic intrusion of Caledonian age. Lateral correlation along 40 km of strike length through *ca.* 6 km of layered cumulates has allowed reconstruction of the form and evolution of the magma chamber. The chamber has the shape of a double-bottomed bowl with an asymmetrical, wide lip which extends towards the south. The northern (Fongen) part of the intrusion occupies the deepest part of the chamber. The southern (Hyllingen) part occupies the lip-like protrusion. Initial $^{87}Sr/^{86}Sr$ ratios (Sr_0) in cumulates range from 0.70308 to 0.70535 and reflect mixing between uncontaminated replenishing magma (with *ca.* Sr_0 0.70308) and resident magma contaminated by partial melts of metapelitic country rocks. Mineral chemistry and Sr_0 show a good correlation through most of the layered series, with more evolved mineral compositions having more contaminated isotopic signatures. Assimilation and fractional crystallization therefore were linked during much of the magma chamber evolution. The intrusion, which has a dioritic bulk composition, is divided into four evolutionary stages based on mineral chemistry.

Stage I comprises a basal reversal. Stage II has fairly constant compositions in any profile normal to modal layering and is characterized by the presence of numerous, large, raft-like, dominantly metabasaltic, inclusions. Stage III consists of a major compositional regression, ending with the most primitive assemblages in the intrusion (with olivine Fo_{75}, plagioclase An_{63}). Stage IV comprises a sequence showing strong normal fractionation, ending with quartz-bearing syenites with low-temperature end-member mineral compositions at the roof. A major feature in the Hyllingen Series is the presence of systematic lateral compositional variations in mineral chemistry, with more evolved compositions along the strike of modal layering approaching the southern margin. These discordant relations between modal and cryptic layering developed as a result of the crystallization of compositionally zoned magma along an inclined floor.

The upper part of the magma chamber became zoned with respect to Sr-isotopes and chemical composition. The lower part remained isotopically homogeneous but chemically zoned until progressive mixing took place with new, uncontaminated magma during the formation of Stage III. Extensive compositional zoning developed both from the roof down (by the generation of roof melts) and from the floor up (by repeated, quiescent emplacement of dense magma along the floor and by mixing as a result of new, dense magma fountaining into buoyant, resident magma).

1. INTRODUCTION

The section which is exposed through a layered intrusion, and the degree of exposure, are both critical features if detailed petrological studies are to be feasible without the aid of extensive drilling. Intrusions that have been subjected to folding offer unique opportunities for study as they can reveal a variety of unusually complete sections. Provided that the deformation can be unravelled, and any accompanying metamorphism is not too severe, three-dimensional aspects of the magma chamber can be reconstructed which are not possible in less deformed intrusions.

Early systematic studies of layered intrusions concentrated on stratigraphic variations in a single profile perpendicular to modal layering. In order to establish cryptic variations it was necessary to separate and analyze the major silicate minerals. This time-consuming process, combined with the fact that the cumulus theory (Wager and Brown, 1968) predicted that modal and cryptic layering are concordant, meant that early studies of layered intrusions were essentially unidimensional.

The advent of the electron microprobe for the routine analyses of rock-forming minerals meant that it became feasible to analyze many more samples. Detailed petrological investigations of the synorogenic Fongen-Hyllingen layered intrusion began in the early 1970s while ideas on the origins of layered igneous rocks were changing dramatically. Sections through the deformed and well-exposed intrusion allow access to the floor, walls, and roof, as well as allowing lateral correlation in a *ca.* 6 km-thick sequence of layered cumulates. The discovery of major, systematic lateral compositional variations in the Fongen-Hyllingen Complex coincided with the increasing consideration of models of *in situ* crystallization, and the application of fluid dynamic theory and experiments to magma-chamber processes. Studies of the Fongen-Hyllingen Complex have contributed significantly to the debate on, for example, mechanisms of magma emplacement, the roles of assimilation and fractional crystallization, and the crystallization of compositionally-zoned magma on a sloping floor.

2. REGIONAL SETTING AND COUNTRY ROCKS

The Fongen-Hyllingen Complex, which occupies an area of about 160 km^2, is the largest mafic intrusion in the central Scandinavian Caledonides (Figure 1). It was emplaced during the later stages of the Caledonian orogeny before the peak of regional metamorphism and the main

---

Figure 1. *(facing page) A. Location of the Fongen-Hyllingen Complex in Norway. B. Simplified geological map of the Fongen-Hyllingen Complex and its country rock envelope. Location of Figure 2A is outlined. C. Idealized subdivision of the layered series into stages on the basis of mineral chemistry (e.g. An% in plagioclase; mg# in mafic minerals) in profiles normal to modal layering. Stages II, III, and IV can be recognized in both northern and southern parts of the complex. D. Lateral correlation of evolutionary stages in the Fongen-Hyllingen Complex. Comparison with Figure 1B shows that this correlation is consistent with the strike of modal layering and the distribution of raft-like inclusions which are concentrated in Stage II. Olivine-rich units are concentrated in the lower part of Stage IV in the Ruten area. Figures 1B and D are based on unpublished work by P. Thy, K.H. Esbensen, S.B. Larsen, J.O. Svane, N.N. Jakobsen, O. Engell-Sørensen, E. Habekost, H.S. Sørensen, B. Paasch, L.G. Knudsen, K.Å. Nielsen, P. Naldal, G.B. Meyer, A. Pedersen, and J. R. Wilson.*

penetrative deformation. The complex is now situated in the uppermost nappe of the Trondheim Nappe Complex.

The present form of the complex is largely due to the effects of the late Caledonian Tydal synform (Wilson and Olesen, 1975), the southerly plunging hinge-zone of which passes through the northern part of the intrusion. Felsic differentiates from the eastern Hyllingen area have yielded a zircon U-Pb age of 426+8/-2 Ma (Wilson *et al.*, 1983) and a Rb-Sr whole rock

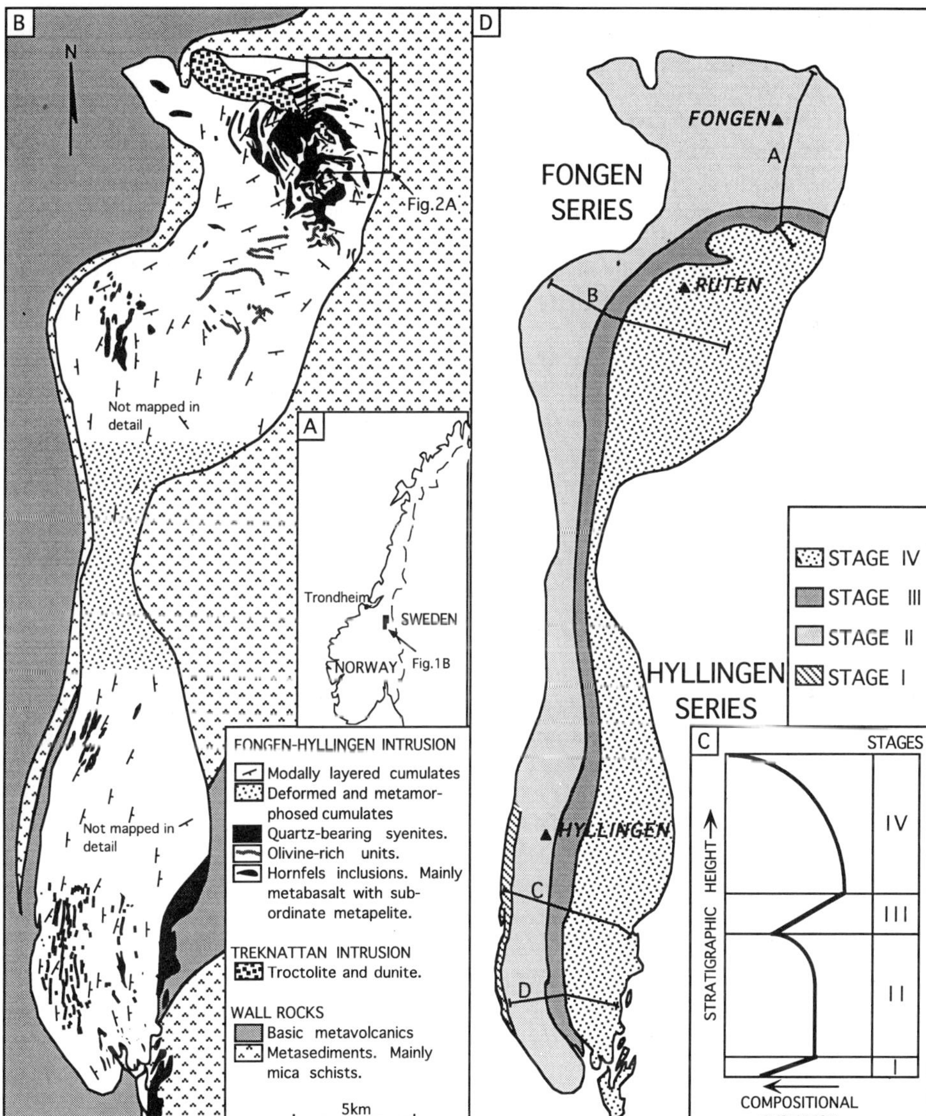

isochron age of 405±9 Ma (Wilson and Pedersen, 1981). These ages are interpreted as recording the times of magmatic crystallization of the final differentiates and post-metamorphic cooling respectively. The ages bracket the main peaks of regional amphibolite facies metamorphism and penetrative deformation in the central Scandinavian Caledonides. The complex therefore belongs to an important group of Caledonian synorogenic mafic (and commonly layered) intrusions in Scotland (*e.g.* the "Newer Gabbros"; Wadsworth, 1982) and Scandinavia (*e.g.* the Honningsvåg Intrusive Suite in northern Norway; Robins *et al.*, 1987).

The Fongen-Hyllingen Complex was intruded roughly along the contact between dominantly metabasic volcanics and metapelites (Figure 1B) that were folded and regionally metamorphosed prior to emplacement of the complex. A large portion of the western margin of the complex is separated from metabasic volcanics by a thin zone of metapelites. Close to the contact with gabbroic rocks belonging to the Fongen-Hyllingen Complex, folded metapelites show signs of partial melting, and cordierite-, sillimanite-, and andalusite-bearing zones are developed, all containing almandine garnet. These parageneses indicate a pressure during crystallization of 5-6 kb (Olesen *et al.*, 1973), equivalent to a depth of 15-20 km. The volcanics have been contact metamorphosed to pyroxene hornfels close to the intrusion and hornblende hornfels further away. A swarm of dominantly plagioclase-phyric basaltic dykes cuts the folded metapelites and metabasic rocks, but pre-dates intrusion of the Fongen-Hyllingen Complex. All these country rock lithologies occur as inclusions in the layered complex.

Regional penetrative deformation and amphibolite facies metamorphism affected the area after intrusion of the complex. Most of the country rocks developed a steep, roughly north-south foliation during this episode, although some areas were partly protected by the contact metamorphism. Kyanite was locally developed at the expense of andalusite in metapelites, indicating an increase in pressure to about 7 kb (Olesen *et al.*, 1973), and metabasic rocks developed amphibolite facies mineralogy. Most of the massive Fongen-Hyllingen Complex was not affected by regional metamorphism and deformation so that primary igneous features can be studied in detail.

3. MAIN FEATURES OF THE FONGEN-HYLLINGEN COMPLEX

This part of Norway has been recently glaciated, and most of the area above the tree line at about 800 m is superbly exposed. Since 1973, a large part of the complex has been mapped on a scale of about 1:10,000 by a total of 14 M.Sc. students under the supervision of the first author.

The Fongen-Hyllingen Complex is named after the major mountains in the northern and southern parts of the area respectively (Figure 1D). A zone of medium- to coarse-grained non-layered dioritic rocks usually occurs along the country rock contact; no chilled margin is developed. A possible feeder to the intrusion may be preserved in the extreme northwestern part of the complex (Wilson *et al.*, 1981a). The northern part of the complex is cut by a minor intrusion - the Treknattan intrusion (Sørensen and Wilson, in press) - that consists dominantly of dunite and troctolite (Figure 1B). The remaining, volumetrically overwhelming part of the complex, consists dominantly of modally layered gabbroic and dioritic rocks which comprise the layered series. Numerous country rock inclusions are enclosed in the layered series.

3.1. Layering features

The rocks comprising the layered series are dominantly medium-grained with an average grain size of 1-3 mm. Small-scale modal layering is the most extensively developed type, with individual layers generally from 2-30 cm thick. Both modally graded layers (generally normally graded) and isomodal layers are widely developed. Monomineralic layers are seldom observed; melanocratic layers generally contain more than one mafic phase and at least 10% plagioclase, whereas leucocratic layers contain >10% mafics. Layers of "average rock", such as those in the Skaergaard Intrusion (illustrated in Figure 10 of Wager and Brown, 1968), are only locally developed. Igneous lamination, defined by the orientation of plagioclase laths and/or tabular pyroxenes in the plane of modal layering, is locally developed. Individual layers can seldom be traced along strike for more than about 30 m before they taper out. In the Ruten area of the northern part of the complex, however, characteristic olivine-rich units can be traced along strike for up to 3 km (Figure 1B).

Magmatic load casts have been described from one of these units (Thy and Wilson, 1980). Trough structures and local discordances are sporadically developed throughout the layered sequence and are evidence for the intermittent action of erosive magmatic currents.

3.2. Inclusions

A major feature of the Fongen-Hyllingen Complex is the presence of abundant country rock inclusions, many of which are raft-like and concordant or subconcordant to modal layering. The majority of the inclusions are metabasaltic hornfelses, but metapelites are also present locally, consistent with the lithology of the wall rocks. These inclusions are most abundant in the lower portions of the layered sequence in both the northern and southern parts of the complex (Figure 1B). In the western Hyllingen area, raft-like inclusions occupy about 22% of the area and measure up to about 1500 x 100 m (average 200 x 25 m). Where exposure and topography allow detailed observations it emerges that the inclusions commonly form a three-dimensional network (Figure 2). Modal layering, identical to that elsewhere in the layered series, is developed between raft-like inclusions and continues right up to the contact of the overlying raft. This feature implies that crystal settling was not responsible for the origin of this modal layering.

The complicated geometrical relationships between layered mafic rocks and inclusions in the Fongen area (Figure 2) illustrate the problem of lateral correlation along modal layering on a local scale. The summit of Fongen mountain is situated in the largest inclusion in the complex which has a total thickness of more than 500 m and covers an area of several square kilometres. The complicated outcrop pattern is partly due to topographic effects, but the interfingering relationship between metabasalt and layered mafic rocks is clear. These raft-like inclusions are believed to represent blocks of the roof that became partially enveloped by magma as the chamber expanded in response to magma replenishment (Habekost and Wilson, 1989).

In some cases the layered rocks show evidence of slumping around and/or depression beneath relatively small (m to dm-sized) country rock inclusions. These xenoliths are believed to have become detached from the local roof and sank through the magma before impacting on the floor where they disrupted the partially consolidated layered mafic rocks.

3.3. Cumulate stratigraphy

Because of the form of the Fongen-Hyllingen Complex and the lateral compositional variations described below, no single profile through the layered series covers the entire

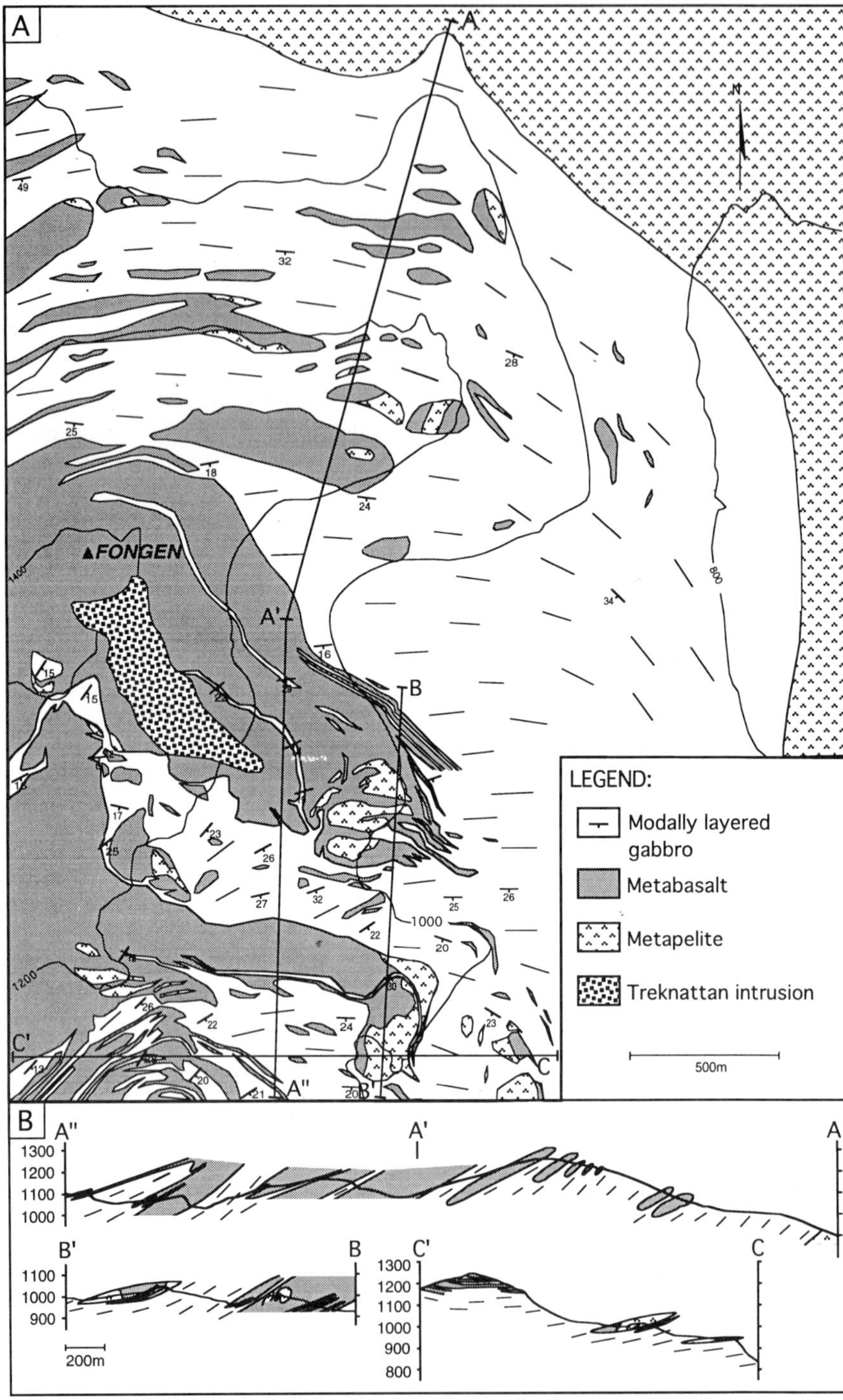
A
FONGEN
A'
A''
B
B'
C
C'
LEGEND:
Modally layered gabbro
Metabasalt
Metapelite
Treknattan intrusion
500m
B
1300
1200
1100
1000
900
800
200m

stratigraphic sequence, and the composite stratigraphic column in Figure 3 has been compiled from several profiles, omitting the raft-like inclusions. The stratigraphically lowest layered mafic rocks are found to the north of Fongen mountain and the highest rocks are developed at the eastern margin of the Hyllingen part of the complex. Lateral correlation between these two parts of the complex is discussed below. Texturally the rocks display typical cumulus/ intercumulus relationships. The large number of cumulus phases, however, commonly results in postcumulus overgrowth playing an important role. A noteworthy exception to this is the presence of intercumulus calcic amphibole, commonly occurring as oikocrysts, through much of the stratigraphy (see below). Biotite, quartz, and K-feldspar occur as intercumulus phases before they adopt cumulus status in the upper part of the Hyllingen Series (Figure 3). The uppermost modal layering roughly coincides with the entry of K-feldspar as an early-crystallizing phase. Above this the quartz-bearing ferrosyenites have granular, non-cumulate textures, but continue to become more evolved upwards, as will be discussed later.

Figure 3 shows that there is a compositional reversal at the base of the profile with highly-evolved rocks (olivine ferrodiorite with An_{38} and Fo_{11}) near the floor. Above this there is a *ca.* 1200 m-thick sequence of olivine ferrogabbros and gabbronorites in which mineral compositions are fairly constant (An_{46-53}; Fo_{35-40}). This is followed by a *ca.* 400 m-thick compositional regression (from An_{47}, Fo_{37} to An_{57}, Fo_{73}) in which apatite and Fe-Ti oxides both cease to be cumulus phases, ending with the most primitive assemblage in the profile. Cr-rich spinel is sporadically present in these primitive cumulates. Above this the assemblage has a fairly constant composition for *ca.* 1000 m (most of the profile B sequence in Figure 3), after which the assemblage becomes progressively more evolved through a thickness of more than 1600 m with the successive entry of cumulus Fe-Ti oxide, calcic amphibole, apatite, biotite, zircon, quartz, K-feldspar, and allanite. The quartz-bearing ferrosyenites (with albite, An_2, and hedenbergite with an *mg#* of 0) which comprise the final differentiates are in contact with country rock amphibolites which form the roof in the Hyllingen part of the complex.

Several important points emerge from Figure 3:

a) The average composition of the layered sequence is broadly dioritic rather than gabbroic. This has implications for the composition of the parental magma.
b) The most primitive rocks in the layered series do not occur at the base but near the middle of the cumulate stratigraphic sequence.
c) There is a reaction relationship between olivine and Ca-poor pyroxene.
d) Hydrous phases are present throughout most of the cumulate stratigraphy. Ca-amphibole is commonly an intercumulus phase and becomes a cumulus phase in evolved assemblages. It is joined by biotite in the latest differentiates.
e) The major solid solution minerals in the final differentiates have low temperature end-member compositions.
f) The main solid solution silicates cover extremely wide compositional ranges: olivine Fo_{73-0};

Figure 2. *(facing page) A. Geological map showing the relationship between layered rocks and inclusions in the eastern Fongen area (area outlined in Figure 1B). Modal layering is generally concordant with the raft-like inclusions (Figure 2B); the apparent discordance in parts of Figure 2A is due to extreme local topographic effects. B. Cross sections through Figure 2A.*

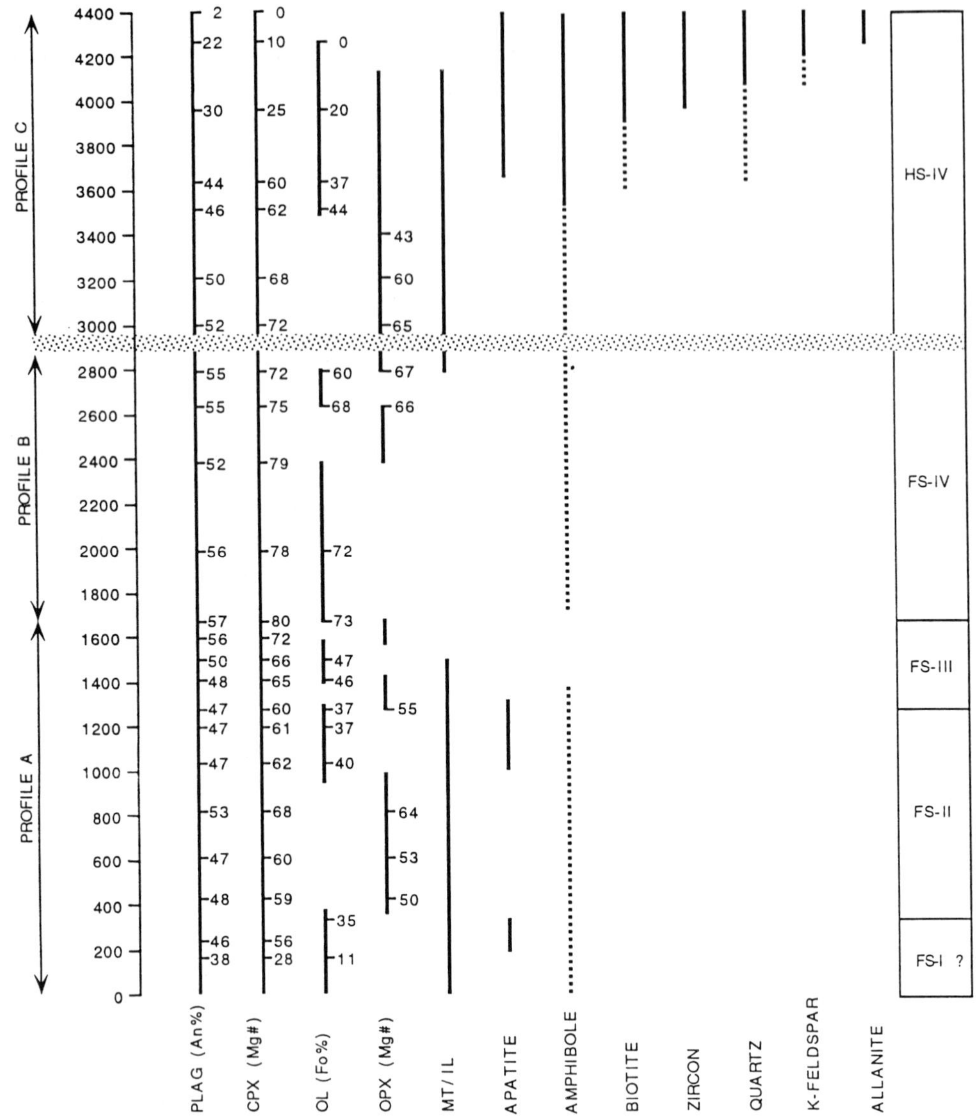

Figure 3. *Cumulate stratigraphy of the Fongen-Hyllingen Complex. The lower part is based on profile A and the upper part of profile B in the Fongen Series (FS). The upper part is based on Stage IV in profile C in the Hyllingen Series (HS). See Figure 1D for location of profiles. Stratigraphic thickness excludes the raft-like inclusions. Lateral compositional variations (Figure 5) result in a slight gap between the top of FS-IV in profile B and the base of HS-IV in profile C.*

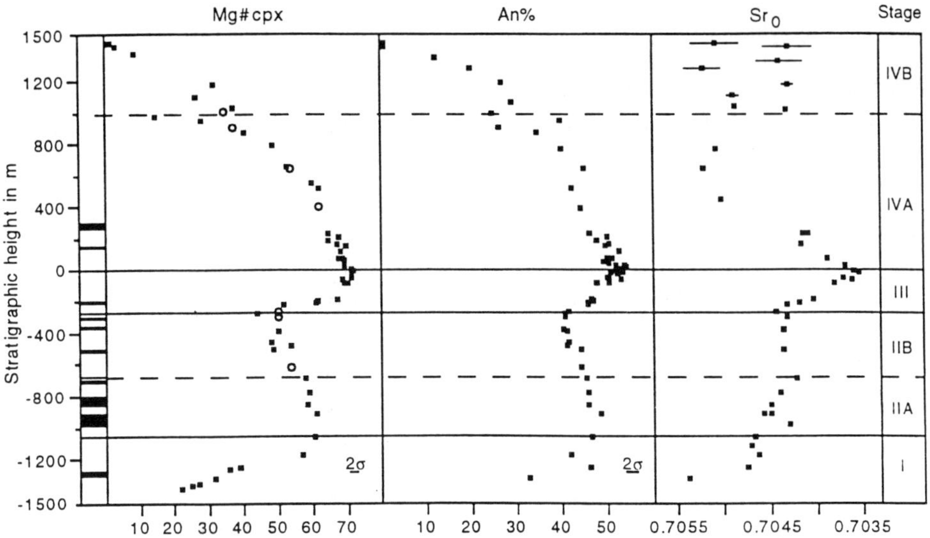

Figure 4. *Compositional variations in profile C (Figure 1D) through the Hyllingen Series. From left to right: stratigraphic thickness (using the boundary between Stages III and IV as datum); inclusions (in black); mg# in Ca-rich pyroxene (open circles - compositions calculated assuming equilibrium with either olivine or Ca-poor pyroxene where Ca-rich pyroxene is not present); An% in plagioclase; initial Sr-isotope ratios, 2σ is contained within the symbols unless indicated; subdivision into evolutionary stages.*

plagioclase $An_{57\text{-}2}$; Ca-rich pyroxene *mg#* 80-0.

g) The fairly constant compositions through thick sequences of cumulates and the major regression in the central part of the sequence, imply that magma replenishment played an important role in the evolution of the Fongen-Hyllingen Complex.

Wilson *et al.* (1981a) considered that the lowest cumulates in the Fongen area comprized ultramafic rocks (with magnesian olivine (Fo_{86}) and Cr-spinel) which they referred to as Fongen Lower Zone. These rocks are now recognized as belonging to the separate Treknattan intrusion (Figure 1B). This reduces the thickness of the layered series in the Fongen area and means that the most primitive rocks in the complex are found in the middle of the layered series and not at the base.

3.4. Subdivision of the layered series into Stages

The southern part of the complex - the Hyllingen Series - was subdivided by Wilson and Larsen (1985) into four evolutionary stages on the basis of mineral chemical variations in a series of profiles normal to modal layering (Figure 1C). These are referred to here as HS-I, HS-II etc. The compositional variation with stratigraphic height outlined in Figure 1C summarizes the mineral chemical data shown in Figure 4. The Hyllingen Series has a maximum thickness (including inclusions) of about 4200 m (Wilson and Larsen, 1985); the profile illustrated in Figure 4 has a thickness of about 3000 m. HS-I comprises a basal reversal, the lower part of which consists of non-layered dioritic rocks; indistinct, wispy modal layering occurs about 100-200 m above the base, followed after a few tens of metres by well-developed modal layering. The basal reversal of HS-I, which is about 340 m thick, therefore continues into the layered sequence. Modal layering is discordant to the western margin of the intrusion and the unlayered diorite by *ca.* 7°. HS-II is defined by a sequence with fairly constant compositions ending with a trend to more evolved rocks. HS-III comprises a gradual regression to more primitive compositions, ending with the most primitive compositions in the profile. In HS-IV the rocks become progressively more evolved upwards, ending with solid-solution end-member compositions at the roof. In addition to these compositional variations normal to the strike of modal layering, the rocks become increasingly evolved along the strike approaching the southern margin of the series. These lateral compositional variations will be considered in more detail below. The subdivision of HS-II and HS-IV into substages (HS-IIA etc.) shown in Figure 4 is based on a combination of mineral and Sr-isotopic data and will be discussed below together with the isotopic variations.

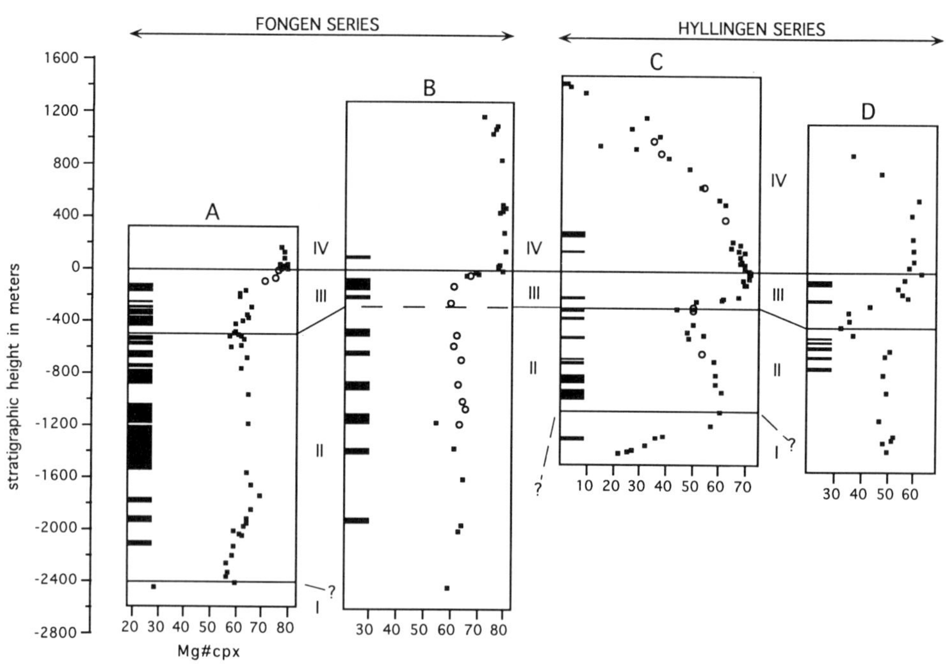

Figure 5. *Cryptic variations in Ca-rich pyroxene in profiles through the Fongen-Hyllingen Complex. (open circles - compositions calculated assuming equilibrium with olivine or Ca-poor pyroxene where Ca-rich pyroxene is not present). Lateral correlation is based on the strike of modal layering, the compositional regression of Stage III, and the distribution of raft-like inclusions. The locations of profiles A-D are shown in Figure 1D.*

On the basis of phase and cryptic layering, the northern part of the complex, which has a maximum thickness (including inclusions) of 6200 m, was divided by Wilson *et al.* (1981a) into the Fongen Sequence, the Transitional Series, and the Ruten Sequence. The entire northern part of the complex is referred to as the Fongen Series and subdivided into Stages (FS-II etc.), using the same criteria as for the Hyllingen Series. Since the layered sequence along the western margin has not been studied in detail in the northern part of the complex the existence of FS-I has not been established, although there is a reversal in the extreme northern part of FS (Figures 3 and 5). The Fongen Sequence, Transitional Series, and the Ruten Sequence of Wilson *et al.* (1981a) comprise FS-II, FS-III and FS-IV respectively.

3.5. Lateral correlation of Stages

The northern and southern parts of the complex are linked by a narrow zone where deformation and metamorphism are extensively developed (Figure 1B). Lateral correlation between the Fongen and Hyllingen Series across this intervening area has been established on the basis of the strike of modal layering, the distribution of raft-like inclusions, and the general lithologies developed, and HS-II, III, and IV are correlated with FS-II, III, and IV respectively (Figures 1D and 5). The boundary between Stages III and IV is used as a datum for the stratigraphic thickness in Figure 5.

Both FS-II and HS-II are characterized by ubiquitous raft-like, dominantly metabasaltic inclusions (Figures 1B and 5). Including the thickness of these inclusions, FS-II has a thickness of 1900-2200 m (Figure 5), decreasing to 800-1000 m in HS-II. Some raft-like inclusions are also present in Stage III in which the gradual compositional regression is developed over a thickness of between 300 and 500 m. The lower part of FS-IV is characterized by several laterally persistent olivine-rich units in the Ruten area (Figure 1B). The uppermost part of FS-IV does not reach such evolved compositions as HS-IV since the roof and uppermost part of the layered series have been eroded away.

The lateral correlation is based on the similarity in the mineral chemical trends in different parts of the complex. One of the major features which emerges from Figure 5 is that the mineral compositions become increasingly evolved along the strike of modal layering from north to south. For example, at the Stage II/III boundary in the Hyllingen Series, Wilson and Larsen (1985) reported that olivine, plagioclase, and Ca-rich pyroxene varies from Fo_{23}, An_{46}, *mg*#49 to Fo_{3}, An_{23}, *mg*#17 over a distance of about 7 km along strike approaching the southern margin. At the Stage III/IV boundary the compositions vary from Fo_{75}, An_{63}, *mg*#79 to Fo_{20}, An_{55}, *mg*#46. These cryptic variations along the strike of modal layering are accompanied by the successive entry of cumulus apatite and zircon (Figure 6). The mineral compositions at the top and bottom of Stage III in the Fongen Series are very similar in profiles A and B in Figure 5 (FS-II/III boundary: Fo_{35}, An_{46}, *mg*#58; FS-III/IV boundary Fo_{72}, An_{60}, *mg*#79). These compositions are considerably more primitive than the northern part of the Hyllingen Series for the Stage II/III boundary, but similar for the Stage III/IV boundary.

The areal compositional variations in the Hyllingen Series shown in Figure 6 illustrate the validity of the subdivision into Stages and the systematic tendency for the rocks to become more evolved along the strike of modal layering approaching the southern margin. Figure 6A also serves to demonstrate the problem of using a zonal subdivision for the Fongen-Hyllingen Complex. For example, if the entry of cumulus apatite were used to define a cumulate zone (which expresses the degree of evolution, *i.e.* fractionation of the assemblage), this zone would

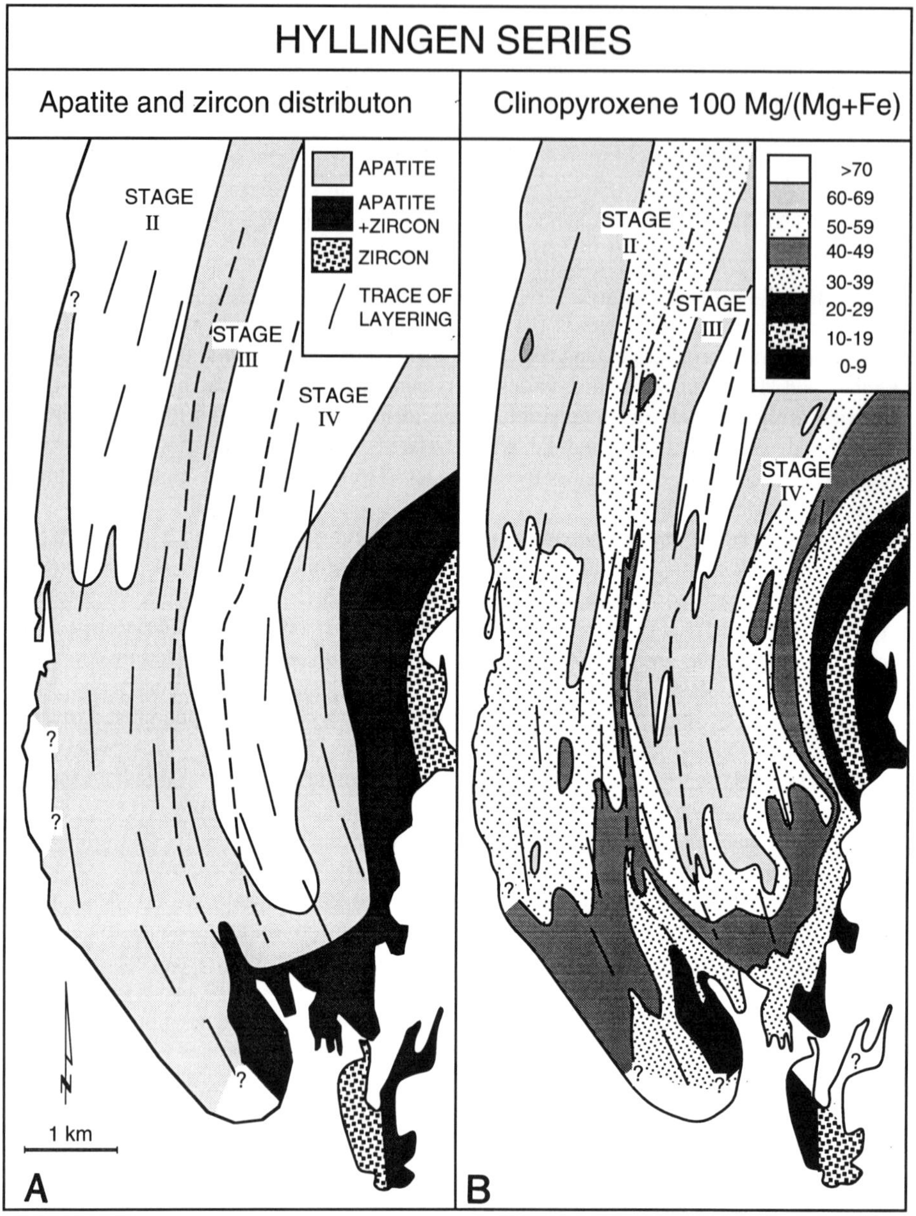

Figure 6. *A. Distributuion of cumulus apatite and zircon in the Hyllingen Series. B. Contoured compositional variation of mg# in Ca-rich pyroxene in the Hyllingen Series. The figures are based on ca. 300 samples from 11 profiles normal to the modal layering (after Figure 11 in Wilson and Larsen, 1985).*

occur three times in the northern part of the Hyllingen Series, but only once in the extreme south.

3.6. Sr-isotopic variation in the Hyllingen Series

Isotopic variations in the Fongen-Hyllingen Complex have recently been studied by Sørensen and Wilson (1995), and some of the results are summarized here. Figure 4 shows that there is generally a very close correlation in most of the Hyllingen Series between mineral chemistry and Sr-isotopic composition, with higher isotopic ratios in more evolved rocks, implying that there is a close relationship between assimilation and fractional crystallization. It is not possible to determine the precise composition of the contaminant end-member. The isotopic data, however, suggest that it was a composite mixture of metabasic and metapelitic wall rocks or partial melts of these, providing a high Sr_0 contaminant. The total range in Sr_0 is from 0.70308 to 0.70535. The highest values are at the base of HS-I and in the upper part of HS-IV, and the lowest value is at the FS-III/IV boundary, consistent with the mineral compositional evolution.

In detail, the basal reversal of HS-I which is so clearly developed for *mg#* in Ca-rich pyroxenes, is only defined by a single Sr_0 value at the base of the profile in Figure 4. This is, however, the highest single Sr_0 value (0.70535) in the entire profile. Above this Sr_0 decreases gradually through the upper part of HS-I and the lower part of HS-II to a value of about 0.7043. In the upper *ca.* 400 m of HS-II, Sr_0 remains fairly constant at 0.7043 whereas the mineral compositions become increasingly evolved. The lower part of HS-II, in which mineral compositions remain constant whereas Sr_0 decreases, is referred to substage HS-IIA. The section above this, where mineral compositions become more evolved and Sr_0 remains constant, defines HS-IIB. The sampling interval in Figure 4 does not allow sufficient resolution to establish whether the point at which *mg#* and An% begin to decrease coincides exactly with the point above which Sr_0 remains constant; there may be an interval up to 200 m thick in which both are essentially constant.

Sr_0 decreases systematically from 0.7044 to 0.7035 through the *ca.* 280 m-thick HS-III, after which it increases to 0.7052 through the lowest *ca.* 900 m of HS-IV, roughly parallel to the mineral compositional trend. This section is referred to as HS-IVA. The correlation between Sr_0 and mineral chemistry breaks down in HS-IVB which comprises the upper *ca.* 600 m of HS-IV; whereas the minerals become increasingly evolved towards the roof, Sr_0 varies erratically from 0.7043 to 0.7052. The scatter in Sr_0 in HS-IVB could to some extent reflect postcumulus disturbance of the Rb-Sr isotopic system in these highly evolved rocks. The Rb/Sr ratio is very high in the uppermost samples (up to 3.8), so that minor redistribution of Rb will give rise to a large error in the age correction.

The ε_{Nd} (at 426 Ma ago) data for the same profile show very similar trends to Sr_0 (Sørensen and Wilson, 1995). The lowest ε_{Nd} value (1.58) occurs at the base of HS-I, the highest (5.27) at the HS-III/IV boundary, and the correlation between mineral chemistry and ε_{Nd} breaks down in HS-IVB.

4. DISCUSSION

4.1. Parental magma, fractionation trend, and conditions of crystallization

In the absence of a chilled margin and contemporaneous dykes the composition of the parental magma to the Fongen-Hyllingen Complex can only be inferred. The coprecipitation of

Ca-rich and Ca-poor pyroxenes, textural evidence for a reaction relationship between olivine and Ca-poor pyroxene, and the development of quartz-bearing late differentiates imply that the parental magma had tholeiitic affinity.

There are several important features, however, that are not typical of tholeiitic fractionation in deep-seated magma chambers. There is no sequence of ultramafic cumulates at the base of the layered series and the average composition of the Fongen-Hyllingen Complex is dioritic rather than gabbroic, suggesting crystallization from an andesitic (or basaltic andesite) parental magma. The fact that plagioclase is albitic in the final differentiates is also an unusual feature. The presence of hydrous minerals (calcic amphibole, biotite) indicates that crystallization took place under elevated p_{H2O} conditions, but the persistence of anhydrous mafic minerals into the most evolved assemblages and the absence of large volumes of granitic differentiates indicate that fractionation did not follow a typical calc-alkaline trend. Fractionation in the Fongen-Hyllingen Complex can be considered to be intermediate between typical tholeiitic and calc-alkaline trends, resulting from crystallization of (basaltic)-andesite magma under elevated and increasing p_{H2O} (Wilson *et al.*, 1981b). The low Sr_0 (0.70308) and high ε_{Nd} (5.84) in the most primitive cumulates point to derivation from a depleted mantle source. Fractionation en route to the magma chamber or during storage in a deeper reservoir could account for the rather evolved characteristic of the Fongen-Hyllingen magma. Alternatively, partial melting of depleted mantle material in the presence of a hydrous fluid may have resulted in the formation of a basaltic or basaltic-andesitic magma consistent with the elevated p_{H2O} (Wyllie, 1981). However, the correlation of Sr_0 with mineral compositions indicates that assimilation was an important parameter in controlling the evolution of the complex.

4.2. The development of compositionally-zoned magma

It will emerge below that compositional zoning of the magma was a vital feature during evolution of the Fongen-Hyllingen Complex. The processes by which compositional zoning can develop will be briefly reviewed here as background for the ensuing discussion.

Roof melting. Buoyant melts released from anatexis of a horizontal roof can lead to an isolated roof melt which has a sharp interface with more primitive, denser magma below. If the roof is inclined, however, the partial melts may mix with the resident magma as they migrate upwards to produce a compositionally-zoned hybrid with the most buoyant, coolest, most contaminated magma at the top (Campbell and Turner, 1987). The heat required for melting the roof is supplied by the release of latent heat of crystallization at the floor and by the addition of new, hot magma to the chamber.

Compositional convection. Evolved magma can also accumulate at the roof of the chamber as a result of compositional convection if buoyant melt is liberated during crystallization. During crystallization on a horizontal floor, buoyant melt released during compositional convection will mix with the overlying magma. However, depending on the shape of the magma chamber, buoyant melt released by crystallization at the margin may be able to reach the roof zone (Huppert *et al.*, 1987).

Magma replenishment by the emplacement of dense magma along the floor. Repeated, quiescent influx of dense magma along the floor can elevate the resident magma column (Huppert and Sparks, 1980); repetition of this process can lead to extensive compositional zoning in the magma with dense, hot, primitive magma at the base and successively more buoyant, cooler, evolved magma upwards. This mechanism does not involve mixing between the new and resident magma.

Forceful emplacement of dense magma into the chamber. The forceful emplacement of dense, primitive magma by fountaining into more-evolved, less-dense resident magma is a very efficient mechanism for the development of compositional zoning in the lower part of the chamber by progressive hybridization (Campbell and Turner, 1989).

The first two processes produce compositional zoning near the roof; processes involving magma replenishment develop compositional zoning near the floor. All the above-mentioned processes may have been involved in the development of extensive compositional zoning in the magma during development of the Fongen-Hyllingen magma chamber.

The elevated Sr-isotope ratios in the stratigraphically highest rocks in HS-IV are interpreted as reflecting contamination by dominantly metapelitic rocks. These highly-evolved rocks with low-temperature end-member mineral compositions and, for example, allanite and zircon as liquidus phases, cannot, however, represent pure roof melts; extreme fractional crystallization must also have been involved in their generation.

In view of the extensive occurrence of cumulus Fe-Ti oxides in the Fongen-Hyllingen Complex (Figure 3), it seems likely that the melt liberated during the early stages of crystallization was buoyant. The extensive network of metabasaltic rafts in the magma chamber may have impeded mixing, but permitted melt liberated from the crystallization front to migrate upwards towards the roof. The presence of a sloping roof could also have been important in this context, allowing buoyant, evolved melt to migrate upwards without extensive mixing with the magma reservoir, but instead generating compositional zoning in the upper portion of the Fongen-Hyllingen magma chamber.

It will be argued below that HS-III crystallized from compositionally-zoned magma produced by mixing between new and resident magma, requiring forceful magma emplacement. The rates of crystallization and replenishment during formation of the interval of Stage II where mineral compositions are fairly constant through a thick interval (the lower part of FS-II and HS-IIA) must have roughly balanced. Replenishment during this interval may not have involved magma mixing and the resident magma column may simply have been repeatedly, or continuously, elevated.

4.3. The origin of discordant relations between modal and cryptic layering

The single most important feature of the Fongen-Hyllingen Complex in terms of the origin of layering is the convincing evidence for systematic, discordant relations between modal and cryptic layering shown schematically in Figure 7A. If modal layering represents the crystallization front (*i.e.* an isochronous surface in the evolving magma chamber), the discordant relations mean that low temperature assemblages were crystallizing near the southern margin at the same time as relatively high temperature assemblages were crystallizing away from the margin. As a first approximation, cryptic layering (defined, for example, by *mg#* in Ca-rich pyroxene) and phase layering (defined, for example, by the entry of cumulus apatite) represent an isochemical and isothermal surface. This means that at any one time the magma in contact with the floor was crystallizing systematically lower temperature mineral assemblages approaching the margin, *i.e.* the magma in contact with the floor was cooler and more evolved towards the magma chamber margin (Figure 7B).

A premise for this is that modal layering represents the crystallization front. This is, in fact, generally assumed in layered intrusions, but is difficult to prove. When modal and cryptic layering are concordant, which was usually taken for granted in earlier studies of layered intrusions, it is logical to assume that they both developed parallel to the magma chamber

Figure 7. (facing page) A. The nature of the discordant relations between modal and cryptic layering in the Hyllingen Series. Cryptic layering is illustrated by the compositional variation of cumulus olivine and the entry of cumulus apatite when olivine reaches about Fo_{37}. B. Modal and cryptic layering are interpreted as representing isochrons and isotherms respectively. C. Development of discordant relations between modal and cryptic layering by crystallization of compositionally stratified magma along an inclined floor. Cryptic layering, which reflects compositional zoning in the magma, probably dips at an angle (α) between that of the crystallization front and the horizontal. D. Crystallization during elevation of the resident zoned magma by influx of dense magma along the chamber floor will give rise to a compositional regression. The angle β will depend on the relative rates of influx and crystallization. Hypothetical compositional variations through profiles X, Y and Z are shown in Figure 7E. E. Hypothetical compositional profiles (X, Y, and Z) through a regression like that in Figure7D. The compositional variation of olivine is shown in three profiles, together with the entry of apatite at about Fo_{37}. Note that the compositional regression develops concordant with modal layering; compositions become more evolved along modal layering from X to Z (up-slope in Figure7D); the Z-shaped pattern of distribution shown by the entry of cumulus apatite (compare with Figure 6).

floor. In the Hyllingen Series they cannot both represent the crystallization front, and the question arises as to whether either of them in fact do so.

There is, however, evidence from the Fongen-Hyllingen Complex that implies that modal layering formed parallel to the crystallization front. As will be discussed in more detail below, the inflection in the mineral chemical trends at the boundary between Stages II and III (*i.e.* the base of the regression which defines Stage III) represents the initiation of a major magma influx event into the chamber. The Stage II/III boundary is entirely concordant with modal layering throughout the Fongen-Hyllingen Complex, not only where modal and cryptic layering are concordant in the Fongen Series, but also in the southern part of the Hyllingen Series where the discordance is strongest. The influx of new, dense, primitive magma will almost simultaneously have an effect on the entire magma chamber floor (the role of mixing and/or elevation of stratified magma are discussed later) and produce higher temperature assemblages. The fact that the Stage II/III boundary (and that between Stages III and IV) is concordant to modal layering is strong evidence in favour of their mutual formation on the crystallization front *i.e.* they both represent isochronous surfaces in the layered intrusion.

The discordance between modal and cryptic layering in the Hyllingen Series rules out crystallization from magma undergoing chamber-scale thermal convection and crystal settling from intermittent currents as suggested for Skaergaard by Wager and Brown (1968). The most likely scenario involves the *in situ* crystallization of compositionally stratified magma along an inclined floor (Figure 7C). Modal layering preserves the advancing crystallization front whereas cryptic layering reflects compositional zoning of the magma. Since the geometry of compositional zoning in the magma is controlled by liquid density, this will be horizontal. Cryptic layering will probably dip at an angle intermediate between the inclination of the crystallization front and the horizontal compositional zonation of the magma (Irvine, 1981). The angle of discordance between modal and cryptic layering in the Hyllingen Series varies systematically from 0° in the north to about 20° approaching the southern margin. The model

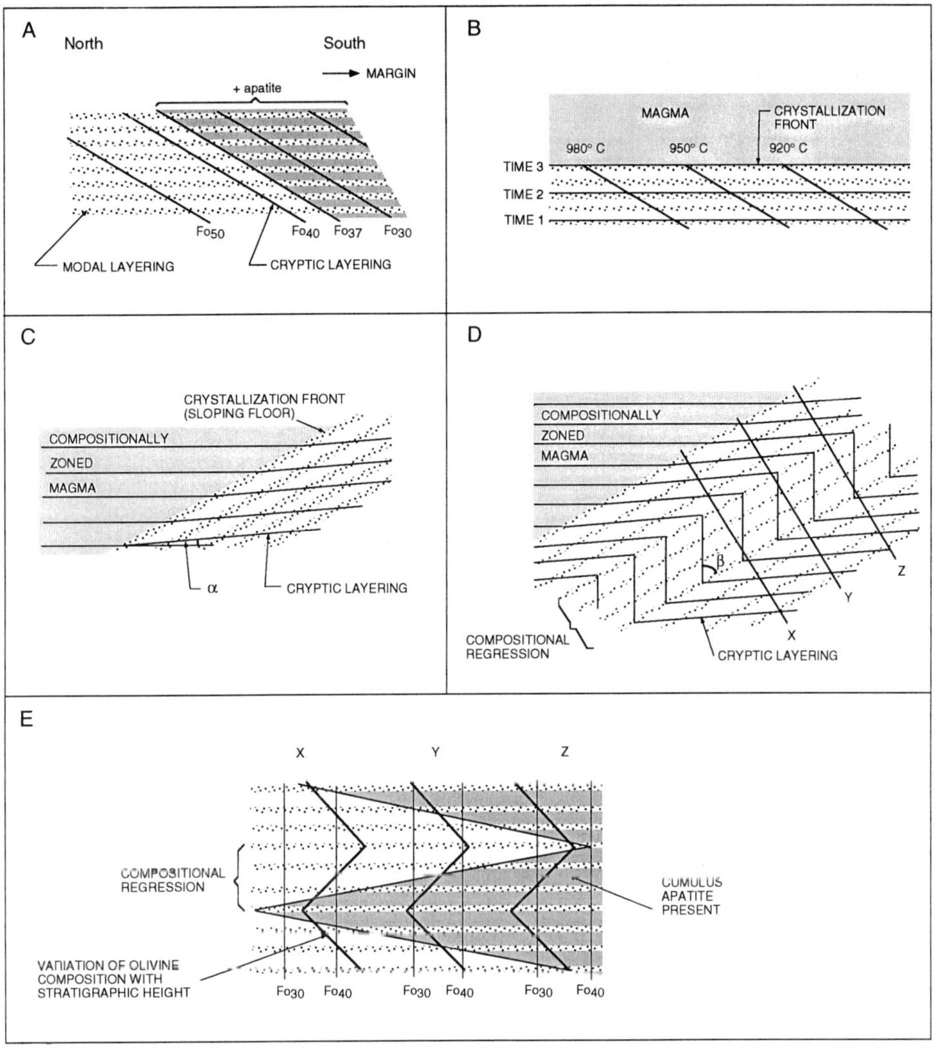

implies that the magma chamber floor sloped slightly more steeply than this towards the margin, *i.e.* the floor was saucer-shaped. This model, which was suggested by Wilson and Larsen (1985) and Wilson *et al.* (1987), is based on the concept of downdip accretion first proposed by Irvine (1981) and Irvine *et al.* (1983).

4.4. The origin of compositional regressions

Compositional regressions in layered intrusions form in response to the influx of new, primitive magma into the chamber. Crystallization along an inclined floor during elevation of a compositionally zoned magma column will produce a regression as progressively denser, more primitive magma comes into contact with the sloping floor (Figure 7D and E). The magma column can be elevated in response to the quiescent addition of new magma at the base of the chamber.

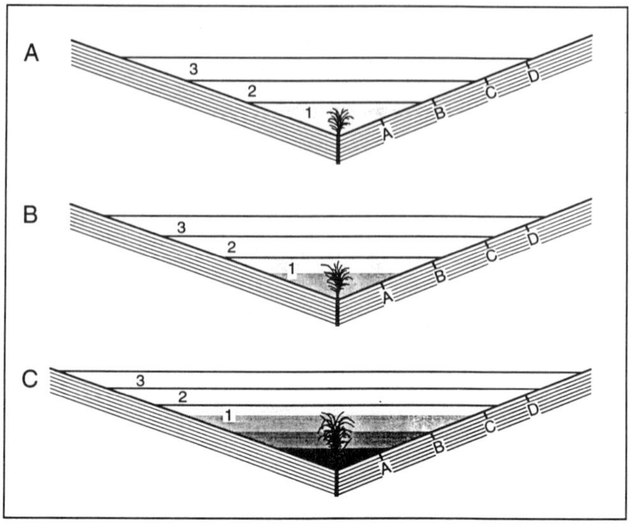

Figure 8 *A. Replenishment by fountaining of dense magma into compositionally zoned resident magma in a broad, funnel-shaped chamber. The floor at A will rapidly be affected by hybrid magma. Points B, C and D are in contact with resident magma. B. Continued magma addition and mixing will result in increasingly primitive magma at A. The overlying, compositionally zoned resident magma above the mixing zone has been elevated so that C and D have been inundated by denser, more primitive layers of resident magma; point B has been inundated by hybrid magma. C. The floor from A-C is covered by hybrid magma which has not reached D. The hybrid is compositionally zoned with dense, primitive, relatively uncontaminated magma at the base and more buoyant, evolved, contaminated magma at the top. The compositional regression develops at A because of magma mixing. That at B and C develops first as a result of the elevation of compositionally zoned magma along the inclined floor and later as a result of mixing. That at D is exclusively a result of the elevation of zoned resident magma.*

As outlined previously, however, magma replenishment can take place forcefully, with the new magma fountaining into the chamber through a narrow conduit, as described by Campbell and Turner (1989). Provided that the new and resident magma have similar viscocities, resident magma becomes entrained into the turbulent fountain of new magma before ponding at the floor (Figure 8). The proportion of new to resident magma in the ponded hybrid gradually increases as influx continues and the efficiency of entrainment decreases. The hybrid magma can thus become compositionally zoned during the fountaining process. If there is an isotopic contrast between new, little contaminated (low Sr_0) magma and resident, contaminated (high Sr_0) magma, progressive mixing will gradually result in lower Sr_0 values towards the base of the hybridized magma column. Crystallization will then give rise to a compositional regression both in terms of mineral chemistry and Sr-isotopic ratio.

The magma which is above the level affected by the mixing will, of course, be elevated during expansion of the chamber in order to accommodate the volume of additional magma emplaced along the base of the magma column. If compositional zoning of the magma is

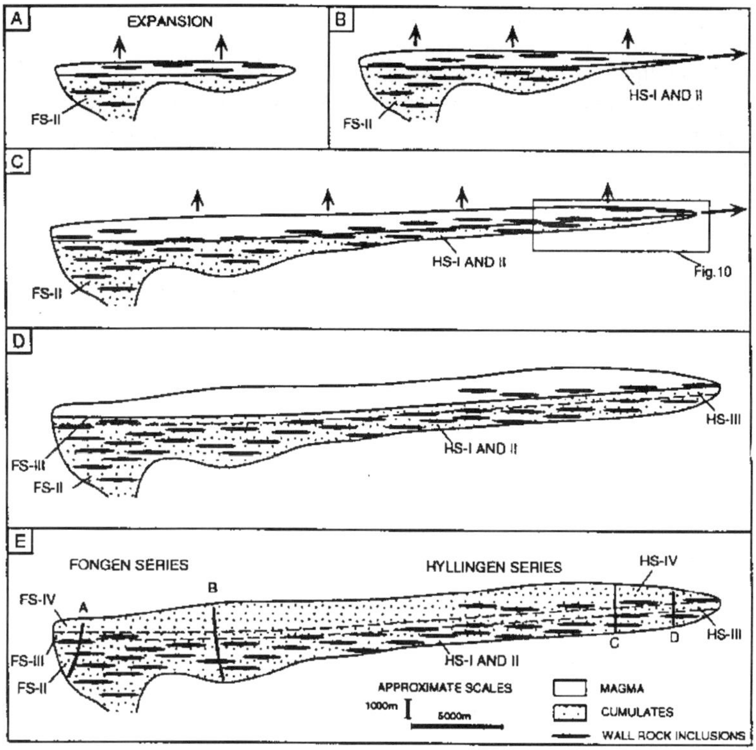

Figure 9. *Schematic illustration of the form and evolution of the Fongen-Hyllingen magma chamber. Stage III is used as datum. A. The magma chamber initially developed in the Fongen area in a period with continuous or repeated magma addition (FS-II). The chamber expanded by roof elevation. Raft-like inclusions of country rocks are interconnected in three dimensions. The "double-bottomed" bowl shape represents that seen at the present level of exposure (Figure 1). The saddle is probably not a persistent feature in the third dimension. B. The chamber expanded to the south in a wedge-shaped protrusion. The Hyllingen Series began to crystallize on the inclined floor of the wedge. C. HS-I formed at the leading edge of the wedge while HS-II (and FS-II) formed away from the wedge during continuous magma addition and magma chamber expansion (the framed area is shown in more detail in Figure 10). Magma addition, and therefore magma chamber expansion, temporarily ceased shortly after the stage of evolution shown in Figure 9C. At this time formation of HS-I ceased. D. Stage III formed during a new period of major magma replenishment. E. When magma addition ceased the resident magma crystallized in a closed system to produce FS-IV and HS-IV. The locations of profiles A-D in Figure 1D are shown schematically. A and B are bent to compensate for folding.*

present in the elevated portion above the mixing zone, this will inflict a compositional regression as the crystallizing floor comes into contact with progressively more primitive magma layers (Figure 7D). This means that compositional regressions can form simultaneously by the two processes at different levels on the magma chamber floor (Figure 8).

4.5. Evolution of the Fongen-Hyllingen magma chamber

The form and evolution of the Fongen-Hyllingen magma chamber based on the combined results of field, petrographical, and chemical studies, is summarized in Figure 9. The main features which lead to this model include: the orientation of modal layering which allows lateral correlation; the distribution of raft-like inclusions; the subdivision into stages in the Fongen and Hyllingen Series; the lateral correlation of the compositional regression defining Stage III; the lowest stage in the Hyllingen Series (HS-I) comprises a major, systematic, basal reversal; FS-II is considerably thicker than HS-II (*ca.* 1800 and 800 m respectively in Figure 5); modal layering at the base of HS is discordant to the western margin and the marginal, unlayered diorites by an angle of *ca.* 7°; the discordant relations between modal and cryptic layering in HS imply crystallization along an inwardly inclined floor. Stage III is used as a datum for Figure 9. A feeder is shown schematically in Figure 9 at the base of the deepest part of the magma chamber. Figure 9E represents the reconstructed *pre-deformational* form of the Fongen-Hyllingen magma chamber; this must be borne in mind when comparison is made with the geological map in Figure 1.

It will emerge below that the basal reversal that comprises HS-I formed from compositionally-zoned magma during lateral expansion of the wedge-shaped protrusion towards the south. This expansion occurred after crystallization of the lower part of FS-II in the north, and after extensive compositional zoning had been established in the magma. In order to deal with the history of the magma chamber, it is therefore necessary to start with Stage-II in the Fongen Series.

Stage II in the Fongen Series (FS-II). The magma chamber initially developed near the boundary between metabasaltic and metapelitic country rocks. Expansion of the chamber took place partly by elevation of its roof, but magma also penetrated the roof along fractures. These fractures were enlarged to form dykes which spread laterally into sill-like bodies. During this stage of development the roof zone consisted of a network of metabasaltic hornfels and magma. The crystallization front moved up to engulf the network of interconnected roof rocks which now appear as *in situ* inclusions (Figure 9A). Some fragments of roof rocks became detached from the roof and sank through the magma to cause impact structures in the partly consolidated floor cumulates.

Whereas the metabasaltic roof rocks became contact metamorphosed to pyroxene hornfels, interleaved metapelites were partially melted and added a buoyant, low-melting component to the magma at the roof. Evolved melt also reached the roof zone by compositional convection. As outlined above, the fairly constant mineral compositions through a thick sequence of cumulates in the lower part of FS-II formed in response to crystallization during repeated or continuous magma addition. Strong compositional zoning therefore developed in the magma during the crystallization of FS-II; from below as a result of the repeated emplacement of dense, primitive magma along the floor, and from above as a of result melting of the roof and compositional convection.

Stages I and II in the Hyllingen Series (HS-I and HS-II). Compositional regressions at the bases of layered intrusions (basal reversals) are fairly common and can develop by a variety of processes:

a) Assimilation of an upwardly decreasing amount of evolved material from the floor (Campbell and Turner, 1987).
b) Reaction between cumulus minerals and an upwardly-decreasing amount of trapped liquid (Raedeke and McCallum, 1984).
c) Fractional crystallization in a feeder channel. This could remove high temperature phases so that the first magma to reach the intrusion is relatively evolved. This is only likely to be important in the early stages of magma emplacement before the feeder system heats up.
d) Upwardly decreasing quenching. Magma emplaced against a cool floor will be chilled so that the initial products of crystallization approach the composition of the parental magma. As the degree of undercooling decreases the minerals will become increasingly primitive until they are in equilibrium with the parental magma. The rocks that crystallized from undercooled magma will become increasingly primitive away from the floor.

None of these processes, however, can explain the basal reversal in HS-I. The low *mg#* of the mafic phases and the presence of apatite and zircon in the lower part of HS-I, the systematic nature of the compositional regression, the fact that there is an isotopic as well as a mineral chemical regression, and the absence of textural evidence for extensive chilling cannot adequately be explained by any of the above processes.

The Hyllingen Series is envisaged as having developed by the formation of a wedge-shaped protrusion towards the south in response to the inflation of the magma chamber in the north, during the crystallization of the lower part of FS-II (Figures 9B and 10). The floor of the wedge sloped upwards towards the south and developed as an extension of the chamber in which the Fongen Series had begun to crystallize *i.e.* the magma chamber expanded both laterally and vertically. This protrusion developed after a thickness of *ca.* 1000 m of layered cumulates had accumulated in the northern, Fongen part of the chamber. Fractionated, residual magma resided in the upper part of the chamber, together with the earliest roof melts. The first magma that came into contact with the new floor at the leading edge of the expanding wedge was the buoyant, evolved, contaminated magma at the chamber roof (Figure 10A). Fractional crystallization of this magma gave rise to the highly evolved rocks at the base of HS-I. Buoyant melt released by compositional convection during crystallization at the leading edge would already be in the roof melt region and would add to it a highly evolved, low-temperature melt fraction.

As the wedge expanded in response to continued magma emplacement (Figure 9C), progressively denser, less evolved and less contaminated magma entered the wedge. Crystallization of this increasingly primitive magma on the floor in the expanding wedge gave rise to the basal reversal of HS-I. Some of the initial heat loss was through the floor, but as the thickness of the basal rocks increased and the floor heated up, heat loss was increasingly through the roof. Modal layering is envisaged as having started to develop at some distance from the floor when the cooling rate decreased sufficiently to allow oscillatory nucleation of slightly supercooled magma (Figure 10A).

While HS-I was crystallizing at the leading edge of the expanding wedge, HS-IIA (and FS-II) were crystallizing towards the centre of the chamber (Figure 10B). The fairly constant mineral compositions in the lower part of Stage II, which are interpreted as evidence for cryst-

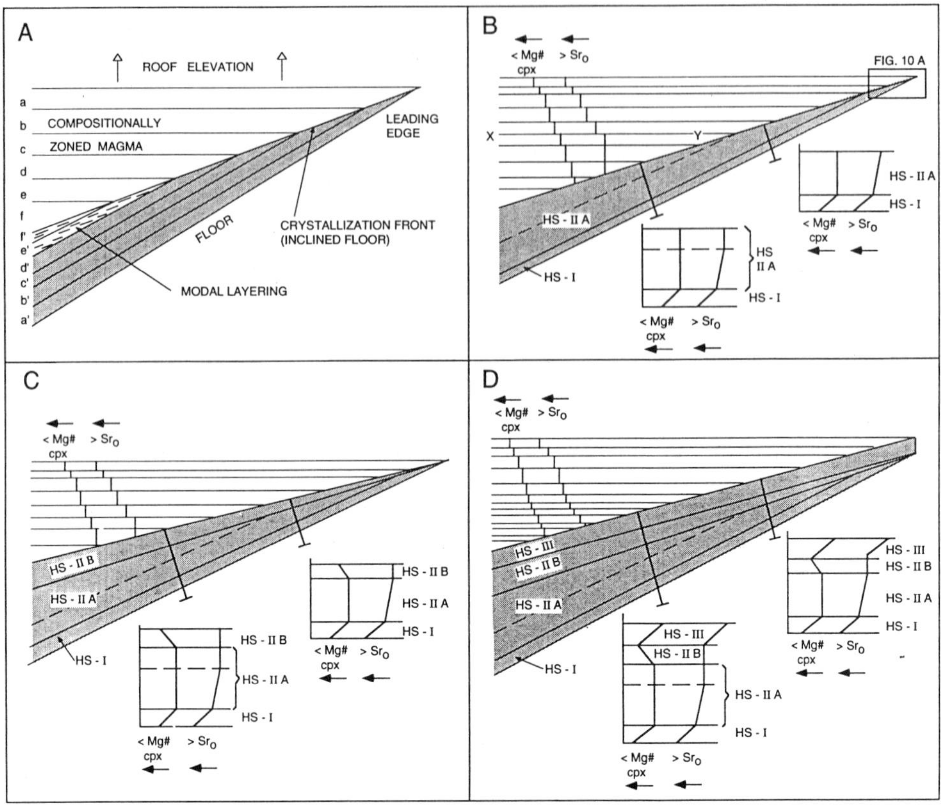

Figure 10. *Schematic illustration of magma stratification, mineral compositions (indicated by mg# in Ca-rich pyroxene) and Sr-isotope trends (Sr_0) during evolution of the Fongen-Hyllingen Complex. A. Formation of the basal reversal of HS-I at the leading edge of the expanding wedge-shaped magma chamber (Figure 10B). Magma layers a, b, c etc. are parental to a′, b′, c′ etc. respectively on the floor. Modal layering develops parallel to the inclined crystallization front at some distance above the floor. B. HS-I and HS-IIA crystallized from compositionally zoned magma during fairly continuous expansion of the wedge-shaped magma chamber. Note that the lower part of the resident magma (below interface XY) was isotopically homogeneous. Magma influx ceased immediately after this stage of development. C. HS-IIB crystallized over the entire sloping floor after magma influx ceased. HS-IIB is therefore discordant to HS-I and HS-IIA. D. HS-III crystallized during magma influx (by fountaining) and mixing at the base of the chamber (to the left in Figure 10C). The lower part of the magma column became compositionally more primitive and isotopically zoned during magma mixing indicated by an increased number of "steps" in the zoned magma. The inflection in isotopic trend at the base of HS-III is delayed in the up-slope profile because this part of the floor was only effected by elevated resident magma for some time until finally being flooded by hybrid magma, as illustrated in Figure 8.*

allization during magma addition, are therefore intimately linked to the formation of the basal reversal of HS-I. Both formed during fairly continuous magma addition. Magma chamber expansion during this period involved the formation of the extensive network of mainly metabasaltic inclusions in the roof zone.

The regression in both mineral chemistry and Sr_0 in HS-I implies that the magma at the roof was both chemically and isotopically zoned. The trends imply that there was not an isolated, homogeneous, buoyant roof melt with a sharp interface to underlying magma, but rather that the zoning was gradual. As discussed above, the interval in which mineral compositions remain constant (HS-IIA) reflects a steady state situation in which the effects of the rate of elevation of the zoned magma column (*i.e.* magma influx) and the rate of crystallization were essentially equal. Whereas mineral compositions remain constant in HS-IIA, Sr_0 continues to decrease. Since the Sr_0 in cumulates reflects that of the parental magma, this must mean that the Sr-isotopic compositional gradient in the liquid extended into this portion of the parental magma (Figure 10B).

The interval in the upper part of HS-II where mineral compositions become increasingly evolved upwards (HS-IIB) resulted from fractional crystallization when the rate of magma influx slowed or ceased. If magma influx stopped at the HS-IIA/B boundary, the magma in the leading edge of the wedge-shaped protrusion stagnated and HS-I ceased to develop. Fractional crystallization occurred from stagnant, stratified magma on the entire sloping floor and mineral compositions became more evolved upwards. HS-IIB is therefore discordant to HS-I and HS-IIA in Figure 10C.

The constant Sr_0 values in HS-IIB in Figure 4 reflect the fact that the Sr-isotopic composition of the relevant part of the parental magma column was constant. This means that whereas there was an isotopic compositional gradient in the magma parental to HS-I and HS-IIA, the magma interval parental to HS-IIB was isotopically homogeneous. This is supported by additional results of Sørensen and Wilson (1995) from the top of HS-IIB. They found that minerals become more evolved along the strike of modal layering approaching the southern margin (*e.g.* *mg#* in Ca-rich pyroxene decreases systematically from 50 to 25 over a distance of 7 km along the strike of modal layering at the HS-II/III boundary), consistent with the crystallization of compositionally zoned magma along an inclined floor, but that Sr- and Nd-isotopic compositions remain essentially constant. This implies that whereas the parental magma to HS-IIB was zoned chemically, it was isotopically homogeneous.

It would be a remarkable coincidence if the interface in the magma separating an isotopically homogeneous portion below from an isotopically zoned portion above (interface XY in Figure 10B) exactly reached the point on the sloping floor represented by profile C in Figures 1D and 4 when magma influx ceased at the HS-IIA/B boundary (*i.e.* profile C would be exactly at point Y in Figure 10B). As mentioned earlier, there may be an interval, up to 200 m thick (Figure 4), in which both mineral chemistry and Sr_0 are constant. Closer sampling in the critical interval and in additional profiles through HS would shed further light on the nature of the compositional zoning of the parental magma.

If the magma chamber floor in the vicinity of profile C in HS had a slope of about 3° during the crystallization of HS-II (as estimated by Wilson and Larsen, 1985), and levelled out to the north, the thickness of the isotopically homogeneous magma column was between about 180 and 350 m. This is consistent with the 250-300 m thickness of HS-IIB in Figure 4 in which Sr_0 is constant.

Stage III. The compositional regression of HS-III is, in many respects, similar to that of HS-I except that it spans a more primitive range of mineral compositions and a less contaminated range of Sr_0 values (Figure 4). It is therefore tempting to interpret HS-III as having formed in a similar way to HS-I, as was done by Wilson and Larsen (1985). It has just been argued above, however, that the part of the magma column parental to HS-IIB was zoned in terms of major element chemistry but isotopically homogeneous (Figure 10B); crystallization of this magma along an inclined floor during elevation could therefore produce the mineral chemical trend but not the isotopic trend of HS-III in Figure 4.

The isotopic trend can best be explained by crystallization during the gradually increasing effect of the addition (by fountaining) of a primitive, isotopically uncontaminated component. Magma emplacement occurred at the base of the magma chamber which was situated to the north of the Hyllingen Series. The start of the regression in mineral chemistry and of that in Sr_0 coincide in Figure 4, implying that hybrid magma rapidly reached this part of the Hyllingen Series. However, Sørensen and Wilson (1995) report the presence of a *ca.* 50 m delay in the start of the regression in ε_{Nd} in profile C, increasing to 130 m in a profile 2 km further to the south. This is consistent with crystallization during elevation of the compositionally zoned, but isotopically homogeneous, resident magma column above the hybrid before the underlying, ponding hybrid magma floods the respective profiles, as illustrated in Figure 8 and Figure 10D.

The most primitive rocks in the entire layered sequence of the Fongen-Hyllingen Complex occur at the top of Stage III in FS and in the northern part of HS and mark crystallization from the most magnesian, uncontaminated magma in the chamber (Sr_0 0.70308 and ε_{Nd} 5.84). This magma was present at the base of the hybrid magma column when influx ceased.

Stage IV. The range in mineral chemistry in Stage IV in Figure 4 (*ca.* 1500 m thick in profile C) extends continuously from the most primitive to the most evolved compositions in this particular profile with no significant breaks or regressions. Recent detailed work (not presented here) has shown evidence for repeated minor magma influx in the lower part of FS-IV in profile B; compositional reversals here are related to laterally continuous olivine-rich units. In the Hyllingen part of the magma chamber the effect of these repeated minor influxes merely periodically slowed the rate of fractional crystallization. The fact that the most evolved compositions occur at the roof implies that an upper border series did not develop below the roof.

The steady increase in Sr_0 through the 900 m-thick HS-IVA implies that there was a Sr-isotopic gradient in the parental resident magma column. Extrapolation on Figure 4 shows that the fairly constant value of Sr_0 in HS-II (0.7044) is intercepted about 300 m above the base of HS-IV, which suggests that the maximum thickness of the hybrid magma column after influx in Stage III was approximately of this magnitude. Consideration of Sr-isotopic ratios reveals that only about 30% of this consists of new influx magma. This in turn means that only about 6% (30% of 300 m = 90 m; 90 m = 6% of the total magma column of 1500 m; assuming as a first approximation that the cumulate thickness is equivalent to that of the parental magma column) of the magma column that was present above the base of stage HS-IV was new influx magma while the rest was variably evolved, contaminated residual magma (Sørensen and Wilson, 1995).

As explained earlier, HS-I crystallized at the leading edge of the expanding Hyllingen Series wedge-shaped protrusion from the highly evolved, most buoyant magma at the roof (Figure 10A). The model implies that this roof magma was also parental to the upper part of HS-IVB, but whereas the solid solution minerals become increasingly evolved upwards and reach low-temperature end-member compositions just below the roof, Sr_0 varies unsystematically over a wide range (Figure 4). Despite their compositional scatter with stratigraphic height, the samples define an isochron (Wilson and Pedersen, 1981), implying that this is at least partly a primary feature. Sr-isotope isochron plots of samples from HS-IVB give an Sr_0 of 0.7047 which is much less contaminated than HS-I (0.70537). It seems that the roof melt must have homogenized isotopically with less contaminated magma before the crystallization of HS-IVB.

Fractional crystallization during the formation of the Fe-rich assemblages of HS would have released buoyant melt. Isotopic homogenization could have taken place by progressive mixing as the density of the crystallizing magma layers decreased by this process until they mixed with overlying magma of equal density. However, the Sr-isotopic ratios fluctuate strongly in HS-IVB, which may be related to intermittent episodes of mixing. It may be significant in this respect that Sr_0 varies between the constant value of HS-IIB (0.7044) and the most contaminated value at the base of HS-I (0.70537).

5. CONCLUSIONS

Several important points emerge from two-dimensional study of the Fongen-Hyllingen Layered Intrusive Complex:

a) Modal layering formed on the crystallization front - the advancing floor of the magma chamber - and therefore, represents isochronous surfaces.
b) Cryptic layering reflects the composition of the parental magma and represents isochemical surfaces in the magma column.
c) Systematic discordant relationships between modal and cryptic layering can develop as a result of crystallization of compositionally zoned magma along an inclined floor.
d) Mineral chemistry and initial Sr-isotopic ratios generally show a strong correlation in the Fongen-Hyllingen Complex, with more evolved cumulates having more contaminated isotopic signatures. Contamination took place largely by hybridization of resident magma with partial melts of pelitic roof rocks.
e) The upper part of the magma column became zoned with respect to Sr-isotopes and chemical composition. The lower part remained isotopically homogeneous (but chemically zoned) until progressive hybridization took place with new, uncontaminated magma during formation of the major regression of Stage III.
f) Extensive compositional zoning in the Fongen-Hyllingen magma chamber therefore developed both from the roof down (by roof melting and as a result of compositional convection) and from the floor up (by the repeated, quiescent emplacement of dense magma along the floor and by mixing as a result of new, dense magma fountaining into buoyant, resident magma).
g) Compositional regressions developed as progressively more primitive magma came into contact with the inclined floor. In the Hyllingen Series, compositionally zoned magma *near the top* of the chamber crystallized *on the sloping floor* at the leading edge of the wedge-shaped chamber during expansion. The basal reversal in the Hyllingen Series retains the crystalline products of compositionally-zoned roof magma *in reverse sequence*. The major, systematic regression in the central part of the layered series in the Hyllingen Series (HS-III) developed both as a result of the elevation of compositionally zoned magma

and as a result of mixing between evolved resident magma and new primitive magma. These processes took place simultaneously at different levels on the inclined floor.

h) It is essential to consider the dynamic development of the magma chamber in order to appreciate *the simultaneous formation of the basal reversal at the margin, and the layered sequence with constant mineral compositions towards the centre of the magma chamber* during expansion of the chamber in response to magma addition.

6. ACKNOWLEDGEMENTS

Financial support for research on Fongen-Hyllingen has been provided by the Danish Natural Science Research Council and the Carlsberg Foundation.

7. REFERENCES

Campbell, I.H., & Turner, J.S., 1987. A laboratory investigation of assimilation at the top of a basaltic magma chamber. *J. Geology* **95**, 155-72.

Campbell, I.H., & Turner, J.S., 1989. Fountains in magma chambers. *J. Petrology* **30**, 885-923.

Habekost, E.M., & Wilson, J.R., 1989. Raft-like metabasaltic inclusions in the Fongen-Hyllingen layered complex, and their implications for magma chamber evolution. *J. Petrology* **30**, 1415-41.

Huppert, H.E., Sparks, R.S.J., Wilson, J.R., Hallworth, M.A., & Leitch, A., 1987. Laboratory experiments with aqueous solutions modelling magma chamber processes. II. Cooling and crystallization at an inclined plane. In: Parsons, I. (ed.) *Origins of Igneous Layering*. D. Reidel Publishing Co., Dordrecht, 539-68.

Huppert, H.E., & Sparks, R.S.J., 1980. The fluid dynamics of a basaltic magma chamber replenished by influx of hot, dense, ultrabasic magma. *Contr. Miner. Petrol.* **75**, 279-89.

Irvine T.N., 1981. A liquid-density controlled model for chromitite formation in the Muskox Intrusion. *Carnegie Inst. Wash. Yrbk.* **80**, 317-24.

Irvine, T.N., Keith, D.W., & Todd, S.G., 1983. The J-M Platinum-Palladium Reef of the Stillwater Complex, Montana: II. Origin by double-diffusive convective magma mixing and implications for the Bushveld Complex. *Econ. Geol.* **78**, 1287-334.

Olesen, N.Ø., Hansen, E.S., Kristensen, L.H., & Thyrsted, T., 1973. A preliminary account on the geology of the Selbu-Tydal area, the Trondheim region, Central Norwegian Caledonides. *Leid. geol. Meded.* **49**, 259-76.

Raedeke, L.D., & McCallum, I.S., 1984. Investigations in the Stillwater Complex: part II. Petrology and petrogenesis of the Ultramafic Series. *J. Petrology* **25**, 395-420.

Robins, B., Haukvik, L., & Jansen, S., 1987. The organization and internal structure of cyclic units in the Honningsvåg Intrusive Suite, north Norway: implications for intrusive mechanisms, double-diffusive convection and pore-magma infiltration. In: Parsons, I. (ed.) *Origins of Igneous Layering*. D. Reidel Publishing Co., Dordrecht, 287-312.

Sørensen, H.S., & Wilson, J.R., 1995. A strontium and neodymium isotopic investigation of the Fongen-Hyllingen layered intrusion, Norway. *J. Petrology* **36**, 161-87.

Sørensen, H.S., & Wilson, J.R., in press. Petrology of the Treknattan Intrusion in the Fongen-Hyllingen complex, Trondheim Region, Norway: a late intrusion into an evolved layered complex. *J. Petrology*.

Thy, P., & Wilson, J.R., 1980. Primary igneous load-cast deformation strutures in the Fongen-Hyllingen layered basic intrusion, Trondheim Region, Norway. *Geol. Mag.* **117**, 363-71.

Wadsworth, W.J., 1982. The basic plutons. In: Sutherland, D. (ed.) *Igneous rocks of the British Isles*. John Wiley, 135-48.

Wager, L.R., & Brown, G.M., 1968. *Layered Igneous Rocks*. Oliver & Boyd, Edinburgh. 588 pp.

Wilson, J.R., & Larsen, S.B. 1985. Two-dimensional study of a layered intrusion - the Hyllingen Series, Norway. *Geol. Mag.* **122**, 97-121.

Wilson, J.R., & Olesen, N.Ø., 1975. The form of the Fongen-Hyllingen gabbro complex, Trondheim region, Norway. *Norsk Geologisk Tidsskrift* **55**, 423-39.

Wilson, J. R., & Pedersen, S., 1981. The age of the synorogenic Fongen-Hyllingen complex, Trondheim region, Norway. *Geologiska Føreningens i Stockholm Førhandlingar* **103**, 429-35.

Wilson, J. R., Hansen, B., & Pedersen, S., 1983. Zircon U-Pb evidence for the age of the Fongen-Hyllingen complex, Trondheim region, Norway. *Geologiska Føreningens i Stockholm Førhandlingar* **105**, 68-70.

Wilson, J.R., Esbensen, K.H., & Thy, P., 1981a. Igneous petrology of the synorogenic Fongen-Hyllingen layered basic complex, south-central Scandinavian Caledonides. *J. Petrology* **22**, 584-627.

Wilson, J.R., Esbensen, K.H., & Thy, P., 1981b. A new pyroxene fractionation trend from a layered basic intrusion. *Nature* **290**, 325-6.

Wilson, J.R., Menuge, J.F., Pedersen, S., & Engell-Sørensen, O. 1987. The southern part of the Fongen-Hyllingen layered mafic complex, Norway: emplacement and crystallization of compositionally stratified magma. In: Parsons, I. (ed.) *Origins of Igneous Layering.* D. Reidel Publishing Co., Dordrecht, 145-84.

Wyllie, P.J., 1981. Plate tectonics and magma genesis. *Geol. Rundsch.* **70**, 128-53.

LAYERED INTRUSIONS
R.G. Cawthorn (editor)

Layered Alkaline Igneous Rocks of the Gardar Province, South Greenland

B.G.J. Upton[a], I. Parsons[a], C.H. Emeleus[b], and M.E. Hodson[c].

[a]Department of Geology and Geophysics, Edinburgh University, Edinburgh, EH9 3JW, U.K.

[b]Department of Geological Sciences, Durham University, Durham, DH1 3LE, U.K.

[c]Macaulay Land Use Research Institute, Aberdeen, AB9 2QJ, U.K.

Abstract

The Gardar Province comprises a Proterozoic suite of intrusive and extrusive igneous rocks, produced in an extensional continental environment between 1350-1130 Ma. The principal magmatic lineages can be related to the evolution of aluminous transitional olivine basalt, *via* trachyte through to comenditic and phonolitic residues. Some of the residual products are peralkaline and occasionally agpaitic. Some twelve major intrusive complexes characterize the province; whereas these are dominantly salic (syenites, nepheline syenites, and granites), intermediate to basic intrusions also occur.

Layered cumulates are prominent within virtually all of the larger intrusions. Layering takes the form of modal layering, phase layering, cryptic layering, and igneous lamination, typically involving parallel orientation of tabular feldspar crystals. Xenolithic or autolithic slabs of roof rocks, or concentrated zones of roof-derived xenoliths, lying concordant with the modal layering and/or igneous lamination, can constitute a fifth element contributing to the layering.

The layering typically dips inwards towards linear or point foci (according to the dyke- or stock-like nature of the intrusion). Marginal border groups of varying relative widths, commonly surround the layered cumulates composing the central portion. Layering can be low-angled 'saucer-shaped' or relatively steep (30-45°), giving V-shaped cross-sections.

Modal layering is generally manifested as subordinate mafic layers within dominantly feldspathic sequences. Some of this layering is inferred to have occurred through *in situ* growth as a consequence of intermittent cessation of feldspar nucleation. Additionally, gravitational sorting of ferromagnesian crystals from less dense felsic minerals produced modal layering. Other processes also appear to have operated to generate modal layering in specific instances. Cumulus minerals (or mineral aggregates) are inferred to have been capable of settling though their host melts. However, in the Ilímaussaq agpaitic magmas, sodalite floated. Additionally, sodic labradorite anorthosite xenoliths and megacrysts also appear to have floated in basaltic and hawaiitic melts.

Structures commonly present which resemble those of clastic sedimentary rocks include normal (modal) grading, cross-bedding, angular unconformities, channels, and load structures. Stacks of channels ("trough stacks") occur in several intrusions. Slump structures, involving both incoherent mafic cumulus mush and consolidated cumulate breccias, are also common. Such features are recognizable, not only in gabbroic cumulates, but in syenogabbroic, syenitic, nepheline syenitic (including agpaitic), and also granitic cumulates. Low viscosity of the mag-

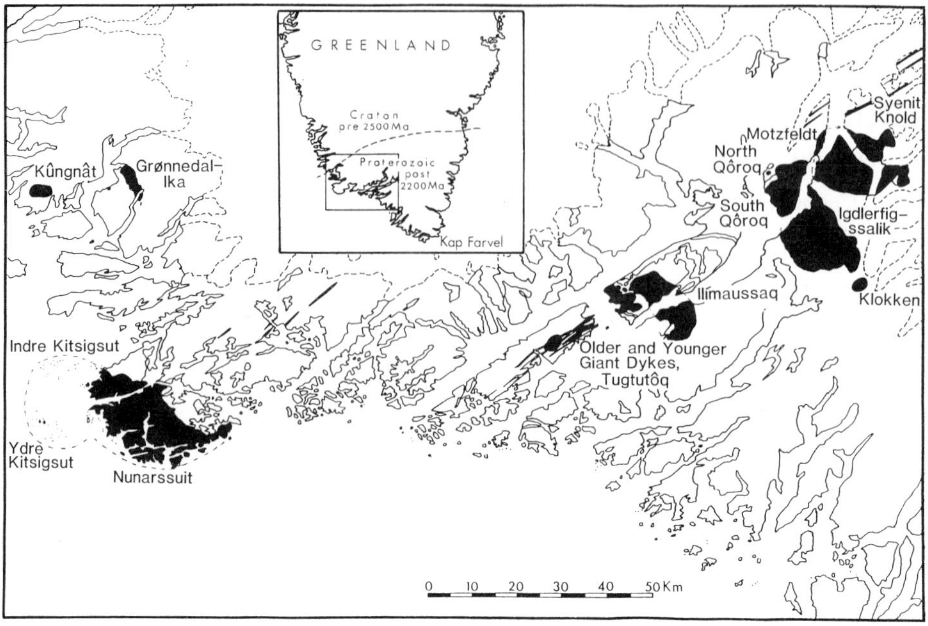

Figure 1. *Sketch map showing principal intrusions exhibiting cumulate layering in the Gardar Province, south Greenland.*

mas is thought to have been of paramount importance in permitting efficient density sorting between melts, crystals, and rocks.

1. INTRODUCTION

The Gardar Province of South Greenland comprises a mid-Proterozoic (*ca.* 1350-1150 Ma), continental intraplate and rift-related assemblage of mildly to strongly alkaline rocks (Emeleus and Upton, 1976; Upton and Emeleus, 1987). The igneous rocks are present as extrusions, hypabyssal intrusions, and plutons. The distribution of the principal plutons is shown in Figure 1. Apart from volumetrically subordinate lamprophyric and carbonatitic rocks, the majority of the Gardar igneous rocks appear to be related to a transitional alkali basalt - trachyte lineage leading either to comenditic or phonolitic residues. The plutonic rocks are mildly alkalic gabbros and syenogabbros which mainly occur as massive dykes (up to 800 m wide) and ring-dykes, and syenites, nepheline syenites, and alkalic granites, which typically compose 'central-type' complexes.

The plutonic intrusions under consideration represent magma chambers that ranged from 3-5 km diameter (*e.g.* Klokken intrusion) to larger bodies with diameters of 10-30 km (*e.g.* Nunarssuit and Igdlerfigssalik intrusions). Some of these were crudely cylindrical chambers, but the aspect ratio (depth to width), changed radically as crystallization advanced. In their initial stages of formation, the so-called giant-dyke intrusions, were tabular vertical-sided chambers with widths between 0.1-1.0 km and lengths and probably vertical extents of tens of kilometres.

The stratigraphic sequences of layered rocks available for study in this region of considerable vertical relief range from a few hundred metres to several kilometres. In the case of the Klokken, Kûngnât, Ilímaussaq, Igdlerfigssalik, and Grønnedal-Ika plutons and the Tugtutôq giant dykes, the accessible statigraphic successions appear to range from 1-5 km thick.

At their present levels of exposure, the plutons are inferred to have consolidated at relatively shallow crustal depths (<10 km and, for several, probably <5 km). Whereas several display well-chilled margins against the country-rocks, the inner facies of the plutons are typically coarse-grained cumulates with poorly compacted ortho- to mesocumulate textures. Cooling of even the largest bodies was sufficiently rapid to have prevented extensive sub-solidus recrystallization and grain-boundary equilibration.

2. LAYERING

Modal layering can be found throughout the entire plutonic compositional spectrum (Ferguson and Pulvertaft, 1963) and generally involves concentration of ferromagnesian minerals relative to felsic minerals. In the gabbros, the modal layering is defined by concentration of olivine (± Ti-magnetite, ilmenite, and apatite) relative to plagioclase. In the syenogabbroic facies, salitic to ferrosalitic clinopyroxene joins the cumulus assemblage and is concentrated, together with olivine, FeTi-oxides, and apatite, in the mafic layers. Feldspars are usually strongly zoned plagioclase with ternary feldspar rims and/or alkali feldspar in the interstices (*e.g.* Upton, 1964; Parsons and Brown, 1988). The same assemblage, but with more evolved compositions, characterizes the melanocratic layers of the Gardar syenites, quartz syenites, and some of the nepheline syenites. Spectacular modal layering is present within the eudialyte-bearing nepheline syenites (kakortokites) at Ilímaussaq, in which the bases of layered units are defined by concentrations of arfvedsonite, accompanied by aegirine, eudialyte, aenigmatite, and alkali feldspar.

Other manifestations of layering in the plutons include cryptic layering, phase layering, due to addition or subtraction of specific cumulus phases, parallel orientation of crystals, most evident with respect to lamination of tabular feldspars, and crudely defined layers ('rafts') of xenoliths. In all of the layered plutons, the layering features dip inwards towards a central focus (in the case of the 'central type' intrusions) or towards a central axis (in the case of the plutonic dykes). In some of the intrusions layering is steep close to the margins, where it parallels the contacts, but serially decreases in inclination towards the interior. Occasionally (as in the plutonic dykes) the layering can be traced down dip to axial zones where it is horizontal. Thus the layering of the Gardar plutons is in some cases (*e.g.* Grønnedal-Ika, Ilímaussaq agpaites and Igdlerfigssalik) saucer-shaped whereas in other instances (Klokken and, probably Kûngnât) the layering does not gradually flatten out but appears to retain a gradient of 30-40° until within two or three hundred metres of the focus. The geometry is like a stack of cones that sometimes steepen to angles of >60° in the marginal facies. In the case of the giant dykes, the internal structures defined by the layering are canoe-like. Where steep marginal layering is present in the Gardar intrusions, it is assumed to be in the same orientation as it was when it formed. However, the question remains open, in most cases, as to whether the presently observed dips in the pluton interiors reflect slopes at the time the cumulus pile was growing, or whether the original layering dipped more shallowly and subsequently became steepened through compaction and/or magma withdrawal.

Modal layering occurs in the Gardar cumulates in a wide variety of styles and scales. Features regarded as analogues of sedimentary structures, such as normally-graded modal layering, cross-bedding, trough or channel structures, load-casts, and sag structures beneath autoliths or xenoliths are common and provide a profusion of way-up criteria. These, together with cryptic layering progressing upwards from higher-temperature to lower-temperature assemblages, show that the majority of the Gardar cumulate suites grew from the 'floors' upwards. The principal exceptions are the unique sequence of sodalite syenites (naujaites) of Ilímaussaq and the relic succession of syenites ('granular syenites') of the Klokken Intrusion, which are inferred to have accreted downwards from the roof.

Commonly, thick sequences (hundreds and sometimes thousands of metres thick) of modally homogeneous, feldspathic cumulates occur, interrupted by very subordinate sequences displaying modal layering. The most striking illustration of this phenomenon is that of the main Nunarssuit syenite where very pronounced layering is confined to stratigraphic units near the base of the visible succession, overlain by a sequence of unlayered syenite several kilometres thick. Such evidence is taken to imply that, whatever factors controlled the development of modal layering (or, more specifically, the factors responsible for increasing the modal ratio of ferromagnesian cumulus to felsic cumulus, from essentially cotectic proportions), these represented infrequent perturbations within the otherwise uniform, steady-state crystallization of large magmatic bodies.

The scale on which layering occurs varies widely. At its least, modal layering involves mafic layers only a few mm thick. More generally it involves mafic mineral-rich layers from 1-20 cm thick. The arfvedsonite-rich units in the Ilímaussaq kakortokite sequence can also be a metre or more thick.

Mafic layers vary very widely with respect to their lateral persistence; the most laterally extensive mappable layers in the Ilímaussaq (Bohse *et al.*, 1971) and Klokken (Parsons and Becker, 1987) intrusions can be traced for a kilometre or more. In Ilímaussaq, individual layered units can be traced over some 5 km and it is not implausible to suppose that they extend across the whole width of the intrusion, for some 17 km (Sørensen and Larsen, 1987). Commonly however, mafic layers vary in thickness and diminish and pass laterally into the normal rock. Olivine-rich layers in the Tugtutôq Younger Giant Dyke thicken down-dip to culminate in dunitic layers several metres thick in the near-horizontal axial zone. Mafic layers can appear cyclically, with upper and lower margins being either diffuse or sharply defined. Normally-graded layering, in which mafic layers have sharply-defined bases but which grade up into feldspathic 'normal' rock, is well displayed in several of the intrusions. Inversely-graded modal layering occurs at Klokken (Parsons and Becker, 1987).

In the following section we present synopses of the features exhibited by some of the more strikingly layered Gardar plutons. The sequence in which the layered intrusions are described is, very roughly, that of increasing degree of fractionation of the parent magma.

3. THE YOUNGER GIANT DYKE COMPLEX

The Younger Giant Dyke Complex (Upton and Thomas, 1980; Upton and Fitton, 1985; Upton, 1987) consists of a system of bifurcating and *en echelon* dyke segments extending over *ca.* 145 km from the inland ice to the Labrador Sea. The entire system is regarded as the prod-

Figure 2. *Troctolitic cumulates in a branch of the Younger Giant Dyke Complex, western Tugtutôq with thin, sharply bounded peridotite (olivine cumulate) layers. These latter are nearly parallel but show lateral discontinuities and bifurcations. (Hammer, centre right, as scale. Hammer shaft ca. 60 cm long).*

uct of a single large-volume intrusion of transitional olivine basalt magma. Widths of the component dyke segments range from <1 m up to *ca.* 800 m. The combined width of the various branches at any one sector is typically between 0.5 and 1.5 km. The dyke margins are sharp, chilled, and essentially vertical against granitoid country rocks.

Whereas much of the intrusion consists of apparently homogeneous troctolitic gabbro, some segments, ranging from a few hundred metres to 3 km long, exhibit moderate to strongly developed modal layering defined by concentrations of mafic minerals. These layered cumulates are symmetrically flanked by marginal unlayered zones that can vary from a few metres width to *ca.* 100 m. The generalized form of the layers, regarded as reflecting the instantaneous morphology of the magma chamber walls and floor, is that of a V-shaped synform, confined within steep to vertical walls. The cumulates are inferred to have grown mainly on the sidewalls of convecting chambers that were elongate (up to 3 km) and narrow (typically <0.5 km). The chamber depth at the time of dyke intrusion may have been several kilometres (or tens of kilometres?), shallowing progressively as the cumulate sequence accumulated from below upwards. Thus the chambers evolved from being initially vertically tabular to horizontally tabular in the late stages of congelation.

In the more primitive layered sectors (seen where the Dyke Complex is inferred to be most deeply eroded), the cumulus minerals are solely plagioclase and olivine, and the residual melt is

***Figure 3**. Horizontal, near-parallel, layering in the axial zone of a branch of the Younger Giant Dyke Complex, western Tugtutôq. In-weathering ol-cumulate layers alternate with thinner, more resistant, pl-ol-cumulate layers. (0.5 m scale subdivided into 5 cm units).*

inferred to have been denser than the parental melt. The layering involves olivine cumulates (feldspathic peridotite) alternating with plagioclase-olivine cumulates (troctolite). Although the troctolite layers are typically thicker by a factor of 10 or more than the peridotite layers (Figure 2), the latter thicken down-dip and, in the axes of the layered dyke portions, peridotite layers several metres thick may be separated by troctolitic layers no more than 10 cm thick (Figure 3).

Remarkable features, regarded as having originated as gullies or troughs, up to 30 m broad, eroded into the inclined chamber floor, with concave-upward profiles, trend down-dip and perpendicular to the dyke walls (Mingard, 1990). These troughs, separated laterally by some hundreds of metres, are filled by layers of peridotite up to 1 m thick (Figure 4). It is deduced that these transverse troughs were eroded by dense olivine-laden magma slurries that descended from the dyke walls and flowed towards the dyke's synformal axis. The channels so formed were then filled as the dense crystals separated from the less dense melt.

In more shallowly eroded levels of the Giant Dyke Complex, ilmenite, magnetite, and apatite are present as cumulus minerals and contribute to the mafic layers and, in sectors where still more evolved cumulates occur, clinopyroxene joins the cumulus assemblage. By this stage the cumulus feldspar is a relatively sodic plagioclase, evolving, with further differentiation, to alkali feldspar (microperthite). The total compositional range for olivine in the complex is Fo_{68}-Fo_{3} (Mingard, 1990). Feldspar ranges from An_{64} through potassic oligoclase to alkali feldspars in the syenites at *ca.* $Or_{42.5}Ab_{55.5}An_2$. Cumulus clinopyroxene appears at *ca.* *mg*#88 persisting to *ca.* *mg*#70 in the syenites (where *mg*# = 100 x Mg/(Mg + total Fe)).

The character of the layering varies widely in different parts of the Dyke Complex. In the troctolite - peridotite sectors, the peridotite layers, which generally range from a few millimetres to *ca.* 15 cm thick, tend to be sharply defined above and below with no indication of normal or reverse grading (*cf.* Figures 2 and 3). Textural analysis suggests that, in these sectors, plagioclase and olivine were growing together as glomerocrysts up to 2 cm diameter and that these crystal aggregates accumulated to give rise to the troctolites. Periodic failure of plagioclase to nucleate left olivine crystallizing alone to produce the intermittent, generally thinner, peridotite layers (Mingard, 1990). In the more evolved sectors, however, plagioclase appears to have crystallized as discrete, non-aggregated, tabular crystals and feldspar lamination is commonly a contributory factor in the layering. Normally-graded layers make their appearance, with sharply defined mafic bases and progressive upward dilution by plagioclase marking the transition up into 'modal average' gabbro or syenogabbro.

Figure 4. *Transverse section across a sharply defined, asymmetric, trough filled with olivine cumulate (peridotite), surrounded by essentially homogeneous pl-ol-cumulate (troctolite), Younger Giant Dyke Complex, west Tugtutôq. The trough measures ca. 9 m across by ca. 2 m thick in the middle. (The scale is 0.5 m long). Some streaky layering, involving plagioclase cumulus, can be seen within the trough to the right of the scale. The long axes of such troughs lie perpendicular to the dyke walls and are hypothesized to spread, down-dip, towards the horizontal zone in the dyke axis. The olivine cumulate has well developed jointing and weathers-in relative to the surrounding troctolite.*

Beyond the stage of fractionation where clinopyroxene attains cumulus status there are situations where feldspar is concentrated to give rise to leucocratic layers. This reaches a climax in the nepheline syenites of the Syenitknold sector (Figure 1) where intermittent rhythmic layering occurs. Rhythmic units, 5-10 cm thick, involving sharply defined mafic bases grading up into leucocratic feldspathic tops, are separated by intervening homogeneous layers of 'modal average' syenite.

Although much of the layering seen in the Younger Giant Dyke Complex is essentially parallel, strong cross-lamination, small-scale angular unconformities and channel structures in some sectors testify to magmatic erosion of partially consolidated cumulates. These are attributed to scour by descending crystal-melt slurries or mushes which appear to have played a role throughout most of the crystallization history. Slump structures are not uncommon. At one locality convolute disturbances among near-vertical layers close to a dyke wall are attributable to steep side-wall cumulates losing cohesion and collapsing. Whereas the slump structures typically indicate participation of incoherent semi-consolidated or mushy cumulate up to several metres thick, slump breccias close to the axial zones at two localities point clearly to break-up and down-slope transport of peridotitic cumulates that were largely consolidated and possibly even jointed at the time of disruption (Figure 5).

On emplacement, the Younger Giant Dyke magma appears to have penetrated and broken up an earlier, very coarse-grained, layered anorthosite body (Upton and Thomas, 1980; Upton, 1987). The resultant anorthosite xenoliths and discrete plagioclase megacrysts were concen-

Figure 5. *Disrupted, and internally deformed autoliths of peridotite in disturbed troctolitic and peridotitic matrix. Detail from a slump breccia, Younger Giant Dyke Complex, western Tugtutôq.*

Figure 6. *Irregularly spaced, modally-graded layering in the main Nunarssuit syenite intrusion. A broad amplitude erosional trough is seen below the hammer. (Hammer head, ca.18 cm long).*

trated by flotation in the upper parts of the intrusion (western Tugtutôq and adjacent mainland). A foundered mass of presumed roofing-facies gabbro, crowded with anorthositic debris is also seen in Syenitknold near the inland ice. Densities of the megacrysts and xenoliths are estimated to have ranged from 2.70-2.82 and to have been buoyant in the parental Giant Dyke magma with a calculated density close to 2.85 (Mingard, 1990).

4. THE NUNARSSUIT COMPLEX

The Nunarssuit Complex (Harry and Pulvertaft, 1963; Parsons and Butterfield, 1981), measures *ca.* 45 x 25 km. It is composed of at least three granitic intrusions, several bodies of augite syenite and part of an early gabbro intrusion. The gabbro is similar to those of the Younger Giant Dyke Complex and possesses similar styles of layering.

The Nunarssuit Syenite itself may be a composite of some five distinct syenite bodies (P. Greenwood, *pers. comm.*) but the most striking development of layering is confined to one major unit. The layering is principally developed around the southwestern margin of the syenite, dipping 15-40° to the northeast and is thus seen at the lowest levels of an otherwise largely homogeneous body. In this layered zone, (traceable for *ca.* 15 km and *ca.* 75 m thick in southwest Nunarssuit), the layering typically involves concentrations of pyroxene and olivine, *ca.* 4-8 cm thick, separated by 10-30 cm of more feldspathic syenite. The mafic layers commonly show normal grading. Discordant features, apparently identical to those seen in cross-

Figure 7. *Complex "soft sediment" deformation in large slump structures, Nunarssuit syenite. Undisturbed, (pale) modally layered syenite cumulates are seen in the foreground, rising (middle distance) into an irregular culmination beneath what appear to be two separate slump deposits, lying on either side. A figure, (lower left) provides scale. The slump deposits appear to be at least 6 m thick. The topographical relief on the chamber floor, as indicated by the irregular bases of the slumps, is also in the region of 5-6 m.*

bedded sediment, are observed (Figure 6). In its most extreme development, the discontinuities take the form of channels, with open U-shaped sections, up to 30 m across and 2 to 3 m deep, that are interpreted as erosional scours into underlying cumulates that were subsequently filled with layered or unlayered cumulate (Harry and Pulvertaft, 1963, Figures 57 and 58; Parsons and Butterfield, 1981).

In places, multiple troughs occur with repetition of the mafic layers stacked one upon the other, sometimes through 20 to 30 cycles. These "trough stacks" may attain vertical thicknesses of ten or more metres and have been observed with faint parallel layering to either side. As in the celebrated trough layering of Skaergaard (Wager and Deer, 1939; Wager and Brown, 1968), the process involved was one that was capable of multiple repetition. However, the relatively structureless 'levees' of unlayered gabbro that occur at Skaergaard (Irvine, 1983) do not have a counterpart at Nunarssuit (nor in any of the other examples of trough layering in the Gardar intrusions).

The lowest facies of some layers are devoid of cumulus feldspar. These ultramafic high-density cumulates are composed mainly of ferroan pyroxene and olivine with alkali feldspar as intercumulus oikocrysts. Harry and Pulvertaft (1963, Figure 59) noted load pouches (casts) where bulbous masses of the dense cumulate had settled down into underlying feldspathic material. Discrete clasts of ultramafic cumulate up to 50 cm length, occurring in association with some of the trough layering are regarded as 'rip-up' clasts of already consolidated material. Occasionally such ultramafic autoliths are so concentrated as to compose slump breccias comparable with those of the Younger Giant Dyke Complex. The autolithic clasts consist only of the most ultramafic facies of the cumulates and, in this respect also resemble those in the slump breccias in the Younger Giant Dyke Complex.

Slump structures, again resembling those in the Giant Dyke gabbros, indicate gravitational 'soft-sediment' deformation (Figure 7). Spherical masses of mafic rock with diffuse comet-like 'tails' trailing up-dip indicate down-slope motion of dense, incompletely consolidated cumulate masses. Juxtaposition of slump-folded material against scarp-like features of undeformed layered syenites suggest that penecontemporaneous faulting may have been responsible for some vigorous topographic relief on the magma chamber floor, triggering gravity slides (Harry and Pulvertaft, 1963).

The many features analogous to sedimentary structures within these syenites suggest that the Nunarssuit cumulates formed by inward and upward accumulation on a magma chamber side-wall or floor. Whereas there is evidence for significant thicknesses of unconsolidated cumulus material on the chamber floor, the slump structures of various kinds point to gravitational instability and the peeling loose of dense masses of cumulate, from sources west of, and structurally higher than, the layered localities along the Nunarssuit coast. It is speculated that these sources may have been located on steep chamber walls that formerly lay south and west of the present outcrop.

Although xenoliths are generally scarce in Nunarssuit, there are two or three zones in the west of the complex where they are abundant. The xenoliths, of meta-basalt and meta-pelite, tend to be lenticular, up to 100 m long and with their long axes parallel to the xenolith zone boundaries. The zones, up to about 100 m thick, have an easterly dip, conformable with the syenite layering. Harry and Pulvertaft (1963) considered the possibility that they represent foundered masses that sank down to the chamber floor in separate episodes of roof collapse. However, they concluded that more probably the xenolith zones are roof pendants, partially disrupted by the intrusion but not far removed from their original positions.

Layering is also seen locally in the Helene Granite unit of the Nunarssuit complex, particularly on the island of Qiartorfiq. Mafic layers rich in clinopyroxene and olivine show non-parallel, concave-up patterns resembling cross-bedded sedimentary structures (Harry and Emeleus, 1960).

5. THE KÛNGNÂT COMPLEX

The Kûngnât complex comprises syenites, ranging from relatively primitive larvikitic types, (composed of alkali feldspar, olivine, clinopyroxene, amphibole, and mica, accompanied by opaque oxides and apatite), grading to more evolved quartz syenites and small bodies of alkali granite, together with a ring-dyke suite of gabbroic and syenogabbroic rocks, (Upton, 1960).

Two syenite stocks, with diameters of *ca.* 3 km, are essentially cylindrical with vertical or steeply outward-dipping walls. Both stocks display inward-dipping modal layering resulting from concentrations of dense cumulus minerals (mainly ferroan olivine and pyroxene), on the centimetre to decimetre scale, within a matrix of essentially homogeneous meso- to leucocratic syenite (Figure 8). Layering is sporadically present throughout the eastern stock. Around the

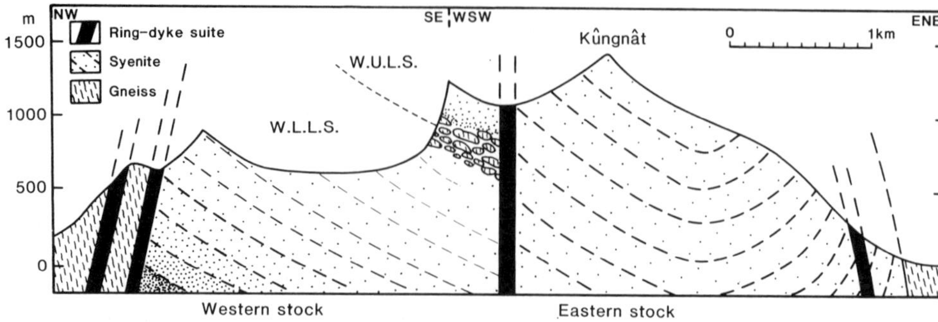

Figure 8. *A diagrammatic cross-section of the Kûngnât Complex (vertical scale = horizontal scale). The attitude of cumulate layering is shown schematically in dashed lines. Intensity of dotted ornamentation in west Kûngnât indicates upward decrease in mafic index in a) the lower layered series (W.L.L.S.) and b) above a zone of gneiss xenoliths, the upper layered series (W.U.L.S.). Intrusive sequence was (1) western syenite stock, (2) eastern syenite stock, (3) syenogabbro-gabbro ring dyke.*

eastern contact wall against Archaean gneiss, thin and impersistent mafic layers dip inwards at angles of up to 80°. Otherwise, however, layering inclination in the syenites is generally between 45 and 15°.

Layered structures are best developed in the western stock, within an exposed section of some 1,800 m divided into a lower and upper series (W.L.L.S. and W.U.L.S., Figure 8) by a diffuse layer of country-rock xenoliths. The lower series cumulates show upward passage from modally layered syenite, through syenites in which modal layering is virtually absent but which display some feldspar lamination, into essentially homogenous and isotropic quartz syenites that persist up to, and probably include, the thick and apparently conformable "raft" of inclusions. The upper layered series consists of relatively highly evolved quartz syenites which show modal layering and some lamination. Although the roof has been stripped by erosion, the distinctly miarolitic character of the uppermost rocks in the upper series suggests a shallow level of emplacement and proximity to the former magma chamber roof. Regular cryptic variation is present in both the lower and upper western layered series (Upton, 1960; Stephenson and Upton, 1982). The re-appearance of modal layering immediately above the xenolith horizon may signify that a roofing collapse promoted the processes necessary for layer formation.

Modal layering in the syenites is generally normally graded; this is particularly well displayed in the western Kûngnât lower series. Although much of the layering in these is essentially parallel, discontinuities, cross-bedding, and channel (trough) structures are common. Symmetrical channels, eroded through regularly layered syenites on the western side of the stock, have widths of up to 5 m and amplitudes of *ca.* 1 m. Concentration of ferromagnesian minerals is most extreme in the channel axes.

An extreme form of trough layering is shown at two horizons, separated vertically by some metres of poorly layered syenite, in the lower layered rocks of west Kûngnât, close to the southern contact zone. Here, modally well-graded layers, crescentic in form and concave upwards, are stacked one above the other in parallel. These trough stacks dip towards the centre

***Figure 9**. Stacked sets of troughs filled with melanocratic (ol-cpx-rich) cumulates, seen in strike section, in the western lower layered series, Kûngnât. A reclining figure, (upper field, centre) provides scale. The trough layers dip towards the centre of the western stock at between 30-40°. The attitude of layering in these western lower layered series syenites can be discerned from the stratification visible in the dark ridge in the background.*

of the intrusion, with their long axes approximately normal to the outer contact wall (Figure 9). Several of these trough stacks lie side by side, separated laterally by *ca.* 10 m of unlayered or poorly layered syenite. In each stack, the broadest and best defined "trough" is that at the base, with widths of *ca.* 30 m and depths of between one and two metres. Some twenty to thirty similar troughs may overlie it with trough widths and intensity of sorting diminishing upwards. Whatever process was responsible for their genesis appears to have been cyclic, to have commenced suddenly with maximum effect and then to have serially diminished. Distinctly leucocratic layers are absent. Despite differences of detail, the trough-like structures in the Kûngnât syenites are regarded as homologues of the trough structures of the Younger Giant Dyke and the Nunarssuit syenites. The consistent "way-upness" of grading, cross-bedding, troughs, and cryptic layering again shows that the western Kûngnât stock accumulated upwards from an inwardly-inclined floor. The layering in the western syenite stock extends virtually to the contact zones leaving no room for a marginal border group of any significant thickness.

The steep layering around the eastern margins of the eastern stock shows undulations or flutings that are mostly concave inwards. Cross-cutting relationships indicate that the layers young inwards from the contact zone. These layers are inferred to have grown *in situ* on the steep boundary layer of the cooling magma chamber and to have constituted a marginal border group, several hundred metres broad, enclosing a less steeply inclined inner layered syenite body.

Kûngnât differs significantly from the Younger Giant Dyke and Nunarssuit intrusions in lacking any evidence for "soft-sediment" slumping or slump breccias. For whatever reason, cumulates formed on any steep surfaces were coherent to the extent that masses of mafic/ultramafic cumulate did not become detached. The absence of side-wall cumulates (specifically in West Kûngnât) suggests that crystals nucleating on, or close to, the boundary layer settled continuously beside the chamber walls to contribute to a steadily accumulating crystal talus at their base.

6. THE KLOKKEN COMPLEX

Klokken is a small, slightly elliptical stock (*ca.* 3 x 4 km; Figure 1), in the east of the Gardar Province, principally composed of syenite but with a surrounding sheath of gabbroic rocks. The syenite core of Klokken (Figure 10) exhibits a unique style of layering (*cf.* Parsons and Becker, 1987 for a review with bibliography; only subsequent papers will be cited here).

***Figure 10**. The upper part of the Klokken layered series viewed from the East. The vertical distance from summit to foreground is about 230 m. The terraces are layers of granular syenite, with pale laminated syenite sometimes visible in scree-covered areas between them. The focus of the layering, which characteristically has 30-40° inward dips, is to the right of the summit.*

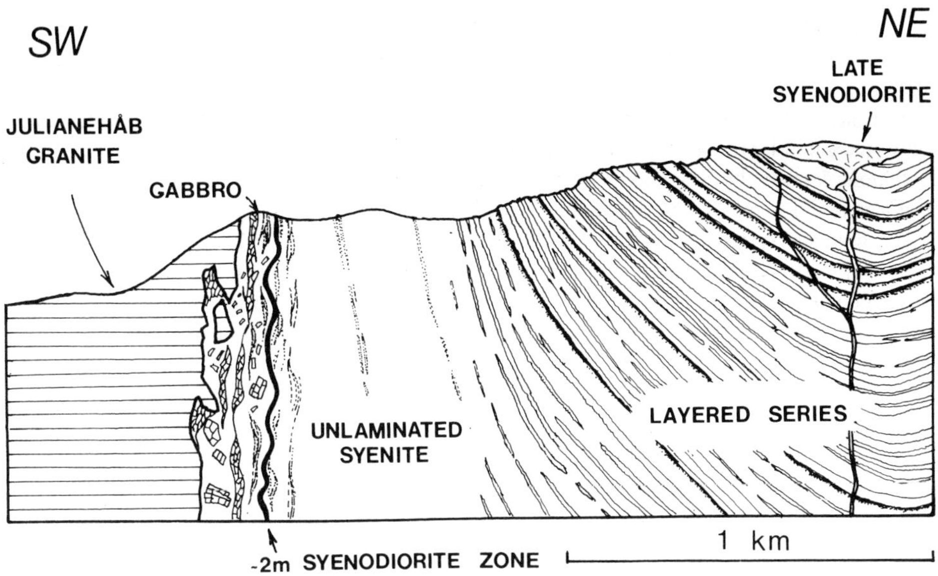

Figure 11. *Stylized cross-section (vertical = horizontal scale) across the SW portion of the Klokken intrusion (after Parsons and Brown, 1988). Granular layers and inversely graded macro-rhythmic mineral layers are shown in a generalized way in the layered series, and the topography is not exact.*

The gabbro sheath is up to 200 m thick; it narrows upwards and is absent around the southern margin. The gabbros contain a variety of wall-parallel structures, including vertical mafic layering, lenses of pegmatite (feldspars *ca.* 1 m long) and units with wavy interlayering of clinopyroxene and plagioclase grown normal to the walls. Inside the gabbro sheath is an annular zone of unlaminated and apparently unlayered syenite, up to 600 m broad (Figure 11). However, this zone shows progressive, inwardly directed, compositional evolution of the olivine, pyroxene, and (ternary) feldspar, providing a good example of a cryptically layered sidewall cumulate (Parsons and Brown, 1988).

Modally layered syenites occupy the core of the complex (Figure 11). Whereas the composition of these is generally similar to those of the augite syenites elsewhere in the Province the style of layering is different and, in certain respects, unique. The layering forms a near-perfect series of stacked cones, with inward dips of 30-50°. These dips are maintained to within 200 m of the focus of the intrusion, where they flatten out to *ca.* 15°. The actual focal point, however, is not exposed. Outcrop is nearly continuous and the total vertical section exposed amounts to *ca.* 600 m.

Two texturally distinct syenite types, granular and laminated, are present in the layered series. Sugary-textured, granular syenite forms sheets that make up around 15% of the layered series. They are most abundant in the highest 100 m, where they compose most of the succession. There is no regularity in thickness of individual sheets, which vary from >10 m in the highest exposures, to only a few centimetres elsewhere. Individual sheets are laterally discontinuous, although the largest extend at least 1 km along strike. The grain-size decreases upwards in a regular manner, from *ca.* 10 mm in the lowest sheets to *ca.* 1-2 mm in the highest. The granular syenite also shows a progressive increase in Mg/(Fe + Mg) in its mafic phases and an increase in An content of its feldspars, from lowest to highest members. Trace elements also follow the same inverted cryptic variation trend. Thus, although now seen as separated, discontinuous layers, the granular syenite bodies collectively show all the hallmarks of having been part of a downwardly accreted upper border group.

The granular syenites are interbedded with coarser, laminated syenites with tabular alkali feldspars (*ca.* 10-25 x 1-3 mm) that formed as an upwardly accreting sequence of inward dipping cumulates. These rocks display macro-rhythmic, inversely graded, layering (Parsons 1979). In each layer the proportion of feldspar decreases steadily upwards, characteristically over *ca.* 2 m, grading into nearly monomineralic hedenbergite-rich horizons, lacking cumulus feldspar. In a few instances the uppermost 20 cm changes from pyroxenite to fayalite olivinite with intercumulus ilmenite. The thickness of the inversely graded layers, irregularly separated by conformable units of "normal" syenite and granular syenite layers, is variable. In a few places normally-graded micro-rhythmic layering on the scale of a few centimetres, sometimes

Figure 12*. Load-pouch (above hammer head)) at base of granular layer which is resting on an inversely graded mafic layer in strongly laminated syenite, with an intermediate degree of sorting. To the left of the pouch is a flame structure. In the laminated syenite the 20 mm alkali feldspars, flattened parallel to (010), flow around the load-pouch and up into the flame with near perfect parallelism. The focus of the intrusion is to the left, as indicated by the dip of the axis of the flame, towards the bottom left.*

showing cross-bedding, is superimposed on the inversely graded rhythm and there is one instance of a normally graded, cross-cutting channel structure (Parsons and Butterfield, 1981).

The mafic/ultramafic rocks of the inversely graded units have orthocumulate textures. The contrast in crystal size between the large alkali feldspar tablets and the smaller pyroxenes (*ca.* 5 x 1 x 1 mm) and olivines (*ca.* 1 mm spheres), and the constancy of these crystal sizes throughout the layered series are striking features. In samples with intermediate contents of mafic phases, the much smaller pyroxenes are enclosed in feldspar and outline a euhedral feldspar primocryst core. Because of the large size of the feldspars, the pyroxenes and feldspars in the layers are hydraulically right-way-up, assuming Stoke's Law settling, but this is not true for the olivines at the tops of layers, which should have sunk more rapidly than the pyroxene. A further notable feature is the high degree of modal sorting. Some pyroxene layers contain >90% hedenbergite and no olivine, whereas the olivine-rich layers may be >90% fayalite. Minor intercumulus amphibole, biotite, titanite, and quartz may occasionally be present (Parsons *et al.*, 1991).

Cryptic variation related to stratigraphic height is not seen in the laminated syenites but the feldspars, pyroxenes, and olivines show compositional changes along strike; the more evolved compositions occurring in the peripheral zones. This has been ascribed to sub-solidus changes brought about by circulating fluids at temperatures above the feldspar solvus. The fluids persisted to sub-solvus temperatures, promoting turbidity and coarse exsolution in the laminated syenites, but not affecting the granular syenites the feldspars of which largely retain their high-temperature features.

The upper surfaces of the inversely graded layers are of two types: either the ultramafic character dies out over a few centimetres or a granular syenite layer rests directly on top of the mafic/ultramafic layer. Such interfaces are sharp but have complex shapes, many features of which can be matched with load structures in sedimentary rocks. Flame structures, penetrating upwards into granular syenite, filled with laminated syenite in which the lamination is parallel with the 'flame' walls, are ubiquitous (Figure 12). The 'flames' are not vertical, but have axes that dip towards the focus of layering, more steeply than the layering itself. Viewed down-dip, they are symmetrical and often sack-shaped. By analogy with similar structures in sediments (Ankatell *et al.*, 1970), the granular layers were sliding relatively up-dip during their formation. The flames are separated by load pouches, or in some cases, detached load balls, of granular syenite and are sometimes (but infrequently) filled with pegmatite, indicating that they formed during the final stages of crystallization. The load pouches and balls differ from those developed in sediments only with respect to their exceptionally large size.

The size of the load pouches and separation of the flame structures is a function of the density contrast between the granular syenite and the laminated syenite beneath. When the latter is leucocratic the flames may extend as much as 4-5 m into the granular layer and the repeat distance may be 2-4 m down-dip. When the intervening syenite is more mafic (*i.e.* much denser than the overlying granular syenite) the flames are small (*ca.* 0.5 m) and closely spaced (0.3-0.5 m). The Klokken load pouches are generally larger than those described from Nunarssuit by Harry and Pulvertaft (1963). The presence of load pouches at the base of granular syenites which are less dense than underlying mafic laminated syenite was explained by Parsons and Becker (1987), who showed that the density relationships would reverse if the mafic minerals in the underlying layer were contained in a slurry containing a few percent of aqueous fluid.

The granular syenites are inferred to have been derived from a cryptically layered upper border group which was subjected to repeated delamination events. Sheets successively spalled

***Figure 13**. Layering in side-wall cumulates of foyaite unit SM5, Motztfeldt. The layering is nearly vertical and is thought to define channels (or flutings) on the mechanical boundary layer of a phonolite-filled magma chamber. The structures are concave towards the interior of the intrusion and young from left to right. (Hammer shaft, ca. 30 cm long).*

off and sank gently to become enveloped in the upwardly accreting coarse laminated syenites. According to this model, the density of all facies of the granular upper border group were greater than those of the underlying residual melts. Furthermore, it implies that spalling was controlled by a developing system of joints lying approximately parallel to the cryptic (and textural) layering in the upper border group.

7. THE IGALIKO COMPLEXES

The Igaliko region comprises four large intrusive complexes principally composed of nepheline syenites, ranging from slightly undersaturated augite syenites to peralkaline foyaites (Emeleus and Harry, 1970). The probable age sequence (oldest to youngest) of the complexes is: 1) North Qôroq (*ca.* 7 x 4 km), 2) Motzfeldt (*ca.* 15 x 20 km), 3) South Qôroq (*ca.* 26 x 10 km) and 4) Igdlerfigssalik (*ca.*16 x 11 km).

7.1. North Qôroq

The foyaite intrusions composing most of this complex display feldspar lamination and scarce, generally thin, discontinuous, steeply inclined to vertical mafic layers. Side-wall cumulates, developed on steep boundary layers, appear to have predominated in this complex.

7.2. Motzfeldt

There are over twenty intrusive syenite units in this complex, grouped into three formations (Bradshaw, 1988). Feldspar lamination and sporadic mafic layering occur in all three. Steep layering (>40°) is particularly prominent in the southern part of the central foyaite; the cumulus minerals in the mafic layers are nepheline, brown amphibole, apatite and aegirine-augite, with alkali feldspar and aenigmatite among the intercumulus phases. Figure 13 illustrates near-vertical, non-parallel, mafic layers defining erosional 'flutings' on an inward-growing chamber wall.

7.3. South Qôroq

Emeleus and Harry (1970) distinguished five separate intrusions which they termed SS.1 to SS.5. Parallel mafic layers, with conformable lamination, are locally well developed in SS.5. Mafic to ultramafic layers, up to 25 cm thick, represent concentrates of cumulus aegirine-augite, fayalite and magnetite. Characteristics of the South Qôroq pyroxenes and olivines are described by Stephenson, (1972, 1974). Faulting and fragmentation of mafic/ultramafic cumulates occurred when the feldspathic cumulates were only partially consolidated, producing disturbed flow patterns in the latter (Figure 14). Steep side-wall layering is well developed in parts of the SS.4b augite syenite, (Stephenson, 1976). This involves (<1 cm) ultramafic layers (olivine, clinopyroxene, magnetite, apatite cumulus) separated by 10-15 cm units of the more leucocratic, nepheline-bearing augite syenite.

7.4. Igdlerfigssalik

This youngest complex in the Igaliko suite comprises seven principal intrusive units. Layering features occur sporadically in these but are best developed within the SI.4 augite syenite unit (Emeleus and Harry, 1970). This shows both lamination and modal layering, the latter reflecting concentration of olivine, clinopyroxene, magnetite, and apatite. West of Igdlerfigssalik, an outer zone, up to 300 m thick, is biotite-rich and well-laminated towards its chilled contact with the country rocks (Emeleus and Harry, 1970, Plate 2). Rare, thin mafic layers occur. Further in, near-vertical mafic layers are present in a 100-200 m zone of pale syenite which, in turn, is succeeded inwards by a 100-150 m zone of finer-grained, darker syenite characterized by sub-parallel vertical, or steeply inward-dipping, mafic layers. The latter are *ca.* 5 mm thick, separated by 10-20 mm thick intervals of leucocratic syenite. Regular, sub-parallel layers are cut by curvi-planar layers which face towards the centre (*i.e.* concave inwards) and which are interpreted as defining steeply inclined erosional surfaces developed on the inward-growing chamber walls (Emeleus and Harry, 1970, Figure 15).

Further north, near Qôroq Fjord, the character of the layering in the darker syenite zone changes progressively. Mafic layers become thicker, and the feldspars more tabular and parallel. The modal layering and the feldspar lamination are conformable and dip at progressively lower angles to the east and southeast, fanning out from near-vertical to between 50 and 30° traced from south to north over *ca.* 800 m of strike. Layering on the south side of Qôroq Fjord, through a sequence *ca.* 200 m thick, involves rhythmically repeated layers, up to 3 m thick which dip towards the intrusion centre at 35-40° (Emeleus and Harry, 1970). The layers have abruptly defined bases, rich in olivine, clinopyroxene, opaque oxides, and apatite, grading up into more felsic, well-laminated syenite. Stacked channel layers occur which resemble those at Nunarssuit and Kûngnât, with axes radial to the intrusion margins.

The excellent exposure and dissection of SI.4 show the changes inwards from steep, side-wall cumulates with erosional features, to lower-angled layering developed on an inclined chamber floor. On these more gently dipping surfaces gravitational accumulation of cumulus

***Figure 14**. Disturbed foyaitic cumulates in South Qôroq. Tabular slices of coherent, ultramafic, pyroxene-rich cumulates are thought to have become detached and incorporated within unconsolidated felsic crystal mush. The lamination of sparse alkali feldspars in the melanocratic cumulates is inferred to have predated the detachment whereas the swirled lamination between the two dark slices is thought to have resulted from differential movement of the ultramafic slices. The lower slice was itself undergoing further break-up prior to congelation of the felsic matrix.*

crystals is inferred, with the orientation of the platy feldspars attributed to alignment by gentle flow. Channel erosion, with subsequent infilling by dense cumulus phases, may reflect sporadic, more powerful, magma currents of crystal-laden melt detached from poorly consolidated side-wall cumulates. Emeleus and Harry (1970) recorded anorthositic and basic xenoliths, up to 2 m across, in the SI.4 syenites, which disturb the underlying layering features, suggesting that they sank and acted as "drop-stones" onto unconsolidated crystal mush. The xenoliths are overlain by undisturbed layers.

Whereas units SI.5 and SI.6 show little layering, SI.7, a foyaite with very tabular feldspar (*ca.* 15 x 3 mm), is characterized both by strong lamination and conformable modal layering. Differential weathering of cumulates with slight variation of felsic and mafic minerals has given rise to a succession of large-scale, gently curving dip-slopes and escarpments. The structures define a stack of saucer-shaped layers, each with a flat base and with edges rising at angles of up to 60° (Emeleus and Harry, 1970, Figure 28).

The evidence suggests that the morphology of the Igaliko syenite magma chambers, for much of their evolution, was basinal. The widespread development of steep side-wall layering implies a high degree of adherence of cumulus crystals to the boundary layers. As in common with most other Gardar plutons, there is no clear distinction between 'marginal border groups' and more shallowly inclined central layered series.

8. THE GRØNNEDAL-IKA COMPLEX

The Grønnedal-Ika Complex (Emeleus, 1964; Bedford, 1989) consists of a layered nepheline syenite intrusion cut by a xenolith-rich, unlayered nepheline syenite and a carbonatite intrusion. Further complexity is added by several generations of dykes and extensive faulting. The following account concerns the early layered nepheline syenite body which was originally *ca.* 6 x 3.5 km across. The least disturbed (by faulting and younger intrusions) and most continuous section is that extending southeast from the northwest contact zone (Emeleus, 1964, Figure 8). The main layered intrusion comprises a Lower and Upper Series of nepheline syenites separated by a thick wedge (or raft) of country-rock gneiss. Layering in the complex is partly defined by feldspar lamination (often perfectly developed), by various lithostratigraphic units including some layers of gneiss xenoliths and by some modal layering. Dips are inward and approximately 1,200 m of cumulates are exposed.

The basal member of the Lower Series is an altered, generally structureless, coarse-grained brown (nepheline) syenite in which alkali feldspar (up to 2 cm) is the only recognizable original mineral. Several gneiss xenoliths, up to 100 m thick, occur in its upper part. There is gradation upwards into well-laminated foyaitic cumulates. Cumulus minerals in these were sanidine (subsequently exsolved and ordered and generally <1.5 cm in length), nepheline, aegirine-augite, and minor apatite. Biotite, cancrinite, amphibole, and calcite are confined to the intercumulus

Figure 15. Side-wall cumulates in the SI.4 augite syenite unit of the Igdlerfigssalik Complex. A channel (fluting, cf. Figure 13) has been eroded through an outer facies showing near-vertical microrhythmic layering and coated with a concentration of mafic cumulus prior to inward growth of a new syenite facies essentially devoid of modal layering.

Figure 16. *Small-scale modal layering in Upper Series foyaites, Grønnedal-Ika complex. The base of a rhythmic unit is marked by a thin mafic layer rich in aegirine-augite, overlain by mesocratic material in which cumulus nepheline (white) is prominent. The downwardly convex lobes at the base of the mafic layer, with their intervening cusps of felsic material, are thought to be small-scale load pouches. (Hammer head ca. 13 cm long).*

matrix.

The Lower Series is overlain by a massive slab of gneiss, *ca.* 270 m thick, which separates it from the Upper Series. The slab is deduced to have detached from the chamber roof and settled to the floor as a coherent tabular mass. No rocks have been identified as belonging to any upper border group at Grønnedal-Íka, in contrast with Klokken and Ilímaussaq. The presumption is that at Grønnedal-Íka, those crystals nucleating on the roof did not remain *in situ* to compose a downward-growing cumulate suite.

The Upper Series, which is much thicker than the Lower Series, consists of finely laminated foyaite within which lies a distinctive unit *ca.* 300 m thick of more pyroxene-rich nepheline syenite. Small-scale mafic layering in the complex appears to be virtually confined to this unit (Figure 16); aegirine-augite-rich streaks and patches are common and there are at least six pyroxene-rich layers, *ca.* 2 cm thick, separated by 5-15 cm thick layers of more leucocratic foyaite, that are traceable for up to 40 m along strike (Emeleus, 1964, Figure 4). These mafic layers generally grade into the more feldspathic rock on both upper and lower surfaces. Cumulus aegirine-augite and apatite are concentrated in the mafic layers. Biotite, alkali amphibole, opaque oxides, aenigmatite, cancrinite, sodalite, and carbonate are intercumulus phases; alkali amphibole frequently appears to have replaced aegirine-augite along very irregular boundaries.

The leucocratic foyaite is similar but with a preponderance of cumulus alkali feldspar (commonly 1-1.5 cm in length) and nepheline. Unlike the rocks of most of the other Gardar intrusions, olivine is virtually absent and magnetite does not occur as a cumulus phase.

The layering elements in the complex generally have steep dips (>30°), steepening to vertical close to the contacts (Emeleus, 1964, Plate 2). It has been argued that the high dips are essentially primary and that low-density contrasts between the magma and the majority of accumulating phases (feldspar and nepheline), together with relatively rapid congelation of trapped melt, stabilized the cumulate, despite the steepness of the slopes. However, some outcrops, especially in the Upper Series, show intense disturbance of lamination which may have resulted from slumping (Bedford, 1989, Plate 2). Dips in the Upper Series are generally greater than in the Lower Series but appear to decrease towards the centre of the complex. It is however, uncertain whether this change is primary, or was imposed later.

No evidence for magmatic erosion has been found in Grønnedal-Ika and it may be that sufficiently vigorous convective circulation never developed within the magma chamber. It would appear that only within the interval during which the relatively pyroxene-rich unit was developing in the Upper Series, did the conditions necessary for development of modal layering exist.

9. THE OLDER GIANT DYKE COMPLEX, TUGTUTØQ

This is a dyke of exceptional size (*ca.* 20 x 0.5 km) which preceded the (still more voluminous) Younger Giant Dyke system (Upton *et al.*, 1985). For most of its length it has a composite character, with gabbroic and syenogabbroic rocks forming a marginal sheath enclosing an axial zone composed of salic rocks. The boundary between the two is gradational over two to three metres. The mafic side-wall zones, symmetrically disposed on either side of the dyke, are up to 100 m broad. The width of the salic central zone varies from *ca.* 300 to 400 m across. The dyke, which trends roughly west-southwest to east-northeast, is believed to be tilted towards the east-northeast about an axis roughly normal to the length of the dyke and, with subsequent erosion, shallower levels are exposed in the east-northeast and deeper levels to the west-southwest (Figure 17).

The basic marginal zones show progressive evolution inwards, olivine compositions changing from Fo_{53} to Fo_{16}. Poorly-defined vertical mafic layering parallel to the contacts is occasionally present. The early mineral phases were olivine, plagioclase, ilmenite, magnetite, and apatite. Residual melts are inferred to have flowed up along the boundary layer to accumulate in the higher levels of the intrusion.

On the hypothesis that the intrusion has been tilted, Upton (1964) suggested that the total vertical sequence presented along the 20 km of outcrop is *ca.* 2 km (*cf.* Figure 17). Whereas the Older Giant Dyke syenitic suite is essentially devoid of modal layering or crystal lamination, it is considered as a layered intrusion because it exhibits cryptic variation and "phase layering" from west-southwest to east-northeast, *i.e.* inferentially up-sequence (Upton *et al.*, 1985). Minerals regarded as cumulus in the "lower" parts of the dyke are alkali feldspar (*ca.* $Or_{38.5}Ab_{61}An_{0.5}$), salitic pyroxene, fayalitic olivine, Ti-magnetite, and apatite, with intercumulus alkali amphibole, biotite, and nepheline. Towards the east-northeast, olivine is lost, the feldspar becomes progressively more potassic and the pyroxene more aegirine-rich. Intercumulus amphibole diminishes to zero and the modal amount of nepheline increases through the pulaskitic facies. In the "upper" (*i.e.* most easterly) foyaite and sodalite foyaite facies of the Giant Dyke, nepheline was a cumulus mineral that co-precipitated with potassic feldspar (>43%) and aegirine. Biotite was joined by sodalite, cancrinite, analcime (and other zeolites) in the intercumulus component.

Although it was proposed that the syenitic core crystallized by side-wall growth in a stratified magma body, (Upton *et al.*, 1985), the regular upward cryptic and phase layering is more simply explicable in terms of progressive congelation upwards in a closed magma body. Assuming that all the rock types of the Older Giant Dyke are cogenetic products of *in situ* fractionation of a parental hawaiitic magma (Upton *et al.*, 1985), it may be inferred that the syenitic rocks are underlain by synformally layered syenogabbros and gabbros analogous to those exposed in axial zones of the Younger Giant Dyke. The sill-like upper portion of the intrusion shown in Figure 17 is also based on evidence from the Younger Giant Dyke.

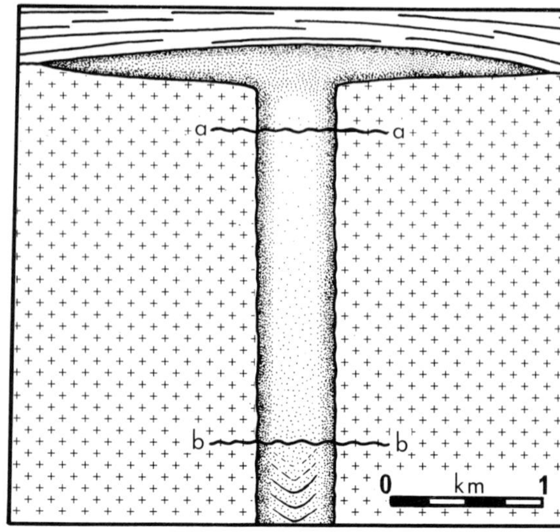

Figure 17. Schematic cross-section through the Older Giant Dyke, Tugtutôq. (Vertical scale = horizontal scale). Lines a and b denote the highest and lowest erosional levels presented by the outcrop. The sill-like culmination exploiting the unconformity plane between the early Proterozoic granitoid country rocks and the supracrustal Eriksfjord Formation, is hypothetical, as is the synformal (modal) layering indicated diagrammatically below b. The dyke exhibits cryptic variation inwards in its side-wall cumulates and upwards in the syenites, pulaskites and foyaites of the interior. Decrease of stipple density diagrammatically reflects increase in differentiation.

The lack of feldspar lamination is noteworthy in that the grain-size and degree of feldspar tabularity in the Giant Dyke foyaites are not very different from those of Grønnedal-Ika in which feldspar parallelism attains a high degree of perfection. The essentially random orientation of the feldspars, together with the absence of modal layering, may denote failure of strong convective flow to develop within this magma body. Furthermore, textures indicate that the rocks are relatively uncompacted cumulates.

10. THE ILÍMAUSSAQ COMPLEX

The Ilímaussaq Complex (Sørensen, 1958; Ferguson, 1964; Bailey *et al.*, 1981a) was intruded into granitoids and the unconformably overlying lavas and sandstones of the Eriksfjord Formation that form part of its roof. The complex, measuring *ca.* 17 x 8 km, involved three intrusive events, the first of which produced silica-undersaturated augite syenite.

Originally the augite syenite may have been part of a large intrusion, occupying most of the area of the whole complex (Nielsen and Steenfelt, 1979), the bulk of which was lost by stoping during emplacement of the younger intrusions. A thin remnant (up to 2 km broad) is, however, retained around the western and southern margins. In the southwestern sector, the augite syenite possesses a chilled margin against the granitic wallrocks, indicating a benmoreitic composition for the parental magma. The chilled zone grades into coarser syenite showing strongly developed rhythmic layering (Hamilton, 1964), generally dipping into the intrusion at 40-50° although locally steepening to nearly vertical. A few of the layered units exhibit grading over thicknesses of 20-30 cm from sharply defined bases rich in ferrosalite, fayalitic olivine, and Ti-magnetite into feldspar-rich upper portions. Steep inwardly inclined layering is traceable to within a few tens of metres of the contact. Whether the layering in cross-section was originally saucer-shaped or cone-shaped is unknown.

After a second intrusive event involving emplacement of quartz syenite and alkali granite sheets in the uppermost part of the augite syenite (Bailey *et al.*, 1981a), a third and culminating event saw the emplacement of a peralkaline, iron-rich phonolitic magma from which an agpaitic rock suit formed (Sørensen and Larsen, 1987; Larsen and Sørensen, 1987). This agpaitic magma is thought to have developed as a low-density, volatile-rich residue from a large underlying (layered) gabbroic - syenitic complex. Ilímaussaq is the only Gardar pluton in which the original roof is (partially) preserved. Beneath the roof, the agpaitic upper border group is also retained. Crystallization of the agpaitic residuum proceeded essentially as a closed system to give rise to one of the most remarkable layered cumulate sequences in the Province and certainly the most exotic in terms of mineralogy and geochemistry (Figure 18).

The nature and origin of the layering have been widely discussed in the literature. The downward-growing upper border group, showing progressive differentiation, consists of pulaskite grading into foyaite (cumulus alkali feldspar, fayalite, hedenbergite, Ti-magnetite, and apatite, with intercumulus nepheline, sodalite, alkali amphibole, aegirine, and aenigmatite). The

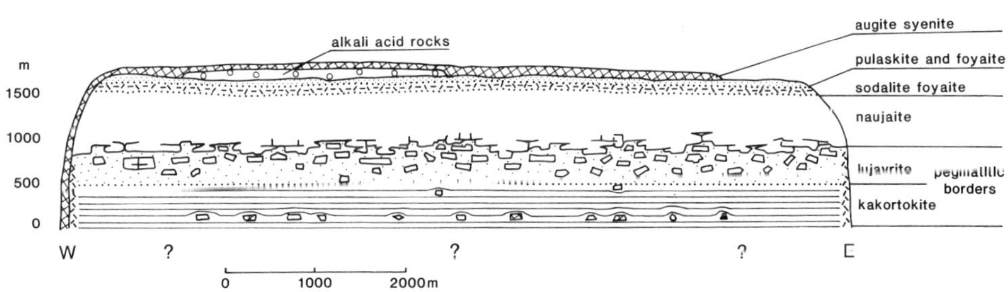

Figure 18. *Schematic cross-section of the Ilímaussaq complex (after Bailey et al., 1981a). Vertical scale = horizontal scale. The early augite syenite intrusion is preserved as a thin marginal sheath around the western flanks, and as a partial roofing zone. Pulaskite and foyaite, sodalite foyaite and naujaite constitute successive stages in the evolution of the down-growing upper border group of the agpaitic intrusion. The kakortokites and (lower part of) the lujavrites represent up-grown floor cumulates. Detached blocks of naujaite have been incorporated within the kakortokite-lujavrite suite. Late-stage lujavrite residues have extensively invaded and brecciated the overlying naujaite. The alkali acid rocks represent a separate intrusion.*

Figure 19. *View of the kakortokite series, Ilímaussaq, from the north. The distant mountain ridge is composed of earlier Proterozoic granitoids, rising above the southeastern contact zone of the intrusion. A large autolith (outlined) of naujaite is seen, centre right, with layered kakortokites draped over it.*

foyaite in turn, passes downwards into sodalite foyaite, by which stage nepheline and sodalite were among the cumulus phases. Beneath this is a sodalite-rich syenite (naujaite) considered to have accreted as a flotation cumulate (Ussing, 1912; Sørensen, 1969). The naujaite comprises 30-40% (and occasionally up to 90%) modally of idiomorphic sodalite typically 2-3 mm across (Sørensen and Larsen, 1987). These are generally enclosed by oikocrysts of alkali feldspar, alkali amphibole, eudialyte, and aegirine and lower-temperature phases, centimetres to decimetres across. Modal layering is weakly developed in the naujaites.

According to Larsen and Sørensen (1987), the magma remaining beneath this upper border zone had probably already developed repeated layering through a double diffusive convection mechanism. Density-graded units *ca.* 7 m thick of eudialyte-bearing nepheline syenite (kakortokite, Figure 19) formed from the layered magma by successive upward crystallization of individual layers. Twenty-nine of these macro-units are exposed (Bohse and Andersen, 1981). According to Sørensen and Larsen (1987), the primary mechanism for the macro-rhythmic layering involved differences in nucleation and growth-rate of the cumulus minerals in relation to the degree of undercooling in a multiply saturated magma. In the kakortokites, this mechanism and density sorting worked in the same direction. The grading in the units was thus enhanced by density sorting during crystal settling. The layers define a saucer- (or bowl-)

shape, with steep to vertical dips at the margins, shallowing abruptly inwards to a generalized slope of 10-20° a few hundred metres in from the contacts. This is taken to denote the original form of the magma chamber floor (Bohse and Andersen, 1981; Figure 20).

The kakortokite sequence involve potassium feldspar, nepheline, arfvedsonite, eudialyte, aegirine, and (at some stages) aenigmatite, as cumulus phases. The idealized layered unit is 10 m thick and consists of a three-layer sandwich (Bohse and Andersen, 1981). Arfvedsonite-rich black kakortokite at the base of each unit grades up into feldspathic white kakortokite, with the prominent intervening presence of eudialyte-rich red kakortokite in some units. Within these kakortokite macro-units small-scale modal layering (Figure 21) and low-amplitude troughs (due to erosion and sorting by magma flow?) can be present. Feldspar lamination, well-developed in the mafic "black kakortokite" unit bases, typically diminishes toward the unit tops (Upton, 1961). Slump structures in three of the macro-units point to thicknesses of over 20 m of unconsolidated crystal mush having been present (Bohse and Andersen, 1981).)

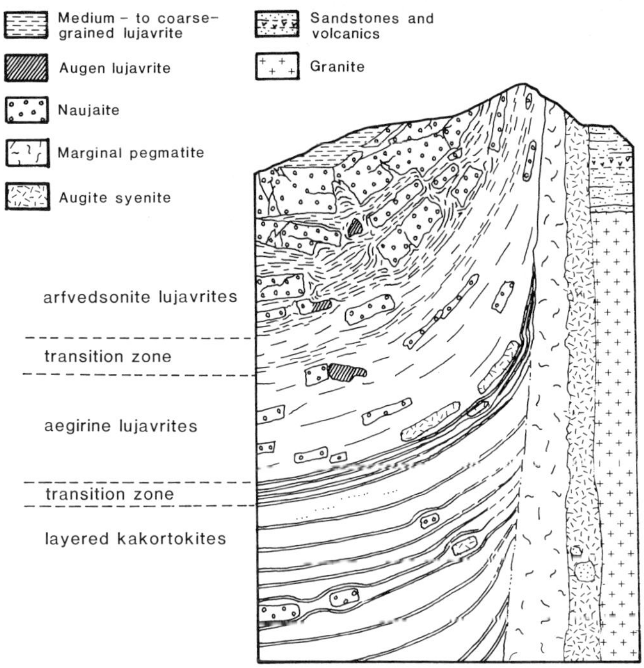

Figure 20. *Schematic section across the southern margin of the Ilímaussaq Complex (from Bohse and Andersen, 1981), showing conformity of the kakortokite - lujavrite cumulate sequence and the banking of the layering from sub-horizontal in the interior of the agpaitic intrusion to steep, and occasionally near-vertical attitudes, close to the marginal (pegmatite) facies. Solid inclusions enveloped during accretion of the cumulate pile included fragments of the pre-agpaite, augite syenite intrusion. The majority of inclusions, however, were derived from the roof involving naujaite and occasionally, a lujavrite variety ("augen lujavrite" cf. Bohse and Andersen, 1981).*

The kakortokites pass upwards into still more highly fractionated eudialyte-poor nepheline syenites (lujavrites) which compose a sandwich horizon between the kakortokites and the overlying naujaite. Graded modal layering is developed in some facies of the lujavrites. Successively higher lujavrite horizons contain more of the incompatible components, leading to crystallization of facies that are potentially economic ores of U, Th, Be, and Nb.

Figure 21. *A modally well graded small-scale unit within one of the larger units in the kakortokite series. A sharply-defined layer base, rich in arfvedsonite cumulus, grades up into increasingly felsic cumulates, with an approximately cm-thick feldspar concentrate at the top. (Hammer shaft ca. 60 cm long).*

The naujaites forming the roof at this stage of the chamber's evolution, congealed early with respect to the accumulation of the presently exposed kakortokite-lujavrite series. Incipient break-up of the consolidated naujaites allowed blocks of this material, up to 10s of metres long, to undergo intermittent detachment (delamination) and sink to become buried in the accreting kakortokite-lujavrite cumulates (Figures 18, 19, and 20). There are clear analogies between this situation and that envisaged at Klokken.

The magma body from which the agpaitic cumulates formed was horizontally tabular, with lateral dimensions of roughly 8 x 17 km but a thickness probably little greater than the observed thickness of the exposed agpaites (*ca.* 1.5 km; *cf.* Figure 18). With a total observed volume of *ca.* 200 km^3, inferred to represent about 2% residue of a parental transitional alkali basalt magma, it is necessary to propose some 10,000 km^3 of such a parent (Larsen and Sørensen, 1987). Bailey *et al.* (1981b) concluded that a continuum may have existed from augite syenite to the agpaites and that extensive fractionation could have taken place beneath the present erosion level. Crystallization in excess of 99% of the augite syenite magma would have been required to produce the final lujavrites.

11. DISCUSSION

Gardar magma viscosities appear to have been sufficiently low throughout a very wide range of magma compositions for even small density differences between melts and solid or semi-solid materials (*viz.* rocks, crystals, or crystal-liquid mushes) to permit gravitational processes to occur. Yield-strengths (*c.f.* McBirney and Noyes, 1979) were apparently low enough for individual crystals to accumulate by flotation (*e.g.* sodalite in Ilímaussaq naujaite), for crystal aggregates to settle, as in the "snow-flake" troctolite cumulates of the Younger Giant Dyke, for buoyant xenoliths and (mega-) xenocrysts to float and produce flotation breccias (anorthositic debris in the Younger Giant Dyke), and for disrupted country rocks to sink in the salic magmas. In at least three instances rocks which crystallized early against a roof (upper border group rocks) were stoped and enveloped by less dense residual magmas that evolved beneath them. At Ilímaussaq the pulaskite - foyaite - sodalite foyaite - naujaite upper border group is generally intact but naujaite autoliths subsided to become enveloped by the uprising kakortokite-lujavrite pile. At Klokken, the upper border group is now seen only as disrupted slabs which sank to be successively overwhelmed by the upgrowing floor cumulates. At Syenitknold in the Younger Giant Dyke system, a single massive autolith of roof-zone gabbro sank within the trachytic residual magma that had evolved beneath it.

There are numerous examples where poorly consolidated cumulate detached and migrated downslope as mass-flows or slumps. In the Younger Giant Dyke these appear to have collapsed from vertical side-walls. In Nunarssuit, it may be inferred that the convoluted masses of "soft sediment" plus cumulate clasts were similarly accelerated down a steep boundary layer, finally coming to rest on slopes (which have a present dip of 45-30°) where they consolidated. In some cases layering that was developing by *in situ* crystal growth was interrupted by mass-flow processes. It is harder to understand why slump features are apparently not present in some of the intrusions (*e.g.* the Kûngnât and Klokken syenites) where nucleation on steep side-walls is also thought to have taken place.

The presence or absence of, and degree of, crystal alignment in these intrusions provides much food for thought. Parallel alignment of feldspar crystals is by far the most obvious manifestation and depends to a large extent on the morphology of the feldspars concerned. Thus, feldspar lamination is well developed in some gabbroic cumulates containing tabular labradorite and andesine grains but the most perfect parallelism is found in some of the nepheline syenite cumulates containing wafer-thin alkali feldspars inferred to have crystallized initially as sanidine. Correspondingly, lamination is weak or undetectable in the syenite cumulates in which the feldspars are more squat and equant. Layering is sometimes indicated by alternation of units that are well and poorly laminated. The latter are less well packed and have correspondingly higher intercumulus contents. Examples are seen in the layered anorthosite xenoliths in the Younger Giant Dyke on Tugtutôq and in the Grønnedal nepheline syenite succession. Variation in degree of feldspar parallelism has been noted (above) for the graded kakortokite units in Ilímaussaq. Compaction effects have undoubtedly operated, but not to the extent where they produce uniformly laminated products. Lamination is notably absent from the Younger Giant Dyke troctolitic cumulates characterized by "snow-flake" plagioclase and olivine aggregates. These are thought to have settled as composite bundles and any subsequent compaction failed to deform these delicate structures.

An enigmatic difference between the intrusions, relates to the degree to which steep marginal border groups are present. In simplest terms this appears to be influenced by the facility with which side-wall cumulus adhered. At Klokken, parts of the Igaliko intrusions, eastern Kûngnât, the Older Giant Dyke, and parts of the Younger Giant Dyke, steep side-wall cumulates formed and remained stable. At west Kûngnât, and in parts of the Younger Giant Dyke however, side-wall cumulates are absent and centrally inclined modal layering extrapolates up to the vicinity of the contact zones. Cumulus crystals in these cases seem to have been incapable of sticking to the walls. For the Nunarssuit syenite partial adherence and intermittent mass-flow collapse is indicated. At Ilímaussaq, the extreme aspect ratio (horizontally tabular) of the agpaitic chamber clearly did not lend itself to extensive side-wall growth although, as indicated by Bohse and Andersen, (1981), (*viz.* Figure 20), the steep to vertical marginal banking of the kakortokite-lujavrite series clearly shows fairly stable adherence of side-wall cumulates. At present there is no clear understanding of the factors governing this very variable ability of the cumulus grains to stick to steep boundary surfaces.

Feldspar lamination, as in the nepheline syenites of the Igaliko and Grønnedal-Ika Complexes, can be present along steeply dipping side walls. Such evidence indicates that the fabric was not brought about by crystal settling. However, there does seem to have been some degree of physical orientation of the feldspars, whether on gently or steeply inclined surfaces, by variable magmatic flow regimes. Whereas some measure of compaction undoubtedly occurred as temperatures approached the solidi, the overall textural features are believed to have been preserved from an early stage in the accumulation of the various cumulate sequences.

12. CONCLUSIONS

The Gardar Province demonstrates, in a remarkable manner, how similar styles of layering involving comparable phenomena are shown by a wide compositional spectrum of rocks. These range, at their simplest, from those involving bi-mineralic cumulus (plagioclase plus olivine) to complex poly-phase assemblages as exemplified by the kakortokites in which up to six cumulus species participated. However, the great bulk of the cumulate rocks involve a "gabbroic" cumulus assemblage of feldspar, olivine, clinopyroxene, ± opaque oxides, and apatite. The principal difference between these rocks and tholeiitic gabbroic sequences is that in the Gardar intrusions, much of the feldspar was alkali feldspar rather than plagioclase.

Crystal nucleation is believed to have taken place principally within the boundary layers. However, gravitational migration of crystals, crystal aggregates, coherent rock masses, crystal + melt "mushes", and crystal-poor melts was ubiquitous and is taken to imply low viscosities and yield strengths. It is suggested that rising concentration of halogens with increasing fractionation was responsible for maintaining low viscosities despite falling temperatures and increasing silica contents.

Processes controlling the presence or absence of modal layering appear to have been complex and to have involved differences in nucleation and growth rates, as well as to crystal sorting mechanisms. Intermittent failure of feldspar to nucleate, allowing mafic cumulates to develop, may have been commonplace across a wide compositional spectrum of the Gardar magmas. Gravity-driven ("sedimentary") processes appear to have played a secondary role, modifying cumulus on which a "primary" density contrast had been imposed *e.g.* by the presence or absence of feldspar. Crystal sorting on grounds of density, within flowing crystal-melt slurries, is believed to have given rise to much of the modal grading observed.

Evidence for magmatic flow capable of eroding previously deposited cumulates, sometimes to a depth of metres, is widespread and provides further support for low viscosities. Conditions for formation of layered cumulates appear to have been optimal in these Gardar magmas in that high-density ferrian species (principally olivine, pyroxene, magnetite and, occasionally amphibole) crystallized together with low-density (calcium-poor) feldspars and feldspathoids, from very fluid alkalic magmas.

13. ACKNOWLEDGEMENTS

The primary mapping of the Gardar complexes was accomplished in association with the Geological Survey of Greenland, to whom we are deeply indebted. Our thanks go also to the many field assistants, research students, boat crews and helicopter pilots without whose unstinted help over the years, these investigations would have been impossible. Financial assistance for field work from the Royal Society, the Natural Environment Research Council, and the Carnegie Trust for Scottish Universities is also gratefully acknowledged. We are grateful also to Y. Cooper, D. Baty, and L. Thorburn for assistance with photographs, text-figures and manuscript preparation. Publication of Figures 18 and 20 is by permission of Grønlands Geologiske Undersøgelse.

14. REFERENCES

Ankatell, J.M., Cegla, J., & Dzulynski, S., 1970. On the deformational structures in systems with reversed density gradients. *Rocznik Polskiego Towartzystwa Geologocznego* **40**, 1-29.

Bailey, J.C., Larsen, L.M., & Sørensen, H., 1981a. Introduction to the Ilímaussaq intrusion with a summary of the reported investigations. In: Bailey, J.C., Larsen, L.M., & Sørensen, H. (eds.) *The Ilímaussaq intrusion, South Greenland. Grønlands Geol. Unders.*, Rap. No. 103, 5-17.

Bailey, J.C., Rose-Hansen, J., Lovberg, L., & Sørensen, H., 1981b. Evolution of Th and U whole-rock contents in the Ilímaussaq intrusion. In: Bailey, J.C., Larsen, L.M., & Sørensen, H. (eds.) *The Ilímaussaq intrusion, South Greenland. Grønlands Geol. Unders.*, Rap. No. 103, 87-98.

Bedford, C., 1989. The mineralogy, geochemistry and petrogenesis of the Grønnedal-Ika Alkaline Igneous Complex, South-West Greenland. Ph.D. thesis (unpubl.), Univ. Durham.

Bohse, H., & Andersen, S., 1981. Review of the stratigraphic divisions of the kakortokite and lujavrite in southern Ilímaussaq. In: Bailey, J.C., Larsen, L.M., & Sørensen, H. (eds.) *The Ilímaussaq intrusion, South Greenland. Grønlands Geol. Unders.* Rap. No. 103, 53-62.

Bohse, H., Brooks, C.K., & Kunzendorf, H., 1971. Field observations on the kakortokites of the Ilimaussaq intrusion, South Greenland. *Grønlands geol. Unders.* Rap. No. 38, 43 pp.

Bradshaw, C., 1988. A petrographic, structural and geochemical study of the alkaline igneous rocks of the Motzfeldt Centre, South Greenland. Ph.D. thesis (unpubl.), Univ. Durham.

Emeleus, C.H., 1964. The Grønnedal-Ika alkaline complex, South Greenland. The structure and geological history of the complex. *Bull. Grønlands geol. Unders.* **45**, (also *Meddr. Grønland* **172**, (3)).

Emeleus, C.H., & Harry, W.T., 1970. The Igaliko syenite complex. General description. *Bull. Grønlands geol. Unders.* **85**, (also *Meddr. Grønland* **186**, (3)).

Emeleus, C.H., & Upton, B.G.J., 1976. The Gardar period in Southern Greenland. In: Escher, A., & Watt, W.S. (eds.) *The Geology of Greenland.* The Geological Survey of Greenland, Copenhagen, 153-81.

Ferguson, J., 1964. Geology of the Ilímaussaq alkaline intrusion, South Greenland. Description of map and structure. *Bull. Grønlands geol Unders.* **39**, 82 pp., (also *Meddr. Grønland* **174**, (4), 82 pp).

Ferguson, J., & Pulvertaft, T.C.R., 1963. Contrasted styles of igneous layering in the Gardar province of South Greenland. *Min. Soc. Amer., Spec. Paper* **1**, 10-21.

Hamilton, E.I., 1964. The geochemistry of the northern part of the Ilímaussaq intrusion, S.W. Greenland. *Bull. Grønlands geol. Unders.* **42**, (also *Meddr. Grønland* **162** (1)).

Harry, W.T., & Emeleus, C.H., 1960. Mineral layering in some granite intrusions of S.W. Greenland. *Int. Geol. Congr. 21st Session. Norden*, 172-81.

Harry, W.T., & Pulvertaft, C.T.R., 1963. The Nunarssuit intrusive complex, South Greenland. *Bull. Grønlands geol. Unders.* **36** (also *Meddr. Grønlands* **169** (1)).

Irvine, T. N., 1983. Skaergaard trough-layering structures. *Carnegie Inst. Wash., Y. Bk.* **82**, 289-94.

Larsen, L.M., & Sørensen, H., 1987. The Ilímaussaq intrusion - progressive crystallisation and formation of layering in an agpaitic magma. In: Fitton, J.G., & Upton, B.G.J. (eds.) *Alkaline Igneous Rocks, Spec. Publ. Geol. Soc. Lond.* **30**, 473-88.

McBirney, A.R., & Noyes, R.M., 1979. Crystallisation and layering of the Skaergaard Intrusion. *J. Petrology* **20**, 487-554.

Mingard, S.C., 1990. Crystallisation and layering of the Younger Giant Dyke Complex, SW Greenland. Ph.D. thesis (unpubl.), Univ. Edinburgh.

Nielsen, B. L., & Steenfelt, A., 1979, Intrusive events at Kvanefjeld in the Ilímaussaq igneous complex. *Bull. geol. Soc. Denmark* **27**, 143-55.

Parsons, I., 1979. The Klokken gabbro - syenite complex, South Greenland: cryptic variation and origin of inversely graded layering. *J. Petrology* **20**, 653-94.

Parsons, I., & Becker, S.M., 1987. Layering, compaction and post-magmatic processes in the Klokken intrusion. In: Parsons, I. (ed.) *Origins of Igneous Layering*. Dordrecht: Reidel Publ. Co., 29-92.

Parsons, I., and Brown, W.L., 1988. Sidewall crystallisation in the Klokken intrusion: zoned ternary feldspars and coexisting minerals. *Contr. Miner. Petrol.* **98**, 431-43.

Parsons, I., & Butterfield, A.W., 1981. Sedimentary features of the Nunarssuit and Klokken syenites, South Greenland. *J. geol. Soc. Lond.* **138**, 289-306.

Parsons, I., Mason, R.A., Becker, S.M., & Finch, A.A., 1991. Biotite equilibria and fluid circulation in the Klokken Intrusion. *J. Petrology* **32**, 1299-333.

Sørensen, H., 1958. The Ilímaussaq batholith. A review and discussion. *Bull. Grønlands geol. Unders.* **19**, 48 pp. (also *Meddr. Grønland* **162**, (3)).

Sørensen, H., 1969. Rhythmic igneous layering in peralkaline intrusions. An essay review on Ilímaussaq (Greenland) and Lovozero (Kola, USSR). *Lithos* **2**, 261-83.

Sørensen, H., & Larsen, L.M., 1987. Layering in the Ilímaussaq alkaline intrusion, South Greenland. In: Parsons, I. (ed.) *Origins of Igneous Layering*. Dordrecht: Reidel Publ. Co., 1-28.

Stephenson, D., 1972. Alkali pyroxenes from nepheline syenites of the South Qôroq Centre, South Greenland. *Lithos* 5, 187-201.

Stephenson, D., 1974. Mn and Ca enriched olivines from nepheline syenites of the South Qôroq Centre, South Greenland. *Lithos* 7, 35-41.

Stephenson, D., 1976. The South Qôroq nepheline syenites, South Greenland: petrology, felsic mineralogy and petrogenesis. *Bull. Grønlands geol. Unders.* **118**.

Stephenson, D., & Upton, B.G.J., 1982. Ferromagnesian silicates in a differentiated alkaline complex: Kûngnât Fjeld, South Greenland. *Miner. Mag.* **46**, 283-300.

Upton, B.G.J., 1960. The alkaline complex of Kûngnât Fjeld, South Greenland. *Bull. Grønlands geol. Unders.* **27**, (also *Meddr. Grønland*, (123)).

Upton, B.G.J., 1961. Textural features of some contrasted igneous cumulates from South Greenland. *Meddr. Grønland* **123**, (6), 1-29.

Upton, B.G.J., 1964. The geology of Tugtutôq and neighbouring islands, South Greenland. Pt.3: Olivine gabbros, syeno-gabbros and anorthosites. Part 4: The nepheline syenites of the Hviddal composite dyke. *Grønlands geol. Unders.* **48**, 80 pp.

Upton, B.G.J., 1987. Gabbroic, syenogabbroic and syenitic cumulates of the Tugtutôq Younger Giant Dyke Complex, South Greenland. In: Parsons, I. (ed.) *Origins of Igneous Layering*. Dordrecht: Reidel Publ. Co., 1-28.

Upton, B.G.J., & Emeleus, C.H., 1987. Mid-Proterozoic alkaline magmatism in southern Greenland: the Gardar province. In: Fitton, J.G., & Upton, B.G.J., (eds.) *Alkaline Igneous Rocks. Spec. Publ. Geol. Soc. Lond.* **30**, 449-71.

Upton, B.G.J., & Fitton, J.G., 1985. Gardar dykes north of the Igaliko Syenite Complex, southern Greenland. *Geol. Surv. Greenland, Report* **127**, (2), 24 pp.

Upton, B.G.J., & Thomas, J.E., 1980. The Tugtutoq Younger Giant Dyke Complex, South Greenland: fractional crystallisation of transitional olivine basalt magma. *J. Petrology* **21**, 167-98.

Upton, B.G.J., Stephenson, D., & Martin, A.R., 1985. The Tugtutôq Older Giant Dyke Complex: mineralogy and geochemistry of an alkali gabbro - augite syenite - foyaite association in the Gardar Province of South Greenland. *Miner. Mag.* **49**, 623-42.

Ussing, N.V., 1912. Geology of the country around Julianehaab, Greenland. *Meddr. Grønland* **38**, 426 pp.

Wager, L.R., & Brown, G.M., 1968. *Layered Igneous Rocks*. Edinburgh: Oliver & Boyd, Ltd., 588 pp.

Wager, L.R., & Deer, W.A., 1939. Geological investigations in East Greenland. III. The petrology of the Skaergaard Intrusion, Kangerdlugssuaq, East Greenland. *Meddr. Grønland* **105**, (4).

LAYERED INTRUSIONS
R.G. Cawthorn (editor)

The Great Dyke of Zimbabwe

A.H. Wilson

Department of Geology and Applied Geology, University of Natal, Private Bag X10, Dalbridge, 4014, South Africa.

Abstract

The Great Dyke of Zimbabwe is unique within the family of large layered intrusions by virtue of its highly elongate form. Apart from the tectonic controls that gave rise to the series of linked magma chambers which together comprise this intrusion, the width to length ratio profoundly affected the layering style, rock-types, mineral compositions and the form of mineralized ore bodies. The intrusion developed as a series of initially isolated chambers which became linked at progressively higher levels during the filling process. The dynamic interplay of crystallization and magma emplacement gave rise to the succession of cyclic units within the ultramafic sequence. The entire length of the Great Dyke (some 550 km) was linked at a level corresponding to the top of the Ultramafic Sequence and at this stage influxes of new magma effectively ceased. The initial magma of the Great Dyke was high magnesian (15.6% MgO), relatively enriched in silica, but with low initial $^{87}Sr/^{86}Sr$ indicating low crustal contamination. The various primary processes of magma mixing resulting from emplacement of new magma into an expanding chamber gave rise to economically important chromitite layers, while fractionation combined with influx of magma caused the formation of base metal sulphides enriched in platinum group elements.

This paper considers the following aspects of the Great Dyke: its tectonic setting, structure, form and development of the magma chambers, initial magma composition, emplacement of magma, crystallization and fractionation, and mineralization.

1. INTRODUCTION AND GENERAL GEOLOGY

Large-scale layered intrusions characterize stable cratonic areas in the late Archaean and early Proterozoic periods. The emplacement of the Great Dyke at 2.46 Ga (Hamilton, 1977) is the only major geological event at this period and it therefore marks the Archaean-Proterozoic boundary in the Zimbabwe Craton. The Great Dyke is one of a group of layered intrusions world wide, that are of approximately the same age and remarkably similar in structure, stratigraphy and composition. The Great Dyke is different from those intrusions in that it is a true dyke at depth in some parts. These intrusions also have economic chromite and platinum group element mineralization.The Great Dyke (Figure 1) is a linear body of mafic and ultramafic rocks 550 km in length and between 4 km and 11 km wide. It trends in a north-northeast direction and intrudes granitoids and greenstone belts of the Archaean Zimbabwe Craton. The northern end of the Great Dyke is bounded by the margin of the Zambezi Province where it underwent deformation, fragmentation and rotation related to the 500 Ma (Pan African) orogeny. The southern limit of the Great Dyke is some 30 km north of the margin of the Limpopo Province. Associated with and parallel to the Great Dyke is a set of major craton-wide fractures and a suite of satellite dykes. Quartz gabbro satellite dykes flank the east and western sides of the Great Dyke (East Dyke and Umvimeela Dyke respectively), whereas ultramafic rocks comprise the southern satellite dyke complex (Figure 1).

The Ultramafic Sequence of the Great Dyke is well layered and is capped at four localities by gabbroic rocks of the Mafic Sequence. The positions of the gabbroic portions represent the centres of up to five discrete magma chamber compartments which make up the Great Dyke. From north to south these are the Mvurodona, Darwendale, Sebakwe, Selukwe and Wedza Subchambers. A 'boatlike' or doubly plunging structure results in the preservation of the remnants of the gabbroic zones in topologically lowest areas.

2. HISTORICAL ASPECTS AND PREVIOUS WORK

The major geological linear feature of the Great Dyke was first recognized between 1865 and 1872 by Carl Mauch in his traverses from Port Natal to the Zambezi River (Harger, 1934). Mennell (1910) noted the continuity of this geological feature and interpreted it as a 'gently inclined sheet' of coarsely crystalline picrite. The first petrological account of the Great Dyke was by Zeally (1915), in which he used the term 'Great Dyke of Norite', and later gave a description of the platinum occurrences (Zeally, 1918). Wagner (1914) was the first to report on the layered form of the Great Dyke and to recognize its synclinal structure.

Early prospecting on the Great Dyke was promoted by the discovery of platinum in the Bushveld Complex and was first reported in the Great Dyke by Maufe (1925), with the economic potential of the chromitite layers having been recognized some years earlier. Keep (1930) described the chromite and asbestos deposits in the northern parts of the Great Dyke.

Lightfoot (1940) summarized the petrography of the Great Dyke rocks and in the same year Weiss (1940) carried out the first gravimetric and magnetic measurements. Following on his interpretation the structure and component rocks of the Great Dyke were discussed by Tyndale-Biscoe (1949) and Hess (1950) carried out the first mineral composition study and concluded that the differentiation was similar to that of the Bushveld Complex.

Worst (1958, 1960) presented the first comprehensive account of the entire body and carried out detailed mapping. Worst (1964) gave accounts of the structure and differentiation and the chromitite resources. Detailed studies of the upper chromitite layers and the sulphide zone in the Darwendale Subchamber were carried out by Bichan (1969, 1970). Wilson (1982, 1992) undertook major investigations on the mineralogical associations, textures, petrology and structure in the Darwendale Subchamber. Detailed studies on the sulphide zones in the Wedza Subchamber were presented by Prendergast (1988, 1990, 1991) and Prendergast and Keays (1989), and in the Darwendale Subchamber by Wilson and Naldrett (1989), Naldrett and Wilson (1989, 1990), Wilson *et al.* (1989) and Wilson and Tredoux (1990).

Figure 1. *(facing page) Geological map of central Zimbabwe Craton showing the Great Dyke, its satellites and associated fractures. Divisions into chambers and subchambers are indicated on right. Circled numbers refer to localities of gravity profiles shown in Figure 7. Abbreviations: MSC, Mvuradona Subchamber; P, Popoteke fault set; GF, Gurungwe fault; MF, Mchingwe fault; M, Mutorashanga. The inset shows the locality of the Great Dyke in relation to the cover rocks and basement in Zimbabwe.*

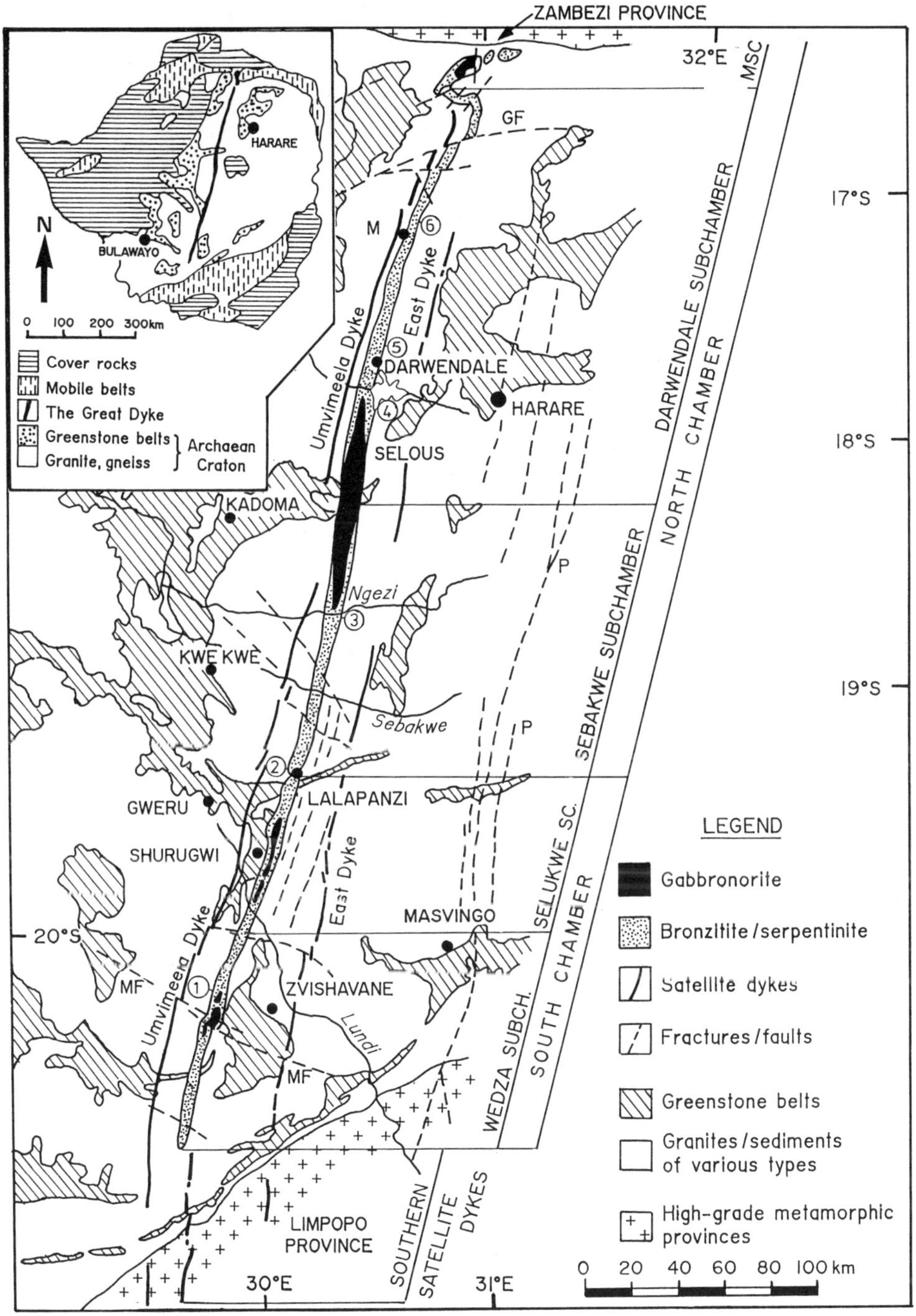

ZAMBEZI PROVINCE
32°E
MSC
GF
17°S
18°S
19°S
20°S
M
East Dyke
Umvimeela Dyke
DARWENDALE
HARARE
SELOUS
KADOMA
Ngezi
KWE KWE
Sebakwe
P
GWERU
LALAPANZI
SHURUGWI
MASVINGO
ZVISHAVANE
Lundi
MF
LIMPOPO PROVINCE
SOUTHERN SATELLITE DYKES
30°E
31°E
DARWENDALE SUBCHAMBER
NORTH CHAMBER
SEBAKWE SUBCHAMBER
SELUKWE SC.
SOUTH CHAMBER
WEDZA SUBCH.
LEGEND
Gabbronorite
Bronzitite/serpentinite
Satellite dykes
Fractures/faults
Greenstone belts
Granites/sediments of various types
High-grade metamorphic provinces
0 20 40 60 80 100 km
N
HARARE
BULAWAYO
0 100 200 300km
Cover rocks
Mobile belts
The Great Dyke
Greenstone belts
Granite, gneiss
Archaean Craton

3. TECTONIC SETTING

The tectonic control of the form of this intrusion is one of the most intriguing problems of Great Dyke geology. Several explanations have been proposed for the structure of the Great Dyke as well as the associated and colinear fracture pattern and the satellite dykes. These include wrench tectonics, the result of an abortive rift system, a failed greenstone belt and vertical tectonics resulting from crustal flexure (Wilson and Prendergast, 1989, and references therein). Wilson (1987) suggested a pure shear model with emplacement of the Great Dyke during a period of crustal extension. The sequence of events relating to the emplacement of the

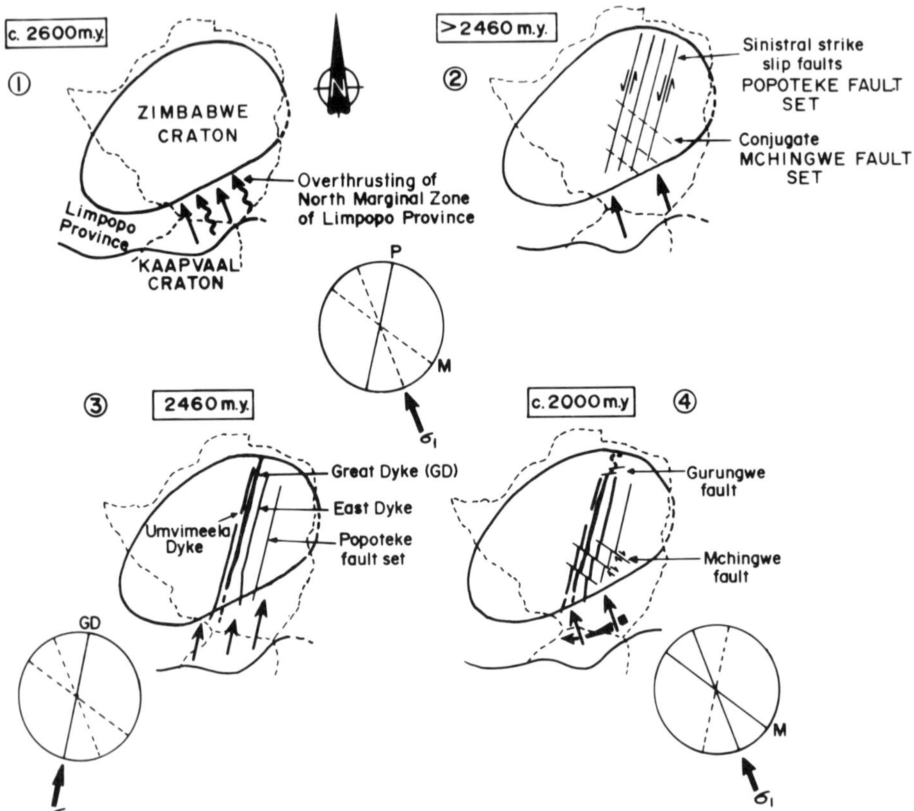

__Figure 2.__ Schematic representation of sequence of events associated with the emplacement of the Great Dyke. (1) collision of Zimbabwe and Kaapvaal Cratons and northward overthrusting of north marginal zone of the Limpopo Province; (2) development of sinistral strike-slip faults of Popoteke fault set (labelled P in Figure 1) together with conjugate Mchingwe fault set (MF in Figure 1); (3) rotation of maximum compressive stress causing extensional conditions and emplacement of Great Dyke and satellites; (4) post-Great Dyke reactivation of Mchingwe fault set resulting in dextral movement.

Great Dyke in this model are as follows (Figure 2):

Stages 1 and 2: A north-northwest-directed maximum compressive stress, as a result of overthrusting of the north marginal zone of the Limpopo Province onto the southern part of the Zimbabwe Craton, induced the major Popoteke fracture system, together with the conjugate Mchingwe fault set. Sinistral strikeslip movement occurred along the faults.

Stage 3: Extension occurred along these faults by rotation of the maximum compressive stress (from north-northwest to north-northeast) with subsequent emplacement of Great Dyke magma into the dilated fracture system, as a series of linked magma chambers. There is strong evidence (see later discussion) to suggest that the magma was emplaced periodically and over an extended period and was concurrent with extensive crystallization. Coeval with the main emplacement event, quartz gabbros were emplaced as flanking satellite dykes that extend almost the entire length of the Great Dyke.

Stage 4: Following emplacement of the Great Dyke, rotation of the maximum compressive stress back to the north-northwest direction caused dextral movement along the Mchingwe fault set together with further dyke emplacement on the north-northwest fracture pattern. These sets of dykes are called the Bubi and Crystal Springs Swarms (Robertson and van Breemen, 1970).

On a broader tectonic scale Hatton and von Gruenewaldt (1990) related the emplacement of the Great Dyke, and similar layered intrusions such as the Widgiemooltha dyke swarm of southwestern Australia and the early Proterozoic layered intrusions of the Fennoscandian shield to rifting, which in turn is part of major orogenic cycles resulting from plate tectonic processes. These intrusions are all characterized by an abundance of orthopyroxene-bearing rocks, reflecting the relatively high silica content of the primary magma. It is generally believed that upwelling asthenosphere in these regions caused subsequent melting of the lithosphere with subsequent contamination of mafic magma resulting in the characteristic high-SiO_2, high-MgO parent magmas. The largest of the Widgiemooltha Suite is the 585 km long Binneringie Dyke (McCall and Peers, 1971). The related Jimberlana Dyke (McClay and Campbell, 1976; Campbell *et al.*, 1970) is 180 km in length and up to 2.5 km wide and is remarkably similar in form, structure and age to the Great Dyke. The Finnish layered intrusions vary greatly in size and degree of preservation and in their undeformed state appear to be elongate bodies with well-defined cyclic units. Initial magmas of these intrusions, as indicated by gabbroic dykes (Alapieti and Lahtinen, 1989) are relatively high in MgO (±16%) and therefore similar to the initial magma of the Great Dyke.

4. INITIAL LIQUID COMPOSITION

As noted previously, the Great Dyke has, in common with many layered intrusion of this type, characteristically high SiO_2 and MgO contents. This is evident from the early crystallization of high magnesium orthopyroxene following extensive olivine crystallization. The proportion of rock types of the Great Dyke are generally more ultramafic when compared with the Bushveld Complex but this does not in itself indicate a more primitive magma composition as it could also reflect repeated emplacement of magma in the earlier stages of the chamber. The compositions of the most magnesian olivine and cumulus orthopyroxene are valid indicators of primary magma composition. For the Great Dyke these are $Fo_{92.0}$ and $En_{91.5}$ respectively. The most magnesian composition of olivine for the Bushveld Complex is Fo_{89} in the Lower Zone and Fo_{90} for the Potgietersrus limb (see Eales and Cawthorn, this volume).

For such magnesian mineral compositions relatively small differences are, however, indicative of significantly different magma compositions. For a basaltic magma with 8.2% FeO this would amount to a magnesium content of approximately 15.5% MgO for the initial liquid of the Great Dyke, and 12.5% MgO for that of the Bushveld Complex. The latter composition is in close agreement with that proposed by Davies *et al.* (1980) as the initial liquid for the Bushveld Complex. The extensive development of orthopyroxenites in both these layered intrusions is also indicative of relatively high SiO_2 contents in the initial magmas. A further indication of the ultramafic nature of the Great Dyke magma is the high Cr_2O_3 contents (up to 0.71%) in orthopyroxene which are significantly higher than the maximum of 0.60% Cr_2O_3 observed in the Bushveld Complex.

The isotopic characteristics of the Great Dyke also provide constraints on the origin of the initial magma. Unlike the Bushveld Complex the initial Sr values for minerals and whole rocks are essentially constant, even for samples widely separated in the stratigraphy and from different subchambers (Hamilton, 1977). This indicates that extensive crustal contamination of basic magma by felsic continental crust did not take place in the Great Dyke. Initial $^{87}Sr/^{86}Sr$ is 0.70261±4 which is a further indication of a primitive and uncontaminated initial magma. It may be concluded that the high SiO_2 content of the Great Dyke magma was therefore a

Table 1
Compositions of some initial liquids which have been proposed for the Great Dyke (Nos. 2-5) in comparison to the initial liquid of the Bushveld Complex (No. 1)

	1	2	3	4	5
SiO_2	55.70	51.91	49.08	52.77	52.07
Al_2O_3	12.74	3.48	9.97	11.04	10.69
Fe_2O_3	1.09	2.14	1.16	1.23	-
FeO	7.80	7.90	9.51	8.20	10.77
MnO	0.09	0.18	0.17	0.14	0.17
MgO	12.44	29.61	20.12	15.60	14.61
CaO	6.96	3.38	7.60	7.60	7.25
Na_2O	2.02	0.32	1.34	1.77	1.54
K_2O	1.03	0.11	0.43	0.69	0.74
TiO_2	0.36	0.14	0.57	0.55	0.51
P_2O_5	-	0.03	0.04	0.11	0.07
Cr_2O_3	0.14	0.79	0.10	0.29	0.34
NiO	0.04	-	-	0.06	0.06
Pt ppb	14*				0.64
Pd	9				4.20
Au	2.8				0.08
Ir	0.22				0.22
Ru	2.3				0.92

1. Bushveld initial liquid (Davies *et al.*, 1980). *Average of PGE data for possible Bushveld liquids from Davies and Tredoux (1985). 2. Suggested liquid for cyclic unit 1 (Bichan, 1970). 3. Chill phase to Peregwe satellite dyke (Robertson and van Breemen, 1970). 4. Chill to East Dyke offshoot (Wilson, 1982). 5. Chill to East Dyke offshoot (Prendergast and Keays, 1989).

primary characteristic derived from silica-enriched subcontinental lithospheric mantle.

A liquid with about 16% MgO (and 53% SiO_2), being the same as that for a chill margin on a dyke considered to be an offshoot of the East Dyke (Wilson, 1982), is in good agreement with observed mineral compositions, and modelling using this composition is consistent with the observed crystallization sequence (see later discussion on order of crystallization). This is, therefore, considered to be the parental magma composition of the Great Dyke and is compared in Table 1 with some previously suggested initial liquid compositions and the proposed initial liquid composition for the Bushveld Complex.

5. STRATIGRAPHIC SUBDIVISIONS AND CYCLIC UNITS

Wilson (1982) suggested that the Great Dyke stratigraphy be formally subdivided into a lower Ultramafic Sequence and upper Mafic Sequence. Worst (1958, 1960) established the

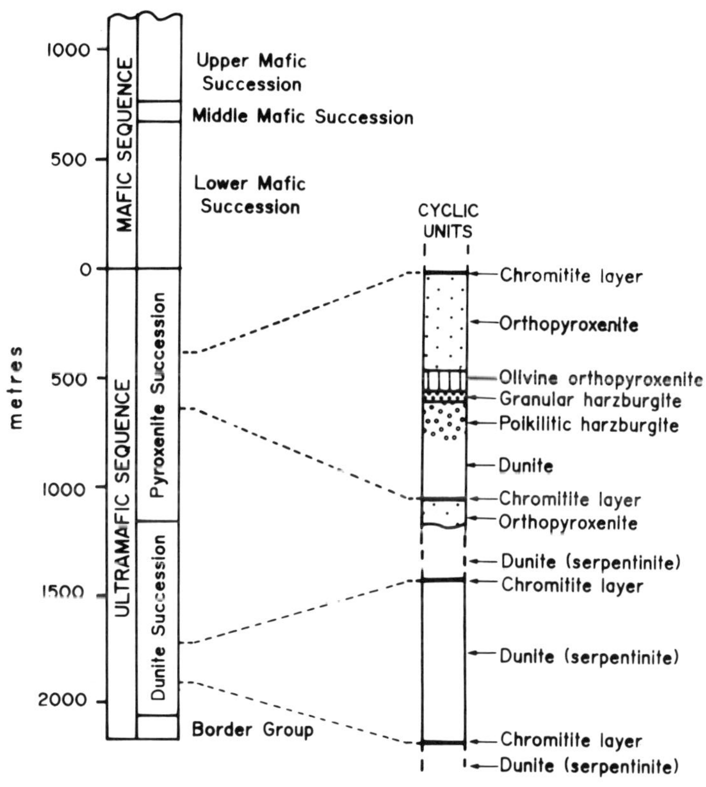

Figure 3. *Subdivision of the Great Dyke stratigraphy into the Dunite and Pyroxenite Successions in the Ultramafic Sequence, and into the Lower, Middle and Upper Successions in the Mafic Sequence. The lithological structure of the cyclic units in the Ultramafic Sequence is also shown (after Wilson and Prendergast, 1989). Schematic cyclic units are shown for the Dunite and Pyroxenite Successions.*

lateral continuity of layers of pyroxenite, serpentinite and chromitite in the Ultramafic Sequence and numbered each lithologically distinct layer downwards from the top of the sequence. In this original work the association of rock types as comprising contiguous units was not recognized and therefore the numbering system showed inconsistencies when considered on a genetic basis. Jackson (1970) recognized distinct cyclic units in the Ultramafic Sequence and Wilson (1982) proposed a numbering system related to cyclic units.

Previous numbering systems were based on the sequence of common rock types with each lithological unit having a different number irrespective of the genetic association with other rock types. In all parts of the Great Dyke the top of the Ultramafic Sequence comprises well-

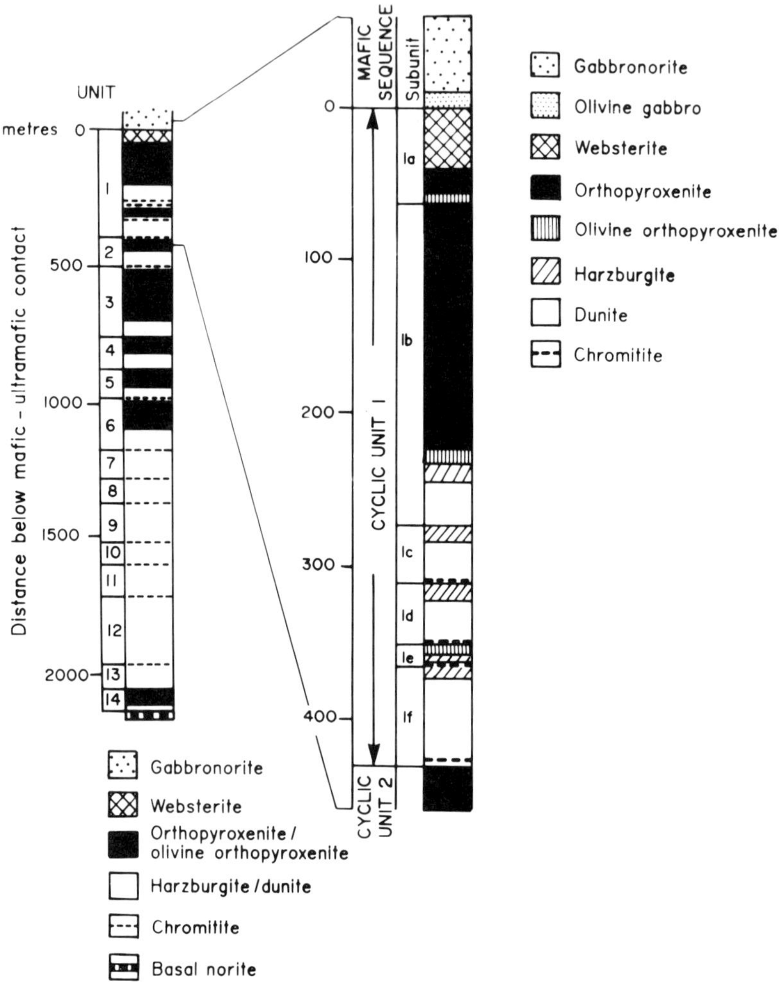

Figure 4. Subdivision of the Ultramafic Sequence in the Darwendale Subchamber into cyclic units with detailed subdivision of Cyclic Unit 1 into subunits (after Wilson, 1992).

developed cyclic units with a lower dunite or harzburgite layer overlain by a pyroxenite layer (Figure 3). In the lower part of the Sequence the cyclic units are defined by chromitite layers within dunite, and pyroxenites are absent. Wilson and Prendergast (1989) suggested on this basis that the Ultramafic Sequence be further subdivided into an upper Pyroxenite Succession and a lower Dunite Succession (Figure 3), each containing easily recognizable cyclic units. In the Darwendale Subchamber 14 cyclic units are recognized (Figure 4). This subdivision of the Ultramafic Sequence of the Great Dyke therefore shows similarities to the subdivision of the Stillwater Complex which has a lower Peridotite Member and an upper Pyroxenite Member (Jackson, 1961). Dunite in the Great Dyke is not preserved in surface outcrop as it has been totally replaced by serpentinite. Deep drilling has shown that the degree of serpentinization decreases with depth and unaltered dunites are encountered in unfractured areas at depths of about 300 m, above which pervasive serpentinization takes place.

Although designation of subdivisions on a finer scale is possible in all cyclic units, this only becomes practically possible in the uppermost and well exposed Cyclic Unit 1 (Figure 4). This cyclic unit represents the topmost portion of the Ultramafic Sequence and is also characterized by the first appearance of cumulus clinopyroxene in the distinctive websterite layer at the very top. Cyclic Unit 1 is subdivided into six subunits on the basis of changes in lithology and narrow chromitite layers. By local convention the pyroxenite layers are referred to on a 'P' notation with that in Cyclic Unit 1 being the P1 pyroxenite.

Reconciliation of the designated cyclic units to mineral compositional variations has important implications for the understanding of layered intrusions. As discussed in a later section, reversal to more magnesian compositions of cumulus phases in the cyclic units in the Pyroxenite Succession takes place within the pyroxenite layer below the lithological boundary of the cyclic unit (i.e. below the chromitite layer, where it occurs, or below the base of the pyroxenite-dunite boundary) indicating that the chemical boundary of the cyclic unit is located some distance below the apparent lithological boundary. In the Dunite Succession the compositional reversal in olivine appears to be coincident with the base of the chromitite layer. For practical purposes the base of the cyclic units in the Ultramafic Sequence of the Great Dyke should be taken at the lithological boundaries even though the mechanistic control may have been initiated some distance below this level.

Table 2
Main Features of the Subchambers of the Great Dyke

	Length (km)	Estimated total thickness (m)	Thickness (m) of P1 layer	Thickness (m) of websterite layer
			Axis 230	38
Darwendale	210	3350		
			Margin 140	12
Sebakwe	120	3350	210	32
Selukwe	96	1900	180	8
Wedza	80	1900	160	11

6. CHAMBERS AND SUBCHAMBERS

A characteristic feature of the Great Dyke is the presence of distinct layering styles in different areas by which Worst (1958, 1960) suggested that four complexes existed along the

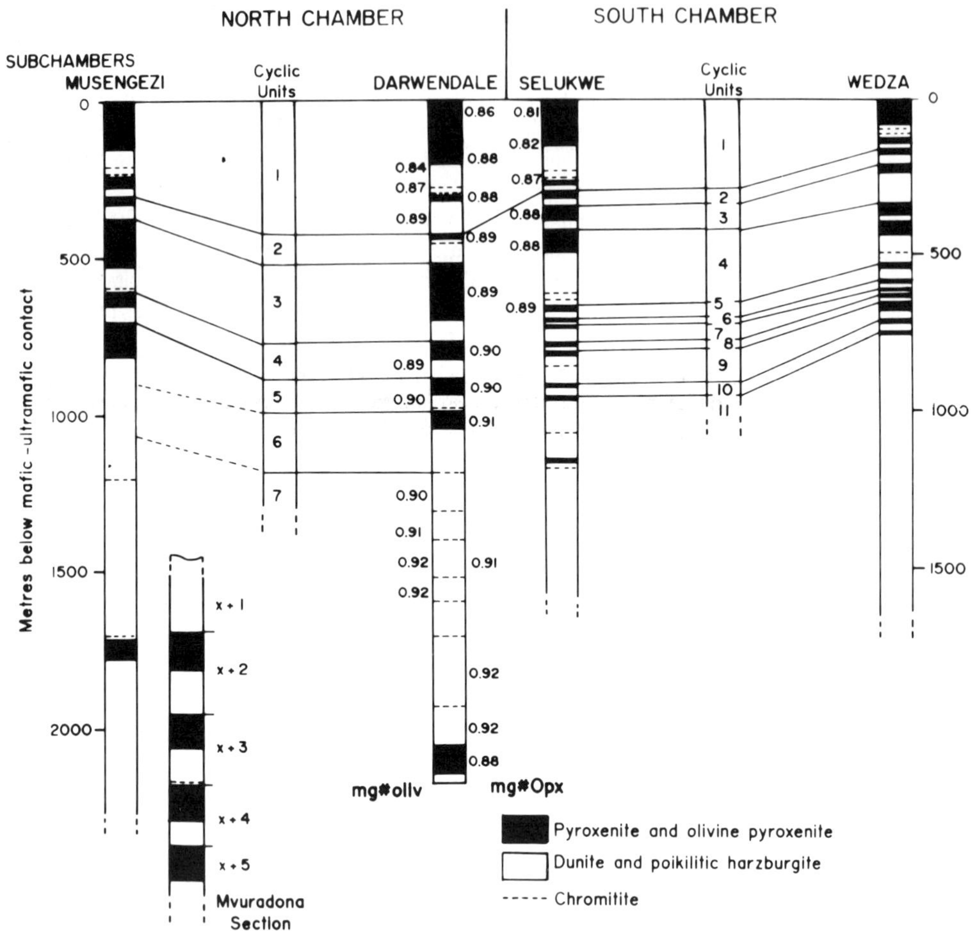

Figure 5. Correlative lithostratigraphy and subdivision of the Ultramafic Sequences in the North and South Chambers. Implied correlation is shown between cyclic units (numbered) within each of the subchambers. Strong correlation exists between subchambers within each of the two chambers, but there is a clear contrast in form and thicknesses of the cyclic units between the North and South Chambers. In contrast Cyclic Unit 1 shows strong correlation between the chambers and indicates that all magma chambers were linked at this stage. The exposed stratigraphy of the Mvuradona section is shown (the "x" indicating the unknown relative position of the observed cyclic units) and this does not correlate with any other section in the North Chamber.

length of the intrusion. From north to south these were called the Musengezi, Hartley, Selukwe and Wedza Complexes. Prendergast (1987) proposed that two main magma chambers constituted the Great Dyke, called the North and South Chambers, and Wilson and Prendergast (1989) suggested that these comprised several smaller subchambers (Figure 1), each of which possessed contiguous and distinct structure, layering pattern, rock types and thickness. The names of the complexes, as originally given by Worst (op. cit), were retained formally in the subchamber nomenclature, except for the Hartley Complex, which now comprises the Darwendale and Sebakwe Subchambers. In addition, Wilson and Prendergast (1989) proposed that a fifth and much smaller subchamber, called the Mvuradona Subchamber, may exist at the northern extremity of the Great Dyke. Each of the four main subchambers has a remnant of the Mafic Sequence at its centre. A summary of the thicknesses, length of chambers and main features is given in Table 2.

The nature of the Ultramafic Sequence is different in different parts of the Great Dyke. In the North Chamber the Ultramafic Sequence is characterized by relatively few, thick cyclic units (100 m thick on average) with well developed pyroxenite layers (Figure 5). In contrast, the South Chamber has a greater number of thinner cyclic units (10-30 m thick) with olivine pyroxenites predominating over pyroxenites in the upper parts of the units. The Ultramafic Sequence is well exposed in both the North and South Chambers with the pyroxenite and olivine pyroxenite layers being more resistant to weathering than the serpentinite layers (Figure 6). In the southern region there is also no indication of the existence of a lower Dunite Succession at surface, although thick continuous intervals of fresh dunite are known to exist in deep boreholes at depths of greater 700 m.

The degree of continuity of the layering and specific lithologies between the chambers and subchambers of the Great Dyke are indicative of the emplacement mechanism. The layering pattern of the lower ultramafic successions is quite different in the various subchambers, but

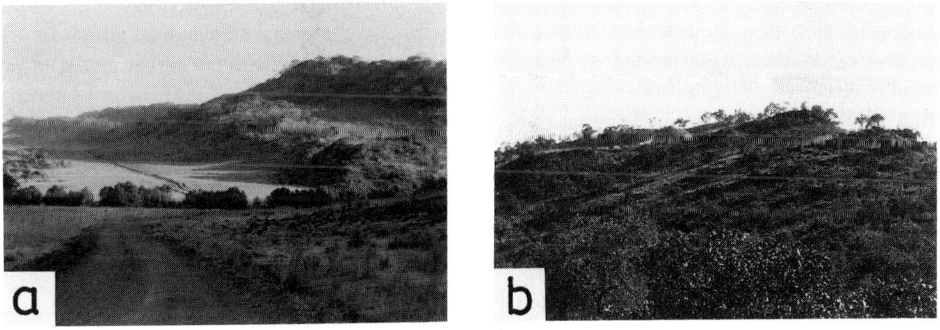

Figure 6. *Layering types in the Ultramafic Sequence. (a) Layering on the east flank of the Darwendale Subchamber (view south) showing the more resistant pyroxenite layers dipping to the west with easily eroded dunite/serpentinite layers between them. Each of the combinations of a dunite layer with a pyroxenite layer comprises a thick cyclic unit approximately 100m thick. (b) Cyclic units in the Sebakwe Subchamber also illustrating the differing resistances of the pyroxenite and dunite layers but in this part the cyclic units are much thinner (10-30 m thick).*

there is remarkable continuity in detailed aspects of the uppermost cyclic unit in the Ultramafic Sequence and in the lowermost Mafic Sequence throughout the length of the Great Dyke, suggesting that the entire magma chamber was linked at the level of Cyclic Unit 1.

7. STRUCTURE OF THE GREAT DYKE AND ITS MAGMA CHAMBERS

The structure and shape of the Great Dyke have been determined from gravity investigations. The earliest study showed the intrusion to have a bell-shaped Bouger gravity anomaly (Weiss, 1940) and this was confirmed by later studies which also indicated major variations along the length of the Great Dyke (Podmore, 1970; 1982). A summary of the gravity profiles and transverse sectional models (Podmore and Wilson, 1987) which are consistent with outcrop patterns and distribution of rock types is shown in Figure 7.

Several important conclusions relating to form, shape and thickness of the Great Dyke arise from the gravity study. The Great Dyke is essentially a symmetrically disposed arrangement of layers dipping toward the central longitudinal axis. A major deep structure is inferred along almost the entire length of the Great Dyke but it is not present where the North and South Chambers abut. This deep structure is interpreted as a continuous feeder dyke by which magma was emplaced into developing magma chambers. Several of the gravity profiles show marked irregularities, such as asymmetry and tilting. The apparent tilt of the Darwendale Subchamber towards the west has implications for the nature of the platinum mineralized sulphide zone and an understanding of its petrology and distribution in this subchamber (see later discussion). In addition, some models require the existence of deep-seated magma chambers or extensions of the present magma chambers at depth. Based on the gravity profiles the relative volumes of the chambers and subchambers may be estimated (Figure 8). The North Chamber is of significantly greater volume compared with the South Chamber and a gradual and progressive change takes place from the Wedza Subchamber in the south to the Darwendale Subchamber in the north.

There have been various explanations for the transverse shape of the Great Dyke. Worst (1960) suggested that the layers were initially subhorizontal in form, abutting at high angle against wall-rocks, and then later subsidence gave rise to the observed synclinal structure. The wall-rock contacts were faulted as part of a graben structure. Wilson and Prendergast (1987, 1989) showed that the layers are essentially flat-lying only towards the axis (Figure 9), and that the attitude of the layers implies that close to the margins the dip may have decreased further to give essentially horizontal sheets beyond the present outcrop limits of the Great Dyke. A further important observation is that for all layers in the Ultramafic Sequence which can be traced from the margin to the axis, and for those where deep drilling data are available, there is a gradual thinning of the rock units towards the margins. Wilson (1982) showed the presence of a Border Group parallel to the margins of the Great Dyke, and suggested that all layers become asymptotic and merge with the Border Group. Pyroxenite layers which are up 50 m thick in the axial zone, thin to less than a few metres in the Border Group. The sequence of chromitites, pyroxenites and harzburgites in the Border Group is lithologically similar to the stratigraphic succession in the Ultramafic Sequence.

8. THE ULTRAMAFIC SEQUENCE

The ideal cyclic unit encountered in the Ultramafic Sequence of the Great Dyke comprises a basal chromitite overlain by a dunite layer which grades upwards through harzburgite and olivine pyroxenite into a pyroxenite which marks the top of the unit. The development of chro-

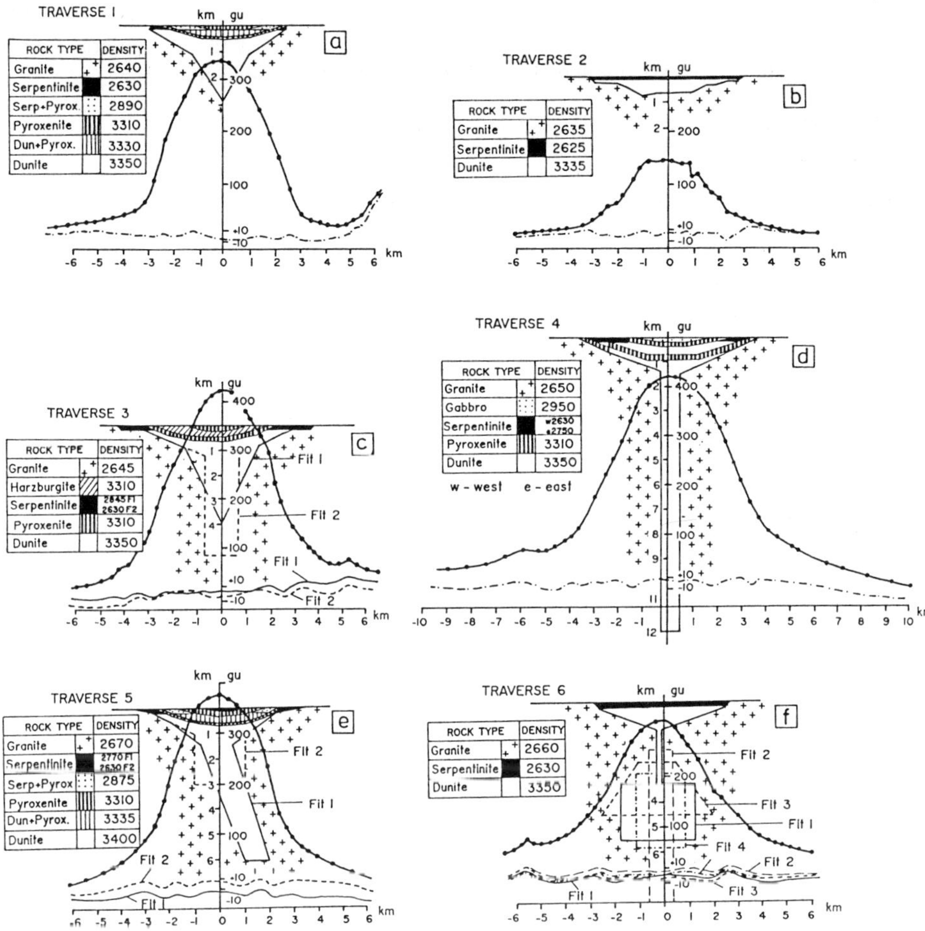

***Figure** 7. Bouger gravity anomaly profiles for six traverses across the Great Dyke (after Wilson and Podmore, 1987). The locality of the traverse number is shown on Figure 1. Transverse sectional models, consistent with surface geology, provide best fit with the gravity data. Sample stations are shown by the dots on the anomaly profile and residuals to the model fit are shown on a scale of +10 to -10 gu. Densities of rock types are given in kg m^3. Each of the traverses provides important information on the structure of the Great Dyke. (a) Typical section of South Chamber. (b) Southern extremity of Sebakwe Subchamber showing thin layered succession and lack of deep root zone. (c) and (d) Deep structure to the North Chamber representing feeder dyke. (e) Tilted structure of the layered sequence which is consistent with field observations. (f) Various fits all showing presence of deep-seated magma chambers beneath the layered succession in the northern Darwendale Subchamber.*

mitite layers may be related to the size of the subchamber, with the thickest layers occurring in the Darwendale and Sebakwe Subchambers. The entire ideal cyclic unit is not always developed; thus olivine pyroxenite may be the most evolved rock-type in some units.

8.1. Chromitites

In the Dunite Succession of the Darwendale Subchamber massive chromitite layers 10 - 15 cm in thickness are coarse-grained with little or no primary silicate material (magnesite, talc and secondary serpentine minerals occur in fractures) and are of economic significance. These are informally termed the lower group chromitites by Prendergast and Wilson (1989). The lower group chromitite layers mark the boundaries between cyclic units in the Dunite Succession. There may also be minor chromitite layers situated 1 to 1.5 m above the main chromitite layer. The contacts of the chromitite layers with the dunite are usually sharp although disseminated upper and lower contacts are also observed. In general the chromitite layers have sharp lower boundaries of coarse-grained chromitite grading upwards into finer-grained, and often microcyclically layered top boundaries. Massive chromitite contains polygonal grains with planar crystal boundaries. With increasing proportions of olivine massive chromitite grades, both vertically and laterally, through semi-massive chromitites into disseminated olivine chromitite and chromite dunite (Figure 10) in which chromite occurs as clusters on the margins and triple junctions of olivine grains. Some of the disseminated chromitites and olivine chromitites show fine-scale (millimetre) layering caused by variations in the proportions of olivine and chromite. There is also a lateral variation in that massive chromitite layers in the centre of the Great Dyke gradually change to disseminated olivine

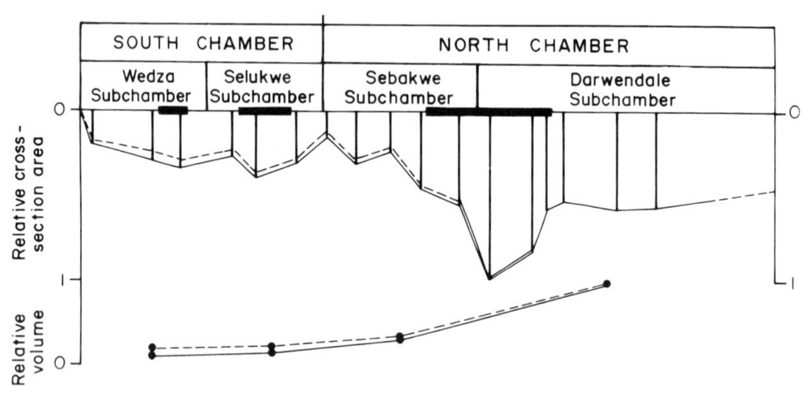

Figure 8. *Cross-sectional area for eighteen gravity traverses (positions identified by vertical lines) across the Great Dyke relative to that section in the Darwendale Subchamber which has the greatest area and taken as unity. Data taken from Podmore and Wilson (1987). Lower part of diagram shows relative volumes for each of the subchambers calculated from the cross-sectional shape and the longitudinal extent of each of the subchambers. Dashed lines represent calculations not taking into account elevation differences (650 m) in the depth of erosion between the north and south ends of the Great Dyke, and solid lines show the corrected variations. The black horizontal bars represent positions of gabbro remnants within the subchambers.*

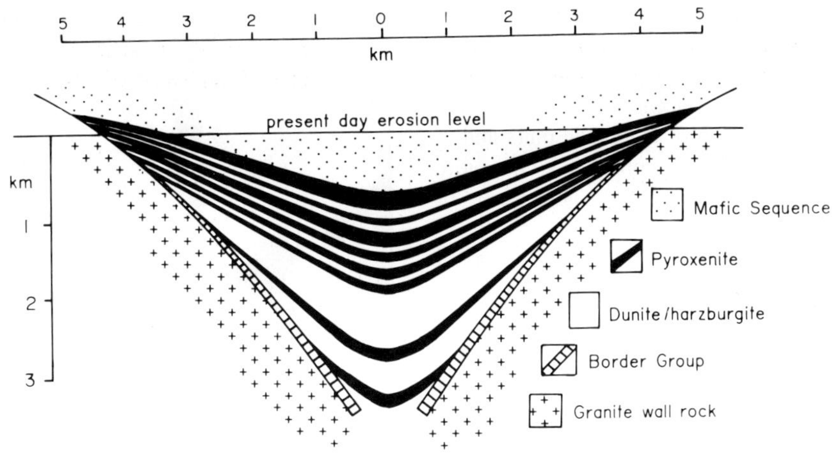

Figure 9. *Transverse section of the layered sequence of the Great Dyke in the Darwendale Subchamber based on borehole intersections, field data and gravity interpretations of attitude of layering and its relationship to wall-rocks. The layering in the Ultramafic Sequence away from the axis is essentially planar with constant dip which contrasts with previous subcircular (Worst, 1960) or curvilinear models of Wilson and Prendergast (1989). Note the small angular decrease and progressive thinning of layers towards the margins. The offlapping arrangement of layers with the wall-rocks causes the uppermost pyroxenite layer (P1) to be in close proximity to the margins of the intrusion. The postulated extension of the layering prior to the present-day erosion level is shown.*

chromitite at the margins. This has been well documented for the chromitite layer C7 of Cyclic Unit 7 of the Darwendale Subchamber (Prendergast and Wilson, 1989).

Chromitite C5 (of Cyclic Unit 5) has been mined extensively south of the village of Mutorashanga (Figure 1) where it comprises a layer 10-20 cm thick with a characteristic hanging-wall of strongly disseminated olivine chromitite up to 1.5 m thick. In the Mutorashanga region other lower group chromitites (C7, C8, and C10) are also mined (Figure 11). In the Darwendale Subchamber there are many thin, discontinuous chromitite layers in the Dunite Succession but it is not known how these relate to the development of cyclic units or subunits within the major cyclic units.

Chromitite C5 of the Darwendale Subchamber is grouped by Prendergast and Wilson (1989) with the lower group chromitites, but several features distinguish it from other chromitites in this group. It is also the only well-developed chromitite layer of the lower group which overlies orthopyroxenite, as the others are contained wholly within dunite. There is typically a layer of harzburgite 1-10 cm thick above the chromitite layer and it is separated from the underlying granular pyroxenite by a narrow zone (1-20 cm thick) of disseminated chromite, layers of very fine-grained chromite or poikilitic harzburgite (Figure 12). The orthopyroxene in the harzburgite shows extensive crystallographic continuity, commonly over several tens of centimetres. These observations indicate that although chromitite layers are situated close to the lower boundaries of cyclic units they do not mark the physical base of the units. Other than in Cyclic Unit 5, chromitite layers are not well developed in the lower part of the Pyroxenite Succession. Wilson (1982) reported concentrations of chromite at the base of each of the cyclic units and in some cases narrow discontinuous chromitite layers are developed but none are of economic importance.

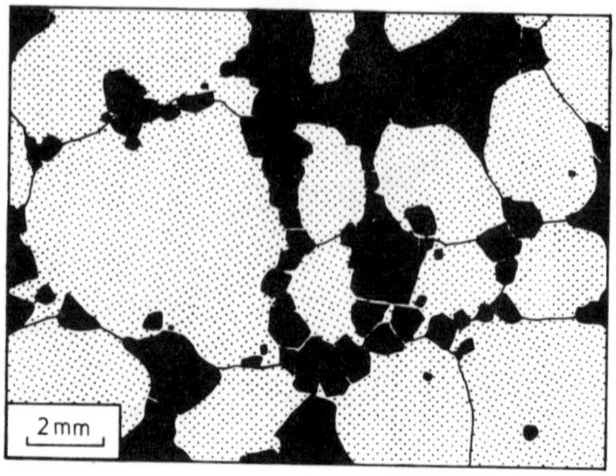

Figure 10. *Drawing of thin section of chromite dunite (olivine, dots; chromite, black) (from Prendergast and Wilson, 1989).*

The upper group chromitites occur in Cyclic Units 1 and 2 at the top of the Ultramafic Sequence in all subchambers and show remarkable consistency in form, structure and composition. Six chromitite layers are described by Wilson and Prendergast (1987) but only two (chromitite layers C1c and C1d) are extensively mined. Mining of these two layers has taken place almost continuously on both sides of the Darwendale and Sebakwe Subchambers. These chromitites are characteristically more disseminated than the lower group chromitites and are generally 1.2-1.8 m in thickness. In most occurrences they are made up of a series of small-scale chromitite layers 5-10 cm thick separated by harzburgite, but in some cases the chromitite occurs as a single massive layer. In general, the upper group chromitites show more complex layering than occurs in the lower group, and there is also more primary silicate material giving rise to a characteristic nodular texture. Although massive chromitites up to 15 cm thick are found in some areas, the structure of these chromitites is generally one of narrow coarse-grained layers of massive chromite seldom more than a few centimetres thick and interlayered with finer-grained chromite poikilitically enclosed by orthopyroxene giving rise to the nodules 1-3 cm in diameter. The core of each nodule comprises one or more grains of relict and highly corroded olivine (Figure 13a) with the distribution of

Figure 11. *Chrome mining operations in the lower group chromitites at Mutorashanga near the northern end of the Darwendale Subchamber. The hills are serpentinite close to the axis of the Great Dyke and the trenches represent surface workings on the chromitite layers. View is north facing and the layering dips gently (at approximately 6°) to the south.*

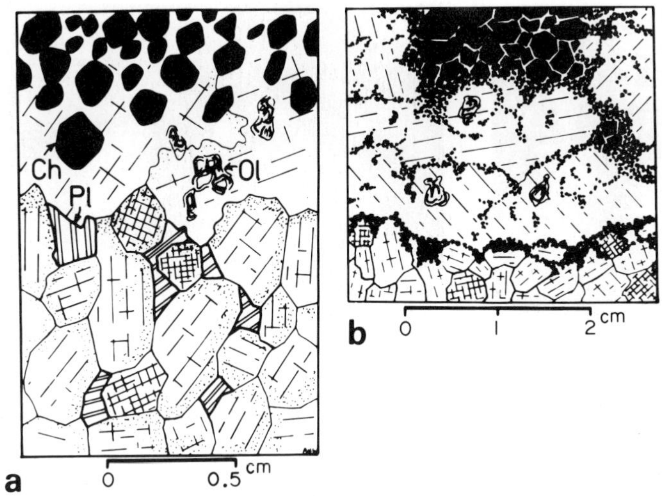

Figure 12. Drawings of photomicrographs of the interface between chromitite C5 and underlying orthopyroxenite of Cyclic Unit No. 6 . (a). Large grains of chromite (Ch) in disseminated footwall of chromitite layer. Relict and corroded olivine (Ol) enclosed by optically continuous orthopyroxene overlying cumulus grains of orthopyroxene. Postcumulus plagioclase (Pl) is interstitial to the pyroxene. (b) Fine-grained chromite crystals draped over cumulus orthopyroxene of the P6 pyroxenite layer. Outlines of original olivine crystals, of which only remnants remain, are marked by the fine-grained chromite. The relict texture also indicates that the fine-grained chromite was draped over the olivine crystals. Olivine crystals have been replaced by optically continuous orthopyroxene (hatching) (from Wilson and Prendergast, 1989).

the fine-grained chromite crys-tals outlining the original olivine grains before corrosion.

There is a significant lateral variation in the upper group chromitites between the axis and the margin of the Great Dyke. In chromitite C1d there is a systematic decrease in grain-size of the chromite towards the margin, together with an increase in olivine/chromite ratio, a decrease in thickness of the layer and an increase in size and abundance of the nodules. At the south end of the Sebakwe Subchamber, chromitite C1d comprises a single layer in the marginal zone but towards the axis splits into two chromite-rich layers, separated by a harzburgite layer. This would indicate that the chromitite layer in the marginal zones comprises two subunits.

8.2. Dunite and poikilitic harzburgite

Dunite is the dominant rock-type of the lower Ultramafic Sequence in both the North and South Chambers. Its texture is one of interlocking olivine grains with typical planar boundaries and triple-point junctions (Figure 13b). Chromite is a ubiquitous primary mineral comprising 1-4% by volume and is generally concentrated at olivine grain margins or at the triple-point junctions. Some chromite is also enclosed by olivine near the margins but none occurs in the centres of olivine indicating enclosure only during the latter stages of olivine growth. Small amounts of pyroxene (both ortho- and clinopyroxene), and zoned plagioclase occur interstitial to the olivine and enclose chromite. The olivine grains typically show strain or dislocation twinning related to the triple-point intersections. This may be explained by grain-coarsening or annealing processes.

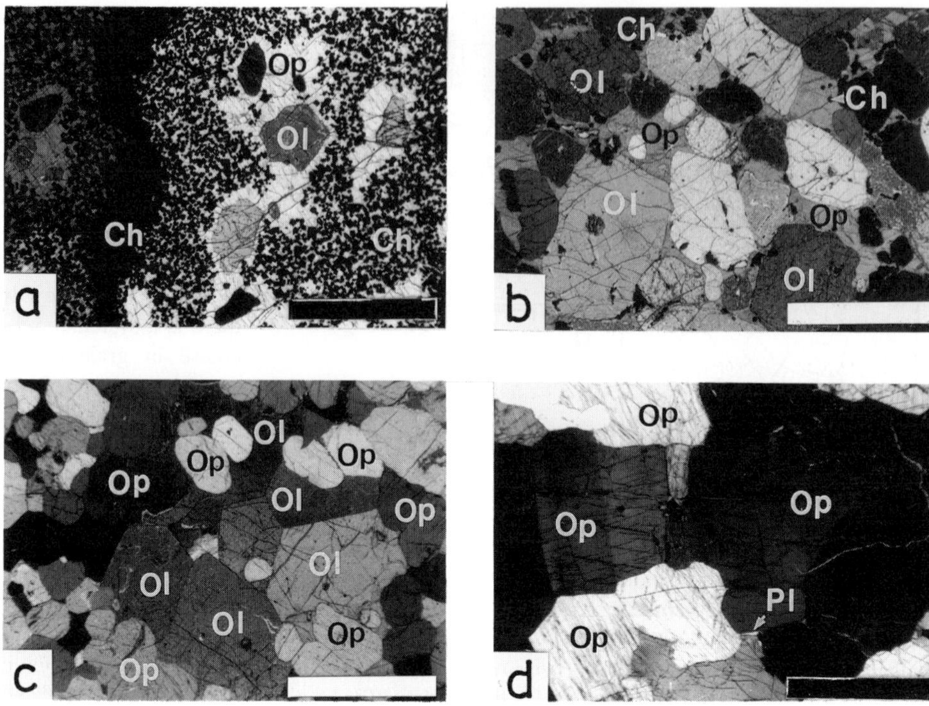

Figure 13. *Photomicrographs of rocks from the Ultramafic Sequence. (a) Olivine in reaction relationship with orthopyroxene in disseminated chromitite C1c. Fine-grained chromite is also enclosed in orthopyroxene. (b) Dunite with interstitial orthopyroxene. Note very fine-grained chromite enclosed in the orthopyroxene and near the margins of the cumulus olivine. (c) Olivine pyroxenite with rounded and embayed pyroxene crystals enclosed by olivine. (d) Coarse-grained orthopyroxenite with very small amounts of clinopyroxene and plagioclase at crystal junctions. Note the glide planes in the orthopyroxene crystals related to marginal discontinuities. All scale bars represent 3 mm. Taken under crossed polars. Nomenclature: Ol - olivine; Op - orthopyroxene; Ch - chromite; Pl - plagioclase.*

Small-scale layering within cyclic units is common on the basis of grain-size and in the proportion of olivine to chromite. In many cyclic units in the Darwendale Subchamber layering on a scale of centimetres can be observed on a continuous basis, giving rise to hundreds or even thousands of layers in a single outcrop. Dunites show marked lateral petrographic variation in places where they can be studied in a single layer (such as in Cyclic Unit 6 in the Darwendale Subchamber). Towards the margin of the Great Dyke there is a reduction in grain-size and an increase in the proportions of interstitial pyroxene. In all subchambers dunite layers in the axis appear to grade into harzburgite towards the margins.

The development of dunite in the various parts of the Ultramafic Sequence of the Great Dyke is dependent on the size of the subchamber and the position within the succession. The

Darwendale Subchamber has extensive dunite in the lower Dunite Succession, whereas poikilitic harzburgite becomes an important component in the Pyroxenite Succession. Poikilitic harzburgites are more extensive than dunite in the Ultramafic Sequence of the smaller subchambers (e.g. Selukwe and Wedza), and where dunites are formed they have more abundant interstitial pyroxene than in the dunites of the North Chamber.

Poikilitic harzburgite is distinctive in the field by its nodular appearance due to the presence of large (1-5 cm in diameter), optically continuous orthopyroxene crystals which are more resistant to weathering than the surrounding olivine grains. Olivine is contained within the orthopyroxene but is highly corroded and irregular in form. That the olivine grains were originally larger and euhedral is indicated by the mantle of chromite in a similar form to that described previously for the upper group chromitites.

8.3. Granular harzburgite and olivine pyroxenite

Granular harzburgite marks the textural transition from poikilitic harzburgite to olivine orthopyroxenite (also referred to here as olivine pyroxenite) by which the pyroxene becomes granular and no longer encloses olivine. Olivine occurs as discrete grains. With increasing proportion of orthopyroxene the rock-type grades into olivine pyroxenite. As the proportion of olivine decreases, its textural form changes from discrete grains to highly irregular crystals interstitial to and partly enclosing rounded orthopyroxene crystals (Figure 13c). This texture contrasts with that of the poikilitic harzburgites where rounded olivine crystals are entirely enclosed by orthopyroxene. In the smaller subchambers olivine pyroxenite predominates over pyroxenite.

Interstitial plagioclase is an important minor constituent (comprising 4-10% by volume) of the harzburgites in Cyclic Unit 1 near the margin of the Darwendale Subchamber and is widely developed in this unit in the smaller subchambers. These rocks were referred to as picrites by Worst (1958), but more correctly they should be termed orthopyroxene-bearing troctolites or feldspathic harzburgites. Phlogopite is also an important minor constituent of these rocks.

8.4. Pyroxenite

Pyroxenite is the dominant rock-type in the Pyroxenite Succession and stratigraphically is the uppermost rock-type in the cyclic units. In the lower cyclic units it is exceptionally coarse grained (crystals lengths up to 10 mm) and consists almost entirely of orthopyroxene. The pyroxene crystals show well-defined glide twins with planes related to knick points in the crystal outline (Figure 13d). Plagioclase and clinopyroxene are minor components and these commonly occur at the well-developed triple-point junctions of the minerals. In general the average grain-size of the pyroxenes in the lower cyclic units is noticeably dependent on the size of the magma chamber with the grain-size largest for the Darwendale Subchamber and smallest for the Wedza Subchamber. This difference is related to rate of cooling which influences the grain-coarsening processes.

The texture of the uppermost pyroxenite of Cyclic Unit 1, also called the P1 layer, is similar in all parts of the Great Dyke where this unit is exposed, and therefore is quite different from the lower cyclic units. This pyroxenite (Figure 14a) is much finer-grained than those of the lower cyclic units and is important because it marks the transition from the Ultramafic Sequence to the Mafic Sequence. Plagioclase and clinopyroxene are postcumulus phases in the orthopyroxenite part of the layer and these become relatively abundant higher in the succession. The top of the unit is marked by the status of the clinopyroxene becoming cumulus in the development of the websterite layer (Figure 14b).

A great deal is known about the P1 pyroxenite because of its generally good exposure and extensive diamond drilling to investigate the important zone of platinum mineralization (see below), which is located near the top of the pyroxenite. The pyroxenite layer shows marked lateral variations in thickness, texture and composition. It is thickest in the axis of the Darwendale Subchamber at some 220 m but thins to about 150 m at the marginal extremity (Wilson and Prendergast, 1989). In other subchambers the axial thickness of the pyroxenite layer is less but the corresponding lateral decrease is also observed. The websterite layer has a maximum thickness of 33 metres in the axis of the Darwendale Subchamber and thins to less

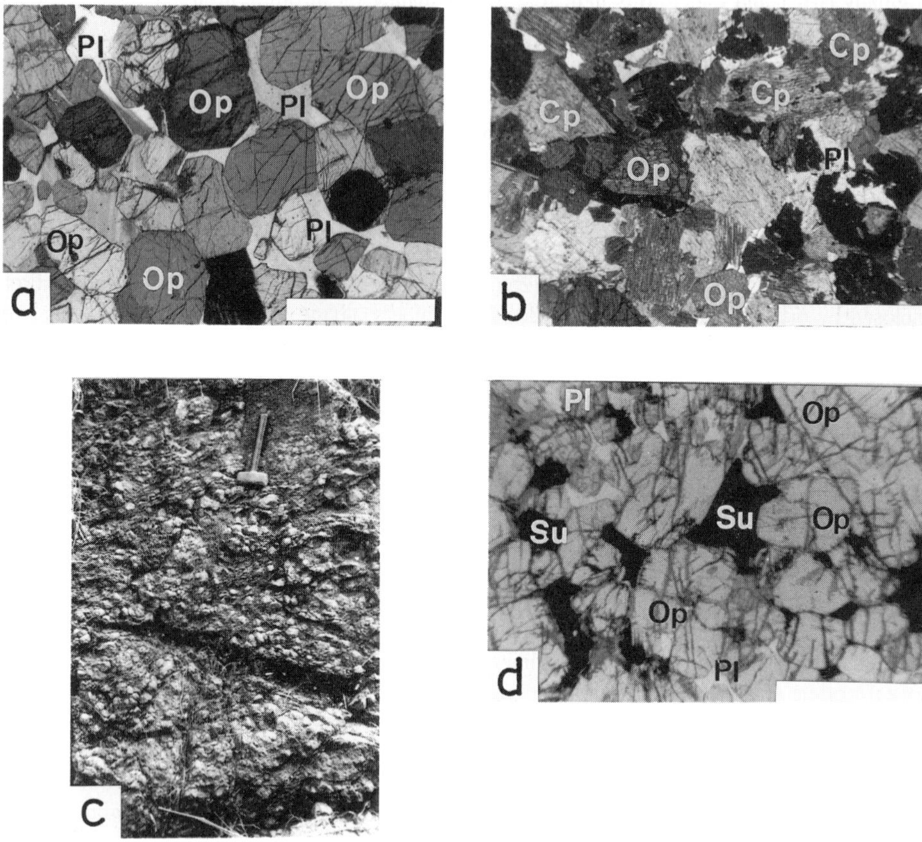

Figure 14. *Photomicrographs and field textures of the P1 pyroxenite. (a) Cumulus orthopyroxene with interstitial postcumulus plagioclase. Crossed polars. (b) Websterite with cumulus clino- and orthopyroxene. Note the highly irregular grain boundaries where in contact with plagioclase. Crossed polars. (c) Field exposure of the nodular pyroxenite. Hammer (25 cm long) for scale. (d) Net-textured sulphide interstitial to cumulus orthopyroxene. Plagioclase is also interstitial. Bar scales represent 3 mm. Nomenclature: Ol - olivine; Op - orthopyroxene; Cp - clinopyroxene; Su - sulphide; Pl - plagioclase.*

than 7 m at the margin. Thicknesses of the orthopyroxenite and websterite layers in the various subchambers of the Great Dyke are summarized in Table 2.

Texturally the P1 pyroxenite comprises elongate orthopyroxene crystals in a matrix of interstitial clinopyroxene and plagioclase. These interstitial phases, for which Wilson and Prendergast (1989) and Wilson (1992) used the term *oikocryst* to describe the textural form, are subspherical or ovoid in shape and give rise to a characteristic nodular pyroxenite (Figure 14c) by virtue of the more resistant nature of these minerals. The nodular pyroxenite is best developed towards the top of the cyclic unit and marks an increase in abundance of clinopyroxene and plagioclase. The association of this field texture with the zone of sulphide mineralization gave rise to the prospecting term 'potato reef' for this part of the succession, and is developed in the P1 layer in all subchambers of the Great Dyke. The texture is best developed in the large subchambers (particularly Darwendale) and is not particularly well-developed in the smaller Wedza Subchamber. The nodular texture persists into the overlying websterite because of the formation of plagioclase oikocrysts. Other minor minerals that occur in the P1 layer are interstitial sulphide (Figure 14d), phlogopite, magnetite, K-feldspar, quartz, sphene, amphibole, apatite and zircon.

Fine-scale layering is well developed in the P1 pyroxenite and is related to grain-size, grain shape, fabric and development of oikocrysts (Wilson, 1992). In the Wedza and Selukwe Subchambers the websterite layer interdigitates with orthopyroxenite towards the margins, whereas this relationship is absent in the axis (Prendergast, 1991). This gives rise to important discordant layering relationships with new phases appearing on the liquidus at progressively lower stratigraphic levels nearer the margin.

Primary erosion features have been observed in this part of the Ultramafic Sequence in both the Darwendale and Wedza Subchambers (Wilson, 1992; Prendergast, 1991). In the Wedza Subchamber the erosion channels are filled with fine-grained mafic rock, and in some parts the websterite layer has been entirely eroded so that the base of the Mafic Sequence lies directly on orthopyroxenite.

9. THE MAFIC SEQUENCE

The Mafic Sequence is best preserved and achieves maximum thickness in the Darwendale Subchamber, but the general characteristics observed in this part apply to the other subchambers. Subdivision of the Mafic Sequence (see Figure 3) is into the Lower, Middle and Upper Mafic Successions (Wilson and Wilson, 1981; Wilson and Prendergast, 1989). The subdivision is on the basis of mappable textural characteristics. Further subdivisions into units are based on chemical reversals and detailed changes in texture. A summary of the rock-types and thicknesses of the subdivisions in the Darwendale Subchamber are as follows:

Lower Mafic Succession (approx. 700 m thick): Medium- to coarse-grained gabbro, norite and gabbronorite containing primary orthopyroxene. These rocks are free of olivine except for a narrow olivine gabbro layer at the base of the unit.

Middle Mafic Succession (approx. 100 m thick): Fine- to medium-grained gabbro and feldspathic orthopyroxenites some of which contain olivine. Many of these rock-types are texturally similar to those in the P1 pyroxenite.

Upper Mafic Succession (approx. 300 m thick): Dominantly norites with iron-rich orthopyroxene derived by inversion of pigeonite. Towards the top of the succession primary magnetite is present.

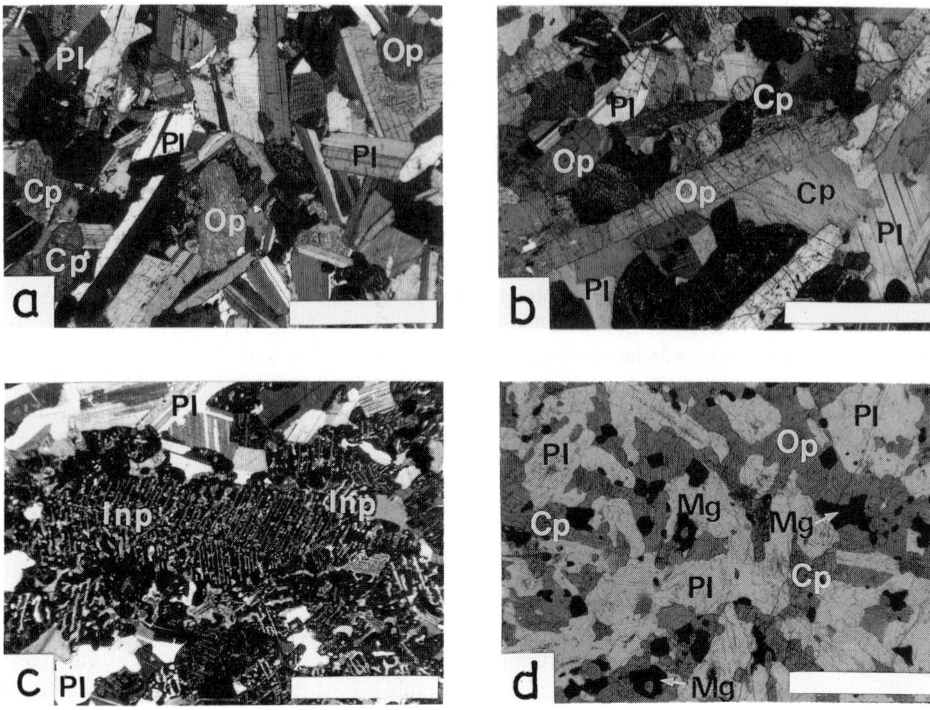

Figure 15. *Photomicrographs of the Mafic Sequence. (a). Gabbro from near the base of the Lower Mafic Succession. (b). Highly elongate orthopyroxene crystals in the Middle Mafic Succession. Plagioclase is interstitial. (c) Inverted pigeonite in norite from the Upper Mafic Succession. Note the relict twin plane preserved by the clinopyroxene exsolution. Plagioclase is cumulus. (d) Magnetite ferro-gabbronorite from the Upper Mafic Succession. All scale bars represent 3 mm. Nomenclature: Op - orthopyroxene; Cp - clinopyroxene; Pl - plagioclase; Inp - inverted pigeonite; Mg - magnetite.*

The base of the Lower Mafic Succession is marked by a thin layer (1-20 m) of olivine gabbro. Preferential weathering gives rise to a distinctive 'pockmarked' weathered outcrop. This unit is overlain by a thick succession of monotonous gabbronorites which show an increasing abundance of orthopyroxene upwards in the succession and a gradual transition from cumulus orthopyroxene (Figure 15a) at the base to large optically continuous postcumulus orthopyroxene at the top of the succession. Fine-scale layering is observed in much of this succession and in the lower parts cross-bedding and erosion structures are indicative of magma density currents. Similar features are observed in the lower gabbroic rocks of the Wedza and Selukwe Subchambers. In the latter two subchambers a narrow chromitite layer is present in some places at the websterite-gabbro contact at the base of the succession. This development is widespread in the Selukwe Subchamber but occurs only in the marginal zones in the Wedza Subchamber. In all parts of the Great Dyke, other than in the Darwendale

Subchamber, the upper portion of the Lower Mafic Succession is the highest part exposed of the Mafic Sequence.

The Middle Mafic Succession is a complex layered package of a variety of more primitive rock-types compared to the Lower Mafic Succession. The basal pyroxenite is characterized by extreme elongation of cumulus orthopyroxene (Figure 15b). Other rock-types include olivine-bearing gabbro, and feldspathic pyroxenites in which the feldspar forms large interstitial and optically continuous crystals. The full extent of this unit in the Darwendale Subchamber is not known because of very poor exposure of the upper gabbroic rocks in the central axis.

The Upper Mafic Succession is characterized by the presence of cumulus pigeonite (with well-developed clinopyroxene herringbone exsolution) now inverted to large plates of optically continuous orthopyroxene (Figure 15c). Magnetite appears as a cumulus phase (Figure 15d), but iron-rich olivine and apatite-rich rocks, characteristic of the upper portions of many large layered intrusions are absent. Based on mineral composition trends Wilson and Prendergast (1989) suggest that approximately 150 m is missing from the top of the Mafic Sequence. Quartz gabbro occurs in the central down-faulted block of the Wedza Subchamber but the relatively magnesian pyroxenes contained in this rock-type indicate that it formed as a hybrid from extensive roof contamination rather than extreme fractionation of mafic magma (Wilson and Prendergast, 1989).

10. THE BORDER GROUP

The marginal zones of the Great Dyke are rarely exposed but a succession of relatively evolved pyroxenites and norites of the Border Group (Wilson, 1982) are recognized at several different localities and at different structural and stratigraphic levels. These range from the base of the Dunite Succession at the north of the Darwendale Subchamber to near the top of the Pyroxenite Succession in the Wedza and Selukwe Subchambers. Observations suggest that the Border Group is variably developed, from being entirely absent to several of tens of metres wide, and varies from a fine-grained, massive zone to a steeply dipping, complexly layered sequence of widely different lithologies. Autolithic fragments of the Border Group are observed in both the Ultramafic and Mafic Sequences in the Darwendale and Selukwe Subchambers. A feature of the Border Group in the Selukwe and Wedza Subchambers is that in some pyroxenites the orthopyroxene crystals are highly elongate and aligned perpendicular to the margins of the intrusion.

11. XENOLITHS

Inclusions of country rock occur as xenoliths in many parts of the Great Dyke. In the upper part of the Mafic Sequence in the Darwendale Subchamber these are fragments of greenstone belt (diorites, magnetite gabbro, serpentinite, quartzite and banded iron formation) and range in size from several metres to hundreds of metres. Extensive recrystallization and partial melting of the xenoliths has resulted in coarse-grained pegmatitic quartz gabbros. Ultramafic inclusions are essentially unmodified, and quartzite, commonly show cross-bedding, and pebble-bearing arkoses have clearly resisted recrystallization. Some fragments of banded iron formation show extensive recrystallization of magnetite to grunerite. Small granite xenoliths are also observed in the marginal zones of the Darwendale Subchamber. In the Musengezi Subchamber highly altered gabbroic rocks of the Mafic Sequence, originally reported by Worst (1958) as hornblende gneiss, was erroneously described by Wilson and Prendergast (1989) as a xenolith.

In the Selukwe Subchamber there are many hundreds of autoliths from the Border Group as well as xenoliths (including chromitite) from the greenstone belt in both the Ultramafic and Mafic Sequences.

12. SATELLITE INTRUSIONS

Satellite intrusions are associated with the Great Dyke and therefore are an important part of the magmatic episode. Broadly these are subdivided into two groups called the Southern and Outer Satellite Dykes. The Southern Satellite Dykes (also called the Main Satellite Dykes (Wilson *et al.*, 1987)) outcrop over a total distance of 80 km at the south end of the Wedza Subchamber. The Outer Satellite Dykes comprise two major dykes associated with the extensive fracture system which runs parallel to the Great Dyke for most of its length.

The Southern Satellite Dykes (see Figure 1) comprise a series of elongate and aligned mafic bodies between 150 and 600 m wide. The dominant rock-types of these dykes are norite and gabbronorite together with layers of websterite (some olivine-bearing) and feldspathic harzburgite. In texture and composition many of these rock-types are similar to those occurring in the Border Group. Layering, where it occurs, is also subvertical and parallel to the dyke margins. One group of dykes has been dated by Rb-Sr at 2545±120 Ma (Robertson and van Breemen, 1970) and is therefore strongly indicated to be part of the Great Dyke magmatic event. The largest of these dykes is postulated to be a feeder or a root zone of a higher subchamber of the Great Dyke, now entirely eroded.

The Outer Satellite Dykes comprise the extensive Umvimeela Dyke (see Figure 1) situated 1-18 km west of the Great Dyke, and the East Dyke, 10-24 km to the east. Space shuttle imagery and aeromagnetic surveys show that the East Dyke is virtually continuous along the entire length of the Great Dyke. Both dykes extend 80 km south of the termination of the Wedza subchamber and intrude the northern zone of the Limpopo Mobile Belt. The Umvimeela and East Dykes are similar in bulk composition and mineralogy. They are essentially quartz gabbros and gabbronorites with subophitic to intersertal textures. The pyroxenes and plagioclase are strongly zoned, generally have similar composition to those of the Border Group and are also hydrothermally altered to a lesser or greater extent. Minor K-feldspar, quartz, and micrographophyric intergrowths are common. These textures and mineral compositions indicate that interaction with wall rocks occurred in the local environment. An offshoot of the East Dyke, situated about 70 km northwest of Masvingo, is considered to be the best representation of the Great Dyke primary magma (Table 1). This rock is a fine-grained olivine basalt with microphenocrysts of olivine and pyroxene of composition $Fo_{91.6}$ and $En_{91.2}$ which correspond closely to the most magnesian compositions observed in the layered sequence.

13. MINERAL COMPOSITIONS

Mineral compositional variations have been investigated in most sections of the Great Dyke (Worst, 1958; Wilson, 1982; Wilson and Prendergast, 1989; Wilson, 1992; Coghill and Wilson, 1993), but are best documented for the Darwendale Subchamber. All show variations consistent with a fractionating tholeiitic (and relatively silica-rich) magma. Compositional reversals and irregularities are important indicators of magmatic processes.

Chromite in chromitite layers shows a trend in the Ultramafic Sequence of initially upward increasing values for MgO (Figure 16) and Cr_2O_3, then decreasing from Cyclic Unit 9 to the top of the sequence. A similar trend is observed for the ratio Cr/Fe in chromite. Interstitial chromite in the dunites and harzburgites shows similar trends between chromitite layers with marked reversals taking place at the boundaries of the chromitite layers. The upper group chromitites (of Cyclic Units 1 and 2) are the most evolved in having the lowest MgO and Cr_2O_3 contents and lowest Cr/Fe ratios. Where chromite is enclosed in silicate minerals the chromite associated with olivine is always more iron-rich than that in pyroxene. This is explained by Wilson (1982) by the greater degree of down-temperature re-equilibration taking place with olivine than pyroxene, and therefore attainment of more iron-rich compositions. Silicate minerals show a frozen diffusion profile against the margins of chromite grains (Wilson, 1982) as a result of the coupled diffusion process of Mg into olivine and Fe^{2+} into chromite with the process becoming progressively more difficult at lower temperatures. Effective diffusion in the olivine was terminated at about 800°C for orthopyroxene and 650°C for olivine.

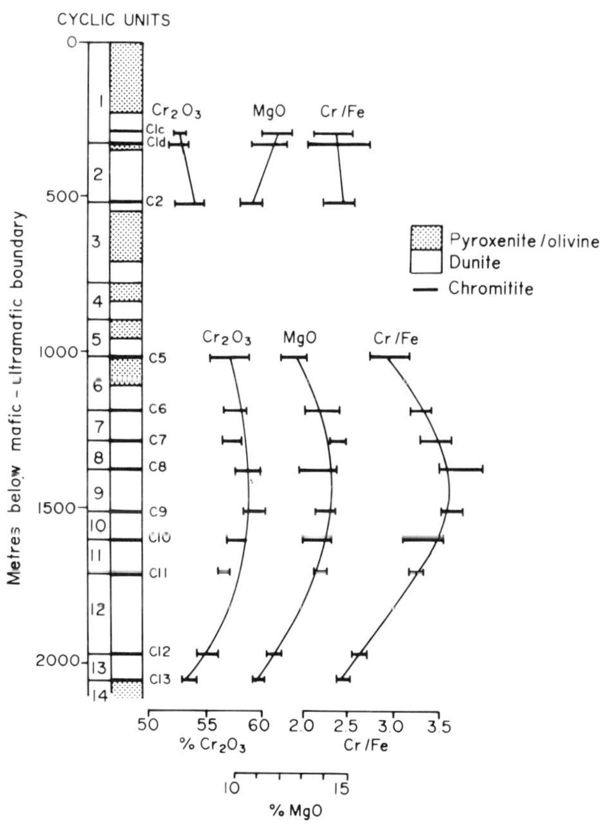

Figure 16. Compositional parameters of chromite in chromitite layers (C1 to C13) in the Darwendale Subchamber. Bars represent the compositional range. Data after Worst (1960), Wilson (1982), and unpublished data.

The ternary parameters (Al/3R, Cr/3R and Fe^{3+}/3R, where 3R represents the cations $Al+Cr+Fe^{3+}$) show discrete fields (Figure 17) and a very restricted range for the chromitite layers, and for chromite in dunite and in olivine pyroxenite. In contrast, the chromite in poikilitic harzburgite shows a very wide range in composition and indicates extensive re-equilibration by reaction with trapped liquid and by solid-state diffusion.

Olivine compositions have not been investigated for all sections of the Great Dyke because of the complete serpentinization of dunites and harzburgites on surface outcrop. Selected units in the Darwendale Subchamber show normal fractionation trends within cyclic units with major reversals taking place coincident with, or immediately above, chromitite layers where very magnesian olivines (Fo_{92}) are observed. Olivine in Cyclic Unit 1 is more evolved and also shows a regular trend of increasing iron upwards in the sequence (Fo_{91-87}).

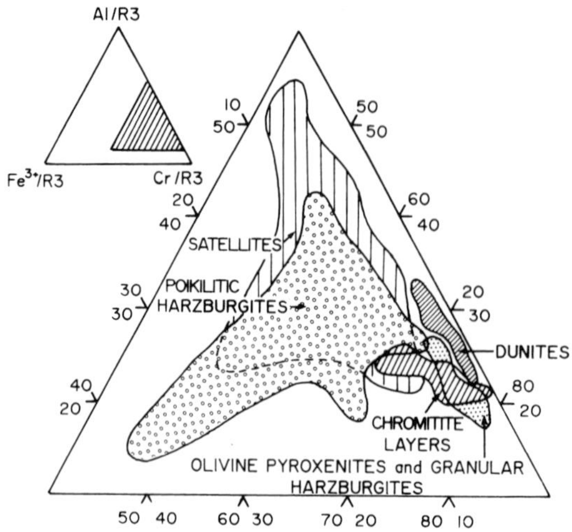

Figure 17. *Ternary diagram showing compositional fields of chromite in the Great Dyke in different lithologic environments. R3 represents summation of cations Cr+Al+Fe^{3+}. The highly restricted field of the chromite in the chromitite layers contrasts with those in other environments, and in particular the satellite dykes. The field of chromite compositions of the granular harzburgites lies within that of the olivine pyroxenites rather than the poikilitic harzburgites (from Wilson, 1982).*

Orthopyroxene compositions have been extensively investigated for all subchambers (Wilson, 1982; Prendergast and Keays, 1989; Wilson, 1992; Prendergast, 1991; Coghill and Wilson, 1993) and give a clear picture of the fractionation trend. The same pattern emerges for all subchambers but is most clearly demonstrated in the thicker cyclic units of the Darwendale Subchamber. There is a steady upwards enrichment in iron through the Pyroxenite Succession with the most magnesian pyroxenes being En_{91}. Near the top of the orthopyroxenite layer of Cyclic Unit 1 the composition is En_{85}. A very clear feature of the pyroxenite layers is that towards the top of each layer there is a progressive reversal to more magnesian compositions. This is explained by the repeated influx of primitive magma mixing with the pre-existing, more evolved magma at the same time that crystallization of pyroxene was taking place. The compositional reversals monitor these mixing events. In the websterite layer the rate of iron enrichment increases dramatically compared to the overall trend of pyroxene compositions in the underlying Ultramafic Sequence and continues into the gabbroic rocks of the Mafic Sequence. The continuous trend of pyroxene composition in the websterite layer and the Lower Mafic Succession is shown by Cr_2O_3 concentration and Mg# of clinopyroxene (Figure 18). One major reversal in orthopyroxene composition takes place in the Middle Mafic Succession (Figure 19) marking influx of primitive magma, but again reverts to the normal trend for the gabbroic rocks in the Upper Mafic Succession. The more iron-rich compositions and the reversed trend observed in the lowermost pyroxenite layer (Cyclic Unit 14 of the Darwendale Subchamber) (Figure 19) are a feature common to many large layered intrusions. Petrogenetic interpretation of orthopyroxene trends are entirely consistent with field observations and place strong constraints on the origin of the Great Dyke magmatic system. The overall low degree of fractionation in the Ultramafic Sequence is consistent with repeated influx of primitive magma into the magma chambers. Apart from the Middle Mafic Succession, there is no evidence of further influxes following the formation of Cyclic Unit 1 thereby allowing the observed continuous fractionation trend.

Trends of clinopyroxene compositions are the same as those of orthopyroxene where these pyroxenes co-exist (in the websterite layer of Cyclic Unit 1 and in the Mafic Sequence).

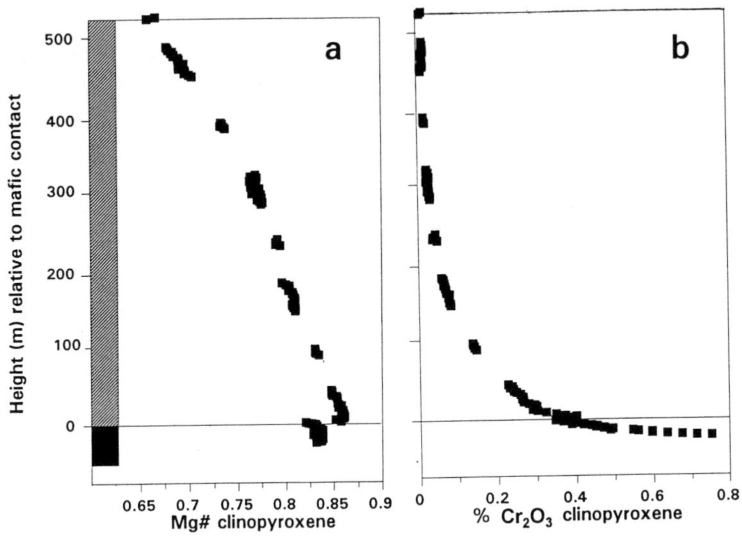

Figure 18. *Compositional variation of clinopyroxene through the websterite layer of Cyclic Unit 1 and the Lower Mafic Succession for (a) Mg# as Mg/(Mg+Fe^{2+}), and (b) Cr_2O_3. Note the compositional reversal at the contact of the websterite and gabbronorite. Ornamentations of stratigraphic column: solid black - websterite; hatched - gabbronorite.*

Chromium contents are high in the most magnesian pyroxenes (up to 1.4% Cr_2O_3 in clinopyroxene in Cyclic Unit 1 and up to 0.71% Cr_2O_3 in orthopyroxene in Cyclic Unit 6). These high chromium contents reflect the primitive nature of the Great Dyke magma (Hughes, 1976).

Plagioclase in the ultramafic rocks is characteristically normally zoned with ranges in composition from An_{88} to An_{64}. In the mafic rocks this mineral shows an overall wide range in composition becoming more sodic upwards in the succession from An_{88} at the base of the Lower Mafic Succession to An_{48} in the Upper Mafic Succession. Zoning is complex with normal, reversed, and oscillatory patterns in evidence. Margins of the crystals commonly exhibit myrmekitic texture as described by Wager and Brown (1968) for plagioclase in the Bushveld Complex.

14. PLATINUM GROUP ELEMENT AND BASE METAL MINERALIZATION

The occurrence of platinum mineralization in the P1 pyroxenite has resulted in this zone of the Great Dyke being most extensively studied, and therefore it merits special consideration, both from the aspect of genesis of the mineralization and also from the contribution of these studies to the origin of layered intrusions in general. Platinum group elements (PGE) are associated with sulphide in the P1 pyroxenite layer as a broad disseminated zone 30-60 m wide in the lower part of the layer (called the lower sulphide zone or LSZ) and as a narrow highly enriched zone 2-8 m wide at the top of the P1 orthopyroxenite (called the main sulphide zone or MSZ) (Figure 20). Only the MSZ is of economic importance and is a major resource of platinum in the western world, ranking second in potential ore reserves to that of the Bushveld

Complex. The platinum horizon was first reported by Zealley in 1918 and occurs in all parts of the Great Dyke where this level is preserved. Of particular significance is the 'potato' reef or nodular pyroxenite (described previously) which is well developed in this part of the succession and was used as a marker horizon for the MSZ by early prospectors. In outcrop it is commonly stained by copper hydroxides and carbonates.

The MSZ is effectively a six-metal ore body comprising Pt, Pd, Rh, Au, Cu, and Ni and is subdivided into a lower platinum-enriched sub-zone and an upper base-metal-enriched sub-zone. This subdivision arises from the clear separation of the base metals and the platinum group elements (Figure 21). The general distribution of the PGE is of metal concentrations rising gradually upwards through the zone but falling to very low levels at the top of the zone over a vertical interval of less than 30 cm. A further fine structure in the distribution of PGE is reported by Wilson and Tredoux (1990) and by Prendergast and Keays (1989). The uniformity of the metal distributions in all subchambers of the Great Dyke is remarkable and points to consistency of processes and conditions which would only be possible if these subchambers were linked at this level (Prendergast and Wilson, 1989). A further deduction is that the mineralization was established as a result of primary magmatic processes associated with the segregation of sulphide from the magma (Prendergast and Keays, 1989; Wilson *et al.*,1989;

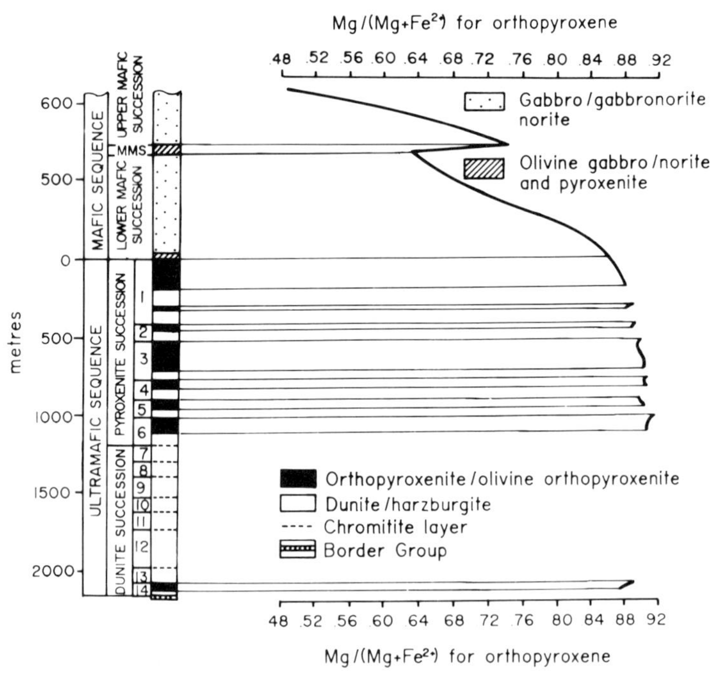

Figure 19. *Variation of orthopyroxene compositions through the Mafic and Ultramafic Sequences. Note the extreme rate of change of composition in the mafic rocks compared with that in the underlying ultramafic rocks. The only major reversal in the Mafic Sequence occurs in the Middle Mafic Succession.*

Prendergast, 1990; Naldrett and Wilson, 1990; Wilson and Naldrett, 1989), rather than by later hydrothermal processes which would inherently impose a variable character on the metal distribution. There also exist systematic lateral and longitudinal variations in the ore body in the Darwendale, Wedza, and Selukwe Subchambers which are related to primary magmatic processes. In the Darwendale Subchamber the form of the mineralized zone is asymmetric across the Great Dyke, with characteristics of the east margin having similarities to the axial zone in a wider mineralized interval of relatively low grade.

The stratigraphic order of enrichment of the metals in association with the sulphides is regarded as the same order as the metal peaks are observed from the base upwards in the MSZ; i.e. from the base to the top of the mineralized zone this is Ir - Pd - Pt - Ni - Au - Cu. This sequence is best explained if the sulphide liquated near the floor of the magma chamber with extraction of the PGE and chalcophile metals from the overlying convecting magma in the order of their apparent partition coefficients into sulphide (Prendergast and Keays, 1989). The validity of employing the concept of partition coefficients for PGE (which are likely to have precipitated by complexation processes) in sulphide is questionable and Coghill and Wilson (1993) suggest that the term 'affinity factor' is more applicable in this respect.

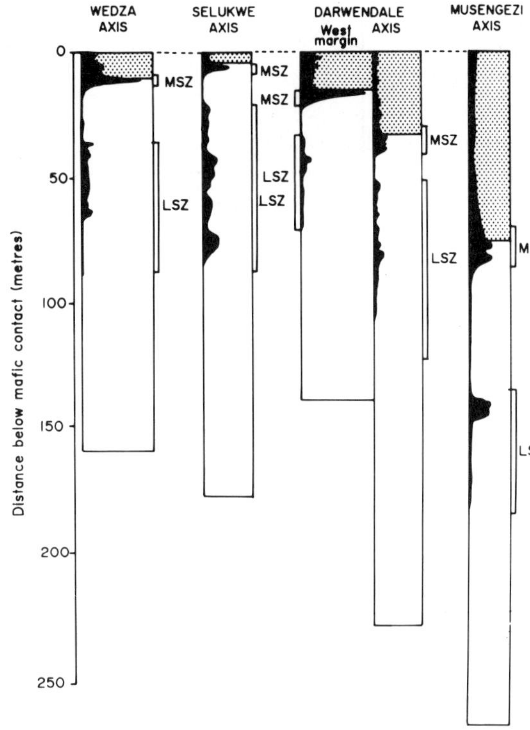

Figure 20. Relative volume distribution of sulphide (solid black ornament) in the P1 pyroxenite in the various subchambers of the Great Dyke on the basis of Ni+Cu assay data. The total thickness of the orthopyroxenite (open) and websterite (stippled ornament) layers are indicated. MSZ - main sulphide zone; LSZ - lower sulphide zone. Note the significant difference in the sulphide distribution between the axis and west margin of the Darwendale Subchamber (from Prendergast and Wilson, 1989).

The mineralogy of the MSZ has been studied in detail in several of the subchambers: Darwendale (Johan *et al.*, 1989); Wedza (Prendergast, 1989; 1990); Selukwe (Coghill and Wilson, 1993). The close association of PGE with base metal sulphides is striking. The main sulphide minerals in this zone are pyrrhotite, pentlandite, chalcopyrite, and pyrite occurring as interstitial phases to the silicate minerals. Where the sulphides are abundant they have a sub-poikilitic net-textured appearance.

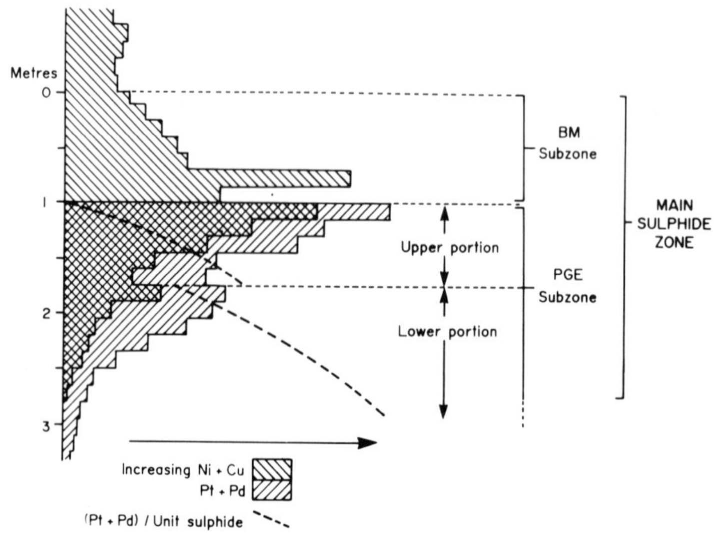

***Figure 21**. Generalized distribution of base metals (Cu+Ni) and the platinum group elements (Pt+Pd) through the main sulphide zone (MSZ) of the Great Dyke. The MSZ is subdivided into the upper base-metal (BM) subzone and the lower platinum group element (PGE) enriched subzone. The latter in turn shows a fine structure on the basis of the stepped distribution of metals into an upper and lower portion. The gentle rise in values upwards in the stratigraphic section, which contrasts with the abrupt decrease in metal concentrations at the top of the PGE subzone, characterizes the mineralized zone and is remarkably similar in all subchambers. The dashed line represents PGE per unit sulphide and this decreases markedly upward in the PGE subzone (from Prendergast and Wilson, 1989).*

In zones of abundant sulphides there is a marked increase in alteration of the silicates to a distinctive assemblage of hydrosilicates which includes tremolite, talc, magnetite, biotite, and chlorite. Prendergast (1990) explained this alteration as resulting from the release of highly reactive fluids arising from the interaction of the sulphides with magmatic fluids exsolving from trapped late-stage liquid. This hot acidic fluid would have been highly reactive towards the silicate minerals as well as causing recrystallization of sulphide. The platinum group minerals (PGM) constitute both high- and low-temperature species. High-temperature phases include primary sulphides such as braggite ((Pt,Pd)S), cooperite (PtS), and laurite (RuS_2). Low-temperature minerals include the bisthmuthotellurides moncheite ($PtTe_2$), maslovite (PtBiTe), michenerite (PdBiTe), kotulskite (PdTe), polarite (PdBi), and arsenides such as sperrylite ($PtAs_2$) and hollingworthite (RhAsS). One textural form of sperrylite is also considered to be a high-temperature primary crystallization phase. The association of high- and low-temperature PGM is the result of a complex multi-stage process involving the preservation of some high-temperature phases, strong incorporation of the metals into early-formed sulphide, with subsequent release of these metals and recrystallization of new phases during the cooling process (Coghill and Wilson, 1993). The presence of small amounts of highly reactive hydrous fluids played an important role in the recrystallization process with redistribution of the PGE taking place on a local scale.

Evans *et al.* (1994) considered the mobilization and secondary dispersion of PGE in the oxidized zone in the outcrop and suboutcrop of the MSZ in the Darwendale Subchamber. The primary PGM assemblage is indicated to have altered to dominantly platinum-iron alloys in the weathering environment which is to a depth of several tens of metres.

15. PETROGENESIS

15.1. Macro-cyclic units

The macro-cyclic layering of the Ultramafic Sequence and the consistent compositional reversals occurring at the base of the cyclic units are most easily explained by repeated injections of primary magma into the chamber. The amount of mixing of the new and resident magma would depend on the fluid dynamics of the system (Huppert and Sparks, 1980) and the relative densities and viscosities of the two magmas (Turner and Campbell, 1986). Mineral composition trends for the Darwendale Subchamber (Wilson, 1982), Selukwe Subchamber (Coghill and Wilson, unpublished data), and Wedza Subchamber (Wilson, unpublished data) reflect the gradually evolving liquid composition throughout the crystallization history but with repeated injections of primary magma terminating the crystallization of initially olivine in the Dunite Succession (by the formation of chromitite layers), and then orthopyroxene in the Pyroxenite Succession. The dynamics of mixing appears to have been different in these two parts of the sequence because the compositional reversals are sharp in the dunites but become slowly reversed in the pyroxenites before the onset of olivine crystallization marking the base of the next cyclic unit (Wilson, 1982).

The greater rate of change of pyroxene composition occurring at the stratigraphic level of the upper P1 pyroxenite, and then proceeding into the websterite layer and the mafic rocks (Figure 19), indicates that the processes in the magma chamber changed from that stage. In contrast to the cyclic units of the Ultramafic Sequence which indicate repeated influxes of magma, there is no evidence of new magma influxes in the Lower Mafic Succession. A further influx of magma, however, gave rise to the more magnesian mineral compositions which characterize the Middle Mafic Succession. The formation of the Mafic Sequence and the first appearance of cumulus plagioclase as resulting from emplacement of a major pulse of more differentiated magma cannot, however, be excluded, but this in itself would not have affected the rate of differentiation within the chamber.

Each subchamber has its own distinctive pattern of layering in the lower Ultramafic Sequence. Higher in the sequence the subchambers within the North and South Chambers became linked at that level so that layering in the upper Pyroxenite Succession was similar in the Wedza and Selukwe Subchambers but different from that in the Darwendale and Sebakwe Subchambers which are also similar to each other. Above Cyclic Unit 1, both the North and the South Chambers were linked with form, style, and structure of the layering being remarkably similar in all parts of the Great Dyke. A synopsis of these observations is that, either, (1) the barriers between the various isolated subchambers, and then later those between the major chambers were breached as magma continued to be emplaced into the chamber system, or (2) ponding of dense magma and subsequent crystallization took place initially within each of the essentially contiguous compartments until these were physically continuous at higher level.

A major problem is the reconciliation of the stratigraphic thickness of the Mafic Sequence compared with that of the Ultramafic Sequence. Modelling of the fractionation trends combined with mass balance considerations (Wilson, 1982), indicates that either the magma chambers were effectively an open system during the period of combined crystallization and magma emplacement which gave rise to the Ultramafic Sequence, or else there was a considerable sill-like lateral extension of the upper Ultramafic Sequence and most of the Mafic Sequence, which has now been entirely eroded (Podmore and Wilson, 1987). It should be noted that this is a problem of mass balance and not one of fluid dynamics pertaining to the evolution of the magma system.

15.2. Order of crystallization

The observed order of crystallization in the Great Dyke, as deduced from the cumulus assemblage in all magma chambers (except where magmatic erosion has taken place) is: chromite - olivine - orthopyroxene - clinopyroxene - plagioclase - pigeonite - magnetite. This is precisely

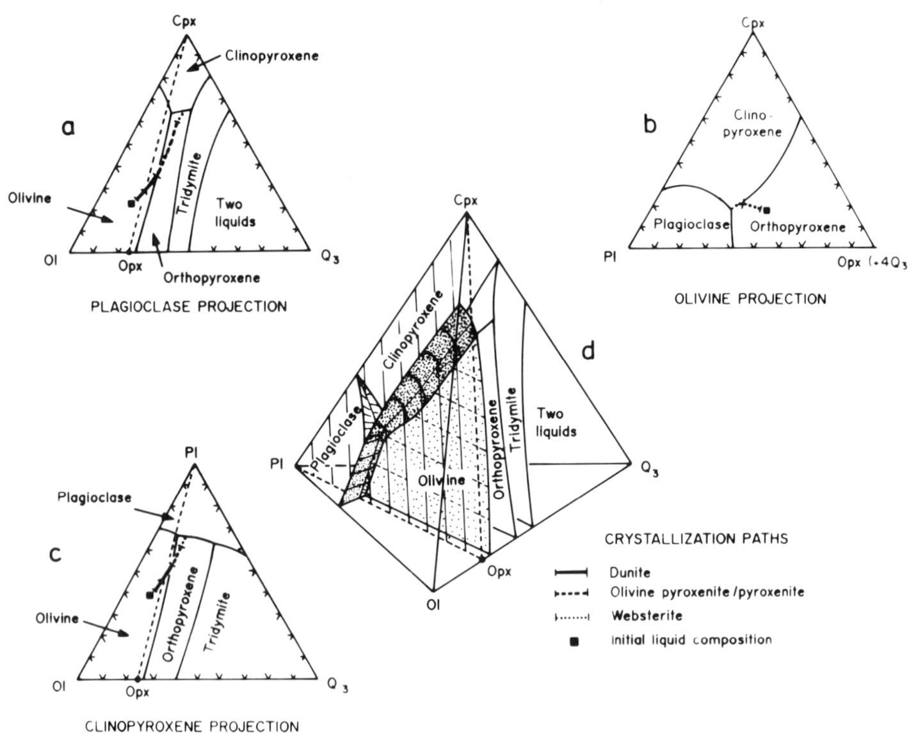

Figure 22. Modelled evolutionary path of the Great Dyke magma using the chill phase of the East Dyke offshoot (composition 4 in Table 2, after Wilson, 1982). The projections and phase diagram planes (a) - (c) are in the basalt tetrahedron (d) after Irvine (1970, 1979). The modelled liquid paths correspond to the different mineral assemblages crystallizing to give the Ultramafic Sequence (dunite, pyroxenite and websterite) (from Wilson, 1982).

the order of crystallization observed in the microphenocryst and groundmass assemblage observed in the chill zone close to the East Dyke and which is accepted as the most likely primary composition of the Great Dyke magma. Furthermore, equilibrium extraction modelling of this composition, shown projected in the basalt tetrahedron (Figure 22), also gives the observed order of silicate crystallization.

As the magma composition approached the phase boundary for each primary (cumulus) mineral phase, that mineral would appear first as abundant postcumulus material, then as an oikocryst and finally as a discrete cumulus phase. This is observed for orthopyroxene in each of the cyclic units of the Ultramafic Sequence in the poikilitic harzburgite, for clinopyroxene as nodules below the websterite, and for plagioclase in the nodular pyroxenite immediately below the first appearance of the mafic rocks.

15.3. Lateral variations and primary structure

Major lateral variations are observed in almost all aspects of the structure, layering, textures, mineral compositions, and mineralization. Individual layers become markedly thinner towards the margins and eventually coalesce in the Border Group. The amount of interstitial material increases away from the axis and the development of oikocryst phases (such as orthopyroxene in the poikilitic harzburgites) occurs at stratigraphically lower levels in the marginal zones compared to the same part of the sequence in the axis. Cumulus minerals also show a continuous progression to more evolved compositions from axis to margin. This is particularly well demonstrated in the P1 pyroxenite of Cyclic Unit 1 but is also observed in lower dunite and pyroxenite layers of the Ultramafic Sequence where these are intersected by drilling. Chromitite C1c in Cyclic Unit 1 shows major lateral variations in texture and composition.

The reason for the lateral variations is considered to be related to the shape of the Great Dyke magma chamber, and the geometry of the layers and how these were influenced by the heat flow through the walls and floor during the crystallization process (Prendergast, 1991; Wilson and Prendergast, 1989). The most marked lateral variations in the preserved succession are observed in Cyclic Unit 1 as this unit shows the greatest proximity to both the floor/wall and the roof of the intrusion (see Figure 9) as result of the flared structure of the Great Dyke. Although the major heat loss would have been from the roof of the intrusion the transverse trumpet shape of the magma chamber would have entailed significant heat loss from the walls. The resulting thermal regime was such that there was a gradational variation in heat flow from the axial environment, where the crystallizing phases were underlain by a thick sequence of hot cumulates, to the marginal environment where the layered sequence was close to the wall/floor of the intrusion. This would have had a profound effect on the types of cumulate textures observed within the Ultramafic Sequence of the Great Dyke and similar controls may also explain the apparent upward displacement of compositional reversals towards the margins observed in many layered intrusions (Wilson and Engell-Sorensen, 1986). Progressive or repeated injection of magma into a chamber with a sloping floor would cause essentially the same magma layer to come into contact with successively higher levels of the floor resulting in an inverted compositional sequence, particularly in the earliest crystallization products observed near the stratigraphic base of the layered succession.

Prendergast (1991) appealed to a compositionally and thermally stratified magma column to explain the discordant relationships between modal and cryptic layering at the margins of the Wedza Subchamber on the basis of the model proposed by Wilson and Larsen (1985) to explain similar phenomena in the Fongen-Hyllingen layered mafic complex in Norway. Wilson

(1992) showed that lateral variations in incompatible trace elements in the P1 pyroxenite could not be explained by a single magma composition and also appealed to a compositionally zoned magma body.

Studies of primary textures and petrofabrics (Wilson, 1992; and unpublished data) indicate that the synclinal attitude of the layering in the Great Dyke is largely primary with minor accentuation due to later downwarping of the axial zone. Lateral thinning of individual layers may therefore be related to crystallization taking place from a more restricted volume of magma nearer the margins with significantly higher cooling rates, rather than by greater degree of compaction in the axial zone (McKenzie, 1987).

15.4. Cumulus history and controls on mineral compositions

The vertical and lateral variations in texture and mineral compositions which may be observed within individual layers in the Great Dyke allow critical evaluation of the cumulus concept in the genesis of layered intrusions (Wilson, 1992). In that paper the terms ortho- and adcumulus were redefined on the basis of end-members of a process continuum, as the original proposers of the nomenclature (Wager and Brown, 1968) intended, and not the purely textural interpretation which has become common usage. Interpretation of textures does not unequivocally lead to classification of a rock as a particular cumulus type. Wilson (1992) showed a strong dependence of orthopyroxene major element composition in the P1 layer on the whole-rock content of incompatible elements. These relationships in the Great Dyke were interpreted as the result of original cumulus crystals undergoing re-equilibration with trapped liquid to more evolved compositions but without significant loss of liquid by compaction of the cumulates, although some degree of compaction cannot be ruled out, and in some cases this may be extreme. Modelling also permitted an estimate of the amount of trapped liquid in the cumulates (at the point at which they were no longer in exchange contact with the overlying body of magma) and this was shown to be relatively small ranging from <1% to a maximum of 13%. The relatively large abundance of interstitial plagioclase and clinopyroxene (up to 30%) in the axial facies of the Great Dyke is indicated to be as oikocrysts of adcumulus status, or heteradcumulus in the terminology of Wager and Brown (1968), and were not derived solely from the trapped liquid. Textures in the sulphide zones of the Great Dyke are strongly influenced by the development of plagioclase oikocrysts and where best developed, this gives rise to the nodular pyroxenite (or 'potato' reef). This texture also influences the distribution of sulphide in that it concentrates around the margins of the oikocrysts while still in its initial liquid state.

16. CONCLUDING REMARKS

Aspects of Great Dyke geology illustrate superbly many fundamental features arising from the emplacement and crystallization of mafic layered intrusions. The shape and dimensions of the intrusion imposed strong controls on cooling properties. The Great Dyke of Zimbabwe is unique among large layered intrusions in its extreme length to width ratio, but at the same time shows clearly that differences in the size and positions within individual subchambers can profoundly affect configurations and styles of layering, textures, and mineral compositions. The distinctly primitive magma of the Great Dyke gave rise to compositions of olivine and pyroxenes that are more magnesian than those observed for most other major layered intrusions. Coupled with this are the extreme monomineralic end-member adcumulates occurring in the larger subchambers.

The structure of the magma chambers is well-established and there is clear indication of a feeder zone for most of the length of the Great Dyke as well as expansion of the magma chamber at depth, or even separate magma chamber compartments. Also significant is the geophysical evidence of tilting of part of the Darwendale Subchamber which is supported by strong asymmetry in rock-types and the nature of the sulphide mineralization in the Darwendale Subchamber on the east and west sides.

The overwhelming geochemical evidence for the origin of macro-layering in the Ultramafic Sequence is one of repeated influxes of magma arresting the normal process of crystallization and causing significant compositional reversals in the mineral compositions. The dynamics of such processes are complex and sharp compositional reversals observed in the cyclic units of the Dunite Succession contrast with the gradual reversals observed in the Pyroxenite Succession. The filling process may have commenced with individual magma chambers (here recognized as subchambers) and then continued with coalescence between them at progressively higher stratigraphic levels. This gave rise to distinct layering patterns in each of the subchambers. Finally, the two all-embracing major chambers (the North and South Chambers) were linked at the level of Cyclic Unit 1. At this stage the magma chamber underwent continuous differentiation until the level of the Middle Mafic Succession where a further influx of more primitive magma took place. A consequence of this fractionation was that the magma achieved sulphide saturation causing the important base-metal and PGE-enriched zones at the top of the Ultramafic Sequence.

A further important aspect relating to the geology of layered intrusions is an appreciation of lateral variations within individual layers of the Great Dyke. This is the result of a strong lateral heat-flow profile, the relative importance of which is dependent on the shape of the magma chamber with its narrow width-to-length aspect ratio. The results of this thermal control are clearly apparent in major lateral changes in rock-types, textures, mineral compositions and style of mineralization.

17. ACKNOWLEDGEMENTS

Martin Prendergast and Bruce Lipin are acknowledged for constructive comments on the manuscript. The University of Natal Research Committee and the Foundation for Research Development (FRD) are thanked for providing funds over the years to facilitate Great Dyke research. Publication of figures 4 and 17, and 2, 3, 8, 9, 10, 12, 16, 19, and 21 is by permission of Oxford University Press and the Institute of Mining and Metallurgy, respectively.

18. REFERENCES

Alapieti, T.T., & Lahtinen, J.J., 1989. Early Proterozoic layered intrusions in the northeastern part of the Fennoscandian Shield. *5th Internat. Platinum Symposium. Guide 29. Geol. Surv. Finland*, 3-41.

Bichan, R., 1969. Origin of chromite seams in the Hartley Complex of the Great Dyke. In: Wilson, H.D.B. (ed.) *Magmatic Ore Deposits. Econ. Geol. Monogr.* **4**, 95-113.

Bichan, R., 1970. The evolution and structural setting of the Great Dyke, Rhodesia. In: Clifford, T.N. and Gass, I.G. (eds.) *African Magmatism and Tectonics.* Edinburgh: Oliver and Boyd, 51-71.

Campbell, I.H., McCall, G.J.H., & Tyrwhitt, D.S., 1970. The Jimberlana Norite, Western Australia - a smaller analogue of the Great Dyke of Rhodesia. *Geol. Mag.* **107**, 1-12.

Coghill, B.M., & Wilson, A.H., 1993. Platinum-group minerals in the Selukwe Subchamber, Great Dyke, Zimbabwe: implications for PGE collection mechanisms and post-formational redistribution. *Miner. Mag.* **57**, 613-33.

Davies, G., Cawthorn, R.G., Barton, J.M., & Morton, M., 1980. Parental magma to the Bushveld Complex. *Nature* **287**, 33-5.

Davies, G., & Tredoux, M., 1985. The platinum-group element and gold contents of the marginal rocks and sills of the Bushveld Complex. *Econ. Geol.* **80**, 838-48.

Evans, D.M., Buchanan, D.L., & Hall, G.E.M., 1994. Dispersion of platinum, paladium and gold from the Main Sulphide Zone, Great Dyke, Zimbabwe. *Trans. Instn. Min. Metall. (Sect. B: Appl. earth sci.)* **103**, 57-67.

Hamilton, J., 1977. Strontium isotope and trace element studies on the Great Dyke and Bushveld mafic phase and their relation to early Proterozoic magma genesis in southern Africa. *J. Petrology* **18**, 24-52.

Harger, H.S., 1934. An early Transvaal geological map, by Carl Mauch. *Trans. Geol. Soc. S. Afr.* **37**, 1-4.

Hatton C.J., & von Gruenewaldt, G., 1990. Early Precambrian layered intrusions. In: Hall, R.P., & Hughes, D.J. (eds.) *Early Precambrian Basic Magmatism.* New York: Blackie, 56-82.

Hess, H.H., 1950. Vertical mineral variation in the Great Dyke of Southern Rhodesia. *Trans. Geol. Soc. S. Afr.* **53**, 159-68.

Hughes, C.J., 1976. Parental magma of the Great Dyke of Rhodesia - voluminous late Archaean high magnesium basalt. *Trans. Geol. Soc. S. Afr.* **79**, 179-82.

Huppert, H.E., & Sparks, R.S., 1980. The fluid dynamics of a basaltic magma chamber replenished by influx of hot, dense, ultrabasic magma. *Contr. Miner. Petrol.* **75**, 279-89.

Irvine, T.N., 1970. Crystallization sequences in the muskox intrusion and other layered intrusions. I. Olivine - pyroxene - plagioclase relations. *Spec. Publ. Geol. Soc. S. Afr.* **1**, 441-76.

Irvine, T.N., 1979. Rocks whose composition is determined by crystal accumulation and sorting. In: Yoder, H.S., (ed.) *The Evolution of Igneous Rocks.* Princeton University Press, 245-306.

Jackson, E.D., 1961. Primary textures and mineral associations in the Ultramafic Zone of the Stillwater complex, Montana. *Prof. Paper U.S. Geol. Surv.* **358**, 106 pp.

Jackson, E.D., 1970. The cyclic unit in layered intrusions - a comparison of repetitive stratigraphy in the ultramafic parts of the Stillwater, Muskox, Great Dyke and Bushveld Complexes. *Spec. Publ. Geol. Soc. S. Afr.* **1**, 391-424.

Johan, Z., Ohnenstetter, D., & Naldrett, A.J., 1989. Platinum group elements and associated oxides and base metal sulphides of the Main Sulphide Zone, Great Dyke, Zimbabwe. In: Papunen, H. (ed.) *Abstr. 5th Intern. Platinum Symposium. Bull. Geol. Soc. Finland* **61**, 53-4.

Keep, F.E., 1930. The geology of the chromite and asbestos deposits of the Umvukwe Range, Lomagundi and Mazoe Districts. *S. Rhod. Geol. Surv. Bull.* **16,** 10 pp.

Lightfoot, B., 1940. The Great Dyke of Southern Rhodesia. *Proc. Geol. Soc. S. Afr.* **43**, 27-54.

Maufe, H.B., 1925. Platinum in Southern Rhodesia – rocks likely to contain it. *Bulawayo Chronicle*, May.

McCall, G.J.H., & Peers, R., 1971. Geology of the Binneringie Dyke, Western Australia. *Geol. Rundsch.* **60**, 1174-263.

McClay, K.R., & Campbell, I.H., 1976. The structure and shape of the Jimberlana intrusion, Western Australia, as indicated by an investigation of the Bronzite Complex. *Geol. Mag.* **113**, 129-39.

McKenzie, D.P., 1987. The compaction of igneous and sedimentary rocks. *J. Geol. Soc. Lond.* **144**, 299-307.

Mennell, F.P., 1910. Geological structure of Southern Rhodesia. *Quart. J. Geol. Soc.* **66**, 353-7.

Naldrett, A.J., & Wilson, A.H., 1989. Distribution and controls of platinum group-element mineralisation in Cyclic Unit 1 of the Great Dyke, Zimbabwe. In: Papunen, H. (ed.) *Abstr. 5th Intern. Platinum Symposium. Bull. Geol. Soc. Finland* **61**, 3.

Naldrett, A.J., & Wilson, A.H., 1990. Horizontal and vertical variations in noble metals in the Great Dyke of Zimbabwe: A model for the origin of the PGE mineralization by fractional segregation. *Chem. Geol.* **88**, 279-300.

Podmore, F., 1970. The shape of the Great Dyke as revealed by gravity surveying. *Spec. Publ. Geol. Soc. S. Afr.* **1**, 610-20.

Podmore, F., 1982. The first Bouguer anomaly map of Zimbabwe. *Trans. Geol. Soc. S. Afr.* **85**, 127-33.

Podmore, F., & Wilson, A.H., 1987. A reappraisal of the structure, geology and emplacement of the Great Dyke, Zimbabwe. In: Halls, H.C., & Fahrig, W.F. (eds.) *Mafic Dyke Swarms. Spec. Paper Geol. Assoc. Can.* **34**, 433-44.

Prendergast, M.D., 1987. The chromite ore field of the Great Dyke, Zimbabwe. In: Stowe, C.W. (ed.) *Evolution of Chromite Ore Fields,* New York: Van Nostrand Reinhold, 89-108.

Prendergast, M.D., 1988. The geology and economic potential of the PGE-rich Main Sulphide Zone of the Great Dyke, Zimbabwe. In: Prichard, H.M., Potts, P.J., Bowles, J.F.W., & Cribb, S.J. (eds.) *Geo-Platinum '87.* Barking, Essex: Applied Science, 281-302.

Prendergast, M.D., 1989. The geology and stratigraphic setting of the Wedza-Mimosa Platinum Deposit, Great Dyke, Zimbabwe. (Abstr.). *Bull. Geol. Soc. Finland* **61**, 13-4.

Prendergast, M.D., 1990. Platinum-group minerals and hydrosilicate 'alteration' in Wedza-Mimosa Platinum Deposit, Great Dyke, Zimbabwe - genetic and metallurgical implications. *Trans. Instn. Min. Metall. (Sect. B: Appl. Earth Sci.)* **99**, B91-105.

Prendergast, M.D., 1991. The Wedza-Mimosa platinum deposit, Great Dyke, Zimbabwe: layering and stratiform PGE mineralisation in a narrow mafic magma chamber. *Geol. Mag.* **128**, 235-49.

Prendergast, M.D., & Keays, R.R., 1989. Controls of platinum-group element mineralization and the origin of the PGE-rich Main Sulphide Zone in the Wedza Subchamber of the Great Dyke, Zimbabwe: Implications for the genesis of, and exploration for, stratiform PGE mineralisation in layered intrusions. In: Prendergast, M.D., & Jones, M. J. (eds.) *Magmatic Sulphides – The Zimbabwe Volume.* Instn. Min. Metall. Lond. 43-69.

Prendergast, M.D., & Wilson, A.H., 1989. The Great Dyke of Zimbabwe II: Mineralisation and mineral deposits. In: Prendergast, M.D., & Jones, M. J. (eds.) *Magmatic Sulphides – The Zimbabwe Volume.* Instn. Min. Metall. Lond. 21-42.

Robertson, I.D.M., & van Breemen, O., 1970. The southern satellites of the Great Dyke, Rhodesia. *Spec. Publ. Geol. Soc. S. Afr.* **1**, 621-44.

Turner, J.S., & Campbell, I.H., 1986. Convection and mixing in magma chambers. *Earth Sci. Rev.* **23**, 255-352.

Tyndale-Biscoe, R., 1949. The geology of the country around Gwelo. *Geol. Surv. S. Rhod. Bull* **39**, 145 pp.

Wager, L.R., & Brown, G.M., 1968. *Layered Igneous Rocks.* San Francisco: W.H. Freeman & Co., 588 pp.

Wagner, P.A., 1914. The geology of a portion of the Belingwe District of Southern Rhodesia. *Trans. Geol. Soc. S. Afr.* **17**, 39-54.

Weiss, O., 1940. Gravimetric and earth magnetic measurements of the Great Dyke of Southern Rhodesia. *Trans. Geol. Soc. S. Afr.* **43**, 143-51.

Wilson, A.H., 1982. The geology of the Great "Dyke", Zimbabwe: the ultramafic rocks. *J. Petrology* **23**, 240-92.

Wilson, A.H., 1992. The geology of the Great Dyke, Zimbabwe: crystallisation, layering and cumulate formation in the P1 Pyroxenite of Cyclic Unit 1 of the Darwendale subchamber. *J. Petrology* **33**, 611-63.

Wilson, A.H., & Naldrett, A.J., 1989. Vertical and lateral variations in the petrology, structure and mineral chemistry of Cyclic Unit 1 of the Darwendale Subchamber of the Great Dyke, and their bearing on PGE and base metal mineralisation. *Bull. Geol. Soc. Finland* **61**, 2.

Wilson, A.H., & Prendergast, M.D., 1987. The Great Dyke of Zimbabwe an overview. In: *Guidebook for the 5th Magmatic Sulphides Field Conference*, Harare, Zimbabwe, 23-55.

Wilson, A.H., & Prendergast, M.D., 1989. The Great Dyke of Zimbabwe: I: Tectonic setting, stratigraphy, petrology, structure, emplacement and crystallisation. In: Prendergast, M.D. and Jones, M.J. (eds.) *Magmatic Sulphides - The Zimbabwe Volume.* Instn. Min. Metall., Lond., 1-20.

Wilson, A.H., & Tredoux, M., 1990. Lateral and vertical distribution of the platinum-group elements, and petrogenetic controls on the sulphide mineralisation, in the P1 Pyroxenite Layer of the Darwendale Subchamber of the Great Dyke, Zimbabwe. *Econ. Geol.* **85**, 556-84.

Wilson, A.H., & Wilson, J.F., 1981. The Great "Dyke". In: Hunter, D.R. (ed.) *Precambrian of the Southern Hemisphere.* Amsterdam: Elsevier, 572-8.

Wilson, A.H., Naldrett, A.J., & Tredoux, M., 1989. Distribution and controls of platinum-group element and base metal mineralisation in the Darwendale Subchamber of the Great Dyke, Zimbabwe. *Geology* **17**, 649-52.

Wilson, J.F., 1987. The tectonic setting of the Great Dyke of Zimbabwe and the Mashonaland Igneous Event. (Abstr.). *14th Colloquium of African Geology, CIFEG Occ. Publ.* **12**, 140-1.

Wilson, J.F., Jones, D.L., & Kramers, J.D., 1987. Mafic dyke swarms in Zimbabwe. In: Halls H.C., & Fahrig W.F. (eds.) *Mafic Dyke Swarms. Spec. Paper. Geol. Assoc. Can.* **34**, 433-44.

Wilson, J.R., & Engell-Sorensen, O., 1986. Basal reversals in layered intrusions are evidence of emplacement of compositionally stratified magma. *Nature* **323**, 616-8.

Wilson, J.R., & Larsen, S., 1985. Two dimensional study of a layered intrusion the Hyllingen Series, Norway. *Geol. Mag.* **122**, 97-124.

Worst, B.G., 1958. The differentiation and structure of the Great Dyke of Southern Rhodesia. *Trans. Geol. Soc. S. Afr.* **61**, 283-354.

Worst, B.G., 1960. The Great Dyke of southern Rhodesia. *S. Rhod. Geol. Surv. Bull.* **47**, 234 pp.

Worst, B.G., 1964. Chromite in the Great Dyke of Southern Rhodesia. In: Haughton, S.H. (ed.) *The Geology of some Ore Deposits in Southern Africa,* Johannesburg: Geol. Soc. S. Afr., **2**, 209-24.

Zeally, A.E.V., 1915. The Great Dyke of norite of Southern Rhodesia: Petrology of the Selukwe portion. *Trans. Roy. Soc. S. Afr.* **5**, 1-24.

Zeally, A.E.V., 1918. The occurrence of platinum in Southern Rhodesia. *S. Rhod. Geol. Surv. Short Report* **3**.

LAYERED INTRUSIONS
R.G. Cawthorn (editor)

The Rum Layered Suite

C.H. Emeleus[a], M.J. Cheadle[b], R.H. Hunter[b], B.G.J. Upton[c], and W.J. Wadsworth[d]

[a]Department of Geological Sciences, University of Durham, South Road, Durham, DH1 3LE, United Kingdom.

[b]Department of Earth Sciences, University of Liverpool, Brownlow Street, Liverpool L69 3BX, United Kingdom.

[c]Department of Geology and Geophysics, University of Edinburgh, West Mains Road, Edinburgh, EH9 3JW, United Kingdom.

[d]Department of Geology, University of Manchester, Oxford Road, Manchester, M13 9PL, United Kingdom.

Abstract

Palaeocene igneous activity in the Rum Central Complex culminated in the formation of the ultrabasic and gabbroic rocks of the Layered Suite. Its three components, the Eastern Layered Series, the Western Layered Series and the Central Series, represent a continuum in time during which replenishments of picritic (MgO 15-20 wt.%) and basaltic magmas ponded in thin sill-like bodies at the Lewisian gneiss - Torridonian sandstone unconformity, each contributing incrementally to a layered cumulate sequence. The magmas were probably guided during ascent by the long-lived Long Loch Fault. Peridotite (olivine - chrome-spinel) cumulates formed from picritic magma. The residual (basaltic) magma mixed with resident residual magma from earlier batches, and with small amounts of siliceous rheomorphic melts from country rocks, forming (isotopically contaminated) allivalitic (= troctolitic), plagioclase - olivine cumulates or, less commonly, gabbroic (plagioclase - olivine - clinopyroxene) cumulates. Residual basaltic magma was probably also intruded as gabbroic sheets and plugs, and extruded as lavas. Widespread slump and shear structures indicate mechanical instability of unconsolidated cumulate mushes, especially in the allivalites. Ultrabasic breccias are common in the Central Series, and are attributed to (i) disruption of earlier cumulates as new batches of magma rose along an elongate, north-south feeder zone and (ii) collapse of cumulates into this zone during episodic magma withdrawal. Equilibrated textures, lack of compositional zoning in olivine and pyroxene, offsets between compositional and modal variation at unit and other lithological boundaries, the occurrence of finger structures and other replacement features, and the compositional modification of ultrabasic rocks adjoining late-stage gabbroic veins, all attest to the pervasive influence of migrating intercumulus liquids during crystallization and consolidation of the cumulates.

1. INTRODUCTION

1.1. Regional setting

The Rum Central Complex lies within the British Tertiary Volcanic Province, an area of mainly Palaeocene igneous activity in NW Scotland, NE Ireland and the adjoining seas. The province formed during the opening of the North Atlantic (Upton, 1988), and coincides broadly with a zone in which the lithosphere had been locally thinned during the late Palaeozoic and the Mesozoic (Meisner *et al.*, 1986) and which may have focussed magma genesis (Thompson and Gibson, 1991). Volcanism occurred between 63 - 51 Ma but was probably restricted to <3 Ma on Rum in the interval 61 - 58 Ma. Almost all the igneous rocks have reversed magnetic polarities, correlated with Chron 26r (Mussett *et al.*, 1988).

1.2. Local geology

Rum lies on a ridge of Precambrian rocks separating two Mesozoic basins (Binns *et al.*, 1974, their figure 4). An outline of the geology is given in Figure 1; the details are described by Emeleus (1996) and depicted on published maps (British Geological Survey, 1994; Emeleus, 1994). Pre-Tertiary rocks consist principally of a thick (>2.5 km) succession of WNW-dipping sandstones of the Proterozoic Torridon Group. Archaean - early Proterozoic (Lewisian) gneisses underlie the intra-basin ridge but are only exposed within the central complex. Mesozoic rocks are limited to ~ 50 m of Triassic sandstones in northwest Rum and faulted slivers of Lower Jurassic sandstones, limestones and shales preserved on the southeast edge of the central complex (λ, Figure 1). There is evidence for faulting of the Torridonian beds, including movement on the Long Loch Fault, before emplacement of the central complex.

The sequence of igneous and tectonic events during the Palaeocene is summarized below (see also Figure 1):

1) Igneous activity commenced with the eruption of basalt lava flows of the Eigg Lava Formation (Smith, 1985; Emeleus, 1994). On Rum, these are restricted to faulted relics associated with the Jurassic sediments (see above).
2) Intrusion of sparse basaltic dykes and sills into the Torridonian beds probably post-dates 1.
3) Early activity in the central complex is manifest in central uplift on the Main Ring Fault, when Lewisian gneiss and basal members of the Torridon Group were brought to high levels within the complex. This was accompanied by tilting of country rocks and by the formation of the Welshman's Rock and Mullach Ard Faults when masses of sandstone slid off a dome that was forming over the central complex. The deformation and uplift are attributed to ascending acid magma (see below).
4) Ensuing subsidence on the Main Ring Fault led to caldera formation, the intrusion of igneous breccias, the intrusion and extrusion (as ash flows) of rhyodacites, and the formation of breccias and mega-breccias, probably by collapse of caldera walls. These features are preserved in the Northern Marginal and Southern Mountains zones. The igneous breccias contain abundant blocks, in some cases exceeding 100 m in diameter, of coarse gabbro, together with rare peridotite clasts, proving that mafic plutonic intrusions

---

Figure 1. *(facing page) Geological sketch map of Rum. Layered Suite : ELS, Eastern Layered Series; WLS, Western Layered Series (AMM = Ard Mheall Member; HBM = Harris Bay Member); CS, Central Series. Stage 1: WG, Western Granite; NMZ, Northern Marginal Zone; SMZ, Southern Mountains Zone. MAF, Mullach Ard Fault; MRF, Main Ring Fault system; LLF, Long Loch Fault; WRF, Welshman's Rock Fault; λ, Lower Jurassic sediments and lavas of the Palaeocene Eigg Lava Formation. The arrows point to small outcrops amongst the Palaeocene igneous rocks and Precambrian sediments and gneisses of the SMZ. The line joining A-A' marks the position of the gravity profile in Figure 20; the solid line, denoted 4C at its NW end, marks the approximate position of section C, Figure 4.*

were already present at this early stage (Emeleus, 1996).

5) Intrusion of the Western Granite and other granitic rocks followed shortly after 4.

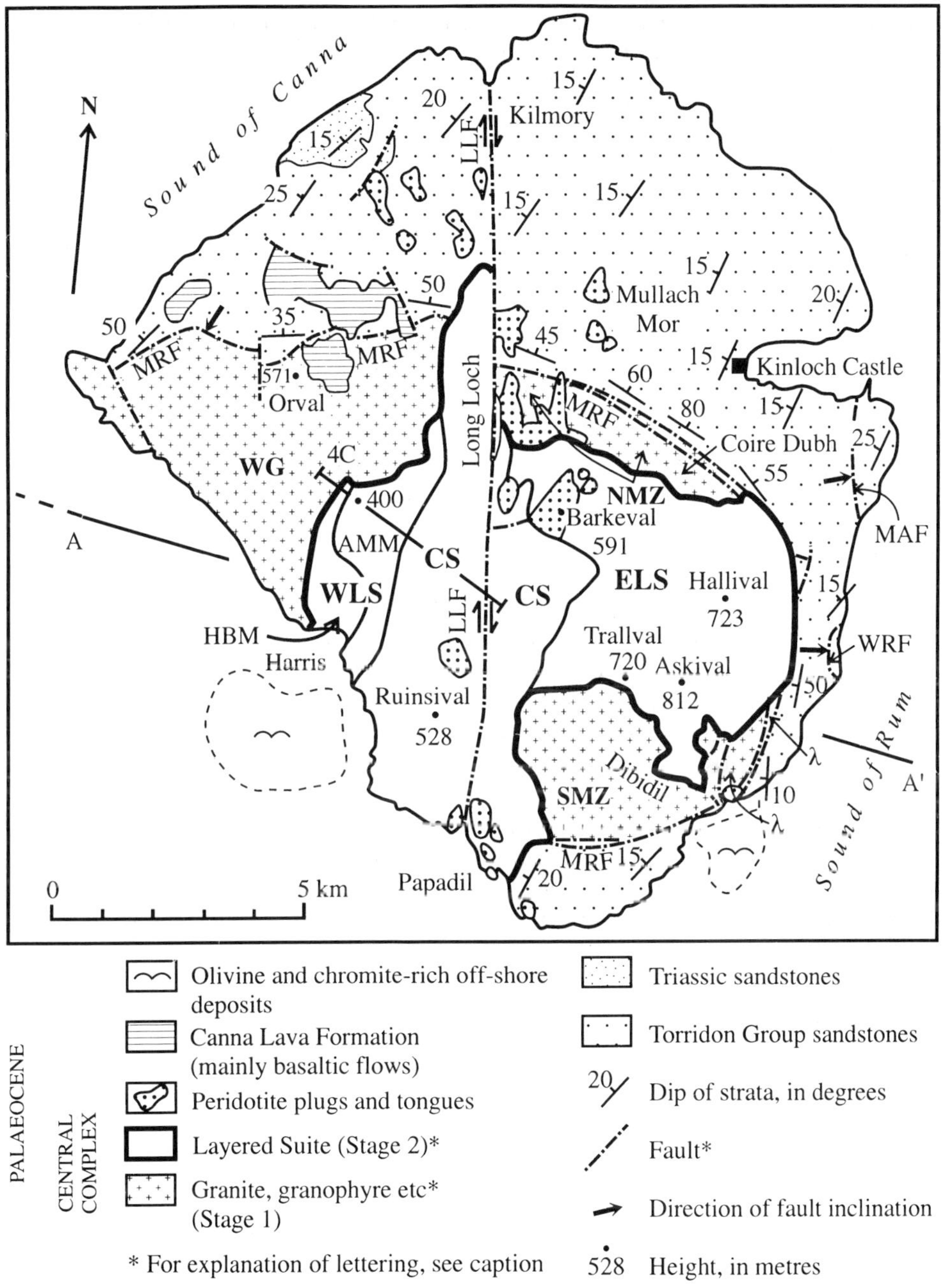

6) Renewed uplift on the Main Ring Fault is recorded in the complex faulting found in the Southern Mountains Zone and the faulted contacts on the north of the Western Granite. (Events 3-6 = Stage 1 of the Central Complex; Emeleus *et al.*, 1985)
7) A change from dominantly silicic to basaltic magmatism was marked by the intrusion of cone-sheets and radial dykes, and NW-trending dykes of the regional dyke swarm. The minor intrusions are predominantly of tholeiitic and transitional to mildly alkaline basalts (Forster, 1980; McClurg, 1982). The cone-sheets focus at a depth of ~2.5 km below central Rum, about 0.5 km southeast of Barkeval (Emeleus, 1996).
8) Near-surface mafic magmatism reached its peak with the emplacement of plutonic basic and ultrabasic rocks that compose the Layered Suite (Eastern and Western Layered Series, and Central Series), together with an array of peridotite, gabbro and dolerite plugs. This involved the intrusion of large volumes of primitive basaltic magma. Rocks of the Layered Suite form the core of the central complex and are excellently exposed over about 35 km^2 (Figures 1, 2, 3, and 6).
(Event 8 = Stage 2 of the Central Complex; Emeleus *et al.*, 1985)
9) Cessation of igneous activity in the central complex was followed by vigorous subaerial erosion of the complex and its surroundings. This was coeval with eruption of basaltic and more evolved flows of the Canna Lava Formation, which were fed from sources outside

Figure 2. *Hallival (south face) with thick, gently west-dipping layered allivalite (pale-weathering cliffs) and peridotite (dark weathering) in the upper part of the Eastern Layered Series. The dark bands in allivalite are subsidiary peridotite layers. This is the type locality for allivalite (Harker, 1908). Scale: the summit is about 220 m higher than foreground.*

Rum. As the lavas flowed into an evolving valley system, they became interbedded with fluviatile conglomerates containing abundant clasts derived from the central complex and its surroundings (Emeleus, 1985).

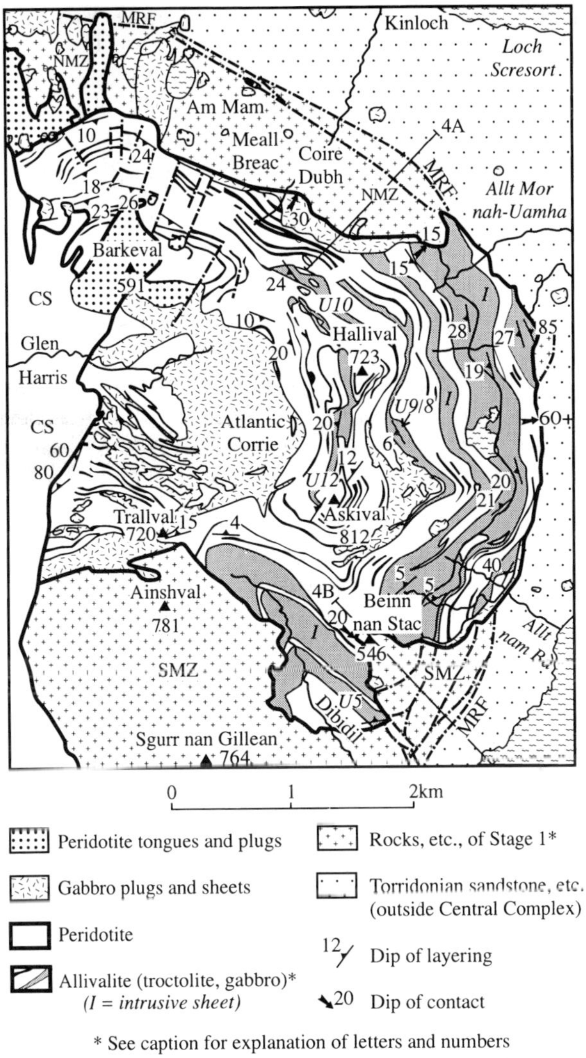

Figure 3. *Sketch map of the Eastern Layered Series. Abbreviations:* I, *intrusive allivalite sheets; U10, U5, etc., layered units; approximate positions of sections A and B, Figure 4, are indicated by solid lines labelled* 4A *and* 4B. *For other abbreviations see Figure 1. (Based in part on published 1:10,000 maps, reproduced by permission of the Director, British Geological Survey: NERC copyright reserved).*

10) A few basalt dykes intrude the Canna Lava Formation. They include the only examples of normally-magnetized Palaeocene rocks from Rum and its surroundings (Mussett *et al.*, 1988, their figure 2).

11) A north-south fault system (the Long Loch Fault) bisects the complex (Figure 1). Dextral movement on this affected the Layered Suite and earlier rocks but the actual date of movement is otherwise poorly constrained.

Regional and local glaciers covered Rum during the Pleistocene and many of the best exposures of the Layered Suite rocks occur in corrie walls and steep hillsides (Figure 2), whereas valley floors are commonly mantled by till and other glacial deposits. Amongst the Recent deposits, the most important in the present context are the olivine- and chromite-rich alluvial sands offshore from Harris and Dibidil (Figure 1; Gallagher *et al.*, 1989).

Rum is the site of a large positive Bouguer gravity anomaly (McQuillin and Tuson, 1963; British Geological Survey, 1994), similar to those over the Skye and Mull central complexes (Bott and Tuson, 1973). The aeromagnetic map of Rum shows a strong negative total force anomaly over the Layered Suite (British Geological Survey, 1994). Palaeomagnetic measurements on the Torridon Group rocks reveal that their primary magnetisation was reset in the vicinity of the central complex (Robinson and McClelland, 1987).

2. RUM LAYERED SUITE

2.1. General structure

The Layered Suite is divisible into three: (i) an Eastern Layered Series (ELS), (ii) a Western Layered Series (WLS) and (iii) a Central Series (CS) (Figure 1). The genetic relationship of these has not been clearly resolved but the current view is that the Eastern and Western Layered Series are lateral equivalents, albeit at slightly different structural levels, and that the Central Series represents a slightly younger structure which developed during successive magma chamber replenishment and withdrawal events (*c.f.* Figure 4C and Figure 21 below).

The margins of the Layered Suite either dip outwards at angles of *ca.* 20°-30° (Figure 4A-C) or are steep, as seen, for example, to the east of Hallival and Askival where the pre-existing Main Ring Fault probably determined the course of the contact with Torridonian rocks. Layering has developed right up to the margins in places but elsewhere the marginal relations are obscure or complicated by intrusion breccias, hybrid rocks and gabbros (Greenwood, 1987).

Within the exposed intrusion, peridotite is the dominant rock type forming ~70% of the outcrop. Allivalite forms <10% and the remainder is composed of gabbro. The term allivalite was originally proposed by Harker (1908, p.71; see Figure 2) for the augite-poor, ultrabasic

---

Figure 4. *(facing page) Cross-sections through the margins of members of the Layered Suite: A. Northeastern margin of the Eastern Layered Series NNE of Hallival; B. Southeastern margin of the Eastern Layered Series at Beinn nan Stac; C. Western Layered and Central Series at Ard Mheall. The approximate positions of A and B are shown in Figure 3 and of C on Figure 1. Symbols: open triangles = ultrabasic breccia; pecked lines = schematic representation of layering; other symbols as on Figures 1 and 2. The allivalite on An Dornabac (Figure 4C; in area C of Figure 6) has probably been detached and downfaulted from the Western Layered Series. (Figures 4A and 4B reproduced by permission of the Director, British Geological Survey: NERC copyright reserved).*

rocks in which plagioclase and olivine are in approximately equal amounts or in which plagioclase predominates. Subsequently, the term has gained wider usage, but in this account it refers to clinopyroxene-poor rocks in which plagioclase (>An_{80}) is modally greater than olivine.

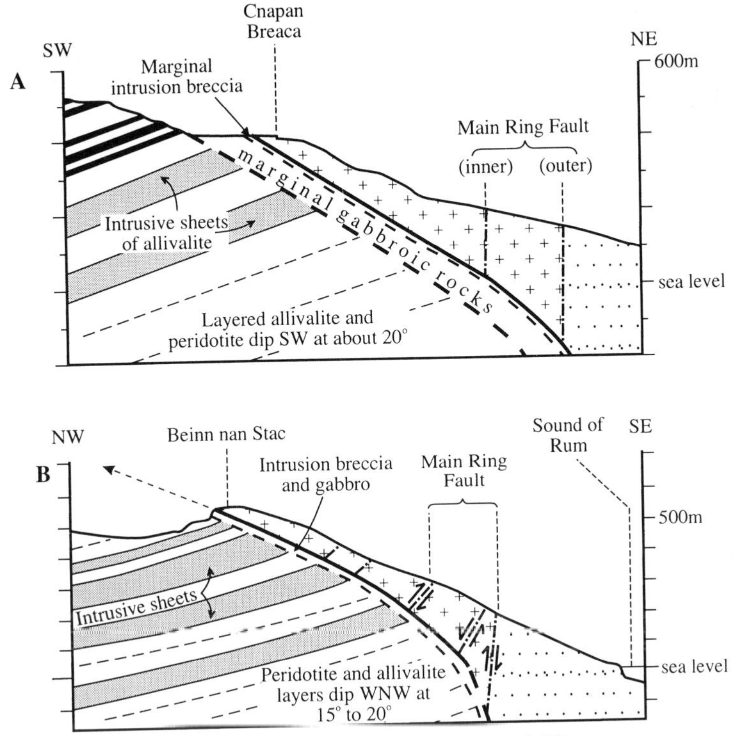

(Vertical and horizontal scales equal on 4A and 4B)

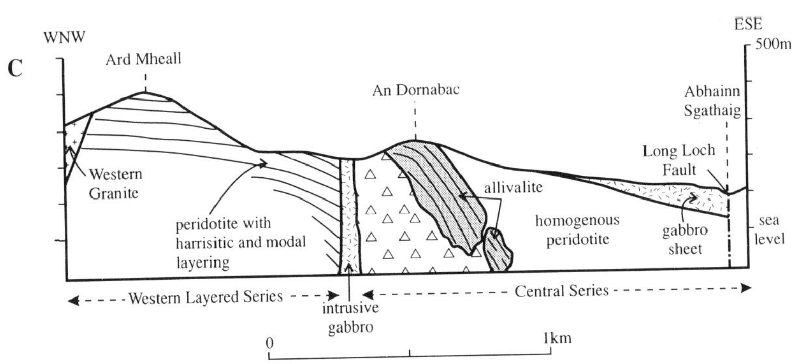

Within the Eastern Layered Series, the exposed sequence has been divided into sixteen macro-rhythmic units, each broadly defined by a feldspathic peridotite base and an allivalitic (or, rarely, gabbroic) top (Figures 3, 5), whose thicknesses range from a few tens of metres to ~100 m. The proportion of peridotite to allivalite ranges from ~95:5 to ~15:85, decreasing from west to east (Volker and Upton, 1990, their figure 5). The macro units attain their maximum thicknesses on the mountains of Askival and Hallival (Figures 2, 3). Layering within the Eastern Layered Series dips generally inwards towards the central focus of the basic-ultrabasic complex in Atlantic Corrie. Dips are typically 10°-15°. However, near the eastern margin they are commonly up to 20° and may locally be as much as 40° (Figure 3). At the western limit of the Eastern Layered Series on Trallval, dips in the allivalite increase westward, towards the Central Series (Figure 3).

The Western Layered Series (Figures 1, 6) consists of two members. The lower, Harris Bay Member, is dominated by layered gabbro (~130 m thick) with well-developed olivine crescumulate textures ('harrisite'; see Figure 9). At the top of the gabbro, there is a smooth compositional change over ~50 m (= the Transitional Series of Wadsworth, 1961) into the 400-m thick layered peridotite of the Ard Mheall Member (Figure 1). There is no *in situ* layered allivalite in the Western Layered Series although detached blocks apparently derived from the Western Layered Series occur within the Central Series (Figure 6, area C). The layering on Ard Meall hill generally dips at 5°-10° to the east and southeast (Figure 6), however, layering is nearly horizontal in gabbros within 2 m of the margin at Harris, and at the eastern limit of the Ard Meall Member, dips increase towards the Central Series, reaching 47° to the north of An Dornabac (Figure 6).

The Central Series comprises a mélange of mega-blocks of layered allivalite some exceeding several 100's m across, surrounded either by heterolithic feldspathic peridotite breccias or relatively massive and homogeneous feldspathic peridotite (see Figures 14, 15). Variable dips for the layering show that the blocks have been variably rotated. In some, the dips approach vertical. Relationships throughout the northern part of the Central Series (Figure 6, area A) suggest that the blocks are disrupted fragments of formerly continuous allivalitic layers with

---

Figure 5. *(facing page) Stratigraphy and comparative scales of layering within the Eastern Layered Series. In A - C, allivalite is white, peridotite is black; in C and D the density of the dots reflects the modal concentration of olivine. A. Macro-rhythmic units (numbers shown on left) of the Eastern Layered Series on Hallival and Askival, based on Brown (1956) and Faithfull (1985). The relative thickness of peridotite increases towards the west (e.g. see Volker and Upton, 1990, their figure 5). These units represent the largest scale mapping stratigraphy. B. Unit 14 allivalite stratigraphy on Hallival. Only subsidiary peridotites >15 cm in thickness are shown. This scale of layering and lithofacies variation is thought to represent the scale of 'event' stratigraphy in the intrusion. C. Lithofacies variation in Unit 14 allivalites. Lithofacies show no systematic ordering and, generally, contacts are diffuse/gradational. Reverse and normal density grading is present. Grain-size differences and sorting are also present, but not illustrated. Some facies are characterized by thin, feldspathic allivalite or anorthositic laminae which are thought to represent deformed (highly attenuated) intraclasts (shown schematically). It should be noted that there is comparable complexity in layering and lithofacies variation in the peridotites as in the allivalites. D. Cm-scale modal variation. Each layer also shows subtle differences in grain size and sorting.*

areal extents of 1.5 x 0.4 km and vertical thickness up to 30 m (see Figure 21). Discrete mappable zones of breccia, up to 400 m in width, extend broadly north-south (Figure 6). In the southern part of the Central Series, on and around Ruinsival mountain, these breccia zones have intrusive relationships with massive peridotite and are radially disposed about a point 1.5 km north of Ruinsival (Volker and Upton, 1990). In the Ruinsival area there is evidence of more than one generation of breccia zones.

Throughout the central part of the intrusion and in the country rocks to the north, steep-sided, intrusive plugs up to a kilometre in diameter involve both massive (homogeneous) and layered peridotite (Wadsworth, 1994). In addition, elongate protrusions (tongues) of peridotite

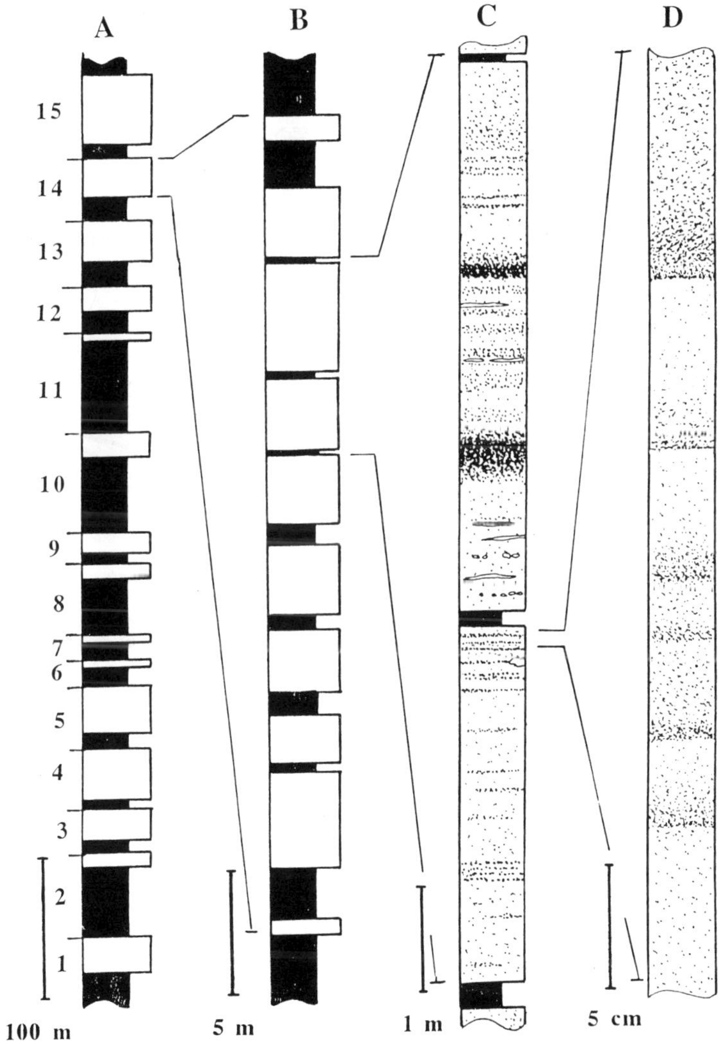

extending northwards from the main complex transgress the Northern Marginal Zone (Figures 1, and 6).

A wide variety of isotropic, layered and heterogeneous gabbros form plugs and semi-concordant sheets within the Eastern Layered Series and the Central Series and are particularly abundant in Atlantic Corrie (Figure 3). Also, modally variable gabbros are an important element of the marginal rocks surrounding the layered cumulates, where their form is less easy to define.

2.2. Lithologies and layering structures

The peridotites typically comprise 60-90% cumulus olivine, poikilitically enclosed by plagioclase and subordinate clinopyroxene. Although some are dunitic (*s.s.*), most are more properly called feldspathic peridotite. Amongst the rocks mapped as allivalites, troctolite (colour index 20-50) is the dominant lithology, although olivine gabbro and gabbro are present in parts of Units 1, 3, 8, 9, 10, and in Unit 14 on Trallval. Olivine gabbros also occur in the upper parts

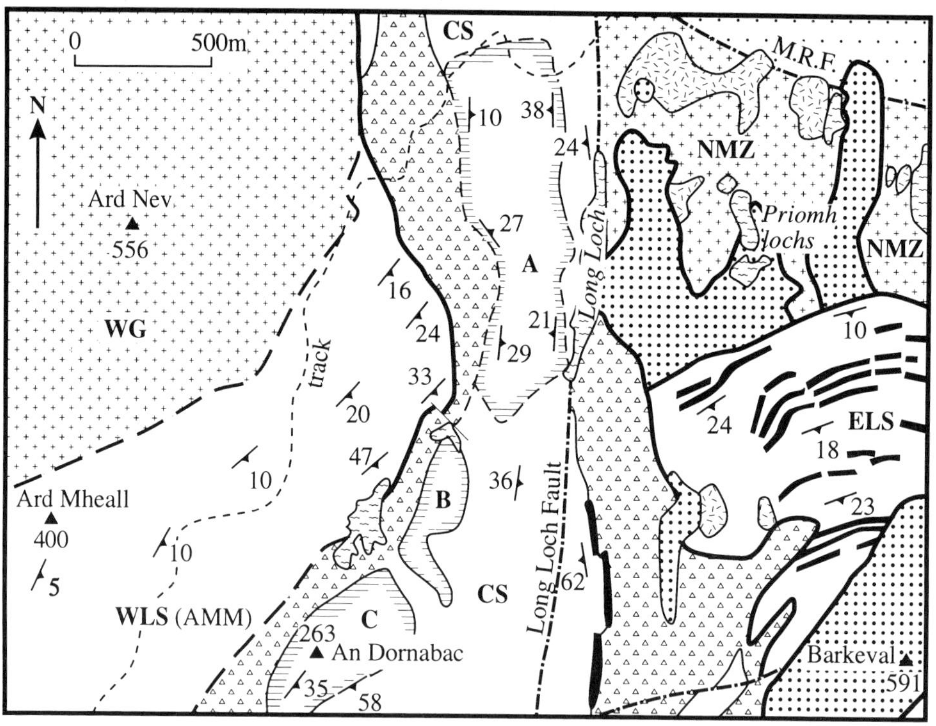

Figure 6. Sketch map of the northern part of the Central Series with the adjoining parts of the Eastern and Western Layered Series. Open triangles = areas of abundant ultrabasic breccia. A, B, and C are areas of disrupted, coherent to semi-coherent, layered rocks (broadly allivalitic) within the northern part of the Central Series. Other symbols and abbreviations as for Figure 1. (Based in part on published 1:10,000 maps, reproduced by permission of the Director, British Geological Survey: NERC copyright reserved).

of Units 4 and 5, where they are interpreted to be either intrusive sheets (Brown, 1956; Emeleus, 1996) or the upper parts of the units (Faithfull, 1985).

The suite is layered on all scales. On the largest scale (Figures 2, 3, 5A) are the macro-rhythmic peridotite-allivalite units which define the general mappable stratigraphy. However, within these units, there is considerable internal variation. The allivalites contain concordant thin layers and lenses of peridotite which range in thickness from a few centimetres to several metres, particularly in Units 12-15. The peridotites and allivalites are internally layered on scales of centimetres to decimetres (Figures 7, and 8) and show considerable lithofacies variation.

Throughout the suite, the layered peridotites (*s.l.*) consist of concordant harrisitic peridotite interlayered with stratified, modally and grain-size graded peridotite. Spectacular development of harrisite occurs in the Western Layered Series (Figure 9) where there are layers of parallel, 'branching' olivine growths up to 12 m thick. Where harrisitic texture is abundant, as for example 600 m ENE of Ard Mheall, some of the component crescumulate olivines have been broken and the fragments re-worked as single crystal clasts and as crystal clusters, and form lags within the grain-size graded interlayers. Within the Eastern Layered Series, the harrisite is often finer-grained and its crescumulate nature only evident in thin section. Stratified peridotite interlayered with the harrisites commonly has bases rich in Cr-spinel. Mineral lineation and lamination are abundant, particularly where re-worked harrisitic olivine is involved.

Figure 7. *Fine-scale layering in peridotite, Western Layered Series, north of Loch An Dornabac. White book = 20 cm high.*

Figure 8. *Small-scale layering in allivalite, Eastern Layered Series, Unit 13, Hallival. The layering is deformed above a small anorthositic intraclast (bottom right); note a small reverse fault. Ruler = 25 cm.*

Grain-size grading may be normal or reverse. Some layers have reversely grain-size and density (modally) graded bases with normally graded upper portions.

The allivalites of the Eastern Layered Series show various lithofacies. Most of the stratification is sub-parallel; layer boundaries may be diffuse or sharp and layer thicknesses vary from centimetres to tens of metres (Figure 8). Bases of individual layers may be slumped (Brown, 1956, his figure 34); layers may be massive and unstratified or show diffuse to strong cm-dm scale stratification defined by differences in mode and grain size (Figure 8). Density and grain-size grading are common (Figures 10, and 11). There appears to be no systematic vertical ordering of the lithofacies and most are continuous along strike for tens to hundreds of metres. Mineral lamination is widespread, particularly with respect to the tabular plagioclase; olivines and the relatively uncommon cumulus pyroxenes can show linear fabrics. Feldspathic laminae, typically of mm-cm thickness and cm-dm in length, are locally common and are interpreted as feldspathic cumulate clasts that were flattened and deformed during deposition and consolidation of the series (Figure 5C).

Figure 9. *Harrisitic peridotite, Central Series, Ard Mheall. Dark olivine crystals, outlined by white, interstitial plagioclase, have grown upwards from a substrate of equigranular olivine. The irregular, upper surface of the harrisite is covered by normal peridotite. Coin = 3 cm.*

Figure 10. *Density-graded layering, Central Series, west of Long Loch (Area "A", Figure 6). Density-graded layers (olivine-rich bases to feldspar-rich tops) contain angular clasts of anorthositic-allivalite. One layer shows slumping and deformation. Two basalt sheets cut the layered rocks. Hammer head = 15 cm.*

Modally- and grain-size-graded layers are well exposed in the Long Loch Member of the Central Series (Figure 10). Layers are of dm-m thickness, with peridotitic bases grading upwards into allivalite, which itself may be modally layered with distinct mineral lamination. Low-angle cross-stratification is present locally. Feldspathic intraclasts are an important element of this sequence (Figure 10) and range from angular to highly attenuated, flattened laminae (*c.f.* Figure 5C). These intraclasts form lags and can show imbrication. These layered rocks grade laterally over a few hundred metres into heterogeneous slumped facies. A 15 m-thick, poorly-stratified layer of feldspathic peridotite caps the sequence; it has coarse lags in basal scours, containing peridotite and allivalite clasts up to 10 cm in size. These confer a conglomeratic appearance to the feldspathic peridotite layer. Elsewhere in the Long Loch Member, blocks of peridotite and allivalite up to a metre diameter have caused deformation of underlying cumulates in the manner of sedimentary drop-stones (Emeleus, 1987, his figure 9).

Figure 11. *Density-graded layering (olivine-rich bases to feldspar-rich tops) in Unit 15 allivalite, Eastern Layered Series, Askival summit area. Hammer shaft = 40 cm.*

2.3. Deformation and replacement structures.

Evidence reminiscent of soft-sediment deformation is abundant throughout the layered sequence (Brown, 1956; Wadsworth, 1961). This occurs on all scales from localized loading to massive slump sheets and mélanges. Load structures are common where peridotite overlies allivalite. These are sometimes detached and sheared out against underlying layers, while the associated flame structures may also be attenuated and sheared into the overlying layer (Figure 12). Often, there is evidence for simultaneous loading and shearing such that the upper portion of the loaded layer is a smeared-out mixture of mafic laminae, recumbent fold limbs and heterogeneous allivalite.

Figure 12. *Attenuated, deformed allivalitic flame-structures in the base of peridotite overlying layered allivalite. Central Series west of Long Loch (Area "A", Figure 6). Hammer shaft = 40 cm.*

Evidence for slumping of layers is widespread but is most obvious where there are contrasted lithologies (*e.g.* peridotite-allivalite). However, thick allivalite layers show widespread evidence of slumping. Slumping and loading commonly appear to have occurred simultaneously. The Unit 7 allivalite is a 2-5 m thick sheet of slumped allivalite/peridotite throughout its exposure of several km lateral extent (Figure 13).

Locally, peridotite layers can be demonstrated to have intrusive relationships with semi-coherent allivalite. At the top of Unit 14 from Hallival to Trallval, peridotite has extensively intruded and become inter-folded with allivalite which was clearly a semi-coherent mush at the time (Renner and Palacz, 1987, fig. 3; Volker and Upton, 1990). Elsewhere, continuous, grain-size graded layers of peridotite can be seen to have intruded and plastically deformed sequences of coherent stratigraphy (*e.g.*, in

Figure 13. *Comprehensive slumping in the allivalite of Unit 7 on Barkeval. The hand on the left gives the scale.*

Figure 14. *Ultrabasic breccia, Central Series (Figure 6, area B). Coarse breccia of angular blocks of peridotite and layered allivalite occur in a (feldspathic) peridotite matrix. The breccia overlies peridotite with weak layering. Hammer shaft = 35 cm.*

the Central Series).

Much of the Central Series is characterised by thick mélanges of blocks of peridotite and layered allivalite in a feldspathic peridotite matrix which commonly is highly deformed (Figure 14; McClurg, 1982; Emeleus, 1987, fig. 11). The blocks show all stages of disaggregation, *in situ* melting and plastic deformation. Clast lithologies in the breccias range from peridotite to allivalite (Figure 15); the hosts are highly variable in composition, both on the outcrop scale and more generally.

Replacement structures are widespread. Generally, these represent replacement of allivalite by peridotite and take the form of finger structures penetrating from the tops of peridotite layers into overlying allivalite (Figure 16; Butcher *et al.*, 1985; Morse *et al*, 1987). The wave-length and amplitude of the finger structures vary. The largest observed are up to a metre in amplitude; more typically they are of cm-dm scale. Generally, these fingers cut across (but do not disturb) any layering or fabric within allivalite and are perpendicular to layer contacts, even where these have significant dip. Pipes of peridotite, up to a metre in diameter and several metres in height, have been described from allivalites on Trallval where they are discordant to allivalite layering (Volker, 1983; Volker and Upton, 1990).

Figure 15. Closer view of breccia, Central Series. Angular to sub-angular blocks of layered allivalite, feldspathic peridotite and peridotite in a feldspathic peridotite matrix. Hammer head = 15 cm.

Figure 16. Finger structures in peridotite of the Central Series, northwest of Long Loch. Fingers project upwards from peridotite into weakly-layered feldspathic peridotite. Hammer shaft = 40 cm.

Throughout the Layered Suite, cross-cutting, intrusive relationships are present. On the largest scale are the gabbro and peridotite plugs and sheets (Emeleus, 1994). Additionally, smaller bodies of homogenous peridotite commonly cross-cut, intrude, thermally erode and replace layered allivalite (Figure 17). Intrusive breccias, as distinct from slump breccias, are an important element of the

Central Series. At the smallest scale, mm-dm-thick veins of gabbro are widespread throughout the Central Series and within the peridotites of the Eastern Layered Series. Thin veins of feldspathic peridotite are also abundant which may be concordant or transgressive to host stratification (Butcher, 1985).

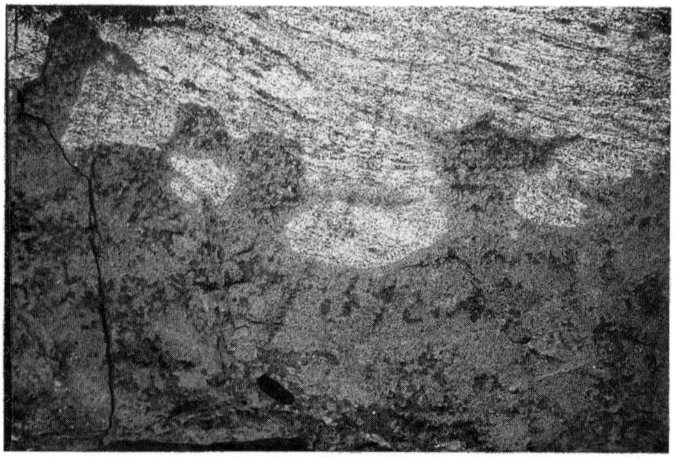

__Figure 17.__ Replacement structures in Unit 7, Eastern Layered Series, Barkeval. Allivalite has been replaced by irregular, lobate protrusions of peridotite from the top of the underlying peridotite sheet. Fine-scale layering with flat-lying folds in the allivalite is undisturbed by the peridotite. Lens cap at base = 5 cm.

2.4. Textures

Most of the layered rocks and intrusive peridotites exhibit classic cumulate textures (Wager *et al.*, 1960). The average grain sizes is in the range 1-3 mm, although size-grading is a common feature of layering. Linear or laminar fabrics are common. The majority of the rocks would be described as adcumulates or heteradcumulates. They are granular to poikilitic, and most show a significant approach to local textural equilibrium.

Feldspathic peridotites have subhedral to rounded (equigranular) olivines cemented by poikilitic feldspar and subordinate clinopyroxene. Equant crystals of Cr-spinel may occur in isomodal layers, as modally-graded layers at the bases of peridotites, or dispersed through the rock. Olivine-rich peridotite (dunite) typically has an equilibrated equigranular texture (Figure 18A) although a subgrain fabric may be present in larger grains. As the modal proportion of olivine decreases to ~50%, plagioclase starts to become sub-poikilitic to equigranular.

Textures in allivalites and anorthosites are commonly equigranular and equilibrated (Figure 18B). Most olivine and clinopyroxene is equigranular but can be poikilitic. Cr-spinel when present is equidimensional. Tabular feldspar may show near-perfect planar lamination with com-

__Figure 18.__ (facing page and overleaf) Photomicrographs of layered rocks from Hallival, Eastern Layered Series. A. Typical granular peridotite with moderate grain boundary equilibration; Unit 10 (scale bar = 1 mm). B. Granular anorthosite showing a high degree of textural equilibration; Unit 14 (scale bar = 1 mm). C. Laminated troctolite with feldspar laths draped around olivine oikocrysts during simultaneous growth and compaction; Unit 10 (scale bar = 0.5 mm). D. Laminated troctolite in which olivines show deformation bands associated with compaction; Unit 10 (scale bar = 1 mm).

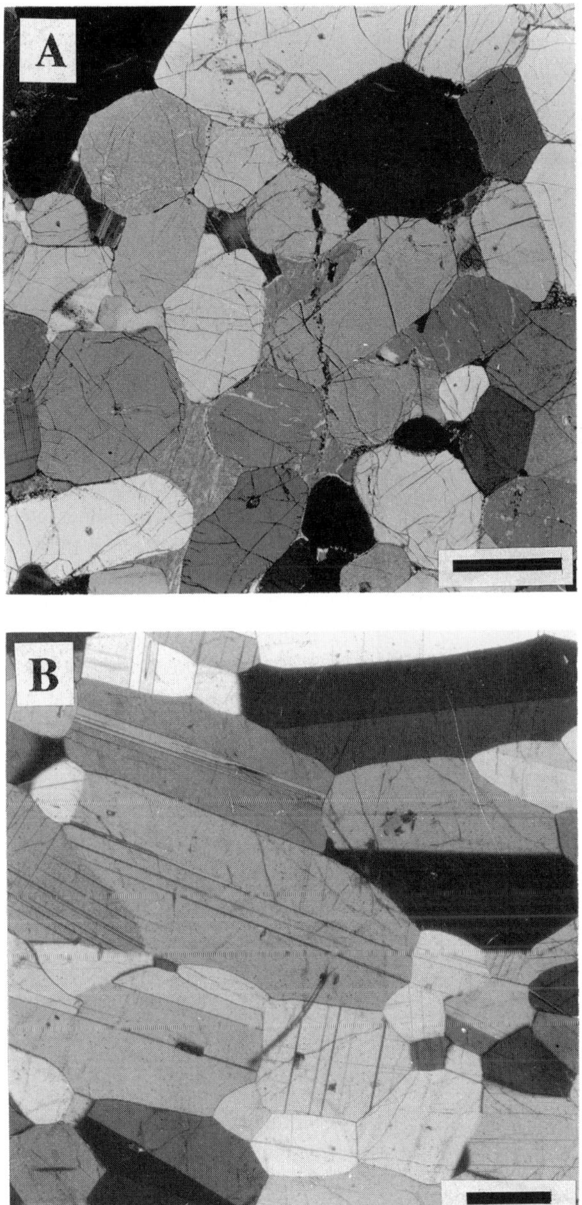

pactional draping around olivine and/or clinopyroxene oikocrysts (Figure 18C). Strained olivine occurs in the allivalites (Figure 18D). Typically the allivalites contain 60-80% cumulus plagioclase. Pure anorthosite layers are common throughout the allivalites, particularly in higher units on Hallival. Wager *et al.* (1960) cited these rocks as type examples of adcumulates.

Crescumulate harrisite is characterized by highly tabular olivine with well-developed parallel growth structures. They can be over 30 cm in length and up to 2 cm in thickness. Interstitial plagioclase feldspar and subordinate clinopyroxene are prominent and phlogopite and amphibole are minor interstitial components. The plagioclase is commonly zoned.

Most of the intrusive olivine gabbros of the sheets and plugs have ophitic textures. The marginal gabbroic rocks have spherulitic intergrowths of plagioclase and clinopyroxene and show a considerable range in grain size even within a thin section. The spherulitic intergrowths are thought to have formed at high degrees of supercooling (Greenwood, 1987). The marginal rocks may contain orthopyroxene, especially where they are cut by rheomorphic acid veins.

2.5. Mineral compositions

There have been numerous individual studies of mineral compositions from the Layered Suite (summarized in Emeleus, 1987, and additionally Renner and Palacz, 1987) but no de-

tailed synthesis for the whole suite has yet been attempted. This is largely because the cryptic variation is complicated, varies from unit to unit and, to a lesser extent, within units. Further, despite the wealth of data available, gaps in the coverage hamper attempts the understand the detailed origin of the variation across the intrusion as a whole.

Olivines from the suite are largely unzoned and have a compositional range of Fo_{92-70}. In general, the allivalites consistently contain olivines with lower Fo contents than those of the underlying peridotites. On a smaller scale, another general feature of the cryptic variation is the compositional 'reversal' that commonly occurs at the tops and bottoms of units. These reversals are generally offset from discordances in the modal layering and from discordances in Ni contents of the olivines, attesting to an origin by the interaction of fluids from cumulate mushes above and below (Tait, 1985). Overall cryptic variation, however, varies dramatically from one unit to another. For example, Unit 14 shows a limited cryptic variation with height ($Fo_{86.5-82}$) (Renner and Palacz, 1987), whereas Units 3 and 4 together show a systematic variation from bottom to top of Fo_{83}-Fo_{71} (Faithfull, 1985). Olivine-rich sands dredged offshore from Harris (Figure 1), containing olivine eroded from the Layered Suite, provide an averaged sample of the compositional variation exhibited by a large area of the intrusion. The olivines have a compositional range of Fo_{92}-Fo_{80} with a mean composition of ~Fo_{87} (Gallagher *et al.*, 1989). Olivines less magnesian than Fo_{80} are

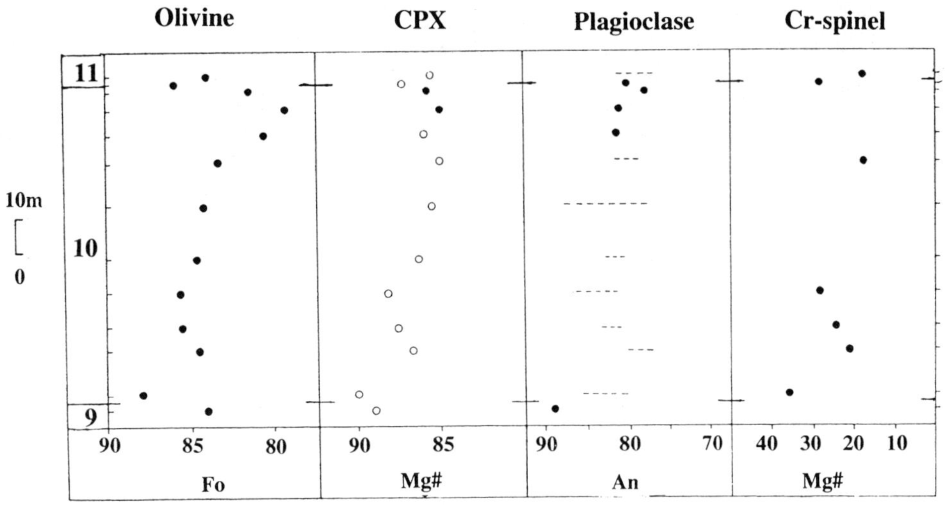

Figure 19. *Cryptic variation in Unit 10 and parts of adjacent units, Eastern Layered Series. Filled circles = averaged analyses of cumulus crystals; open circles = averaged analyses of intercumulus pyroxene; dashed lines = apparent range of intercumulus plagioclase compositions; Mg# = 100Mg/(Mg+Fe); levels at which samples were taken are indicated by ticks at the margins. After Dunham and Wadsworth (1978, their figure 1). Obvious marginal zoning is excluded in averages of pyroxene and plagioclase.*

absent from this sample, suggesting that the least magnesian olivines ($<Fo_{80}$) found in, for example, Units 1, 3, 4, 5 and 10 (Faithfull, 1985; Tait, 1985) were not sampled by the source of these sands.

Clinopyroxene is diopsidic with a restricted compositional range in the layered rocks ($Mg_{46}Fe_5Ca_{49}$ - $Mg_{42}Fe_{18}Ca_{40}$). Strong compositional zoning is absent, but marginal zonation has been reported by Faithfull (1985). Tait (1985), Faithfull (1986) and Renner and Palacz (1987) all found good correlation between the Mg# of clinopyroxene and the Fo content of olivine (see also Figure 19). The fact that olivine would have crystallized before pyroxene in the peridotites and troctolites supports the suggestion that olivine and clinopyroxene were re-equilibrated by post-cumulus processes.

Plagioclase core compositions range between An_{90}-An_{79}, but both cumulus and intercumulus plagioclase are complexly zoned to An_{50}. Reverse zonation is as common as normal zonation (McClurg, 1982; Volker, 1983), but unfortunately there has been little detailed study of the plagioclase zonation.

Chromite is concentrated in thin chromitite seams (generally less that 10 mm thick), at peridotite-allivalite boundaries (*e.g.* Brown, 1956; Young, 1984), as discontinuous thin seams within peridotites between harrisitic and non-harrisitic layers (Wadsworth, 1961) and as dispersed crystals within the peridotites. Compositions are very variable, especially in the chromitite seams (Henderson, 1975; Henderson and Suddaby, 1971; Dunham and Wilkinson, 1985) with MgO ranging from 5-15%, Al_2O_3 from 10-40%, and Cr_2O_3 from 21-45%.

Chromite from the offshore dredged sands (Gallagher *et al.*, 1989) has a range of Cr_2O_3 from 23-45 wt.% with a mean of 31-33 wt.% and a marked negative correlation between Mg and Cr contents. Several ideas for the origin of the seams have been suggested. These include preferential settling of chromite grains (Brown, 1956; Wadsworth, 1961), and that they are the product of mixing and reaction of newly emplaced picritic magma with either remnant magma in the chamber (Dunham and Wilkinson, 1985) and/or with floor melt (Huppert and Sparks, 1980), or with upward-moving interstitial liquid from the underlying cumulate mush (Young, 1984).

Minor amounts of other silicate minerals (intercumulus orthopyroxene, kaersutite, phlogopite) are found in rocks from the Layered Suite. The hydrous phases are especially present in the well-developed harrisites of the Western Layered Series supporting the suggestion that most harrisites grew as a consequence of supersaturation due to changing water content of the magma (Donaldson, 1974). Late-stage teschenite veins and segregations in peridotites of the Central Series contain sphene, epidote, analcime, zircon and chlorite (Kitchen, 1985). Various sulphides (for example, chalcopyrite and pyrrhotite) are conspicuous in the peridotites of Units 1-6 (Faithfull, 1986), trace amounts of platinum-group elements occur in sulphides low in this part of the succession (Hulbert *et al.*, 1992) and electrum (46.4% Au, 49.9% Ag) is present in sulphide-rich droplets above a chromitite band at the Unit 11/12 junction north of Hallival (Dunham and Wilkinson, 1985).

Figure 19 shows the cryptic variation exhibited by olivine, clinopyroxene, plagioclase and Cr-spinel from the type unit, Unit 10, as recorded by Dunham and Wadsworth (1978). These profiles are shown as an example of the cryptic variation exhibited by a unit, but they are not necessarily typical of the variation shown by other units. Indeed, these profiles are subtly different to profiles from elsewhere in Unit 10 (Tait, 1985; Butcher, 1985).

The lack of compositional zonation in the olivines and pyroxenes, together with the constancy of composition sometimes shown within individual unit peridotites, supports arguments that the olivines (and clinopyroxenes) grew by equilibrium crystallization and/or have re-equilibrated during the formation of the units. Exceptions include rare Fo_{92} olivines (Gallagher *et al.*, 1989), which probably represent primary olivines which crystallized from parental magmas (see later) in rapidly cooled dykes. The presence of both mineral compositional and textural re-equilibration was probably caused by long cooling times and re-equilibration with migrating fluids. The lack of extreme mineral compositional variation is largely a consequence of the constant replenishments of the Rum magma chamber by picritic magmas. The overall complexity shown by cryptic variation between and within the different units reflects the complex interplay of the various primary and post-cumulus processes responsible for their formation and illustrates the danger of assuming that all the units formed in the same way. The apparent ability of olivine and clinopyroxene major-element chemistry to re-equilibrate reduces their usefulness as indicators of the primary processes responsible for the formation of the units. More studies of plagioclase zonation and minor-element olivine and pyroxene chemistry (for example nickel contents; Tait, 1985; Renner and Palacz, 1987) are required to further elucidate the origin of the Layered Suite.

2.6. Isotope Analyses

Studies of radiogenic isotopes have concentrated on Units 8 - 15 in the Eastern Layered Intrusion on Hallival (Palacz, 1984; 1985; Palacz and Tait, 1985; Renner and Palacz, 1987). In the lower part of the Unit 10 peridotite, $^{87}Sr/^{86}Sr$ varies from 0.7036 to 0.7043 and $^{143}Nd/^{144}Nd$

is 0.51281 (Initial values at ~ 60 Ma). Progressive changes occur through the overlying feldspathic peridotites ($^{87}Sr/^{86}Sr$ between 0.7049 and 0.7053 and $^{143}Nd/^{144}Nd$ from 0.51271 to 0.51253) to the allivalite which has $^{87}Sr/^{86}Sr$ ~0.706 and $^{143}Nd/^{144}Nd$ from 0.51249 to 0.5123. These and similar data have been interpreted to show that the Eastern Layered Intrusion was built up from batches of uncontaminated picritic magma (from which peridotite crystallized) injected into a magma chamber containing crustally-contaminated and relatively evolved basaltic magma (from which the allivalite and some gabbro crystallized). The Sr, Nd and Pb isotopic data suggest that the contaminant was about 7% of upper crustal amphibolite-facies Lewisian gneiss, rather than granulite-facies gneiss, as in the Skye Main Lava Series (Dickin, 1981), or Torridonian sandstone (Palacz, 1985). Profiles of isotopic variation through the layered units (Palacz, 1985, his figure 3; Palacz and Tait, 1985, their figure 3; Renner and Palacz, 1987, their figure 2) show major isotopic changes across unit boundaries, and discontinuities and reversals within units that commonly, but not invariably, mirror lithological variations. These perturbations were attributed to the varying affects of new influxes of primitive magma, mixing of primitive magma with resident contaminated magma, and migrating, contaminated, residual magma reacting with the cumulates. In a study of Unit 14, Renner and Palacz (1987) identified a number of abrupt decreases in $^{87}Sr/^{86}Sr$ within the allivalite, which they attributed to fresh influxes of uncontaminated basaltic magma. A similar influx could explain the low $^{87}Sr/^{86}Sr$ values of the gabbros overlying troctolites in Unit 9 (Palacz, 1984; $^{87}Sr/^{86}Sr$ ~0.704 compared with 0.7050 - 0.7055 in the troctolite). Greenwood (1987) found that Sr and Pb isotopes in the problematical marginal gabbro of the Layered Suite supported the view, gained from field and elemental geochemical studies, that the marginal gabbros were hybrid rocks formed by the contamination of peridotite and allivalite by anatectic melts from the country rocks, including the Torridonian arkosic sandstones.

From oxygen isotope studies, it has been shown that heated meteoric waters have reacted to varying degrees with the Layered Suite and its surroundings, producing a large range of $\delta^{18}O$ values (-6 to +10.7‰) (Forster and Harmon, 1983; Greenwood *et al.*, 1992). Greenwood *et al.* (1992) considered that the isotopic evidence identified the contact of the Layered Suite as the principal zone along which hot meteoric fluids flowed as the pluton cooled. This zone coincides with considerable hydrothermal alteration of the marginal gabbros.

3. PETROGENESIS

3.1. Development of concepts

Harker provided the first detailed maps and comprehensive account of Rum (Harker, 1903; 1908). He suggested that the layered peridotitic and gabbroic rocks were generated by separate sheet-like injections of already differentiated olivine-rich and feldspar-rich magma, which built the succession from the roof downwards (Harker, 1908, pp.74-75). Bailey (1945; 1956) recognized that much of the Rum Central Complex was bounded by an arcuate fault system, the Main Ring Fault, and he demonstrated that the Lewisian gneiss (recognized by Tilley, 1944) and basal Torridonian sandstone only occurred within the ring fault, proving central uplift *of the Precambrian rocks* by as much as two kilometres. Bailey (1945) thought it unlikely that the layered ultrabasic rocks had originated by crystal fractionation and sedimentation, as had recently been proposed for the Skaergaard Intrusion (Wager and Deer, 1939). Wager and Brown (1951) however, discerned a rhythmic layering in the ultrabasic rocks and found similarities with layering in the Skaergaard. They suggested that the Rum rocks also had accumulated from the bottom upwards.

Brown's (1956) study of Hallival and Askival proved a landmark in understanding the geology of Rum as well as the processes producing igneous layering. He established a stratigraphy for the layered rocks, in which fifteen of the sixteen major major units (the 'Eastern Layered Series') were recognized. Brown demonstrated that, compared with Skaergaard, there was negligible cryptic variation through the >700 m of layered rocks. He concluded that the layered rocks had built from the bottom upwards through fractionation of successive batches of new magma, each of which initially precipitated magnesian olivine crystals ($\pm$ chromite), succeeded by olivine + calcic plagioclase (and sometimes diopsidic clinopyroxene), and finally by calcic plagioclase alone. Petrographic analysis combined with mineral compositional data (largely based on the optical properties of the minerals) allowed discrimination between early formed, high-temperature mineral phases and those that had crystallized from trapped liquid (the 'cumulus' and 'intercumulus' minerals of later studies, *e.g.* Wager *et al.*, 1960). In the absence of chilled margins or a convincing marginal border group, estimation of the composition of the Rum parental magma proved difficult. Using the compositions and proportions of intercumulus minerals in the first attempt to model the parental magma in the absence of a chilled marginal facies, Brown deduced that the layered rocks formed from the accumulation of high-temperature phases precipitated by aluminous tholeiitic basalt magma (Brown, 1956, his table 7), the residual magma being extruded during contemporaneous volcanism. Thus, in contrast to Skaergaard, the Rum Complex was envisaged to have functioned as an open magmatic system, episodically replenished from a deeper source and venting to the surface. The smaller-scale (cm to mm) layering (*e.g.* Figure 8) was attributed to crystal winnowing by gentle magmatic currents. Few examples of gravity-stratified layering were noted and strongly erosive features, such as cross-bedding or trough-banding are virtually absent. Localized, complex folding of the layering (*c.f.* Figure 13) was compared with slump folds in sediments and the suggestion made that slumping had occurred in unconsolidated cumulates on gentle slopes (< 5°). The thickness of slumped layers (rarely >4 m) indicated the maximum order of the thickness of crystal mush on the magma chamber floor. In the absence of a marginal chilled facies or border group, it was suggested that the Rum layered rocks had not crystallized at their present level, but had consolidated at depth and been been elevated as a solid, piston-like block, lubricated by basaltic magma which crystallized to form the marginal gabbro.

Wadsworth (1961) recognized several layered units in southwest Rum and deduced that they probably represented a lower stratigraphic level than the Eastern Layered Series. The distinctive harrisitic textures in these rocks were interpreted to have formed when tranquil magmatic conditions allowed upward growth of olivine from the magma chamber floor, whereas peridotite breccias formed when unstable cumulates collapsed down fault scarps in the magma chamber. Both Brown and Wadsworth emphasized the distinctive textures of rocks formed by bottom accumulation of crystals and crystallization of the trapped magma; from these and other studies evolved the concepts and terminology of igneous cumulates (Wager *et al.*, 1960). These researches, which are described and discussed in some detail in "Layered Igneous Rocks" (Wager and Brown, 1968, p. 46-97), provided the basic framework for numerous subsequent studies of the Rum layered rocks and, despite modifications and refinements, many of the original interpretations remain fundamental to our understanding of this complex, and of processes in layered intrusions in general.

The silicic rocks of the Northern Marginal Zone and Southern Mountains Zone (Figure 1; Hughes 1960; Dunham 1968; Smith 1985) include both shallow crustal intrusions and extrusive sequences (Williams, 1985, Emeleus *et al.*, 1985). Since they and other members of the Phase 1 activity are intruded by the Layered Suite, and in the absence of significant down-faulting within the Main Ring Fault *after* emplacement of the acid rocks, it follows that the Layered Suite must have been intruded close to the contemporary land surface. Greenwood (1987) confirmed partial melting of the varied country rocks with the generation of rheomorphic acid melts and deduced that the Eastern Layered Series marginal gabbros were hybrids between these melts and the Layered Suite magmas, rather than crystallized from a separate intrusion. These observations reinforced the view that the layered rocks had formed *in situ* rather than being forced upwards as a solid block (Emeleus, 1987; Young *et al.*, 1988). Further supporting evidence came from (i) the presence of undisturbed layered structures to within a few metres of the contacts, (ii) structures in the marginal zone at Harris that were clearly controlled by the near-vertical contact with the Western Granite, and (iii) the rare examples of chilled marginal rocks at Harris and Beinn nan Stac (Greenwood, 1987; Greenwood *et al.*, 1990).

The chilled marginal rocks provided evidence that picritic magma had been involved in formation of the suite (Greenwood *et al.*, 1990). Support for this concept stems from the fact that olivines range to compositions as magnesian as Fo_{92} and also from observations that textures and whole-rock and mineral compositions of some late-stage dykelets cutting the Eastern Layered Series demonstrate that highly magnesian melts did attain high levels in the crust (McClurg, 1982). These observations, together with Gibb's (1976) proposal for a 'eucritic' magma with suspended olivine and Donaldson's (1975) suggestion that hydrous feldspathic peridotite magma had been involved in formation of ultrabasic breccias, started the move away from postulating a basaltic parent (*e.g.* Brown, 1956) to parental magmas of more magnesian compositions (see below).

Mapping by McClurg (1982) and Volker (1983; Volker and Upton 1990) extended Brown's Eastern Series stratigraphy westward, adding much detail about the small-scale layering and providing abundant compositional data on the layered rocks and their minerals. McClurg recognized that a wide, north-south-trending swathe of ultrabasic breccias and layered rocks occurs near the Long Loch and Volker's mapping extended this to the south coast. These rocks, which were interpreted to intrude both the Eastern and Western Layered Series, were termed the Central Series (Figures 1 and 6). Butcher (1984) and Faithfull (1985) added considerable detail to Brown's original mapping of the Eastern Layered Series. They also showed that late-stage gabbroic veins, derived from fractionated intercumulus liquids, metasomatically altered their surroundings (Butcher, 1985) and that the peridotite finger structures (Figure 16) in allivalite overlying peridotite were of replacement origin, caused by pore magma migrating up from the crystallizing peridotite (Butcher *et al.*, 1985; Robins, 1982; Morse *et al.*, 1987). The complexity of peridotite-allivalite relationships within units and their variability along strike had been noted in Unit 14 (Maaløe, 1978) and was subsequently emphasized by other workers (*e.g.* Faithfull, 1985).

Most of the research on Rum in the last 25 years has concentrated on the Eastern Layered Series. A high proportion of the studies were on the well-exposed units high in the succession, with the emphasis on their mineral, bulk rock, and isotope compositions. The normal cryptic variation in Unit 10, and discrepancies between cryptic and phase layering across the base and top of the unit were discovered (see above) and attributed by Dunham and Wadsworth (1978) to mixing between residual resident magma and fresh magma of the succeeding episode. Tait (1985) envisaged the thick basal peridotite in Unit 10 as having formed by settling of suspended olivine from a new pulse of picritic magma, followed by crystallization of allivalite from resident, isotopically contaminated magma. Heterogeneity in the upper part of the peridotite and lateral olivine compositional variation in the allivalite were attributed to differing degrees of re-equilibration with migrating intercumulus liquids. Bédard *et al.* (1988) found that Unit 9 peridotite intruded the Unit 9 allivalite and that a small peridotite in Unit 10 also intruded allivalite (Bédard *et al.*, 1988, their figure 15). They suggested the layered succession was composed of a picritic sill complex emplaced into layered troctolite. They also described structures where peridotite has replaced allivalite, and suggested that other features such as the pyroxene-rich upper part of Unit 9 allivalite showed that gabbro had formed by sub-solidus metasomatic replacement of layered allivalite. They reinterpreted the evidence presented by Young and Donaldson (1985) who had argued that the wavy structures at and near the base of the pyroxene-rich part of the layer resulted from loading caused by the deposition of dense (pyroxene-rich) crystal mush on top of a less dense (feldspar-rich) mush. Similar undulatory layered structures in Unit 14 on Trallval were attributed to loading (Volker and Upton, 1990; Bédard and Sparks, 1991; Volker and Upton, 1991). Donaldson (1982) suggested that bifurcating layers of harrisite and small harrisite pods and lenses might have formed when segregated intercumulus melt became trapped beneath impervious layers in the cumulates, and Young (1984) described chromite concentrations at an allivalite-peridotite boundary (Units 7/8), ascribing these to reaction between expelled pore liquids and overlying primitive magma. These observations highlighted the important role played by migrating intercumulus liquids in producing 'diagenetic' modifications to the cumulates, with strong overprinting and even obliteration of primary compositional, textural and structural features (*c.f.* Irvine, 1980; Hunter, 1987).

Fluid dynamical studies of basaltic magma chambers were applied to the Rum magma chamber (Huppert and Sparks, 1980; Tait, 1985; Sparks *et al.* 1985). These studies emphasized the importance of melt density in determining the evolution of the magma chamber, of compositional convection of melts within the porous crystal pile and of compaction in cumulates. Huppert and Sparks thought that dense picritic replenishments would have ponded at the base of the chamber below any resident basaltic magma. However, suggestions that crystals, olivine for example, would be kept in suspension within the convecting magma until some *en masse* settling event are now thought to be incorrect, because dense crystals should progressively settle from the boundary layers of convecting magma (Martin and Nokes, 1988).

Small-scale layering has generally been attributed to magmatic sedimentation processes under tranquil conditions, with slumping causing sporadic disturbances. An examination by Elias (1989, 1991) of structural and other evidence from the Unit 12 allivalites in the Hallival-Askival area convinced him that these rocks are commonly highly deformed and he proposed that most of the cumulates had been transported by mass flow from the margins towards the centre of the intrusion. Density flows redistributed olivine cumulates and thick, feldspar-rich cumulates fed debris flows from which the allivalite layers formed. Instability of a different sort had already been recognized where faulting of early-consolidated cumulates caused fragmentation and the resultant debris was transported to form ultrabasic breccias (magmatic screes; Wadsworth, 1961, 1992), although others stress that many of the breccias resulted from intrusion of ultrabasic melts (Volker, 1983; Volker and Upton, 1990), possibly of hydrous feldspathic peridotite composition (Donaldson, 1975). Significantly, the breccias are commonest in the proposed Layered Suite feeder zone (*e.g.* McClurg, 1982; see below).

3.2. Magma composition

The exact nature of the Rum parental magma(s) has been debated (see above) and suggested compositions have ranged from basaltic (Brown, 1956) to ultrabasic (Donaldson, 1975; Gibb, 1976; Forster, 1980; McClurg, 1982) to both (Renner and Palacz, 1987). However, the existence of chilled picrite dykes with magnesian olivine (Fo_{92}) (McClurg, 1982) and chilled contacts (Greenwood *et al.*, 1990) with 13-20 wt.% MgO shows that very magnesian liquids *must have been* present during the formation of the Layered Suite. It is this evidence, when combined with the mineralogy, the bulk composition of the Layered Suite rocks (*e.g.* McClurg, 1982; Volker, 1983; Butcher, 1984; Faithfull, 1986) and with the forsterite contents of offshore detrital olivine sands (Gallagher *et al.*, 1989), which suggests that the composition of the mean parental magma must have contained approximately 18+/-2 wt.% MgO. It is likely that this bulk composition was supplied directly to the Rum magma chamber, although the magma would invariably have carried some suspended olivine (Gibb, 1976) and according to its ascent rate and opportunity for olivine fractionation, reduced the MgO content of the melt portion of the magma during its ascent from the mantle. It is extremely unlikely that the bulk composition of the melt supplied to the intrusion had a MgO content much above 18-20 wt.%, because melts with greater than 20 wt.% MgO are largely confined to the Archaean (komatiites). Study of segregations within some of the peridotites suggests the magma had transitional to mildly alkaline affinities (Kitchen, 1985; Faithfull, 1986), in harmony with the observation that low-Ca pyroxene was not precipitated. The magma may also have been hydrous, with resident magmas in the chamber containing up to 1% water (Donaldson, 1975; Kitchen, 1985; Tait, 1985) although water in the primary magma would likely have been significantly less (~ 0.1 wt.%). A mean parental composition of 18-20 wt.% MgO provides three important constraints on the origin and petrogenesis of the Layered Suite:

1) The temperature of the mantle source region: this must have been anomalously hot regardless of whether melting in the source region was anhydrous or hydrous. If the source region was essentially dry, the mantle would have had a potential temperature of approximately 1600°C, similar to the present day source region for Hawaii (Watson and McKenzie, 1991). If it was wet, the potential temperature would have been slightly lower.
2) The total volume of magma involved in the evolution of the suite: simple mass balance calculations based on the crystallization of a 18 wt.% MgO liquid (able to crystallize approximately 30-35 vol.% olivine, corresponding to an equilibrated olivine composition at Fo_{86}), and using map-derived estimates of the bulk composition of the present day intrusion, suggest that the intrusion represents approximately one half of the total volume of supplied magma and therefore that the equivalent of the intrusion volume may have been erupted as basaltic lava. Gravity modelling (Figure 20) suggests that the intrusion volume is of the order of 700 km^3. This would require 700 km^3 of erupted basalt (approximately half the volume of the Skye Main Lava Series). However, Rum-derived lava fields have yet to be identified. Either they have been eroded away, are concealed beneath later fields and/or the sea, or have simply not been recognized. There is also the possibility that some of the residual magma formed the gabbroic and doleritic intrusions in and around the Rum Complex.

3) The thickness of the shallow-level magma chamber: if each unit of the Eastern Layered Series represents one replenishment event of the magma chamber, then peridotite layers with 60-80 vol.% olivine may have formed from a body of 18 wt.% MgO magma which was approximately two to three times the present day thickness of the peridotite layers. These calculations suggest that replenishments of the Rum magma chamber produced a 100-200 m thick sill-like sheet of magma, overlying a considerable thickness of crystal mush (*c.f.* Dunham, 1970). The sheet of magma progressively decreased in thickness due to crystallization and venting, only to be re-inflated by the next replenishment event.

3.3. Origin of layering

Although some of the peridotites are inferred to have originated as discrete intrusions or as metasomatic replacement bodies, most of the exposed Layered Suite is considered to have accumulated on the floor of the magma chamber. The common occurrence of interbedded crescumulate harrisite in Layered Suite peridotites indicates tranquil *in situ* floor crystallization, while large (>5cm) intraclasts, the ubiquitous presence of grain-size grading and soft-sediment deformation all suggest reworking and floor accumulation. Roof cumulates are absent in the few localities where the roof of the magma chamber is exposed (*e.g.* Figures 4A and B). At Beinn nan Stac (Figure 4B), the floor cumulates of the Eastern Layered Series extend to the marginal gabbro which is only a few metres thick. This observation raises the possibility that, although much crystallization must have occurred near the roof, most of the crystals subsequently were redeposited on the floor of the chamber. The mechanism by which crystals were transported to the floor from either the roof or from interior of the magma chamber remains unclear. Deposition may at times have been effected by simple crystal settling. However, the common presence of flow-derived features in the cumulates requires at least some current activity and hence the likely operation of laterally spreading density currents that may have formed from roof-derived crystal-laden plumes.

The Rum magma chamber was an open system and the 16 macro-rhythmic units of the Eastern Layered Series (Figure 5A) clearly represent major replenishment events. Each has a major peridotite unit at its base and as such this boundary represents a major change in composition of the accumulating crystal pile. However the existence of subsidiary peridotites within the allivalites (*e.g.* Butcher *et al.,* 1985; Faithfull, 1985) and the study of a single unit from the Eastern Layered Series (Renner and Palacz, 1987) suggest that individual units may be the product of several subordinate replenishment events rather than one single major event. Furthermore, the complexity of the peridotites, with interbedded harrisites and stratified peridotites (Wadsworth, 1961) and of the allivalites themselves (*e.g.* Faithfull, 1985; Bédard *et al.*, 1988; Volker and Upton, 1990) reveal that the fundamental cumulate event stratigraphy is represented by lithologically coherent sub-units on the metre to decimetre scale (Figure 5B). Each of these represents an accumulation event, by either *in situ* harrisite crystallization, by a depositional event on the floor of the magma body, or an intrusion of magma into the cumulate pile. Individual sub-units may comprise modal and grain-size variation often defining layering on a cm-mm scale. This small-scale layering may be due to changes in the fluid dynamics of depositing gravity currents, variations in the concentration of entrained crystals and/or by sorting during crystal deposition.

3.4. Intrusive peridotites and finger structures

The significance of intrusive peridotite bodies among the predominantly accumulative peridotites of the Rum Layered Suite has generated considerable debate (Bédard *et al.*, 1988;

Volker and Upton, 1991; Bédard and Sparks, 1991). Such bodies clearly exist as peripheral tongues and plugs (Wadsworth, 1994), but clear intrusive relationships within the Layered Suite can also be found, for example on the west slopes of Barkeval, and it is possible that some of the subhorizontal peridotites elsewhere in the Eastern Layered Series are also intrusive, as suggested by Bédard *et al.* (1988). However, unequivocal field evidence is scarce, and in any event the distinction between simple cumulates, re-worked crystal mushes (some of which may be locally invasive to earlier but incompletely consolidated parts of the Layered Suite if sufficiently energetic mass-flows are generated) and injections of crystal-rich magma, is inevitably blurred.

Many of the peridotites have finger-like protrusions (Brown, 1956; Robins, 1982) which cut through the layering of the overlying allivalite without deforming it. These protrusions have been interpreted as (i) replacement of the overlying allivalite by ascending intercumulus melt after it had been deposited on the peridotite (Butcher *et al.*, 1985) or (ii) dissolution of the allivalite by hot intrusive peridotite magma (Morse *et al.*, 1987). The two crucial observations that can be used to determine their origin are finger mineralogy and underlying peridotite thickness. Some fingers contain significant, often oikocrystic, clinopyroxene, indicating that clinopyroxene was a liquidus phase of the finger magma. Ascending magma of this composition derived from underlying compacting peridotite, would have been too cold to dissolve the overlying allivalite and therefore these fingers indicate intrusive peridotites. Other fingers do not contain significant clinopyroxene and thus may be replacive in origin. The occurrence of extensive but very thin (few cm.) subsidiary peridotites with fingers also argues against an intrusive origin for these finger-bearing peridotites. A comprehensive study of the finger mineralogy through the whole spectrum of fingered peridotites is essential to determine the importance of intrusive peridotites.

3.5. Postcumulus processes

The importance of post-cumulus processes in the Rum intrusion cannot be over emphasized. Solidification rates would have been comparable to migrating melt velocities (1-0.1 m/yr) (Sparks *et al.*, 1985) and cooling time-scales (1°-10°C/year) would have been longer than these required for mineral re-equilibration. Crystals deposited on the floor of the magma chamber would have formed touching frameworks of 40-60% crystals. The melt that filled the pore space between the crystals may have partially crystallized *in situ*, but would have been much more likely to move, due either to compaction of the cumulate pile or convection of the less dense residual fluid resulting from *in situ* crystallization (*c.f.* Irvine, 1987). Movement of the fluid would have led to mineral and textural re-equilibration, including the development of adcumulates. The uniformity of olivine forsterite contents within individual peridotites which were generated by both *in situ* bottom growth (harrisites) and by crystal sedimentation from the roof of the magma chamber, and the common occurrence of re-equilibrated mineral textures are evidence for the pervasiveness of postcumulus re-equilibration within the cumulate pile. Further, compaction and compositional convection in cumulates leading to the expulsion of cotectic and eutectic melts may provide an explanation for many of the compositional complexities of the allivalites (Bédard *et al.*, 1988), the numerous late-stage veins in the intrusion (Butcher, 1985; Kitchen, 1985) and possibly for the extensive transgressive gabbros present in the centre of the intrusion (Figure 3).

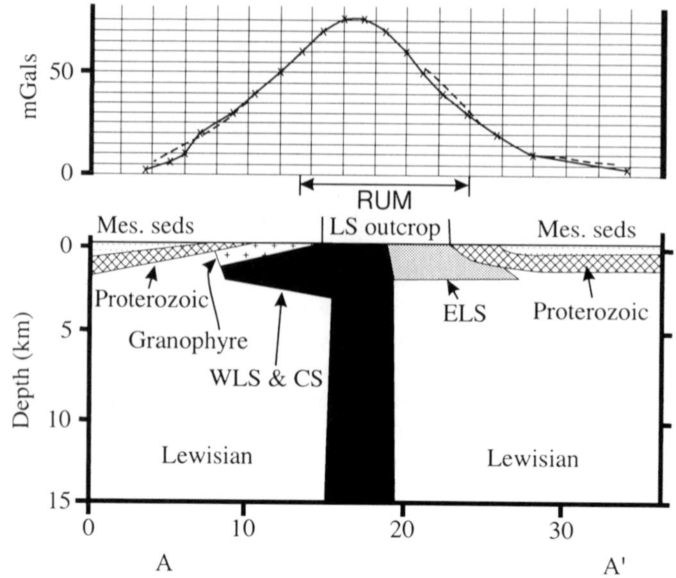

Figure 20. *Gravity model of the Rum Central Complex. The measured Bouguer gravity anomaly is shown on the upper graph and the model presented in this paper in the lower figure. The gravity data are shown by crosses joined by a continuous line and the anomaly resulting from the model by a dotted line (where this departs from the measured profile). Gravity data abstracted from the 1:250,000 British Geological Survey UK and Continental Shelf Series, Bouguer Gravity Anomaly Map, Tiree, Sheet 56°N-08°W. Densities: Western Layered Series and Central Series = 3.2 g/cm^3, Eastern Layered Series = 3.05 g/cm^3, Lewisian gneisses = 2.79 g/cm^3, Western Granophyre = 2.7 g/cm^3, Proterozoic (Torridonian) = 2.65 g/cm^3 and the Mesozoic sedimentary rocks = 2.5 g/cm^3.*

3.6. Geometry of emplacement

The Layered Suite is the culmination of prolonged igneous activity during which pulses of transitional basaltic and picritic magmas built up a body of predominantly dense mafic rocks. The pronounced positive Bouguer gravity anomaly over the Central Complex has previously been interpreted in terms of a steep-sided, approximately cylindrical, body of dense rock extending down to many kilometres, the depth estimate depending on assumptions about the relative proportions of peridotite and olivine gabbro (McQuillin and Tuson, 1963). Broadly similar models have been proposed for the Bouguer gravity anomalies over Palaeocene central complexes elsewhere in the province, for example Mull and Skye (Bott and Tuson, 1973). For the Rum model, subsurface cross-sections of the complex were taken to be similar to its present day exposed limits. If, however, the basic and ultrabasic rocks extend laterally some distance beyond present surface limits, as suggested by circumstantial evidence north and south of the complex (Figure 1), then a tabular or disc-shaped body with relatively narrow feeders may be a more appropriate solution (Figure 20).

Such a mushroom-shaped body is our preferred model, with the head of the mushroom having formed at the Lewisian-Torridonian unconformity. We suggest that the magma may have intruded laterally at this level partly because this boundary may have represented a neutral buoyancy level for the magma. The magma density would have been between 2.75-2.77 g/cm^3 (assuming that the magma contained only a small proportion of crystals) which would have been greater than the density of the overlying Torridonian and Mesozoic sedimentary rocks (<2.65 g/cm^3), but possibly less than that of the underlying Lewisan gneisses (2.75-2.80 g/cm^3). The presence of low-density acid magmas would also have constituted a further trap.

The magma supply was maintained through narrow, approximately north-south, dyke-like conduits whose siting was controlled by a precursor to the right-lateral Long Loch Fault. It is possible that similar movements along this line of weakness resulted in some pull-apart, facilitating the entry of magma into this area. Pulses of magma were fed through the conduits, which had a history of intermittent opening. We envisage episodes of magma ascent and chamber inflation with concomitant cumulate growth and lava expulsion, as well as episodes of deflation and magma withdrawal, as evidenced by structures in and adjacent to the Central Series. Conduits were continually being opened, breaking through earlier cumulates, contributing to the complex relationships found in the Central Series (Figure 21) which represents the consolidated remains of the final input of magma. Active conduits now appear as intrusive breccias, whereas collapse breccias represent times when magma supply waned, or magma was withdrawn, with concomitant subsidence and faulting. The pulses of magma flowed out across the floor of a persistently reconstituted thin, sill-like magma chamber. At the present level of exposure, their lateral extent was partly constrained by the Main Ring Fault. However, from time to time, magma was intruded beyond these limits, especially at the northern and southern ends of the feeder zone. Dense olivine-rich magma also intruded the poorly consolidated substrate, as sheets guided by differentially lithified feldspathic cumulates. Consolidation of individual pulses was from the base upwards, giving the distinctive pattern of compositionally limited cryptic layering, repeated on the unit scale in the layered successions. Crystallization was principally in the roof zone where olivine (+ chromite), was joined by feldspar and finally clinopyroxene. The cumulus crystals were carried to the floor under conditions that are considered to have been relatively tranquil; although they gave rise to some current sorting, the absence of cross-bedding and trough structures is stressed (*c.f.* Upton *et al.*, this volume).

As cumulates crystallized from each batch of magma, a complementary, more evolved, residual liquid fraction formed. The compositions of these residues are uncertain, as is the volume, although in *any* model this must be large (see above). Both the composition and volume of the residua will have been controlled by the compositions of the pulses of new magma and by the amount and composition of residua from earlier cycles resident in the magma chamber. A proportion of these residua may have been retained within the system, contributing to the Layered Suite and the gabbroic sheets, but in view of the volume of residual basaltic material generated, it is most likely that complementary surface eruptions occurred. Field evidence for contemporaneous extrusion is circumstantial. Fine-grained basic hornfelses are locally common as xenoliths, especially in the east of the suite, and some contain relics of amygdales (Faithful, 1985). The xenoliths are likely to have been derived from basalts in the roof zone of the intrusion. Although it is possible that they were derived from lavas cogenetic with the Layered Suite, an equally likely source would have been the much earlier basalts of the Eigg Lava Formation which occur nearby (Figure 1). Clasts of highly altered amygdaloidal (olivine) basalt, common in the oldest conglomerates of the Canna Lava Formation on Rum, may also have been derived from lavas erupted from the developing Central Complex.

As the Layered Suite grew, space for successive pulses was probably achieved by a combination of uplift in the roof zone and subsidence of the floor. Whereas it is clear that uplift and country-rock deformation accompanied emplacement of the acid rocks in Stage 1, evidence for uplift directly related to Stage 2 is less obvious. Circumstantial evidence from the palaeogeography of Rum and its surroundings indicates that the growth of the Rum Central Complex was accompanied by the formation of a sizeable mountain mass, which was vigorously eroded at the end of Stage 2 (Emeleus 1985; Williamson and Bell, 1994). This edifice would largely have been built up by successive acidic and basaltic effusions, but uplift may also have contributed. On a smaller scale, some of the faulting on Beinn nan Stac (Figure 4B) is reversed and may have occurred in response to uplift during early stages of growth of the Layered Suite.

Assuming that the layering was essentially horizontal (*c.f.* Jackson, 1961, p.82), or dipped

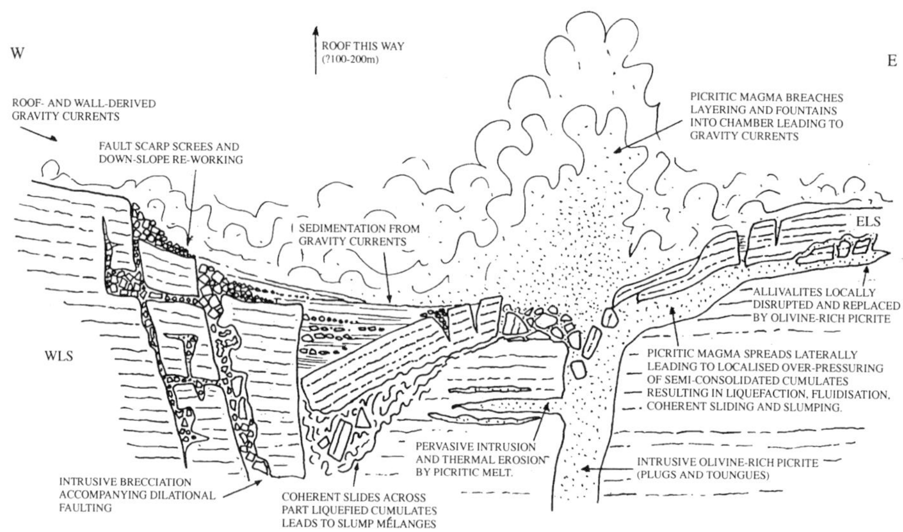

Figure 21. *Schematic reconstruction of possible events leading to the formation of the Central Series. Replenishment or rejuvenation of the magma chamber with picritic magma (with or without crystals) causes disruption, sliding and slumping of earlier formed layered sequences. Events may breach the magma chamber floor or be intrusive, locally overpressuring floor cumulates. Replenishment may be associated with faulting and subsidence of the floor of the chamber. Deposition and reworking of cumulate debris and primary crystals may occur from gravity currents associated with replenishments, or from catastrophic instabilities of mushes formed at the walls and roof possibly triggered by volcano-tectonic disturbances. Accommodation space on the floor is generated by faulting or during subsidence or collapse associated with replenishment events. No real scale is implied but the west to east section is vertically exaggerated. Layering is depicted schematically.*

centrally at low angles (<5°), when it formed, evidence of subsequent subsidence may be inferred from the centrally-directed dips of 10°- 25°. These may result from a combination of cumulate compaction and foundering of the chamber floor near an unstable feeder zone. Deformation of the floor would then have led to extensive slumping of dense, variably consolidated cumulates. Since compaction will have been a continuous process, considerable thicknesses (? >5 m) of unconsolidated cumulates will only rarely have been able to form. Steeper, centrally-directed dips close to the outer and inner margins of the Eastern Layered Series may have resulted from fault drag as central parts subsided *en bloc*. East of Hallival and Askival, drag occurred against the Torridonian rocks as the Eastern Layered Series subsided; on the western end of Trallval, layering in the Eastern Layered Series deformed as solidified parts of the Central Series subsided.

Although the Layered Suite is made up of three distinct series of rocks, its evolution is envisaged to have been continuous. Parts of the Eastern Layered Series west of Barkeval may be correlated with the upper Western Layered Series, from which they are virtually indistinguishable in the field. Relict masses of both Eastern and Western Series rocks occur within the Central Series. Large, steeply inclined masses of layered allivalite members in breccias south of the Long Loch (Figure 6) closely resemble allivalite in higher units in the Eastern Layered Series and the An Dornabac Member of the Central Series is probably complementary to the Ard Mheall Member of the Western Layered Series, from which it was probably separated during syn-magmatic faulting and subsidence (*c.f.* Figure 21). Parts of the Long Loch Member in area A, Figure 6, may correlate with allivalites high in the Eastern Layered Series.

Whereas Stages 1 and 2 are very different in terms of rock types and their modes of occurrence, xenoliths of coarse gabbro and peridotite in intrusive igneous breccias of Stage 1 prove that there was significant earlier deep-seated mafic plutonism, and occurrences of mixed (basic and acid) magma rocks in Stage 1 demonstrate that basaltic magma was available alongside the acid magmas. The igneous activity in the Central Complex is therefore seen as a continuum in time, driven mainly by picritic and basaltic magmas. Plutonic mafic rocks formed during both the earliest and latest stages, bracketing the acid rocks which are regarded as the combined products of fractionation and rheomorphism.

4. CONCLUSIONS

1) The Layered Suite of the Palaeocene Rum Central Complex consists of three complementary members each with well-developed successions of ultrabasic and basic cumulates. These are the Eastern Layered Series (700 m thick), the Western Layered Series (520 m thick) and the Central Series (~2000 m thick?).
2) The presence of chilled picrite dykes and rare chilled contacts, and the mineralogy and bulk composition of many of the Layered Suite rocks provide compelling evidence that high-Mg melts (13 - 20 wt.% MgO) were available throughout the formation of the suite.
3) The magmas were of transitional character and ranged from picrite to basalt in composition. The cumulus phases are forsteritic olivine, calcic plagioclase, diopsidic augite and chromite; cumulus Ca-poor pyroxene is absent.
4) Magma entered the Complex through a central feeder zone, parallel to the Long Loch Fault and probably controlled by this long-lived fracture. Initiation of the complex may have been facilitated by pull-apart on a precursor of the present fault. The conduits through which

magma entered the complex are now occupied by zones of intrusive ultrabasic breccias; ultrabasic breccia zones with collapse features mark the sites of subsidence attendant on magma withdrawal.

5) Repetitive batches of magma were intruded at a shallow crustal level, forming thin, tabular sill-like bodies whose limits were locally constrained by the pre-existing Main Ring Fault. Space for each batch of magma entering the chamber was made by a combination of uplift of the roof and subsidence of the floor.
6) The Layered Suite is the exposed, upper part of a flattened disc-like body of dense, mafic rocks, at least 2 km in thickness and extending some distance beneath country rocks around the complex. The overall shape is mushroom-like.
7) The layered rocks built from the base upwards as batches of magma, crystallizing at or near the cooler roof zone, forming bottom cumulates and a pool of residual, basaltic magma. Whereas much of this residual magma was probably erupted during contemporary surface volcanism, some formed intrusive gabbroic sheets.
8) Major layered units typically consist of olivine-chromite basal cumulates with overlying olivine-plagioclase, plagioclase-olivine, plagioclase-olivine-clinopyroxene and (rarely) clinopyroxene-plagioclase-(olivine) cumulates; there is small-scale vertical and horizontal variation within this simple succession. The scale of the layering varies from >50 m to < 1 cm. Large- and small-scale layers are generally conformable, large-scale layers are laterally continuous, and the layering rarely dips at angles >25°. The original dips are inferred to have been low (<5°) and to have been subsequently modified by compaction and slumping. The layering on the 1-5 m scale is considered to represent the important increments in building the Layered Suite.
9) Limited cryptic layering is recorded within individual macro-rhythmic units. The overall compositional variation, however, is little different from that found in individual units. The pronounced breaks in compositional and modal layering at or near unit boundaries may mark significant time breaks in cumulate formation, and therefore in magma input.
10) Slow cooling of the Layered Suite favoured the formation of thick piles of porous, partly crystallized cumulates. Melt migration through these is believed to have played a major role in formation of the ultimate products. Compaction processes were attended by ascent of low-density residual intercumulus melts. Conversely, high-density magnesian melts were able to percolate downwards. The migrating melts reacted with the cumulate matrices. Evidence of these migrations is seen in the large- and small-scale replacements of allivalite by peridotite (and possibly by gabbro). Cumulate fabrics matured and equilibrated as melts were expelled upwards and mineral compositions were modified across peridotite-allivalite/gabbro boundaries, giving rise to the discrepancies between cryptic and modal layering at unit boundaries.
11) Repetitive admission of primitive magmas along the Long Loch feeder system maintained the complex at an overall high temperature. The cumulate body, estimated to be 2-3 km thick and 20 km diameter, would have had a high thermal capacity. Absence of inter-layer quenched facies, such as have been reported from the Kap Edward Holm Complex (Tegner *et al.*, 1993), implies high magma supply rates and the cumulate pile provided a major, shallow crustal heat store as crystallization proceeded. These factors allowed relatively unfractionated picritic melts to attain near-surface levels. Thus, a high-temperature body of cumulates, at shallow crustal level, annealed sufficiently slowly for development of textures

reminiscent of those expected in larger, deeper-seated complexes, as well as extensive thermal metamorphism and anatexis of adjacent crustal rocks.

12) The absence of cross-bedding and trough-layering suggests that much of the laterally continuous, small- and large-scale layering was formed under relatively tranquil conditions which were periodically punctuated by catastrophic collapses and slumping of unstable cumulate piles towards the Central Series feeder system. Vigorous convective circulation of magma and down-wall currents were inhibited by the thin, horizontally-tabular form of the episodically replenished magma chamber. Layering continues to within a metre or less of the contacts which were presumably hot. Chilled marginal facies are uncommon and may have been rarely formed.

13) Slumping occurred when unconsolidated cumulates became unstable on the floor of a dynamic chamber experiencing repeated influxes and withdrawal of magma. Slump structures are most easily discernible in the layered allivalitic rocks.

14) The three series composing the Layered Suite represent a continuum of cumulate formation. The Western Layered Series may correlate with the lower parts of the Eastern Layered Series, the An Dornabac Member of the Central Series probably originally overlay the Ard Mheall Member of the Western Layered Series, and part of the Long Loch Member may be equivalent to the higher layers in the Eastern Layered Series. Other parts of the Central Series, however, are probably younger than any of the rocks preserved within the Eastern and Western Layered Series.

15) Isotopic studies show that picritic magma and some of the basaltic magma entering the chamber were uncontaminated, whereas the resident magma had been contaminated by about 7% of upper crustal, amphibolite-facies Lewisian gneiss. Contamination by Torridonian sedimentary rocks is slight, and was limited to rocks at the margins of the complex.

5. ACKNOWLEDGEMENTS

The technical assistance of K.L. Atkinson, G. Dresser, and A. Carr during the preparation of this account is gratefully acknowledged, as is the permission of the Director of the British Geological Survey to include material used in several of the figures.

6. REFERENCES:

Bailey, E.B., 1945. Tertiary igneous tectonics of Rhum (Inner Hebrides) *Quart. J. Geol. Soc. Lond.* **100**, 165-91.

Bailey, E.B., 1956. Hebridean notes: Rhum and Skye. *Liverpool Manchester Geol. J.* **1**, 420-26.

Bédard, J.H., Sparks, R.S.J., Renner, R., Cheadle, M.J., & Hallworth, M.A., 1988. Peridotite sills and metasomatic gabbros in the Eastern Layered Series of the Rhum complex. *J. Geol. Soc.* **145**, 207-24.

Bédard, J.H., & Sparks, R.S.J., 1991. Comments on 'The structure and petrogenesis of the Trallval and Ruinsival areas of the Rhum ultrabasic pluton' by J.A. Volker, & B.G.J. Upton. *Trans. Roy. Soc. Edin.: Earth Sci.* **83**, 389-90.

Binns, P.E., McQuillin, R., & Kenolty, N., 1974. The geology of the Sea of the Hebrides. *Rep. Inst. Geol. Sci.* **73/14**, 43 pp.

Bott, M.H.P., & Tuson, J., 1973. Deep structure beneath the Tertiary volcanic regions of Skye, Mull and Ardnamurchan, north-west Scotland. *Nature* **242**, 114-6.

British Geological Survey, 1994. 1:50,000 Series Sheet 60 (Scotland), Rum Solid and Drift Geology, 3rd Edition.

Brown, G.M., 1956. The layered ultrabasic rocks of Rhum, Inner Hebrides. *Phil. Trans. Roy. Soc. Lond.* **240B**, 1-53.

Butcher, A.R., 1984. A study of postcumulus structures in the Rhum layered intrusion, Inner Hebrides. Ph.D. Thesis, University of Manchester.

Butcher, A.R., 1985. Channelled metasomatism in Rhum layered cumulates - evidence from late-stage veins. *Geol. Mag.* **122**, 503-18.

Butcher, A.R., Young, I.M., & Faithfull, J.W., 1985. Finger structures in the Rhum Complex. *Geol. Mag.* **122**, 491-502.

Dickin, A.P., 1981. Isotope geochemistry of Tertiary igneous rocks from the Isle of Skye, N.W. Scotland. *J. Petrology* **22**, 155-89.

Donaldson, C.H., 1974. Olivine crystal types in harrisitic rocks of the Rhum pluton and in Archean spinifex rocks. *Geol. Soc. Am. Bull.* **85**, 1721-6.

Donaldson, C.H., 1975. Ultrabasic breccias in layered intrusions - the Rhum complex. *J. Geol.* **83**, 33-45.

Donaldson, C.H., 1982. Origin of some of the Rhum harrisite by segregation of intercumulus liquid. *Miner. Mag.* **45**, 201-9.

Dunham, A.C., 1968. The felsites, granophyre, explosion breccias and tuffisites of the north-eastern margin of the Tertiary igneous complex of Rhum, Inverness-shire. *Quart. J. Geol. Soc. Lond.* **123**, 327-52.

Dunham, A.C., 1970. The emplacement of the Tertiary igneous complex of Rhum. In: Newall, G., & Rast, N. (eds.). Mechanisms of Igneous Intrusion. *Spec. Issue Geol. J.* **2**, 23-32.

Dunham, A.C., & Wadsworth, W.J., 1978. Cryptic variation in the Rhum layered intrusion. *Miner. Mag.* **42**, 347-56.

Dunham, A.C., & Wilkinson, F.C.F., 1965. Sulphide droplets and the U11/12 chromite band, Rhum: a mineralogical study. *Geol. Mag.* **122**, 539-48.

Elias, R.T., 1989. The origin of cyclic layering in the Eastern Layered Series of the Rhum Intrusion. *Geol. Soc. Lond. Newsletter* **18**, 38.

Elias, R.T., 1991. Cumulus processes and melt-migration in layered intrusions and the use of image analysis to quantify microscopic textures in cumulates. Ph.D. Thesis, University of Liverpool.

Emeleus, C.H., 1985. The Tertiary lavas and sediments of northwest Rhum, Inner Hebrides. *Geol. Mag.* **122**, 419-37.

Emeleus, C.H., 1987. The Rhum layered complex, Inner Hebrides, Scotland. In: Parsons, I. (ed.). *Origins of Igneous Layering.* Dordrecht: Reidel, 263-86.

Emeleus, C.H., 1994. 1:20,000 Solid Geology map of Rum (2nd Edition). Edinburgh: Scottish Natural Heritage.

Emeleus, C.H., 1996. Geology of Rum and the adjoining islands. *Mem. Brit. Geol. Surv.*, Sheet 60 with parts of sheets 61 and 70 (Scotland).

Emeleus, C.H., Wadsworth, W.J., & Smith, N.J., 1985. The early igneous and tectonic history of the Rhum Tertiary Volcanic Centre. *Geol. Mag.* **122**, 451-7.

Faithfull, J.W., 1985. The Lower Eastern Layered Series of Rhum. *Geol. Mag.* **122**, 459-68.

Faithfull, J.W., 1986. Petrology and geochemistry of gabbroic and ultrabasic rocks from eastern Rhum. Ph.D. Thesis, University of Durham.

Forester, R.W., & Harmon, R.S., 1983. Stable isotope evidence for deep meteoric-hydrothermal circulation: Isle of Rhum, Inner Hebrides, Scotland. *Proc. 3rd Internat. Conference on Water-Rock Interactions* (Abstr.), 135-6.

Forster, R.M., 1980. A geochemical and petrological study of the Tertiary minor intrusions of Rhum, northwest Scotland. Ph.D. Thesis, University of Durham.

Gallagher, M.J., & 11 others, 1989. Marine deposits of chromite and olivine, Inner Hebrides of Scotland. Brit. Geol. Surv. Tech. Rep. WF/89/13. *Brit. Geol. Surv. Miner. Reconnaissance Programme Rep.* **106**, 20 pp.

Gibb, F.G.F., 1976. Ultrabasic rocks of Rhum and Skye: nature of the parent magma. *J. Geol. Soc. Lond.* **132**, 209-22.

Greenwood, R.C., 1987. Geology and petrology of the margin of the Rhum ultrabasic intrusion, Inner Hebrides, Scotland. Ph.D. Thesis, University of St. Andrews.

Greenwood, R.C., Donaldson, C.H., & Emeleus, C.H., 1990. The contact zone of the Rhum ultrabasic intrusion: evidence of peridotite formation from magnesian magmas. *J. Geol. Soc. Lond.* **147**, 209-12.

Greenwood, R.C., Fallick, A.E., & Donaldson, C.H., 1992. Oxygen isotope evidence for major fluid flow along the contact zone on the Rhum ultrabasic intrusion. *Geol. Mag.* **129**, 243-6.

Harker, A., 1903. One-Inch Geological Map of the Small Isles of Inverness-shire, Sheet 60 (Scotland). Geological Survey of Great Britain, Edinburgh: H.M.S.O.

Harker, A., 1908. The geology of the Small Isles of Inverness-shire. *Mem. Geol. Surv. Scot.* Sheet 60. Edinburgh: H.M.S.O.

Henderson, P., 1975. Reaction trends shown by chrome-spinels of the Rhum layered intrusion. *Geochim. Cosmochim. Acta* **39**, 1035-44.

Henderson, P., & Suddaby, P., 1971. The nature and origin of the chrome-spinel of the Rhum layered intrusion. *Contr. Miner. Petrol.* **33**, 21-31.

Hughes, C.J., 1960. The Southern Mountains Igneous Complex, Isle of Rhum. *Quart. J. Geol. Soc. Lond.* **96**, 111-38.

Hulbert, L.J., Duke, J.M., Eckstrand, O.R., Scoates, R.F.J., Theriault, R.J., LeCheminant, G.M., Williamson, B., Kyser, T.K., Gallagher, M.J., Gunn, A.G., & Grinenko, L.N., 1992. Metallogenesis and geochemical evolution of Cyclic Unit 1, Lower Eastern Layered Series, Rhum. (Abstr.) In: Foster, R.P. (ed.) *Mineral deposit modelling in relation to crustal reservoirs of the ore-forming elements.* Institution of Mining and Metallurgy.

Hunter, R.H., 1987. Textural equilibrium in layered igneous rocks. In: Parsons, I. (ed.) *Origins of Igneous Layering.* Dordrecht: Reidel, 473-503.

Huppert, H.E., & Sparks, R.S.J., 1980. The fluid dynamics of a basaltic magma chamber replenished by influx of hot, dense ultrabasic magma. *Contr. Miner. Petrol.* **75**, 279-89.

Irvine, T.N., 1980. Magmatic infiltration metasomatism, double-diffusive fractional crystallization, and adcumulus growth in the Muskox and other layered intrusions. In: Hargraves, R.B. (ed.) *Physics of Magmatic Processes.* Princeton: Princeton University Press, 325-83.

Irvine, T.N., 1987. Appendix 2. Processes involved in the formation and development of layered igneous rocks. In: Parsons, I. (ed.) *Origins of Igneous Layering.* Dordrecht: Reidel, 649-56.

Jackson, E.D., 1961. Primary textures and mineral associations in the Ultramafic Zone of the Stillwater Complex, Montana. *U.S. Geol. Surv. Prof. Paper* **358**, 106 pp.

Kitchen, D.E., 1985. The parental magma on Rhum: evidence from alkaline segregations and veins in the peridotite from Salisbury's Dam. *Geol. Mag.* **122**, 529-37.

McClurg, J.E., 1982. Petrology and evolution of the northern part of the Rhum ultrabasic complex. Ph.D. Thesis, University of Edinburgh.

McQuillin, R., & Tuson, J., 1963. Gravity measurements over the Rhum Tertiary plutonic complex. *Nature,* **199**, 1276-7.

Maaløe, S., 1978. The origins of rhythmic layering. *Miner. Mag.* **42**, 337-45.

Martin, D., & Nokes, R.I., 1988. Crystal settling in a vigorously convecting magma chamber. *Nature* **332**, 534-6.

Meisner, R., Matthews, D., & Weaver, Th., 1986. The 'Moho' in and around Great Britain. *Annal. Geophy.* **4 (B.6)**, 659-64.

Morse, S.A., Owen, B.E., & Butcher, A.R., 1987. Origin of finger structures in the Rhum Complex: phase equilibrium and heat effects. *Geol. Mag.* **124**, 205-10.

Mussett, A.E., Dagley, P., & Skelhorn, R.R., 1988. Time and duration of activity in the British Tertiary Igneous Province. In: Morton, A.C., & Parson, L.M. (eds.). *Early Tertiary Volcanism and the Opening of the North Atlantic.* Spec. Publ. Geol. Soc. Lond. **39**, 337-48.

Palacz, Z.A., 1984. Isotopic and geochemical evidence for the evolution of a cyclic unit in the Rhum intrusion, north-west Scotland. *Nature,* **307**, 618-20.

Palacz, Z.A., 1985. Sr-Nd-Pb isotopic evidence for crustal contamination in the Rhum intrusion. *Earth Planet. Sci. Lett.* **74**, 35-44.

Palacz, Z.A., & Tait, S.R., 1985. Isotopic and geochemical investigations of cyclic unit 10 from the Eastern Layered Series of the Rhum intrusion. *Geol. Mag.* **122**, 485-90.

Renner, R., & Palacz, Z.A., 1987. Basaltic replenishment of the Rhum magma chamber: evidence from unit 14. *J. Geol. Soc.* **144**, 961-70.

Robins, B., 1982. Finger structures in the Little Kufjord layered intrusion, Finnmark, Northern Norway. *Contr. Mineral. Petrol.* **81**, 290-5.

Robinson, M.A., & McClelland, E.A., 1987. Palaeomagnetism of the Torridonian of Rhum, Scotland: evidence for limited uplift of the Central Intrusive Complex. *Earth Planet. Sci. Lett.* **85**, 473-87.

Smith, N.J., 1985. The age and structural setting of limestones and basalts on the Main Ring Fault in southeast Rhum. *Geol. Mag.* **122**, 439-45.

Sparks, R.S.J., Huppert, H.E., Kerr, R.C., McKenzie, D.P., & Tait, S.R., 1985. Postcumulus processes in layered intrusions. *Geol. Mag.* **122**, 555-68.

Tait, S.R. 1985. Fluid dynamic and geochemical evolution of cyclic unit 10, Rhum, Eastern Layered Series. *Geol. Mag.* **122**, 469-84.

Tegner, C., Wilson, J.R., & Brooks, C.K., 1993. Intraplutonic quench zones in the Kap Edvard Holm Layered Gabbro Complex, East Greenland. *J. Petrology* **34**, 681-710.

Thompson, R.N., & Gibson, S.A., 1991. Subcontinental mantle plumes, hotspots and pre-existing thinspots. *J. Geol. Soc. Lond.* **148**, 973-8.

Tilley, C.E., 1944. A note on the gneisses of Rum. *Geol. Mag.* **81**, 129-31.

Upton, B.G.J., 1988. History of Tertiary igneous activity in the N Atlantic borderlands. In: Morton, A.C., & Parson, L.M. (eds.) *Early Tertiary volcanism and the opening of the NE Atlantic.* Spec. Publ. Geol. Soc. Lond. **39**, 429-54.

Volker, J.A., 1983. The geology of the Trallval area, Rhum, Inner Hebrides. Ph.D. Thesis, University of Edinburgh.

Volker, J.A., & Upton, B.G.J., 1990. The structure and petrogenesis of the Trallval and Ruinsival areas of the Rhum ultrabasic complex. *Trans. Roy. Soc. Edin: Earth Sci.* **81**, 69-88.

Volker, J.A., & Upton, B.G.J., 1991. Reply to comments by J.H. Bédard, & R.S.J. Sparks. *Trans. Roy. Soc. Edin: Earth Sci.* **82**, 391.

Wadsworth, W.J., 1961. The layered ultrabasic rocks of south-west Rhum, Inner Hebrides. *Phil. Trans. Roy. Soc. Lond.* **244B**, 21-64.

Wadsworth, W.J., 1992. Ultrabasic igneous breccias of the Long Loch area, Isle of Rhum. *Scott. J. Geol.* **28**, 103-13.

Wadsworth, W.J., 1994. The peridotite plugs of northern Rum. *Scott. J. Geol.* **30**, 167-74.

Wager, L.R., & Brown, G.M., 1951. A note on rhythmic layering in the ultrabasic rocks of Rhum. *Geol. Mag.* **88**, 166-8.

Wager, L.R., & Brown, G.M., 1968. *Layered Igneous Rocks.* Edinburgh: Oliver and Boyd, 588 pp.

Wager, L.R., Brown, G.M., & Wadsworth, W.J., 1960. Types of igneous cumulates. *J. Petrology* **1**, 73-85.

Wager, L.R., & Deer, W.A., 1939. Geological investigations in East Greenland: Part III - The petrology of the Skaergaard intrusion, Kangerdlugssuaq. *Medd. Grønland* **105** (4), 323 pp.

Watson, S., & Mackenzie, D., 1991. Melt generation by plumes: a study of Hawaiian Volcanism. *J. Petrology* **32**, 501-37.

Williams, P.J., 1985. Pyroclastic rocks in the Cnapan Breaca felsite, Rhum. *Geol. Mag.* **122**, 447-50.

Williamson, I.T., & Bell, B.R., 1994. The Palaeocene lava field of west-central Skye, Scotland: Stratigraphy, palaeogeography and structure. *Trans. Roy. Soc. Edin.: Earth Sci.* **85**, 39-75.

Young, I.M., 1984. Mixing of supernatant and interstitial fluids in the Rhum layered intrusion. *Miner. Mag.* **48**, 345-50.

Young, I.M., & Donaldson, C.H., 1985. Formation of granular textured layers and laminae within the Rhum crystal pile. *Geol. Mag.* **122**, 519-28.

Young, I.M., Greenwood, R.C., & Donaldson, C.H., 1988. Formation of the Eastern Layered Series of the Rhum Complex, northwest Scotland. *Can. Miner.* **26**, 225-33.

LAYERED INTRUSIONS
R.G. Cawthorn (editor)

The Stillwater Complex

I.S. McCallum

Department of Geological Sciences, University of Washington, Seattle, WA, 98195, U.S.A.

Abstract

The Stillwater Complex was emplaced at 2.7 Ga at a depth of 10-15 km. Laramide deformation and subsequent erosion have exposed the basal hornfels and a thick, but incomplete, section of the complex. The complex is divided into a Basal Series, an Ultramafic Series and a Banded Series. The Basal Series forms an irregular sheet-like mass composed of early-formed cumulate rocks and coeval sills of diabase and sulphide-rich mafic norite. The Ultramafic Series is made up of cycles of harzburgite and bronzitite with conformable layers of chromitite in the lower parts of each cycle. The cyclic units and absence of cryptic variation in the Ultramafic Series indicate repeated influxes of magma and venting of fractionated magma from the chamber. The Lower and Upper Banded Series are composed primarily of norite and gabbronorite while the Middle Banded Series is predominantly anorthosite, troctolite and olivine gabbro. Cryptic variation in the Lower and Upper Banded Series is consistent with crystal fractionation and accumulation. In contrast, plagioclase in the Middle Banded Series has a uniform composition throughout which is consistent with a sorting mechanism in which plagioclase crystals were suspended in a convecting magma for extended periods. In the Lower Banded Series, a PGE-rich sulphide zone (J-M Reef) is associated with the reappearance of olivine, most likely due to a magma influx. Two models have been proposed for the origin of the reef. The magmatic model holds that sulphides became enriched in PGE during batch segregation of an immiscible sulphide melt which came into contact with a large volume of silicate melt during magma mixing. In the hydromagmatic model, it is proposed that PGE enrichments are the result of leaching of underlying cumulates by Cl-rich hydrous fluids exsolved during the late stages of intercumulus crystallization. Crystallization sequences reveal that at least two compositionally distinct magmas were involved in the formation of the complex. The magma that formed the Ultramafic Series was rich in both MgO and SiO_2 and appears to be related to the mafic norite sills. The magma that formed the Middle Banded Series was tholeiitic in character. Isotopic and trace element compositions preserve a record of crustal interaction. Addition of older continental crust to the mantle source regions *via* subduction is the most plausible contamination mechanism.

1. INTRODUCTION

The Stillwater Complex crops out on the northern edge of the Beartooth Range, one of the major exposed blocks in the Wyoming Archean Province (Figure 1). The complex is separated from the main Beartooth block by the Mill Creek-Stillwater Fault Zone and from the North Snowy Block by the West Boulder Fault (Figure 1, inset). The intrusive contact between the complex and the underlying metasedimentary rocks is locally exposed between the Boulder River and Chrome Mountain and in the Mountain View area (Figure 1). Between Chrome Mountain and the West Fork of the Stillwater River, the lower part of the complex is in fault contact with the hornfels. From the West Fork to the main Stillwater River, the complex is in fault contact with a younger quartz monzonite along the Bluebird Thrust, while east of the Stillwater River the complex has been intruded by the same quartz monzonite. Along its northern margin, the complex is overlain by Paleozoic and Mesozoic sedimentary rocks. For most of its length this contact is an angular unconformity except for the area between the West Fork and the Stillwater River where the contact is marked by the Horseman Thrust. The exposed part of the complex covers an area of ~180 km^2 with a maximum length of 47 km and a maximum width of 8 km (Figure 1). Gravity measurements north of the Beartooth front are consistent with a northward subsurface extension of the complex as a relatively flat sheet for approximately 30 km (Bonini, 1982).

Evidence for a major crust-forming episode that extended from ~3000 to 2740 Ma is preserved in the main Beartooth Range (Wooden and Mueller, 1989). This episode culminated in the production of voluminous granodiorites and granites of the Long Lake Suite between 2780 and 2740 Ma and isotopic data indicate that the intrusion of the Stillwater mafic magma at 2700 Ma was related to this same event. During the Proterozoic, mafic dykes were emplaced throughout the Beartooth Range and the complex and surrounding rocks were subjected to a low-grade regional metamorphism. The area was uplifted, tilted towards the north, and eroded during late Proterozoic. Subsidence and sedimentation from Middle Cambrian through Lower Cretaceous covered the complex with a sequence of sedimentary rocks up to 3000 metres thick. Laramide deformation during the late Cretaceous-early Tertiary resulted in uplift, tilting, and erosion which exhumed the late Proterozoic erosional surface.

A precise Sm-Nd isochron on mineral separates from a gabbronorite from the West Fork Adit portal gave a crystallization age of 2701±8 Ma (DePaolo and Wasserburg, 1979). Nunes (1981) determined an age of 2713±3 Ma on zircons from the Basal Series and Premo *et al.* (1990) determined a U-Pb zircon age of 2705±4 Ma for the dyke/sill suite that is associated with the Basal Series, indicating that the sills are coeval with the main complex.

The complex contains important reserves of base and noble metals. Sulphide-rich rocks associated with the Basal Series, adjacent hornfels, and lowermost Ultramafic Series have been extensively explored as a source of copper and nickel since the late nineteenth century. Chromite-rich seams associated with peridotites of the Ultramafic Series have also been extensively explored. During wartime periods when the demand for chromium was high, these deposits were mined in the Benbow, Mountain View and Gish areas. Stillwater chromites represent about 80% of the identified chromium reserves in the United States. The occurrence of platinum and palladium minerals in the complex has been known since the thirties but it was not until 1973 that the major platinum-group element-enriched zone, the J-M Reef, was discovered. The reef, which is composed of disseminated sulphides in a narrow zone within the lower part of the Banded Series, is presently being mined on both sides of the Stillwater Valley.

Figure 1. *(facing page) Map of the Stillwater Complex showing the major subdivisions, major faults, mineralized zones and locations referred to in the text. Inset shows the Stillwater Complex in relation to the major blocks which make up the Beartooth Mountains.*

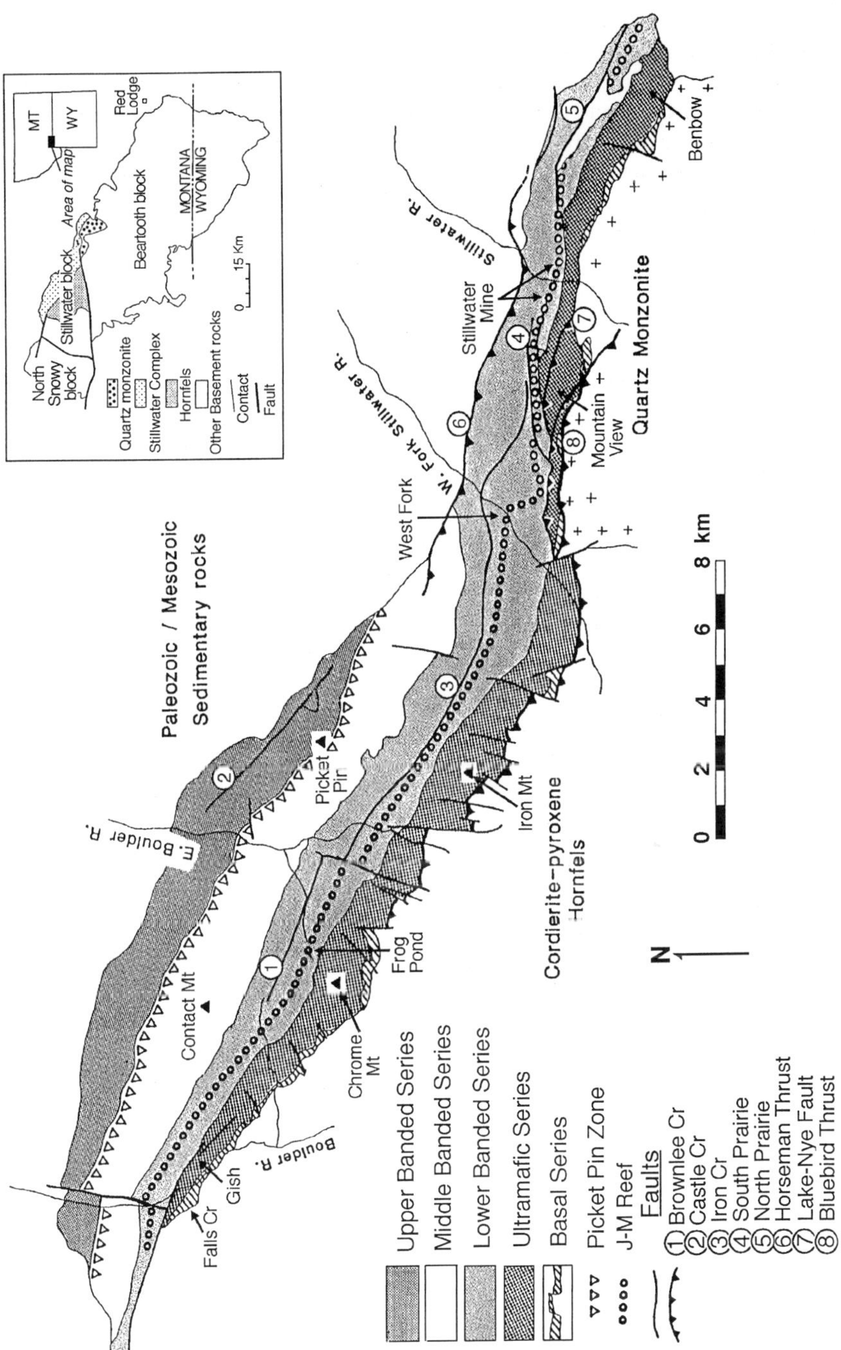

Paleozoic / Mesozoic Sedimentary rocks
North Snowy block
Stillwater block
Beartooth block
Area of map
Red Lodge
MT
WY
MONTANA
WYOMING
Quartz monzonite
Stillwater Complex
Hornfels
Other Basement rocks
Contact
Fault
0 15 Km
E. Boulder R.
Boulder R.
W. Fork Stillwater R.
Stillwater R.
Contact Mt
Picket Pin
Chrome Mt
Frog Pond
Iron Mt
Cordierite–pyroxene Hornfels
West Fork
Stillwater Mine
Mountain View
Quartz Monzonite
Benbow
Gish
Falls Cr
0 2 4 6 8 km
N
Upper Banded Series
Middle Banded Series
Lower Banded Series
Ultramafic Series
Basal Series
Picket Pin Zone
J-M Reef
Faults
1 Brownlee Cr
2 Castle Cr
3 Iron Cr
4 South Prairie
5 North Prairie
6 Horseman Thrust
7 Lake-Nye Fault
8 Bluebird Thrust

2. GEOLOGY OF THE COMPLEX

2.1. Structure and layering

Approximately 6 km of Laramide uplift has exposed the pre-Middle Cambrian erosional surface at a mean elevation of ~3000 metres (Jones *et al.,* 1960). Five major high-angle reverse faults of Laramide age with strikes subparallel to the layering and steep northeast dips have affected the Banded Series (faults 1-5, Figure 1). On each of these faults, the hanging wall has been elevated preserving remnants of Cambrian sedimentary rocks adjacent to the fault in the footwall block. Much of the movement along these faults appears to be along planes coincident with the igneous layering, *e.g.*, the South Prairie Fault (Figure 2).

A set of south-dipping thrust faults occurs in the eastern half of the complex (Page and Nokleberg, 1974). The northern fault of the Bluebird Thrust system has juxtaposed Ultramafic Series rocks against Banded Series rocks in the West Fork area (Figure 1). Further to the east, movement along this fault system has rotated a large wedge of the Ultramafic Series to form the Mountain View Block which is bounded on the north by the Lake-Nye Fault, which merges with the Bluebird Thrust toward the west. This fault has truncated the chromitite deposits of the Mountain View block and removed about 1000 metres of the Ultramafic Series in the Nye Creek area. The Horseman Thrust, which forms the northern boundary of the complex from Picket Pin Creek to Little Rocky Creek, has thrust slices of the complex over the Paleozoic strata (Figure 2).

Closely spaced, steeply dipping transverse faults with displacements ranging from less than a metre to several hundred metres are common in the Basal and Ultramafic Series but seldom extend far into the Banded Series rocks (Figure 1). Restriction of these faults to the lowermost units of the complex suggests that many of these faults may represent reactivated basin margin

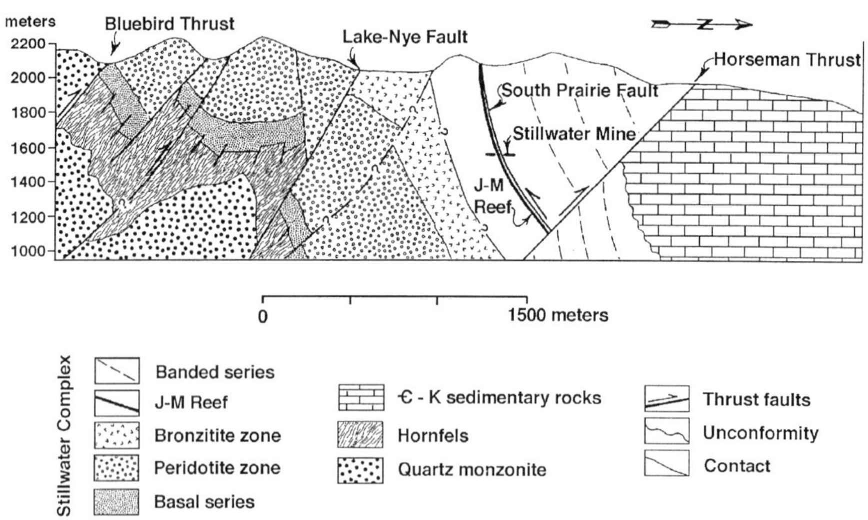

Figure 2. Structural section through the Stillwater Complex in the Mountain View area (after Turner et al., 1985).

growth faults that developed during the formation of the complex. The latest movement along these transverse faults postdated that along the south-dipping thrusts.

The fraction of Stillwater rocks which display layering is relatively small. The typical outcrop is modally uniform although many rocks show an igneous lamination defined by preferred orientation of plagioclase and augite. Anorthosites and bronzitites form megalayers up to several hundred metres thick, which can be traced across the entire complex. Thinner layers can be traced for distances of a few tens to a few hundreds of metres. Modally graded layers and rhythmic layering (repetitive modally graded layers) occur but are not common and size-graded and cross-bedded layers are rare. The most spectacular example of rhythmic layering is the inch-scale layering composed of alternating plagioclase-rich and pyroxene-rich layers. Macrorhythmic layers, which grade upwards over a distance of several metres from a pyroxene-rich base to a plagioclase-rich top, are locally preserved in gabbronorites on Contact Mountain.

2.2. Contact aureole

In the East Boulder Plateau, pelitic rocks have been thermally metamorphosed to pyroxene hornfels near the contact. A distinctive blue quartzite occurs as thin layers within the hornfels and banded iron formations form extensive outcrops at Iron Mountain and south of Chrome Mountain. Hornfels occurs as quartz-bearing and quartz-free varieties (Page, 1977; Barker, 1975). Quartz-bearing hornfels in the vicinity of the contact consists mainly of quartz-hypersthene-plagioclase-cordierite assemblages. At some distance from the contact, cummingtonite takes the place of hypersthene (Labotka, 1985). Quartz-free hornfelses are less common and restricted to contact zones and xenoliths within the complex. They consist dominantly of hypersthene and cordierite with minor plagioclase and locally contain abundant sulphide. Barker (1975) suggested that the quartz-free rocks might represent residues from partial melts whereas Page (1977) suggested that these two hornfels types reflect differences in original bulk composition. Assemblages in the iron-formation are consistent with peak metamorphic temperatures around 825°C (Labotka, 1985) and pressures between 300 and 400 MPa.

2.3. Major subdivisions of the complex

The complex has been subdivided into five major units: Basal Series, Ultramafic Series, Lower Banded Series, Middle Banded Series, and Upper Banded Series (Figures 1 and 3). Each series has been further subdivided into a number of zones and subzones (Figure 3). Also shown in Figure 3 is the stratigraphic distribution of the cumulus minerals and a compressed version of the variation in mode of cumulus minerals as a function of stratigraphic height for sections through the Banded Series in the Contact Mountain area and the Ultramafic Series in the Mountain View area. Each of the series and zone boundaries is based on the appearance or disappearance of one or more cumulus minerals. An additional unit, the Sill/Dyke Suite, is associated with the Basal Series.

2.4. Mineralogy

The most abundant primary minerals are olivine, orthopyroxene, clinopyroxene, inverted pigeonite, plagioclase, and chrome spinel. Minor primary minerals are quartz, phlogopite, amphibole, apatite, magnetite, ilmenite, and sulphides. A large variety of secondary minerals is present, the most abundant being serpentine and talc (after olivine and orthopyroxene), zeolites

Figure 3. Composite stratigraphic section compiled from the detailed sections of McCallum et al. (1980) and Raedeke and McCallum (1984) showing the zones and modal stratigraphy (cumulus minerals only). "S" denotes sulphide-enriched zones, "A - K" denote the major chromitites, and "P" denotes podiform concentrations of sulphide of limited lateral extent.

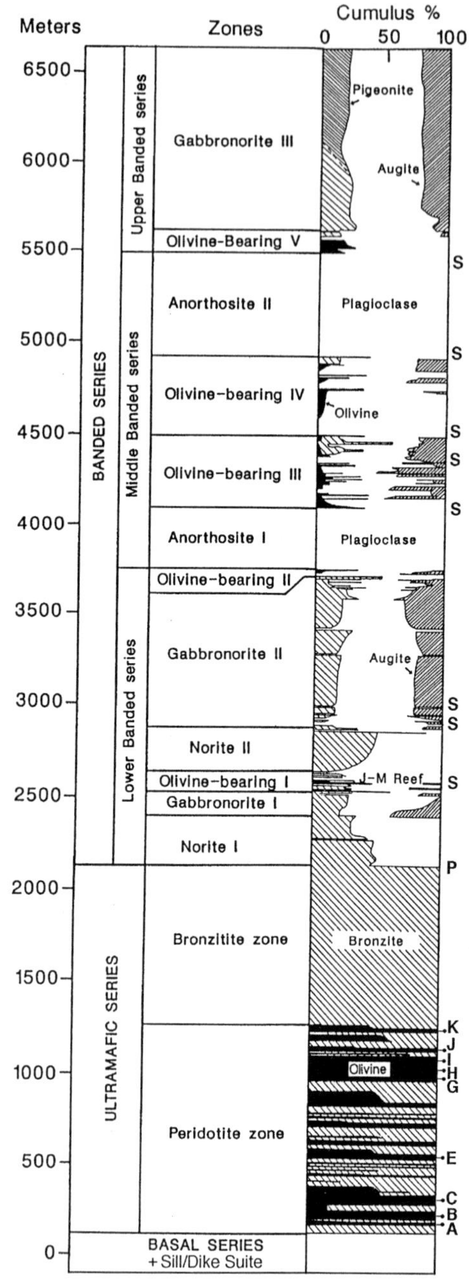

and zoisite (after plagioclase) and chlorite and actinolite (after pyroxene). The mineral assemblages are summarized in Table 1.

2.4.1. Olivine

Olivine occurs as a cumulus mineral in peridotites, harzburgites, troctolites and olivine gabbros and is the major constituent in "discordant dunite" masses. In the Ultramafic Series, olivine ranges from Fo_{90} to Fo_{79}. The more Fe-rich olivines are from the lowermost cycles of the Peridotite Zone while the most Mg-rich olivines are associated with chromitites. In all other Ultramafic Series samples the olivines show a very restricted compositional range (Fo_{86-84}). In troctolites and olivine gabbros from the Banded Series, olivine ranges from Fo_{79} to Fo_{64}. The lack of a compositional overlap between Ultramafic Series and Banded Series olivines is consistent with the stratigraphic gap between the last occurrence of olivine in the Peridotite Zone and its reappearance in Lower Banded Series. Alteration of olivine in the Ultramafic Series varies from the formation of a few veins of serpentine (+ magnetite) to complete replacement of entire outcrops with serpentine + magnetite ± talc ± calcite. In troctolites of the Banded Series, olivine is commonly altered to a pale brown amphibole which is surrounded by a rim of pale green chlorite adjacent to plagioclase.

Table 1
Rock names and terminology

Rock name	Cumulus Minerals	Postcumulus Minerals	Notation*
Peridotite	Ol (±Chr)	Opx, Cpx, Plag, Phl, (Amph, Ap)	oC/ocC
Harzburgite	Ol, Opx (±Chr)	Cpx, Plag, (Phl, Amph, Ap)	obC
Chromitite	Chr (±Ol)	Opx, Cpx, Plag, (Phl, Amph, Ap)	cC
Bronzitite	Opx	Plag, Cpx, (Qtz, Phl, Ap)	bC
Norite	Plag, Opx/Pig	Cpx, (Ap, Qtz)	pbC
Olivine gabbro	Plag, Cpx, Ol	Opx (Ap)	paoC
Gabbronorite	Plag, Opx/Pig, Cpx	(Qtz, Ap, Mt)	pbaC
Troctolite	Plag, Ol	Opx, Cpx (Ap)	poC
Olivine gabbronorite	Plag, Opx, Cpx, Ol	(Ap)	pbaoC
Anorthosite	Plag	Opx/Pig, Cpx, Qtz, (Mt)	pC

*Abbreviations: C = cumulate, p = plagioclase (plag); o = olivine (ol); c = chromite (chr); b = orthopyroxene/pigeonite (opx/pig); a = augite (cpx); qtz = quartz; ap = apatite; amph = amphibole; phl = phlogopite, mt = magnetite. Parentheses indicate minor phase that is not always present. Sulphides may occur as interstitial minerals in any assemblage.

2.4.2. Pyroxenes

Orthopyroxene occurs as a cumulus mineral in bronzitites, harzburgites, norites, and gabbronorites and as a postcumulus mineral in all other rocks. In virtually all samples containing coexisting olivine and orthopyroxene, the orthopyroxene is slightly more magnesian than the olivine indicating a close approach to equilibrium between these two minerals. The main charge-balanced substitutions are $^{[6]}$[Al,Cr]$^{[4]}$Al ⇔ $^{[6]}$Mg$^{[4]}$Si and $^{[6]}$Ti$^{[4]}$Al$_2$ ⇔ $^{[6]}$Mg$^{[4]}$Si$_2$, where [6] and [4] refer to octahedral and tetrahedral sites, respectively. Orthopyroxenes are unzoned and contain fine lamellae of augite along (100). The rims of orthopyroxenes generally have many fewer augite lamellae and are depleted in Ca and REE relative to the cores due to the exsolution of augite components out of the grain and their reprecipitation as blebby augite along grain boundaries.

Clinopyroxene (augite) is present in all zones, but in the complex as a whole it is less abundant than orthopyroxene. It occurs as a cumulus mineral in gabbronorites and olivine gabbros and as an intercumulus mineral in all other rock types. Fe and Mg distribution between coexisting pyroxenes suggests a close approach to equilibrium at high temperature. Element substitutions are the same as those outlined above for orthopyroxene with the addition of a minor $^{[8]}$Na$^{[6]}$Al ⇔ $^{[8]}$Ca$^{[6]}$Mg substitution. Cumulus augites tend to be elongated along the *c* axis and in most gabbros and gabbronorites the long axes of augite grains are randomly oriented in the plane of lamination. Postcumulus augites reach dimensions of 20 cm in some anorthositic samples and may show a decrease in Mg/(Mg+Fe) of ~8 mol% from center to edge. In Mg-rich augite, fine orthopyroxene and pigeonite lamellae have exsolved on (100) of the augite host, whereas more Fe-rich augites from the Banded Series contain both (001) and (100) lamellae of low-Ca pyroxene. These lamellae were initially exsolved at high temperature as pigeonite and during slow cooling, the lamellae coarsened with the (001) lamellae growing much faster due to the more rapid diffusion of Ca, Mg and Fe along the *c* axis. At some point during the cooling cycle, transformation of pigeonite to orthopyroxene was initiated in the (001) lamellae with eventual complete transformation to orthopyroxene.

Cumulus pigeonite (now inverted to orthopyroxene) is restricted to gabbronorites in the Upper Banded Series. The inversion process has produced an unusual poikilitic texture in which each orthopyroxene grain contains multiple domains of (001) exsolution lamellae, each domain delineating an original cumulus pigeonite. The relict (001) and (100) augite lamellae which exsolved prior to inversion commonly form a herring bone pattern consistent with a precursor pigeonite twinned on (100). In many samples the regular lamellae are accompanied by blebby augite. After inversion, the orthopyroxene exsolved fine augite lamellae on (100). Oikocrysts of inverted pigeonite, commonly in epitaxial intergrowths with augite, are common in anorthosites.

2.4.3. Plagioclase

Plagioclase is the most abundant cumulus mineral throughout the Banded Series and it occurs as an intercumulus mineral throughout the Ultramafic Series. Grain sizes of cumulus plagioclase vary widely from <0.1 cm to ~ 1 cm even within a single thin section. In the Middle Banded Series, the average grain size of plagioclase is 2 to 3 times that of plagioclase in the Lower and Upper Banded Series (McCallum *et al.*, 1980). Sharp grain-size discontinuities also occur within anorthosites within a few metres of the contacts (Boudreau and McCallum, 1986). In most norites, gabbros, and gabbronorites, tabular plagioclase crystals define a distinct igneous lamination. Lamination is less pronounced in anorthosites and troctolites.

Plagioclase ranges from An_{88} to An_{60} in the Banded Series. A similar range is observed in the intercumulus plagioclase of the Ultramafic Series. Systematic decreases in *average* An content with stratigraphic height occur in the Lower and Upper Banded Series, but no such systematic variation is observed in the Middle Banded Series. FeO contents range up to 0.52 wt% and correlate with FeO contents in coexisting pyroxenes. Cumulus plagioclase is relatively homogeneous in norites and gabbronorites while plagioclase in anorthosites and troctolites shows more extensive zoning with normal, reversed and patchy zoning patterns, often within the same grain. Within the Banded Series, alteration of plagioclase to zoizite may affect only part of a grain or it may be pervasive throughout entire outcrops.

2.4.4. Chromite

In the Ultramafic Series the highest concentrations of chromite occur in the peridotite member of each cyclic unit. Chromite is present in minor amounts in harzburgites and bronzitites and in the olivine-bearing rocks of the J-M Reef. Within peridotite, chromite occurs as massive seams from a few cm to ~ 1 metre thick, as irregular patches of chromitite, and as disseminated grains. In massive chromitites, chromite reaches its maximum MgO and Cr_2O_3 contents and minimum Al_2O_3 and Fe_2O_3 contents while in rocks with sparsely disseminated chromites, the reverse is the case (Campbell and Murck, 1993). The primary compositions of chromites are retained in massive layers whereas disseminated chromites have undergone extensive subsolidus exchange with silicates. Chromites in the J-M Reef are more Fe-rich than those in the Ultramafic Series.

2.4.5. Apatite

Apatite is an important minor mineral throughout the complex (Boudreau *et al.,* 1986). Cl-rich apatite (>6.0% Cl) is characteristic of the lower third of the complex and a change to more F-rich apatite (>1.4% F) occurs within OB-I (olivine-bearing unit I - see Figure 3) just above the J-M Reef (Figure 4). Within the J-M Reef, chlorapatite, which is associated with pegmatitic olivine- and phlogopite-bearing rocks, contains >2.0% total rare earth elements (REE) with a typical LREE-enriched pattern. Apatites with such high chlorine contents are rare in igneous rocks and available evidence indicates that Stillwater chlorapatite is a product of high-temperature hydrothermal activity (Boudreau and McCallum, 1989).

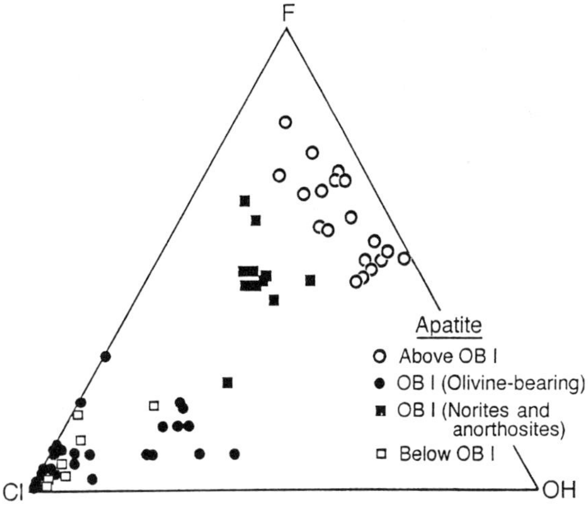

Figure 4. *Variations in Cl-F-OH of apatite from different levels of the complex (after Boudreau et al., 1986).*

2.4.6. Phlogopite and amphibole

Phlogopite is a minor intercumulus mineral in peridotites from the Ultramafic Series and occurs as an interstitial mineral in the olivine-bearing rocks of OB-I. Compositional inhomogeneities are common but, in general, the major element compositions indicate a close approach to equilibrium with the other major silicates. Page and Zientek (1987) showed that phlogopite in peridotite contains 74% to 80% of the phlogopite end-member and roughly equal amounts of the annite and siderophyllite end-members. Cr and Ti occupy octahedral sites in the phlogopite structure. For Cr, the substitution mechanism is $^{[6]}Mg^{[4]}Si \Leftrightarrow {}^{[6]}Cr^{[4]}Al$ while for Ti, two substitutions are involved, *i.e.*, $^{[6]}Mg_2 \Leftrightarrow {}^{[6]}Ti^{[6]}\square$ and $^{[6]}Mg_2{}^{[4]}Si \Leftrightarrow {}^{[6]}Ti^{[4]}Al_2$. Stillwater phlogopites are enriched in Cl (up to 0.5%) and F (up to 0.5%) compared to those from other layered intrusions, with the exception of the Bushveld Complex.

Primary amphibole is rare and restricted to the lowermost peridotite members of the Ultramafic Series. Phlogopite and amphibole tend to be mutually exclusive. Brown amphibole, which is invariably a late crystallizing, interstitial mineral, occurs as rims around chromites and as interstitial material replacing postcumulus augite indicative of a melt reaction relationship. Amphibole compositions are somewhat variable but most are pargasite or pargasitic hornblende with up to 4.5% TiO_2 and 1.8% Cr_2O_3 (Page and Zientek, 1987).

2.4.7. Sulphides, tellurides, arsenides, alloys

In the Basal Series, Sill/Dyke Suite, and adjacent hornfels, sulphides are common and locally form massive concentrations. Mafic norite dykes are particularly rich in sulphides with pyrrhotite being the most abundant followed by chalcopyrite and pentlandite. In the Ultramafic Series, sulphides are not abundant although most samples contain minute amounts of interstitial sulphide. The largest concentrations of sulphide minerals in the Ultramafic Series are associated with chromite-rich layers. In the Banded Series, 10 sulphide-bearing zones have been mapped by McCallum *et al.* (1980). By far the most important of these is the J-M Reef in which sulphides occur as disseminated grains and patches in troctolite and/or anorthosite in a

1-2 metre thick mineralized zone. The most abundant minerals in the reef are pyrrhotite, chalcopyrite, and pentlandite with much smaller amounts of moncheite, braggite, cooperite, kotulskite, Pt-Fe alloy, cubanite, sperrylite, vysotskite, telluropalladinite, keithconnite, stillwaterite and several other rare PGE-rich species. Pentlandite is the major host for Pd.

3. BASAL SERIES AND SILL/DYKE SUITE

3.1. Basal Series

The Basal Series forms an irregular sheet-like mass from ~60 m to ~400 m (average ~160 m) thick separating the cumulates of the Ultramafic Series from the hornfelses. The upper contact of the Basal Series is placed at the base of the first cyclic unit of the Ultramafic Series. The lower contact is more irregular and cuts across stratigraphic units suggesting emplacement of the Stillwater magma along an unconformity. Page (1979) included within the Basal Series all igneous rocks below the first appearance of cumulus olivine whereas Zientek (1983) considered the sill/dyke suite as a separate unit. The Basal Series has been cut out by thrust faults in the central part of the complex (Figure 1).

The dominant lithology is a uniform bronzitite cumulate which forms a laterally extensive upper layer. This unit is underlain by a variety of cumulate-textured rocks which are most commonly norites but also include anorthosites, gabbros, peridotites, and sulphide-bearing assemblages. Xenoliths of cordierite-pyroxene hornfels are fairly common. Although the same rock types occur in all areas where the Basal Series has been examined, there is a lack of continuity along strike and each area is unique in terms of lithologic sequences and thicknesses.

Page (1979) has documented an upward decrease in intercumulus mineral content, and a general upward increase in Mg/(Mg+Fe) through the basal bronzitite. Orthopyroxene shows a wide range in composition from En_{30-90} with the bulk of the grains in the range En_{60-80}, while plagioclase ranges from An_{60-83}. Major Fe-Ni-Cu sulphide concentrations occur in Basal Series rocks and adjacent hornfels in the Benbow, Nye Basin, and Mountain View areas. Sulphides occur as massive accumulations, interconnected matrix sulphide, and as isolated blebs and aggregates. Textures are suggestive of crystallization from a sulphide melt although, in some samples, there is evidence for subsolidus remobilization of sulphide. Ni/Cu ratios are variable but average around 1.0 with an average Cu+Ni content of 0.5%.

3.2. Sill/Dyke Suite

Zientek (1983) described two petrographically distinct types of sills and dykes which intrude the rocks at the base of the complex. One type has a diabasic texture, with compositions ranging from norite to gabbronorite, and is generally sulphide-poor. The other type is a mafic norite in which sulphides make up from 2 to 40% of the rock. Mafic norites are spatially associated with the Basal Series rocks while the diabases are associated with hornfels. The non-porphyritic texture and chilled margins of both types indicate that they were intruded as liquids. Field relations and age determinations indicate that the intrusion of the sills and dykes was contemporaneous with or slightly predated the formation of the main layered series (Premo *et al.*, 1990), lending credence to the suggestion that some of these intrusions may have formed from the same magmas that gave rise to the layered series.

Helz (1985) distinguished five chemical groups, four of which correspond to Zientek's diabase group and the remaining group corresponds to the mafic norite. The different groups cannot be related to each other by any simple fractionation process and the relative abundances of each type varies along strike. The mafic norite and the Mg-gabbronorite groups are the most promising candidates for a Stillwater parental magma. These suites have Mg/(Mg+Fe) appropriate for the crystallization of Mg-rich olivines and pyroxenes of the Ultramafic Series, their high SiO_2 and MgO contents are consistent with the early crystallization of orthopyroxene, and their low alkali contents are consistent with the crystallization of calcic plagioclase.

4. ULTRAMAFIC SERIES

The base of the Ultramafic Series is placed at the first appearance of significant quantities of cumulus olivine, while the upper boundary is placed at the horizon where plagioclase appears as a cumulus phase. The basal contact of the Ultramafic Series is preserved in the western part of the complex and locally in the Mountain View and Benbow areas whereas the upper contact is exposed at intervals along the length of the complex (Figure 1). The Ultramafic Series is subdivided into a lower **Peridotite Zone [PZ]** in which olivine ± orthopyroxene ± chromite are cumulus phases and an upper **Bronzitite Zone [BZ]**, in which orthopyroxene is the cumulus phase (Jackson, 1961). The Peridotite Zone comprises a repetitive sequence of cyclic units; a complete cyclic unit consists of peridotite-harzburgite-bronzitite. The Bronzitite Zone is relatively uniform except for a few thin layers containing olivine ± chromite. Variations in thickness of the Ultramafic Series range from 2000 m at Mountain View to 840 m at Chrome Mountain and most likely reflect topographic relief on the floor of the complex during initial stages of emplacement. The stratigraphic section and cryptic variation through the Ultramafic Series at Mountain View is shown in Figure 5.

4.1. Peridotite Zone

Raedeke and McCallum (1984) described 21 cyclic units at Mountain View and 20 at Chrome Mountain. Of the 21 cyclic units at Mountain View, 15 have the complete sequence, five are missing the peridotite member and one does not contain the bronzitite member. The number of cycles depends on the status accorded textural and modal variations within otherwise uniform lithologies. For example, Page *et al.* (1972) subdivided the peridotite member in cyclic unit 2 from the Nye Basin into nine subunits on the basis of abrupt changes in olivine grain sizes and abundance of chromite. The lower contacts of the peridotite units are sharp. Olivine, commonly accompanied by a small amount of chromite, forms the framework of cumulus grains and orthopyroxene occurs as oikocrysts enclosing partially resorbed olivines. Intercumulus plagioclase makes up between 2 and 15% of the peridotites at Mountain View. Late-crystallizing augite, phlogopite, and amphibole in decreasing order of abundance are minor interstitial constituents. Apatite and sulphides are present in trace amounts.

Layers of massive and disseminated chromite occur in the peridotite member of many of the cyclic units. The main chromitite seams are traditionally referred to as **A** (lowermost) through **K** (uppermost). Mining has been restricted to the **G** and the **H**, which are well-exposed at Mountain View (Figure 6). Individual layers of chromitite range from a single crystal thick to a metre or more. Layers thicken and thin over short distances and a single layer may bifurcate into two or more sublayers. In the units below and between massive chromitites, pegmatites are common. In the G chromitite, some pegmatites contain up to 10% of coarse-grained phlogopite and higher than normal abundances of sulphides. The most abundant interstitial mineral in chromitites is chrome-rich augite, stabilized by the high Cr_2O_3 activity.

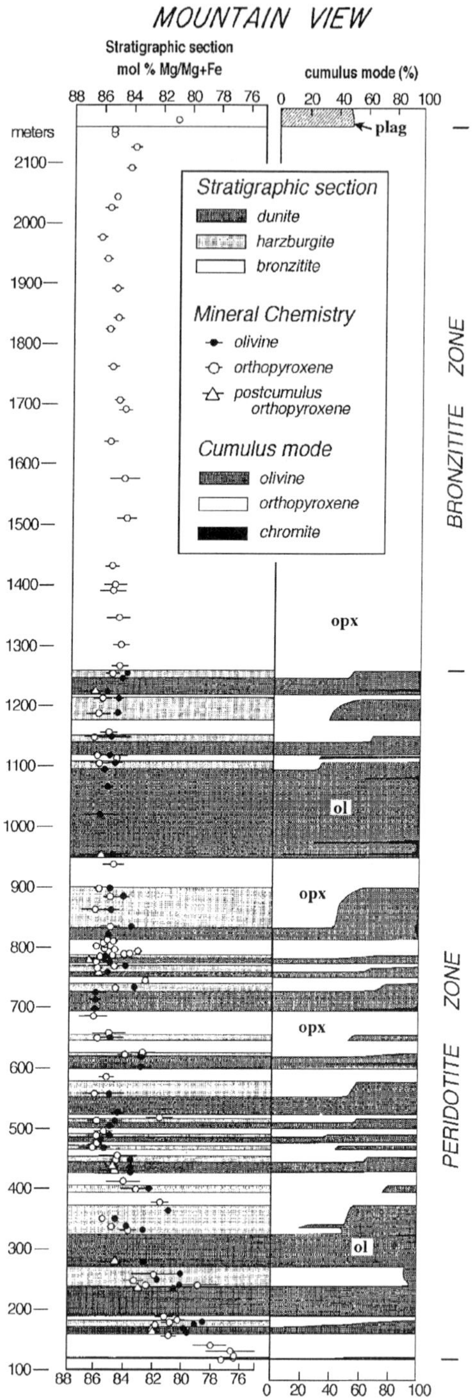

Figure 5. *Stratigraphy, cryptic variation and modal variation in the Ultramafic Series at Mountain View. Thicknesses are measured from the base of the Basal Series (after Raedeke and McCallum, 1984).*

The contact between peridotite and overlying harzburgite is marked by an abrupt increase in the amount of orthopyroxene and a gradational change in texture from poikilitic to equigranular. Layering in harzburgite occurs on a centimetre scale in the form of layers alternately rich in olivine or orthopyroxene. Olivine makes up, on average, ~35% of the harzburgite and, with few exceptions, the olivine/orthopyroxene ratio decreases up section. In some harzburgites, both olivine and orthopyroxene show evidence of secondary enlargement, while in a few others, olivine shows evidence of resorption and reaction with interstitial melt. Augite forms sparse oikocrysts, and plagioclase occurs as space-filling grains. Chromite, phlogopite, amphibole, apatite and sulphides are rare.

The contact of harzburgite with bronzitite is marked by the abrupt disappearance of olivine. Bronzite crystals adjacent to plagioclase are generally euhedral while those enclosed by poikilitic augite are highly embayed which suggests the reaction: bronzite + liquid $\rightarrow$ augite. Chromite and interstitial quartz are present in small amounts while phlogopite, apatite and sulphides are very rare. Irregular patches of pegmatite occur randomly throughout the bronzitites. These pegmatitic patches are modally similar to the "normal" bronzitite, suggesting an isochemical grain-coarsening mechanism.

Textures and abundances of minerals in the Peridotite Zone are consistent with

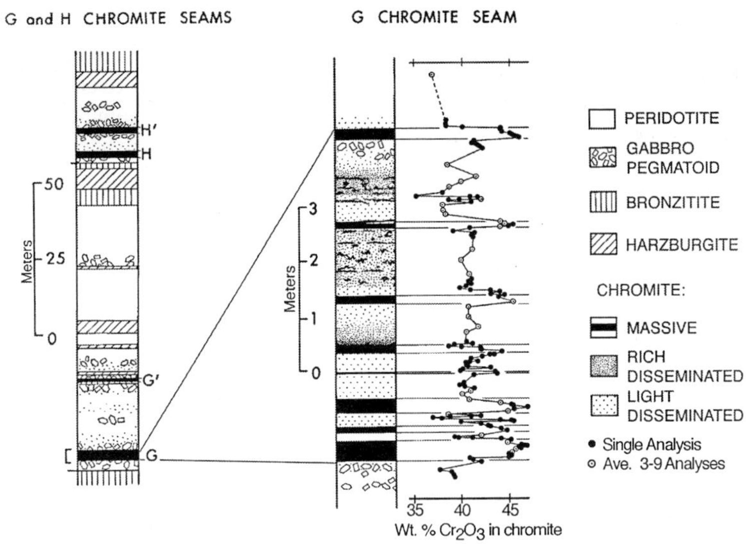

Figure 6. *Stratigraphy in the vicinity of the G and H chromitites at Mountain View and variations of chromite compositions through the G chromitite (after Campbell and Murck, 1993).*

an overall crystallization sequence of olivine (±chromite) → orthopyroxene → plagioclase → clinopyroxene → phlogopite → amphibole with reaction relationships between olivine and orthopyroxene, orthopyroxene and clinopyroxene, and clinopyroxene and amphibole.

4.2. Bronzitite Zone

The Bronzitite Zone at Mountain View appears to be uniform. However, exposure is not complete and minor amounts of other lithologic units may be present. At Chrome Mountain, a narrow harzburgite outcrops near the bottom of the zone and a thin chromite-bearing harzburgite occurs approximately 30 metres below the upper contact. Intercumulus minerals in the Bronzitite Zone are the same as those in bronzitites in the cyclic units. In the uppermost 20 metres of this zone, intercumulus plagioclase and augite become more abundant.

Textures range from orthocumulate to adcumulate. In orthocumulates, most bronzite grains have morphologies and shapes that are consistent with homogeneous nucleation and crystal settling. In some adcumulates, coalescence of grains, grain growth, and subsolidus annealing have modified the primary textures. Zoning in bronzites is minor although some grains contain concentrically arranged arrays of exsolved lamellae of Fe-Ti oxides. Interstitial plagioclase grains show a range of composition and zoning patterns as might be expected for crystallization from an intercumulus melt.

4.3. Discordant dunites

At Chrome Mountain, Iron Mountain and the Gish area, discordant dunites crop out over large areas. These rocks have a smooth, tan, weathered surface which is distinct from that of the "normal" cumulates. In most outcrops, the dunites appear to *replace* previously formed

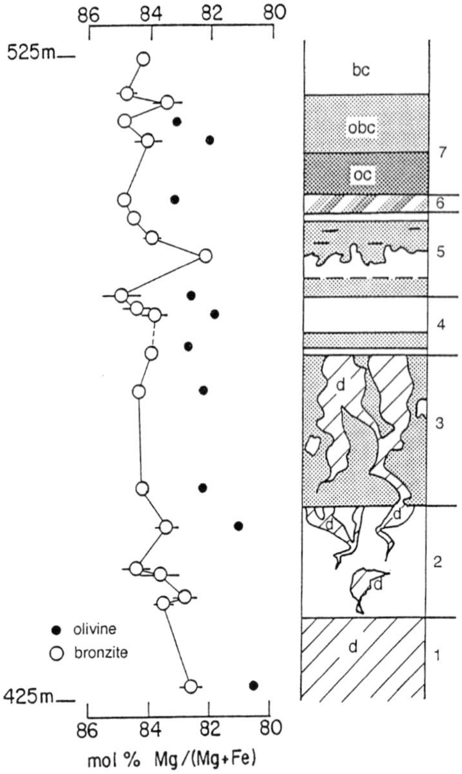

***Figure** 7. Detailed stratigraphy through a 100 m section of the Peridotite zone on the south flank of Chrome Mountain. (1) serpentinized dunite, (2) interlayered bronzitite and pegmatitic bronzitite with irregular veins and masses of dunite, (3) harzburgite with irregular masses of dunite, (4) bronzitite-harzburgite-pegmatitic bronzitite sequence, (5) harzburgite-bronzitite sequence with sinuous contact marked by a thin chromitite, (6) fine-scale interlayered peridotite, harzburgite, and bronzitite, (7) normal cyclic unit.*

cumulates while in a few locations, the dunites *crosscut* the cumulate layers. In the former, contacts between the discordant dunite and primary cumulates tend to be irregular or sinuous while in the latter the contacts tend to be planar. In both cases the contacts are sharp. The dunites are extensively serpentinized, olivine-rich rocks containing minor chrome spinel and sparse orthopyroxene oikocrysts. While the discordant dunites occur in all cyclic units, they are most abundant in the lower units where they form veins and irregular masses which both cross and are crossed by the cumulate layering, indicating formation at the same time as the cumulates (Figure 7). The compositions of olivine and orthopyroxene in the discordant dunites are not significantly different from those in the "normal" cumulates. The pyroxene-rich pegmatites are commonly associated with the discordant dunites. Contacts between these pegmatites and surrounding cumulates are sinuous and coarse-grained chromite is concentrated along the contact. The observations are consistent with a process in which dissolution of bronzite releases Cr which is precipitated as chrome spinel.

4.4. Mineral compositions

The change in mineral composition with stratigraphic height through the Peridotite Zone shows a systematic trend of *upward* Mg-enrichment (from En_{76} to En_{86}) through the lowermost 400 metres of the complex followed by little compositional variation through the remainder of the series (Figure 5). Minor elements in olivines and pyroxenes follow a similar pattern. Throughout the Bronzitite Zone, orthopyroxene compositions remain constant ($En_{85\pm2}$). Minor and trace elements show a larger range but such variation as exists again shows no systematic variation with stratigraphic position: Cr contents are high (average Cr_2O_3 = 0.6%) and rare earth element (REE) abundances show a typical HREE-enriched pattern with $[Ce/Yb]_n < 0.15$ (Papike *et al.*, 1995).

The total range in REE abundances is small, the patterns are subparallel, and there is a significant negative Eu anomaly (Lambert and Simmons, 1987).

Intercumulus plagioclase in ultramafic samples shows some zoning ranging in one sample of bronzitite from An_{69} to An_{86}; however, all but a few points lie between An_{75} and An_{80}. Clinopyroxene oikocrysts, which increase in abundance towards the top of the Bronzitite Zone, are largely unzoned and have compositions indicating equilibrium with coexisting orthopyroxenes at near-solidus temperatures. During subsolidus cooling, element redistribution is limited to grain boundaries and to intracrystalline exsolution lamellae.

Data of Campbell and Murck (1993) on coexisting minerals from the G and H chromitite show that the maximum Cr_2O_3 content in chromite occurs in the massive layers (Figure 6) and there is a positive correlation among X_{Mg}, X_{Cr} and the modal abundance of chromite. In chromite-rich layers, both olivine and chromite have higher X_{Mg} than in chromite-poor layers which has been attributed to extensive subsolidus exchange of Fe and Mg between coexisting chromite and silicates *within* layers (Irvine, 1967). With the exception of Mg-rich olivines associated with massive chromitites, the olivines are relatively constant in composition ($Fo_{84\text{-}87}$).

5. BANDED SERIES

The Banded Series comprises all rocks in which *cumulus* plagioclase is a major constituent. On the basis of detailed sections in the Contact Mountain and Picket Pin areas, McCallum *et al.* (1980) split the Banded Series into the Lower Banded Series (LBS), Middle Banded Series (MBS) and Upper Banded Series (UBS) and further subdivided these units into twelve zones. The boundaries between units are placed at distinctive lithologic breaks which are easily identified in the field. In their map of the Banded Series, Segerstrom and Carlson (1982) identified most of the same zones. The Lower Banded Series is composed of norites and gabbronorites with minor amounts of olivine-bearing cumulates which host the platiniferous J-M Reef. The Middle Banded Series is made up of anorthosites, olivine gabbros and troctolites and the Upper Banded Series comprises gabbronorites with minor troctolite and norite.

The most complete sections through the Banded Series are exposed from Contact Mountain to the Picket Pin area (Figure 1); an unknown thickness of stratigraphically higher cumulates lie beneath the sedimentary cover. East of Picket Pin, Paleozoic sediments progressively onlap the uppermost cumulates covering the Upper Banded Series and eventually the Middle Banded Series. In the eastern third, only the Lower Banded Series is exposed south of the Horseman Thrust. A simplified stratigraphic section and compositions of the primary minerals in the Banded Series are shown in Figure 8.

5.1. Lower Banded Series (LBS)

The lower contact of the LBS is placed at the horizon marking the first appearance of cumulus plagioclase and the upper contact is placed at the base of the first thick anorthosite unit (Figure 3). The Lower Banded Series has been subdivided into six zones: Norite-I (N-I), Gabbronorite-I (GN-I), Olivine-bearing-I (OB-I), Norite-II (N-II), Gabbronorite-II (GN-II), and Olivine-bearing-II (OB-II).

5.1.1. Norite-I Zone (N-I) and Gabbronorite-I Zone (GN-I)

Cumulus orthopyroxene and plagioclase are in approximately cotectic proportions in N-I whereas clinopyroxene occurs as oikocrysts making up between 1 and 10% of the rock. Minor amounts of sulphide, apatite and quartz are commonly present. Modally graded and

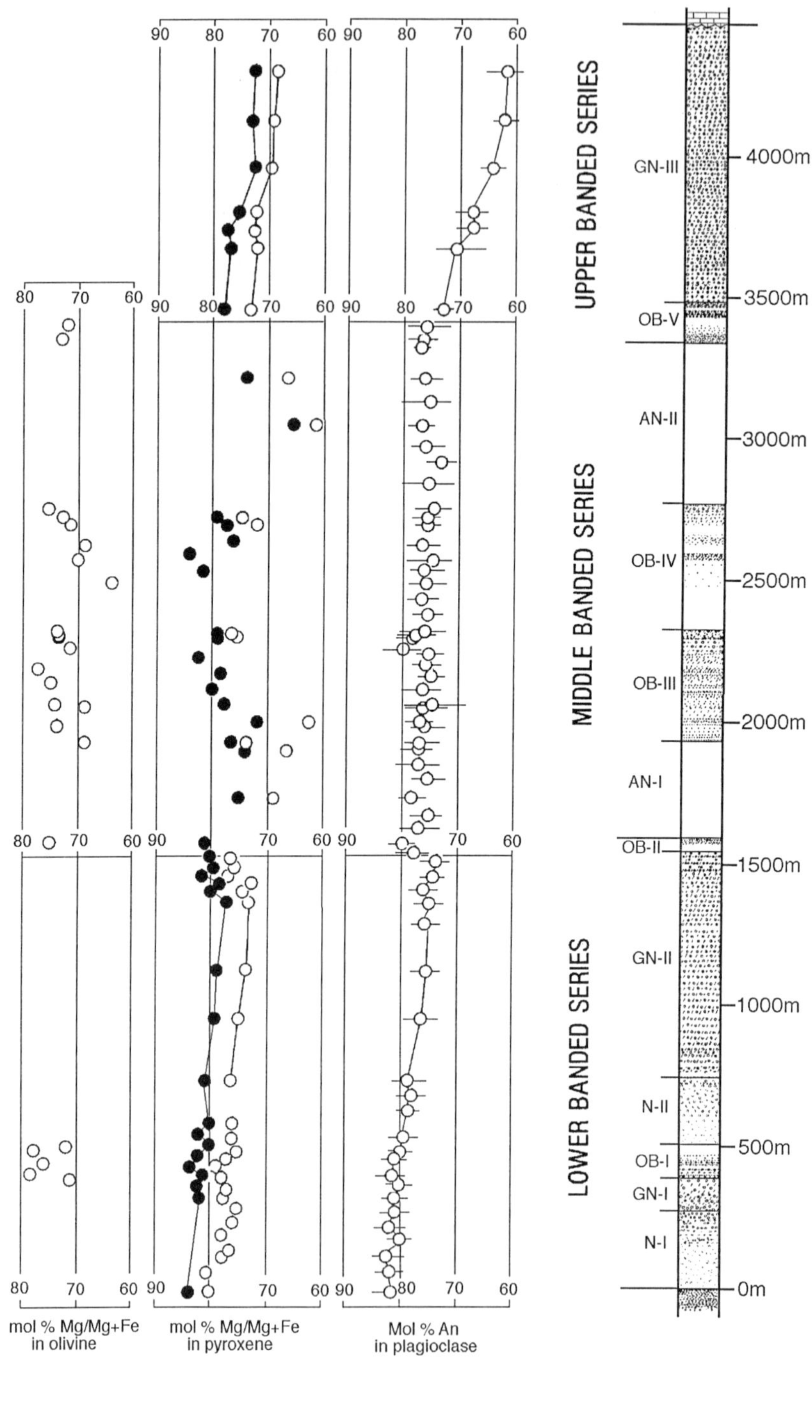

UPPER BANDED SERIES
MIDDLE BANDED SERIES
LOWER BANDED SERIES
GN-III
OB-V
AN-II
OB-IV
OB-III
AN-I
OB-II
GN-II
N-II
OB-I
GN-I
N-I
4000m
3500m
3000m
2500m
2000m
1500m
1000m
500m
0m
mol % Mg/Mg+Fe in olivine
mol % Mg/Mg+Fe in pyroxene
Mol % An in plagioclase

Figure 8. (facing page) Stratigraphic variations in compositions of plagioclase, pyroxene, and olivine through the Banded Series. Data sources: McCallum et al. (1980), Criscenti (1984), Page and Moring (1987), Meurer and Boudreau (1996), Salpas et al. (1983), Haskin and Salpas (1992), Czamanske and Scheidle (1985), Boudreau and McCallum (1986). In pyroxene column, filled circles are clinopyroxene, open circles are orthopyroxene.

rhythmic layering are common, particularly in the more leucocratic members. A prominent anorthosite about 2 metres thick occurs about midway through the N-I Zone. This sharply-bounded layer of anorthosite has no complementary mafic layer indicating that *localized* crystal sorting is not the mechanism responsible for the concentration of plagioclase. The contact between N-I and GN-I is placed at the first appearance of cumulus augite. In the upper part of GN-I, there is a complex, laterally extensive unit characterized by layers of norite, gabbronorite, and anorthosite which are locally disturbed and associated with abundant pyroxenite xenoliths which are commonly surrounded by a narrow rim of chrome spinel.

Page and Moring (1987) have identified seven subzones within N-I and GN-I on the basis of distinctive modal changes in outcrops located close to the west portal of the Stillwater Mine. These outcrops display spectacular rhythmic layering with many of the layers showing modal grading, cross-bedding, channel structures, onlap and offlap structures, and slump structures which are clearly syndepositional. Orthopyroxene ranges from En_{83} to En_{75} and plagioclase from An_{83} to An_{78} (Figure 8).

5.1.2. Olivine-bearing Zone I (OB-I)

The basal contact of OB-I, which was placed by McCallum *et al.* (1980) at the first appearance of cumulus olivine in the Banded Series, is well-defined but irregular and may represent an unconformity. The upper contact of OB-I is placed at a horizon marked by a distinctive textural change from mottled anorthosite to layered norite. The reappearance of olivine in a series of cyclic units, the unconformable lower contact, and the occurrence of bronzitite xenoliths are consistent with multiple injections of olivine-saturated magma followed by a prolonged period of mixing before the magma returned to a relatively uniform composition represented by the overlying norite zone.

Surface mapping, logging of drill cores, and mapping of exploration adits and mine exposures have revealed a remarkable degree of lateral variation in OB-I. Stratigraphic sections through OB-1 in the Frog Pond/Dead Tree, West Fork, and Stillwater River areas are shown in Figure 9, although these sections may not be representative of the entire zone. The Frog Pond/Dead Tree area, which represents the most complete section through OB-I, is approximately 120 metres thick. Ten olivine-bearing members (O_1-O_{10}) composed of coarse-grained to pegmatitic peridotites and/or troctolites have been recognized by Todd *et al.* (1982). In the part of OB-I below the J-M Reef, these units are interlayered with norites, gabbronorites, and minor anorthosites. Above the reef, anorthosite predominates (Figure 9). Troctolitic layers grade into norite layers along strike, individual layers commonly pinch out, and there are local unconformities and onlapping sequences. Todd *et al.* (1982) noted the existence of cyclic units within the upper part of OB-I with a typical cycle composed of peridotite, troctolite and anorthosite. In the West Fork area, OB-I is well exposed in the West Fork Cliffs where the first discovery of the J-M Reef in outcrop was made in 1974 (Mann *et al.*, 1985). Here, olivine zones O_1 through O_4 are absent although they may be represented by

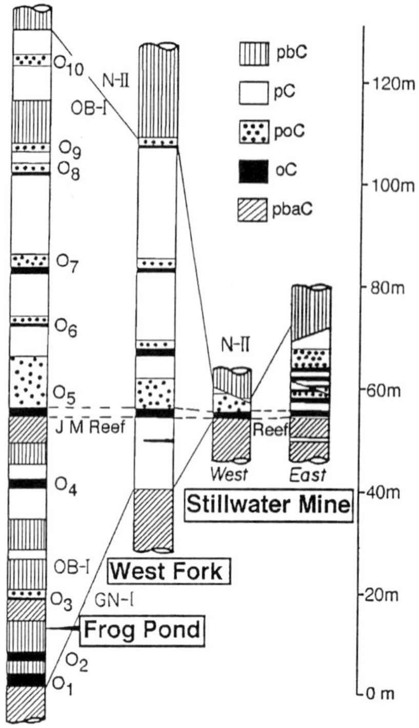

Figure 9. *Stratigraphic sections through OB-I at the Frog Pond adit (Todd et al., 1982), West Fork adit (Mann et al., 1985) and Stillwater Mine (Turner et al., 1985; Barnes and Naldrett, 1986). Note the presence of faults in the Stillwater Mine sections.*

pyroxene-rich layers (Mann and Lin, 1985). The lowermost olivine layer at West Fork is correlated with O_5 from Frog Pond since both are associated with the mineralized J-M Reef. The section above the reef at West Fork is similar to that at Frog Pond (Figure 9).

In the eastern part of the complex, where the reef is being mined, stratigraphy in OB-I is less regular, in part because of Laramide faults and in part because of thinning of units across basement highs. Mapping in the Stillwater Mine reveals that GN-I and OB-I become progressively thinner as they are traced west from the Stillwater Valley until GN-I disappears and OB-I is reduced in thickness. The South Prairie reverse fault has disrupted OB-I in this region. The main strand of this fault is confined to the hanging wall norites but numerous splays affect the reef package in the Mountain View area. Towards the west and east the South Prairie Fault cuts progressively higher into the hanging wall. With the exception of the olivine-bearing member that hosts the J-M Reef, the olivine-bearing units, which are prominently developed in the Frog Pond-West Fork areas, are absent, or poorly developed, in the Stillwater Mine.

5.1.3. J-M Reef

The J-M Reef is not restricted to a single stratigraphic position within OB-I. At Frog Pond and West Fork, the reef, which contains 1-2% disseminated sulphides through 1-3 metres, is associated with the O_{5B} unit which consists of a 1-1.5 m thick pegmatitic peridotite overlain by a troctolite up to 3.5 metres thick which is the host of the main PGE mineralization (Figure 9). The reef is generally confined to the troctolite but it varies considerably in thickness and grade and in some localities it is absent (LeRoy, 1985). The most common sulphides are pyrrhotite, pentlandite (containing up to 5% Pd), and chalcopyrite with minor moncheite, cooperite, braggite, kotulskite, Pt-Fe alloy and various arsenides. The reef averages 20-25 ppm Pt + Pd over a thickness of ~2 metres with a Pd/Pt ratio of ~3.6 (LeRoy, 1985). In the West Fork and Frog Pond adits, localized downwarps in the stratigraphy in which the mineralized zone is significantly thickened, have been compared to the pothole structures of the Merensky Reef.

In the Stillwater Mine, OB-I (commonly referred to as the reef package) is quite different from that at Frog Pond and West Fork. The mineralized zone is correlated with the O_{5B} unit at Frog Pond but the underlying, and several of the overlying, olivine-bearing units are not present (Figure 9). The base of the reef package is placed at the first stratigraphically continuous olivine-rich layer which lies *discordantly* on a rhythmically layered sequence of gabbronorites, norites and anorthosites. A typical reef package is composed of a basal pegmatitic olivine-rich rock overlain by a variety of coarse-grained to pegmatitic assemblages containing ameboidal olivine in a matrix of plagioclase and pyroxene, informally referred to as "mixed rock" (Bow *et al.*, 1982). The mixed rock is overlain by a sequence of troctolite, mottled anorthosite, and norite. The upper contact of the reef package is placed at the point where the olivine-bearing norite grades into olivine-free norite. PGE mineralization in the mine occurs at four levels relative to the base of the reef package: (1) Footwall zone in GN-I just below the lower contact of the reef package, (2) Basal zone which straddles the basal contact, (3) Main zone, and (4) Upper zone (Raedeke and Vian, 1986). Mineralized zones are generally less than 3 metres thick except where several of the zones have coalesced to form thickened zones, referred to as "ballrooms" by mine geologists. Ore is patchily developed; areas of high grade ore are separated by low grade areas up to 100 metres wide. In the eastern part of the Stillwater Mine, the highest grade PGE-sulphides occur in Upper and Main zones. As the reef package is traced west, the rocks become progressively richer in pyroxene at the expense of olivine and the highest ore grades in the reef progressively cut down section and occur primarily in the Main, Basal and Footwall zones. Turner *et al.* (1985) suggested that the westward progression from olivine-rich to pyroxene-rich reef rocks appears to be related to pothole-like structures.

To constrain the source of the metals and to evaluate isotopic equilibrium, Bosch *et al.* (1991) and McCallum *et al.* (1992) measured Pb isotopic compositions of multiply leached sulphides and plagioclases from the J-M Reef and vicinity. The *least radiogenic* Pb isotopic composition of each plagioclase and sulphide is plotted in Figure 10. The plagioclase data define a tight cluster close to a 2.7 Ga "source" isochron, reinforcing the assumption that the most primitive values represent the initial Pb isotopic composition of

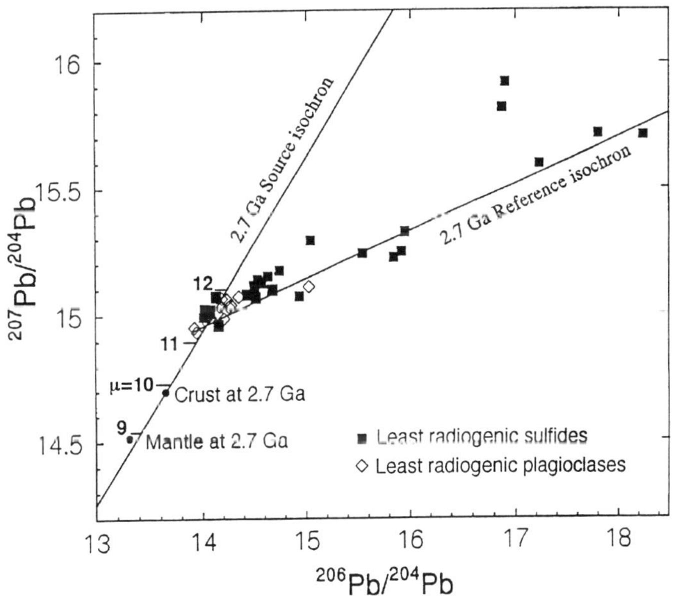

Figure 10. *A Pb-Pb isotope plot of the least radiogenic data from plagioclase and sulphide leaches. A 2.7 Ga source isochron and range of μ values (based on the Stacey and Kramers (1975) crustal evolution model) are shown along with a 2.7 Ga reference isochron. Note the restricted range of plagioclase, the wide range of sulphide values and the high μ values.*

the complex. In contrast, the least radiogenic sulphides show a much wider range indicating that a component of post-emplacement Pb has been incorporated into their structures during low-temperature recrystallization. The sulphide data are consistent with mixing of initial lead and radiogenic lead derived from a younger hydrothermal source but the scatter in the data precludes assigning a unique age or identifying the source of the hydrothermal fluids. It appears that the bulk of the Pb and other chalcophile elements were derived from the mantle around 2.7 Ga, a conclusion reached independently by Marcantonio *et al.* (1993) on the basis of Os isotopic studies.

In summary, the J-M Reef is broadly continuous across the entire complex but remarkably variable when viewed in detail. PGE-bearing sulphides are restricted to stratigraphically narrow zones and in some areas mineralized zones may be stacked. The basal contact of the reef, while regionally conformable, is marked by local depressions and highs which are not necessarily correlated with ore grade. Pegmatoids are abundantly developed in all lithologies, hydrous minerals are common and evidence for remobilization and recrystallization is widespread. Chlorapatite, phlogopite and chromite are distinctive accessory phases. Petrologic evidence supports the hypothesis that OB-I initially formed in response to the injection of olivine ($\pm$ plagioclase) saturated magma into a chamber containing fractionated magma, but this mechanism alone cannot explain all the features observed in this zone.

5.1.4. Norite-II Zone (N-II) and Gabbronorite-II Zone (GN-II)

N-II is a fairly uniform norite although modal proportions are somewhat variable. Drilling has confirmed the occurrence of several olivine-bearing members in N-II although none are exposed. The boundary between N-II and GN-II is marked by the transition from intercumulus to cumulus augite. In the Contact Mountain area, the lowermost 100 metres of GN-II contain numerous layers of anorthosite within gabbronorite. The anorthosites range from ~10 cm to ~15 m in thickness and two of them contain conformable layers of sulphides.

GN-II is the host for the spectacular outcrops of inch-scale rhythmic layering which occurs in a variety of forms. The most common type is composed of pyroxene-rich layers approximately 1 to 2 cm thick separated by plagioclase-rich layers 2 to 4 cm thick. Doublet layers, in which two closely spaced pyroxene layers are separated by a thicker layer of plagioclase, are a relatively common variant. A third variation involves two superimposed layering patterns involving different repeat distances. The layering is best developed in norites and gabbronorites which contain plagioclase in excess of cotectic proportions suggesting that the magma may have been slightly undersaturated in pyroxene. Toward the top of the sequence, the layering becomes progressively diffuse and the rock grades into a homogeneous norite.

Boudreau (1987) observed that, in many instances, the layer spacing is proportional to crystal grain size and suggested that layering develops in response to crystal aging within crystal suspensions. Because larger crystals have a lower surface free energy per mole than small crystals, the smaller crystals are more easily dissolved. As the crystal + liquid system evolves, initial minor perturbations in crystal size distribution are enhanced as larger crystals tend to grow at the expense of smaller crystals to diminish the total surface free energy of the system. The aging process leads to a long-range order similar to that observed in inch-scale layering. Boudreau (1987) has successfully modeled the formation of both singlet and doublet layers in a computer simulation in which layer growth by crystal aging was simulated by solving the differential equations for diffusion/reaction and crystal growth.

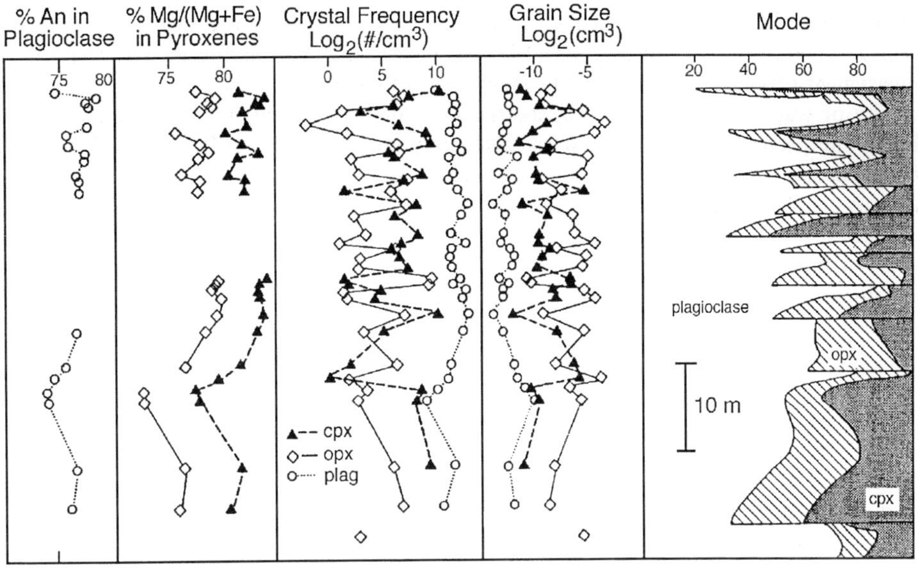

Figure 11. *Stratigraphic section through the macrorhythmic units at the top of GN-II showing modal variations, grain sizes, crystal frequency, and mineral compositions.*

Layering of a different type occurs in the uppermost 50 metres of GN-II as a series of macrorhythmic units which are primarily defined by variations in the plagioclase/pyroxene ratio (Figure 11). Criscenti (1984) mapped 10 macrorhythmic units in the Cairn Ridge-East Boulder Peak areas ranging from 18 metres to 2 metres in thickness; the thickest units can be traced for several kilometres along strike. Units 1 to 8 show a sharp lower contact and a upward increase in plagioclase content and pyroxene grain size. Units 8 to 10 show large modal variations but no sharp internal contacts. The frequency of pyroxenes (number of grains per cm^3) varies by several orders of magnitude throughout the macrorhythmic units whereas plagioclase frequency fluctuates only slightly (Figure 11) indicating that an oscillatory nucleation and growth mechanism was involved in the formation of these units.

5.1.5. Olivine-Bearing II Zone (OB-II)

The lower contact of OB-II is placed at the base of an anorthosite which overlies the uppermost macrorhythmic unit. The anorthosite is overlain by a laminated *gabbro*. The uppermost 5-8 m of OB-II is a remarkable assemblage in which irregular masses of troctolite up to football size are set in a matrix of anorthosite producing a distinctive pattern in outcrop, referred to as "pillow troctolite" by Hess (1960). The contact between the troctolite/anorthosite and the underlying gabbro is sinuous and highly discordant. Irregular patches of gabbro are isolated within troctolite/anorthosite and patches of troctolite are spotted throughout the gabbro close to the contact. Pegmatites containing plagioclase and pyroxene megacrysts up to 20 cm across are common. Olivine in troctolite occurs as large *ameboidal* grains containing abundant inclusions of small plagioclases. The upper contact of OB-II with the overlying anorthosite of the Middle Banded Series is sharp.

The origin of the unusual assemblages in OB-II remains one of the unsolved problems of Stillwater petrology. McCallum *et al.* (1977) suggested that troctolite formed by metasomatic replacement of gabbro by incongruent dissolution of pyroxene induced by an increase in the activity of water with the overlying anorthosite acting as an impermeable cap to migrating fluids. The presence of a volatile phase also promoted grain coarsening. Problems with this model include the source of the fluid, which has not been identified, and the absence of hydrous phases and fluid inclusions. Alternatively, the relations in OB-II could be explained by thermal erosion in response to the influx of magma. In this model, the sinuous contacts could be an erosional effect and the troctolite could represent the first "cumulate" formed from the new magma. This model, however, has difficulty in explaining the unique textures, the presence of pegmatoids and the apparently isolated gabbro blocks within troctolite.

5.2. Middle Banded Series (MBS)

The Middle Banded Series lithologies are distinctly different from those of both the Lower and Upper Banded Series. Plagioclase makes up 82 vol% of the MBS, olivine and augite are the most common *cumulus* mafic minerals, and cumulus orthopyroxene is relatively minor. The major lithologies are anorthosite, troctolite, and olivine gabbro; gabbronorites are rare and norites are absent. The sequence of crystallization is olivine/plagioclase $\rightarrow$ augite $\rightarrow$ orthopyroxene. The *average* grain size of plagioclase crystals in anorthosites is 2 to 3 times that in the cumulates of LBS and UBS and the plagioclases show complex zoning patterns. The most plausible explanation for these features is crystallization of the MBS from a magma significantly different in composition from that which produced the Ultramafic and Lower Banded Series, as suggested by Raedeke (1982) and Irvine *et al.* (1983).

5.2.1. Anorthosite-I and II Zones (AN-I and AN-II)

AN-I and II are relatively uniform anorthosites containing oikocrysts of augite, orthopyroxene or inverted pigeonite and quartz (1-5%) along with minor intercumulus magnetite, sulphides (pyrrhotite, pentlandite, chalcopyrite $\pm$ pyrite), and rare fluorapatite. Prominent zones of disseminated sulphides occur just below the upper contacts of both AN-I and AN-II. The thickness of AN-I is variable from a maximum of 400 metres in the eastern part of Contact Mountain to less than 200 metres west of the Boulder River. AN-II varies from 600 metres thick in the Contact Mountain area to less than 200 metres at Picket Pin.

In most samples, plagioclase shows a range in grain sizes with almost all grains falling between 1 and 10mm. Preferred orientation of plagioclase is rare. Plagioclase grains enclosed within pyroxene and adjacent to quartz are commonly subhedral and tabular while those in pyroxene-poor areas are anhedral with textures characteristic of annealed rocks. Anorthosites contain pyroxene-rich and pyroxene-poor domains on a scale of decimetres to dekametres which led Haskin and Salpas (1992) to conclude that the anorthosites are built up by the coalescence of metre-sized masses (rockbergs) of partially consolidated plagioclase cumulates.

Average compositions of plagioclase in AN-I (An_{77}) and AN-II (An_{76}) are virtually constant (Figure 9) and there is no systematic variation in major and trace element compositions with stratigraphic height (McCallum *et al.*, 1980; Salpas *et al.*, 1983). This large-scale homogeneity contrasts sharply with the heterogeneous nature of plagioclases on a cm scale. A single thin section may show grains with normal, reversed, patchy, convoluted and asymmetric zoning patterns (Czamanske and Scheidle, 1985). In a sample from AN-II, 80% of points analyzed lie between An_{74} and An_{78}, 90% between An_{73} and An_{79}, 99% between An_{69} and An_{83} and it is just as common for rims to be reversely zoned as normally zoned. Most of the zoning appears to be

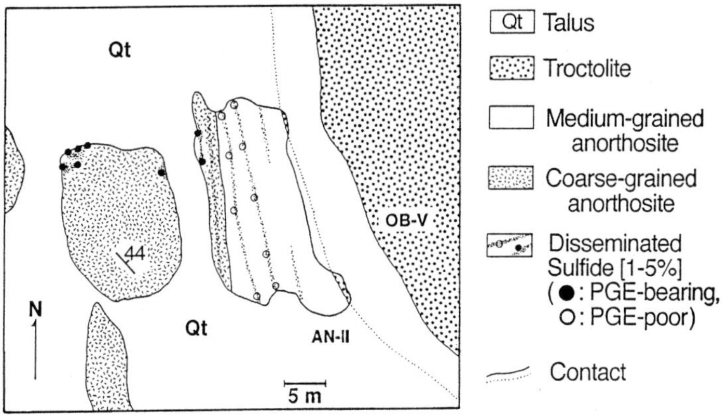

Figure 12. *Geologic map of part of the Picket Pin PGE deposit ~75 metres northwest of the summit of Picket Pin. "Up' is to the right. The PGE-bearing sulphides are located in the coarse-grained anorthosite.*

a primary feature although compaction, dissolution and reaction with migrating interstitial fluid have also modified plagioclase compositions. Pyroxene oikocrysts are also zoned; two large pyroxene oikocrysts analyzed in detail show a range in Mg/(Mg+Fe) from 0.66 to 0.60 which is much smaller than the range predicted for fractional crystallization of pyroxene from a trapped intercumulus liquid (Salpas *et al.*, 1996).

5.2.2. The Picket Pin Pt-Pd Zone

The Picket Pin deposit is a zone of disseminated, PGE-bearing sulphide which occurs in the upper 150 metres of AN-II (Boudreau and McCallum, 1986). The sulphide zone is traceable at the same stratigraphic position over 22 km. Sulphides are concentrated in a zone ~10 metres below the top of AN-II at a contact between a coarse-grained anorthosite containing up to 20% intercumulus pyroxene and an overlying medium-grained anorthosite adcumulate (Figure 12). Sulphides occur as podiform and discontinuous lenticular accumulations which are grossly conformable. Discordant sulphide-bearing pipes, which occur to a depth of 150 m in the footwall anorthosite, lead directly to the stratabound lenses. Boudreau and McCallum (1992) have suggested that the anorthosite in the upper 10 m of AN-II is the result of infiltration of interstitial liquids into the overlying troctolite during compaction of the thick anorthosite pile. The pipe-like nature of the footwall mineralization strongly supports a model of sulphide concentration by migration of a residual melt and/or fluid upward through footwall anorthosite.

5.2.3. Olivine-bearing Zones III and IV (OB-III and OB-IV)

The stratigraphy of these zones was defined by McCallum *et al.* (1980) for a section across Contact Mountain and extended to the east and west by Meurer and Boudreau (1996). The major stratigraphic units are traceable laterally over considerable distances but the thickness of these units varies significantly. Troctolites and anorthosites, which are predominant in the lower parts of each zone, are overlain by olivine gabbro. The uppermost unit in both OB-III and OB-IV is a unique olivine gabbronorite in which *both* olivine and orthopyroxene appear to be cumulus minerals in addition to plagioclase and clinopyroxene (Figure 3). Olivine gabbros and gabbronorites are generally cotectic and well laminated. Cyclic units in OB-III consist of a troctolite - anorthositic troctolite - olivine gabbro sequence. The sharp but irregular basal contact of the cyclic units most likely represents an erosional unconformity. Troctolites come in two types. *Banded troctolites* contain olivine as equant grains which define a wispy layering and are commonly cross-bedded. *Discordant troctolites*, which occur as isolated blobs and as finger-like protrusions *within* laminated gabbro, contain ameboidal grains of olivine up to 2 cm with small inclusions of plagioclase, similar to those at the top of OB-II.

There is no *systematic* variation of plagioclase, olivine, or cumulus pyroxene compositions as a function of stratigraphic position or lateral position (Figure 8). Plagioclase grains show complex zoning patterns and compositional ranges identical to those in AN-I and II. The average value of all plagioclase is An_{77}. Plagioclase in anorthositic layers tends to have a blocky habit whereas plagioclase in gabbro is generally tabular which led Meurer and Boudreau (1996) to suggest that anorthosites formed by the coalescence of plagioclase grown in a stress-free environment whereas the cotectic gabbros were subjected to a uniaxial stress in a compacting cumulate pile. Intercumulus pyroxenes tend to be more variable in composition and generally richer in FeO than the nearest adjacent cumulus pyroxene.

Many petrographic and geochemical features suggest a petrogenetic link between OB-III and IV and AN-I and AN-II which together make up the Middle Banded Series. These include plagioclase in excess of cotectic proportions, the similarity of grain size in plagioclase throughout the MBS, the complex zoning patterns, the constancy of mineral compositions and the similarity of Pb and Nd isotopic ratios (Wooden *et al.,* 1991; Lambert *et al.,* 1994).

5.3. Upper Banded Series (UBS)

5.3.1. Olivine-bearing Zone V (OB-V)

The basal member of this zone, which is very well exposed on Picket Pin Peak, is a banded troctolite similar to that at the base of OB-III (Figure 3). Olivine-rich lenses occur locally along the basal contact. Modally graded layers, cross-bedding, and scour-and-fill structures are pervasive in this troctolite indicating strong current action during its formation. The banded troctolite is overlain by an 80 m thick anorthosite which passes upward into a repetitive sequence of anorthosite - norite - gabbronorite. The gabbronorites exhibit a unique texture in which large blocky orthopyroxene, up to 1 cm, are associated with small acicular augite which, along with the plagioclase, define a strong lamination. The gabbronorite also shows evidence of current action in the form of rip-up clasts and xenoliths of anorthosite and unusual gabbronoritic "snowballs" within a layered gabbronorite.

5.3.2. Gabbronorite-III Zone (GN-III)

The lower contact of this zone is placed at the base of a thick sequence of uniform, laminated gabbronorite. Throughout this zone, plagioclase, augite and low-Ca pyroxene are present in cotectic proportions. In the lower part of GN-III, orthopyroxene is clearly a cumulus mineral while in the central and upper parts of the zone, orthopyroxene occurs as poikilitic crystals formed by the inversion of cumulus pigeonite. Magnetite becomes an important postcumulus mineral in this zone. In the central part of GN-III, irregular zones of pegmatite, some discordant, some conformable, are common as are xenoliths of anorthosite within fine-grained gabbro. Veins and dykes of pegmatitic hornblende-plagioclase-quartz up to 50 cm wide, formed during a late magmatic stage, are common in the upper part of GN-III.

Mineral compositions in GN-III show smooth variations with stratigraphic height consistent with fractional crystallization (Figure 8). Plagioclase ranges from An_{75} at the base to An_{62} at the top of GN-III while the Mg/(Mg+Fe) ratio in low-Ca pyroxene ranges from 0.75 to 0.67 over the same interval (Raedeke, 1982). There is a continuum in mineral compositions from GN-II to GN-III even though the Middle Banded Series is sandwiched between (Figure 3).

6. SOME PETROGENETIC CONSIDERATIONS

6.1. Cryptic variation and fractionation trends

6.1.1. Ultramafic Series

The Ultramafic Series, which shows minimal cryptic variation above the 400 m horizon, was formed by repeated injections of olivine-saturated magma (± intratelluric olivine) into a chamber located at a depth of about 10-15 km (Raedeke and McCallum, 1984). Injection of new magma could have occurred at any point during the formation of a cyclic unit. The dynamical behavior of these injected batches depended primarily on their density, viscosity, and momentum relative to that of the melt resident in the chamber (Campbell and Turner, 1989). Given the *slight* decreases in melt densities accompanying olivine fractionation, it is likely that injected batches, even though hotter, would have been slightly more dense than the magma in the chamber.

If the injected batches had sufficient momentum they would form turbulent fountains which would entrain some of the resident magma and eventually form a gravity-stratified mixed layer at the floor of the chamber containing suspended crystals. At the base of the hybrid layer, crystals are transferred to a stagnant layer through which they settle. No sorting is anticipated and olivine and/or orthopyroxene would accumulate on the floor in their order of saturation. As the thermal gradients decayed, the convective velocity decreased and the zones of crystal retention diminished in size. If the interval between magma influxes was long, the hybrid layer and the resident magma would mix and homogenize. By analogy with modern systems, it appears likely that venting of the chamber would have been triggered by magma influx.

As the blanket of insulating cumulates thickened, the fraction of heat lost through the floor was steadily reduced resulting in an *upward* decreasing residual porosity which reached a steady state around 400 metres. Below this horizon, reaction of cumulus minerals with interstitial liquid produced the Fe-enriched zone. Above this horizon, heat loss through the roof became the dominant method of cooling in the large aspect-ratio chamber and there is no significant variation in the compositions of olivines and pyroxenes since the net effect of influx, mixing, and venting buffered the magma at a relatively constant composition (McCallum and Raedeke, 1984). As the chamber matured, the frequency of magma injections apparently decreased although the sporadic reappearance of olivine-bearing cumulates in the Bronzitite zone indicates a continuing influx of fresh magma.

6.1.2. Banded Series

Fractionation trends in the Banded Series are summarized in Figure 13. The oblique trend, which is defined by norites and gabbronorites from the Lower and Upper Banded Series, is continuous even though the MBS is sandwiched between the LBS and UBS. The main vertical trend is defined largely by anorthosites from the Middle Banded Series whereas anorthositic rocks from OB-I define a subsidiary vertical trend and Ultramafic Series samples define a

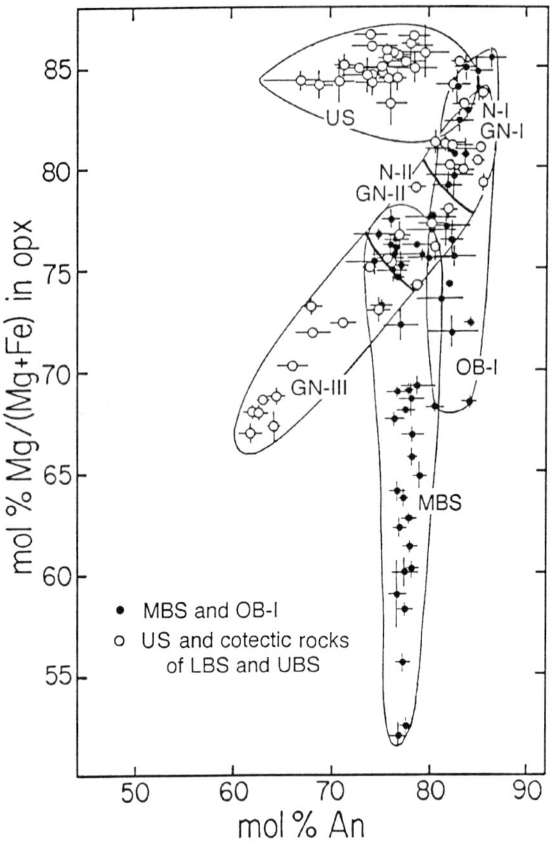

Figure 13. *Mol% An in plagioclase versus mol% Mg/(Mg+Fe) in orthopyroxene for Ultramafic Series (US) and Banded Series (BS) samples. The Fe-rich parts of the vertical MBS and OB-I trends are defined by intercumulus pyroxenes.*

broad horizontal trend. Plagioclases in the MBS have a restricted compositional range and plot close to the intersection of the LBS and UBS.

The oblique trend can be modeled by a system undergoing progressive fractional crystallization with minerals crystallizing in roughly cotectic proportions. The vertical trend, which is also observed in lunar anorthosites, has been modeled by Raedeke and McCallum (1980) by equilibrium crystallization of a mixture of plagioclase (well in excess of cotectic proportions), minor "cumulus" pyroxene plus the products of crystallization of a variable amount of trapped melt. The abundance of plagioclase relative to intercumulus liquid buffers the plagioclase composition at a near-constant value while the pyroxene composition is determined by the relative amounts of "cumulus" pyroxene and trapped melt. The vertical trend can be reproduced using trapped liquid contents from 1 to 20%. The horizontal trend for the Ultramafic Series can be similarly modeled by orthopyroxene fractionation with the variation in plagioclase composition resulting from crystallization of trapped melt.

6.2. Migration and trapping of intercumulus melts

Migration of melts and fluids through the pores of partially solidified cumulates during compaction and crystallization, and the trapping of some portion of these melts are recognized as fundamental processes in the formation of cumulates. Trace-element abundance in anorthosites (Salpas *et al.*, 1983, 1996; Haskin and Salpas, 1992) shed some light on the scale of these processes. If anorthosites are considered to be mixtures of cumulus plagioclase plus the products of *trapped* intercumulus liquid, cumulate theory would predict that samples with the highest amounts of trapped liquid should have the highest modal abundances of pyroxene and should be enriched in plagioclase-incompatible elements. A convenient element pair with which to test this prediction is La and Sc since both have very low $D^{plag/liq}$ values while Sc is compatible in pyroxene ($D^{cpx/l} > 1$). On

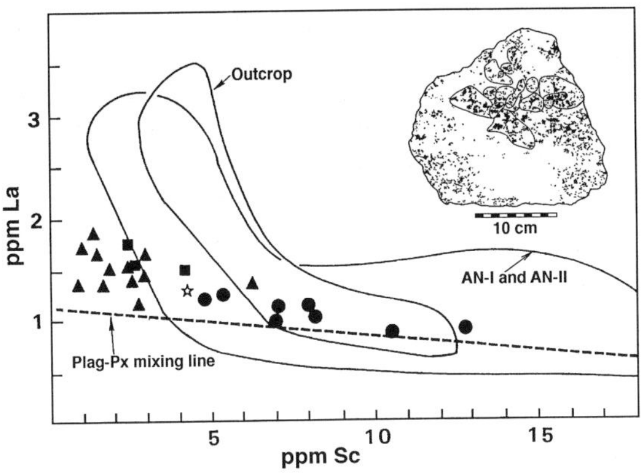

Figure 14. La vs. Sc in anorthosites. Field labeled "AN-I and AN-II" denotes samples from traverses through these zones and field labeled "outcrop" denotes samples from a 100 m^2 outcrop within AN-II. Individual points are for subsamples from the boulder shown in inset. Triangles: pyroxene-free, magnetite-free. Squares: pyroxene-free, magnetite-bearing. Circles: pyroxene-bearing, magnetite-free. Star: average of all samples. The dashed line is for a mixture of plagioclase and pyroxene. Inset shows the distribution of pyroxene in the boulder. Lines delineate individual oikocrysts.

a La *versus* Sc plot, mixtures of plagioclase + trapped liquid should lie on a line with a positive slope. In fact, regardless of the scale of sampling (traverse, single outcrop, or single boulder), La-Sc data on anorthosites show *negative* slopes which are particularly pronounced at low Sc values (Figure 14).

The pore space between the plagioclase grains filled in large part with adcumulus plagioclase and heteradcumulus pyroxene that crystallized *in equilibrium with the bulk liquid* and in small part with interstitial plagioclase and pyroxene derived by crystallization of trapped liquid *which in no case exceeded 9%*. Given that the boulder sample is free of trace minerals and has an average trapped liquid of only 1.3% (Salpas *et al.*, 1995), it is clear that intercumulus liquid was able to migrate over distances well in excess of boulder dimensions (decimetres). Traverse and outcrop samples with low modal pyroxene (low Sc) have the highest concentrations of trace minerals (quartz, Fe-Ti oxides, apatite, sulphides), incompatible trace elements (Figure 14), and deuteric minerals indicating migration of intercumulus liquid over considerably longer distances (metres to dekametres) during the growth of intercumulus minerals. Finally, the distribution of sulphides in the Picket Pin deposit indicate that, after vapour-saturation was reached, late-stage, fluid-saturated melts migrated over distances of several hundred metres.

6.3. Parental magmas

Documentation of crystallization sequences in the complex as a whole (McCallum *et al.*, 1980) and OB-I (Todd *et al.*, 1982) revealed that a single parental magma was inadequate to explain the data. The sequence peridotite → harzburgite → bronzitite → norite → gabbronorite in the Ultramafic Series and the Lower Banded Series required a different magma composition from that which formed the sequence troctolite → olivine gabbro → olivine gabbronorite → gabbronorite in the Middle Banded Series (and OB-I). The former magma has been called the U-type and the latter the A-type by Irvine *et al.* (1983). It is noteworthy that gabbronorites could have crystallized from either parent.

Trace-element and isotopic data have been used to provide geochemical tests of the two-magma hypothesis. Plagioclases from OB-I show a wide range in *absolute* REE abundances and significantly different *relative* REE abundances even within a single cyclic unit consistent with the addition of batches of a *new* magma to the chamber during the formation of OB-I (Lambert and Simmons, 1988). These new magmas had *relative* REE concentrations, *e.g.* lower Nd/Sm, distinct from the magmas that formed the Ultramafic Series. The influx of a new magma occurred initially in small volumes and the new magma and resident magma retained separate identities for some time prior to mixing. With repeated influxes, the effect of new magma gradually became more pronounced. The most significant change within OB-I occurs within cyclic unit 5 which coincides with the J-M Reef suggesting that the reef-forming event was associated with a *major* influx of new magma.

Nd isotopic ratios are potentially the most useful in distinguishing magma types since, unlike Sr, Pb and to a lesser extent Os, they have not been disturbed by post-crystallization processes. DePaolo and Wasserburg (1979) reported an $\varepsilon_{Nd(2701)}$ of -1.6 ± 0.6 for six samples from a wide stratigraphic range (Figure 15). Lambert *et al.* (1989, 1994) observed a wider spread of initial ratios (ε_{Nd} = +1.9 to -5.2) and concluded that at least two isotopically distinct magmas were required. Examination of their data, however, reveals that four of the samples analyzed were collected from the sulphide-rich zone at the base of the complex and two from the lowermost chromitite; these samples showed the most negative values (ε_{Nd} = -2.7 to -5.2), which are comparable to those of the footwall hornfels (ε_{Nd} = -3.7) and mafic norite sill (ε_{Nd} = -3.4), indicating that the initial magma batches had suffered significant contamination from a *local*

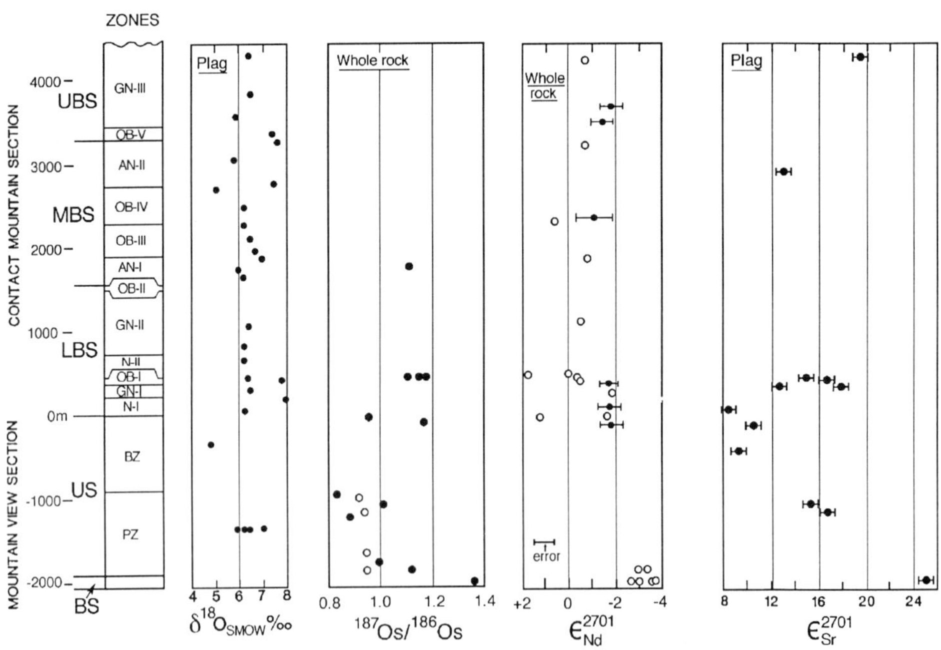

Figure 15. Stratigraphic variation of $\delta^{18}O$ in plagioclase (Dunn, 1986), initial ratios of Os isotopes (Marcantonio et al., 1993; Lambert et al., 1989), Nd isotopes (DePaolo and Wasserburg, 1979; Lambert et al., 1989) and Sr isotopes (Stewart and DePaolo, 1987). The O and Sr data were determined on plagioclase separates and the Os and Nd on whole rocks. Note that Basal Series samples show evidence of localized contamination.

source. When these samples are omitted, the remaining samples show an ε_{Nd} range from -1.9 to +1.9 with both the lowest and highest values coming from samples believed to have been derived from the A magma.

Initial $^{187}Os/^{186}Os$ ratios for the Ultramafic Series were measured on A, C, H and J chromitites by Marcantonio *et al.* (1993) and their values (0.92±0.02 at 2.7 Ga) fall within the range of chondritic (mantle) values (Figure 15). Additional Os isotopic data from Lambert *et al.* (1994) on the G, H, I and K chromitites also showed near-chondritic values. However, samples from the J-M Reef and chromites from the B chromitite and the Bronzitite zone have consistently higher initial values (average of 1.15±0.04). The reef samples have much higher Re/Os ratios than the chromitites and require a much larger age correction. To further complicate the issue, Marcantonio *et al.* (1993) documented rhenium mobilization by hydrothermal fluids and suggested that some of the osmium isotopic variability may be due to this effect.

In summary, the strongest evidence for multiple magmas is the variable crystallization sequences and the range of compositions encountered in the coeval sill/dyke suite. Trace-element and isotopic evidence for two distinct magmas, while suggestive, is not compelling.

6.3.1. Parental magma compositions for the UMS and LBS (U-type)

The first estimates of parental magma compositions were based on "chilled margin" samples (Hess, 1960; Jackson, 1971). However, both compositions (Table 2) belong to the Group 1 gabbronoritic dykes as defined by Helz (1985), who pointed out that members of this group are poor choices for U-type parental liquids because of their differentiated compositions, high REE abundances and inappropriate crystallization sequences.

Longhi *et al.* (1983) addressed the parental magma problem by determining the crystallization sequence of a Stillwater bronzite diabase dyke which has an age and phenocryst assemblage appropriate for Stillwater parental magmas. At pressures in the range of 300 to 500 MPa, olivine is followed by orthopyroxene, then augite and finally plagioclase (Figure 16), which differs slightly from the observed sequence of olivine → orthopyroxene → plagioclase → augite. The compositions of liquidus olivine and orthopyroxene are very close to the most primitive compositions observed in the Ultramafic Series but the cotectic field boundary between olivine and orthopyroxene at low pressure (Figure 16A) is difficult to reconcile with the evidence for olivine reaction. In addition, the bronzite diabase is enriched in incompatible elements and alkalis relative to those computed for the Stillwater parental magma resulting in plagioclase ($An_{\sim70}$) which is more albitic than the most primitive plagioclase in the complex.

Helz (1985) noted that mafic norite and magnesian gabbronorite of the basal Sill/Dyke Suite have geochemical characteristics comparable to those inferred for melts parental to the Ultramafic Series (Table 2). Experimental data at 150 and 300 MPa and low f_{O_2} (CCO buffer) on the crystallization sequences in these two compositions and a 50-50 mix have been reported by Helz (1995). At both pressures, the mafic norite composition crystallized in the order orthopyroxene → plagioclase → clinopyroxene (no olivine) while the magnesian gabbronorite crystallized in the order olivine → plagioclase → clinopyroxene → orthopyroxene. The 50-50 mix has the requisite crystallization sequence (olivine [Fo_{79-80}] → orthopyroxene [En_{79-81}] → plagioclase → clinopyroxene) at 150 MPa, but olivine and orthopyroxene are reversed in the sequence at 300 MPa which again implies a cotectic relationship between olivine and orthopyroxene at this pressure. On the basis of these results, Helz (1995) suggested that the complex crystallized at a relatively low (100-200 MPa) pressure with the Basal Series

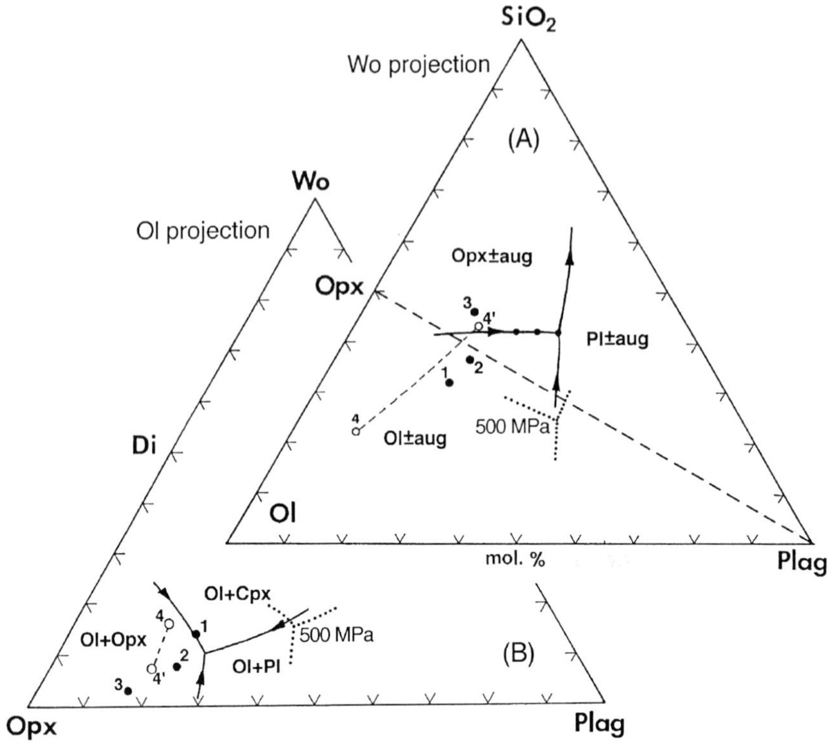

Figure 16. *Projections of liquidus boundaries and rock compositions at 1 bar in the system olivine-plagioclase-wollastonite-silica (after Longhi et al., 1983). (A) Projection from Wo on to the olivine-plagioclase-silica plane. The phase boundaries are for liquids at or below augite saturation. (B) Projection from olivine on to the wollastonite-orthopyroxene-plagioclase plane. (1): WSD-14 bronzite diabase (Longhi et al., 1983), (2) CC2-813 mafic norite sill (Helz, 1985), (3) calculated parental magma (McCallum, 1988), (4-4') AFC (assimilation - fractional crystallization) path for komatiite (McCallum, 1988).*

cumulates forming from a mafic norite liquid, the cyclic units of the Peridotite Zone from a mix of the two magmas and the Bronzitite Zone cumulates representing a reversion to the mafic norite magma. These are intriguing suggestions but more information on the compositions and relative abundances of the phases in the experiments is required to assess the viability of this model. Indirect support for the Helz model has been provided by Papike *et al.* (1995) who have shown that the REE pattern calculated for the Stillwater U-magma has a slope similar to that of the mafic norite and gabbronorite dykes (Figure 17).

McCallum (1988) used the MELTS program (Ghiorso and Sack, 1995) to compute a parental magma composition that is consistent with (1) the observed crystallization sequence, (2) the compositions of the most primitive cumulus minerals, *and* (3) the relative proportions of the cumulus minerals. The third constraint is particularly important since an infinite range of magma compositions can satisfy the first two constraints. The calculations were carried out in a fractional crystallization mode at a pressure of 300 MPa with oxygen fugacities constrained to follow the QFM buffer. A composition which provides the best fit to all constraints is listed in Table 2 (column 6).

Although the compositions of the U-type parental magma determined using these different approaches differ in detail, it is clear that they share some common characteristics, specifically, high MgO, relatively high SiO_2, low alkalis, CaO, Al_2O_3, and TiO_2. In many, but not all, respects they are comparable to modern boninites. At first glance, it appears that the pressure of 100-200 MPa inferred by Helz is inconsistent with the pressure of 300-400 MPa recorded by hornfels assemblages. However, it should be kept in mind that the crystallization sequences will record pressures during formation whereas the hornfels may record peak pressures after the entire complex was emplaced.

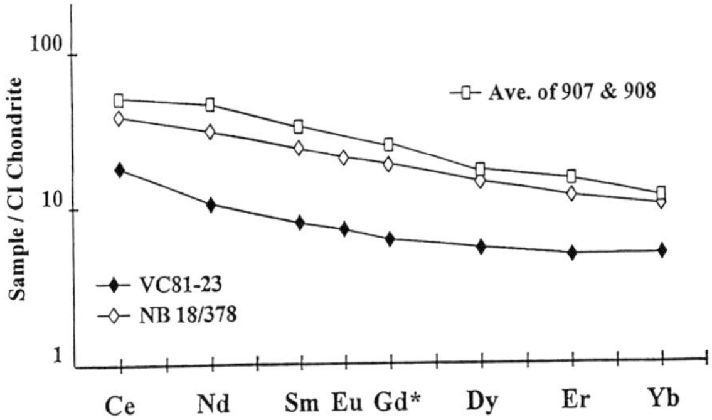

Figure 17. *CI chondrite-normalized REE plot of calculated parental U-magma based on SIMS analyses of bronzite from bronzitite samples 907 and 908 (from Papike et al., 1995) compared to REE in mafic norite (VC81-23) and Mg-gabbronorite NB18/378 (from Lambert and Simmons, 1988). The asterisk indicates that Gd values are interpolated in the SIMS data.*

6.3.2. Parental magma composition for the MBS

The crystallization sequence in the Middle Banded Series is typical of tholeiitic magma and a range of compositions can satisfy the known constraints. Irvine *et al.* (1983) have suggested that this magma (A-type) was hyper-aluminous since, in their model, the thick anorthosites are formed by crystallization of this magma. However, such aluminous compositions are conspicuously absent from the Sill/Dyke Suite. The most abundant members of this suite are gabbronorites (Group 1 of Helz, 1985). The crystallization sequence of a typical gabbronorite (Table 2) which plots near the center of Helz's Group 1 was predicted using MELTS. This composition crystallizes in the same order as inferred for the MBS cumulates (olivine → plagioclase → clinopyroxene → orthopyroxene) and produces olivine and plagioclase of approximately the correct composition.

6.3.3. Evidence for crustal assimilation

Evidence for a crustal component in Stillwater parental magmas comes mainly from isotopic data. Simmons and Lambert (1982) reported initial Sr isotopic ratios with a range of $\varepsilon_{Sr(2701)}$ from +1.4 to +31.3 while Stewart and DePaolo (1987) reported a range of $\varepsilon_{Sr(2701)}$ = -2.0 to +25 (average = +14.3) (Figure 15). These ranges are larger than expected for a homogeneous magmatic system and indicate some post-crystallization disturbance, but important conclusions can still be extracted from the data. First, the maximum ε_{Sr} occurs in a sample from the Basal Series suggesting that the first influxes of magma suffered the maximum amount of

Table 2
Compositions of proposed parental magmas

	1	2	3	4	5	6	7	8
SiO_2	50.68	49.41	52.20	51.20	48.2	54.1	54.4	48.3
TiO_2	0.45	1.20	0.69	0.30	1.15	0.6	0.5	1.52
Al_2O_3	17.64	15.79	9.80	12.6	16.20	12.7	10.8	14.5
Fe_2O_3	0.26	2.11	1.04	-	2.30	-	1.3	1.09
FeO	9.88	10.25	9.36	12.0	9.30	9.5	9.2	13.6
MnO	0.15	0.20	0.20	0.19	0.19	-	0.2	0.24
MgO	7.71	7.36	16.70	13.0	9.64	14.5	13.2	7.00
CaO	10.47	10.88	8.18	8.8	10.82	7.6	7.8	11.0
Na_2O	1.87	2.19	0.58	1.1	1.56	0.6	1.2	1.47
K_2O	0.24	0.16	1.68	0.29	0.25	0.2	1.0	0.13
P_2O_5	0.09	0.11	-	0.06	0.11	-	-	0.14
	99.92	99.95	100.00	99.51	100.43	100.0	99.6	98.99

1. Parental magma proposed by Hess (1960). 2. Parental magma proposed by Jackson (1971). 3. Stillwater bronzite diabase (WSD-14) (Longhi *et al.,* 1983). 4. Mafic norite, Sill/Dyke Suite (Helz, 1985). 5. Magnesian gabbronorite, Sill/Dyke Suite (Helz, 1985). 6. Computed parental magma (McCallum, 1988). 7. Magma formed by assimilation of granodiorite by komatiite (McCallum, 1988). 8. Gabbronorite, Sill/Dyke Suite (Helz, 1985).

contamination, most likely by *localized* interaction with the country rock. Second, the positive ε_{Sr} values are consistent with pre-emplacement contamination of magmas.

The negative ε_{Nd} (-2 to -5) values of samples from the Basal Series, lowermost Ultramafic Series and the Sill/Dyke Suite are also consistent with localized contamination during emplacement. The slightly negative to slightly positive ε_{Nd} values (-2 to +2) of the main series cumulates suggest derivation of the parent magma from a mantle source with a slight long-term Nd/Sm enrichment (relative to depleted mantle at 2.7 Ga) or one contaminated by LREE-enriched ancient crustal material. The Os isotopic data are difficult to interpret unambiguously. The roughly chondritic values of most (but not all) Ultramafic Series chromitites limit the amount of crustal contamination of U-magmas whereas the elevated Os isotopic ratios of the reef samples may be the result of incorporation of a crustal component into A-magmas. However, a *later* remobilization of the reef sulphides and incorporation of radiogenic Os from an external source cannot be discounted.

Pb isotopic compositions of leached plagioclase provide the clearest evidence for the addition of a crustal component (Wooden *et al.,* 1991; McCallum *et al.,* 1992). On a $^{207}Pb/^{204}Pb$ *vs* $^{206}Pb/^{204}Pb$ plot, the data define a broad trend roughly parallel to a 2.7 Ga isochron. Samples from the Basal Series and lowermost Ultramafic Series lie slightly above the main trend defined by the Banded Series samples again indicating *local* contamination of the lower part of the complex during emplacement. The initial Pb isotopic compositions of the main trend are unusually radiogenic (μ ~11-12) (Figure 10) and are identical to those of the late Archean (2.73-2.79 Ga) granitoid suite of the eastern and central Beartooth Mountains (Wooden and Mueller, 1988). Wooden *et al.* (1991) rejected a model in which primitive mantle melts assimilated late Archean granitoids on the grounds that all Stillwater magmas would have to be contaminated to the same degree to produce the observed uniformity of Pb isotopic and Th/U values, which is unlikely given the highly variable Th/U of the granitoids. Wooden *et al.* (1991) suggested that subduction of Archean crust around 2.8 Ga formed an enriched and relatively homogeneous mantle source which was later melted to produce magmas which were parental to the granitoid suite and later to the Stillwater Complex

The $\delta^{18}O$ composition of the Stillwater magma(s), calculated from plagioclase-basalt fractionation factors, ranges from 4.7 to 6.7 per mil with most values lying close to the average value of 5.9 per mil, *i.e.*, within the range of values for mantle-derived melts (Dunn, 1986). With the exception of sulphides from the Basal Series, Zientek and Ripley (1990) documented uniform $\delta^{34}S$ values throughout the complex indicating that the complex crystallized from a very uniform sulfur reservoir, most likely derived from the mantle. The stable isotope data do not support models calling on *large* amounts of crustal contamination.

6.3.4. Sources of parental magmas

Longhi *et al.* (1983) suggested that U-type magmas might have formed by assimilation of continental crustal material by primary komatiitic magmas. MELTS computations confirm that magmas with the major and trace element characteristics of U-magmas can be generated by such an AFC process (compare analyses 6 and 7, Table 2). However, large amounts of assimilation are required; under isenthalpic conditions Ma/Mc (mass assimilated/mass crystallized) ~ 1 (McCallum, 1988). This would most likely result in Os isotopic ratios much higher than those observed in the chromitites. Further tests of this model must await better constraints on the composition of crust which might have been assimilated.

By analogy with boninites, partial melting of subcontinental harzburgitic mantle, enriched in incompatible elements and possibly water, has been suggested as a mechanism for the production of U-magmas (Wooden *et al.*, 1991). Addition of older continental crust to the mantle source *via* subduction would elevate Pb and Sr isotopic ratios and lower Nd isotopic ratios but have little effect on Os ratios since the high Os content of the mantle renders it essentially immune to crustal contamination. This model satisfies most of the known constraints.

Relative to U-type magmas, A-type magmas have Nd isotopic ratios that are variable but slightly higher and Os isotopic ratios that are also variable but significantly higher. For this reason, it is unlikely that the same mantle source was involved since most mantle sources have chondritic to sub-chondritic Os ratios. A-type magmas are geochemically evolved and may have developed the higher Os ratios by assimilation of Archean crustal rocks by mafic magmas derived from partial melting of a depleted mantle lherzolite (Lambert *et al.*, 1994). Thus, at least three sources (enriched subcontinental lithospheric mantle, depleted mantle, continental crust) are required to explain the geochemical features of the Stillwater rocks. There is also abundant evidence for mixing of magmas during the crystallization of the complex.

6.4. Origin of Stillwater anorthosites

The Middle Banded Series contains 82 vol% plagioclase, a value which is well in excess of cotectic proportions involving plagioclase and pyroxene, and the key to understanding the anorthosite problem lies in finding a source of the excess plagioclase. Three sources have been suggested: (1) a hyper-aluminous magma, (2) a magma containing abundant intratelluric plagioclase, and (3) plagioclase which failed to accumulate on the floor during the crystallization of cumulates below the MBS.

Hess (1960) proposed that the anorthosites crystallized from an aluminous melt formed by *resorption* of earlier-formed plagioclase. This model requires excessive superheat to resorb the plagioclase needed to form 1000 metres of anorthosite. McCallum *et al.* (1980) also called on early crystallization of plagioclase but to circumvent the superheat problem and the absence of cumulus plagioclase in the Ultramafic Series they proposed crystallization in a pressure gradient in which the melt was saturated in plagioclase in the upper, low-pressure, region while pyroxene was saturated at the base. Since the effect of pressure on the cotectic composition is small, for this model to have any validity the chamber must be large, very well mixed and the degree of crystallization small.

Irvine *et al.* (1983) proposed that the anorthosite layers and the overlying troctolite and olivine gabbro formed sequentially from a hyper-aluminous magma that remained saturated in plagioclase over an *extended* crystallization interval before reaching the plagioclase-olivine cotectic. This model cannot easily explain the thick anorthosites of uniform composition, the coarse grain size of the plagioclase, and the complex zoning patterns. In addition, such aluminous compositions are absent from the Sill/Dyke Suite. The MBS magma was geochemically evolved, presumably by fractionation of mafic phases at some deeper level, and it was likely to be close to olivine saturation at the time of its emplacement. However, anorthosites contain intercumulus quartz and this model requires that *all* olivine, which must have initially crystallized from the intercumulus liquid, reacted out. In a variant of this model, Barnes and Naldrett (1986) suggested that aluminous magmas could have formed by fractionation of orthopyroxene from the U magma at pressures up to 1 GPa. However, it has not been demonstrated that such magmas would have the requisite low-pressure crystallization sequence.

Czamanske and Bohlen (1990) suggested that the major anorthositic zones formed from the "accidental" injection of a mafic magma containing intratelluric plagioclase which had formed by fractionation in a lower crustal chamber. This model implicitly assumes that the quartz-bearing anorthosites and overlying troctolites were derived from the same magma and therefore it is subject to the criticism regarding olivine discussed above. In one sense, this model is no different from the *in situ* fractionation models discussed above, except that it relegates the separation of plagioclase from mafic minerals to a deeper, hidden chamber.

6.4.1. An alternative model of anorthosite formation

The Stillwater Complex provides incontrovertible evidence for crystal sorting on a grand scale and it is instructive to evaluate the evidence for derivation of the excess plagioclase by fractionation and sorting *within* the magma chamber. Could the excess plagioclase represent that which did not accumulate on the floor during the crystallization of the Ultramafic Series and/or the Lower Banded Series but was "stored" at some other location in the chamber? The similarity of *average* plagioclase compositions in the anorthosites, bronzitites, norites and gabbronorites supports this idea. The magnitude of the Eu anomaly bears on this question since co-crystallization and removal of plagioclase would be reflected in increasingly negative Eu anomalies in cumulus bronzites of the Bronzitite Zone. Negative Eu anomalies *are present* in bronzites from all levels of the Ultramafic Series (Lambert and Simmons, 1987); $(Eu/Eu^*)_{opx}$ values lie between 0.5 and 0.9 with the lowest values in the uppermost few metres of the series. However, orthopyroxene has an intrinsic negative Eu anomaly (McKay *et al.*, 1990) and at oxygen fugacities appropriate for the Stillwater Complex, $(Eu/Eu^*)_{opx} \sim 0.9$. The data are consistent with some plagioclase fractionation during the *latest* stage of crystallization of bronzitites. A more likely source of plagioclase is that which failed to accumulate during the formation of the norites and gabbronorites of the LBS, as originally suggested by Hess (1960). Additional support for this model is provided by the data of Loferski *et al.* (1994) who observed that plagioclase in AN-I and AN-II had absolute and relative REE abundances very similar to those in N-I and quite different from those in the Ultramafic Series.

Given the importance of sorting, it is necessary to evaluate mechanisms by which this might be achieved. The key parameter in controlling the fate of crystals in a cooling, crystallizing, and convecting magma chamber is S, the ratio of crystal settling or flotation rate (V_g) to the convective velocity (U_o) (Marsh and Maxey, 1985). The lower the value of S, the greater the extent of crystal retention. Neutrally buoyant crystals (S=0) are completely retained and simply follow fluid streamlines, dense crystals are concentrated in upwelling regions while buoyant crystals are concentrated in the downflow regions. At constant S values, different fluid flow patterns result in different retention volumes.

In addressing the same problem, Martin and Nokes (1989) used an experimental approach to simulate crystal motions within a convecting fluid under different thermal regimes. In the experiments most relevant to the Stillwater, the system was cooled from above only and the mean velocity of convection was much greater than the particle settling rate. Despite the low S values and high crystal retention, Martin and Nokes showed that particle removal occurs at the lower boundary of their system where the vertical component of the convective velocity is zero. An important conclusion from this work is that, *under steady state conditions*, a convecting magma which is crystallizing minerals in cotectic proportions will eventually deposit these minerals in near-cotectic proportions yet have retention zones in which the minerals are present in non-cotectic proportions.

At the point of plagioclase saturation (~1200°C, ~300 MPa), the density of the Stillwater melt was 2.68 g cm^{-3} (the presence of 0.5% H_2O would lower these values by ~0.02 g cm^{-3}) whereas plagioclase An_{85} has a density of 2.70 g cm^{-3} (Lange and Carmichael, 1987). It is clear that the retention zone for plagioclase grains in the convecting magma is much larger than that for mafic phases which have densities in the range of 3.25 to 3.35 g cm^{-3}. The coarse grain size and the complex zoning patterns of plagioclases are consistent with their remaining suspended in a convecting magma for an extended period of time.

Since plagioclase grains tend to concentrate in downwelling regions whereas the mafic phases concentrate in upwelling regions, there should be sorting in a lateral sense as well as a vertical sense and some of the lateral variation in thickness of monomineralic layers may be due to such a mechanism. Size sorting within the zone of crystal retention is also predicted. Thus, abrupt changes in crystal size at horizons conformable to the regional layering, which are particularly common in the bronzitites, could simply reflect fluctuations in convective velocity. A likely cause of variations in U_o or convective flow patterns, is addition of new magma batches, so it is probably no coincidence that variations in layering styles, textures and modes occur in the vicinity of lithologic boundaries.

At the low S values characteristic of plagioclase, the zone of retention of plagioclase is large and an entrained plagioclase might circulate hundreds of times, enhancing the probability of large crystals with complex zonation and resorption patterns. Under conditions of multiple saturation, S values of pyroxene are likely to be >10 times those of plagioclase, with a correspondingly large difference in retention volumes. Regardless of cotectic proportions, concentrations (well in excess of cotectic proportions) of "low S phases" such as plagioclase would build up in the convecting magma until a steady state was achieved at which point cotectic norites or gabbronorites would accumulate on the floor. While the uncertainties in most parameters render a quantitative analysis impossible, the effect of different S values is real. It appears inescapable that the crystallization of the norites and gabbronorites (and possibly upper bronzitites) was accompanied by a retention zone rich in plagioclase.

The model outlined above has some interesting implications. In the first place a large volume of retention implies that the magma was maintained at, or slightly below, its liquidus. Since the liquidus dT/dP gradient is superadiabatic, suspended plagioclase crystals would tend to suffer some resorption during the upwelling part of the convective cycle and enhanced growth during the downwelling part. The extent of resorption is limited since the heat of dissolution must be extracted from the surrounding melt, but it is likely that the resorption textures commonly observed in the anorthosites may be due to dissolution during convective transport.

While a fraction of the plagioclase in the anorthosites might have been derived from intratelluric crystals in magmas injected into the chamber, there is no reason to believe that this is the primary source. The compositions and textures of plagioclase in the Middle Banded Series are consistent with the internal sorting mechanism described above. It is also clear that the conditions in the chamber were periodically perturbed by the addition of batches of olivine-saturated magma which incorporated some of the suspended plagioclase by mixing. As discussed earlier, anorthosites apparently formed by coalescence of plagioclase mushes to form anorthositic rockbergs. An increase in density of suspensions due to the crystallization of pyroxene and the expulsion of most of the interstitial liquid may have initiated the accumulation of the rockbergs. In any event, rapid accumulation is indicated by the lack of fractionation in the anorthosites.

6.5. Origin of the PGE deposits

Two markedly different petrogenetic models have been proposed for the PGE-rich J-M Reef. In the **orthomagmatic model**, advocated by Barnes and Naldrett (1986) and Campbell *et al.* (1983), the sulphides accumulated with their high PGE tenors *at the same time* as the rocks enclosing them by a process of batch segregation of an immiscible sulphide liquid formed during magma mixing. A key aspect of this model is the entrainment of fractionated magma resident in the chamber into turbulent plumes of injected magma thereby permitting sulphide droplets to come in contact with a large volume of silicate melt (Figure 18). Barnes and Naldrett (1986) suggested that magmas ranging in composition from olivine-saturated to plagioclase-saturated, which formed by fractionation of pyroxene in a lower crustal chamber, were injected into the Stillwater chamber in a series of pulses where they mixed with a resident magma. Lateral and vertical stratigraphic variations in OB-I are the results of different volumes and different compositions of injected magma and variable distance from the feeder system (Figure 18).

In this model, factors controlling the PGE content of the magmatic sulphide liquid are the PGE and S contents of the silicate magmas which mixed, the distribution coefficients (D) of the PGE between coexisting sulphide and silicate liquids, and the mass ratio of (silicate magma)/(sulphide magma), referred to as the "R factor". To achieve PGE concentrations of the magnitude observed, both R and D must be large. While models of this type can be adjusted to accommodate most of the observations, some problems remain. In the first place, a *range* of magma compositions from olivine-saturated to plagioclase-saturated is required to explain stratigraphic variations. Second, the common occurrence of pegmatites, along with

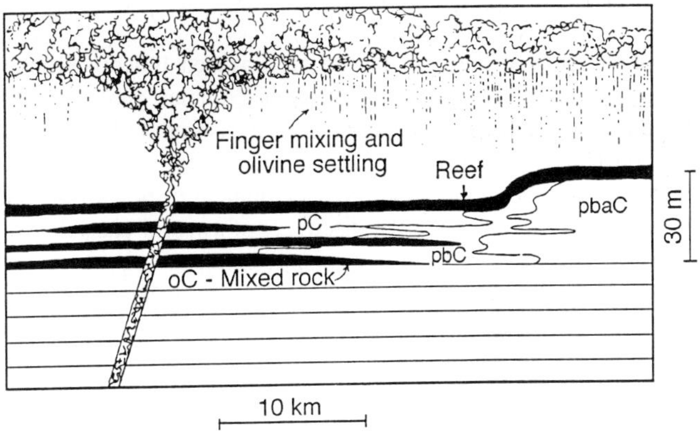

Figure 18. *Model of Barnes and Naldrett (1986) for the origin of OB-I and the J-M Reef. Small magma influxes produced the lower olivine-bearing layers of limited lateral extent. A larger pulse was involved in the formation of the reef package and its associated sulphides. Sulphide droplets acquired high PGE concentrations due to the large R factors (details in text). The horizontal and vertical scales are approximate.*

relatively high proportions of phlogopite, chlorapatite and other hydrous minerals, is difficult to explain in a strictly magmatic model as is the occurrence of ore in stratabound zones *below* the main reef. Third, distribution coefficients of ~10^7 and a silicate liquid column up to 7 kilometres thick are required to explain the observed enrichments in Pt and Pd in the reef, whereas experimentally determined D values for Pd are generally much lower (Peach and Mathez, 1993).

In the **hydromagmatic model**, most recently discussed by Boudreau and McCallum (1989, 1992), it is proposed that the PGE-enrichments are a consequence of leaching by Cl-rich hydrous fluids exsolved from intercumulus magma during the latter stage of crystallization (Figure 19). These fluids leached PGE (+ S and other soluble components) from intercumulus sulphide in cumulates *below* the reef and transported them upwards to be redissolved or deposited where fluid-saturation fronts encounter discontinuities marked by changes in composition of the intercumulus silicate liquids. Boudreau and McCallum (1989) refer to this process as *vapour-driven constitutional zone refining*. Two variants of this model exist. In one variant, the entire reef package is believed to be the product of the metasomatic process (Boudreau, 1988). The other variant postulates that sulphides and silicates in the reef package were formed by magmatic influxes in a manner similar to that described in the orthomagmatic model, and the sulphides were *later* enriched in PGE by the high-temperature metasomatic fluid infiltrating from below. Boudreau and McCallum (1992) presented numerical models of the degassing of intercumulus liquids which suggest that the thick cumulus sequences below the reef can act as chromatographic columns to separate PGE and S during the degassing. The PGE are enriched at a sulphide-dissolution front as upward migrating sulphide-undersaturated fluids resorb cumulus sulphides. Arguments against the hydromagmatic model include the restricted distribution of mineralized rock to narrow zones within OB-I, the absence of pipes which might represent fossil fluid channelways, and lack of experimental support for the required high solubility of PGE in high temperature Cl-rich fluids.

7. SUMMARY

The quarter century since the publication of the first edition of "Layered Igneous Rocks" has witnessed a wealth of new information on all aspects of Stillwater Complex geology. New maps have been completed, detailed stratigraphic sections have been measured and described, and a precise age has been established. A greatly expanded geochemical data base has compelled the reassessment of old petrogenetic models and the development of new ones. The single most important event of the past two decades was the discovery of a world-class deposit of platinum-group elements (J-M Reef) associated with the reappearance of olivine in the Banded Series rocks. Although the J-M Reef is similar in many respects to the Merensky Reef of the Bushveld Complex there are significant differences in the tenor of the ore and the relative abundances of platinum and palladium. An abundance of information has been obtained on the ore zone and the rocks in its immediate vicinity, much of which remains to be interpreted.

Field and geochemical evidence for multiple magma injections into an evolving magma chamber is very strong but much remains to be learned about the physics of the processes of magma influx and mixing. The realization that at least two chemically distinct parental magmas were involved has spurred effort to determine the compositions, sources, and frequency of injection of these magmas. There is a growing body of evidence that samples of the parental magmas have been preserved in the coeval dyke/sill sequence at the base of the complex. The magma that formed the Ultramafic Series had major element characteristics similar to those of

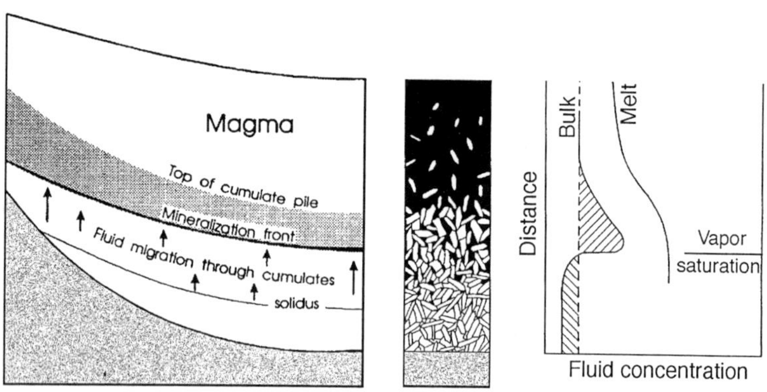

__Figure 19.__ Model of Boudreau and McCallum (1992) for the origin of PGE-enriched sulphide zones. Crystallization of interstitial liquid deep in the cumulate pile leads to fluid saturation. This Cl-rich fluid migrates upwards and carries with it the fluid-compatible elements (S, PGE, Cu, Ni, As, Te) which were originally contained in a minor sulphide phase. The upward migration of fluid is limited to the level at which the interstitial liquid is fluid-saturated, since fluid reaching this level must redissolve in the fluid-undersaturated liquids. As the crystal pile thickens, the fluid saturation boundary moves upwards until it encounters a lithologic discontinuity or a sulphide-rich zone. The line labeled "Bulk" represents the volatile concentration in the bulk system (crystals + liquid). The line labeled "Melt" represents the concentration of volatile in the melt only.

modern boninites whereas the magma that formed the olivine-bearing rocks of the Banded Series had tholeiitic affinities. Trace elements and radiogenic isotopes have proven useful in distinguishing these different magma types and indicate that two mantle sources and a crustal contaminant are required. However, there is no consensus on whether the crustal component was incorporated into the mantle source *via* subduction or was added by assimilation during storage and transit in the crust.

Anorthosites, which are abundant in the Stillwater Complex, continue to attract interest, in part because lunar anorthosites are believed to have formed in a similar manner to those in layered intrusions. Evidence has accumulated that anorthosites have formed by the coalescence of plagioclase-rich suspensions (rockbergs) which themselves formed by large-scale sorting in a convecting magma. Anorthosites also provide evidence for large-scale migration of intercumulus melts and fluids.

A result of first-order importance was the discovery of the critical role of fluids during the crystallization of the complex. This has led to development of a hypothesis that transport of ore-forming components in chlorine-rich hydromagmatic fluids was the mechanism for producing enrichments in platinum group elements. However, it is safe to say that such fluid-based models are not universally accepted and models involving a strictly magmatic origin for the ore zones have considerable support.

8. ACKNOWLEDGEMENTS

This work has been supported by the National Science Foundation (Grant EAR-9406243) and the National Aeronautics and Space Administration (Grant NAGW-3352). I thank Linda Raedeke, Ed Mathez, Alan Boudreau, Peter Salpas, Todd Dunn, Louise Criscenti and Hugh O'Brien for their time and effort devoted to the Stillwater project. Without their contributions, this work would not have been possible. I thank Barbara Murck and Steve Barnes for stimulating discussions and for making sure we heard an alternative viewpoint. I also thank Dick Vian, chief geologist at the Stillwater Mine, and mine geologists Jim Dahy, Rad Langston, and Ennis Geraghty for providing much useful information and insightful discussions. I am grateful to Manville Corporation geologists Bob Mann, Stan Todd, Lynn LeRoy and Sam Corson for their assistance. I am also indebted to many other geologists who have generously shared their time and ideas, particularly John Longhi, Jim Papike, Don DePaolo, Brian Stewart, Mike Zientek, Bob Carlson, Ken Segerstrom, and Bruce Lipin. Finally, all of us who have worked on the complex in the past quarter century owe a major debt of gratitude to the pioneering Stillwater geologists, most notably, Joe Peoples, Art Howland, Dale Jackson, Bill Jones, and Harry Hess.

9. REFERENCES

Barker, R.W., 1975. Metamorphic mass transfer and sulphide genesis, Stillwater Intrusion, Montana. *Econ. Geol.* **70**, 275-98.

Barnes, S.J., & Naldrett, A.J., 1986. Geochemistry of the J-M Reef of the Stillwater Complex, Minneapolis Adit area. II. Silicate mineral chemistry and petrogenesis. *J. Petrology* 27, 791-825.

Bonini, W.E., 1982. The size of the Stillwater Complex: An estimate from gravity data. In: Walker, D. & McCallum, I.S. (eds.) *Magma Oceans and Stratiform Layered Intrusions*. L.P.I. Technical Report 82-01, 53-5.

Bosch, D., Nelson, B.K., & McCallum, I.S., 1991. Initial lead composition of feldspars from the Stillwater Complex, Montana. *EOS* **72**, 298.

Boudreau, A.E., 1987. Pattern formation during crystallization and the formation of fine-scale layering. In: Parsons, I. (ed.) *Origins of Igneous Layering.* Dordrecht: Reidel, 453-71.

Boudreau, A.E., 1988. Investigations of the Stillwater Complex. IV. The role of volatiles in the petrogenesis of the J-M Reef, Minneapolis Adit section. *Can. Miner.* **26**, 193-208.

Boudreau, A.E., & McCallum, I.S., 1986. Investigations of the Stillwater Complex. Part III. The Picket Pin Pt-Pd deposit. *Econ. Geol.* **81**, 1953-75.

Boudreau, A.E., & McCallum, I.S., 1989. Investigations of the Stillwater Complex: Part V. Apatites as indicators of evolving fluid composition. *Contr. Miner. Petrol.* **102**, 138-53.

Boudreau, A.E., & McCallum, I.S., 1992. Concentration of Platinum Group Elements by magmatic fluids in layered intrusions. *Econ. Geol.* **87**, 1830-48.

Boudreau, A.E., Mathez, E.A., & McCallum, I.S., 1986. Halogen geochemistry of the Stillwater and Bushveld Complexes: Evidence for the transport of the platinum-group elements by Cl-rich fluids. *J. Petrology* **27**, 967-86.

Bow, C, Wolfgram, D., Turner, A., Barnes, S., Evans, J., Zdepski, M., & Boudreau, A., 1982. Investigations of the Howland reef of the Stillwater Complex, Minneapolis Adit area: Stratigraphy, structure and mineralization. *Econ. Geol.* **77**, 1481-92.

Campbell, I.H., & Murck, B.W., 1993. Petrology of the G and H chromitite zones in the Mountain View area of the Stillwater Complex, Montana. *J. Petrology* **34**, 291-316.

Campbell, I.H., & Turner, J.S., 1989. Fountains in magma chambers. *J. Petrology* **30**, 885-923.

Campbell, I.H., Naldrett, A.J., & Barnes, S.J., 1983. A model for the origin of the platinum-rich sulphide horizons in the Bushveld and Stillwater Complexes. *J. Petrology* **24**, 133-65.

Criscenti, L.J., 1984. The origin of macrorhythmic units in the Stillwater Complex. Unpubl. M.Sc. thesis, University of Washington, 109 pp.

Czamanske, G.K., & Bohlen, S.R., 1990. The Stillwater Complex and its anorthosites: An accident of magmatic underplating. *Am. Miner.* **75**, 37-45.

Czamanske, G.K., & Scheidle, D.L., 1985. Characteristics of the Banded series anorthosites. In: Czamanske, G.K., & Zientek, M.L. (eds.) *The Stillwater Complex, Montana: Geology and Guide.* Spec. Publ. Montana Bureau of Mines and Geology **92**, 334-45.

Czamanske, G.K., & Zientek, M.L., 1985 (eds.) *The Stillwater Complex, Montana: Geology and Guide.* Spec. Publ. Montana Bur. Mines and Geology **92**, 396 pp.

DePaolo, D.J., & Wasserburg, G.J., 1979. Sm-Nd age of the Stillwater Complex and the mantle evolution curve for neodymium. *Geochim. Cosmochim. Acta* **43**, 999-1008.

Dunn, T., 1986. An investigation of the oxygen isotope geochemistry of the Stillwater Complex. *J. Petrology* **27**, 987-97.

Ghiorso, M.S., & Sack, R.O., 1995. Chemical mass transfer in magmatic processes IV. A revised and internally consistent thermodynamic model for the interpolation and extrapolation of liquid-solid equilibria in magmatic systems at elevated temperatures and pressures. *Contr. Miner. Petrol.* **119**, 197-212.

Haskin, L.A., & Salpas, P.A., 1992. Genesis of compositional characteristics of Stillwater AN-I and AN-II thick anorthosite units. *Geochim. Cosmochim. Acta* **56**, 1187-212.

Helz, R.T., 1985. Composition of fine-grained mafic rocks from sills and dikes associated with the Stillwater Complex. In: Czamanske, G.K., & Zientek, M.L. (eds.) *The Stillwater Complex, Montana: Geology and Guide.* Spec. Publ. Montana Bureau of Mines and Geology **92**, 97-117.

Helz, R.T. 1995. The Stillwater Complex, Montana: a subvolcanic magma chamber? *Am. Miner.* **80**, 1343-6.

Hess, H.H., 1960. *Stillwater Igneous Complex, Montana.* Geol. Soc. Am. Mem. **80**, 230 pp.

Irvine, T.N., 1967. Chromian spinel as a petrogenetic indicator. Part 2. Petrologic applications. *Can. J. Earth Sci.* **4**, 71-103.

Irvine, T.N., Keith, D.W., & Todd, S.G., 1983. The J-M platinum palladium reef of the Stillwater Complex, Montana: II. Origin by double-diffusive convective magma mixing and implications for the Bushveld Complex. *Econ. Geol.* **78**, 1287-334.

Jackson, E.D., 1961. Primary textures and mineral associations in the Ultramafic Zone of the Stillwater Complex, Montana. *U.S. Geol. Surv. Prof. Paper* **358**, 106 pp.

Jackson, E.D., 1967. Ultramafic cumulates in the Stillwater, Great Dyke, and Bushveld Intrusions. In: Wyllie, P.J. (ed.) *Ultramafic and Related Rocks*, New York: Wiley, 20-38.

Jackson, E.D., 1971. The origin of ultramafic rocks by cumulus processes. *Fortsch. Mineral.* **48**, 128-74.

Jones, W.R., Peoples, J.W., & Howland, A.L., 1960. Igneous and tectonic structures of the Stillwater Complex, Montana. *U.S. Geol. Surv. Bull.* **1071-H**, 281-340.

Labotka, T.C., 1985. Petrogenesis of metamorphic rocks beneath the Stillwater Complex: Assemblages and conditions of metamorphism. In: Czamanske, G.K., & Zientek, M.L. (eds.) *The Stillwater Complex, Montana: Geology and Guide*. Spec. Publ. Montana Bureau of Mines and Geology **92**, 70-6.

Lambert, D.D., & Simmons, E.C., 1987. Magma evolution in the Stillwater Complex, Montana: I. Rare earth element evidence for the formation of the Ultramafic series. *Am. J. Sci.* **287**, 1-32.

Lambert, D.D., & Simmons, E.C., 1988. Magma evolution in the Stillwater Complex, Montana: II. Rare earth element evidence for the formation of the J-M Reef. *Econ. Geol.* **83**, 1109-26.

Lambert, D.D., Morgan, J.W., Walker, R.J., Shirey, S.B., Carlson, R.W., Zientek, M.L., & Koski, M.S., 1989. Rhenium-osmium and samarium-neodymium isotopic systematics of the Stillwater Complex. *Science* **244**, 1169-74.

Lambert, D.D., Walker, R.J., Morgan, J.W., Shirey, S.B., Carlson, R.W., Zientek, M.L., Lipin, B.R., Koski, M.S., & Cooper, R.L., 1994. Re-Os and Sm-Nd isotope geochemistry of the Stillwater Complex, Montana: Implications for the petrogenesis of the J-M Reef. *J. Petrology* **35**, 1717-53.

Lange, R.A., & Carmichael, I.S.E., 1987. Densities of Na_2O-K_2O-CaO-MgO-FeO-Fe_2O_3-Al_2O_3-TiO_2-SiO_2 liquids: new measurements and derived partial molar properties. *Geochim. Cosmochim. Acta* **51**, 2931-46.

LeRoy, L.W., 1985. Troctolite-anorthosite zone I and the J-M Reef: Frog Pond Adit to Graham Creek area. In: Czamanske, G.K., & Zientek, M.L. (eds.) *The Stillwater Complex, Montana: Geology and Guide*. Spec. Publ. Montana Bureau of Mines and Geology **92**, 325-33.

Loferski, P.J., Arculus, R.J., & Czamanske, G.K., 1994. Rare earth element evidence for the petrogenesis of the Banded series of the Stillwater Complex, Montana, and its anorthosites. *J. Petrology* **35**, 1623-49.

Longhi, J., Wooden, J.L., & Coppinger, K.D., 1983. The petrology of high-Mg dikes from the Beartooth Mountains, Montana: A search for the parent magma of the Stillwater Complex. *J. Geophys. Res.* **88**, Suppl., B53-69.

Mann, E.L., & Lin, C.-P., 1985. Geology of the West Fork adit. In: Czamanske, G.K., & Zientek, M.L. (eds.) *The Stillwater Complex, Montana: Geology and Guide*. Spec. Publ. Montana Bureau of Mines and Geology **92**, 247-52.

Mann, E.L., Lipin, B.R., Page, N.J., Foose, M.P., & Loferski, P.J., 1985. Guide to the Stillwater Complex exposed in the West Fork area. In: Czamanske, G.K., & Zientek, M.L. (eds.) *The Stillwater Complex, Montana: Geology and Guide*. Spec. Publ. Montana Bureau of Mines and Geology **92**, 231-46.

Marcantonio, F., Zindler, A., Reisberg, L., & Mathez, E.A., 1993. Re-Os isotopic systematics in chromitites from the Stillwater Complex. *Geochim. Cosmochim. Acta* **57**, 4029-37.

Marsh, B.D., & Maxey, M.R., 1985. On the distribution and separation of crystals in convecting magma. *J. Volcanol. Geotherm. Res.* **24**, 95-150.

Martin, D., & Nokes, R., 1989. A fluid-dynamical study of crystal settling in convecting magma. *J. Petrology* **30**, 1471-500.

McCallum, I.S., 1988. Evidence for crustal recycling during the Archean: The parental magmas of the Stillwater Complex. *Lunar Planet. Inst. Tech. Rep.* **88-02**, 92-4.

McCallum, I.S., Raedeke, L.D., & Mathez, E.A., 1977. Stratigraphy and petrology of the Banded zone of the Stillwater Complex, Montana. *EOS* **58**, 1245.

McCallum, I.S., Raedeke, L.D., & Mathez, E.A., 1980. Investigations in the Stillwater Complex: Part I. Stratigraphy and structure of the Banded zone. *Am. J. Sci.* **280A**, 59-87.

McCallum, I.S., Thurber, M.W., Bosch, D., & Nelson, B.K., 1992. Lead isotopic compositions of plagioclases and sulphides in the Stillwater Complex: Evidence for isotopic disequilibrium and remobilization. *Lunar Planet. Sci.* **XXIII**, 867-8.

McKay, G.A., Wagstaff, J., & Le, L., 1990. REE distribution coefficients for pigeonite: Constraints on the origin of the mare basalt europium anomaly. *Lunar Planet. Sci. Conf.* **XXI**, 773-4.

Meurer, W.P., & Boudreau, A.E., 1996. The petrology and mineral compositions of the Middle Banded series of the Stillwater Complex, Montana. *J. Petrology,* in press.

Nunes, P.D., 1981. The age of the Stillwater Complex; a comparison of U-Pb zircon and Sm-Nd isochron systematics. *Geochim. Cosmochim. Acta* **45**, 1961-3.

Page, N.J., 1977. Stillwater Complex, Montana: Rock succession, metamorphism and structure of the complex and related rocks. *U.S. Geol. Surv. Prof. Paper* **999**, 79 pp.

Page, N.J., 1979. Stillwater Complex, Montana: Structure, mineralogy, and petrology of the Basal Zone with emphasis on the occurrence of sulphides. *U.S. Geol. Surv. Prof. Paper* **1038**, 69 pp.

Page, N.J., & Nokleberg, W.J., 1974. Geologic map of the Stillwater Complex, Montana. *U.S. Geol. Surv. Miscellaneous Investigations Series* **I-797**.

Page, N.J., Shimek, R., & Huffman, C., 1972. Grain-size variations within an olivine cumulate, Stillwater Complex, Montana. *U.S. Geol. Surv. Prof. Paper* **800-C**, 29-37.

Page, N.J., & Moring, B.C., 1987. Petrology of the noritic and gabbronoritic rocks below the J-M Reef in the Mountain View area, Stillwater Complex, Montana. *U.S. Geol. Surv. Bull.* **1674-C**, 1-47.

Page, N.J., & Zientek, M.L., 1987. Composition of primary postcumulus amphibole and phlogopite within an olivine cumulate in the Stillwater Complex, Montana. *U.S. Geol. Surv. Bull.* **1674-A**, 1-35.

Papike, J.J., Spilde, M.N., Fowler, G.W., & McCallum, I.S., 1995. SIMS studies of planetary cumulates: Orthopyroxene from the Stillwater Complex, Montana. *Am. Miner.* **80**, 1208-21.

Peach, C.L., & Mathez, E.A., 1993. Sulphide melt-silicate melt distribution coefficients for nickel and iron and implications for the distribution of other chalcophile elements. *Geochim. Cosmochim. Acta* **57**, 3013-21.

Premo, W.R., Helz, R.T., Zientek, M.L., & Langston, R.B., 1990. U-Pb and Sm-Nd ages for the Stillwater Complex and its associated dikes and sills, Beartooth Mountains, Montana: Identification of a parent magma. *Geology* **18**, 1065-8.

Raedeke, L.D., 1982. Petrogenesis of the Stillwater Complex. Unpubl. Ph.D. thesis. University of Washington, 212 pp.

Raedeke, L.D., & McCallum, I.S., 1980. A comparison of fractionation trends in the lunar crust and the Stillwater Complex. *Proc. Conf. Lunar Highlands Crust* 133-53.

Raedeke, L.D., & McCallum, I.S., 1984. Investigations of the Stillwater Complex: Part II. Petrology and petrogenesis of the Ultramafic series. *J. Petrology* **25**, 395-420.

Raedeke, L.D., & Vian, R.W., 1986. A three-dimensional view of mineralization in the Stillwater J-M Reef. *Econ. Geol.* **81**, 1187-1195.

Salpas, P.A., Haskin, L.A., & McCallum, I.S., 1983. Stillwater anorthosites: A lunar analog? *Proc. Lunar Planet. Sci. Conf. 14th, J. Geophys. Res.* **88**, Suppl., B27-39.

Salpas, P.A., Haskin, L.A., Gitlin, E.C., & McCallum, I.S., 1996. A formational model for Stillwater anorthosites: further information from detailed sampling and analysis of a small AN-II boulder. *Geochim. Cosmochim. Acta*, in revision.

Segerstrom, K., & Carlson, R.W., 1982. Geologic map of the banded upper zones of the Stillwater Complex and adjacent rocks, Stillwater, Sweet Grass, and Park Counties, Montana. *U.S. Geol. Surv.* map **I-1383**.

Simmons, E.C., & Lambert, D.D. (1982) Magma evolution in the Stillwater Complex: A preliminary evaluation using REE data for whole rocks and cumulate feldspars. In: Mueller, P.A., & Wooden, J.L. (eds.) *Precambrian geology of the Beartooth Mountains, Montana and Wyoming*. Spec. Publ. Montana Bureau of Mines and Geology **84**, 91-106.

Stacey, J.S., & Kramers, J.D., 1975. Approximation of terrestrial lead isotope evolution by a two stage model. *Earth Planet. Sci. Lett.* **26**, 207-21.

Stewart, B.M., & DePaolo, D.J., 1987. Sr isotopic stratigraphy of the Stillwater Complex, Montana. Evidence for multiple magma injection. *EOS* **68**, 429.

Todd, S.G., Keith, D.W., LeRoy, L.W., Shissel, D.J., Mann, E.L., & Irvine, T.N., 1982. The J-M Pt-Pd reef of the Stillwater Complex, Montana: I. Stratigraphy and petrology. *Econ. Geol.* **77**, 1454-80.

Turner, A.R., Wolfgram, D., & Barnes, S.J., 1985. Geology of the Stillwater County sector of the J-M Reef, including the Minneapolis adit. In: Czamanske, G.K., & Zientek, M.L. (eds.) *The Stillwater Complex, Montana: Geology and Guide*. Spec. Publ. Montana Bureau of Mines and Geology **92**, 210-30.

Wooden, J.L., & Mueller, P.A., 1988. Pb, Sr, and Nd isotopic compositions of a suite of late Archean igneous rocks, eastern Beartooth Mountains: implications for crust-mantle evolution. *Earth Planet. Sci. Lett.* **87**, 59-72.

Wooden, J.L., Czamanske, G.K., & Zientek, M.L., 1991. A lead isotopic study of the Stillwater Complex, Montana: constraints on crustal contamination and source regions. *Contr. Miner. Petrol.* **107**, 80-93.

Zientek, M.L., 1983. Petrogenesis of the Basal zone of the Stillwater Complex, Montana. Unpubl. Ph.D. thesis, Stanford University, 246 pp.

Zientek, M.L., & Ripley, E.M., 1990. Sulfur isotope studies of the Stillwater Complex and associated rocks, Montana. *Econ. Geol.* **85**, 376-91.

LAYERED INTRUSIONS
R.G. Cawthorn (editor)

The Windimurra Complex, Western Australia

C.I. Mathison[a] and A.L. Ahmat[b]

[a]Key Centre for Teaching and Research in Strategic Mineral Deposits, Department of Geology and Geophysics, The University of Western Australia, Nedlands 6907, Western Australia, Australia.

[b]Ashton Mining Limited, 100 Jersey Street, Jolimont 6014, Western Australia, Australia.

Abstract

The stratiform 2.8 Ga Windimurra Complex (2300 km^2) has a total thickness between 13 km (field stratigraphy) and 5 km (gravity modelling). The complex is surrounded by younger granitoids with sheared contact zones, and the roof is not exposed. Phase layering and cumulus mineral compositions show upwards fractionation, and allow recognition of the following subdivisions:

1) Ultramafic Series (UMS, 0.5 km thick), serpentinized olivine (Fo_{90}?) - chromite cumulates;
2) Lower Series (LS, 6-11 km thick), mainly anorthositic leucogabbronorites with olivine gabbroids increasingly abundant upwards, with cumulus plagioclase (An_{85-64}, 76 vol%), augite (*mg* 87-67), orthopyroxene (*mg* 85-61), olivine (Fo_{80-50}), and cumulus magnetite 2 km below the top, where $mg = 100xMg/(Mg+Fe_{Total})$;
3) Middle Series (MS, 1.5 km thick), mainly magnetite gabbronorites with cumulus plagioclase (An_{58}, 57 vol%), augite (*mg* 64), inverted pigeonite (*mg* 56), magnetite, and ilmenite;
4) Upper Series (US, <1 km thick), mainly magnetite olivine gabbros with cumulus plagioclase (An_{57}, 37 vol%), augite (*mg* 56), olivine (Fo_{35}), magnetite, and ilmenite.

The extensively explored and researched Lower Series dominates the complex, and poses many problems. Stratigraphic correlation of the various outcrop areas is uncertain due to scarcity of markers, lateral changes and discordances, non-outcrop, and faulting. Thickness estimates vary from 3 to 11 km depending on the method. Cumulus olivine increases upwards and orthopyroxene decreases, but both tend to appear and disappear irregularly. Anorthositic leucogabbronorite (80-90 vol% plagioclase) is the main rock type, and troctolite, olivine gabbro, and gabbro are common in the upper parts. Overall, mineral compositions show slight reverse fractionation initially, then remain roughly constant and show slight normal fractionation, but fluctuate irregularly in detail. Modal and phase macro-layering are well developed in the LS. Olivine-free modal macrolayers (gabbronorite - anorthositic leucogabbro-norite) are much more planar, uniform, and laterally extensive than olivine-bearing macrolayers. Anorthositic leucogabbronorite layers are commonly more mesocumulate compared with the adjacent adcumulate mafic gabbronorite layers. Imperfect cyclic layering, commonly reflecting the crystallization order plagioclase + olivine first, then augite, and orthopyroxene last with olivine disappearing, becomes more obvious upwards in the LS. Two 500 m thick intervals of magnetite-rich leucogabbroids are present in the LS - the Canegrass Magnetite Zone (CMZ), which is partly discordant to the upper LS, and the enigmatic Shephards Discordant Zone which may be equivalent to CMZ.

The Windimurra Complex is probably one of the most discordantly layered stratiform intrusions, and shows angular discordances between layers, structures resembling unconformities and troughs, non-planar layers, layer terminations, poorly layered intervals with heterogeneous "mixed rocks", lateral changes in modal proportions (olivine particularly) and mineral compositions, and angular discordance between the magnetite-in horizon and modal layering.

1. GENERAL GEOLOGY

The Windimurra Complex, about 500 km northeast of Perth and 50 km east of Mount Magnet, lies near the eastern margin of the Murchison Province of the Archaean granite-greenstone

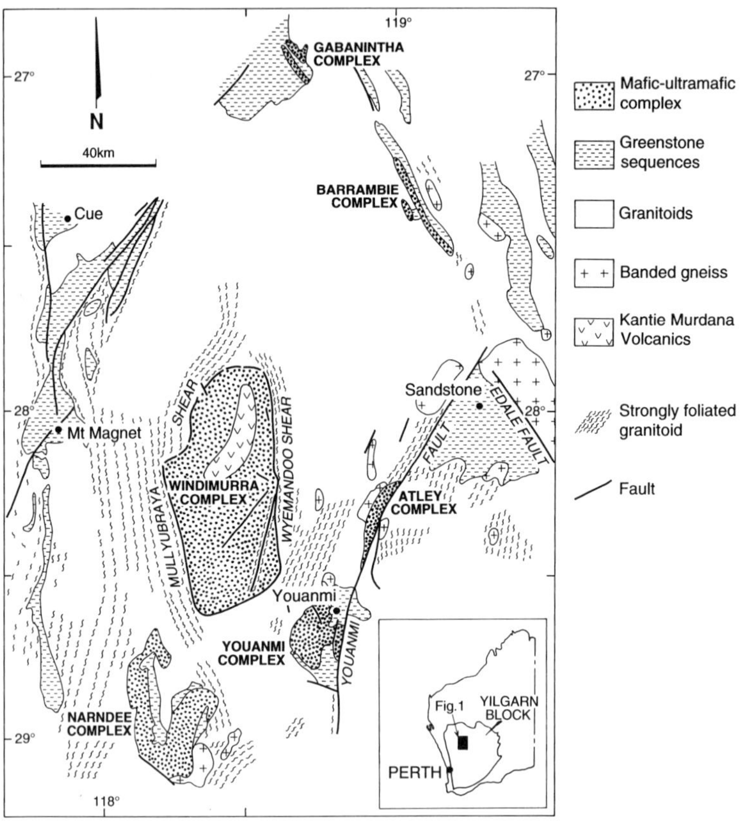

Figure 1. *Regional geological setting of the Windimurra Complex showing adjacent layered intrusions.*

Figure 2. *(facing page) Geological map of the Windimurra Complex showing the main areas of outcrop and the distribution of the main stratigraphic subdivisions (inset).*

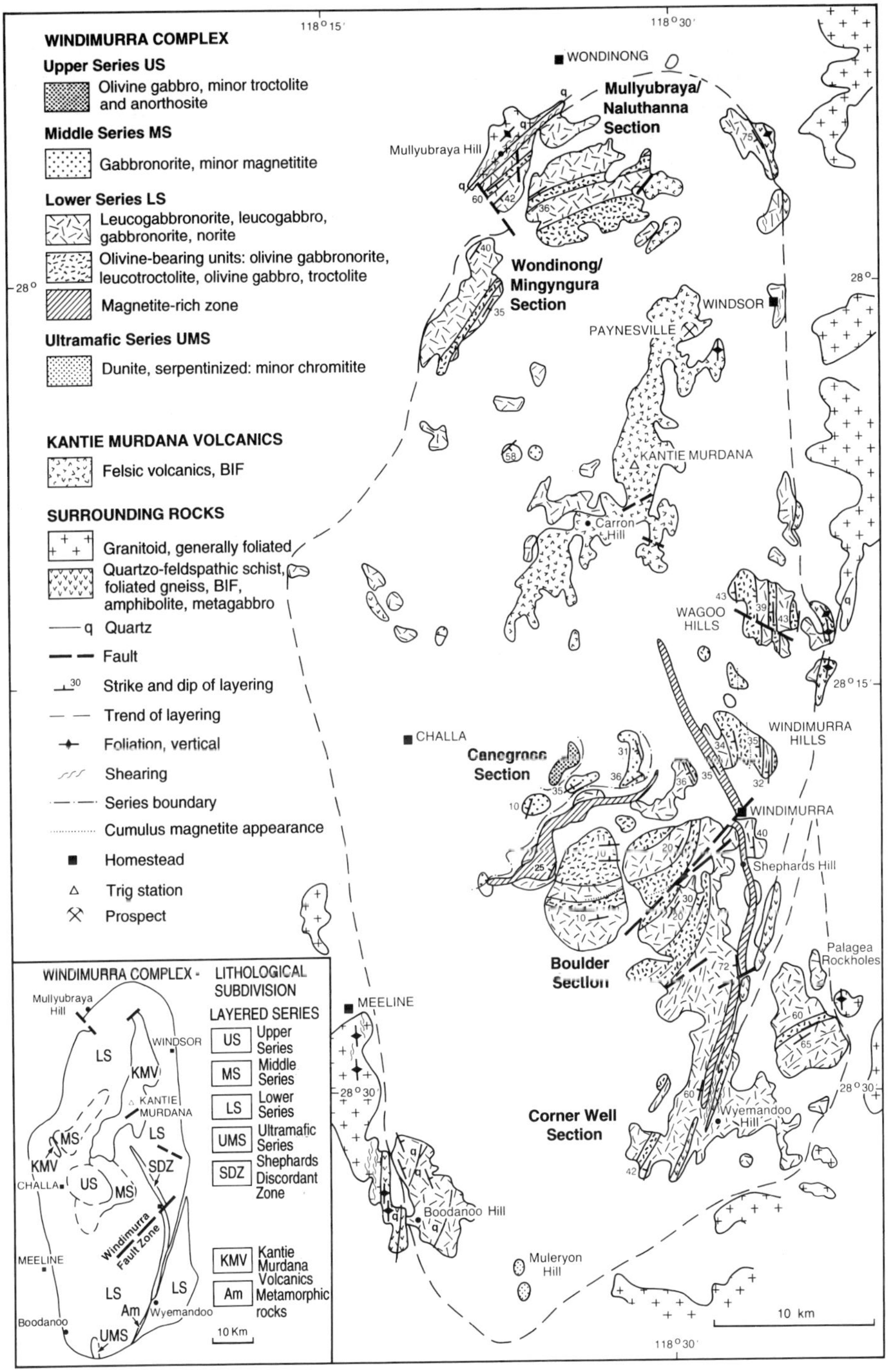
WINDIMURRA COMPLEX
Upper Series US
Olivine gabbro, minor troctolite and anorthosite
Middle Series MS
Gabbronorite, minor magnetitite
Lower Series LS
Leucogabbronorite, leucogabbro, gabbronorite, norite
Olivine-bearing units: olivine gabbronorite, leucotroctolite, olivine gabbro, troctolite
Magnetite-rich zone
Ultramafic Series UMS
Dunite, serpentinized: minor chromitite
KANTIE MURDANA VOLCANICS
Felsic volcanics, BIF
SURROUNDING ROCKS
Granitoid, generally foliated
Quartzo-feldspathic schist, foliated gneiss, BIF, amphibolite, metagabbro
q Quartz
Fault
Strike and dip of layering
Trend of layering
Foliation, vertical
Shearing
Series boundary
Cumulus magnetite appearance
Homestead
Trig station
Prospect
WINDIMURRA COMPLEX - LITHOLOGICAL SUBDIVISION
LAYERED SERIES
US Upper Series
MS Middle Series
LS Lower Series
UMS Ultramafic Series
SDZ Shephards Discordant Zone
KMV Kantie Murdana Volcanics
Am Metamorphic rocks
Windimurra Fault Zone
10 Km
Mullyubraya/ Naluthanna Section
Wondinong/ Mingyngura Section
Canegrass Section
Boulder Section
Corner Well Section
WONDINONG
WINDSOR
PAYNESVILLE
KANTIE MURDANA
Carron Hill
WAGOO HILLS
WINDIMURRA HILLS
CHALLA
WINDIMURRA
Shephards Hill
Palagea Rockholes
MEELINE
Wyemandoo Hill
Boodanoo Hill
Muleryon Hill
Mullyubraya Hill
118° 15′
118° 30′
28°
28° 15′
28° 30′
10 km

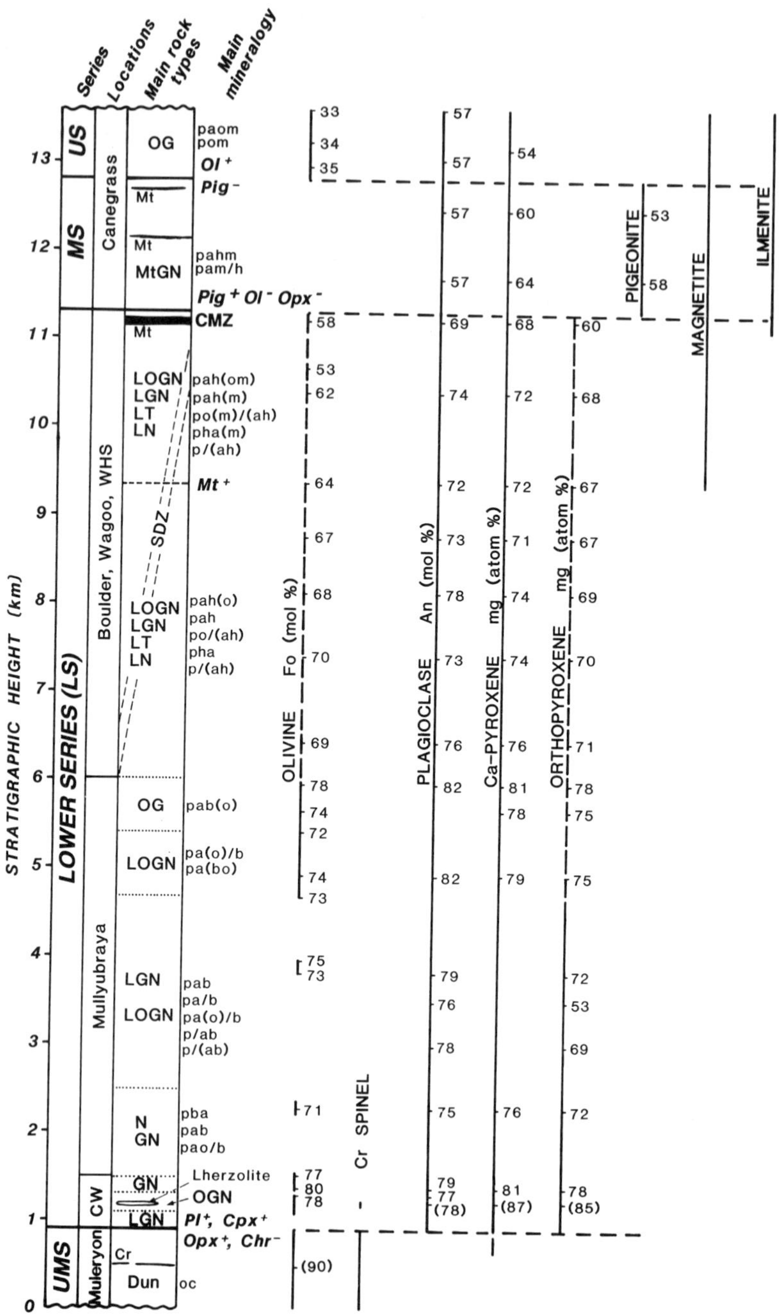
Series
Locations
Main rock types
Main mineralogy
STRATIGRAPHIC HEIGHT (km)
US
MS
LOWER SERIES (LS)
UMS
Canegrass
Boulder, Wagoo, WHS
Mullyubraya
CW
Muleryon
OG
Mt
MtGN
CMZ
LOGN
LGN
LT
LN
SDZ
GN
N
Lherzolite
OGN
Cr
Dun
paom
pom
Ol +
Pig -
pahm
pam/h
Pig + Ol - Opx -
pah(om)
pah(m)
po(m)/(ah)
pha(m)
p/(ah)
Mt +
pah(o)
pah
po/(ah)
pha
pab(o)
pa(o)/b
pa(bo)
pab
pa/b
p/ab
p/(ab)
pba
pao/b
Pl+, Cpx+
Opx+, Chr-
oc
OLIVINE Fo (mol %)
Cr SPINEL
PLAGIOCLASE An (mol %)
Ca-PYROXENE mg (atom %)
ORTHOPYROXENE mg (atom %)
PIGEONITE
MAGNETITE
ILMENITE

Yilgarn Craton, near the western margin of the Southern Cross Province. It is the largest of six layered intrusions (Figure 1). The complex is surrounded and intruded by 2.69 Ga monzogranites (U-Pb single zircon age of Wiedenbeck and Watkins, 1993), and wrapped by shear zones so that the original tectonic setting and host rocks are unknown. Only one contact with greenstone country rocks is preserved. The Kantie Murdana Volcanics (age unknown and up to 1-1.5 km thick) overlies about 15% of the middle of the complex and is a younger sequence or a tectonic slice, not the roof to the complex.

The Windimurra Complex is a stratiform basin-like body, 85 km north-south and 37 km east-west, covering an elliptical area of 2300 km^2, mostly with less than 50 m of topographic relief and only about 10% outcrop (Figure 2). Modal layering dips inwards at angles of about 70° marginally to 10° centrally. Unpublished data of Ahmat and Fletcher give an age of 2.80 ± 0.04 Ga (Sm - Nd mineral isochron). Two east-west gravity traverses 20 km apart with 800 m station spacing (Peters, 1972) indicate steep inward-dipping contacts at 80°. Gravity modelling (Peters, 1972) suggests the sheet-like complex is mostly about 3 km thick, and has a 5 km thick central axial zone (using densities of 2.9 g/cm^3 for the LS - probably slightly high, 3.1 g/cm^3 for the MS and US, and 2.6 g/cm^3 for the surrounding granitoids). These gravity-based estimates differ strongly from the stratigraphic thickness estimated from outcrop sections (6-13 km).

The Windimurra Complex is enveloped by steep inward-dipping shear zones in the surrounding granitoids, 1-1.5 km wide and up to 100 km long with unknown displacements, destroying the original margins and contact relations. The Wyemandoo shear zone on the east side is a complex 1-2 km wide tectonic-metamorphic belt containing andalusite-mica schists derived from felsic volcanic rocks, banded iron formation, amphibolite, and some metagabbro. It penetrates the complex in the southeast, separating it from a rotated block of the complex, the Palagea Block. The complex is also cut by major dextral northeast to southwest faults or shear zones and by sinistral northwest to southeast faults with apparent displacements up to about 1 km. Numerous other smaller shears and faults are present.

The following summary of the geological history of the area is based on studies of the Murchison Province (Ahmat and de Laeter, 1982, Watkins and Hickman, 1990):

1) deposition of the lower Murchison greenstones (Luke Creek Group, 3.0 Ga),
2) intrusion of monzogranites (2.9 Ga, now gneisses),
3) deposition of the upper Murchison greenstones (Mount Farmer Group, 2.8 Ga),
4) intrusion of the Windimurra Complex (2.8 Ga),
5) intrusion of monzogranites (2.69 Ga), and
6) Windimurra Complex intruded by several sets of dykes (e.g. feldspar porphyry 2.67 Ga,

---

Figure 3. *(facing page) Stratigraphic column for the Windimurra Complex, compiled from correlation of the successions in six main areas of outcrop (from Ahmat, 1986). Vertical lines show the stratigraphic distribution of cumulus minerals. UMS = Ultramafic Series, LS = Lower Series, MS = Middle Series, US = Upper Series; LT, LN, LGN, LOGN, OG = leucotroctolite, -norite, -gabbronorite, -olivine gabbronorite, and olivine gabbro respectively, Dun = serpentinized dunite, Cr = chromitite, Mt = magnetitite, p = plagioclase, a = augite, o = olivine, b = Ca-poor pyroxene, m = magnetite, c = chromite, cumulus minerals separated from intercumulus minerals by /; SDZ = Shephards Discordant Zone, CMZ = Canegrass Magnetite Zone, CW = Corner Well.*

dolerite 2.4-2.5 Ga), and affected by regional metamorphism and several phases of deformation (2.68-2.4 Ga).

2. STRATIGRAPHIC SUCCESSION

Figure 3 shows one version of the stratigraphic column for the layered series, compiled using the assemblages and compositions of cumulus minerals assuming negligible lateral change (Ahmat, 1986). The resulting stratigraphic thickness is 13 km - very large for a layered intrusion. Four major subdivisions of the layered series can be recognized on the basis of cumulus mineral assemblages as follows:

Ultramafic Series (UMS) olivine - chromite; Lower Series (LS) plagioclase - orthopyroxene - augite - olivine ± magnetite ± ilmenite; Middle Series (MS) plagioclase - augite - inverted pigeonite - magnetite - ilmenite; Upper Series (US) plagioclase - olivine - magnetite - ilmenite. This upwards progression in cumulus mineral assemblages and the following overall changes in mineral compositions show strong fractionation: plagioclase An_{82-56}, augite *mg* 87-54, Ca-poor pyroxene *mg* 88-53, and olivine Fo_{90}? (UMS), then Fo_{80-50} (LS) and Fo_{37-32} (US).

Ultramafic Series. The UMS is represented by a small, isolated occurrence at Muleryon Hill (1 km^2, max. 900 m thick, min. 300 m) near the southern margin of the complex (Figure 2). Serpentinized dunites were olivine - chromite mesocumulates with completely chloritized intercumulus ? plagioclase and pyroxene. Olivine composition is inferred to have been Fo_{90} from Mg-Fe partitioning with coexisting chromite (method of Irvine, 1965). Thin chromitites, mostly less than 2 cm thick, dip southwest at 45°. The chromite is Al- and Fe-rich, with 23 wt% Al_2O_3, 30 wt% Cr_2O_3, 28 wt% Fe_2O_3, 11 wt% FeO, 6 wt% MgO, and about 2 wt% TiO_2, broadly similar to some of the chromite in the Rhum and Fiskenaesset intrusions. The Windimurra chromite differs from typical chromites in PGE-bearing layered intrusions in having higher Al, Ti, Fe^{2+} and Fe^{3+} and lower Cr and Mg. Muleryon Hill rocks are assumed to be a basal ultramafic series beneath the LS because chromite with identical composition occurs in rare lherzolitic mesocumulates in the lower LS in the Corner Well and Wondinong areas. However, the restricted occurrence of the UMS is problematic.

Lower Series. The LS is the dominant part of the Windimurra layered series making up 90% of outcrops, with a total thickness ranging from 11 km (Figure 3) to 6 km (estimate given later). Dips of modal layering generally decrease inwards from 70° near the margins to 15°. The UMS - LS contact is not exposed, and the gap in olivine composition between the UMS (Fo_{90}?) and the LS (Fo_{80}) suggests that a significant thickness of UMS and LS is not exposed.

Anorthositic leucogabbronorite (about 85 vol% plagioclase) is the commonest rock type and alternates with prominent modal layers of more mafic gabbroids. Cumulus phases have the following typical composition ranges and average abundances: plagioclase mainly An_{82-69} (76 vol%), augite *mg* 81-68 (13 vol%), orthopyroxene *mg* 78-60 (8 vol%), olivine Fo_{80-58} (2 vol%), magnetite (1.4 to 0.8 wt% V_2O_3, 7-18 wt% TiO_2) and ilmenite (total Fe-Ti oxides 1 vol%). Not all these phases usually occur together in the one rock. Olivine abundance increases upwards, and a cumulus magnetite-in horizon occurs about 2 km below the top, providing the only potential stratigraphic marker. Cumulus ilmenite appears about 0.5 km below the top. Two 0.5 km thick magnetite-rich successions are present - the locally discordant Canegrass Magnetite Zone CMZ at the top of the LS, and the Shephards Discordant Zone SDZ which appears to be discordant to some of the host LS rocks.

Table 1
Calculated overall composition of the Windimurra Complex and its major subdivisions, and analyses of typical macrorhythmically layered gabbronorites

wt%	UMS Ultramafic Series	LS Lower Series	MS Middle Series	US Upper Series	Windimurra Complex	anorthositic gabbronorite	mafic gabbronorite
SiO_2	42.50	49.40	51.10	41.80	48.80	49.10	50.70
TiO_2	0.10	0.16	1.40	3.60	0.50	0.41	0.23
Al_2O_3	4.40	24.30	14.80	10.40	21.30	25.60	17.50
FeO(t)	13.80	4.60	11.10	25.70	7.00	5.20	8.10
MnO	0.10	0.11	0.20	0.32	0.13	0.07	0.15
MgO	38.90	4.80	7.00	8.10	7.30	3.00	8.50
CaO	0.14	14.30	12.20	8.50	12.90	14.00	12.90
Na_2O	0.02	2.20	2.10	1.40	2.00	2.50	1.80
K_2O	0.02	0.04	0.06	0.07	0.04	0.08	0.04
P_2O_5	0.01	0.04	0.03	0.08	0.04	0.08	0.04
Total	100.00	100.00	100.00	100.00	100.00	100.00	100.00
mg number	83	65	53	36	65	54	65

This chapter on the Windimurra Complex concentrates on the LS because of its extent, and the extensive research and exploration. The lower parts of the LS were explored for platinum, and the SDZ and CMZ for vanadium.

Middle Series. The Middle Series (MS) and Upper Series (US) have restricted lateral extent in the middle of the complex, and the stratigraphic succession is not known in detail because of poor exposure. The MS and US rocks show typical tholeiitic iron enrichment, and differ from LS rocks in having a finer grain size (2-3 mm), consistent lamination of plagioclase, higher colour index (50-60), less modal macrolayering, and less modal variety of rock types including fewer anorthositic rocks.

The MS is about 1.5 km thick with dips varying from 10° to 35°, and is defined by the presence of inverted pigeonite and the absence of olivine, though the contact with the LS is not exposed. Pigeonite appeared at more evolved compositions in the Windimurra crystallization sequence (Ca-poor pyroxene composition *mg* 60) than in typical layered intrusions (*mg* 70). Fe-Ti oxide gabbronorites dominate, and maximum composition ranges and average modal proportions of cumulus minerals are as follows: plagioclase An_{60-56} (47 vol%), augite *mg* 67-61 (33 vol%), inverted pigeonite *mg* 59-53 (10 vol%), magnetite (0.65 wt% V_2O_3, 13.5 wt% TiO_2), and ilmenite (total oxides 10 vol%). Magnetite is locally concentrated in rhythmically layered oxide gabbronorites.

Upper Series. The change from MS to US rocks is marked by the disappearance of inverted pigeonite and the reappearance of olivine. Rock types in the US are mainly Fe-Ti oxide-bearing olivine gabbros, with some troctolite and minor anorthosite. Typical mineral assemblages, maximum composition ranges and proportions are as follows: plagioclase An_{58-56} (38 vol%), augite *mg* 56 (33 vol%), olivine Fo_{37-32} (13 vol%), magnetite (0.25 wt% V_2O_3, 13-

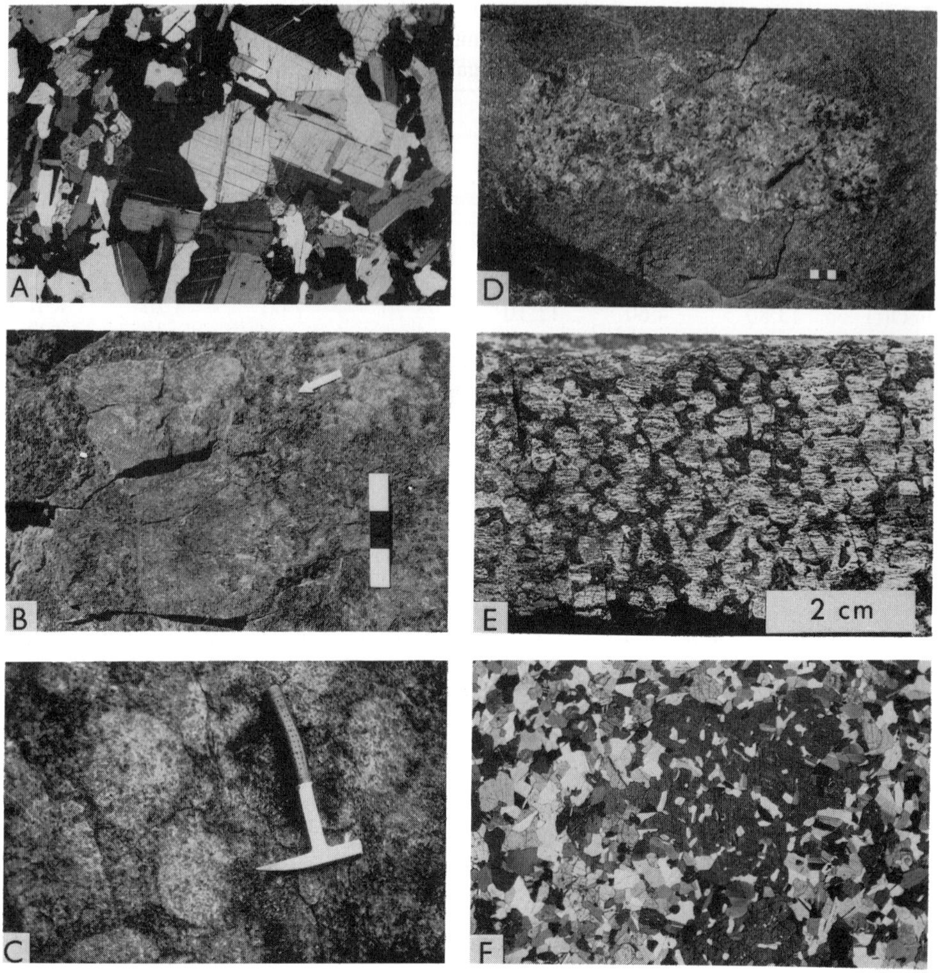

Figure 4. *Textures and structures in some Lower Series rocks.* ***A:*** *gabbronorite with cumulophyric cluster of pure plagioclase (width of field = 14mm).* ***B:*** *"mixed rock" olivine leucogabbro showing anorthosite inclusions and irregular clusters of coarse olivine grains (arrowed), from the main troctolite interval in Wagoo Hills (scale bar = 15 cm).* ***C:*** *"boulder anorthosite" with spheroidal anorthosite inclusions in a more mafic matrix, Mullyubraya section.* ***D:*** *inclusion of coarse pegmatitic leucogabbro in finer grained olivine gabbro, Wagoo Hills, similar stratigraphic level to B (scale = 5 cm).* ***E:*** *olivine (serpentinized) - magnetite cumulate from base of cyclic unit about 50 m above the base of the Canegrass Magnetite Zone.* ***F:*** *"chill" gabbronorite with oikocrysts of orthopyroxene enclosing plagioclase but not augite, with plagioclase and augite in the matrix (width of field = 14 mm).*

16 wt% TiO_2) and ilmenite (total oxides 16 vol%). An upper contact with the roof is unknown, and the compositions and lack of cumulus apatite suggest that a considerable volume of more fractionated rocks could be missing.

Calculated average compositions of the UMS, LS, MS, and US are given in Table 1, together with an estimate of the bulk composition of the layered series in the Windimurra Complex assuming 6 vol% UMS, 77 vol% LS, 11 vol% MS, and 6 vol% US. The compositions demonstrate the plagioclase-rich nature of the LS, and of the layered series as a whole.

3. PETROLOGY OF LOWER SERIES ROCKS

3.1. Composition, texture, and microstructure

Cumulus minerals are mainly plagioclase, augite, orthopyroxene, and olivine, and include magnetite and ilmenite near the top of the LS. The number of cumulus phases in a simple rock tends to increase from mainly three in the lower LS to four or five higher up. Intercumulus minerals include pyroxenes and Fe-Ti oxides, with minor lower temperature minerals such as quartz, apatite, and hornblende, together with minor sulphides (pyrrhotite > chalcopyrite > pentlandite > pyrite). Rocks are mostly adcumulates as shown by lack of strong normal zoning in plagioclase, small amounts of intercumulus minerals, and low contents of incompatible elements in whole-rock analyses (e.g. K_2O, P_2O_5 <0.08 wt% even in the most fractionated US rocks). Table 1 lists chemical compositions for an anorthositic leucogabbronorite (a typical LS rock with about 85 vol% plagioclase) and an adjacent mafic gabbronorite (65 vol% plagioclase), commonly found as alternating macrolayers in modally layered sequences.

Textures of LS rocks are typical of cumulates in layered intrusions, and show extensive textural equilibration. Lamination of plagioclase is generally weak, and grain size is commonly 5-10 mm, coarser than in most layered intrusions. Large oikocrysts (5-15 cm) of augite or orthopyroxene occur in anorthositic rocks. Photomicrographs of common rock types appear in Mathison and Ahmat (1991, their figures 5.3 and 5.4).

Plagioclase grains show a common tendency to segregate or to cluster, forming anorthositic inclusions from 1 to 50 cm across. Phenocrysts of plagioclase may be present, typically 2-5 mol% An more calcic than in the host rock. Small adcumulate clusters or microxenoliths of anorthosite are common in mafic gabbroids (Figure 4A) as in the "cumulophyric" clusters of Rutherford and Hermes (1984).These features suggest the presence of more than one generation of plagioclase in a given rock. Larger angular to rounded inclusions of pure anorthosite (10-50 cm) occur in heterogeneous leucogabbroids called "mixed rocks", which commonly have sporadic clusters of coarser olivine and coarse pyroxene (Figure 4B). "Boulder anorthosite" (Figure 4C) is a type of mixed rock with abundant, well-sorted spheroidal anorthosite fragments in a more pyroxene-rich matrix, and occurs mainly in the lower LS. Xenoliths of coarse anorthositic leucogabbroids also occur in mafic gabbroids (Figure 4D), but the reverse is virtually unknown.

Most LS gabbroids, particularly the anorthositic varieties, show the effect of alteration or metamorphism, mainly the amphibolization of pyroxenes which tends to be more extensive within about 4 km of the margins of the complex.

3.2. Common rock types

Windimurra LS rock types can be broadly categorized as anorthositic gabbroids with greater than about 80 vol% plagioclase, or mafic gabbroids with less than about 75 vol%

Table 2
Comparison of macrorhythmic modal layers of mafic gabbronorites (MGN, 1-2 m thick) and anorthositic leucogabbronorites (LGN, 5-10 m thick) in Wagoo Hills

	MGN	LGN
volume % plagioclase	65	85
intercumulus minerals	traces of magnetite	quartz, hornblende, apatite, magnetite and ilmenite
volume % alteration	<5	7-25
cumulate type	adcumulate	mesocumulate
plagioclase zoning	reverse: core An_{68}, rim An_{72}	normal: core An_{75}, rim An_{68}
orthopyroxene *mg*	69	62
augite *mg*	76	70
whole-rock *mg*	69	56

plagioclase, and also as olivine-free or olivine-bearing. The following rocks and cumulus assemblages are the most common in the LS: leucogabbronorite / gabbronorite (plagioclase - augite - orthopyroxene), leuco-olivine gabbronorite / olivine gabbronorite (plagioclase - augite - orthopyroxene - olivine), leuco-olivine gabbro / olivine gabbro (plagioclase - augite - olivine), leucotroctolite (plagioclase - olivine), and leucogabbro / gabbro (plagioclase - augite). Cumulus magnetite is also present above the magnetite-in horizon, but mainly in the mafic rocks.

Ultramafic rocks occur rarely in the LS, mostly in the CMZ and SDZ as magnetitites and magnetite-rich rocks. The latter contain virtually any combination of cumulus plagioclase, augite and olivine with magnetite and minor ilmenite. Olivine-magnetite cumulates are common in the lower part of the CMZ (Figure 4E). Lower parts of the LS contain rare lherzolitic cumulates with traces of chromite, most of which appear to be discordant sheets or pods rather than layers.

3.3. Other rock types

Fine-grained, somewhat chill-like gabbronorites and gabbros occur as 1-2 m thick layers extending laterally for at least 1 km mainly in the lower half of the LS (Figure 4F). The texture is unusual as orthopyroxene oikocrysts have inclusions of plagioclase, but not of augite, which is restricted to the surrounding matrix. In contrast, some gabbronorites and gabbros, also mainly in the lower sections of the LS, have very coarse to pegmatitic texture and could be termed pegmatoids (grain size 1-5 cm or greater). These pegmatoids are either stratiform or transgressive, and the latter type may have formed by replacement.

3.4. Comparison of mafic and anorthositic rocks

Detailed study by Mathison and Booth (1990) of modal macrorhythmic layering (1-10 m thick modal layers) in gabbronorites 3.8 km above the base of the Wagoo Hills section showed several important differences between adjacent mafic and anorthositic layers (Tables 1 and 2). The cumulus mineral assemblages are the same in each rock, but most other aspects differ.

The mafic gabbronorites, with a cumulus assemblage of 65 vol% plagioclase, 20 vol% orthopyroxene, and 15 vol% augite are adcumulates with virtually no intercumulus minerals. In contrast, the anorthositic gabbronorites have much more intercumulus material and alteration, and whole-rock analyses show higher concentrations of incompatible elements such as K, P,

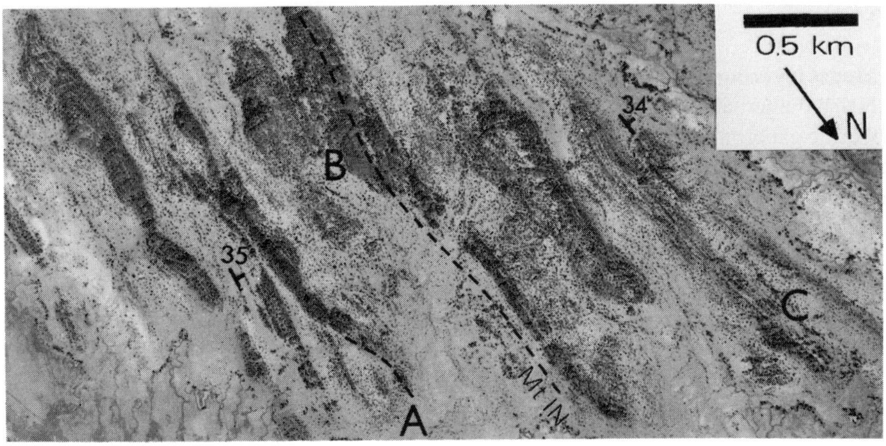

Figure 5. Aerial photograph of Windimurra Hills, showing a large scale discordance resembling an unconformity (A). A trough-like structure (B) is associated with the magnetite-in horizon (mt in), and strike trends are nonparallel (C). (dark signature = mafic gabbroids, light = anorthositic rocks. Dark layers appear to terminate to south and north due to limit of outcrop).

and Zr. Mafic mineral compositions and whole-rock compositions have significantly lower mg numbers by about 7 and 15% respectively in the anorthositic layers, and the compositions of plagioclase and pyroxenes thus decouple across the layer contacts. This is interpreted to be a result of the trapped liquid shift effect of Barnes (1986) due to a higher mass ratio of trapped liquid relative to mafic crystals, coupled with the formation of a denser residual liquid in the anorthositic layers (the compositional convection effect of Tait *et al.*, 1984). The difference in plagioclase core compositions cannot be due to this, and must reflect some difference during the early crystallization of each layer.

This finding suggests that interpretation of mg numbers for mineral and whole-rock data requires caution. Because of the partial dependence of mg numbers on modal proportions, rocks with similar modal proportions should be used to establish stratigraphic composition trends. This has been largely done for Figure 3, but not always in the more detailed columns. Petrographic study of several hundred LS rocks confirms that at least some of these differences between adjacent mafic and anorthositic layers are common, particularly for gabbronorites. However, olivine-bearing leucogabbroids, and some leucogabbros and relatively pure anorthosites commonly lack these mesocumulate features, and are more adcumulate in character.

4. LAYERING

4.1. Modal layering

Rhythmic alternation of relatively mafic gabbroids (plagioclase generally <75 vol%) with anorthositic gabbroids (mostly >80 vol% plagioclase) is a prominent feature of the LS, and occurs at the following scales:

a) megascale: Up to 1 km thick sequences of anorthositic gabbroids extend laterally for at least 3 to 5 km, and alternate with thinner intervals of more mafic gabbroids (100-400 m thick), as shown in Figure 5.
b) mesoscale: The mafic intervals in (a) consist of macrorhythmic alternations of mafic and anorthositic gabbroids mostly 1-50 m thick, and the thick anorthositic sequences typically contain sporadic 1-10 m thick mafic layers (lateral extent typically 0.5-3 km). Mafic layers in the LS outcrop better than anorthositic layers and support more vegetation, so appear darker on aerial photographs (Figure 6).
c) small scale: Typical outcrop-scale rhythmic layering (<1 m) with layers mostly 1-5 cm thick is stratigraphically and laterally intermittent. Angular discordances, mostly less than 20°, are common. Rhythmic layering is not well developed in the LS, but it becomes more common upwards.

Mesoscale macrorhythmic modal layering (terminology of Irvine, 1982) is the dominant type of modal layering visible in LS outcrops, and its regularity and lateral extent relate to rock type: gabbronorite intervals have uniform, regularly repeated, planar mafic layers only 1-3 m thick extending laterally for 2-3 km. In contrast, olivine-bearing layers (particularly olivine-rich troctolite) show much less regular layering, lateral change, and commonly contain mixed rock features as in Figure 4B.

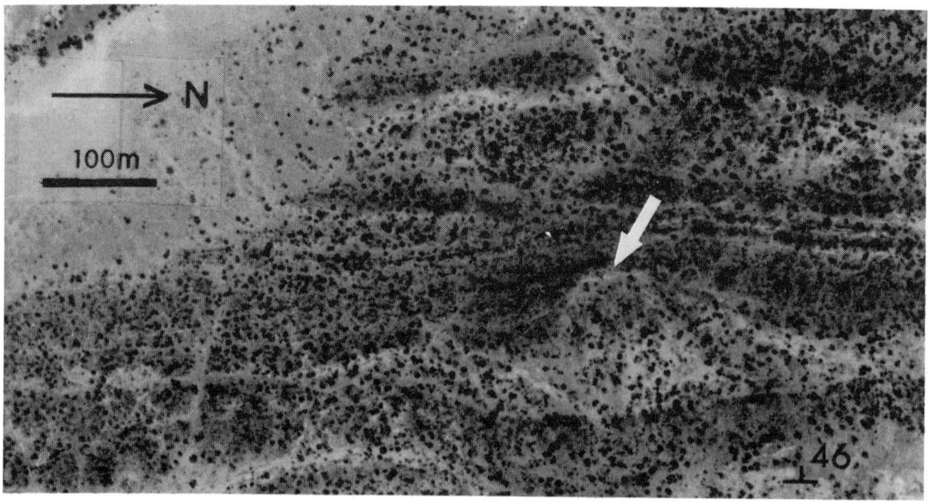

Figure 6. *Dome-shaped upper contact (arrowed) of pegmatitic leucogabbro layer in Wagoo Hills. Dark signature in aerial photograph represents olivine gabbro layers, which appear to terminate to the south due to lack of outcrop.*

4.2. Phase layering

Phase layering shows a wide range of scales and patterns as follows:

1) The whole layered series and the LS show this phase layering at the megascale, and the lower LS cumulus mineral assemblages below about 4.6 km in Figure 3 are mainly plagioclase - orthopyroxene - augite (except for the Corner Well succession), the middle portions also contain olivine, and the upper portions contain mainly plagioclase - augite - olivine - magnetite.
2) Cumulus olivine and orthopyroxene repeatedly appear and disappear in the LS, typically in conjunction with macrorhythmic modal layering. However, obvious cyclic units or fractionation sequences are lacking in the lower LS. In the middle and upper parts of the LS, imperfect cyclic units occur in which troctolite is commonly overlain by olivine gabbro, olivine gabbronorite, and gabbronorite, probably reflecting the crystallization order plagioclase + olivine first, then augite, orthopyroxene, with olivine disappearing before or after the appearance of orthopyroxene. Mineral compositions may show a small, usually gradual reversal of up to 5 atom% across the bases of these cyclic units, and slight normal fractionation of a similar amount. Cyclic units appear to be present at two scales, less than 50 m thick, and several hundred metres thick. The latter typically contain smaller cyclic units.

Cyclic units in the lower 150 m of the CMZ consist of olivine - magnetite cumulate (Figure 4E) overlain by magnetite troctolite then leuco(olivine)gabbro, corresponding to a different crystallization order: olivine + magnetite first, then plagioclase, and augite last. These cyclic units resemble those described by Umeji (1975) in the Freetown layered intrusion.

5. DISCORDANT STRUCTURES

The Lower Series is notable for the typical lack of lateral continuity, and the presence of a wide range of discordant or unconformity-like structures, terminations of layers, and lateral changes in modal proportions (particularly for olivine-bearing layers). The main discordant structures are described below. Mineral compositions generally show negligible changes across the contact zones of the first four examples of discordances.

Shephards Discordant Zone (SDZ). This is an enigmatic 47 km long and 500 m thick layered sequence of magnetite-bearing anorthositic gabbroids and magnetites, similar to the CMZ (Figures 2 and 7). Locally, the SDZ appears to be concordant with the host LS succession below it, and dips west at 35° in the north to 70° in the south. Regionally, the SDZ seems to truncate the magnetite-in horizon. Also, the Corner Well area in the southeast, probably equivalent to the lowest part of the LS, lies to the west of and apparently above the SDZ. The extent of discordancy of the SDZ is unclear, and the contacts are generally not exposed.

Canegrass Magnetite Zone (CMZ). The olivine-rich lower part of the CMZ (dip about 30°) occurs in a trough structure cutting across the strike and dip of the upper part of the Boulder Section (Figure 7). The strike of the CMZ becomes more concordant to the LS further to the northeast. The base of the CMZ marks the beginning of a series of cyclic units.

Windimurra Hills section. A structure resembling an erosional unconformity or part of a large pothole is present (A in Figure 5). The basal contact of a trough-shaped olivine gabbronorite sequence truncates a 200 m thick sequence of modally macrolayered gabbronorites. Removal of such a large thickness by erosion seems unlikely, so perhaps the termination of the gabbronorite layers was a primary feature such as an undulating surface on the crystal mush, followed by draping of subsequent layers after a break in crystallization. Lack of exposure in the contact zone prevents better understanding of this structure. Another discordant feature is the trough-like structure associated with the magnetite-in horizon (B in Figure 5). Some underlying layers appear to terminate against this trough, and the magnetite-in horizon is discordant to the modal and phase layering. The strike directions of adjacent mafic layered sequences are commonly not parallel (C in Figure 5).

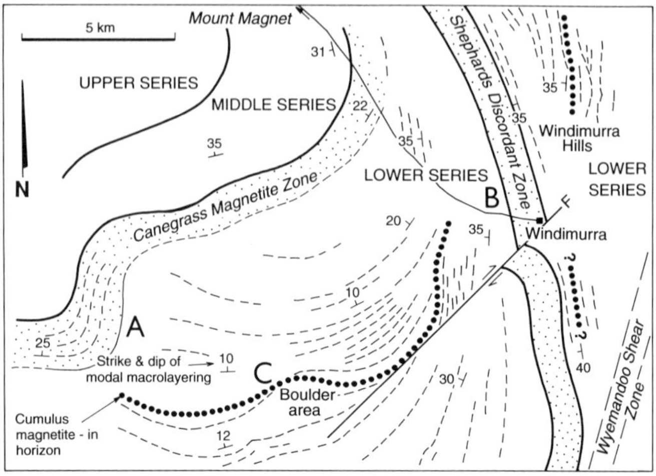

Figure 7. *Structural discordances in the Lower Series, central eastern Windimurra Complex. A: trough-like, olivine-rich lower portion of the Canegrass Magnetite Zone cutting across the upper Boulder Section. B: Shephards Discordant Zone, possibly discordant to the magnetite-in horizon. C: Magnetite-in horizon cutting across modal macrolayering.*

Wagoo Hills section. A heterogeneous coarse leucogabbro layer thickens abruptly to the north along strike from 20 m to 100 m (Figure 6). The lower contact of this layer is roughly planar but the upper contact is domal. A distinctive 1 m thick olivine gabbro layer with prismatic olivine at the base of the overlying mafic sequence follows most of this domal contact as though it was draped over it. This unusual structure could be a primary irregularity in the crystallizing front, or a later formed feature due to compaction or slumping.

Other discordances and layering irregularities. The magnetite-in horizon is discordant to the modal layering in the Boulder Section (Figure 7). In the LS, strikes of macrolayers commonly converge or diverge at up to 30°, and small-scale rhythmic layering shows angular truncations. Macrolayers commonly change laterally or terminate, and examples of these irregularities follow. A 20 m thick olivine gabbro layer about 100 m stratigraphically above the domal feature in Wagoo Hills terminates abruptly to the south. The main troctolite zone in Wagoo Hills changes along strike to the north from melatroctolite to leucotroctolite to olivine gabbro, and is irregularly layered. The main magnetitite layer in CMZ thickens from 2 m to 15 m over 10 km along strike to the northeast.

6. STRATIGRAPHIC SUCCESSIONS IN THE LOWER SERIES

6.1. Generalized sequence

Olivine is rare in the lower parts, except in the Corner Well sequence, and appears with increasing frequency upwards, whereas cumulus orthopyroxene is common in the lower parts,

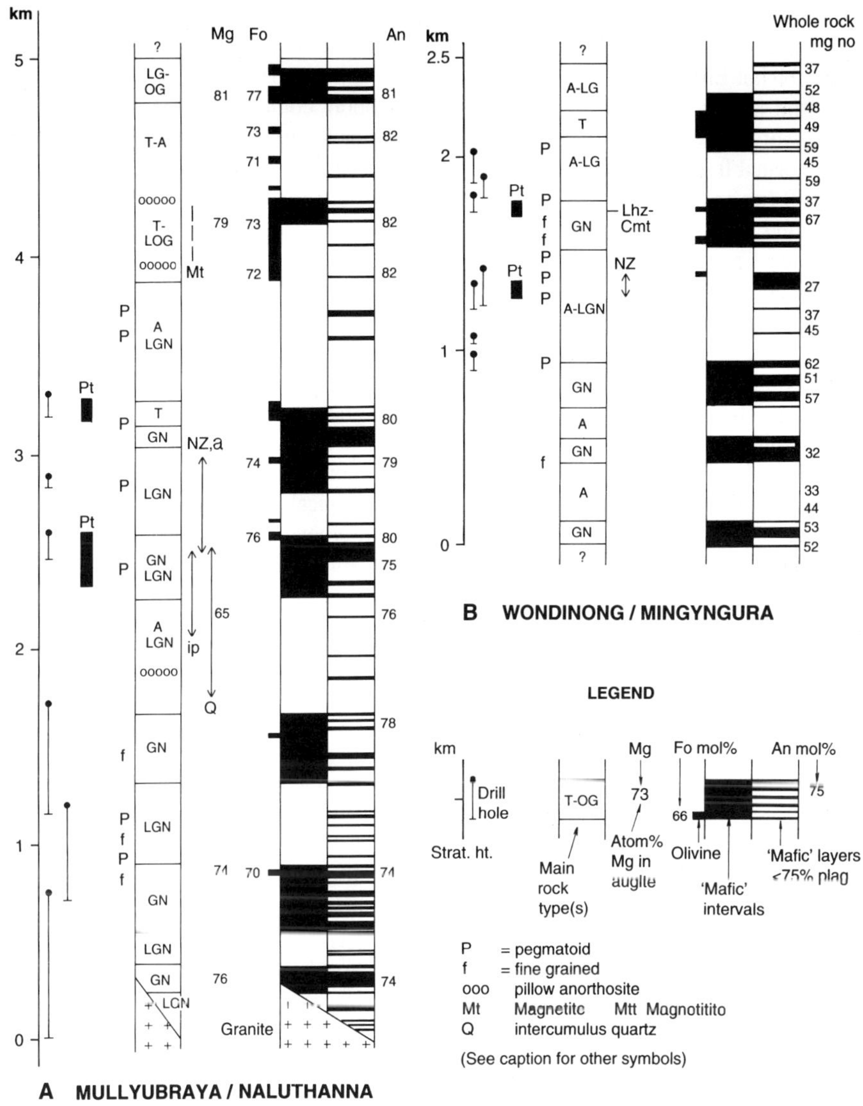

Figure 8. Summary stratigraphic columns for the Lower Series in the north-western Windimurra Complex. The base of section B appears to be equivalent to 1 km above the base of section A. Explanation of symbols for Figures 8, 9, 10, and 11 (also see legend): GN = gabbronorite, OGN = olivine gabbronorite, OG = olivine gabbro, T = troctolite, A = anorthosite, L (as prefix) = anorthositic; Lhz = lherzolitic olivine cumulate, Cmt = chromitite, Pt = main PGE anomalies, a = intercumulus apatite, NZ = strongly normally zoned plagioclase, ip = inverted pigeonite, MIH = cumulus magnetite-in horizon, upward-pointing arrow = cyclic or megacyclic units. Individual mafic layers are shown in the right-hand column where thickness and spacing allows, and the left-hand column shows broadly mafic and anorthositic intervals.

and becomes less common upwards. Cumulus olivine and cumulus orthopyroxene commonly occur together in the middle parts of the LS in olivine gabbronorite. Individual macrolayers tend to be thicker lower down, and cyclic units and small-scale rhythmic layering become more common and better defined upwards. Initially, mineral compositions show slight reverse changes upwards, then are relatively constant with minor fluctuations of uncertain significance, and show normal fractionation near the top. The LS can be roughly subdivided as follows:

- **lower LS** (up to about 4.6 km in Figure 3): mainly leucogabbronorite, initially very coarse grained and interlayered with anorthosite with large pyroxene oikocrysts; contains an interval of more fractionated leucogabbronorite near the top.
- **middle LS** (up to about 9.4 km in Figure 3): first main appearance of olivine (± anomalous PGE), ± plagioclase megacrysts, mixed rocks, and boulder anorthosite; slightly more mafic than lower portion, modal augite starts to exceed orthopyroxene.
- **upper LS**: above the magnetite-in horizon, cumulus orthopyroxene less common, olivine common, gabbronorites rarer, culminating in the magnetite-rich CMZ.

The stratigraphic succession in each of five outcrop areas is summarized below using simplified stratigraphic columns showing individual mafic and anorthositic layers (where their spacing allows) and broader mafic and anorthositic intervals, typical plagioclase, olivine, and augite compositions, distribution of olivine and the main rock types, and several other features given in the explanation for Figure 8. The sections illustrate the lateral variability in this suite of rocks.

6.2. Mullyubraya - Wondinong section

This area in the northwest of the Windimurra Complex has been explored extensively for platinum. Dips of layering decrease inwardly from about 50° to 35°. The Mullyubraya or Naluthanna section (5 km thick, Figure 8A) is the most complete section through the lower half of the LS, and contains at least 50 mafic macrolayers. Anorthositic leucogabbronorite dominates lower down, and leucotroctolite and leuco-olivine gabbro are common higher up. A more fractionated interval of anorthositic leucogabbronorites with more mesocumulate character occurs at 2 to 2.5 km. Minor inverted pigeonite is present, together with more intercumulus quartz and Fe-Ti oxides than usual and in the overlying 0.5 km, strongly normally zoned plagioclase and intercumulus apatite. Augite is *mg* 65 (typically *mg* 75 in this section overall) and orthopyroxene *mg* 54 (typically *mg* 68), yet plagioclase is An_{74}, a marked example of decoupling of pyroxene and plagioclase compositions. The interval from the base to about 2.5 km could be a fractionation cycle. Olivine then becomes more abundant, associated with a zone of anomalous platinum (mainly 20-100 ppb Pt + Pd) between 2.5 and 3.5 km. Mineral compositions show the following typical ranges and trends: plagioclase An_{74-82}, augite *mg* 74-81, orthopyroxene *mg* 64-75 (all initially showing a slight reverse trend upwards), olivine Fo_{70-75} (roughly constant). Whole-rock *mg* numbers (Perring and Vogt, 1991) show at least six major reversals which appear to correlate with the bases of the main mafic intervals.

The Wondinong or Mingyngura section 7 km to the west spans the middle 2.5 km of the main section (Figure 8B), but individual layers and even thick mafic intervals do not correlate with the main section, and the fractionated zone is absent or weakly developed. At 1.75 km, a 20 cm thick 100 m long chromitite with 1-8 ppm Pt + Pd occurs at the contact of a lherzolitic cumulate overlying an anorthosite (Perring and Vogt, 1991).

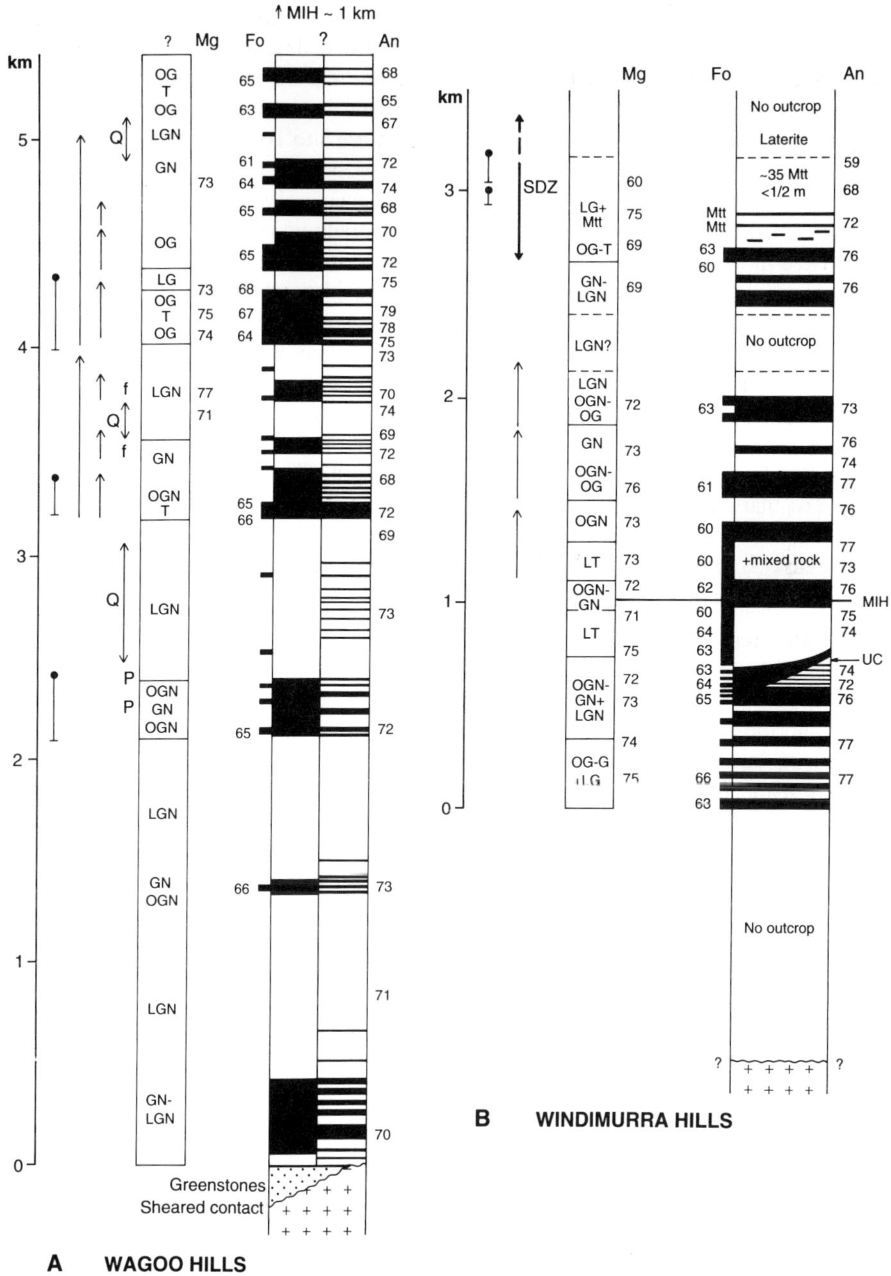

***Figure 9.** Summary stratigraphic columns for the Wagoo Hills and Windimurra Hills outcrop areas. For explanation, see legend and caption in Figure 8.*

6.3. Wagoo Hills section

This section extends inwards from the only exposed, but tectonized, contact with host greenstones. The 5 km thick section with an average dip of 45° contains at least 90 mafic macrolayers (Figure 9A). The section is broadly similar to the Mullyubraya - Wondinong section, although the Wagoo Hills mineral compositions are 5 to 10 mol% more fractionated.

The lower 4 km is mainly anorthositic leucogabbronorite. The lowest 2 km is poorly exposed and preserved, and lacks reliable dips, and is mainly oikocrystic anorthosite alternating with coarse leucogabbronorite, intruded by sheets of granitoid near the margin. A slightly more fractionated interval occurs at 2.3-2.5 km with traces of inverted pigeonite, and intercumulus quartz, apatite, and magnetite, despite the typical plagioclase composition of An_{73}. In contrast, the upper 1.5 km is well preserved and exposed, and is mainly leucogabbro with some olivine gabbro and troctolite and minor gabbronorite. Cyclic units with olivine-bearing bases occur above 3 km. The olivine-rich lower 200 m of the thickest mafic sequence (base at 4 km) is complex and poorly layered, containing mixed rocks and showing lateral variation from melatroctolite to leucotroctolite to olivine gabbro from south to north, with olivine changing composition from Fo_{65} to Fo_{68} over 2 km of strike length. This layering style contrasts strongly with the regular planar macrolayering in leucogabbronorites 300 m below.

Little overall change occurs in mineral compositions which have the following ranges: plagioclase $An_{65\text{-}79}$, olivine $Fo_{61\text{-}68}$, augite *mg* 71-77, orthopyroxene *mg* 60-72. The thickest olivine - bearing interval (base at 4 km) has the least fractionated mineral compositions.

6.4. Windimurra Hills section

This well-exposed and preserved 2.5 km thick section is the most studied part of the LS (Parks and Hill, 1986), and dips west at about 35°. Eight mafic strike ridges are separated by thicker anorthositic intervals with poorer outcrop (Figure 5). At least 110 mafic macrolayers are present, but the summary column (Figure 9B) cannot show this detail. The section is relatively mafic for the LS, but averages about 76 vol% plagioclase. Cumulus olivine and orthopyroxene commonly reappear throughout the sequence. Cumulus magnetite appears about halfway up. Poorly-defined fractionation sequences or cyclic units are present at various scales of thickness from 50 to 500 m. Several discordances and lateral changes are present, as described earlier. Above the magnetite-in horizon, gabbronorite changes to olivine gabbronorite to the north, and troctolite about 200 m above changes laterally to leuco-olivine gabbronorite. Up to 4 mol% Fo variation in olivine composition occurs over 2 km along strike. Generally, no obvious change in mineral compositions occurs across the magnetite-in horizon. Typical mineral compositions ranges are plagioclase $An_{72\text{-}77}$, olivine $Fo_{60\text{-}67}$ mainly (but down to Fo_{56} in some anorthositic troctolites), augite *mg* 71-76, and orthopyroxene *mg* 65-70.

6.5. Boulder section

This is the largest outcrop area in the LS and provides the best section through the upper part (Figure 2). At least 50 mafic macrolayers occur in a 5 km section (Figure 10A). Dips decrease from 65° to 10° but vary along strike, and structural complexities exist (faulting, dip variation, and strike discordance). Curiously, a "magnetite-out" horizon occurs near the base of the section, above the SDZ. The typical pattern of upward increase in olivine and decrease in orthopyroxene occurs. Intercumulus quartz is common around 1-1.5 km and 2.5 km. Boulder anorthosite occurs locally, but “chill” rocks and pegmatoids are rare. The magnetite-in horizon is 300 m higher 6 km to the west of the main section, so is discordant to the modal layering (Figure 7). Olivine compositions vary laterally along this horizon from Fo_{63} in the southwest to

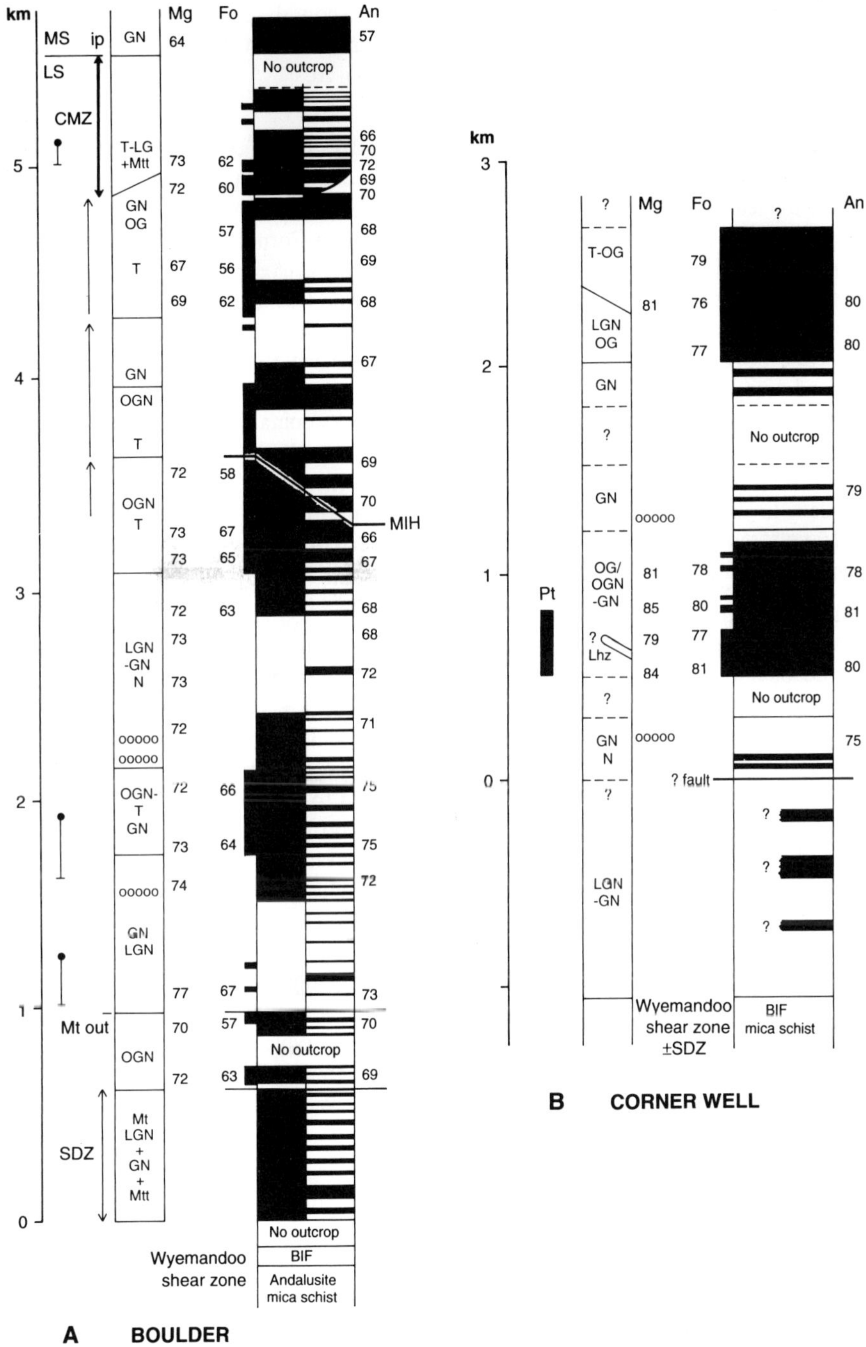

Figure 10. Summary stratigraphic columns for the Boulder and Corner Well outcrop areas. For explanation, see legend and caption in Figure 8.

Fo_{67} over 5 km to the northeast. Mineral compositions show the typical LS trend upwards, and ranges are as follows: plagioclase An_{66-75}, olivine mainly Fo_{60-67} (but Fo_{56} in leucotroctolite at 4.5 km), augite mainly *mg* 70-74, orthopyroxene *mg* 65-70.

6.6. Corner Well section

This area of poor outcrop in the southeast (Figure 2) contains about 2.5 km of presumed LS rocks dipping northwest at 40°, though the section is complicated by faulting. The Corner Well rocks are the least fractionated in the LS (*e.g.* An_{79}, Fo_{78}), but occur along strike from Boulder section rocks (An_{73}, Fo_{67}) to the northeast. However, the Corner Well sequence seems to be extensively faulted.

Two main rock types are present (Figure 10B): olivine-free anorthositic leucogabbronorites with minor mafic gabbronorite macrolayers typical of the lower LS, and olivine-bearing rocks (O'Sullivan, 1987). The latter are unlike typical LS rocks and are finer grained (1-3 mm), laminated, and compositionally relatively uniform with little macrolayering. Mineral compositions are similar to those in the olivine-free rocks but the rocks contain high Ni (100-200 ppm) and Cr (up to 1000 ppm) and have higher mg numbers. Two alternations of the olivine-free and olivine-bearing rock types occur (Figure 10B). The lowermost olivine-bearing interval contains concordant to discordant lenses or sheets of lherzolitic mesocumulate with 70 vol% olivine Fo_{84}, 15 vol% augite *mg* 87, 10 vol% plagioclase An_{79}, orthopyroxene *mg* 85, chromite with the same composition as in the UMS at Muleryon Hill, and 20 - 50 ppb Pt + Pd.

6.7. Shephards Discordant Zone

The SDZ succession is about 2 km west of Windimurra Hills. The lower 150 m consists of magnetite olivine gabbro and magnetite troctolite (Figure 9B, top), overlain by magnetite leucogabbro with podiform lenses of magnetite-olivine rock (2-10 cm thick). Two prominent laterally extensive magnetitites about 50 m apart follow, and minor cumulus ilmenite appears at this level. The remaining sequence (about 400 m) is poorly preserved and exposed, and is altered, metamorphosed, sheared, and deeply weathered to 45 m depth. This main part of the SDZ consists of alternating magnetite anorthositic gabbros and magnetitite layers and lenses (at least 35 magnetitites <0.5 m thick). The saprolitic weathered zone over this sequence is a potential vanadium deposit (Habteselassie, 1994). Cumulus orthopyroxene is absent from this part of the SDZ, but is commonly present 20 km to the south, suggesting lateral change.

Overall, the SDZ shows rapid fractionation from the olivine-bearing base upwards, and the following changes occur over 300 m: plagioclase An_{76-60}, augite *mg* 69 to 75 then 60, magnetite TiO_2 7 to 18 wt%, V_2O_3 1.4 to 0.8 wt%, and Cr_2O_3 2 to 0.1 wt%. Magnetite and whole-rock V_2O_3/TiO_2 ratios continuously decrease upwards, a typical fractionation trend. Olivine composition is Fo_{60-63}. No obvious change in mineral compositions occurs across the base of the SDZ (plagioclase An_{75} and augite *mg* 72 in the underlying gabbronorite).

6.8. Canegrass Magnetite Zone

The CMZ is a 25 km long 600 m thick strongly modally layered sequence of magnetite gabbroids and magnetite-rich layers containing at least 50 mafic and ultramafic macrolayers (Figure 2). A simplified column is shown at the top of the Boulder section (Figure 10A). The CMZ is probably thicker in an olivine-rich trough structure about 4 km wide and discordant to the upper Boulder section (Figure 7). The lower 150 m of this trough has prominent and complexly inter-related modal and cyclic layering reflecting the crystallization order olivine + magnetite, plagioclase, then augite last. Magnetite-rich rocks and magnetite troctolites commonly occur as lenses or troughs about 100 m wide and 1-3 m thick. Strong lateral variation may be present as the sequence appears to be much thinner and impoverished in olivine to the northeast. A prominent plagioclase magnetitite layer extends 15 km to the northeast and thickens from 2 to 15 m. The upper part of the CMZ is poorly exposed and the top is hidden under laterite. In the lower 300 m of the CMZ, plagioclase changes from An_{72} to An_{66}, and this trend may continue in the upper CMZ as plagioclase is An_{57} in the MS. Augite is mainly *mg* 70 to 73. Magnetite compositions show similar changes upwards to those in the SDZ: TiO_2 increases from 7 to 15 wt%, and Cr_2O_3 decreases from about 2 to 0.3 wt%. Slightly anomalous PGE occur in some magnetite-rich rocks (10 - 40 ppb Pt + Pd).

SDZ and CMZ are broadly similar, as are the underlying rocks. The relationship of the SDZ to the LS and to the CMZ is most puzzling. Ahmat (1986) interpreted the SDZ as a late-stage differentiated intrusion into the LS, possibly a feeder for the CMZ, MS, and US. It is possible that the SDZ and the CMZ represent the same magnetite-rich zone at the top of the LS, and that the SDZ is a structural repetition of the CMZ due to tectonism, the mechanism of which is not understood.

Cumulus magnetite and magnetitites appear at an earlier fractionation stage in the Windimurra Complex (plagioclase An_{70}, augite *mg* 72) compared with the Bushveld (plagioclase An_{50-55}, augite *mg* 53), possibly due to higher oxygen fugacity in the former.

6.9. Other areas

The Palagea Block is a 25 km long rotated block in the southeast of the Windimurra Complex with roughly 5 km of poorly exposed and preserved probable LS rocks (Figure 2). The succession does not correlate readily with the other LS sequences. The Boodanoo Hill area in the southwest (Figure 2) contains weakly layered, fine-grained, apatite-bearing magnetite gabbro or gabbronorite (An_{48}), similar to some rocks at the southern end of the SDZ. The Boodanoo Hill rocks may be a separate intrusion.

7. STRATIGRAPHIC CORRELATION FOR THE LOWER SERIES

Stratigraphic correlation of the main outcrop areas is difficult because of lack of sufficient markers and distinctive intervals, lateral variation and discordances, and faulting. The first attempt based on olivine compositions is similar to that used by Ahmat (1986), and assumes progressive upwards fractionation and negligible lateral change (Figure 11A). A second attempt assumes lateral equivalence of the main troctolite sequence and other lithologically similar intervals in the LS, and results in a more reasonable thickness of about 6 km for the LS (Figure 11B). Olivine and plagioclase compositions change laterally along the main troctolite zone from Fo_{80} and An_{82} to Fo_{57} and An_{68} over a strike distance of about 40 km, suggesting lateral discordance between lithological layering and mineral compositions. However, the pattern of change in mineral compositions is not as regular as in the example documented by Wilson and Larsen (1985). The magnetite-in horizon is discordant to the modal layering, and this fits the work of Wilson and Larsen (1986) and Nielsen and Wilson (1991) that shows phase layers are not time horizons, and can cut across modal layers. Whatever criterion is used for lateral equivalence in the LS, significant discordance is present, and the LS probably cannot be represented by a single stratigraphic column. The SDZ is discordant in both attempts, probably due to tectonism.

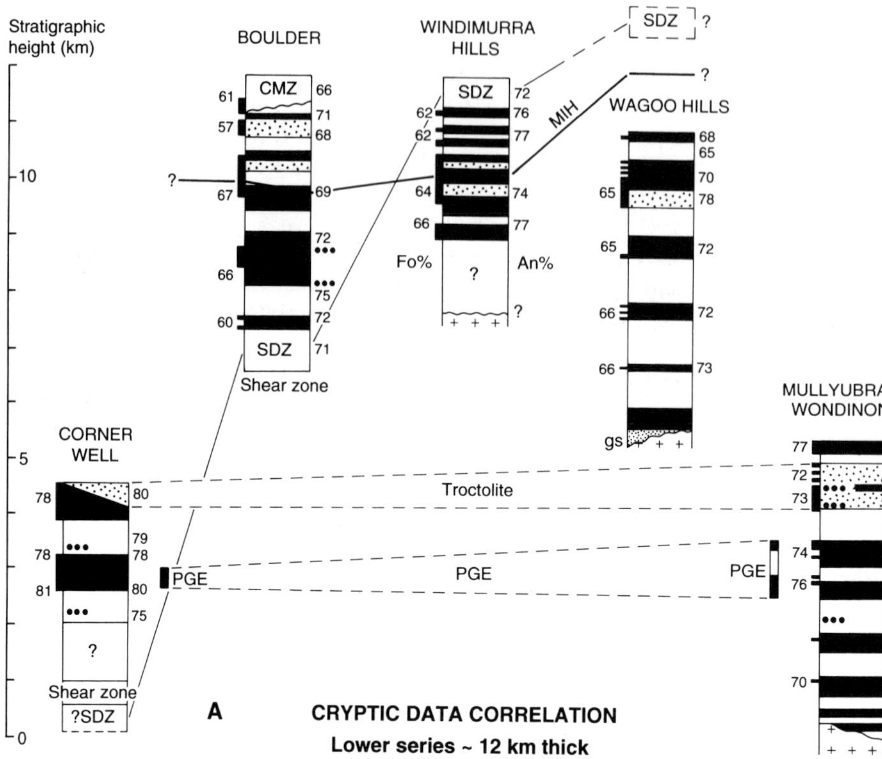

Figure 11. *Two models for stratigraphic correlation of the five main outcrop areas in the Lower Series.* ***A:*** *correlation based mainly on mineral composition data assuming insignificant lateral variation.* ***B:*** *(facing page) lithostratigraphic correlation. Other symbols and abbreviations are as used in Figures 8, 9, and 10, except heavy dots = boulder anorthosite, gs = greenstone country rock.*

The overall thickness of the Windimurra Complex still conflicts with the gravity model, though the latter could be in error because of the small density contrast. Differences could also arise as stratigraphic thickness measured normal to the plane of modal layering may not be the overall thickness of the stratiform intrusion, perhaps because the latter is truncated underneath by thrusting.

8. PETROGENESIS

The Windimurra Complex, particularly the LS, poses several major problems regarding:

1) the parent magmas and their compositions, the limited extent of the UMS and its relation to the LS, and the plagioclase-rich composition of the LS,
2) stratigraphic correlation in the LS, the significance and origin of the various discordances and lateral variations, and the conflicting thickness estimates, and

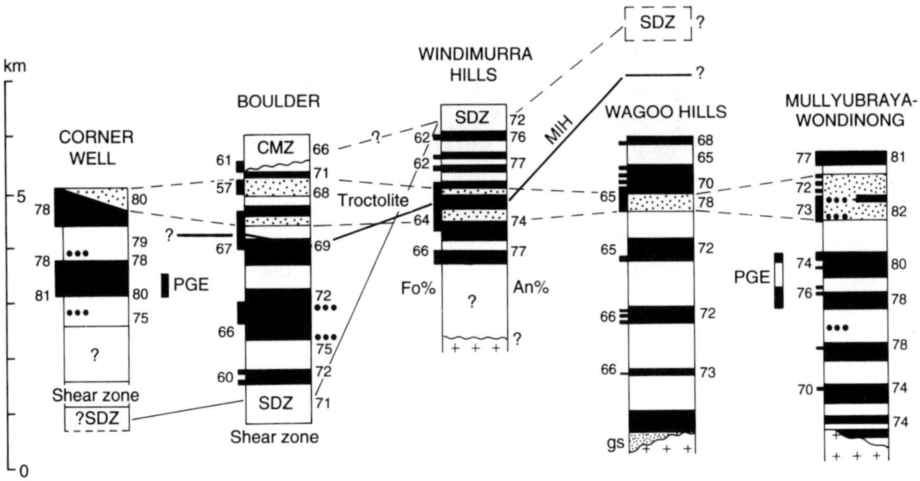

B **LITHOSTRATIGRAPHIC CORRELATION**
Lower Series ~ 6 km thick

Figure 11. *(continued) See previous page for caption.*

3) the significance of the SDZ, and its relation to the LS and the CMZ.
Other problems include the origin of the modal macrolayering, the origins of "chill" rocks and pegmatoids, and why so little of the original country rocks is preserved.

8.1. Parent magmas

The existence of the vastly different UMS and LS suggests at least two parent magmas. The UMS and the rare lherzolitic cumulates in the lower LS at Corner Well and in the Wondinong section provide evidence for an early high-Mg magma. Even in the LS, the overall order of appearance of cumulus phases (plagioclase + orthopyroxene + augite, then olivine, followed later by magnetite and ilmenite) may not fit a single magma. The lower LS suggests the presence of a silica-oversaturated, high-alumina basalt magma probably with the crystallization sequence plagioclase first, then orthopyroxene + augite. Silica oversaturation fits the widespread occurrence of intercumulus quartz. The increasing influence of a more silica-undersaturated, high-alumina basaltic magma is indicated around the middle of the LS where cyclic units with the probable order of crystallization plagioclase + olivine, augite, then orthopyroxene start to become apparent. A further change is indicated around the top of the LS in the CMZ where cyclic units suggest yet another crystallization order: olivine + magnetite, plagioclase, then augite, possibly reflecting absolute iron enrichment and oxygen fugacity increase. Above the CMZ, the iron enrichment in the MS and US, and the appearance of pigeonite in the gap in olivine crystallization demonstrate a typical tholeiitic iron enrichment trend.

Whatever the magma types were in the LS, the initial $^{87}Sr/^{86}Sr$ ratios for plagioclases (Mathison and de Laeter, 1994) and for whole-rocks (Ahmat and de Laeter, 1982) and ε_i Nd = 1.3 (Fletcher *et al.*, 1984) suggest a primitive source and lack of contamination, at least with older evolved crust.

8.2. Plagioclase enrichment

Average compositions of the LS and the whole Windimurra Complex (Table 1) are strongly enriched in calcic plagioclase components and it is likely that neither represents a closed system. The estimated average plagioclase content in the LS (76 vol%) is well above the cotectic proportion crystallizing with pyroxene from a basaltic magma - the "anorthosite problem". This excess of plagioclase could reflect partly the magma composition, and partly the crystallization and accumulation conditions. The presence of phenocrysts and anorthositic inclusions indicates preferential accumulation, or rafting or synneusis of cumulus plagioclase, perhaps because of near-neutral buoyancy which could contribute to the plagioclase-enriched cumulates. However, the fate of the complementary mafic-enriched fraction is a problem, and gravity data provide no evidence. Intrusions of a plagioclase-enriched mush from a deeper body as proposed by Czamanske and Bohlen (1990) for Stillwater anorthosites is possible, but difficult to reconcile with the intimate modal macrolayering.

8.3. Multiple injections and dynamic behaviour

From work on other large layered intrusions, multiple injections could be expected during the formation of the LS. Some evidence for these injections comes from repeated appearances of olivine, presence of cyclic units, mixed rocks, and poorly layered intervals, and reversals in mineral compositions and whole-rock mg numbers. Despite the great thickness of the LS, the extent of differentiation from bottom to top, as recorded in the cumulus mineral compositions, is no greater than that observed in short vertical sections, again suggesting multiple replenishment by less differentiated magma.

Initially, with so much plagioclase accumulating, the resident, fractionating magma would probably be denser than new magma, so that any new injections could have risen up well above the crystallization front without any immediate effect on the cumulates, explaining the lack of obvious cyclic units in the lower LS. However, later injections of a mafic magma could have been denser and spread across the floor with an immediate effect on cumulates, forming cyclic units. The occurrence of identifiable cyclic units increases upwards in the LS, suggesting such an evolution in the density relation between new injections and resident magma.

8.4. Lateral variation and discordance

The Windimurra Complex is notable because of the common occurrence of various types of discordance and lateral variation, and the difficulties of stratigraphic correlation. New magma injections may not have spanned the width of the magma chamber due to factors such as feeder location and geometry, slope and shape of the floor, and limited volume of injections. For example, "proximal" cumulates near a feeder might have more olivine than "distal" cumulates. Presence of an axial feeder zone is supported by the shape from gravity modelling. This fits the features unique to the northern and southern ends of the intrusion such as less fractionated mineral compositions, sporadic lherzolitic cumulates, anomalous PGE values, and the position and restricted occurrence of the UMS at Muleryon Hill. Nielsen and Wilson (1991) show that crystallization on an inclined floor from a zoned magma is likely to produce discordances. The discordances and lateral variations in the LS do not seem to have consistent direction of lateral asymmetry, so perhaps the magma chamber geometry was complicated (*e.g.* series of separate feeders along an axial zone).

Other origins are possible for some of the angular discordances in the Windimurra Complex. Autotectonism of the cumulates before consolidation due to crustal loading from the mass of the intrusion could be involved (Carr *et al.*, 1994). Also, later tectonism could explain some of the discordances. The Windimurra Complex is probably too large, too disrupted, and too poorly exposed in critical areas to be certain about the origin of the discordances. Lack of lateral continuity and presence of discordant structures have probably been under-estimated in studies of layered intrusions, though awareness of these is increasing *e.g.* the olivine-bearing zones in the Stillwater Complex (Czamanske and Zientek, 1985) and the "gap" areas in the Bushveld Complex (Wilson *et al.*, 1994).

9. ACKNOWLEDGEMENTS

The authors wish to gratefully acknowledge the assistance of R.J. Perring, J. Bunting, I. Martin, and other exploration geologists, and Dr R.E.T. Hill (CSIRO) for access to unpublished data. S.J. Barnes, I.H. Campbell, J.R. Wilson, and others provided constructive advice on this and earlier manuscripts.

10. REFERENCES

Ahmat, A.L., 1986. *Petrology, structure, regional geology and age of the gabbroic Windimurra Complex, Western Australia.* Unpublished Ph.D. Thesis, University of Western Australia.

Ahmat, A.L., & de Laeter, J.R., 1982. Rb-Sr isotopic evidence for Archaean-Proterozoic crustal evolution of part of the central Yilgarn Block, Western Australia: constraints on the age and source of the anorthositic Windimurra Gabbroid. *J. Geol. Soc. Aust.* **29**, 177-90.

Barnes, S.J., 1986. The effect of trapped liquid crystallisation on cumulus mineral compositions in layered intrusions. *Contr. Miner. Petrol.* **93**, 524-31.

Carr, H.W., Groves, D.I., & Cawthorn, R.G., 1994. The importance of synmagmatic deformation in the formation of the Merensky Reef potholes in the Bushveld Complex. *Econ. Geol.* **89**, 1398-410.

Czamanske, G.K., & Zientek, M.L., (eds.) 1985. *The Stillwater Complex Montana: Geology and Guide.* Montana Bureau of Mines. Geol. Spec. Publ. **92**, 396 pp.

Czamanske, G.K., & Bohlen, S.R., 1990. The Stillwater Complex and its anorthosites: an accident of magmatic underplating? *Am. Miner.* **75**, 37-45.

Fletcher, I.R., Rosman, K.J.R., Williams, I.R., Hickman, A.H., & Baxter, J.L., 1984. Sm-Nd geochronology of greenstone belts in the Yilgarn Block, Western Australia. *Precamb. Res.* **26**, 333-61.

Habteselassie, M.M., 1994. Vanadium distribution in the Shephards Discordant Zone, Windimurra Complex, Western Australia. *Minerals and Energy Research Institute of Western Australia,* report no. **123**, 162pp.

Irvine, T.N., 1965. Chromian spinel as a petrogenetic indicator. Part I. Theory. *Can. J. Earth Sci.* **2**, 648-72.

Irvine, T.N., 1982. Terminology for layered intrusions. *J. Petrology* **23**, 127-62.

Mathison, C.I., & Ahmat, A.L., 1991. Overview of the Windimurra Complex. *In:* Barnes, S.J., & Hill, R.E.T. (eds.) Mafic-ultramafic complexes of Western Australia. *Geological Society of Australia, Excursion Guidebook* No. **3**, 45-52.

Mathison, C.I., & Booth, R.A., 1990. Macrorhythmically layered gabbronorites in the Windimurra gabbroid complex, Western Australia. *Lithos* **24**, 171-80.

Mathison, C.I., & de Laeter, J.R., 1994. Geological Note: Strontium isotope initial ratios in the Windimurra layered gabbroid complex, Western Australia: a test for Bushveld-type magma mixing. *Aust. J. Earth Sci.* **41**, 281-4.

Nielsen, F.M., & Wilson, J.R., 1991. Crystallisation processes in the Bjerkreim-Sokndal layered intrusion, South Norway: evidence from the boundary between two macrocyclic units. *Contr. Miner. Petrol.* **107**, 403-14.

O'Sullivan, A.P., 1987. Geology and geochemistry of the Corner Well sequence of the Windimurra Complex, Yilgarn Block, Western Australia. Unpublished B.Sc. Hons. Thesis, University of Western Australia.

Parks, J., & Hill, R.E.T., 1986. Phase compositions and cryptic variation in a 2.2 km section of the Windimurra layered gabbroic intrusion, Yilgarn Block, Western Australia - a comparison with the Stillwater Complex. *Econ. Geol.* **81**, 1196-202.

Perring, R.J., & Vogt, J.H., 1991. The Wondinong - Mullyubraya section. *In:* Barnes, S.J., & Hill, R.E.T. (eds) Mafic-ultramafic complexes of Western Australia. *Geological Society of Australia, Excursion Guidebook* No. **3**, 65-76.

Peters, W.S., 1972. Geology and geophysics of the Mt. Walter area and gravity traverses at Windimurra and Mt. Keith. Unpublished B.Sc. Hons. Thesis, University of Western Australia.

Rutherford, M.J., & Hermes, O.D., 1984. Melatroctolite - anorthosite gabbro complex, Cumberland, Rhode Island: petrology, origin, and regional setting. *Bull. Geol. Soc. Am.* **95**, 844-54.

Tait, S.R., Huppert, H.E., & Sparks, R.S.J., 1984. The role of compositional convection in the formation of adcumulate rocks. *Lithos* **17**, 139-46.

Umeji, A.C., 1975. Gravity stratification in the Freetown basic igneous layered complex, Sierra Leone, West Africa. *Geol. J.* **10**, 107-30.

Watkins, K.P., & Hickman, A.H., 1990. Geological evolution and mineralisation of the Murchison Province, Western Australia. *Geol. Surv. West. Aust. Bull.* **137**, 267pp.

Wiedenbeck, M., & Watkins, K.P., 1993. A time scale for granitoid emplacement in the Archaean Murchison Province, Western Australia, by single zircon geochronology. *Precamb. Res.* **61**, 1-26.

Wilson, J.R., Cawthorn, R.G., Kruger, F.J., & Grundvig, S., 1994. Intrusive origin for the unconformable Upper Zone in the Northern Gap, Western Bushveld Complex. *S. Afr. J. Geol.* **97**, 462-72.

Wilson, J.R., & Larsen, S.B., 1985. Two dimensional study of a layered intrusion: the Hyllingen Series, Norway. *Geol. Mag.* **122**, 97-121.

Author Index

Subject Index